Emergence, Complexity and Computation

Volume 22

About this Series

The Emergence, Complexity and Computation (ECC) series publishes new developments, advancements and selected topics in the fields of complexity, computation and emergence. The series focuses on all aspects of reality-based computation approaches from an interdisciplinary point of view especially from applied sciences, biology, physics, or chemistry. It presents new ideas and interdisciplinary insight on the mutual intersection of subareas of computation, complexity and emergence and its impact and limits to any computing based on physical limits (thermodynamic and quantum limits, Bremermann's limit, Seth Lloyd limits…) as well as algorithmic limits (Gödel's proof and its impact on calculation, algorithmic complexity, the Chaitin's Omega number and Kolmogorov complexity, non-traditional calculations like Turing machine process and its consequences,…) and limitations arising in artificial intelligence field. The topics are (but not limited to) membrane computing, DNA computing, immune computing, quantum computing, swarm computing, analogic computing, chaos computing and computing on the edge of chaos, computational aspects of dynamics of complex systems (systems with self-organization, multiagent systems, cellular automata, artificial life,…), emergence of complex systems and its computational aspects, and agent based computation. The main aim of this series it to discuss the above mentioned topics from an interdisciplinary point of view and present new ideas coming from mutual intersection of classical as well as modern methods of computation. Within the scope of the series are monographs, lecture notes, selected contributions from specialized conferences and workshops, special contribution from international experts.

More information about this series at http://www.springer.com/series/10624

Andrew Adamatzky
Editor

Advances in Unconventional Computing

Volume 1: Theory

 Springer

Editor
Andrew Adamatzky
Unconventional Computing Centre
University of the West of England
Bristol
UK

ISSN 2194-7287   ISSN 2194-7295   (electronic)
Emergence, Complexity and Computation
ISBN 978-3-319-81633-3   ISBN 978-3-319-33924-5   (eBook)
DOI 10.1007/978-3-319-33924-5

Printed on acid-free paper

This Springer imprint is published by Springer Nature
The registered company is Springer International Publishing AG Switzerland

Preface

Unconventional computing is a science in flux. What is unconventional today will be conventional tomorrow. Designs being standard in the past are seen now as a novelty. Unconventional computing is a niche for interdisciplinary science, a cross-breed of computer science, physics, mathematics, chemistry, electronic engineering, biology, materials science and nanotechnology. The aims are to uncover and exploit principles and mechanisms of information processing in, and functional properties of, physical, chemical and living systems to develop efficient algorithms, design optimal architectures and manufacture working prototypes of future and emergent computing devices.

I invited world's leading scientists and academicians to describe their vision of unconventional computing and to highlight the most promising directions of future research in the field. Their response was overwhelmingly enthusiastic: over 50 chapters were submitted spanning almost all the fields of natural and engineering sciences. Unable to fit over one and half thousands pages into one volume, I grouped the chapters as "theoretical" and "practical". By "theoretical" I mean constructs and algorithms which have no immediate application domain and do not solve any concrete problems, yet they make a solid mathematical or philosophical foundation to unconventional computing. "Practical" includes experimental laboratory implementations and algorithms solving actual problems. Such a division is biased by my personal vision of the field and should not be taken as an absolute truth.

The first volume brings us mind-bending revelations from gurus in computing and mathematics. The topics covered are computability, (non-)universality and complexity of computation; physics of computation, analog and quantum computing; reversible and asynchronous devices; cellular automata and other mathematical machines; P-systems and cellular computing; spatial computation; chemical and reservoir computing. As a dessert we have two vibrant memoirs by founding fathers of the field.

The second volume is a tasty blend of experimental laboratory results, modelling and applied computing. Emergent molecular computing is presented by enzymatic logical gates and circuits, and DNA nano-devices. Reaction-diffusion chemical computing is exemplified by logical circuits in Belousov–Zhabotinsky medium and geometrical computation in precipitating chemical reactions. Logical circuits realised with solitons and impulses in polymer chains show advances in collision-based computing. Photo-chemical and memristive devices give us a glimpse into the hot topics of novel hardware. Practical computing is represented by algorithms of collective and immune-computing and nature-inspired optimisation. Living computing devices are implemented in real and simulated cells, regenerating organisms, plant roots and slime moulds. Musical biocomputing and living architectures make the ending of our unconventional journey non-standard.

The chapters are self-contained. No background knowledge is required to enjoy the book. Each chapter is a treatise of marvellous ideas. Open the book at a random page and start reading. Abandon all stereotypes, conventions and rules. Enter the stream of unusual. Even a dead fish can go with the flow. You can too.

Bristol, UK   Andrew Adamatzky
March 2016

Contents

Chapter 1
Nonuniversality in Computation: Fifteen Misconceptions Rectified

Selim G. Akl

Abstract In January 2005 I showed that universality in computation cannot be achieved. Specifically, I exhibited a number of distinct counterexamples, each of which sufficing to demonstrate that no finite and fixed computer can be universal in the sense of being able to simulate successfully any computation that can be executed on any other computer. The number and diversity of the counterexamples attest to the general nature of the nonuniversality result. This not only put to rest the "Church–Turing Thesis", by proving it to be a false conjecture, but also was seen to apply, in addition to the Turing Machine, to all computational models, past, present, and future, conventional and unconventional. While ten years have now passed since nonuniversality in computation was established, the result remains largely misunderstood. There appear to be at least two main reasons for this state of affairs. As often happens to new ideas, the nonuniversality result was confronted with a stubborn entrenchment in a preconceived, deeply held, and quasi-religious belief in the existence of a universal computer. This was exacerbated by a failure to read the literature that demonstrates why such a belief is unfounded. Behavior of this sort, sadly not uncommon in science, explains the enduring mirage of the universal computer. The purpose of this chapter is to rectify the most notorious misconceptions associated with nonuniversality in computation. These misconceptions were expressed to the author in personal communications, orally, by email, and in referee reports. Each misconception is quoted verbatim and a detailed response to it is provided. The chapter concludes by inviting the reader to take a computational challenge.

1.1 Introduction

For a long time people believed that the Sun orbits our Earth. We now know better. And so it is with universality in computation. Consider the following statement:

> It can also be shown that any computation that can be performed on a modern-day digital computer can be described by means of a Turing Machine. Thus if one ever found a procedure

S.G. Akl (✉)
School of Computing, Queen's University, Kingston, ON K7L 3N6, Canada
e-mail: akl@cs.queensu.ca

© Springer International Publishing Switzerland 2017
A. Adamatzky (ed.), *Advances in Unconventional Computing*,
Emergence, Complexity and Computation 22,
DOI 10.1007/978-3-319-33924-5_1

1

that fitted the intuitive notions, but could not be described by means of a Turing Machine, it would indeed be of an unusual nature since it could not possibly be programmed for any existing computer [35], p. 80.

The first sentence in the preceding quote is clearly false, and well known counterexamples abound in the literature (see, for example, [36, 63], and for further references [4, 8]). The second sentence, on the other hand, is only half true. Indeed, it is perfectly justified to say that a computation that cannot be performed on a Turing Machine must be of an *unusual nature*. However, this does not mean that such a computation cannot be programmed on any existing computer. Rather, the existence of such a computation would instead mean that the Turing Machine is simply not universal and that the "Church–Turing Thesis" (whose validity is assumed implicitly in the opening quote of this chapter) is in fact invalid.

In January 2005 I showed that the concept of a Universal Computer cannot be realized [4]. Specifically, I exhibited instances of a *computable* function $\mathcal{F}$ that cannot be computed on any machine $\mathcal{U}$ that is capable of only a finite and fixed number of operations per step (or time unit). This remains true even if the machine $\mathcal{U}$ is endowed with an infinite memory and the ability to communicate with the outside world while it is attempting to compute $\mathcal{F}$. It also holds if, in addition, $\mathcal{U}$ is given an indefinite amount of time to compute $\mathcal{F}$.

1.1.1 The Main Theorem

Formally, my result is stated as follows:

> **Theorem 1 Nonuniversality in computation:** *Given n spatially and temporally connected physical variables, $X_1, X_2, \ldots, X_n$, where n is a positive integer, there exists a function $F^n(X_1, X_2, \ldots, X_n)$ of these variables, such that no computer can evaluate F^n for any arbitrary n, unless it is capable of an infinite number of operations per time unit.* ∎

A constructive proof of Theorem 1 seeks to define a function F^n that obeys the following property: F^n is readily computable by a machine M_n capable of exactly n operations per time unit; however, this machine cannot compute F^{n+1} when the number of variables is $n + 1$. While a second machine M_{n+1} capable of $n + 1$ operations per time unit can now compute the function F^{n+1} of $n + 1$ variables, M_{n+1} is in turn defeated by a function F^{n+2} of $n + 2$ variables. In principle, the escalation continues without end.

This point deserves emphasis. While the function $F^{n+1}(X_1, X_2, \ldots, X_{n+1})$ is easily computed by M_{n+1}, it cannot be computed by M_n. Even if given infinite amounts of time and space, machine M_n is incapable of simulating the actions of

M_{n+1}. Furthermore, machine M_{n+1} is in turn thwarted by $F^{n+2}(X_1, X_2, \ldots, X_{n+2})$, a function computable by a third machine M_{n+2}. The process repeats indefinitely. Therefore no computer is universal if it is capable of exactly $T(i)$ operations during time unit i, where i is a positive integer, and $T(i)$ is finite and fixed once and for all, for it will be faced with a computation requiring $V(i)$ operations during time unit i, where $V(i) > T(i)$ for at least one i. (Note that this last statement is, in fact, a more general form of the result in Theorem 1, as it invokes neither the physical variables nor the function F^n.)

Therefore, in order to actually prove Theorem 1 constructively, it suffices to exhibit (at least) one concrete instance of the function F^n satisfying the property described in the previous two paragraphs. Indeed we offer several examples of such a function occurring in:

1. *Computations with time-varying variables:* The variables, over which the function is to be computed, are themselves changing with time.
2. *Computations with time-varying computational complexity:* The computational complexity of the function to be computed is itself changing with time.
3. *Computations with rank-varying computational complexity:* Given several functions to be computed, and a schedule for computing them, the computational complexity of a function depends on its position in the schedule.
4. *Computations with interacting variables:* The variables of the function to be computed are parameters of a physical system that interact unpredictably when the system is disturbed.
5. *Computations with uncertain time constraints:* There is uncertainty with regards to the input (when and for how long are the input data available), the calculation (what to do and when to do it), and the output (the deadlines are undefined at the outset); furthermore, the function that resolves each of these uncertainties, *itself* has demanding time requirements.
6. *Computations with global mathematical constraints:* The function to be computed is over a system whose variables must collectively obey a mathematical condition at all times.

It should be clear at this point that all computable functions used in this work are in fact generalizations of the definition of a mathematical function in that they may enjoy one or more of the following properties: they have real (that is, physical) time as a variable, their variables interact with one another, there exists a condition that must be satisfied while a function is being evaluated, and so on.

1.1.2 A Simple Example: Time-Varying Variables

For instance, suppose that the X_i are themselves functions that vary with time. It is therefore appropriate to write the n variables as $X_1(t), X_2(t), \ldots, X_n(t)$, that is, as functions of the time variable t. Further assume that, while it is known that the X_i

change with time, the actual functions that effect these changes are not known (for example, X_i may be a true random variable). Let

$$F^{n+1}(X_1, X_2, \ldots, X_{n+1}) = F_1(X_1), F_2(X_2), \ldots, F_{n+1}(X_{n+1}), \qquad (1.1)$$

where each F_i is a simple function of one variable that takes one time unit to compute. The problem calls for computing $F_i(X_i(t))$, for $i = 1, 2, \ldots, n$, at time $t = t_0$. Specifically, let $F_i(X_i(t))$ simply represent the reading of $X_i(t)$ from an external medium. The fact that $X_i(t)$ changes with the passage of time means that, for $k > 0$, not only is each value $X_i(t_0 + k)$ different from $X_i(t_0)$, but also the latter cannot be obtained from the former. No computer capable of fewer than n read operations per time unit can solve this problem.

We note in passing that the example in the previous paragraph is deliberately simple in order to convey the idea. Reading a datum, that is, acquiring it from an external medium, is the most elementary form of information processing. Any computer must be able to perform such operation. This simplest of counterexamples suffices to establish nonuniversality in computation. Of course, if one wishes, the computation can be made more complex, at will. While our main conclusion remains unchanged, for some, a more complex argument may sound more convincing. Thus, for example, we may add arithmetic by requiring that $F_i(X_i(t))$ call for reading $X_i(t)$ and incrementing it by one, for $i = 1, 2, \ldots, n$, at time $t = t_0$. Reading $X_i(t)$, incrementing it by one, and returning the result takes one time unit.

In any case, a computer M_n capable of n operations per time unit (for example, one with n processors operating in parallel) can compute all the $F_i(X_i(t))$ at $t = t_0$ successfully. A computer capable of fewer than n operations per time unit fails to compute all the $F_i(X_i(t))$ at $t = t_0$. Indeed, consider a computer M_{n-1} capable of $n - 1$ operations per time unit. M_{n-1} would compute $n - 1$ of the $F_i(X_i(t))$ at $t = t_0$ correctly. Without loss of generality, assume that it computes $F_i(X_i(t))$ at $t = t_0$ for $i = 1, 2, \ldots, n - 1$. Now one time unit would have passed, and when M_{n-1} attempts to compute $F_n(X_n(t_0))$, it would be forced to incorrectly compute $F_n(X_n(t_0 + 1))$.

Is computer M_n universal? Certainly not. For when the number of variables is $n + 1$, M_n fails to perform the simple computation presented in this section and involving time-varying variables. As stated earlier in Sect. 1.1.1, a succession of machines $M_{n+1}, M_{n+2}, \ldots$, each succeeds at one level only to be foiled at the next.

1.1.3 Consequences

The implication of this result to the theory of computation is akin to that of Gödel's incompleteness theorem to mathematics. In the same way as no finite set of axioms A_i can be complete, no computer C_i is universal that can perform a finite and fixed number of operations per time unit. This is illustrated in Fig. 1.1: For every set of axioms A_i there exists a statement G_{i+1} not provable in A_i, but provable in A_{i+1};

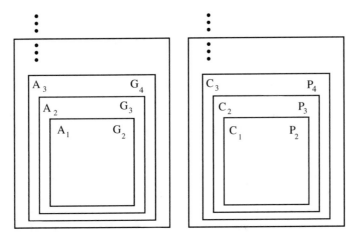

Fig. 1.1 Incompleteness in mathematics and nonuniversality in computation

similarly, for every machine C_i there is a problem P_{i+1} not solvable on C_i, but solvable on C_{i+1}.

The nonuniversality in computation result applies to any computer that obeys the *finiteness condition*, that is, a computer capable of only a finite and fixed number of operations per step (that is, per time unit). It should be noted that computers obeying the finiteness condition include all "reasonable" models of computation, both theoretical and practical, such as the Turing Machine, the Random Access Machine, and other idealized models [53], as well as all of today's general-purpose computers, including existing conventional computers (both sequential and parallel), and contemplated unconventional ones such as biological and quantum computers [5]. It is true for computers that interact with the outside world in order to read input and return output (unlike the Turing Machine, but like every realistic general-purpose computer). It is also valid for computers that are given unlimited amounts of time and space in order to perform their computations (like the Turing Machine, but unlike realistic computers). Even accelerating machines that increase their speed at every step (such as doubling it, or squaring it, or any such fixed acceleration) at a rate that is set in advance, cannot be universal! (Divine intervention, in the form of the Oracle Machine [61], for example, is clearly beyond the scope of the present discussion.)

The only constraint that we have placed on the computer (or model of computation) that claims to be universal is that the number of operations of which it is capable per time unit be finite and fixed once and for all. In this regard, it is important to note that:

1. The requirement that the number of operations per time unit, or step, be *finite* is necessary for any "reasonable" model of computation; see, for example, [56], p. 141.

2. The requirement that this number be *fixed* once and for all is necessary for any
   model of computation that purports to be "universal"; see, for example, [25], p.
   210.

Without these two requirements, the theory of computation in general, and the theory
of algorithms, in particular, would be totally irrelevant.

The consequences to theoretical and practical computing are significant. Thus
the conjectured "Church–Turing Thesis" is false. It is no longer true that, given
enough time and space, any single general-purpose computer, defined a priori, can
perform all computations that are possible on all other computers. Not the Turing
Machine, not a personal computer, not the most powerful of supercomputers. In view
of the computational problems mentioned in Sect. 1.1.1 and presented more fully in
Sect. 1.3 (for details, see [4–15, 18–20, 29, 45–50]), the only possible universal
computer would be one capable of an infinite number of operations per time unit.

In fact, this work has led to the discovery of computations that can be performed
on a quantum computer but not on any classical machine (even one with infinite
resources), thus showing for the first time that the class of problems solvable by
classical means is a true subset of the class of problems solvable by quantum means
[47]. Consequently, the only possible universal computer would have to be quantum
(as well as being capable of an infinite number of operations per time unit).

1.1.4 Motivation

The results described in the previous paragraphs were obtained and published
between 2005 and 2015. They appeared in technical reports, were presented at conferences, and published as journal papers and book chapters, as documented in the
bibliography section. Yet, ten years after they were first announced, they remain
generally misunderstood, and consequently controversial. This is not surprising: Old
habits die hard. It is difficult to abandon ideas that one has believed all of one's
life. This applies to inherited scientific ideas, that people persist in defending, even
if these ideas are demonstrably wrong. Disruptive innovations usually disturb and
destabilize. To announce that the long-held belief in universality in computation is,
in fact, in error was met with great skepticism, to say the least. Most of the animosity
towards the author came from scientists who had not bothered to read his work. The
title of a paper alone was sufficient to ignite their anger.

It started with hostile email. Then came the uninspired referee reports. And finally
Wikipedia removed the entry on Nonuniversality in Computation, created by a student
of the author, after one reader objected. It was clear that these people had not read
any of my work on the subject. This is when I decided to write this chapter, based
on a web page [16] I had created to reply to unfounded criticisms.

The remainder of this chapter is organized as follows. Section 1.2 provides definitions for many of the terms used in the discussion to follow (some of the more
common of these terms were already introduced in this section without definition, in

order not to interrupt the flow of the opening paragraphs). Section 1.3 summarizes the work in my previous papers, culminating in the fundamental result that the 'Universal Computer' is a myth. Section 1.4 is a dialog with an imaginary debater who embodies all of my critics and voices all their misconceptions. Section 1.5 offers a computational challenge to the readers.

1.2 Preliminaries

A *time unit* is the length of time required by a processor to perform a *step* of its computation, consisting of three elementary operations: a *read* operation in which it receives a constant number of fixed-size data as input, a *calculate* operation in which it performs a fixed number of constant-time *arithmetic* and *logical* calculations (such as adding two numbers, comparing two numbers, and so on), and a *write* operation in which it returns a constant number of fixed-size data as output.

A *sequential computer*, consists of a single processor. A *parallel computer* has n processors, numbered 1 to n, where $n \geq 2$. Both computers use the same type of processor and that processor is the fastest possible [2]. The assumption that the computers on hand, whether sequential or parallel, use the fastest conceivable processor is an important one. This is because the speed of the processor used is what specifies the duration of a time unit, as defined in the previous paragraph; the faster the processor, the smaller the time unit. In the time-varying variables computation of Sect. 1.1.2, for example, the value of each input variable $X_i(t)$ changes at the same speed as the processor in charge of evaluating $F_i(X_i(t))$.

The concept of *computational universality* is one of the dogmas in Computer Science, stated as follows: Given enough time and space, any general-purpose computer can, through simulation, perform any computation that is possible on any other general-purpose computer. Statements such as this are commonplace in the computer science literature, and are served as standard fare in undergraduate and graduate courses alike. Sometimes the statement is restricted to the Turing Machine (TM), and is referred to as the "Church–Turing Thesis" (CTT), which stipulates that there exists a Universal Computer, namely, the TM; thus, the following quote is typical:

A Turing Machine can do everything that a real computer can do [56], p. 125.

Other times, the statement is made more generally about a *Universal Computer*, that is, a computer capable of executing any computation that can be performed by any other computer; thus, we are told:

It is possible to build a universal computer: a machine that can be programmed to perform any computation that any other physical object can perform [25], p. 134.

As its name indicates, the CTT is a *conjecture*, whose proof has remained elusive due to the difficulty in defining what it means *to compute*. A few points are worth making in this regard:

1. The *Universal Turing Machine* (UTM) defined by Alan Turing, is 'universal' in the sense of being able to simulate any computation performed by any other special-purpose TM. In other words, the UTM can be programmed to execute any TM computation. This is a provable property, and is not in question here or elsewhere.
2. However, the UTM is not known to be 'universal' in the more general sense of being able to simulate by its own means any computation performed by means of any other computational device (not necessarily a TM). It is this universality of the UTM, in the more general sense, that is expressed by the CTT.
3. My work has shown the CTT, and more generally computational universality, to be false. Specifically, I proved the stronger result that no 'Universal Computer' is possible, thus eliminating, not only the UTM, but also all other models of computation as contenders for the title.

1.3 Inherently Parallel Computations

The term *inherently parallel computation* refers to a computation that can be performed successfully only on a parallel computer with an appropriate number of processors. Examples are computations that involve: time-varying variables, time-varying computational complexity, rank-varying computational complexity, interacting variables, uncertain time constraints, and variables obeying mathematical constraints. These computations also provide counterexamples to the existence of a universal computer [4–15, 18–20, 29, 45–50].

1.3.1 Time-Varying Variables

This computation was described in Sect. 1.1.2. It is reproduced briefly here for completeness in the context of inherently parallel computations. For a positive integer n larger than 1, we are given n functions, each of one variable, namely, $f_1, f_2, \ldots, f_n$, operating on the n physical variables $x_1, x_2, \ldots, x_n$, respectively. Specifically, it is required to compute $f_i(x_i)$, for $i = 1, 2, \ldots, n$. For example, $f_i(x_i)$ may be equal to x_i^2. What is unconventional about this computation, is the fact that the x_i are themselves (unknown) functions $x_1(t), x_2(t), \ldots, x_n(t)$, of the time variable t. It takes one time unit to evaluate $f_i(x_i(t))$. The problem calls for computing $f_i(x_i(t))$, $1 \leq i \leq n$, at time $t = t_0$. Because the function $x_i(t)$ is unknown, it cannot be inverted, and for $k > 0$, $x_i(t_0)$ cannot be recovered from $x_i(t_0 + k)$.

A sequential computer fails to compute all the f_i as desired. Indeed, suppose that $x_1(t_0)$ is initially operated upon. By the time $f_1(x_1(t_0))$ is computed, one time unit would have passed. At this point, the values of the $n - 1$ remaining variables would have changed. The same problem occurs if the sequential computer attempts to first read all the x_i, one by one, and store them before calculating the f_i.

By contrast, a parallel computer consisting of n independent processors may perform all the computations at once: For $1 \leq i \leq n$, and all processors working at the same time, processor i computes $f_i(x_i(t_0))$, leading to a successful computation.

1.3.2 Time-Varying Computational Complexity

Here, the computational complexity of the problems at hand depends on *time* (rather than being, as usual, a function of the problem *size*). Thus, for example, tracking a moving object (such as a spaceship racing towards Mars) becomes harder as it travels away from the observer.

Suppose that a certain computation requires that n independent functions, each of one variable, namely, $f_1(x_1)$, $f_2(x_2)$, ..., $f_n(x_n)$, be computed. Computing $f_i(x_i)$ at time t requires $C(t) = 2^t$ operations, for $t \geq 0$ and $1 \leq i \leq n$. Further, there is a strict deadline for reporting the results of the computations: All n values $f_1(x_1)$, $f_2(x_2)$, ..., $f_n(x_n)$ must be returned by the end of the third time unit, that is, when $t = 3$.

It should be easy to verify that no sequential computer, capable of exactly one constant-time operation per step (that is, per time unit), can perform this computation for $n \geq 3$. Indeed, $f_1(x_1)$ takes $C(0) = 2^0 = 1$ time unit, $f_2(x_2)$ takes another $C(1) = 2^1 = 2$ time units, by which time three time units would have elapsed. At this point none of $f_3(x_3), \ldots, f_n(x_n)$ would have been computed. By contrast, an n-processor parallel computer solves the problem handily. With all processors operating simultaneously, processor i computes $f_i(x_i)$ at time $t = 0$, for $1 \leq i \leq n$. This consumes one time unit, and the deadline is met.

1.3.3 Rank-Varying Computational Complexity

Suppose that a computation consists of n stages. There may be a certain precedence among these stages, or the n stages may be totally independent, in which case the order of execution is of no consequence to the correctness of the computation. Let the *rank* of a stage be the order of execution of that stage. Thus, stage i is the ith stage to be executed. Here we focus on computations with the property that the number of operations required to execute stage i is $C(i)$, that is, a function of i only.

When does rank-varying computational complexity arise? Clearly, if the computational requirements grow with the rank, this type of complexity manifests itself in those circumstances where it is a disadvantage, whether avoidable or unavoidable, to being ith, for $i \geq 2$. For example, the precision and/or ease of measurement of variables involved in the computation in a stage s may decrease with each stage executed before s.

The same analysis as in the previous section applies by substituting the rank for the time.

1.3.4 Interacting Variables

A physical system has n variables, $x_1, x_2, \ldots, x_n$, each of which is to be measured or set to a given value at regular intervals. One property of this system is that measuring or setting one of its variables modifies the values of any number of the system variables unpredictably.

A sequential computer measures *one* of the values (x_1, for example) and by so doing it disturbs the equilibrium, thus losing all hope of recording the state of the system within the given time interval. Similarly, the sequential approach cannot update the variables of the system properly: Once x_1 has received its new value, setting x_2 disturbs x_1 unpredictably.

A parallel computer with n processors, by contrast, will measure *all* the variables $x_1, x_2, \ldots, x_n$ simultaneously (one value per processor), and therefore obtain an accurate reading of the state of the system within the given time frame. Consequently, new values $x_1, x_2, \ldots, x_n$ can be computed in parallel and applied to the system simultaneously (one value per processor).

1.3.5 Uncertain Time Constraints

In this paradigm, we are given a computation each of whose components, namely, the input phase, the calculation phase, and the output phase, needs to be computed by a certain deadline. However, unlike the standard situation in conventional computation, the deadlines here are not known at the outset. In fact, and this is what makes this paradigm truly unconventional, we do not know at the moment the computation is set to start, *what* needs to be done, and *when* it should be done. Certain physical parameters, from the external environment surrounding the computation, become spontaneously available. The values of these parameters, once received from the outside world, are then used to evaluate two functions, call them f_1 and f_2, that tell us precisely *what* to do and *when* to do it, respectively.

The difficulty posed by this paradigm is that the evaluation of the two functions f_1 and f_2 is itself quite demanding computationally. Specifically, for a positive integer n, the two functions operate on n variables (the physical parameters). Only a parallel computer equipped with n processors can succeed in evaluating the two functions on time to meet the deadlines.

1.3.6 Computations Obeying Mathematical Constraints

There exists a family of computational problems where, given a mathematical object satisfying a certain property, we are asked to transform this object into another which also satisfies the same property. Furthermore, the property is to be maintained throughout the transformation, and be satisfied by every intermediate object, if any. More generally, the computations we consider here are such that every step of the computation must obey a certain predefined constraint. (Analogies from popular culture include picking up sticks from a heap one by one without moving the other sticks, drawing a geometric figure without lifting the pencil, and so on.)

1.3.6.1 Rewriting Systems

An example of such transformations is provided by *rewriting systems*. From an initial string ab, in some formal language consisting of the two symbols a and b, it is required to generate the string $(ab)^n$, for $n > 1$. Thus, for $n = 3$, the target string is $ababab$. The rewrite rules to be used are: $a \rightarrow ab$ and $b \rightarrow ab$. Throughout the computation, no intermediate string should have two adjacent identical characters. Such rewrite systems (also known as $\mathcal{L}$-systems) are used to draw fractals and model plant growth [51]. Here we note that applying any *one* of the two rules at a time causes the computation to fail (for example, if ab is changed to abb, by the first rewrite rule, or to aab by the second).

A sequential computer can change only one symbol at once, thereby causing the adjacency condition to be violated. By contrast, for a given n, a parallel computer with n processors can easily perform a transformation on all the inputs collectively. The required property is maintained leading to a successful computation. Thus, the string $(ab)^n$ is obtained in $\log n$ steps, with the two rewrite rules being applied simultaneously to all symbols in the current intermediate string, in the following manner: $ab, abab, abababab$, and so on. It is interesting to observe that a successful generation of $(ab)^n$ also provides an example of a rank-varying computational complexity (as described in Sect. 1.3.3). Indeed, each legal string (that is, each string generated by the rules and obeying the adjacency property) is twice as long as its predecessor (and hence requires twice as many operations to be generated).

1.3.6.2 Sorting Variant

A second example of computations obeying a mathematical constraint is provided by a variant to the problem of sorting. For a positive even integer n, where $n \geq 8$, let n distinct integers be stored in an array A with n locations $A[1], A[2], \ldots, A[n]$, one integer per location. Thus $A[j]$, for all $1 \leq j \leq n$, represents the integer currently stored in the jth location of A. It is required to sort the n integers in place into increasing order, such that:

1. After step i of the sorting algorithm, for all $i \geq 1$, no three consecutive integers satisfy:

$$A[j] > A[j + 1] > A[j + 2] , \tag{1.2}$$

   for all $1 \leq j \leq n - 2$.

2. When the sort terminates we have:

$$A[1] < A[2] < \cdots < A[n]. \tag{1.3}$$

This is the standard sorting problem in computer science, but with a twist. In it, the journey is more important than the destination. While it is true that we are interested in the outcome of the computation (namely, the sorted array, this being the *destination*), in this particular variant we are more concerned with *how* the result is obtained (namely, there is a condition that must be satisfied throughout all steps of the algorithm, this being the *journey*). It is worth emphasizing here that the condition to be satisfied is germane to the problem itself; specifically, there are no restrictions whatsoever on the model of computation or the algorithm to be used. Our task is to find an algorithm for a chosen model of computation that solves the problem exactly as posed. One should also observe that computer science is replete with problems with an inherent condition on how the solution is to be obtained. Examples of such problems include: inverting a nonsingular matrix without ever dividing by zero, finding a shortest path in a graph without examining an edge more than once, sorting a sequence of numbers without reversing the order of equal inputs, and so on.

An *oblivious* (that is, input-independent) algorithm for an $n/2$-processor parallel computer solves the aforementioned variant of the sorting problem handily in n steps, by means of predefined pairwise swaps applied to the input array A, during each of which $A[j]$ and $A[k]$ exchange positions (using an additional memory location for temporary storage) [2]. A sequential computer, and a parallel computer with fewer than $(n/2) - 1$ processors, both fail to solve the problem consistently, that is, they fail to sort all possible $n!$ permutations of the input while satisfying, at every step, the condition that no three consecutive integers are such that $A[j] > A[j + 1] > A[j + 2]$ for all j. In the particularly nasty case where the input is of the form

$$A[1] > A[2] > \cdots > A[n] , \tag{1.4}$$

any sequential algorithm and any algorithm for a parallel computer with fewer than $(n/2) - 1$ processors fail after the first swap.

It is interesting to note here that a Turing Machine with $n/2$ heads succeeds in solving the problem, yet its simulation by a standard (single-head) Turing Machine fails to satisfy the requirements of the problem. Indeed, suppose that the standard Turing Machine is presented with the input sequence $A[1] > A[2] > \cdots > A[n]$:

1. It will either use the given representation of the input, and proceed to perform an operation (a swap, for example), in which case it would fail after one step of the algorithm,
2. Or it will transform the given representation into a different encoding (perhaps one intended to capture the behavior of the Turing Machine with $n/2$ heads) in preparation for the sort, in which case it would again fail since the transformation itself will consist of more than one algorithmic step.

This is a surprising result as it goes against the common belief that any computation by a variant of the Turing Machine can be effectively simulated by the standard model [40].

1.3.7 The Universal Computer Is a Myth

The Principle of Simulation is the cornerstone of computer science. It is at the heart of most theoretical results and practical implements of the field such as programming languages, operating systems, and so on. The principle states that *any* computation that can be performed on *any* one general-purpose computer can be equally carried out through simulation on *any* other general-purpose computer [25, 32, 44]. At times, the imitated computation, running on the second computer, may be faster or slower depending on the computers involved. In order to avoid having to refer to different computers when conducting theoretical analyses, it is a generally accepted approach to define a model of computation that can simulate *all* computations by other computers. This model would be known as a Universal Computer $\mathcal{U}$. Thus, Universal Computation, which clearly rests on the Principle of Simulation, is also one of the foundational concepts in the field [23].

Our purpose here is to prove the following general statement: There does not exist a *finite* computational device that can be called a Universal Computer. Our reasoning proceeds as follows. Suppose there exists a Universal Computer capable of n elementary operations per step, where n is a finite and fixed integer. This computer will fail to perform a computation *requiring* n' operations per step, for any $n' > n$, and consequently lose its claim of universality. Naturally, for each $n' > n$, another computer capable of n' operations per step will succeed in performing the afore-mentioned computation. However, this new computer will in turn be defeated by a problem requiring $n'' > n'$ operations per step.

This reasoning is supported by each of the computational problems presented in Sects. 1.3.1–1.3.6. As we have seen, these problems *can* easily be solved by a computer capable of executing n operations at every step. Specifically, an n-processor parallel computer led to a successful computation in each case. However, *none* of these problems is solvable by any computer capable of at most $n - 1$ operations per step, for any integer $n > 1$. Furthermore, the problem size n itself is a variable that changes with each problem instance. As a result, *no* parallel computer, regardless of how many processors it has available, can cope with a growing problem size, as long as the number of processors is finite and fixed. This holds even if the computer purporting to be universal is endowed with an unlimited memory and is allowed to compute for an indefinite amount of time.

Therefore, the Universal Computer $\mathcal{U}$ is clearly a myth. As a consequence, the Principle of Simulation itself (though it applies to most *conventional* computations) is, in general, a fallacy. In fact, the latter principle is responsible for many other myths in computing, such as the *Speedup Theorem*, the *Slowdown Theorem*, and *Amdahl's Law*. Counterexamples for dispelling these and other myths are presented in [3, 21].

1.4 Misconceptions and Replies

> Almost all the other fellows do not look from the facts to the theory but from the theory to the
> facts; they cannot get out of the network of already accepted concepts; instead, comically,
> they only wriggle about inside [27].

What follows are fifteen misconceptions relating to my nonuniversality result, and
responses to them. They are presented as a dialog with an interlocutor who, I assume
(perhaps wishfully), has read my papers listed in the bibliography on the myth of
universal computation.

1.4.1 Misconception 1

Interlocutor: So, you describe a number of functions that are uncomputable. What's
new about that? Uncomputable functions have been known since the time of Turing.

Response: This is incorrect. The functions that I describe are eminently computable. In my papers listed in the bibliography, every function F^n of n variables can
be easily evaluated by a computer capable of at least n elementary operations per
time unit (an n-processor parallel computer [10], for example). However, a computer
capable of only $n - 1$ or fewer elementary operations per time unit cannot compute
F^n. Nonuniversality in computation immediately follows by simple induction.

1.4.2 Misconception 2

Interlocutor: You are proposing computations that cannot be simulated efficiently.
What's new about that? Your own book [2] and your earlier papers presented such
computations that can be performed in constant time by n processors, but require
time exponential in n when executed on $n - 1$ or fewer processors.

Response: The error in the preceding statement is in the phrase "cannot be simulated efficiently". Indeed, my nonuniversality result does not follow from computations that cannot be simulated efficiently. It follows from computations that *cannot
be simulated at all*. Thus, for each of the functions F^n of n variables that I describe
in my papers listed in the bibliography, no computer capable of fewer than n elementary operations per time unit can simulate the actions of a computer capable of
n elementary operations per time unit. The latter computer is capable of evaluating F^n successfully; the former is not capable of doing so, even if given an infinite
amount of time and memory space. It is this impossibility of simulation that leads to
nonuniversality.

In every one of my examples, the job simply cannot be done by simulation.
The world does not stand still while the simulator is taking its sweet time. In some
examples (the time-varying variables, say), physical time is the enemy: If a moment
is not grasped, it is gone forever, and no amount of simulation will help. In other

examples (the sorting variant, say), if a given condition is violated in the course of the simulation, the computation by the simulator is considered to have failed, regardless of whether the correct solution is eventually reached (simulating a plane landing after it has crashed is not useful to the unfortunate passengers, nor is simulating a surgery after the patient has died of any help to the deceased, and so on).

Simulation is no more than a "mathematical exercise" in such cases.

1.4.3   Misconception 3

Interlocutor: On the issue of simulation. You have added a dimension of time deadlines that is not present in the Turing Machine model. The "Church–Turing Thesis" (I thought) says that if a Turing Machine can do it in any finite amount of time, it is a computable function. However, if the job is to get the function computed by some deadline, then some machines are not up to the job. Is this the essence of what you are saying?

Response: I will start by saying that the Turing Machine confuses the issue because it is, by all measures, an inadequate model to capture what happens in the real world. The claims of universality that I aim to disprove have been made about much more powerful physical devices (e.g. modern computers, robots that move about their environment, sensing and control devices, and so on).

Now that the Turing Machine is out of the way, I will answer your question with a yes and a no:

1. *Yes*: I am describing unconventional computations, in some of which physical time plays a role. However, these are unlike the problems encountered in the field of real time computing, where deadlines are artificial ones imposed by the designer of the application (sometimes they are soft, sometimes they are hard, and they invariably refer to internal machine time). In my computations, nature essentially is in control; for example, a physical variable takes on a value at time unit t which is lost forever at time unit $t + 1$; two physical variables may interact causing one or both to lose their value unpredictably, unavoidably, and irreversibly; computational complexity grows with the passage of time; and so on.
2. *No*: Not all of my counterexamples to universality use time deadlines. See, for example, the problems described in Sect. 1.3.6 in which a mathematical constraint must be satisfied at every step of the computation.

1.4.4   Misconception 4

Interlocutor: What do you think about this way of attacking your time-varying variables problem:

First of all, we should separate the tasks into sensing and computation tasks. That means, I assume that there are sensors equipped with memory which read the values of the variables at time t and write them into appropriate memory locations. So, for every t we have memory locations $x_1(t)$, $x_2(t)$, ..., $x_n(t)$. Furthermore, assume that the values of n and t are stored in some other variables. Then, a universal machine would read the value of n (and t), read the values from memory, and perform the requested computations on these values, which is a rather simple task.

So, the main argument of this is that we allow for this separation. The sensors are peripheral components, which do not perform any computations, but 'just' provide the input for the computation, that is, they sense their values simultaneously at time t and store them in their respective memory locations, and the computations can access these values later on.

I guess that you will argue that this requires n values to be stored in the memory of the machine at the same time, which would contradict a specification of a machine which is independent of n. But if we can reduce the problem you posed to this separation of concerns, we would have the consistence with the traditional theory of computation except for the addition of these sensors, which are not considered to be a part of the universal computing device but of the input specification.

Response: Surely, you cannot "separate" part of the universal computer (in this case the input unit) from the rest of the computer just to fit the problem. The universal computer is one unit, and a computational step is: [read, calculate, write].

The definition of 'to compute' as 'the process of transforming information' applies to all three phases of computation, namely,

1. The input phase, where data such as keystrokes, mouse clicks, or temperatures are reduced, for example, to binary strings;
2. The calculation phase, where data are manipulated through arithmetic and logic operations, as well as operations on strings, and so on; and
3. The output phase, where data are produced as numbers on a screen, or rendered as images and sounds, for instance.

Each of the three phases represents information transformation; each is an integral part of every computation, and no computation worthy of that name can exclude any of them. In particular, input and output are fundamental in the design and analysis of algorithms. Input-sensitive and output-sensitive computations often play an especially important role in deriving lower and upper bounds on algorithmic complexity.

One of the most dramatic illustrations of the interweaving of input and output with calculation is the well-known linear-time planarity testing algorithm [34]. In order to determine whether a graph with V vertices is planar, Step 1 of that algorithm avoids a potentially quadratic-time computation, by reading in the edges of the graph *one at a time* from the outside world; if there is one more edge than $3V - 6$, the graph is declared nonplanar.

But let's assume you have set up n sensors and succeeded in solving the problem. What happens when you discover, the next morning, that there are now, not n, but $n + 1$ inputs? Do you think it is a fair solution for you to go to the sensor shop, buy one more sensor and rush back to attach it to the extra input source and to the computer? And even if you did, what if by the time you return, the time t_0 at which the result is needed would have passed?

1.4.5 Misconception 5

Interlocutor: Your argument is great. It is simple and effective. However, it does not show that there is something *theoretically* flawed with the concept of a universal computer; it shows that a universal computer could never be physically realized. So the "Church–Turing Thesis" is all right if we take it that the space/time needed is infinite. Having said that, I imagine that the distinction between the theoretical and the implementation claim is often overlooked, which makes the proof integral in making that distinction very sharp indeed.

Response: Thank you for reading. But, sorry, I beg to differ. Even if given infinite space and infinite time (which we allow the Turing Machine anyway), no computer that you define once and for all (and are not allowed to change ever again) can solve the problems that I define.

The issue is not with infinite space and infinite time. The issue is: How many operations can you do in one slice of time? You have to define this for every theoretical (and of course practical) computer. Otherwise, analysis of algorithms becomes meaningless, and the running time of a program a void concept. For example, the Turing Machine can do one, two, or three operations per slice of time, depending on your definition, but it is always a finite number. It can read a symbol, then change state, or write a symbol, or move left or right or stay put. Then a new iteration starts. It cannot do an arbitrary number of these fundamental operations per slice of time (you may call the latter a step, an iteration, a time unit, and so on). In passing, I should say that it is this finiteness of the number of operations per time unit that caused the great success of computer science, making the Computer the greatest invention of the 20th century: A machine designed once and for all that can simulate any operation by another machine. In theory, you should be able to buy a computer and use it for the rest of your life, for it will never be faced with a computable function that it cannot compute. Or so we thought ...

Suppose you have defined your machine once and for all (it can be a Turing Machine or a Supercomputer, or the chip in your digital camera or that in your toaster). I will now give you a computable function that it fails to compute. One example of such a computation is the problem of operating on variables that change with time (as described in Sect. 1.3.1). Even if given all the time in the world from here to eternity, even if given all the space in the known Universe and beyond, you could not solve the problem, not even in theory with pen and paper. Why? Because you defined your machine and fixed it once and for all (as one should if one claims to have a Universal Computer that can simulate anything that another computer can compute). And herein lies the Achilles heel of the Universal Computer: I, as a designer of the problem, ask you for a definition of your purported Universal

Computer; I then concoct a computation that thwarts it. This computation, of course, can be easily carried out by another "Universal Computer", but then I'll give you another problem, and so on.

There is nothing here about implementation. I put no limits on what your machine can do, except for one: It must do a finite number of operations per time unit. This number can be as big as you want, it can even be defined by a function of time, but it must be fixed the moment you define your model of computation, and it cannot be infinite. Once you define it, you cannot change it. We can play the game with paper and pencil. I will always win.

I believe that the result is perfectly *theoretical* as well as being perfectly *practical*.

1.4.6 Misconception 6

Interlocutor: While I could not make much sense of your "proof" (right off the bat, I do not think 'spacially and temporally connected variables' is well defined, although I also think it suffers from many other flaws) I decided to take your challenge seriously, within the bounds and the scope of your restrictions, language and assumptions, in an attempt to define a reasonable sounding computational device that would solve the time-varying variables computation. I will assume that the variables appear each on a luminous display of some sort.

I am going to assume relativity (please let me know if spacially and temporally connected variables is supposed to mean something else?). There is space k between displays so that, when they update their values, the light from their screens arrive at my computational device at nearly, but not precisely the same time (determined by k and where my device is in the 'room'). I will move my computational device close to the displays so that I can exploit the hypotenuse the signals from each display must travel first to reach the sensor of my device. Then, it is simply a matter of setting the processing speed to a finite value around 'a constant $\times k$ meters / the speed of light.' In this way, my computational device is able to compute the function F^n as desired.

My device has one sensor. It only needs one sensor, since the signals from the displays come in sequentially. Remember that the displays are separated by a distance and the signal from each display, after they update, must travel this distance (at the speed of light). So my device uses its one sensor to read in a display and then it reuses that sensor some time later after the signal from the next display has reached it (some time later).

Response: Right off the bat, as is clear from my papers, the spatial and temporal relationship among the variables is such that:

1. The variables occupy the same region in physical space; specifically, the distance separating any two variables is no larger than a small constant (whose magnitude depends on the general paradigm under consideration);

2. The variables are constrained by a unique parameter representing true physical time.

Now turning to your solution, I would say it is original; unfortunately, I am afraid you are missing the point of the exercise.

The problems I pose are to be treated independently of specific physical properties. For example, the time-varying variables could be anything one would want them to be (atmospheric pressures for instance, or humidity readings, etc.), and they should not necessarily be on display (they would need to be acquired, that is, measured, first). Light rays, the basis of your argument, should play no role in the solution (for they may not help in general).

However, let's consider your one-sensor solution, assuming the variables are indeed displayed. The difficulty with such setups is that they inevitably break down at some point as the problem scales up. Specifically,

1. The dimensions of the sensor are fixed.
2. As the number of variables n, and hence displays, grows, the angle with the horizontal formed by a line from the furthest display to the sensor approaches 0. There will not be enough real estate on the sensor for a light ray from the furthest display to impinge upon.
3. Thus, the problem poser can make n sufficiently large so as to render the sensor useless.

And one more thing: How does your sensor handle multiple simultaneous (or perhaps overlapping) inputs from displays all equidistant from it?

1.4.7 Misconception 7

Interlocutor: Your definition of computation is unconventional.

Response: Absolutely, if one considers as *unconventional* the passage of time, or the interactions among the constituents of the universe subject to the laws of nature, or for that matter any form of information processing. In any case, my definition of computation may be unconventional, but it is not unrealistic.

Besides, it is important to realize that defining "to compute" as "what the Turing Machine does", which is quite commonly done (see, for example, [28]), leads to a logical reasoning that is viciously circuitous. If computation is what the Turing Machine does, then clearly the Turing Machine can compute anything that is computable. Furthermore, the "Church–Turing Thesis" would move immediately from the realm of "conjecture" to that of "theorem". The fact that it has not done so to this day, is witness to the uncertainty surrounding the widespread definition of "computation". As stated in Sect. 1.1.3, my counterexamples, by contrast, *disprove* the "Church–Turing Thesis".

1.4.8 Misconception 8

Interlocutor: But the Turing Machine was not meant for this kind of computation.

Response: Precisely. Furthermore, the nonuniversality result covers all computers, not just the Turing Machine.

1.4.9 Misconception 9

Interlocutor: Abstract models of computation do not concern themselves with input and output.

Response: This opinion is held by those who believe that computation is the process that goes from input to output, while concerning itself with neither input nor output (see, for example, [28] cited in Sect. 1.4.7). By analogy, one might say that eating is all about digestion, and a Moonlight Sonata interpretation is nothing but the hitting of piano keys. Perhaps.

Perhaps not. A model of computation is useful to the extent that it is a faithful reflection of reality, while being mathematically tractable. In computer science, input and output are not cumbersome details to be ignored; they are fundamental parts of the computation process, which must be viewed as consisting of three essential phases, namely, input, calculation, and output. A computer that does not interact with the outside world is useless. In that sense, the thermostat in your house is more powerful than the Turing Machine. Please remember that my result goes beyond the Turing Machine (which, by the way, is a very primitive model of computation). To ask computer scientists to stick to the Turing Machine and not to look beyond, would be as if physics stopped progressing beyond the knowledge of the Ancient Greeks.

In fact, looking ahead, if computers of the future are to be *quantum* in nature, their main challenge is expected to be, not calculation but, input and output.

1.4.10 Misconception 10

Interlocutor: Classical computability theory does not include variables that change with time.

Response: This is a serious lacuna in classical theory. My work shows that there are in fact many such lacunae. Their result is to severely restrict the definition of computation. Indeed, to define computation merely as function evaluation, with fixed input and fixed output, is unrealistic and naive. It also trivializes the "Church–Turing Thesis" (turning it into a tautology), because it necessarily leads to the kind of sterile circular reasoning mentioned in the response to Sect. 1.4.7.

Having said that, the time-varying variables counterexample is only one of many such refutations of universality in computation. Other examples arise in the computations described in Sect. 1.3 (for details, see [4–15, 18–20, 29, 45–50]).

1.4.11 Misconception 11

Interlocutor: The Church–Turing Thesis applies only to classical computations.

Response: This is certainly not the case. As the multitude of examples listed in [18] amply demonstrate, the commonly accepted statement of the "Church–Turing Thesis" [39] is essentially this:

> There exists a Universal Computer capable of performing any computation that is possible on any other computer.

There are no restrictions, exceptions, or caveats whatsoever on the definition of computation. In fact, a typical textbook definition of computation is as follows:

> A computation is a process that obeys finitely describable rules [52].

What's more, it is suggested in every textbook on the subject that, thanks to the fundamental and complementary notions of simulation and universality, every general-purpose computer is universal: A Turing Machine, a Random-Access Machine, a Personal Computer, a Supercomputer, the processing chip in a cell phone, are all universal. (My result shows this claim to be *false*.)

Going a little further, many authors consider all processes taking place in the Universe as computations. Interested readers may consult [13, 22, 25, 26, 30, 38, 41, 42, 52, 54, 55, 59, 60, 62, 64–68].

I happen to agree with the authors listed in the previous paragraph on the pervasive nature of computation. However, in order to reach my conclusion about nonuniversality in computation, I do not in fact need to go that far. My counterexamples use very simple computations to refute universality: computations whose variables change with time, computations whose variables affect one another, computations the complexity of whose steps changes with the passage of time, computations the complexity of whose steps depends (not on the time but) on the order of execution of each step, computations with uncertain time constraints, computations that involve variables obeying certain mathematical conditions, and so on. These are not uncommon computations. They may be considered unconventional only when contrasted with the standard view of computation as a rigid function evaluation, in which, given a variable x, it is required to compute $f(x)$, in isolation of the rest of the world.

1.4.12   Misconception 12

Interlocutor: Actually, Alan Turing knew that computations involving physical time would cause problems for his machine.

Response: This is the back-pedaling argument *par excellence*. Perhaps Turing knew about the computations that I describe here, but I doubt it.

Indeed, Turing did propose the Oracle Machine (o-machine) [61] as an (unconventional) extension to his a-machine (today known as the Turing Machine). However, the o-machine has nothing to do with the problems I propose to counter universality in computation. Furthermore, the o-machine is a fanciful creation that appeals to some form of divine intervention in order to solve the problems it faces, while the computations that I use to prove nonuniversality are eminently executable on everyday computers, provided the latter are capable of performing the proper number of basic operations per time unit.

The "Church–Turing Thesis" [39] is proof that both Turing and Church were convinced of the universality of the Turing Machine. In any case, I am not aware of any writings by Turing, von Neumann, or anybody else, that hint to nonuniversality in computation, prior to my January 2005 paper on the myth of the universal computer.

This type of argument, exemplified by Misconception 12, reminds me of the words of the American philosopher William James [37]:

> First, you know, a new theory is attacked as absurd; then it is admitted to be true, but obvious and insignificant; finally it is seen to be so important that its adversaries claim that they themselves discovered it.

1.4.13   Misconception 13

Interlocutor: The problem I see with the nonuniversality claim is that it does not appear to be falsifiable. It might be improved by thinking further about how to make the claim more specific, clear, testable, well-defined and worked out into detail. The entire claim would ideally be described in a few very worked out, specific, sentences—needing little external exposition or justification.

Response: Have you had a chance to read any of my papers? There is no lack of mathematical formalism there. For example in one of my papers, M. Nagy and I used quantum mechanics to illustrate five of the six aforementioned paradigms [48]. I am certain you will enjoy that one.

Or how about the best understood problem in computer science, arguably that of sorting a sequence of numbers, but with a twist? See Sect. 1.3.6.2 for a description.

But let me try this since you would like something simple and crisp. Consider a computation C_2 consisting of two processes P_0 and P_1 and two global variables x and y, both initially equal to 0, as shown below:

C_2:
P_0: **if** $x = 0$ **then** $y \leftarrow y + 1$; return y **else** loop forever
P_1: **if** $y = 0$ **then** $x \leftarrow x + 1$; return x **else** loop forever ■

Computation C_2 is time critical and specifies that the **then** part of each **if** statement must be executed immediately when x is found to be 0, and immediately when y is found to be 0, respectively, and no later. Similarly, for the **else** part. This condition can be formalized by introducing global time as a variable in the above formulation, a detail omitted here for simplicity.

A two-processor machine M_2 does the job and returns x and y. A single-processor machine M_1 completes P_0 (alternatively, P_1) but loops forever when attempting P_1 (alternatively, P_0). As well, when trying to simulate M_2's behavior through an interleaved execution (such as for example: $\{x = 0?; y = 0?; y \leftarrow y + 1; x \leftarrow x + 1\}$), M_1 again fails since at least one assignment statement is executed more than one time unit too late. Incidentally, $\{x = 0?; x \leftarrow x + 1; y = 0?; y \leftarrow y + 1\}$, is obviously not a legitimate simulation, as x is incremented without first checking the value of y.

It is clear that M_1 cannot be a universal machine, for it is not capable of properly simulating the actions of M_2. But neither is M_2 universal, for it cannot execute the time-critical computation C_3 below and terminate:

C_3:
P_0: **if** $x = 0$ **then** $y \leftarrow y + 1$; return y **else** loop forever
P_1: **if** $y = 0$ **then** $z \leftarrow z + 1$; return z **else** loop forever
P_2: **if** $z = 0$ **then** $x \leftarrow x + 1$; return x **else** loop forever ■

It is easy to see that C_3 is carried out successfully by a three-processor machine M_3. The latter, however, is in turn defeated by a computation C_4 of the same type as C_2 and C_3.

In general, simple induction shows that no machine M_i is universal for finite i.

1.4.14 Misconception 14

Interlocutor: A fine student just presented your NonUniversality in Computation work in my graduate theory course. Going simply on her presentation, I am not impressed. You start by quoting Hopcroft and Ullman and saying that their statement is clearly false. You took this quote out of the context of all of theoretical computer science which clearly defines that a 'computation' is to start with the entire input presented and is given as much time as it wants to read this input and do its computation. It is true that since Turing, the nature of computation has changed to require real time interactions with the world. But you should not misrepresent past work. Having not studied your arguments at length, the only statement that I have gotten is that a machine is unable to keep up if you only allowed it to read in a fixed number of bits of information per time step and you throw at it in real time an arbitrarily large number of bits of information per time step. In itself, this statement is not very deep.

Response: I am sure the student did a wonderful job, but somehow the message did not get across. I will take your remarks one by one. They are repeated below as Remark A, Remark B, and Remark C.

Remark A: You took this quote out of the context of all of theoretical computer science which clearly defines that a 'computation' is to start with the entire input presented and is given as much time as it wants to read this input and do its computation.

Response to Remark A:

1. I did not find your definition of 'computation' anywhere. It is certainly not in Hopcroft and Ullman's book [35], from which the quote in question was taken, namely:

   > It can also be shown that any computation that can be performed on a modern-day digital computer can be described by means of a Turing Machine. Thus if one ever found a procedure that fitted the intuitive notions, but could not be described by means of a Turing Machine, it would indeed be of an unusual nature since it could not possibly be programmed for any existing computer [35], p. 80.

2. Your definition of 'computation' applies narrowly to some primitive computational models, such as the Turing Machine, Cellular Automata, and so on.

3. Because of (2), your definition trivializes the "Church–Turing Thesis", rendering it a tautology: By defining 'computation' as 'what the Turing Machine does', it obviously follows that 'the Turing Machine can compute everything that is computable'. As mentioned earlier (see the response to Misconceptions 7 and 10 in Sects. 1.4.7 and 1.4.10, respectively), this is a typical example of circular reasoning.

4. Your definition is not sufficiently general to capture the richness and complexity of the notion of 'computation'. Others have proposed more encompassing definitions of 'computation'. Here are a few quotes:

   > In a sense Nature has been continually computing the 'next state' of the universe for billions of years; all we have to do - and actually, all we can do - is 'hitch a ride' on this huge ongoing computation [60].

   > Computation is physical; it is necessarily embodied in a device whose behaviour is guided by the laws of physics and cannot be completely captured by a closed mathematical model. This fact of embodiment is becoming ever more apparent as we push the bounds of those physical laws [57] (see also [58]).

   > Information sits at the core of physics ... every particle, every field of force, even the space-time continuum itself derives its function, its meaning, its very existence entirely ... from answers to yes-or-no questions ... all things physical are information-theoretic in origin [65].

   > Think of all our knowledge-generating processes, our whole culture and civilization, and all the thought processes in the minds of every individual, and indeed the entire evolving

biosphere as well, as being a gigantic computation. The whole thing is executing a self-motivated, self-generating computer program [25].

5. And one more thing about your point that a computation is to "start with the entire input presented". Since *all* of my counterexamples indeed require the entire input to be "present", I will assume that you mean "start with the entire input residing in the memory of the computer". (Incidentally, some of my counterexamples assume the latter as well.) The relevant point here is this: Hopcroft himself has a well-known algorithm that makes no such assumption about the entire input residing in memory. As stated in response to Misconception 4 in Sect. 1.4.4, one of the most dramatic illustrations of the interweaving of input and output with calculation is the well-known linear-time planarity testing algorithm of Hopcroft and Tarjan [34]. In order to determine whether a graph with n vertices is planar, Step 1 of that algorithm avoids a computation that could potentially run in time at least quadratic in n, by reading in the edges of the graph one at a time from the outside world; if there is one more edge than the absolute maximum stipulated by Euler's formula, namely $3n - 6$ (the paper actually uses the loose bound $3n - 3$ in order to allow for $n < 3$), the graph is declared nonplanar [34].

Remark B: It is true that since Turing, the nature of computation has changed to require real time interactions with the world. But you should not misrepresent past work.

Response to Remark B:

1. I am glad to see that you agree with me about the nature of computation. You should know, however, that my counterexamples to universality are not all about "real time interaction with the world". There is a list of such counterexamples in Sect. 1.1, a brief description of some in Sect. 1.3, and a list of references to my papers in the bibliography section. One counterexample involving mathematical constraints (it is a variant of sorting, in which the entire input is available in memory at the outset of the computation) is described in Sect. 1.3.6.2. Note also that my nonuniversality result applies to putative 'universal computers' capable of interaction with the outside world. These 'universal computers' are endowed with unlimited memories and are allowed an indefinite amount of time to solve the problems they are given. They all still fail the test of universality.

2. However, I do take exception to the claim that I misrepresented past work. Below are quotes from famous computer scientists asserting, without caveat, exception, or qualification that a universal computer is possible (often explicitly stating that such a universal computer is the Turing Machine, and essentially taking for granted the unproven "Church–Turing Thesis"):

> Church's thesis: The computing power of the Turing Machine represents a fundamental limit on the capability of realizable computing devices [24], p. 2.

> Anything which can be computed can be computed by a Turing Machine [1], p. 123.

It is theoretically possible, however, that Church's Thesis could be overthrown at some future date, if someone were to propose an alternative model of computation that was publicly acceptable as fulfilling the requirement of 'finite labor at each step' and yet was provably capable of carrying out computations that cannot be carried out by any Turing Machine. No one considers this likely [40], p. 223.

If we have shown that a problem can (or cannot) be solved by any TM, we can deduce that the same problem can (or cannot) be solved by existing mathematical computation models nor by any conceivable computing mechanism. The lesson is: Do not try to solve mechanically what cannot be solved by TMs! [43], p. 152.

Any algorithmic problem for which we can find an algorithm that can be programmed in some programming language, any language, running on some computer, any computer, even one that has not been built yet but can be built, and even one that will require unbounded amounts of time and memory space for ever-larger inputs, is also solvable by a Turing Machine [32], p. 233.

It is possible to build a universal computer: a machine that can be programmed to perform any computation that any other physical object can perform. Any computation that can be performed by any physical computing device can be performed by any universal computer, as long as the latter has sufficient time and memory [33], pp. 63–64.

Hundreds of such statements can be found in the literature; for a sample see [18].

3. Finally, please note that to correct previous mistakes is not to misrepresent the past. This is how science advances. Newton, Darwin, and Einstein were giants who built great scientific edifices. Each edifice, magnificent yet incomplete.

Remark C: Having not studied your arguments at length, the only statement that I have gotten is that a machine is unable to keep up if you only allowed it to read in a fixed number of bits of information per time step and you throw at it in real time an arbitrarily large number of bits of information per time step. In itself, this statement is not very deep.

Response to Remark C:

1. As mentioned in Sect. 1.4.10, the "time-varying variables" computation is but one of many counterexamples to universality.
2. Surely you must know that a reasonable model of computation must be finite. The Turing Machine has a finite alphabet, a finite set of states, and a finite set of elementary operations per step. To assume otherwise would render the fields of complexity theory and algorithm design and analysis useless.
3. Your characterization of my result is erroneous. I describe *computable functions* that no machine that claims to be universal can compute.
4. Let's try the following: *You* define a "universal computer" fulfilling the requirement of 'finite labor at each step' (to quote [40] again), and I will give you a computable function that it cannot compute.

1.4.15 Misconception 15

Interlocutor: I still find myself puzzled by things you say. I have no dispute that there are computations with various constrains such as real time constraints, time-varying aspects, global math constraints, and the like. I know that we have to deal with such computations already on existing machines. But since existing machines are Turing computable, I am puzzled why these constraints force us outside the limits of Turing computability. Can you help me understand?

Response: Thank you for your continued interest and for engaging me in this constructive exchange. This is a privilege that most computer scientists not often afford me on this subject. I will try to address your question.

The Turing Machine of the 1930s as well as the 2015 machine on which I am typing this reply, operate in isolation of their environment. In particular, time for the problems we solve on these machines always means running time, and space means memory space. I dream of computations in which physical time and physical space play a central role, where the environment in which the computations take place affects the computations, and in turn is affected by them. Some of these computations, as you say, are starting to show up already. Many more will come as we expand the realm of computing.

It is important to note here that I am not just talking about interaction. Even if equipped with the ability to interact with the outside world, unlike the Turing Machine but like all computers today, a machine that claims to be universal will fail to solve the problems that I propose as counterexamples to universality.

How is this possible? Simply because physical time and physical space overwhelm any machine that satisfies the fundamental requirement of being fixed once and for all–the very definition of universality. Every problem I describe has a condition that must be satisfied in order for the computation to be said to have been carried out successfully. If this condition is not satisfied, the computation is judged to have failed, regardless of whether a correct output is later produced (by, for example, correcting the error, or by restarting the computation, or by simulation, and so on).

Assume that a computer is supposed to control the reconfiguration of a physical structure of size n. It is required that the structure maintain its integrity at every step, otherwise the structure will collapse, and the operation will be declared to have failed. A universal computer will fail inevitably for a certain structure of a certain size.

In particular, a standard Turing Machine cannot solve the problem in the previous paragraph. By contrast, a Turing Machine with n tapes (and n heads) can. Interestingly, this contradicts the general belief that any computation by any variant of the standard Turing Machine can be simulated on the latter.

1.5 Conclusion

Facts do not go away when scientists debate rival theories to explain them [31].

In conclusion, I offer the reader the following challenge.

Computational Challenge:

Anyone who still does not accept my result, namely, that universality in computation is a myth, has but one option: to prove it wrong. In order to do this, one must exhibit a universal computer capable of a finite and fixed number of operations per time unit, on which each one of the computations in the following classes, and described in [4–15, 18–20, 29, 45–50], can be performed:

1. Computations with time-varying variables
2. Computations with time-varying computational complexity
3. Computations with rank-varying computational complexity
4. Computations with interacting variables
5. Computations with uncertain time constraints
6. Computations with global mathematical constraints.

   It is important that the purported universal computer be able to execute successfully all aforementioned computations, since each one of them, by itself, is a counterexample to computational universality. For a simplified (and perhaps more colorful) version of the challenge, please see [17].

References

1. Abramsky, S., et al.: Handbook of Logic in Computer Science. Clarendon Press, Oxford (1992)
2. Akl, S.G.: Parallel Computation: Models and Methods. Prentice Hall, Upper Saddle River (1997)
3. Akl, S.G.: Superlinear performance in real-time parallel computation. J. Supercomput. **29**, 89–111 (2004)
4. Akl, S.G.: The myth of universal computation. In: Trobec, R., Zinterhof, P., Vajteršic, M., Uhl, A. (eds.) Parallel Numerics, pp. 211–236. University of Salzburg, Salzburg and Jozef Stefan Institute, Ljubljana (2005)
5. Akl, S.G.: Three counterexamples to dispel the myth of the universal computer. Parallel Proc. Lett. **16**, 381–403 (2006)
6. Akl, S.G.: Conventional or unconventional: is any computer universal? In: Adamatzky, A., Teuscher, C. (eds.) From Utopian to Genuine Unconventional Computers, pp. 101–136. Luniver Press, Frome (2006)
7. Akl, S.G.: Gödel's incompleteness theorem and nonuniversality in computing. In: Nagy, M., Nagy, N. (eds.) Proceedings of the Workshop on Unconventional Computational Problems, Sixth International Conference on Unconventional Computation, pp. 1–23. Kingston (2007)
8. Akl, S.G.: Even accelerating machines are not universal. Int. J. Unconv. Comp. **3**, 105–121 (2007)
9. Akl, S.G.: Unconventional computational problems with consequences to universality. Int. J. Unconv. Comp. **4**, 89–98 (2008)

10. Akl, S.G.: Evolving computational systems. In: Rajasekaran, S., Reif, J.H. (eds.) Parallel Computing: Models, Algorithms, and Applications, pp. 1–22. Taylor and Francis, Boca Raton (2008)
11. Akl, S.G.: Ubiquity and simultaneity: the science and philosophy of space and time in unconventional computation. Keynote address, Conference on the Science and Philosophy of Unconventional Computing, The University of Cambridge, Cambridge (2009)
12. Akl, S.G.: Time travel: a new hypercomputational paradigm. Int. J. Unconv. Comp. **6**, 329–351 (2010)
13. Akl, S.G.: What is computation? Int. J. Parallel, Emerg. Distrib. Syst. **29**, 337–345 (2014)
14. Akl, S.G.: Nonuniversality explained. Int. J. Parallel Emerg. Distrib. Syst. (To appear in)
15. Akl, S.G., Nagy, M.: Introduction to parallel computation. In: Trobec, R., Vajteršic, M., Zinterhof, P. (eds.) Parallel Computing: Numerics, Applications, and Trends, pp. 43–80. Springer, London (2009)
16. Akl, S.G.: Non-Universality in Computation: The Myth of the Universal Computer. School of Computing, Queen's University. http://research.cs.queensu.ca/Parallel/projects.html
17. Akl, S.G.: A computational challenge. School of Computing, Queen's University. http://www.cs.queensu.ca/home/akl/CHALLENGE/A_Computational_Challenge.htm
18. Akl, S.G.: Universality in computation: Some quotes of interest. Technical Report No. 2006-511, School of Computing, Queen's University. http://www.cs.queensu.ca/home/akl/techreports/quotes.pdf
19. Akl, S.G., Nagy: M.: The future of parallel computation. In: Trobec, R., Vajteršic, M., Zinterhof, P. (eds.) Parallel Computing: Numerics, Applications, and Trends, pp. 471-510. Springer, London (2009)
20. Akl, S.G., Salay, N.: On computable numbers, nonuniversality, and the genuine power of parallelism. Int. J. Unconv. Comput. **11**, 283–297 (2015)
21. Akl, S.G., Yao, W.: Parallel computation and measurement uncertainty in nonlinear dynamical systems. J. Math. Model. Alg. **4**, 5–15 (2005)
22. Davies, E.B.: Building infinite machines. Br. J. Phil. Sci. **52**, 671–682 (2001)
23. Davis, M.: The Universal Computer. W.W. Norton, New York (2000)
24. Denning, P.J., Dennis, J.B., Qualitz, J.E.: Machines, Languages, and Computation. Prentice-Hall, Englewood Cliffs (1978)
25. Deutsch, D.: The Fabric of Reality. Penguin Books, London (1997)
26. Durand-Lose, J.: Abstract geometrical computation for black hole computation. Research Report No. 2004-15, Laboratoire de l'Informatique du Parallélisme, École Normale Supérieure de Lyon, Lyon (2004)
27. Einstein, A.: Letter to Erwin Schrödinger. In: Gilder, L., The Age of Entanglement, p. 170. Vintage Books, New York (2009)
28. Fortnow, L.: The enduring legacy of the Turing machine. http://ubiquity.acm.org/article.cfm?id=1921573
29. Fraser, R., Akl, S.G.: Accelerating machines: a review. Int. J. Parallel, Emerg. Distrib. Syst. **23**, 81–104 (2008)
30. Gleick, J.: The Information: A History, a Theory, a Flood. HarperCollins, London (2011)
31. Gould, S.J.: Evolution as fact and theory. Discover **2**, 34–37 (1981)
32. Harel, D.: Algorithmics: The Spirit of Computing. Addison-Wesley, Boston (Reading (1992))
33. Hillis, D.: The Pattern on the Stone. Basic Books, New York (1998)
34. Hopcroft, J., Tarjan, R.: Efficient planarity testing. J. ACM **21**, 549–568 (1974)
35. Hopcroft, J.E., Ullman, J.D.: Formal Languages and their Relations to Automata. Addison-Wesley, Boston, Reading (1969)
36. Hypercomputation. http://en.wikipedia.org/wiki/Hypercomputation
37. James, W.: Pragmatism, A New Name for Some Old Ways of Thinking. Longman Green and Co., New York (1907)
38. Kelly, K.: God is the machine. Wired **10** (2002)
39. Kleene, S.C.: Introduction to Metamathematics. North Holland, Amsterdam (1952)
40. Lewis, H.R., Papadimitriou, C.H.: Elements of the Theory of Computation. Prentice Hall, Englewood Cliffs (1981)

41. Lloyd, S., Ng, Y.J.: Black Hole Comput. Sci. Am. **291**, 53–61 (2004)
42. Lloyd, S.: Programming the Universe. Knopf, New York (2006)
43. Mandrioli, D., Ghezzi, C.: Theoretical Foundations of Computer Science. Wiley, New York (1987)
44. Minsky, M.L.: Computation: Finite and Infinite Machines. Prentice-Hall, Englewood Cliffs (1967)
45. Nagy, M., Akl, S.G.: On the importance of parallelism for quantum computation and the concept of a universal computer. In: Calude, C.S., Dinneen, M.J., Paun, G., Pérez-Jiménez, M. de J., Rozenberg, G. (eds.) Unconventional Computation, pp. 176-190. Springer, Heildelberg (2005)
46. Nagy, M., Akl, S.G.: Quantum measurements and universal computation. Int. J. Unconv. Comput. **2**, 73–88 (2006)
47. Nagy, M., Akl, S.G.: Quantum computing: Beyond the limits of conventional computation. Int. J. Parallel Emerg. Distrib. Syst. **22**, 123–135 (2007)
48. Nagy, M., Akl, S.G.: Parallelism in quantum information processing defeats the Universal Computer. Par. Proc. Lett. **17**, 233–262 (2007)
49. Nagy, N., Akl, S.G.: Computations with uncertain time constraints: effects on parallelism and universality. In: Calude, C.S., Kari, J., Petre, I., Rozenberg, G. (eds.) Unconventional Computation, pp. 152–163. Springer, Heidelberg (2011)
50. Nagy, N., Akl, S.G.: Computing with uncertainty and its implications to universality. Int. J. Parallel Emerg. Distrib. Syst. **27**, 169–192 (2012)
51. Prusinkiewicz, P., Lindenmayer, A.: The Algorithmic Beauty of Plants. Springer, New York (1990)
52. Rucker, R.: The Lifebox, the Seashell, and the Soul. Thunder's Mouth Press, New York (2005)
53. Savage, J.E.: Models of Computation. Addison-Wesley, Boston, Reading (1998)
54. Seife, C.: Decoding the Universe. Viking Penguin, New York (2006)
55. Siegfried, T.: The Bit and the Pendulum. Wiley, New York (2000)
56. Sipser, M.: Introduction to the Theory of Computation. PWS Publishing Company, Boston (1997)
57. Stepney, S.: Journeys in non-classical computation. In: Hoare, T., Milner, R. (eds.) Grand Challenges in Computing Research, pp. 29–32. BCS, Swindon (2004)
58. Stepney, S.: The neglected pillar of material computation. Phys. D **237**, 1157–1164 (2004)
59. Tipler, F.J.: The Physics of Immortality: Modern Cosmology. God and the Resurrection of the Dead. Macmillan, London (1995)
60. Toffoli, T.: Physics and Computation. Int. J. Theor. Phys. **21**, 165–175 (1982)
61. Turing, A.M.: Systems of logic based on ordinals. Proc. Lond. Math. Soc. **2**(45), 161–228 (1939)
62. Vedral, V.: Decoding Reality. Oxford University Press, Oxford (2010)
63. Wegner, P., Goldin, D.: Computation beyond Turing Machines. Commun. ACM **46**, 100–102 (1997)
64. Wheeler, J.A.: Information, physics, quanta: The search for links. In: Proceedings of the Third International Symposium on Foundations of Quantum Mechanics in Light of New Technology, pp. 354-368. Tokyo (1989)
65. Wheeler, J.A.: Information, physics, quantum: the search for links. In: Zurek, W. (ed.) Complexity, Entropy, and the Physics of Information. Addison-Wesley, Redwood City (1990)
66. Wheeler, J.A.: At Home in the Universe. American Institute of Physics Press, Woodbury (1994)
67. Wolfram, S.: A New Kind of Science. Wolfram Media, Champaign (2002)
68. Zuse, K.: Calculating space. MIT Technical Translation AZT-70-164-GEMIT, Massachusetts Institute of Technology (Project MAC). Cambridge (1970)

Chapter 2
What Is Computable? What Is Feasibly Computable? A Physicist's Viewpoint

Vladik Kreinovich and Olga Kosheleva

Abstract In this chapter, we show how the questions of what is computable and what is feasibly computable can be viewed from the viewpoint of physics: what is computable within the current physics? what is computable if we assume—as many physicists do—that no final physical theory is possible? what is computable if we consider data processing, i.e., computations based on physical inputs? Our physics-based analysis of these questions leads to some unexpected answers, both positive and negative. For example, we show that under the no-physical-theory-is-perfect assumption, almost all problems are feasibly solvable—but not all of them.

2.1 What Is Computable? What Is Feasibly Computable? Different Aspects of These Questions

The two main questions of theoretical computer science. One of the main objectives of theoretical computer science is to answer the following two fundamental questions:

- The first question is: which tasks are computable in principle?
- Once we learned that a task is, in principle, computable, a natural next question is: is this task *feasibly* computable, i.e., can we perform the corresponding computations in reasonable time?

These questions are usually considered from the viewpoint of a computer scientist. From the viewpoint of a computer scientist, computation is a solution to a well-defined task, performed on a well-defined computational devices. As a result, when formulating and analyzing the above problems, computer scientists usually

V. Kreinovich (✉) · O. Kosheleva
University of Texas at El Paso, 500 W. University, El Paso, TX 79968, USA
e-mail: vladik@utep.edu

O. Kosheleva
e-mail: olgak@utep.edu

© Springer International Publishing Switzerland 2017
A. Adamatzky (ed.), *Advances in Unconventional Computing*,
Emergence, Complexity and Computation 22,
DOI 10.1007/978-3-319-33924-5_2

consider well-defined tasks and computations which consist of a sequence of well-defined elementary steps.

A physicist's understanding is somewhat different. Computer science is, after all, an applied discipline. From the practical viewpoint, we need computations to process data from the real world—so that we will be able to predict the future state of the world and, in situations when we can control this future state, to come up with actions that would result in the best possible outcome.

From this viewpoint, we can distinguish between two different types of computations:

- traditional computations, when we are trying to find a solution to a well-defined (= mathematical) problem, and
- data processing computations, when we process the data coming from the physical world.

Similarly, based on what computational devices we can use, we can distinguish between two possible approaches:

- a "purist" approach, when we are only allowed step-by-step computations on a well-defined computational device, and
- a pragmatic approach, when, in addition to computations, we can set up physical models of the analyzed systems, analog computations—whatever helps.

Thus, each of the two fundamental questions—what is computable? what is feasibly computable?—can be formulated in three different ways. Namely, in addition to the traditional formulation, when we consider computing well-defined mathematical tasks on well-defined computers, we can also consider:

- a pragmatic formulation, when, in addition to well-defined computers, we can use physical processes to help with computations, and, finally
- a data processing formulation when we are interested in processing physical data.

What we do in this chapter. In this chapter, we consider all these three approaches one by one, and we show that they lead to somewhat different answers to the fundamental questions of what is computable and what is feasible computable.

The structure of this chapter is as follow. The pragmatic formulation is discussed in Sects. 2.2 and 2.3, and the data processing formulation is discussed in Sect. 2.4.

2.2 Non-Standard Physical Processes Can Help Computations: Examples Based On Specific Physical Models

Solving NP-complete problems is important. In practice, we often need to find a solution that satisfies a given set of constraints–or at least check that such a solution is possible. Once we have a candidate for the solution, we can feasibly check whether this candidate indeed satisfies all the constraints. In theoretical computer science, "feasibly" is usually interpreted as computable in polynomial time, i.e., in time bounded by a polynomial of the length of the input.

A problem of checking whether a given set of constraints has a solution is called a problem of the class NP if we can check, in polynomial time, whether a given candidate is a solution; see, e.g., [26].

Examples of such problems include checking whether a given graph can be colored in 3 colors, checking whether a given propositional formula—i.e., formula of the type

$$(v_1 \lor \neg v_2 \lor v_3) \& (v_4 \lor \neg v_2 \lor \neg v_5) \& \ldots,$$

is satisfiable, i.e., whether this formula is true by some combination of the propositional variables v_i, etc.

Each problem from the class NP can be algorithmically solved by trying all possible candidates. For example, we can check whether a graph can be colored by trying all possible assignments of colors to different vertices of a graph, and we can check whether a given propositional formula is satisfiable by trying all 2^n possible combinations of true-or-false values $v_1, \ldots, v_n$. Such exhaustive search algorithms require computation time like 2^n, time that grows exponentially with n. For medium-size inputs, e.g., for $n \approx 300$, the resulting time is larger than the lifetime of the Universe. So, these exhaustive search algorithms are not practically feasible.

It is not known whether problems from the class NP can be solved feasibly (i.e., in polynomial time): this is a famous open problem $P \overset{?}{=} NP$. It is known, however, that there are problems in the class NP which are *NP-complete* in the sense that every problem from the class NP can be reduced to this problem. Reduction means, in particular, that if we can find a way to efficiently solve one NP-complete problem, then, by reducing other problems from the class NP to this problem, we can thus efficiently solve all the problems from the class NP.

So, it is very important to be able to efficiently solve even one NP-complete problem. (By the way, both above example of NP problems—checking whether a graph can be colored in 3 colors and whether a propositional formula is satisfiable—are NP-complete.)

Can the use of non-standard physics speed up the solution of NP-complete problems? NP-completeness of a problem means, crudely speaking, that the problem may take an unrealistically long time to solve—at least on computers based on the usual physical techniques. A natural question is: can the use of non-standard physics speed up the solution of these problems?

To answer this question, let us start the analysis of the corresponding physics.

Parallelization: a natural idea. If a person faces a task that would take too much time for him or her working alone—e.g., building a house—this person asks for help. Similarly, when a problem takes too much time to solve on a single computer, a natural idea is to have several computers working on this problem in parallel.

Physical limitations to parallelization speed-up. At first glance, potentially, by dividing the original problem into smaller and smaller pieces and using more and more processors to process these pieces, we can speed up the computation as much as possible.

In reality, however, there are physical limitations on the possible speed-up; see, e.g., [24]. Indeed, let us assume that we have a parallel algorithm that, for all inputs of bit length $\leq n$, solves the original problem in time $T_{par}(n)$.

The user is located at some point in space. The user inputs the problem at this spatial location, and the user expects the result of the computation to be delivered to the same spatial location. Each processor that participates in the desired computation must:

- get this signal (directly or indirectly) from the user's location, and then
- start some other signals that will eventually reach the user at his or her spatial location.

So, if a processor is located at distance r from the user, then the signal going from the user to the processor and back must cover the distance of at least $2r$.

According to modern physics, the speed of all communications is limited by the speed of light c. Thus, the smallest amount of time for this signal transmission is $\dfrac{2r}{c}$. If this time exceeds $T_{par}(n)$, this means that this processor is unable to contribute to the computation result. Thus, only processors for which $\dfrac{2r}{c} \leq T_{par}(n)$, i.e., for which $r \leq R(n) \overset{\text{def}}{=} \dfrac{1}{2} \cdot c \cdot T_{par}(n)$, contribute to the computation. So, we only need to consider processors which are located inside the sphere of radius $R(n)$ centered at the user.

How many processors can fit inside this sphere? A physical bound of the number $N_{proc}(n)$ of these processors can be obtained if we divide the volume $V(R(n))$ of the inside of this sphere by the smallest possible volume ΔV of a processor:

$$N_{proc} \leq \frac{V(R(n))}{\Delta V}.$$

In the Euclidean space, $V(R) = \dfrac{4}{3} \cdot \pi \cdot R^3$, so we conclude that

$$N_{proc}(n) \leq \frac{1}{\Delta V} \cdot \frac{4}{3} \cdot \pi \cdot (R(n))^3 = \frac{1}{\Delta V} \cdot \frac{4}{3} \cdot \pi \cdot \frac{1}{8} \cdot c^3 \cdot (T_{par}(n))^3,$$

i.e., that $N_{proc}(n) \leq \text{const} \cdot (T_{par}(n))^3$, where the multiplicative constant does not depend on the size n of the input.

We can always simulate parallel computations by $N_{par}(n)$ processors on a sequential machine: for this, for each original cycle of the parallel machine, we need to emulate how the state of each of $N_{proc}(n)$ processors change. In this simulation, one step of the original parallel machine requires $N_{proc}(n)$ steps of the simulating sequential machine. Thus, the overall time $T_{\mathrm{seq}}(n)$ of the corresponding sequential machine can be obtained by multiplying the original parallel time $T_{\mathrm{par}}(n)$ by the number of processors: $T_{\mathrm{seq}}(n) \leq T_{\mathrm{par}}(n) \cdot N_{\mathrm{proc}}(n)$. By using the bound $N_{\mathrm{proc}}(n) \leq \mathrm{const} \cdot (T_{\mathrm{par}}(n))^3$, we conclude that

$$T_{\mathrm{seq}}(n) \leq \mathrm{const} \cdot (T_{\mathrm{par}}(n))^4.$$

So, if a problem is difficult to solve on a sequential machine, and there is a huge lower bound on $T_{\mathrm{seq}}(n)$, then we can conclude that there is a related lower bound on the parallel time as well: $T_{\mathrm{par}}(n) \geq \mathrm{const} \cdot (T_{\mathrm{seq}}(n))^{1/4}$. In particular, if—as most computer scientists believe—an NP-complete problem cannot be solved faster than in exponential time $T_{\mathrm{seq}}(n) \geq 2^n$, then we get similar exponential lower bounds on the parallel time as well: $T_{\mathrm{par}}(n) \geq (\sqrt[4]{2})^n$. This is faster than 2^n, but still not feasible.

Important observation: these limitations depend on physics. The above limitations are based on the usual physics, where the space is Euclidean (so that the volume grows as a cube of the radius), and the speeds of all physical processors are limited by the speed of light.

However, it is well known that the actual space-time is different from Euclidean, it is *curved*; see, e.g., [5]. Also, physicists are seriously considering space-time models in which it is possible to exceed the speed of light; see, e.g., [29]. This leads to the possibility of potentially physically realistic situations in which we can solve NP-complete problems in polynomial time. Let us briefly enumerate such situations.

Case of curved space-time. Already in the historically very first non-Euclidean geometry—the hyperbolic Lobachevsky space—the volume $V(R)$ of the inside of the sphere grows exponentially with radius. Thus, in principle, we can fit exponentially many processors within a radius that grows linearly with n. On the resulting parallel machine, if we ask each processor to check one of 2^n Boolean vectors, we can thus solve the NP-complete propositional satisfiability problem in linear time; see, e.g., [21, 24]. So, if the proper physical space is hyperbolic, we can solve NP-complete problems in polynomial time.

Another possible scheme is related to the "almost" black holes [24]. One of the well-known consequences of general relativity is the existence of the "black hole" solutions. A black hole is an area from which nothing comes out (in particular, light cannot escape it, hence it looks black). It is proved that if an object (e.g., a star) is massive enough, it will eventually be crushed by its own gravitational force and form a black hole. If the object is smaller, or if it has a significant electric charge, then it forms an "almost" black hole, i.e., an area from which it is possible but difficult to escape. If you enter this "almost" black hole, you go into the narrow throat; see, e.g., [23], Chaps. 31 and 44. From the outside, it looks like a small particle. So, a natural hypothesis (described in Chap. 44 of [23]) is that all charged elementary particles *are* actually such "almost" black holes.

Each of these throats is gateway to a different space-time. So, to solve a propositional satisfiability problem with n variables $v_1, \ldots, v_n$, we can pick up two particles in our world—which are gateways to different worlds—and:

- ask the folks from the first of these worlds to check the propositional satisfiability of a formula obtained when we plug in $v_n =$ "true", and
- ask the second world to do the same with $v_n =$ "false".

Participants living in each of these world will thus be given a formula with $n - 1$ Boolean variables. To check the satisfiability of each of these formulas, they will repeat the same procedure: find two particles-gateways in their world, and ask the corresponding creatures to solve a problem with $n - 2$ variables, etc. In n steps, we reduce the problem to checking a formula with a single variable, and we need n steps to send the results back. Thus, in linear time, we also get a solution to the propositional satisfiability problem (because, as one can see, in this world with almost black holes, the volume $V(R)$ also grows exponentially with the radius R).

Possibility of velocities exceeding the speed of light. If we allow processes exceeding the speed of light, then we have acausal processes, i.e., the possibility to go back to the past [29]. The simplest thing that we can do in this case is to let a slow computer solve the problem for as long as it takes—and then send the result back in time, so that the user will get it right after he or she requested the solution. More sophisticated schemes are also possible; see, e.g., [10, 11].

Other possible schemes of using non-standard physics to speed up computations. To speed up computations, we can also use the fact that, according to relativity theory, time slows down when one travels at a speed close to the speed of light or in a strong gravitational field (e.g., near the black hole). So, if the whole civilization starts going around at a speed close to the speed of light and/or moves close to the black hole, then, by performing computations on stationary planets far away from the black hole, we get the result much faster—in terms of our time. For example, if 1 year for us will be 10 years for the outside world, then a problem that takes 10 years to compute will be solved after 1 year of our time.

Other possible schemes include the use of quantum effects, etc.; see, e.g., [1, 28].

2.3   What if No Final Theory Is Possible?

In the previous section, we analyzed how specific physical phenomena affect computability. In this analysis, we considered several specific physical models, such as cosmological solutions with wormholes and/or casual anomalies, etc. However, many physicists believe that no physical theory is perfect, i.e., that no matter how many observations support a physical theory, inevitably, new observations will come which will require this theory to be updated. In this section, following [13, 14, 20, 31], we prove that if such a no-perfect-theory principle is true, then the use of physical data

- can enhance computations, and
- can drastically speed up the solution of NP-complete problems: namely, we can feasibly solve almost all instances of each NP-complete problem.

2.3.1 No Physical Theory Is Perfect: How to Formalize the Widely Spread Physicists' Belief

No physical theory is perfect: a widely spread physicists' belief. If we prove that, within a given physical theory, we can speed up the solution to NP-complete problems, will this answer be fully satisfactory?

So far, in the history of physics, no matter how good a physical theory, no matter how good its accordance with observations, eventually, new observations appear which are not fully consistent with the original theory—and thus, a theory needs to be modified. For example, for several centuries, Newtonian physics seems to explain all observable facts—until later, quantum (and then relativistic) effects were discovered which required changes in physical theories.

Because of this history, many physicists believe that every physical theory is approximate—no matter how sophisticated a theory, no matter how accurate its current predictions, inevitably new observations will surface which would require a modification of this theory; see, e.g., [5].

How does this belief affect computations? At first glance, the fact that no theory is perfect makes the question of possible computation of non-computable sequences and of possible speed-up rather hopeless: no matter how good results we achieve within a given physical theory, eventually, this theory will turn out to be, strictly speaking, false—and thus, our computation or speed-up schemes will not be applicable.

In this section, we show, however, that in spite of this seeming hopelessness, an important non-standard computations and speed-up results can be deduced simply from the fact no physical theory is perfect.

How to describe, in precise terms, that no physical theory is perfect: discussion. The statement that no physical theory is perfect means that no matter what physical theory we have, eventually there will be observations which violate this theory. To formalize this statement, we need to formalize what are observations and what is a theory.

What are observations? Each observation can be represented, in the computer, as a sequence of 0s and 1s; actually, in many cases, the sensors already produce the signal in the computer-readable form, as a sequence of 0s and 1s.

An exact description of each experiment can also be described in precise terms, and thus, it will be represented in a computer as a sequence of 0s and 1s. An experiment should specify how long we wait for the result; in this way, we are guaranteed that we get the result.

In each experiment, we can specify which bit of the result we are interested in; for convenience, we can consider producing different bits as different experiments.

Each such experiment is represented as a sequence of 0s and 1s; by appending 1 at the beginning of this sequence, we can view this sequence as a binary expansion of a natural number i. This natural number will serve as the "code" describing the experiment. For example, a sequence 001 is transformed into $i = 1001_2 = 9_{10}$.

(We need to append 1, because otherwise two different sequences 001 and 01 will be represented by the same integer.)

For natural numbers i which correspond to experiment descriptions, let ω_i denote the bit result of the experiment described by the code i.

Let us also define ω_i for natural numbers i which do not correspond to a syntactically correct description of experiments. For example, we can fix a scheme of an experiment that uses a natural number i as a parameter (e.g., repeating a certain procedure i times), and define ω_i as the result of this scheme.

In these terms, all past and future observations form a (potentially) infinite sequence $\omega = \omega_1 \omega_2 \ldots$ of 0s and 1s, $\omega_i \in \{0, 1\}$.

Comment. To make sure that the resulting algorithm is feasible, we need to define experiment descriptions in which a way that the time needed to complete the i-th experiment does not exceed a polynomial of $\log(i)$. From this viewpoint, an experiment in which there is no explicit time limit should be described as a sequence of experiments with a cutoff time 1, 2, ..., t, ...; the allocated time can be indicated, e.g., by adding t special time symbols to the original description of the experiment.

What is a physical theory from the viewpoint of our problem: a set of sequences. A physical theory may be very complex, but all we care about is which sequences of observations ω are consistent with this theory and which are not. In other words, for our purposes, we can identify a physical theory T with the set of all sequences ω which are consistent with this theory.

Not every set of sequences corresponds to a physical theory: the set T must be non-empty and definable. Not every set of sequences comes from a physical theory. First, a physical theory must have at least one possible sequence of observations, i.e., the set T must be *non-empty*.

Second, a theory—and thus, the corresponding set—must be described by a finite sequence of symbols in an appropriate language. Sets which are uniquely described by (finite) formulas are known as *definable*. Thus, the set T must be definable.

Since at any moment of time, we only have finitely many observations, the set T must be closed. Another property of a physical theory comes from the fact that at any given moment of time, we only have finitely many observations, i.e., we only observe finitely many bits. From this viewpoint, we say that observations $\omega_1 \ldots \omega_n$ are consistent with the theory T if there is a continuing infinite sequence which is consistent with this theory, i.e., which belongs to the set T.

The only way to check whether an infinite sequence $\omega = \omega_1 \omega_2 \ldots$ is consistent with the theory is to check that for every n, the sequences $\omega_1 \ldots \omega_n$ are consistent with the theory T. In other words, we require that for every infinite $\omega = \omega_1 \omega_2 \ldots$,

- if for every n, the sequence $\omega_1 \ldots \omega_n$ is consistent with the theory T, i.e., if for every n, there exists a sequence $\omega^{(m)} \in T$ which has the same first n bits as ω, i.e., for which $\omega_i^{(m)} = \omega_i$ for all $i = 1, \ldots, n$,
- then the sequence ω itself should be consistent with the theory, i.e., this infinite sequence should also belong to the set T.

From the mathematical viewpoint, we can say that the sequences $\omega^{(m)}$ converge to ω: $\omega^{(m)} \to \omega$ (or, equivalently, $\lim \omega^{(m)} = \omega$), where convergence is understood in terms of the usual metric on the set of all infinite sequences $d(\omega, \omega') \stackrel{\text{def}}{=} 2^{-N(\omega, \omega')}$, where $N(\omega, \omega') \stackrel{\text{def}}{=} \max\{k : \omega_1 \ldots \omega_k = \omega'_1 \ldots \omega'_k\}$.

In general, if $\omega^{(m)} \to \omega$ in the sense of this metric, this means that for every n, there exists an integer ℓ such that for every $m \geq \ell$, we have $\omega_1^{(m)} \ldots \omega_n^{(m)} = \omega_1 \ldots \omega_n$. Thus, if $\omega^{(m)} \in T$ for all m, this means that for every n, a finite sequence $\omega_1 \ldots \omega_n$ can be a part of an infinite sequence which is consistent with the theory T. In view of the above, this means that $\omega \in T$.

In other words, if $\omega^{(m)} \to \omega$ and $\omega^{(m)} \in T$ for all m, then $\omega \in T$. So, the set T must contain all the limits of all its sequences. In topological terms, this means that the set T must be *closed*.

A physical theory must be different from a fact and hence, the set T must be nowhere dense. The assumption that we are trying to formalize is that no matter how many observations we have which confirm a theory, there eventually will be a new observation which is inconsistent with this theory. In other words, for every finite sequence $\omega_1 \ldots \omega_m$ which is consistent with the set T, there exists a continuation of this sequence which does not belong to T. The opposite would be if all the sequences which start with $\omega_1 \ldots \omega_m$ belong to T; in this case, the set T will be *dense* in the open set of all the sequences starting with $\omega_1 \ldots \omega_m$. Thus, in mathematical terms, the statement that every finite sequence which is consistent with T has a continuation which is not consistent with T means that the set T is *nowhere dense*.

Resulting definition of a theory. By combining the above properties of a set T which describes a physical theory, we arrive at the following definition.

Definition 2.3.1 By a *physical theory*, we mean a non-empty closed nowhere dense definable set T.

Mathematical comment. To properly define what is definable, we need to have a consistent formal definition of definability. In this chapter, we follow a natural definition from [15, 16, 19, 20, 31]—which is reproduced in the Appendix.

Formalization of the principle that no physical theory is perfect. In terms of the above notations, the no-perfect-theory principle simply means that the infinite sequence ω (describing the actual results of all observations) is not consistent with any physical theory, i.e., that the sequence ω does not belong to any physical theory T. Thus, we arrive at the following definition.

Definition 2.3.2 We say that an infinite binary sequence ω is *consistent with the no-perfect-theory principle* if the sequence ω does not belong to any physical theory (in the sense of Definition 2.3.1).

Comment. Are there such sequences in the first place? Our answer is yes. Indeed, by definition, we want a sequence which does not belong to a union of all definable physical theories. Every physical theory is a closed nowhere dense set. Every definable set is defined by a finite sequence of symbols, so there are no more than countably many definable theories. Thus, the union of all definable physical theories is contained in a union of countably many closed nowhere dense sets. Such sets are knows as *meager* (or *Baire first category*); it is known that the set of all infinite binary sequences is not meager. Thus, there are sequences who do not belong to the above union—i.e., sequences which are consistent with the no-perfect-theory principle; see, e.g., [9, 25].

2.3.2 The Use of Physical Computations Can Enhance Computations

How to describe general computations. Each computation is a solution to a well-defined problem. As a result, each bit in the resulting answer satisfies a well-defined mathematical property. All mathematical properties can be described, e.g., in terms of Zermelo–Fraenkel theory ZF, the standard formalization of set theory. For each resulting bit, we can formulate a property P which is true if and only if this bit is equal to 1. In this sense, each bit in each computation result can be viewed as the truth value of some statement formulated in ZF. Thus, our general ability to compute can be described as the ability to (at least partially) compute the sequence of truth values of all statements from ZF.

All well-defined statements from ZF can be numbered, e.g., in lexicographic order. Let α_n denote the truth value of the nth ZF statement, and let $\alpha = \alpha_1 \ldots \alpha_n \ldots$ denote the infinite sequence formed by these truth values. In terms of this sequence, our ability to compute is our ability to compute the sequence α.

Kolmogorov complexity as a way to describe what is easier to compute. We want to analyze whether the use of physical observations (i.e., of the sequence ω analyzed in the previous section) can simplify computations. A natural measure of easiness-to-compute was invented by A. N. Kolmogorov, the founder of modern probability theory, when he realized that in the traditional probability theory, there is no formal way to distinguish between:

- finite sequences which come from observing from truly random processes, and
- orderly sequences like $0101 \ldots 01$.

Kolmogorov noticed that an orderly sequence $0101 \ldots 01$ can be computed by a short program, while the only way to compute a truly random sequence $0101 \ldots$ is to have a print statement that prints this sequence. He suggested to describe this differences by introducing what is now known as *Kolmogorov complexity* $K(x)$ of a finite sequence x: the shortest length of a program (in some programming language) which computes the sequence x.

- For an orderly sequence x, the Kolmogorov complexity $K(x)$ is much smaller than the length $\text{len}(x)$ of this sequence: $K(x) \ll \text{len}(x)$.
- For a truly random sequence x, we have $K(x) \approx \text{len}(x)$; see, e.g., [22].

The smaller the difference $\text{len}(x) - K(x)$, the more we are sure that the sequence x is truly random.

Relative Kolmogorov complexity as a way to describe when using an auxiliary sequence simplifies computations. The usual notion of Kolmogorov complexity provides the complexity of computing x "from scratch". A similar notion of the *relative Kolmogorov complexity* $K(x \mid y)$ can be used to describe the complexity of computing x when a (potentially infinite) sequence y is given. This relative complexity is based on programs which are allowed to use y as a subroutine, i.e., programs which, after generating an integer n, can get the nth bit y_n of the sequence y by simply calling y. When we compute the length of such programs, we just count the length of the call, not the length of the auxiliary program which computes y_n—just like when we count the length of a C++ program, we do not count how many steps it takes to compute, e.g., $\sin(x)$, we just count the number of symbols in this function call. The relative Kolmogorov complexity is then defined as the shortest length of such a y-using program which computes x.

Clearly, if x and y are unrelated, having access to y does not help in computing x, so $K(x \mid y) \approx K(x)$. On the other hand, if x coincides with y, then the relative complexity $K(x \mid y)$ is very small: all we need is a simple for-loop, in which we call for each bit y_i, $i = 1, \ldots, n$, and print this bit right away.

Resulting reformulation of our question. In terms of relative Kolmogorov complexity, the question of whether observations enhance computations is translated into checking whether $K(\alpha_1 \ldots \alpha_n \mid \omega) \approx K(\alpha_1 \ldots \alpha_n)$ (in which case there is no enhancement) or whether $K(\alpha_1 \ldots \alpha_n \mid \omega) \ll K(\alpha_1 \ldots \alpha_n)$ (in which case there is a strong enhancement). The larger the difference $K(\alpha_1 \ldots \alpha_n) - K(\alpha_1 \ldots \alpha_n \mid \omega)$, the larger the enhancement.

Enhancement is possible. Let us show that under the no-perfect-theory principle, observations do indeed enhance computations.

Proposition 2.3.1 *Let α be a sequence of truth values of ZF statements, and let ω be an infinite binary sequence which is consistent with the no-perfect-theory principle. Then, for every integer $C > 0$, there exists an integer n for which $K(\alpha_1 \ldots \alpha_n \mid \omega) < K(\alpha_1 \ldots \alpha_n) - C$.*

Comment. In other words, in principle, we can have an arbitrary large enhancement.

2.3.3 The Use of Physical Observations Can Help in Solving NP-Complete Problems

Towards the main result of this section: that the use of physical observations can help in solving NP-complete problems. In this section, we prove that under the no-perfect-theory principle, it is possible to drastically speed up the solution of NP-complete problems.

How to represent instances of an NP-complete problem. For each NP-complete problem $\mathcal{P}$, its instances are sequences of symbols. In the computer, each such sequence is represented as a sequence of 0s and 1s. Thus, as in the previous section, we can append 1 in front of this sequence and interpret the resulting sequence as a binary code of a natural number i.

In principle, not all natural numbers i correspond to instances of a problem $\mathcal{P}$; we will denote the set of all natural numbers which correspond to such instances by $S_{\mathcal{P}}$.

For each $i \in S_{\mathcal{P}}$, the correct answer (true or false) to the ith instance of the problem $\mathcal{P}$ will be denoted by $s_{\mathcal{P},i}$.

Easier-to-solve and harder-to-solve NP-complete problems. We will show that our method works on "harder-to-solve" NP-complete problems, harder-to-solve in the following sense. By definition, for all NP-complete problems, unless $P = NP$, there is no feasible algorithm for solving all its instances. However, for some easier-to-solve problems, there are feasible algorithms which solve "almost all" instances, in the sense that for each n, the proportion of instances $i \leq n$ for which the problem is solved by this algorithm tends to 1. In this case, while the worst-case complexity is still exponential, in practice, almost all problems can be feasibly solved.

A more challenging case is that of harder-to-solve NP-complete problems, for which no feasible algorithm is known that would solve almost all instances.

In this section, we show that our method works on all NP-complete problems, both easier-to-solve and harder-to-solve ones.

What we mean by using physical observations in computations. In addition to performing computations, our computational device can produce a scheme i for an experiment, and then use the result ω_i of this experiment in future computations. In other words, given an integer i, we can produce ω_i.

In precise theory-of-computation terms, the use of physical observations in computations thus means computations that use the sequence ω as an *oracle*; see, e.g., [26].

Definition 2.3.3 By a *ph-algorithm* $\mathcal{A}$, we mean an algorithm which uses, as an oracle, a sequence ω which is consistent with the no-perfect-theory principle.

Notation The result of applying an algorithm $\mathcal{A}$ using ω to an input i will be denoted by $\mathcal{A}(\omega, i)$.

Definition 2.3.4 Let $\mathcal{P}$ be an NP-complete problem. We say that a feasible ph-algorithm $\mathcal{A}$ *solves almost all instances of* $\mathcal{P}$ if for every $\varepsilon > 0$, and for every natural number n, there exists an integer $N \geq n$ for which the proportion of the instances $i \leq N$ of the problem $\mathcal{P}$ which are correctly solved by $\mathcal{A}$ is greater than $1 - \varepsilon$:

$$\forall \varepsilon > 0 \, \forall n \, \exists N \left(N \geq n \, \& \, \frac{\#\{i \leq N : i \in S_{\mathcal{P}} \, \& \, \mathcal{A}(\omega, i) = s_{\mathcal{P},i}\}}{\#\{i \leq N : i \in S_{\mathcal{P}}\}} > 1 - \varepsilon \right).$$

Comment. The restriction to sufficiently long inputs $N \geq n$ makes perfect sense: for short inputs, NP-completeness is not an issue: we can perform exhaustive search of all possible bit sequences of length 10, 20, and even 30. The challenge starts when the length of the input is high.

Proposition 2.3.2 *For every NP-complete problem $\mathcal{P}$, there exists a feasible ph-algorithm $\mathcal{A}$ that solves almost all instances of $\mathcal{P}$.*

Comments. In other words, we show that the use of physical observations makes all NP-complete problems easier-to-solve (in the above-described sense).

It turns out that this result is the best possible, in the sense that the use of physical observations cannot solve *all* instances.

Proposition 2.3.3 *If $P \neq NP$, then no feasible ph-algorithm $\mathcal{A}$ can solve all instances of $\mathcal{P}$.*

Comment. Another possible idea of strengthening Proposition 2.3.2 is to require that the property

$$\frac{\#\{i \leq N : i \in S_{\mathcal{P}} \,\&\, \mathcal{A}(\omega, i) = s_{\mathcal{P},i}\}}{\#\{i \leq N : i \in S_{\mathcal{P}}\}} > 1 - \varepsilon$$

hold not only for *infinitely many N*, but for *all N* starting with some N_0. It turns out that in this formulation, the use of physical observation does not help.

Definition 2.3.5 Let $\mathcal{P}$ be an NP-complete problem. Let $\delta > 0$ be a real number. We say that a feasible ph-algorithm $\mathcal{A}$ δ-*solves* $\mathcal{P}$ if

$$\exists N_0 \, \forall N \left(N \geq N_0 \rightarrow \frac{\#\{i \leq N : i \in S_{\mathcal{P}} \,\&\, \mathcal{A}(\omega, i) = s_{\mathcal{P},i}\}}{\#\{i \leq N : i \in S_{\mathcal{P}}\}} > \delta \right).$$

Proposition 2.3.4 *For every NP-complete problem $\mathcal{P}$ and for every $\delta > 0$, if there exists a feasible ph-algorithm $\mathcal{A}$ that δ-solves $\mathcal{P}$, then there exists a feasible algorithm $\mathcal{A}'$ (not using physical observations) which also δ-solves $\mathcal{P}$.*

2.4 What if We Take into Account that We Are only Interested in Processing Physical Data

Many physical theories accurately predict which events are possible and which are not, or—in situations where probabilistic (e.g., quantum) effects are important—predict the probabilities of different possible outcomes. At first glance, it may seem that this probabilistic information is all we need.

In this section, we show, however, that to adequately describe physicists' reasoning, it is important to also take into account additional physical knowledge—about what is possible and what is not. We show that, if we limit ourselves to objects which are physically possible, then many seemingly undecidable problems become algorithmically decidable.

How physicists make their conclusions: why probabilities are sometimes not enough. Modern physics makes many very accurate predictions of different events. In situations when quantum effects are important and thus, deterministic predictions are not possible, physics predicts probabilities of different events; see, e.g., [5]. There are still many problems where we cannot accurately predict the events and/or their probabilities, but in many other situations, the accuracy of predictions is truly amazing.

At first glance, once we know the probabilities, we are done: we can thus predict the frequencies with which the corresponding events will occur in real life. In many situations, probabilities are indeed all we need. For example, when we predict that the probability of a coin falling heads is 1/2, this means that in half of the cases, the coin will fall heads, in half, tails, and there is no other information that we can extract from observing the results of an actual coin toss: these results should be random.

This is true not only for coin tossing, but also for other predictions in which the predicted probability is "reasonable", i.e., not too small and not too close to 1. However, the situation is somewhat different when it comes to events with a very small probability. Let us give a few simple examples.

According to statistical physics, entropy of a closed system can only increase. This means, for example, that if we place a cold kettle on a cold stove, it is not possible that the kettle will start boiling by itself, while the stove will get colder—although this transfer of energy from the stove to the kettle does not contradict to the energy conservation law.

How do physicists conclude that this is not possible? They estimate the probability of such an event and conclude that this probability is extremely small. From the purely mathematical viewpoint, the fact that this probability is not zero means that if we wait long enough, then we will still see a kettle boiling on a cold stove. However, this is not what the physicists claim. What they claim is that the kettle *cannot* boil. In other words, they claim that the corresponding event is simply not possible [3–5].

Another example is the impossibility of spontaneous human levitation. The fact that a body has a non-zero temperature means that all the atoms and all the molecules in the body are randomly oscillating. Again, since all the molecules are going in random directions, there is a non-zero probability that they will all go into the same direction and a person will be spontaneously lifted above ground. What the physicists claim is not that such a possibility is rare, they claim is that it is simply not possible.

Physicists make similar conclusions about all irreversible events. For example, if we place a gas in one half of the box and leave another half—separated by a door—empty, then, when we open the door, gas will spread evenly through both halves of the box. From the purely mathematical viewpoint, it is also possible that, vice versa, if we start with a gas which is uniformly spread through the box, then at some future moment of time, all the molecules will concentrate in one half of this box, while the other half will remain empty: the probability of this event is small but still positive.

However, what the physicists claim is that such a spontaneous separation is simply not possible.

Need to go beyond probabilities is in good accordance with common sense. The impossibility of events with very low probability may sound counter-intuitive, but it is actually in good accordance with common sense. Suppose that you flip a coin—which you believe to be fair—several times, and every time it falls heads. If this happens two, three, even ten times in a row, you may still continue believing that the coin is fair and that the actual probability of heads is indeed 1/2. However, what if this happens 30 times in a row? 100 times in a row? Different people may have different thresholds, but for any person, there is some number after which the person will be absolutely sure that this coin is not fair.

Let us give another example. In each state lottery, usually, someone wins the big prize. If the same person wins the big prize two years in a row, one may still claim that this was a random coincidence. But what if the same person wins three years in a row? four years in a row? No matter how much you originally believe in the fairness of the state lottery process, if this continues year after year, eventually, every person will be convinced that the lottery is rigged.

Events with very small probability are not possible: can we describe this physical idea in purely probabilistic terms? We have mentioned that both physics and common sense use a rule that events with very small probability cannot happen. How can we describe this rule in precise terms?

At first glance, it may seem that we can describe this rule in purely probabilistic terms: namely, we can set up some threshold small value $p_0 \ll 1$, and we can claim that any event with probability $\leq p_0$ is not possible. However, a simple example of coin tossing shows that proposal does not work. Indeed, what we want to claim is that after tossing a coin a large number (N) of times, we cannot have a sequence HH...H of all heads. The probability of this event is 2^{-N}, which, for large N, is indeed a very small number. So, at first glance, it may seem that if we take $p_0 \geq 2^{-N}$, then we will be able to make the desired conclusion.

But the situation is not so easy. The problem is that *any* sequence of N heads and tails—including the actual sequence that we will get after tossing a coin N times—has the exact same probability 2^{-N}. So, if we require that no event with probability $\leq p_0$ is possible, we come up with a strange conclusion that no sequence of heads and tails is possible at all—which makes no sense, since, of course, we can flip the coin N times and record the results.

Comment. It is worth mentioning that there is a direct relation between this discussion and the notions of Algorithmic Information Theory, such as algorithmic randomness and Kolmogorov complexity; see, e.g., [22]. The main difference, however, is that the notion of algorithmic randomness is based on the assumption that events with probability 0 cannot occur, while we are trying to describe a more general statement: that not only events with zero probability cannot occur, but events with a sufficiently small positive probability cannot occur either.

Additional information is needed. The above simple example shows that we cannot separate possible from impossible events by only using the known probabilities of different events. Thus, to properly describe physicists' reasoning (and our common sense), we need to supplement the probabilistic information with an additional information about what is possible.

How to describe what is possible. Let U be the set of all theoretically possible events. We assume that we know the probabilities of different events, i.e., that for some subsets $S \subseteq U$, we know the probability $p(S)$ that the actual event will be in S.

From all possible events, a physicist selects a subset T of all events which are possible. The main idea that we want to formalize is that if the probability is very small, then the corresponding event is not possible. What constitutes "very small" depends on the situation, but it is clear that if we have a definable sequence of events $A_1 \supseteq A_2 \supseteq \ldots \supseteq A_n \supseteq \ldots$, with $p(A_n) \to 0$, then for some sufficiently large N, the probability of the corresponding event A_N becomes so small that this event is impossible, i.e., $T \cap A_N = \emptyset$.

This is what we trying to describe for the case of coin tossing: A_n is the event when the heads appear in the first n coin tosses; then, $p(A_n) = 2^{-n} \to 0$.

In general, we arrive at the following formalization:

Definition 2.4.1 [8] Let U be a set with a probability measure p. We say that a subset $T \subseteq U$ is *a set of possible elements* if for every definable sequence A_n for which $A_n \supseteq A_{n+1}$ and $p(A_n) \to 0$, there exists an N for which $T \cap A_N = \emptyset$.

Need to go beyond probabilities. Sometimes, physicists use similar arguments even in situations when we do not know the probabilities. For example, physicists often expand a dependence in Taylor series $f(x) = a_0 + a_1 \cdot x + a_2 \cdot x^2 + \ldots$ When x is small, i.e., when $|x| \leq \delta$ for some small δ, they argue that we can safely ignore quadratic (and higher order) terms in this expansion and assume that $f(x) \approx a_0 + a_1 \cdot x$; see, e.g., [5].

This conclusion is definitely justified if we know the value a_2, or, at least, if we know some a priori bound C on this value. Then, $|a_2 \cdot x^2| \leq C \cdot \delta^2$, so when δ is sufficiently small, this term can indeed be safely ignored. However, physicists make this conclusion even when we do not know of any a priori bound on a_2. Their idea is that values which are too large are highly improbable.

In this case, we also have a series of events $A_n \supseteq A_{n+1}$: namely, A_n is the set of situations in which $|a_2| > n$. Here, we do not have probabilities, but we know that $\cap A_n = \emptyset$. Thus, no matter what is the (unknown) probability measure p, we have $p(A_n) \to 0$. As a result, we can use Definition 2.4.1 and conclude that for a sufficiently large N, events from A_N are impossible—hence $|a_2| \leq N$.

Such situations lead to the following alternative definition that can be used even when we do not know probabilities; see, e.g., [6–8, 15–20]:

Definition 2.4.2 Let U be a set. We say that a subset $T \subseteq U$ is *a set of possible elements* if for every definable sequence A_n for which $A_n \supseteq A_{n+1}$ and $\cap A_n = \emptyset$, there exists an N for which $T \cap A_N = \emptyset$.

What we do in this section. In this section, we show that many problems become algorithmically decidable if we restrict ourselves to physically possible objects.

In general, many problems are not algorithmically decidable. In general, many computational problems are not algorithmically decidable; see, e.g., [2, 30]. As a simple example, let us consider the problem of deciding whether two given real numbers are equal or not.

In this problem, the input consists of two real numbers, and the desired output is "yes" or "no", depending on whether these numbers are equal or not.

To describe this problem in precise terms, we need to formulate how exactly we present the input to a computer. In practice, real numbers come from measurements, and measurements are never absolutely accurate. In principle, we can measure a real number x with higher and higher accuracy (if not now, then in the future). For example, for any integer n, we can measure this number with the accuracy of n binary digits, i.e., with the accuracy of 2^{-n}. As a result of each such measurement, we get a rational number r_n for which $|x - r_n| \leq 2^{-n}$. This is exactly the usual definition of a computable real number: it is a process (maybe algorithmic, maybe involving measurements) that enables us, given an integer n, to generate a rational number r_n for which $|x - r_n| \leq 2^{-n}$ [2, 30].

Computing with computable real numbers means that, in addition to usual computational steps, we can also generate some n, get the corresponding value r_n, and then use this value in computations.

Some things can be computed this way. For example, if we know computable real numbers x and y, then their sum $z = x + y$ is also a computable real number. Indeed, to compute the 2^{-n}-approximation t_n to the sum z, it is sufficient to take the sum $s_n = r_{n+1} + s_{n+1}$ of $2^{-(n+1)}$-approximations r_{n+1} and s_{n+1} to x and y. Indeed, from $|x - r_{n+1}| \leq 2^{-(n+1)}$ and $|y - s_{n+1}| \leq 2^{-(n+1)}$, we can then conclude that

$$|z - s_n| = |(x + y) - (x_{n+1} + y_{n+1})| = |(x - x_{n+1}) + (y - s_{n+1})| \leq$$

$$|x - x_{n+1}| + |y - s_{n+1}| \leq 2^{-(n+1)} + 2^{-(n+1)} = 2^{-n}.$$

However, it is not possible to algorithmically check whether the two computable numbers x and y are equal or not. Indeed, if this was possible, then, e.g., for equal real numbers $x = y = 0$ for which $r_n = s_n = 0$ for all n, our procedure will return the answer "yes". This procedure consists of finitely many steps, thus it can only ask for finitely many values r_n and s_n. Let N be the smallest number which is larger than all the requests n. Then, we can keep the same x, but take instead a different $y' = 2^{-N} \neq 0$ for which $s'_1 = \ldots = s'_{N-1} = 0$ (so our equality-checking procedure will not notice the difference), but $s'_N = s'_{N+1} = \ldots = 2^{-N}$. Since our procedure cannot notice the difference between y and y', it will still produce the same answer – that yes, the inputs are equal—while in reality, the new inputs $x = 0$ and $y' = 2^{-N} \neq 0$ are different.

Similar negative results are known for many other problems [2, 30].

If we restrict ourselves to possible pairs of real numbers, then equality becomes decidable. Let us show, following [17], that if we restrict ourselves to possible pairs (x, y), then it is algorithmically possible to check whether $x = y$ or $x \neq y$.

Indeed, the fact that we consider possible pairs of real numbers means that on the set $U = \mathrm{IR} \times \mathrm{IR}$ of all possible pairs of real numbers, we have a subset T of possible numbers that satisfied Definition 2.4.2. In particular, we can consider the following definable sequence of sets $A_n \overset{\text{def}}{=} \{(x, y) : 0 < |x - y| \leq 2^{-n}\}$.

One can easily see that $A_n \supseteq A_{n+1}$ for all n and that $\cap A_n = \emptyset$. Thus, by Definition 2.4.2, there exists a natural number N for which $T \cap A_N = \emptyset$, i.e., for which no element $s \in T$ belongs to the set A_N. This, in turn, means that for every pair $(x, y) \in T$, either $|x - y| = 0$ (i.e., $x = y$) or $|x - y| > 2^{-N}$.

So, to check whether $x = y$ or not, it is sufficient to compute both x and y with accuracy $2^{-(N+2)}$, i.e., to compute values r_{N+2} and s_{N+2} for which $|x - r_{N+2}| \leq 2^{-(N+2)}$ and $|y - s_{N+2}| \leq 2^{-(N+2)}$. Then:

- if $x = y$, then, due to the triangle inequality, we have

$$|r_{N+1} - s_{N+2}| \leq |x - r_{N+2}| + |x - s_{N+2}| \leq 2^{-(N+2)} + 2^{-(N+2)} = 2^{-(N+1)};$$

- on the other hand, if $x \neq y$, then from $|x - y| > 2^{-N}$, we conclude that

$$|r_{N+1} - s_{N+2}| \geq |x - y| - |x - r_{N+2}| - |y - s_{N+2}| >$$

$$2^{-N} - (2^{-(N+2)} + 2^{-(N+2)}) = 2^{-N} - 2^{-(N+1)} = 2^{-(N+1)}.$$

Thus, by checking whether $|r_{N+1} - s_{N+2}| \leq 2^{-(N+1)}$ or whether $|r_{N+1} - s_{N+2}| > 2^{-(N+1)}$, we can decide whether $x = y$ or $x \neq y$.

Here, we compare rational numbers, i.e., ratios of integers, and for rational numbers, we can indeed algorithmically tell whether one is smaller or the other one is smaller.

Towards a general description of similar properties. To generalize the above result, let us come up with a general description of similar properties [12].

Let us start with reformulating the question of whether $x = y$ in generalizable terms. Specifically, we would like to describe the corresponding property in terms of the observable sequences r_n and s_n describing the real numbers x and y.

The equality between real numbers can indeed be described in these terms. Indeed, if $x = y$, then, for every n, we have

$$|r_n - s_n| \leq |r_n - x| + |x - s_n| \leq 2^{-n} + 2^{-n} = 2^{-(n-1)}.$$

Vice versa, let us assume that we have two computable real numbers x and y for which $|r_n - s_n| \leq 2^{-(n-1)}$ for all n. In this case, due to $|x - r_n| \leq 2^{-n}$ and $|y - s_n| \leq 2^{-n}$, we have

$$|x - y| \leq |x - r_n| + |r_n - s_n| + |s_n - y| \leq 2^{-n} + 2^{-(n-1)} + 2^{-n} = 2^{-(n-2)}.$$

Since this holds for every n, for $n \to \infty$, we get $x = y$.

Thus, the equality between computable real numbers has the form

$$\forall n \, (|r_n - s_n| \le 2^{-(n-1)}).$$

In general, as shown, e.g., in [27, 30], many properties involving limits, differentiability, etc., can be described in similar terms, namely as an *arithmetic formula*

$$Qn_1 \, Qn_2 \ldots Qn_k \, F(r_1, \ldots, r_\ell, n_1, \ldots, n_k), \qquad (2.4.1)$$

where:

- each Qn_i is either a universal quantifier $\forall n_i$ or an existential quantifier $\exists n_i$,
- $r_1, \ldots, r_\ell$ are corresponding sequences, and
- the property F is simply a propositional ("and", "or", and "not") combination of equalities and inequalities between the explicitly computable rational-valued expressions.

In the above example of checking whether two given real numbers are equal:

- we have two sequences $\ell = 2$,
- we only have one quantifier $k = 1$,
- this quantifier is a universal quantifier $Q_1 = \forall$, and
- the property F has (in these terms) the form $|r_1(n_1) - r_2(n_1)| \le 2^{-(n_1-1)}$.

Let us show that for all such arithmetic expressions, the information on what is possible and what is not leads to algorithmic decidability.

Proposition 2.4.1 *For every arithmetic formula of type (2.4.1) and for every set T of possible tuples $r = (r_1, \ldots, r_\ell)$, there exists an algorithm that, given a tuple $r = (r_1, \ldots, r_\ell) \in T$, checks whether or not the given formula holds for this tuple.*

Conclusion. In this section, we have shown that in order to adequately describe physical reasoning, we need to supplement the usual probabilistic information with an additional knowledge describing what is physically possible and what is not. We have also shown that if we restrict ourselves to physically possible objects, then many problems become algorithmically decidable.

2.5 Conclusions

In this chapter, we showed that what is computable and what is feasibly computable depends:

- on what physical processes we allow, and
- on whether we are interested in:

- general computations, in particular, solving mathematical problems, or
- only processing physical data (in which case inputs must satisfy some physics-motivated constraints).

Somewhat surprisingly, the possibility to enhance computations comes not only when we consider specific physical models, but also when we take into account that, according to many physicists, no physical theory is perfect—i.e., no matter how well a theory fits the experimental data, it will eventually have to be modified.

Acknowledgments This work was supported in part by the National Science Foundation grants HRD-0734825 and HRD-1242122 (Cyber-ShARE Center of Excellence) and DUE-0926721. The authors are greatly thankful to the anonymous referees for valuable suggestions.

Appendix: Proofs

A precise definition of definability.

Definition A1 Let $\mathcal{L}$ be a theory, and let $P(x)$ be a formula from the language of the theory $\mathcal{L}$, with one free variable x for which the set $\{x \mid P(x)\}$ is defined in the theory $\mathcal{L}$. We will then call the set $\{x \mid P(x)\}$ $\mathcal{L}$-definable.

Crudely speaking, a set is $\mathcal{L}$-definable if we can explicitly *define* it in $\mathcal{L}$. The set of all real numbers, the set of all solutions of a well-defined equation, every set that we can describe in mathematical terms: all these sets are $\mathcal{L}$-definable.

This does not mean, however, that *every* set is $\mathcal{L}$-definable: indeed, every $\mathcal{L}$-definable set is uniquely determined by formula $P(x)$, i.e., by a text in the language of set theory. There are only denumerably many words and therefore, there are only denumerably many $\mathcal{L}$-definable sets. Since, e.g., in a standard model of set theory ZF, there are more than denumerably many sets of integers, some of them are thus not $\mathcal{L}$-definable.

Our objective is to be able to make mathematical statements about $\mathcal{L}$-definable sets. Therefore, in addition to the theory $\mathcal{L}$, we must have a stronger theory $\mathcal{M}$ in which the class of all $\mathcal{L}$-definable sets is a set—and it is a countable set.

Denotation *For every formula F from the theory $\mathcal{L}$, we denote its Gödel number by $\lfloor F \rfloor$.*

Comment. A Gödel number of a formula is an integer that uniquely determines this formula. For example, we can define a Gödel number by describing what this formula will look like in a computer. Specifically, we write this formula in LaTeX, interpret every LaTeX symbol as its ASCII code (as computers do), add 1 at the beginning of the resulting sequence of 0s and 1s, and interpret the resulting binary sequence as an integer in binary code.

Definition A2 We say that a theory $\mathcal{M}$ is *stronger* than $\mathcal{L}$ if it contains all formulas, all axioms, and all deduction rules from $\mathcal{L}$, and also contains a special predicate $\mathrm{def}(n, x)$ such that for every formula $P(x)$ from $\mathcal{L}$ with one free variable, the formula $\forall y\, (\mathrm{def}(\lfloor P(x) \rfloor, y) \leftrightarrow P(y))$ is provable in $\mathcal{M}$.

The existence of a stronger theory can be easily proven: indeed, for $\mathcal{L}=$ZF, there exists a stronger theory $\mathcal{M}$. As an example of such a stronger theory, we can simply take the theory $\mathcal{L}$ plus all countably many equivalence formulas as described in Definition A2 (formulas corresponding to all possible formulas $P(x)$ with one free variable). This theory clearly contains $\mathcal{L}$ and all the desired equivalence formulas, so all we need to prove is that the resulting theory $\mathcal{M}$ is consistent (provided that $\mathcal{L}$ is consistent, of course). Due to the compactness principle, it is sufficient to prove that for an arbitrary finite set of formulas $P_1(x), \ldots, P_m(x)$, the theory $\mathcal{L}$ is consistent with the above reflection-principle-type formulas corresponding to these properties $P_1(x), \ldots, P_m(x)$.

This auxiliary consistency follows from the fact that for such a finite set, we can take

$$\mathrm{def}(n, y) \leftrightarrow (n = \lfloor P_1(x) \rfloor\ \&\ P_1(y)) \vee \ldots \vee (n = \lfloor P_m(x) \rfloor\ \&\ P_m(y)).$$

This formula is definable in $\mathcal{L}$ and satisfies all m equivalence properties. The statement is proven.

Important comments. In the main text, we will assume that a theory $\mathcal{M}$ that is stronger than $\mathcal{L}$ has been fixed; proofs will mean proofs in this selected theory $\mathcal{M}$.

An important feature of a stronger theory $\mathcal{M}$ is that the notion of an $\mathcal{L}$-definable set can be expressed within the theory $\mathcal{M}$: a set S is $\mathcal{L}$-definable if and only if

$$\exists n \in \mathrm{I\,N}\, \forall y (\mathrm{def}(n, y) \leftrightarrow y \in S).$$

In the paper, when we talk about definability, we will mean this property expressed in the theory $\mathcal{M}$. So, all the statements involving definability become statements from the theory $\mathcal{M}$ itself, *not* statements from metalanguage.

Proof of Proposition 2.3.1. Let us fix an integer C. To prove the desired property for this C, let us prove that the set T of all the sequences which do not satisfy this property, i.e., for which $K(\alpha_1 \ldots \alpha_n \mid \omega) \geq K(\alpha_1 \ldots \alpha_n) - C$ for all n, is a physical theory in the sense of Definition 1. For this, we need to prove that this set T is non-empty, closed, nowhere dense, and definable. Then, from Definition 2, it will follow that the sequence ω does not belong to this set and thus, that the conclusion of Proposition 1 is true.

The set T is clearly non-empty: it contains, e.g., a sequence $\omega = 00 \ldots 0 \ldots$ which does not affect computations. The set T is also clearly definable: we have just defined it.

Let us prove that the set T is closed. For that, let us assume that $\omega^{(m)} \to \omega$ and $\omega^{(m)} \in T$ for all m. We then need to prove that $\omega \in T$. Indeed, let us fix n, and let us prove that $K(\alpha_1 \ldots \alpha_n \mid \omega) \geq K(\alpha_1 \ldots \alpha_n) - C$. We will prove this by contradiction. Let us assume that $K(\alpha_1 \ldots \alpha_n \mid \omega) < K(\alpha_1 \ldots \alpha_n) - C$. This means that there exists a program p of length $\mathrm{len}(p) < K(\alpha_1 \ldots \alpha_n) - C$ which uses ω to compute $\alpha_1 \ldots \alpha_n$. This program uses only finitely many bits of ω; let B be the largest index of these bits. Due to $\omega^{(m)} \to \omega$, there exists an M for which, for all $m \geq M$, the first B bits of $\omega^{(m)}$ coincide with the first B bits of the sequence ω. Thus, the same program p will work exactly the same way—and generate the sequence $\alpha_1 \ldots \alpha_n$—if we use $\omega^{(m)}$ instead of ω. But since $\mathrm{len}(p) < K(\alpha_1 \ldots \alpha_n) - C$, this would means that the shortest length $K(\alpha_1 \ldots \alpha_n \mid \omega^{(m)})$ of all the programs which use $\omega^{(m)}$ to compute $\alpha_1 \ldots \alpha_n$ also satisfies the inequality $K(\alpha_1 \ldots \alpha_n \mid \omega^{(m)}) < K(\alpha_1 \ldots \alpha_n) - C$. This inequality contradicts to our assumption that $\omega^{(m)} \in T$ and thus, that $K(\alpha_1 \ldots \alpha_n \mid \omega^{(m)}) \geq K(\alpha_1 \ldots \alpha_n) - C$. The contradiction proves that the set T is indeed closed.

Let us now prove that the set T is nowhere dense, i.e., that for every finite sequence $\omega_1 \ldots \omega_m$, there exists a continuation ω which does not belong to the set T. Indeed, as such a continuation, we can simply take a sequence $\omega = \omega_1 \ldots \omega_m \alpha_1 \alpha_2 \ldots$ obtained by appending α at the end. For this new sequence, computing $\alpha_1 \ldots \alpha_n$ is straightforward: we just copy the values α_i from the corresponding places of the new sequence ω. Here, the relative Kolmogorov complexity $K(\alpha_1 \ldots \alpha_n \mid \omega)$ is very small and is, thus, much smaller than the complexity $K(\alpha_1 \ldots \alpha_n)$ which—since ZF is not decidable—grows fast with n.

The proposition is proven.

Proof of Proposition 2.3.2. 1°. As the desired ph-algorithm, we will, given an instance i, simply produce the result ω_i of the ith experiment. Let us prove, by contradiction, that this algorithm satisfies the desired property.

2°. We want to prove that for every $\varepsilon > 0$ and for every n, there exists an integer $N \geq n$ for which

$$\#\{i \leq N : i \in S_{\mathcal{P}} \,\&\, \omega_i = s_{\mathcal{P},i}\} > (1 - \varepsilon) \cdot \#\{i \leq N : i \in S_{\mathcal{P}}\}.$$

The assumption that this property is not satisfied means that for some $\varepsilon > 0$ and for some integer n, we have

$$\#\{i \leq N : i \in S_{\mathcal{P}} \,\&\, \omega_i = s_{\mathcal{P},i}\} \leq (1 - \varepsilon) \cdot \#\{i \leq N : i \in S_{\mathcal{P}}\} \text{ for all } N \geq n.$$
$$\tag{2.A.1}$$

Let T denote the set of all the sequences x that satisfy the property (3.1), i.e., let

$$T \overset{\text{def}}{=} \{x : \#\{i \leq N : i \in S_{\mathcal{P}} \,\&\, x_i = s_{\mathcal{P},i}\} \leq (1 - \varepsilon) \cdot \#\{i \leq N : i \in S_{\mathcal{P}}\} \text{ for all } N \geq n\}.$$

We will prove that this set T is a physical theory in the sense of Definition 2.3.1.

Then, due to Definition 2.3.2 and the fact that the sequence ω satisfies the no-perfect-theory principle, we will be able to conclude that $\omega \notin T$, and thus, that the property (3.1) is not satisfied for the given sequence ω. This will conclude the proof by contradiction.

3°. By definition of a physical theory T, it is a set which is non-empty, closed, nowhere dense, and definable. Let us prove these four properties one by one.

3.1°. Non-emptiness comes from the fact that the sequence x_i for which $x_i = \neg s_{\mathcal{P},i}$ for $i \in S_{\mathcal{P}}$ and $x_i = 0$ otherwise clearly belongs to this set: for this sequence, for every N, we have

$$\#\{i \leq N : i \in S_{\mathcal{P}} \,\&\, x_i = s_{\mathcal{P},i}\} = 0$$

and thus, the desired property is satisfied.

3.2°. Let us prove that the set T is closed, i.e., that if we have a family of sequences $x^{(m)} \in T$ for which $x^{(m)} \to \omega$, then $x \in T$.

Indeed, let us take any $N \neq n$, and let us prove that

$$\#\{i \leq N : i \in S_{\mathcal{P}} \,\&\, x_i = s_{\mathcal{P},i}\} \leq (1 - \varepsilon) \cdot \#\{i \leq N : i \in S_{\mathcal{P}}\}$$

for this N. Due to $x^{(m)} \to x$, there exists an M for which, for all $m \geq M$, the first N bits of $x^{(m)}$ coincide with the first N bits of the sequence x: $x_i^{(m)} = \omega_i$ for all $i \leq N$. Thus,

$$\#\{i \leq N : i \in S_{\mathcal{P}} \,\&\, x_i = s_{\mathcal{P},i}\} = \#\{i \leq N : i \in S_{\mathcal{P}} \,\&\, x_i^{(m)} = s_{\mathcal{P},i}\}.$$

Since $x^{(m)} \in T$, we have

$$\#\{i \leq N : i \in S_{\mathcal{P}} \,\&\, x_i^{(m)} = s_{\mathcal{P},i}\} \leq (1 - \varepsilon) \cdot \#\{i \leq N : i \in S_{\mathcal{P}}\},$$

thus

$$\#\{i \leq N : i \in S_{\mathcal{P}} \,\&\, x_i = s_{\mathcal{P},i}\} \leq (1 - \varepsilon) \cdot \#\{i \leq N : i \in S_{\mathcal{P}}\}.$$

So, the set T is indeed closed.

3.3°. Let us now prove that the set T is nowhere dense, i.e., that for every finite sequence $x_1 \ldots x_m$, there exists a continuation x which does not belong to the set T.

Indeed, as such a continuation, we can simply take a sequence

$$x = x_1 \ldots x_m x_{m+1} x_{m+2} \cdots$$

where for $i > m$, we take $x_i = s_{\mathcal{P},i}$ if $i \in S_{\mathcal{P}}$ and $x_i = 0$ otherwise. For this new sequence, for every N, at most m first instances may lead to results different from $s_{\mathcal{P},i}$, so we have

$$\#\{i \leq N : i \in S_{\mathcal{P}} \,\&\, x_i = s_{\mathcal{P},i}\} \geq \#\{i \leq N : i \in S_{\mathcal{P}}\} - m.$$

When $N \to \infty$, then $\#\{i \leq N : i \in S_{\mathcal{P}}\} \to \infty$, so for sufficiently large N, we have

$$\#\{i \leq N : i \in S_{\mathcal{P}}\} - m > (1 - \varepsilon) \cdot \#\{i \leq N : i \in S_{\mathcal{P}}\},$$

thus,

$$\#\{i \leq N : i \in S_{\mathcal{P}} \,\&\, x_i = s_{\mathcal{P},i}\} > (1 - \varepsilon) \cdot \#\{i \leq N : i \in S_{\mathcal{P}}\},$$

and we cannot have

$$\#\{i \leq N : i \in S_{\mathcal{P}} \,\&\, x_i = s_{\mathcal{P},i}\} \leq (1 - \varepsilon) \cdot \#\{i \leq N : i \in S_{\mathcal{P}}\}.$$

Therefore, this continuation does not belong to the set T.

3.4°. Finally, since the formula (2.A.1) explicitly defines the set T, this set T is clearly definable.

So, T is a physical theory, hence $\omega \notin T$, and the proposition is proven.

Proof of Proposition 2.3.3. Let us assume that P$\neq$NP. We then need to prove that for every feasible ph-algorithm $\mathcal{A}$, it is not possible to have

$$\#\{i \leq N : i \in S_{\mathcal{P}} \,\&\, \mathcal{A}(\omega, i) = s_{\mathcal{P},i}\} = \#\{i \leq N : i \in S_{\mathcal{P}}\}$$

for all natural numbers N.

To prove this impossibility, let us consider, for each feasible ph-algorithm $\mathcal{A}$, the set

$$T(\mathcal{A}) \stackrel{\text{def}}{=} \{x : \#\{i \leq N : i \in S_{\mathcal{P}} \,\&\, \mathcal{A}(x, i) = s_{\mathcal{P},i}\} = \#\{i \leq N : i \in S_{\mathcal{P}}\} \text{ for all } N\}.$$

Similarly to the proof of Proposition 2.3.2, we can show that this set $T(\mathcal{A})$ is closed and definable.

Let us prove that the set $T(\mathcal{A})$ is nowhere dense, i.e., that for every finite sequence $x_1 \ldots x_m$, there exists a continuation x which does not belong to the set $T(\mathcal{A})$. Indeed, we can simply extend the original finite sequence $x_1 \ldots x_m$ by 0s. In this case, when the oracle has only finitely many nonzero bits, we can incorporate these bits into an algorithm and get a feasible non-oracle algorithm $\mathcal{A}'$ which produces the same results: $\mathcal{A}'(i) = \mathcal{A}(x, i)$ for all i.

Let us prove, by contradiction, that $x \notin T(\mathcal{A})$. Indeed, if $x \in T(\mathcal{A})$, this would mean that

$$\#\{i \leq N : i \in S_{\mathcal{P}} \,\&\, \mathcal{A}'(i) = s_{\mathcal{P},i}\} = \#\{i \leq N : i \in S_{\mathcal{P}}\}$$

for all N. Thus, the feasible non-oracle algorithm $\mathcal{A}'$ solves all the instances of the original NP-complete problem $\mathcal{P}$, which contradicts to our assumption that P$\neq$NP. This contradiction proves that $x \notin T(\mathcal{A})$ and thus, the set $T(\mathcal{A})$ is indeed nowhere dense.

We have thus proven that the set $T(\mathcal{A})$ is closed, nowhere dense, and definable. The only property which is still missing from the definition of a physical theory (Definition 2.3.1) is non-emptiness. We do not know whether the set $T(\mathcal{A})$ is non-empty or not, but we can prove the desired impossibility in both cases.

If the set $T(\mathcal{A})$ is non-empty, then this set is a theory in the sense of Definition 1, and thus, since the sequence ω satisfies the no-perfect-theory principle, we have $\omega \notin T(\mathcal{A})$. This means that the ph-algorithm $\mathcal{A}$ is not solving all instances of the problem $\mathcal{P}$.

If the set $T(\mathcal{A})$ is empty, this also means that the ph-algorithm $\mathcal{A}$ does not solve all instances of the problem $\mathcal{P}$—no matter what oracle we use.

The proposition is proven.

Proof of Proposition 2.3.4. Let us assume that no non-oracle feasible algorithm δ-solves the problem $\mathcal{P}$. We then need to prove that for every feasible ph-algorithm $\mathcal{A}$, it is not possible to have N_0 for which

$$\#\{i \leq N : i \in S_{\mathcal{P}} \ \& \ \mathcal{A}(\omega, i) = s_{\mathcal{P},i}\} > \delta \cdot \#\{i \leq N : i \in S_{\mathcal{P}}\}$$

for all natural numbers $N \geq N_0$.

To prove this impossibility, let us consider, for each feasible ph-algorithm $\mathcal{A}$ and for each natural number N_0, the set

$$T(\mathcal{A}, N_0) \stackrel{\text{def}}{=}$$

$$\{x : \#\{i \leq N : i \in S_{\mathcal{P}} \ \& \ \mathcal{A}(x, i) = s_{\mathcal{P},i}\} > \delta \cdot \#\{i \leq N : i \in S_{\mathcal{P}}\} \text{ for all } N \geq N_0\}.$$

Similarly to the proof of Proposition 2.3.2, we can show that this set $T(\mathcal{A}, N_0)$ is closed and definable.

Let us prove that the set $T(\mathcal{A}, N_0)$ is nowhere dense, i.e., that for every finite sequence $x_1 \ldots x_m$, there exists a continuation x which does not belong to the set $T(\mathcal{A}, N_0)$. Indeed, similarly to the proof of Proposition 2.3.3, we can extend the original finite sequence $x_1 \ldots x_m$ by 0s. In this case, when the oracle has only finitely many nonzero bits, we can incorporate these bits into an algorithm and get a feasible non-oracle algorithm $\mathcal{A}'$ which produces the same results: $\mathcal{A}'(i) = \mathcal{A}(x, i)$ for all i.

Let us prove, by contradiction, that $x \notin T(\mathcal{A}, N_0)$. Indeed, if $x \in T(\mathcal{A}, N_0)$, this would mean that

$$\#\{i \leq N : i \in S_{\mathcal{P}} \ \& \ \mathcal{A}'(i) = s_{\mathcal{P},i}\} > \delta \cdot \#\{i \leq N : i \in S_{\mathcal{P}}\}$$

for all $N \geq N_0$. Thus, the feasible non-oracle algorithm $\mathcal{A}'$ δ-solves the original NP-complete problem $\mathcal{P}$, which contradicts to our assumption that no such feasible non-oracle algorithm is possible. This contradiction proves that $x \notin T(\mathcal{A}, N_0)$ and thus, the set $T(\mathcal{A}, N_0)$ is indeed nowhere dense.

We have thus proven that the set $T(\mathcal{A}, N_0)$ is closed, nowhere dense, and definable. The only property which is still missing from the definition of a physical theory (Definition 2.3.1) is non-emptiness. We do not know whether the set $T(\mathcal{A}, N_0)$ is non-empty or not, but we can prove the desired impossibility in both cases.

For each N_0, if the set $T(\mathcal{A}, N_0)$ is non-empty, then this set is a theory in the sense of Definition 2.3.1, and thus, since the sequence ω satisfies the no-perfect-theory principle, we have $\omega \notin T(\mathcal{A}, N_0)$, i.e.,

$$\#\{i \leq N : i \in S_{\mathcal{P}} \& \mathcal{A}(\omega, i) = s_{\mathcal{P}, i}\} \leq \delta \cdot \#\{i \leq N : i \in S_{\mathcal{P}}\} \text{ for some } N \geq N_0. \quad (2.A.2)$$

If the set $T(\mathcal{A}, N_0)$ corresponding to a given N_0 is empty, then also $\omega \notin T(\mathcal{A}, N_0)$, i.e., we also have the property (2.A.2).

Since the property (2.A.2) holds for all N_0, this means that the ph-algorithm $\mathcal{A}$ does not δ-solve the problem $\mathcal{P}$.

The proposition is proven.

Proof of Proposition 2.4.1. If the formula (2.4.1) had no quantifiers, then we could simply plug in the corresponding values into this formula and check whether the corresponding formula holds or not. The problem is with the quantifiers: while we can easily check whether some property holds for a specific value n_i, it is not possible to directly check whether this property holds for all infinitely many natural numbers $n_i = 0, 1, 2, \ldots$ The situation would be different if we could have a bound N on possible values of n_i, i.e., if the quantifier had the form $\forall n_i \leq N$ or $\exists n_i \leq N$: in this case, we can simply test all possible values $n_i \leq N$.

Let us show that for tuples from the set T, we can indeed have such bounds on the variables n_i. Let us start with a bound on n_1. For the variable n_1, there are two possible cases: when Q_1 is a universal quantifier and when Q_1 is an existential quantifier. Let us consider these two cases one by one.

In the first case, the formula (2.4.1) has the form $\forall n_1 \, G(n_1)$, for some expression $G(n_1)$ (with one fewer quantifier). Let us take

$$A_n = \{r : \forall n_1 \, (n_1 \leq n \rightarrow G(n_1)) \& \neg \forall n_1 \, G(n_1)\}.$$

One can easily check that $A_n \supseteq A_{n+1}$ and $\cap A_n = \emptyset$. Thus, there exists a natural number N for which $T \cap A_N = \emptyset$. So, for $r \in T$, if $\forall n_1 \, (n_1 \leq N \rightarrow G(n_1))$, we cannot have $\neg \forall n_1 \, G(n_1)$, so we must have $\forall n_1 \, G(n_1)$. Clearly, $\forall n_1 \, G(n_1)$ always implies $\forall n_1 \, (n_1 \leq N \rightarrow G(n_1))$. Thus, for $r \in T$, $\forall n_1 \, G(n_1)$ with an unlimited quantifier is equivalent to a formula $\forall n_1 \, (n_1 \leq N \rightarrow G(n_1))$ with a bounded quantifier.

In the second case, the formula (2.4.1) has the form $\exists n_1 \, G(n_1)$, for some expression G (with one fewer quantifier). Let us take

$$A_n = \{r : \neg \exists n_1 \, (n_1 \leq n \& G(n_1)) \& \exists n_1 \, G(n_1)\}.$$

One can easily check that $A_n \supseteq A_{n+1}$ and $\cap A_n = \emptyset$. Thus, there exists a natural number N for which $T \cap A_N = \emptyset$. So, for $r \in T$, if $\neg \exists n_1 \, (n_1 \leq N \& G(n_1))$, we cannot have $\exists n_1 \, G(n_1)$ we must have $\neg \exists n_1 \, G(n_1)$. Clearly, $\neg \exists n_1 \, G(n_1)$ always implies $\neg\exists n_1\,(n_1 \leq N\ \&\ G(n_1))$. Thus, for $r \in T$, $\neg\exists n_1\,G(n_1)$ is equivalent to $\neg\exists n_1\,(n_1 \leq N\ \&\ G(n_1))$. So, by taking negations, we conclude that the original formula $\exists n_1\,G(n_1)$ with an unlimited quantifier is equivalent to a formula $\exists n_1\,(n_1 \leq N\ \&\ G(n_1))$ with a bounded quantifier.

Now, we have reduced the original formula with k quantifiers to a formula in which the first quantifier is bounded. This bounded-quantifier formula is equivalent to, correspondingly, $G(0)\ \&\ G(1)\ \&\ \ldots\ \&\ G(N)$ or to $G(0) \vee G(1) \vee \ldots \vee G(N)$, where the corresponding formulas $G(n_1)$ have $k - 1$ quantifiers. So, if we can find the truth values of each of these (finitely many) formulas $G(n_1)$, we could be able to check the truth value of the original formula (2.4.1).

For each of these formulas $G(n_1)$ with $k - 1$ quantifiers, we can apply the same reduction to reduce them to formulas with $k - 2$ quantifiers, etc., until we get formulas with no quantifiers at all – which can be therefore directly checked.

This reduction proves that it is indeed algorithmically possible to check whether a given formula (2.4.1) holds or not for a given tuple r. The proposition is proven.

References

1. Aaronson, S.: NP-complete problems and physical reality. SIGACT News **36**, 30–52 (2005)
2. Bishop, E.: Foundations of Constructive Analysis. McGraw-Hill, New York (1967)
3. Boltzmann, L.: Bemrkungen über einige Probleme der mechanischen Wärmtheorie. Wiener Ber. **II**(75), 62–100 (1877)
4. Feynman, R.P.: Statistical Mechanics. W. A. Benjamin, New York (1972)
5. Feynman, R., Leighton, R., Sands, M.: The Feynman Lectures on Physics. Addison Wesley, Boston (2005)
6. Finkelstein, A.M., Kosheleva, O., Kreinovich, V., Starks, S.A., Nguyen, H.T.: To properly reflect physicists' reasoning about randomness, we also need a maxitive (possibility) measure, In: Proceedings of the 2005 IEEE International Conference on Fuzzy Systems FUZZ-IEEE'2005, 22–25 May 2005, pp. 1044–1049. Reno, Nevada (2015)
7. Finkelstein, A.M., Kosheleva, O., Kreinovich, V., Starks, S.A., Nguyen, H.T.: Use of maxitive (possibility) measures in foundations of physics and description of randomness: case study, In: Proceedings of the 24th International Conference of the North American Fuzzy Information Processing Society NAFIPS'2005, 22–25 June 2005, pp. 687–692. Ann Arbor, Michigan (2005)
8. Finkelstein, A.M., Kreinovich, V.: Impossibility of hardly possible events: physical consequences, In: Abstracts of the 8th International Congress on Logic, Methodology and Philosophy of Science, **5**(2), pp. 25–27. Moscow (1987)
9. Jalal-Kamali, A., Nebesky, O., Durcholz, M.H., Kreinovich, V., Longpré, L.: Towards a "generic" notion of genericity: from 'typical' and 'random' to meager, shy, etc. J. Uncertain Syst. **6**(2), 104–113 (2012)
10. Koshelev, M., Kreinovich, V.: Towards computers of generation omega–non-equilibrium thermodynamics, granularity, and acausal processes: a brief survey, In: Proceedings of the International Conference on Intelligent Systems and Semiotics (ISAS'97), pp. 383–388. National Institute of Standards and Technology Publications, Gaithersburg, Maryland (1997)
11. Kosheleva, O.M., Kreinovich, V.: What can physics give to constructive mathematics, In: Mathematical Logic and Mathematical Linguistics, pp. 117–128. Kalinin (1981, in Russian)
12. Kosheleva, O., Kreinovich, V.: Adding possibilistic knowledge to probabilities makes many problems algorithmically decidable, Proceedings of the World Congress of the International Fuzzy Systems Association IFSA'2015, joint with the Annual Conference of the European

Society for Fuzzy Logic and Technology EUSFLAT'2015, Gijon, Asturias, Spain, June 30 –
July 3 2015, to appear (2015)

13. Kosheleva, O.M., Soloviev, S.V.: On the logic of using observable events in decision making,
    In: Proceedings of the IX National USSR Symposium on Cybernetics, pp. 49–51. Moscow
    (1981, in Russian)

14. Kosheleva, O., Zakharevich, M., Kreinovich, V.: If many physicists are right and no physical
    theory is perfect, then by using physical observations, we can feasibly solve almost all instances
    of each NP-complete problem. Math. Struct. Model. **31**, 4–17 (2014)

15. Kreinovich, V.: Toward formalizing non-monotonic reasoning in physics: the use of Kol-
    mogorov complexity and Algorithmic Information Theory to formalize the notions "typically"
    and "normally", In: Sheremetov, L., Alvarado, M. (eds.) Proceedings of the Workshops on Intel-
    ligent Computing WIC'04 associated with the Mexican International Conference on Artificial
    Intelligence MICAI'04, pp. 187–194. Mexico City, Mexico, 26–27 April 2004

16. Kreinovich, V.: Toward formalizing non-monotonic reasoning in physics: the use of Kol-
    mogorov complexity. Rev. Iberoam. de Intel. Artif. **41**, 4–20 (2009)

17. Kreinovich, V.: Negative results of computable analysis disappear if we restrict ourselves to
    random (or, more generally, typical) inputs. Math. Struct. Model. **25**, 100–113 (2012)

18. Kreinovich, V.: Towards formalizing non-monotonic reasoning in physics: logical approach
    based on physical induction and its relation to Kolmogorov complexity. In Erdem, E., Lee,
    J., Lierler, Y., Pearce, D. (eds.) Correct Reasoning: Essays on Logic-Based AI in Honor of
    Vladimir Lifschitz. Lectures Notes in Computer Science, vol. 7265, pp. 390–404. Springer,
    Heidelberg (2012)

19. Kreinovich, V., Finkelstein, A.M.: Towards applying computational complexity to foundations
    of physics, Notes of Mathematical Seminars of St. Petersburg Department of Steklov Institute
    of Mathematics, **316**, 63–110 (2004); reprinted in J. Math. Sci. **134**(5), 2358–2382 (2006)

20. Kreinovich, V., Kosheleva, O.: Logic of scientific discovery: how physical induction affects
    what is computable, In: Proceedings of the International Interdisciplinary Conference Philos-
    ophy, Mathematics, Linguistics: Aspects of Interaction 2014 PhML'2014, pp. 116–127. St.
    Petersburg, Russia, 21–25 April 2014,

21. Kreinovich, V., Margenstern, M.: In some curved spaces, one can solve NP-hard problems in
    polynomial time, Notes of Mathematical Seminars of St. Petersburg Department of Steklov
    Institute of Mathematics **358**, 224–250 (2008); reprinted in J. Math. Sci. **158**(5), 727–740
    (2009)

22. Li, M., Vitányi, P.M.B.: An Introduction to Kolmogorov Complexity. Springer, Berlin (2008)

23. Misner, C.W., Thorne, K.S., Wheeler, J.A.: Gravitation, W. H. Freeman and Company, San
    Francisco (1973)

24. Morgenstein, D., Kreinovich, V.: Which algorithms are feasible and which are not depends on
    the geometry of space-time. Geombinatorics **4**(3), 80–97 (1995)

25. Oxtoby, J.C.: Measure and Category: A Survey of the Analogies between Topological and
    Measure Spaces. Springer, Heidelberg (1980)

26. Papadimitriou, C.: Computational Complexity. Addison Welsey, Reading, Massachusetts
    (1994)

27. Rogers Jr., H.: The Theory of Recursive Functions and Effective Computability. MIT Press,
    Cambridge (1987)

28. Srikanth, R.: The quantum measurement problem and physical reality: a computation theoretic
    perspective, In: Goswami, D. (ed.) AIP Conference Proceedings on Quantum Computing: Back
    Action, **864**, pp. 178–193. IIT Kanpur, India, March 2006

29. Thorne, K.S.: From Black Holes to Time Warps: Einstein's Outrageous Legacy. W.W. Norton
    & Company, New York (1994)

30. Weihrauch, K.: Computable Analysis. Springer, Berlin (2000)

31. Zakharevich, M., Kosheleva, O.: If many physicists are right and no physical theory is perfect,
    then the use of physical observations can enhance computations. J. Uncertain Syst. **8**(3), 227–
    232 (2014)

Chapter 3
The Ideal Energy of Classical Lattice Dynamics

Norman Margolus

Abstract We define, as local quantities, the least energy and momentum allowed by quantum mechanics and special relativity for physical realizations of some classical lattice dynamics. These definitions depend on local rates of finite-state change. In two example dynamics, we see that these rates evolve like classical mechanical energy and momentum.

3.1 Introduction

Despite appearances to the contrary, we live in a finite-resolution world. A finite-sized physical system with finite energy has only a finite amount of distinct detail, and this detail changes at only a finite rate [1–3]. Conversely, given a physical system's finite rates of distinct change in time and space, general principles of quantum mechanics define its minimum possible average energy and momentum. We apply these definitions to classical finite-state lattice dynamics.

3.1.1 Ideal Energy

It was finiteness of distinct state, first observed in thermodynamic systems, that necessitated the introduction of Planck's constant h into physics [4]. Quantum mechanics manages to express this finiteness using the same continuous coordinates that are natural to the macroscopic world. Describing reality as superpositions of waves in space and time, finite momentum and energy correspond to effectively finite bandwidth; hence finite distinctness. For example [3], the average rate v at which an isolated physical system can traverse a long sequence of distinct states is bounded by the average (classical) energy E:

N. Margolus (✉)
Massachusetts Institute of Technology, Cambridge, MA, USA
e-mail: nhm@mit.edu

© Springer International Publishing Switzerland 2017
A. Adamatzky (ed.), *Advances in Unconventional Computing*,
Emergence, Complexity and Computation 22,
DOI 10.1007/978-3-319-33924-5_3

$$v \leq 2E/h, \tag{3.1}$$

taking the minimum possible energy to be zero. Here E/h is the average frequency of the state, which defines a half-width for the energy frequency distribution. If we compare (3.1) in two frames, we can bound the average rate μ of changes not visible in the rest frame, and hence attributable to overall motion:

$$\mu \leq 2pv/h. \tag{3.2}$$

Here p is the magnitude of a system's average (classical) momentum, which is also a half-width for a (spatial) frequency distribution; v is the system's speed.

These kinds of constraints are sometimes referred to as uncertainty bounds, but they in no way preclude precise finite-state evolution. Given rates of change, these bounds define ideal (minimum achievable) average energy and momentum for finite state systems, emulated as efficiently as possible (with no wasted motion or state) by perfectly-tailored quantum hamiltonians [3, 5].

Clearly there can never be more *overall spatial change* μ than *total change* v in a physical evolution: this is reflected in $pv/E = (v/c)^2$. From this and (3.2),

$$E \geq (h\mu/2)/(v/c)^2. \tag{3.3}$$

Thus for a given rate μ of overall motional change, E can only attain its minimum possible value if the motion is at the speed of light; then no energy is invested in rest-frame dynamics (rest energy). In a finite-state dynamics with several geometrically related signal speeds, to minimize all energies (3.3) the fastest signals must move at the speed of light. If we then want to realize the dynamics running faster, we must put the pieces of the system closer together: we can increase p in (3.2), but not v. Of course in finite-state models of particular physical systems, realistic constraints on speeds and separations may require higher energies.

These bounds can be used to define ideal local energies and momenta for some invertible lattice dynamics, determined by rates of distinct change.

3.1.2 Local Change

We restrict our attention to finite-state lattice dynamics that emulate the locality, uniformity and microscopic invertibility of physical law: invertible cellular automata (CA). We assume the dynamics is defined as a regular arrangement of invertible interactions (logic gates), repeated in space and time, each of which independently transforms a localized set of state variables.

This kind of CA format, where the state variables are always updated in independent groups, has sometimes been called partitioning CA, and encompasses a variety of lattice formats that have been used to model physical dynamics [6–13]. It is interesting that all globally invertible CA can be recast in this physically realistic format, as a composition of independent invertible interactions, even if the CA was originally defined as a composition of non-invertible operations on overlapping neighborhoods [14–16]. Historically, CA originated as physics-like dynamics *without* invertibility [17–19].

Now, in the energy bounds above, only rates of change matter, not the amount of state updated in a single operation. This is unrealistic. We can define a large-scale synchronous dynamics, where the global rate of state change is independent of the size of the system. Physically, total energy must be bounded by the total rate of local changes, since each independent local update also obeys an energy bound. We resolve this conflict by allowing synchronous definition, but counting the global average rate of distinct change as if local updates were non-synchronous—which would in fact be true in most relativistic frames.

There is also an issue of what not to count. For a dynamics defined by a set of gate operations, it might seem natural to include, in the minimum, energy required to construct the gates and to turn them on and off. This is the energy needed to construct a perfectly-tailored hamiltonian. Here we ignore this construction energy, and discuss the ideal case where the hamiltonian is given for free (as part of nature), and we only need to account for energy required by state change within the dynamics.

3.1.3   Two Examples

In the remainder of this paper, we introduce and discuss two 2×2 block partitioning CA (cf. [20]). These dynamics are isomorphic to classical mechanical systems, and are simple enough that it is easy to compare energetic quantities, defined by local rates of state change, with classical ones.

The first example is a scalable CA version of the Soft Sphere Model [21], which is similar to Fredkin's classical mechanical Billiard Ball Model [22]. This digital system emulates the integer time behavior of an idealized classical mechanical system of elastically colliding balls, and is computation universal. The CA is scalable in that square blocks of ones (balls) of any size can be collided to simulate a billiard ball computation. This model has not been published before.

The second example is a CA model of an elastic string that exhibits simple-harmonic motion and exactly emulates the continuum wave equation at integer times, averaged over pairs of adjacent sites. This model has been discussed before [7, 23–25], but the analysis of overall translational motion, ideal energy, and their relativistic interpretation, have not been previously published.

3.2   Scalable Soft Sphere CA

Many CA dynamics can be interpreted as the integer-time behavior of a continuous classical mechanical system, started from an exactly specified initial state. This is true, for example, for lattice gas models of fluids. Such *stroboscopic* classical mechanical CA inherit, from their continuous counterparts, conserved quantities such as energy and momentum that we can compare to ideal quantities determined by local rates of state change. Of course the continuum models we have in mind would be numerically unstable if actually run as continuous dynamics, but this issue is not inherited by the finite-state CA [26].

A famous stroboscopic dynamics of this sort is Fredkin's billiard ball model of computation, in which hard spheres moving in a plane, each with four possible initial velocities, are restricted to a square lattice of initial positions. At each integer time, the system is again in such a configuration. To guarantee this property without additional restrictions on initial states, we let billiard balls pass through each other in some kinds of collisions, without interacting.

Figure 3.1 shows a variant of this model in which the balls are much more compressible, so collisions deflect paths inward rather than outward. This variant has the advantage that it is more directly related to a simple partitioning CA (cf. [6]). In the collision illustrated in Fig. 3.1a balls enter from the left with a horizontal component of velocity of one column per time unit, so consecutive moments of the history of a collision occur in consecutive columns.

The collision shown is energy and momentum conserving, and compression and rebound take exactly the time needed to displace the colliding balls from their original paths onto the paths labeled AB. If a ball had come in only at A with no ball at B, it would have left along the path labeled $A\bar{B}$: the collision acts as a universal logic gate.

Figure 3.1b shows a realization of the collision as a simple partitioning dynamics. Each time step in (a) corresponds to two in (b), and again particles are shown at each integer time—drawn dark at even times and light at odd. The rule (c) is inferred from (b), interpreting that diagram as showing the positions of two streams of colliding particles at one moment (dark), and their positions at the next moment

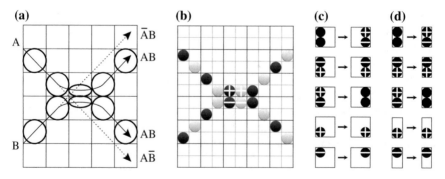

Fig. 3.1 Scalable Soft Sphere dynamics. **a** Stroboscopic view of continuous classical collision, one time step per column. **b** Finite state collision, with particles drawn lighter at odd time-steps. **c** 2D partitioning rule. Only these cases (and their rotations) interact. Otherwise, all particles move diagonally, unchanged. **d** 1D version of the rule

(light). All particles move diagonally across a block, unchanged, in the cases not defined explicitly in (c). If this rule is applied to just the dark particles in each of the dark-bordered 2×2 blocks in (b), ignoring the light particles, it moves them to the light positions; applied to just the light particles in each of the light-bordered blocks, it moves them to the dark positions. The dynamics alternately applies the rule to these two partitions. To also allow collisions like (b) for streams of balls arriving from the right, top, or bottom, we define the rule (c) to have discrete rotational symmetry: in each of the cases shown in (c), each of the four $90°$ rotations of the pattern on the left turns into the corresponding rotation of the pattern on the right.

Note that (b) can also be interpreted as showing a time history of a collision of two particles in a one-dimensional partitioning dynamics (the center of mass dynamics). Then we get the rule (d), with the cases not explicitly shown interchanging the two cell values. Three dimensional versions of the dynamics can be constructed as in [21].

It is not surprising that a time-independent continuous dynamics turns into a time-dependent discrete partitioning. In the continuous model, balls approach a locus of possible collision, interact independently of the rest of the system, and then move away toward a new set of loci. The partitions in the continuous case are just imaginary boxes we can draw around places within which what happens next doesn't depend on anything outside, for some period of time. Thus it is also not surprising that we can assign a conserved energy to partitioning dynamics.

3.2.1   Ideal Energy and Momentum

For a physical realization of the SSS dynamics, let τ be the time needed for gate operations to update all blocks of one partition, and let v_0 be the average speed at which the physical representation of a fastest-moving particle travels within the physical realization of the CA lattice (assuming discrete isotropy).

Equations (3.2) and (3.3) define an ideal (minimum) momentum and energy for a block in which there is a distinct overall spatial change and direction of motion. Clearly these ideal quantities are conserved overall in collisions, since freely moving particles move diagonally at v_0 before and after. Are they also conserved in detail during collisions?

When two freely moving particles enter a single block in the collision of Fig. 3.1b, the number of block changes is reduced: one instead of two. The ideal magnitude of momentum for each freely moving particle before the collision is $p_1 = (h/2\tau)/v_0$. For two colliding particles moving horizontally within a block the ideal is $p_2 = (h/2\tau)/(v_0/\sqrt{2}) = 2p_1/\sqrt{2}$, which is the same as the net horizontal momentum before the collision. Ideal energy is similarly conserved.

Note, however, that the separate horizontal motions of the $+$ and $-$ particles during the next step of the collision of Fig. 3.1b imply an increase in the minimum energy and momentum for that step. This effect becomes negligible as we enlarge the scale of the objects colliding.

3.2.2 Rescaling the Collision

If two columns of k particles are collided in the SSS dynamics, then the resulting collision just shifts the output paths by k positions along the axis of the collision. This is illustrated in Fig. 3.2a for $k = 3$. Thus $k \times k$ blocks of particles collide exactly as in the classical collision of Fig. 3.1a: the SSS CA can perform logic with diagonally-moving square "balls" of any size. When two balls of equal size meet "squarely," moving together along a horizontal axis, each pair of columns evolves independently of the rest; colliding along a vertical axis, pairs of rows evolve independently. Square balls can participate in both kinds of collisions.

During such a collision, from the blocks that change we can infer a net momentum and hence a velocity for the motion of each colliding ball: Fig. 3.2b illustrates this for $k = 100$, with the time unit being the time for a freely moving $k \times k$ ball to travel its length (and width). Looking at just the changes in the top half of (a), we determine the magnitude and direction of minimum average momentum for each block that changes using (3.2), and hence determine a total momentum. Half of the conserved total energy is associated with each ball, so $v/v_0 = v E_{\text{ball}}/v_0 E_{\text{ball}} = p/p_0$ gives the magnitude of velocity of a ball as a function of time, as number and type of changes evolve. This is plotted in (b).

The fraction $1/\gamma$ of the total energy E that is mass energy depends on $(v/c)^2 = (v/v_0)^2(v_0/c)^2$. Thus given (b), it depends on an assumption about the value of v_0/c. The fraction $1/\gamma$, as a function of time in the $k = 100$ collision, is plotted in Fig. 3.2c under different assumptions. The bottom case, $v_0 = c$, has the greatest range but the smallest value at all times. The top case is $v_0 = .2c$. As expected, the faster the speed of the fastest signals, the less the energy tied up in mass, hence the smaller the total energy. Ideally, $v_0 = c$.

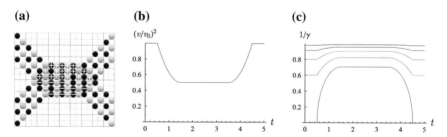

Fig. 3.2 Multiple collisions in the SSS dynamics. **a** Colliding columns of particles are displaced horizontally by the height of the column. **b** Each column is slowed down by the collision. **c** The fraction of energy that is mass during a collision decreases with increasing initial particle speed v_0 (from top, 20% of c, 40%, 60%, 80%, 100%)

3.3   Elastic String CA

In this second example we discuss a classical finite state model of wave motion in an elastic string. This stroboscopic classical mechanical model exactly reproduces the behavior of the time-independent one-dimensional wave equation sampled at integer-times and locations. As in the SSS example, a continuous model is turned into a finite state one by restricting the initial state (in this case the initial wave shape) to a perfect discrete set of possible initial configurations, and this constraint reappears at each integer time. In the continuum limit the discrete constraint on the wave shape disappears; the exactness of the wave dynamics itself (at discrete times) is independent of this limit.

The elastic string CA uses partitioning, but in a different way than the SSS CA: here the partitioning actually constrains the continuous classical dynamics used to define the CA, but in a way that never affects the classical energy. In the SSS case, the time dependence associated with the partitioning completely disappears in the continuous classical-mechanical version of the dynamics.

3.3.1   Discrete Wave Model

Consider an ideal continuous string for which transverse displacements exactly obey the wave equation. Figure 3.3a shows an initial configuration with the string stretched between equally spaced vertical bars. The set of initial configurations we're allowing are periodic, so the two endpoints must be at the same height.[1] Any configuration is allowed as long as each segment running between vertical bars is straight and lies at an angle of $\pm 45°$ to the horizontal.

Initially the string is attached at a fixed position wherever it crosses a vertical bar. We start the dynamics by releasing the attachment constraint at all of the gray bars. The attachment to the black bars remains fixed. In Fig. 3.3b the segments that are about to move are shown with dotted lines: the straight segments have no tendency to move. Under continuum wave dynamics, the dotted segments all invert after some time interval τ. This will be our unit of time for the discrete dynamics. The new configuration at the end of this interval is shown in Fig. 3.3c, with segments that have just moved dotted. At this instant in time all points of the string are again at rest and we are again in an allowed initial configuration. Now we interchange the roles of the black and gray bars and allow the segments between adjacent gray bars to move for a time interval τ. The dynamics proceeds like this, interchanging the roles of the black and gray bars after each interval of length τ. Since attachments are always changed at instants when all energy is potential and the string is not moving, the explicit time dependence of the system doesn't affect classical energy conservation.

---

[1]Unless the right and left edges of the space itself are joined with a vertical offset.

(a)   **(b)**   **(c)**

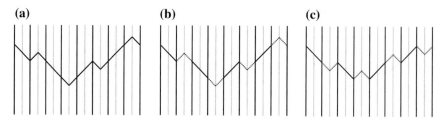

Fig. 3.3 Discrete wave dynamics. Elastic string is held fixed where it crosses black bars

(a)   **(b)**   **(c)**

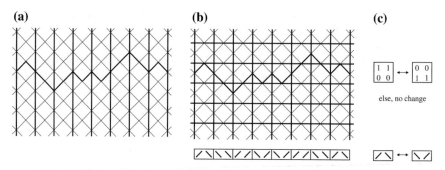

Fig. 3.4 Discrete wave dynamics. **a** A wave configuration. Possible wave paths are indicated by *dotted lines.* **b** *Horizontal and vertical lines* indicate one of two partitions used for discrete update rule. A 1D array summarizing wave gradients is shown below (not part of the 2D dynamics). **c** Top, dynamical rule for 2D wave. Presence of wave-path segments is indicated by 1's. Bottom, equivalent 1D dynamical rule for gradients

We express this dynamics as a purely digital rule in Fig. 3.4. In Fig. 3.4a we show a wave with the black bars marking the attachments for the next step. To simplify the figure we have suppressed the gray bars—they are always situated midway between the black bars and so don't need to be shown. We have also added a grid of 45° dotted lines that shows all of the segments that the string could possibly follow. In Fig. 3.4b we add in horizontal black bars, in order to partition the space into a set of 2 × 2 blocks that can be updated independently. Note that in all cases the segments that are allowed to change during this update step, as well as the cells that they will occupy after the update, are enclosed in a single block. The long box below Fig. 3.4b contains just the slope information from the string. This array of gradients is clearly sufficient to recreate the wave pattern if the height at one position is known. This is not part of the 2D dynamics: it will be discussed as a related 1D dynamics.

Figure 3.4c shows the dynamical rule for a block. Since the dotted lines indicate the direction in which segments must run if they appear in any cell, the state information for each segment is only whether it is there or not: this is indicated with a 1 or a 0. The only segments that change are peaks /\ or valleys \/, and these are represented by two 1's at the top of a block or at the bottom of a block respectively. The rule is that peaks and valleys turn into each other, and nothing else changes. We apply the rule alternately to the blocks shown, and to a complementary partition shifted half a block horizontally and vertically.

3.3.2   Exact Wave Behavior

At the bottom of Fig. 3.4c we've presented a dynamics for the *gradients* of the wave. The full 2D dynamics just turns peaks into valleys and vice versa, leaving straight segments unchanged: we can do that equally well on the array of gradients. As the 2D dynamics interchanges which blocking to use, the dynamics on the gradients also alternates which pairs of gradients to update together. In all cases, the dynamics on the gradients duplicates what happens on the string: if the two dynamics are both performed in parallel, the gradient listed below a column will always match the slope of the string in that column.

The dynamics on the gradients has an interesting property. Turning a peak into a valley and vice versa is exactly the same as swapping the left and right elements of a block. Leaving a // or \\ unchanged is also exactly the same as swapping the left and right elements of a block. In all cases, the dynamics on the gradients is equivalent to a swap.

This means that the left element of a block will get swapped into the right position, and at the next update it will be the left element of a new block and will again get swapped into the right position, and so on. Thus all of the gradients that start off in the left side of a block will travel uniformly to the right, and all that start in the right side of a block will travel uniformly to the left.

This shows that the system obeys a discrete version of the wave equation. Half of the gradients constitute a right-going wave, and half constitute a left-going wave. At any step of the dynamics, the 2D wave in the original dynamics is just the sum of the two waves: it is reproduced by laying gradients end to end.

If we refine the lattice, using more and more cells to represent a wave of given width, smoother and smoother waves can be represented. Of course even without going to a large-scale limit, the CA dynamics is already exactly equivalent to a continuous wave equation with constrained initial wave shapes, sampled at integer times: simply stretch the rightgoing and leftgoing waves constructed out of gradients to the full width of the lattice. This just amounts to drawing the wave shape corresponding to each block of the current partition a little differently.

3.3.3   Overall Transverse Motion

Assume the string carrying a discrete wave wraps around the space. We've discussed the horizontal motion of waves along such a string, but the string itself can move vertically. For example, a pattern such as \/\/\/ ... \/ all the way around the space reproduces itself after two partition update steps, but shifted vertically by two lattice units. This is clearly the maximum rate of travel for a string: one position vertically per update step. Call this v_0.

We can express the net velocity of the string in terms of the populations of right-going and leftgoing gradient segments. Let R_+ be the number of rightgoing segments with slope $+1$ (rightgoing $/$'s), and similarly for R_-, L_+ and L_-. If the width of the space is B blocks, then there are $B = R_+ + R_-$ segments forming the rightgoing wave, and $B = L_+ + L_-$ forming the leftgoing one.

For the rightgoing or leftgoing wave, periodically repeating its sequence of gradients corresponds to an unbounded wave with the same average slope. When both waves have shifted horizontally the width of one period (after $2B$ partition update steps), the net vertical shift is the sum of the slopes of the leftgoing gradients, minus the sum for the rightgoing ones: $(L_+ - L_-) - (R_+ - R_-)$. We can compute this by summing the differences for each pair of slopes grouped together in the columns of one partition. Only columns containing $\backslash/$ or $/\backslash$ contribute a non-zero difference, and so we only need to count the numbers of blocks $B_{\backslash/}$ and $B_{/\backslash}$ that are about to change, to compute the constant velocity

$$\frac{v}{v_0} = \frac{(L_+ - L_-) - (R_+ - R_-)}{2B} = \frac{B_{\backslash/} - B_{/\backslash}}{B}. \tag{3.4}$$

3.3.4 Ideal Energy and Momentum

Only blocks that change have overall motion, and with τ the time taken to update one partition, the frequencies of positive and negative motion are $B_{\backslash/}/\tau$ and $B_{/\backslash}/\tau$. Thus from (3.2), attributing a momentum to each changing block, the total ideal momentum up is $hB_{\backslash/}/2\tau v_0$, and down is $hB_{/\backslash}/2\tau v_0$, so the net ideal momentum $p = (h/2\tau v_0)(B_{\backslash/} - B_{/\backslash})$. From (3.4), the corresponding relativistic energy is $E = c^2 p/v = (hB/2\tau)/(v_0/c)^2$. Letting $v_0 \to c$ to minimize energy, and choosing units with $h = 2$ and $c = 1$ and $\tau = 1$, this becomes

$$E = B \quad \text{and} \quad p = B_{\backslash/} - B_{/\backslash}. \tag{3.5}$$

Energy is the constant width (in blocks) of the string, and momentum is the constant net number of blocks moving up.

There is an interesting subtlety involved in letting $v_0 \to c$ in the 2D dynamics. We interpret all gradient segments as always moving, swapping in pairs in each update in order to recover the wave equation—even though some paired segments are in different blocks when they "swap" identical values. If all block motion forward or backward is at the speed c, each segment must be interpreted as traveling at the speed $c\sqrt{2}$ as it swaps diagonally. If instead we interpret segments as moving up and down (or not moving), none travel faster than light, but the interaction is non-local at the scale of an individual block.

3.3.5   Rest Frame Energy

For the transverse motion of the string to approach the maximum speed, almost all of the block updates must contribute to overall motion, and almost none to just internally changing the string. This slowdown of internal dynamics is a kind of time dilation, which is reflected in the rest frame energy $\sqrt{E^2 - p^2}$. From (3.5),

$$E_r = \sqrt{B^2 - (B_\/ - B_\wedge)^2} \,. \tag{3.6}$$

The energy E_r available for rest-frame state-change decreases as more blocks move in the same direction. In this model total energy E is independent of v, hence rest energy $E_r = E/\gamma$ must approach 0 as $1/\gamma \to 0$. This contrasts with a normal relativistic system that can never attain the speed of light, which has a constant rest energy E_r and a total energy E that changes with v.

   The analysis up to here applies equally well to both the 1D and 2D versions of the dynamics of Fig. 3.4. In 2D, however, there is an additional constraint: there must be an equal number B of positive and negative slopes, so that the string meets itself at the periodic boundary. Since there are also an equal number B of right and left going gradients, $R_+ = L_-$ and $R_- = L_+$. Thus from (3.4) and (3.6),

$$E_r = 2\sqrt{R_+ L_+} \,. \tag{3.7}$$

If R_+/B were the probability for a walker to take a step to the right, and L_+/B the probability to the left, then (3.7) would be the standard deviation for a $2B$-step random walk. Related models of diffusive behavior that make contact with relativity are discussed in [24, 27, 28]. None of these define relativistic objects that have an internal dynamics, however.

3.4   Discussion

Given the definition of a finite-state dynamics, we could try to assign intrinsic properties to it based on the best possible implementation. For example, programming it on an ordinary computer, a basic property is the minimum time needed, on average, to simulate a step of the dynamics. It would be hard, though, to be sure we've found the most efficient mapping onto the computer's architecture, and the minimum time would change if we used a different computer, or built custom hardware using various technologies. The true minimum time would correspond to the fastest possible implementation allowed by nature! Such a definition seems vacuous, though, since we don't know the ultimate laws of nature, and even if we did, how would we find the best possible way to use them?

Surprisingly, a fundamental-physics based definition of intrinsic properties is not in fact vacuous, if we base it on general principles. Assuming the universe is fundamentally quantum mechanical, we couldn't do better than to simply *define* a hamiltonian that exactly implements the classical finite-state dynamics desired at discrete times, with no extra distinct states or distinct state change. This ideal hamiltonian identifies the fastest implementation that is *mathematically possible*, with given average energy.

This procedure assigns to every invertible finite-state dynamics an ideal energy that depends only on the average rate of distinct state change. This is generally not much like a physical energy, though, since we haven't yet included any realistic constraints on the dynamics. For example, each state change might correspond to a complete update of an entire spatial lattice, as in the synchronous definition of a CA. Then the energy would be independent of the size of the system. We can fix this by constraining the finite-state dynamics to be local and not *require* synchrony: defining it in terms of gates that are applied independently.

We expect the ideal energy, and distinct portions of it, to become more realistic with additional realistic constraints. For this reason, we studied invertible lattice dynamics derived from the integer-time behavior of idealized classical mechanical systems. In the examples we looked at, ideal energies and momenta defined by local rates of state change evolve like classical relativistic quantities.

It seems interesting and novel to introduce intrinsic definitions of energy and other physical quantities into classical finite-state systems, and to use these definitions in constructing and analyzing finite-state models of physical dynamics. Since all finite-energy systems in the classical world actually have finite state, and since classical mechanics doesn't, this may be a productive line of inquiry for better modeling and understanding that world. Moreover, inasmuch as all physical dynamics can be regarded as finite-dimensional quantum computation, finite-state models of classical mechanics may play the role of ordinary computation in understanding the more general quantum case.

Acknowledgments I thank Micah Brodsky and Gerald Sussman for helpful discussions.

References

1. Bekenstein, J.D.: Universal upper bound on the entropy-to-energy ratio for bounded systems. Phys. Rev. D **23**, 287 (1981)
2. Margolus, N., Levitin, L.B.: The maximum speed of dynamical evolution. Phys. D **120**, 188 (1998)
3. Margolus, N.: The finite-state character of physical dynamics. arXiv:1109.4994
4. Planck, M.: On the law of distribution of energy in the normal spectrum. Ann. Phys. (Berlin) **309**, 553 (1901)
5. Margolus, N.: Quantum emulation of classical dynamics. arXiv:1109.4995
6. Margolus, N.: Physics like models of computation. Phys. D **10**, 81 (1984)
7. Margolus, N.: Crystalline Computation. In: Hey, A. (ed.) Feynman and Computation. Perseus Books, 267 (1998). arXiv:comp-gas/9811002

8. Toffoli, T., Margolus, N.: Cellular Automata Machines: A New Environment for Modeling. MIT Press, Cambridge (1987)
9. Chopard, B., Droz, M.: Cellular Automata Modeling of Physical Systems. Cambridge University Press, Cambridge (2005)
10. Rothman, D., Zaleski, S.: Lattice Gas Cellular Automata: Simple Models of Complex Hydrodynamics. Cambridge University Press, Cambridge (2004)
11. Rivet, J.P., Boon, J.P.: Lattice Gas Hydrodynamics. Cambridge University Press, Cambridge (2005)
12. Fredkin, E.: A computing architecture for physics. In: CF '05 Proceedings of the 2nd conference on computing frontiers. p. 273. ACM (2005)
13. Wolfram, S.: A New Kind of Science. Wolfram Media, Champaign (2002)
14. Toffoli, T., Margolus, N.: Invertible cellular automata: a review. Phys. D **45**, 229 (1990)
15. Kari, J.: Representation of reversible cellular automata with block permutations. Math. Syst. Theory **29**, 47 (1996)
16. Durand-Lose, J.: Representing reversible cellular automata with reversible block cellular automata. Discrete Math. Theor. Comp Sci. Proc. AA, 145 (2001)
17. Ulam, S.: Random Processes and Transformations. In: Proceedings of the International Congress on Mathematics, 1950, Vol. 2, p. 264. (1952)
18. von Neumann, J.: Theory of Self-Reproducing Automata, University of Illinois Press, Champaign (1966)
19. Zuse, K.: Calculating Space. MIT Tech. Transl. AZT-70-164-GEMIT (1970)
20. Margolus, N.: Mechanical Systems that are both Classical and Quantum. arXiv:0805.3357
21. Margolus, N.: Universal cellular automata based on the collisions of soft spheres. In: Griffeath, D., Moore, C. (eds.) New Constructions in Cellular Automata. p. 231. Oxford University Press, Oxford (2003). arXiv:0806.0127
22. Fredkin, E., Toffoli, T.: Conservative logic. Int. J. Theor. Phys. **21**, 219 (1982)
23. Hrgovčić, H.: Discrete representations of the n-dimensional wave equation. J. Phys. A: Math. Gen. **25**, 1329 (1992)
24. Toffoli, T.: Action, or the fungibility of computation. In: Hey, A (ed.) Feynman and Computation, p. 349. Perseus Books (1998)
25. Margolus, N.: Physics and computation. Ph.D. Thesis, Massachusetts Institute of Technology (1987)
26. Toffoli, T.: Cellular automata as an alternative to (rather than an approximation of) differential equations in modeling physics. Phys. D **10**, 117 (1984)
27. Smith, M.: Representation of geometrical and topological quantities in cellular automata. Phys. D **45**, 271 (1990)
28. Ben-Abraham, S.I.: Curious properties of simple random walks. J. Stat. Phys. **73**, 441 (1993)

Chapter 4
An Analogue-Digital Model of Computation: Turing Machines with Physical Oracles

Tânia Ambaram, Edwin Beggs, José Félix Costa, Diogo Poças and John V. Tucker

Abstract We introduce an abstract analogue-digital model of computation that couples Turing machines to oracles that are physical processes. Since any oracle has the potential to boost the computational power of a Turing machine, the effect on the power of the Turing machine of adding a physical process raises interesting questions. Do physical processes add significantly to the power of Turing machines; can they break the Turing Barrier? Does the power of the Turing machine vary with different physical processes? Specifically, here, we take a physical oracle to be a physical experiment, controlled by the Turing machine, that measures some physical quantity. There are three protocols of communication between the Turing machine and the oracle that simulate the types of error propagation common to analogue-digital devices, namely: infinite precision, unbounded precision, and fixed precision. These three types of precision introduce three variants of the physical oracle model. On fixing one archetypal experiment, we show how to classify the computational power of the three models by establishing the lower and upper bounds. Using new techniques and ideas about timing, we give a complete classification.

T. Ambaram (✉) · J. Félix Costa
Department of Mathematics, Instituto Superior Técnico, Universidade de Lisboa,
Lisboa, Portugal
e-mail: taniambaram@gmail.com

J. Félix Costa
e-mail: fgc@math.ist.utl.pt

E. Beggs · J.V. Tucker
College of Science, Swansea University, Singleton Park, Swansea,
SA2 8PP, Wales, UK
e-mail: e.j.beggs@swansea.ac.uk

J.V. Tucker
e-mail: j.v.tucker@swansea.ac.uk

D. Poças
Department of Mathematics and Statistics, McMaster University, Hamilton,
ON L8S 4K1, Canada
e-mail: diogopocas1991@gmail.com; pocasd@math.mcmaster.ca

© Springer International Publishing Switzerland 2017
A. Adamatzky (ed.), *Advances in Unconventional Computing*,
Emergence, Complexity and Computation 22,
DOI 10.1007/978-3-319-33924-5_4

73

4.1 Introduction

Loosely speaking, by an analogue-digital system we mean a system that makes physical measurements in the course of a digital computation; equivalently, the term hybrid system may be used. The digital computation can be of any complexity, though some embedded systems can be modelled by hybrid systems based upon finite automata. Actually, many processes perform an analogue measurement that is used in a digital computation of some kind. For example, models of analogue computation also involve digital elements: this is found in the construction of the analogue recurrent neural net model (ARNN) (see [1])[1]; in the optical computer model (see [2]); and in the mirror system model (see [3]).

Computational dynamical systems that are able to read approximations to real numbers also behave as analogue-digital systems: they perform digital computations, occasionally accessing some external values. At the moment of access, the "computer" executes some task on the analogue device, such as a test for a given value of a quantity. In the perfect platonic world, this test cab be performed with infinite or arbitrary precision, in the sense that the machine can obtain as many bits of the real number as needed; or, less ideally, with a fixed finite precision provided by the equipment in use.

To analyse the computational capabilities of analogue-digital systems, we introduce an abstract analogue-digital model of computation that couples Turing machines to oracles that are physical processes. Thus, we consider Turing machines with the ability of making measurements. The Turing machines considered are deterministic, but in fact they can use the oracle both to get advice and to simulate the toss of a coin.

Interacting with physical processes takes time. There are cost functions of the form $T : \mathbb{N} \to \mathbb{N}$ that gives the number of time steps to perform the measurements that the Turing machine allows. In weighing using a balance scale—as, indeed, in most measurements—the pointer moves with an acceleration that depends on the difference of masses placed in the pans. It does so in a way such that the time needed to detect a mass difference increases exponentially with the precision of the measurement, no matter how small that difference can be made. This means that the measurement has an exponential cost that should be considered in the overall complexity of the analogue-digital computation.[2]

A possible objection to such a model is that it is not limited in precision; for even if it is sufficiently precise, it sooner or later finds the obstacle of the atomic structure of matter. However, measurement has its own theoretic domain and can only be conceived as a limiting procedure; see [4–6, 8, 9]. It means that measurement is like

---

[1] In the ARNN case, a subsystem of about eleven neurones (variables) performs a measurement of a *real-valued* weight of the network up to some precision and resumes to a computation with advice simulated by a system of a thousand rational neurones interconnected with integer and a very few rational weights.

[2] In contrast, in the ARNN model, the time of a measurement is linear due to the fact that the activation function of each neuron is piecewise linear instead of the common analytic sigmoid.

complexity and can only be conceived asymptotically. Once we limit, in absolute terms, space or time resources of a Turing machine, its complexity vanishes, since all (now finite) sets can be decided in linear time and constant space.[3]

A measurement can be fundamental or derived. Measuring distance is fundamental, but measuring velocity is derived. Commonly, according to Hempel [8], fundamental measurement is based on a partial order of comparisons that, taken to the limit, generates a real number. Comparisons in the sense of Hempel are based on events in the experimental setup. The most common measurement of some concept y—position, mass, electric resistance, etc.—consists of performing the experiment with a test value z, for which we could test one or both of the comparisons "$z < y$" and "$y < z$"; experiments allowing such comparisons are called two-sided.

Experiments in physics provide intuitions about abstract measurements, namely (a) that they result from comparisons, (b) that they have a cost, (c) that they come with errors, and (d) that they are stochastic. Coupling a Turing machine with a physical experiment, we construct a specific type of analogue-digital machine.

Since any oracle has the potential to boost the computational power of a Turing machine, the effect on the power of the Turing machine of adding a physical process is an interesting area to investigate.

Do experiments add significantly to power to the Turing machine, and break the Turing Barrier? Does the power of the Turing machine vary with different experiments?

We classify computational power. In Sect. 4.2, from a number of physical experiments, the so-called *smooth scatter machine*, first seen in [13], is selected as representative of the analogue-digital of machines of interest. In Sect. 4.3, we summarise the analogue-digital model. Sect. 4.4, we begin by summarising the theory of non-uniform complexity classes, which allows us to formalise a real number value for a parameter, by encoding an advice function of some class $\mathcal{F}\star$, and reading it into the memory of the machine. The section continues with lower bounds (starting Sect. 4.4.3) and upper bounds (starting Sect. 4.4.7).

The role of precision of data has been explicated in our earlier papers, but the role of precision in time has been less clear. The notion is subtle: there is time in the physical theory, time that manages the oracle queries, and the runtime of the computation. In addition to explaining the model and its properties, we develop some new techniques and results to explore the nature of timing. In Sect. 4.5, we use a technique, based on knowing time exactly, to lower the upper bounds.

All the concepts will be defined in due course. Among the results proved in this paper are:

Theorem 1 *The classification of the computational power of the analogue-digital model in the absense and presence of errors is as follows:*

---

[3]In the same way, we could also say that tapes of Turing machines can have as many cells as the number of particles in the universe, but in such a case no interesting theory of computability can be developed.

1. *If B is decidable by a smooth scatter machine with infinite precision and exponential protocol, clocked in polynomial time, then $B \in P/\log^2\star$.*
2. *B is decidable by a smooth scatter machine with infinite precision and exponential protocol $T(k) \in \Omega(2^k)$, clocked in polynomial time, if, and only if, $B \in P/\log\star$.*
3. *If B is decidable by a smooth scatter machine with unbounded or fixed precision and exponential protocol, clocked in polynomial time, then $B \in BPP//\log^2\star$.*
4. *B is decidable by a smooth scatter machine with unbounded or fixed precision and exponential protocol, clocked in polynomial time, and having access to exact physical time if, and only if, $B \in BPP//\log\star$.*

Moreover, we will argue that such bounds are, to a large extent, really independent of the analogue system considered.

This paper offers an overview of the development of the analogue-digital model with technical details and new results. Here we have the first complete analysis of the analog-digital machine having access to two-sided measurements, including the full proofs of the lower and upper computational bounds for the polynomial time case.

4.2 Physical Oracles

The substrate of a real number computation is a physical process executed by a machine; the input data is in many cases obtained by a physical measurement process. During a computation, the machine may have access to external quantities. For example, an analogue-digital device controlling the temperature in a building "solves" differential equations while "calling" thermometers to check the values of the temperature through time.

This idea that both computation and data are in some way physical processes motivates the theoretical quest for the limits of computational power of a physical process. We propose the coupling of a Turing machine with a physical oracle. The idea is to replace the standard oracle, which is just a set, by a physical experiment that allows the machine to perform a measurement.

A Turing machine coupled with such an oracle becomes a hybrid computation model, having an analogue component—the experiment—combined with a digital component—the Turing machine. As the standard Turing machine with oracle, the Turing machine with physical oracle will use the information given by the physical oracle in its computations.

4.2.1 Types of Physical Oracles

The physical oracles that we will be considering are physical processes that enable the Turing machine to measure quantities. As far as we have investigated (see [7]), measurements can be classified in one of the three types: *one-sided*—also called *threshold*—*measurement*, *two-sided measurement* and *vanishing measurement*.

Fig. 4.1 *Threshold* Can only measure $y < z$. Can give a sequence of tests approximating z from below

Fig. 4.2 Can measure both $a < x$ and $x < a$. A bisection method can be used to find a

Fig. 4.3 *Vanishing* Can only measure ($a < x$ or $x < a$). A modified bisection method will work. Assume monotonicity on each side of a

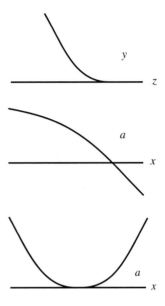

A one-sided experiment is an experiment that approximates the unknown value a just from one side, i.e., it approximates an unknown value a either with values x from above or with values x from below, checking either if $a < z$ or if $z < a$, but not both (see Fig. 4.1).

A two-sided experiment is an experiment that approximates the unknown value a from two sides, i.e., it approximates the unknown value a with values x from above and with values x from below, checking if $a < x$ and $x < a$ (see Fig. 4.2).

A vanishing experiment is an experiment that approximates the unknown value y from the physical time taken by the experiment (see Fig. 4.3).

Note that this type of experimental classification is neither in Hempel's original work in [8], nor in the developed theory synthesised in [9], since, in a sense, these authors only consider the two-sided experiments.

In this paper we deal only with the two-sided measurement. Threshold experiments were considered in [10] and vanishing experiments in [11].

We now illustrate this category with a gallery of two-sided experiments of measurement, emphasising the experimental time.

4.2.1.1 Balance Experiment

The balance experiment is intended to measure the (gravitational) mass of some physical body. To perform the experiment we put the unknown mass m_A of body A in one of the pans and we approximate its value by placing another body, B, in the other pan. Body B can have a known mass, m_B, bigger or smaller than m_A. After

Input: The masses m_A (unknown) and m_B (test mass)
if $m_A > m_B$ then the pan containing A goes down;
if $m_A < m_B$ then the pan containing B goes down
      else both pans will stand side by side;

Fig. 4.4 The balance experiment

placing the body B, the balance can display one of the behaviors explained in the right hand side of Fig. 4.4.

Accordingly to the behavior of the balance, we can change the mass of B in order to approximate m_A, using linear search. Repeating the experiment a number of times and considering bodies with masses each time closer to m_A we get the desired approximation of the unknown value. The time needed to detect a move in the pointer of the balance scale is exponential in the number of bits of precision of the measurement.

4.2.1.2 Elastic Collisions

The collision experiment can be used to measure the (inertial) mass of a body. This experiment was already studied as a physical oracle in [12]. To perform the experiment, we project a test body B with known mass m_B and velocity v_B along a line towards the body A at rest with unknown mass m_A. The mass of B, m_B, can be bigger or smaller than m_A. After the collision the bodies acquire new speeds, accordingly to their relative masses. After the collision we can have one of the behaviors explained in the right hand side of of Fig. 4.5. According to the behavior of the particles, we can change the mass of B in order to approximate m_A, using linear search. Repeating the experiment a number of times and considering masses each time closer to m_A, we get an approximation of the unknown value. The time needed to detect a possible motion of body A is exponential in the number of bits of precision of the measurement.

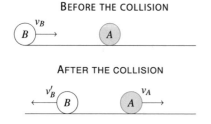

BEFORE THE COLLISION

AFTER THE COLLISION

Input: The masses m_A (unknown) and m_B (test mass)
if $m_A > m_B$ then the test body B moves backwards;
if $m_A < m_B$ then the test body B moves forwards;
      else the test body B comes to rest and A is projected
          forward with speed v_B;

Fig. 4.5 The collision experiment

4.2.1.3 The Foucault's Pendulum

The construction of this *gedankenexperiment* is based on the principles of classical mechanics (see Fig. 4.6). The motion of the pendulum exhibits two coupled harmonic components:

1. A periodic motion of period

$$T = 2\pi \sqrt{\frac{\ell}{g}},$$

   corresponding to the classical period of the pendulum of length ℓ for small amplitude oscillations.
2. Another periodic motion of period

$$\tau = \frac{24\,\text{h}}{|\sin \lambda|},$$

   corresponding to a complete rotation of the plane of oscillation at latitude λ. Moreover, the rotation of the plane of oscillation of the pendulum is *clockwise* to the north of equator and *counterclockwise* to the south of equator.

This two-sided experiment can be designed to locate the equator. The time needed for the pendulum to cross the angular distance of 1^s of arc is given by

$$t_\lambda = \frac{1}{15|\sin \lambda|}\,s,$$

that, for small values of the angle λ, is of the order of

$$t_\lambda = \frac{1}{|15\lambda|}\,s,$$

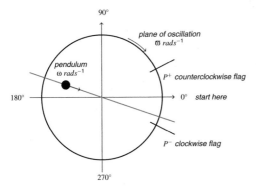

Input: The latitudes $\lambda_0 = 0°$ (equator) and λ (test)
if $\lambda > 0$ then plane of oscillation crosses the flag P^+;
if $\lambda < 0$ then plane of oscillation crosses the flag P^-;
else the plane of oscillation does not rotate ;

Fig. 4.6 Foucault's machine experiment

going to infinity as the angle λ approaches 0. In the case of a dyadic value λ, this time is exponential in the size of λ. The time needed to detect the equator is exponential in the number of bits of precision.

4.2.2 The Smooth Scatter Experiment

We now introduce the measurement experiment that we take as a generic example throughout this paper.

The smooth scatter experiment (SmSE for short), considered in [13], is a variation of the sharp scatter experiment introduced in [14, 15]. As the sharp scatter experiment, it measures the position of a vertex but it has a slightly different experimental apparatus. In the sharp case we considered a vertex in a sharp wedge and in the smooth case we are considering a vertex in a rounded wedge (see Fig. 4.7), where the shape of the wedge is given by a smooth symmetric function.

In order to measure the vertex position y, the smooth scatter experiment sets a cannon at some position z, shoots a particle from the cannon, and then waits some time until the particle is captured in a detection box. We consider that y can take any value in $]0, 1[$. The fire and detection phases define one run of the experiment. After the shooting, we can have one of the behaviours explained in the algorithm of Fig. 4.7 (right). By analyzing in which box the particle is detected, it is possible to conclude the relative position of y and z, i.e., whether the vertex is in the right or left side of the cannon. Then, repeating this procedure, by resetting the cannon position and executing one more run of the experiment, we can better approximate the vertex position.

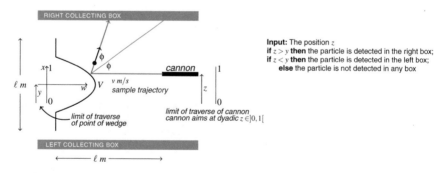

Fig. 4.7 The smooth scatter experiment

4.2.2.1   The Physical Theory

The $SmSE$ is governed by a fragment of Newtonian mechanics, consisting in the following laws and assumptions:

1. Particles obey Newton's laws of motion in the two dimensional plane;
2. Collisions between barriers and particles are perfectly elastic, that is, kinetic energy is preserved;
3. Barriers are rigid and do not deform upon impact;
4. Cannon projects a particle with a given velocity and direction;
5. Detectors are capable of telling if a particle has crossed them; and
6. A clock measures the time.

The experimental clock that measures the physical time $\tau : [0, 1] \rightarrow \mathbb{R}$ is very important in analysing the cost of accessing the oracle. When the Turing machine accesses the oracle it must wait until the end of the experiment in order to continue with its computation, so this means that accessing the oracle no longer takes one computation step but "t steps", where t is a function of the precision of the query.

4.2.2.2   The Time

The access to the $SmSE$ will cost more than one computation step as the physical experiment takes some intrinsic time t to be performed, designated as physical time. As explained in [13, 16], the time required to run the $SmSE$ is given by the following proposition:

Proposition 1 *Consider that $g(x)$ is the function describing the shape of the wedge of a $SmSE$. Suppose that $g(x)$ is n times continuously differentiable near $x = 0$, all its derivatives up to $(n - 1)$th vanish at $x = 0$, and the nth derivative is nonzero. Then, when the SmSE, with vertex position y, fires the cannon at position z, the time needed to detect the particle in one of the boxes is $t(z)$, where:*

$$\frac{A}{|y - z|^{n-1}} \leq t(z) \leq \frac{B}{|y - z|^{n-1}}, \tag{4.1}$$

for some $A, B > 0$ and for $|y - z|$ sufficiently small.

Looking at Eq. 4.1, we conclude that we cannot bound the physical time, as it goes to infinity when z and y become close. Without loss of generality, we will assume from now on that $n = 2$, that is, $g(x) \in C^2$, since the results are essentially the same for values of $n > 2$.

4.3   The Smooth Scatter Machine

We now fix the smooth scatter experiment as paradigmatic of two-sided measurement experiments and will proceed with the developing of our theory of analogue-digital computation.

A Turing machine coupled with the smooth scatter experiment as oracle is called a smooth scatter machine ($SmSM$). This machine was introduced in [13]. The $SmSM$ communicates with the $SmSE$ using the query tape. During a computation, the $SmSM$ performs its standard transitions and, whenever necessary, it consults the oracle. The oracle is an experiment that allows the Turing machine to measure an approximation to the vertex position of the scatter and use this value in the computation. In order to initialize the experiment, the Turing machine writes in the query tape the parameters for the experiment, e.g., the position of the cannon. After some time, the machine will be in other state according to the outcome of the experiment. To guarantee that our machine does not wait unbounded time for the answer, we will add to the Turing machine a *time constructible schedule* $T : \mathbb{N} \to \mathbb{N}$. The schedule depends on the size of the query.

Therefore, after some time not exceeding $T(n)$, where n is the size of the query, if the particle is detected in the right box, then the next state of the Turing machine is q_r; if the particle is detected in the left box, then the next state is q_l; and if the particle is not detected in any box, then the next state is q_t. The machine resumes the computation. The states q_r, q_l, and q_t replace the standard states *yes* and *no* of an oracle Turing machine. Note that, during the consultation of the oracle, the Turing machine waits for the oracle answer but keeps counting the running time in parallel.

4.3.1   Communicating with the Smooth Scatter Experiment

The Turing machine communicates with the $SmSE$ when necessary. The communication is made through the query tape, where the Turing machine writes the parameters of the experiment, in this case the position of the cannon. We consider that the position provided by the Turing machine is written in binary and denotes a number of the form $n/2^k$, where n is a non-negative integer in $]0, 2^k[$ and k is a positive integer. The query z corresponds to the dyadic rational

$$z = \sum_{i=1}^{|z|} 2^{-i} z[i], \tag{4.2}$$

where $z[i]$ denotes the ith bit of z. After processing z, the apparatus should set the position of the cannon and execute the experiment. Note that by Eqs. 4.1 and 4.2, with the vertex at $y \in]0, 1[$, the experimental time grows exponentially with the size of the query, for values of z approaching y.

We consider that our experiment can set the cannon's position in three ways, inducing different computational powers. The three protocols are:

Protocol 1 *Given a dyadic rational z in the query tape, the experiment sets the position of the cannon to $z' = z$. In this case we are working with an error-free $SmSE$ or with an infinite precision $SmSE$—the protocol is $Prot\_IP(z)$. A Turing machine coupled with this oracle is called an* error-free smooth scatter machine.

Protocol 2 *Given a dyadic rational z in the query tape, the experiment sets the position of the cannon to $z' \in [z - 2^{-|z|}, z + 2^{-|z|}]$, chosen uniformly. In this case we are working with an error-prone $SmSE$ with unbounded precision—the protocol is $Prot\_UP(z)$. A Turing machine coupled with this oracle is called an* error-prone smooth scatter machine with unbounded precision.

Protocol 3 *Given a dyadic rational z in the query tape, the experiment sets the position of the cannon to $z' \in [z - \varepsilon, z + \varepsilon]$, chosen uniformly, for a fixed $\varepsilon = 2^{-q}$, for some positive integer q. In this case we are working with an error-prone $SmSE$ with fixed precision—the protocol is $Prot\_FP(z)$. A Turing machine coupled with this oracle is called an* error-prone smooth scatter machine with fixed precision.

We can describe all the protocols in the Algorithm 1.

The procedure $Prot$ used in Algorithm 1 assigns a value to z' according with the protocol, i.e., using infinite precision ($Prot(z, inf) = z$), unbounded precision ($Prot(z, unb) \in [z - 2^{-|z|}, z + 2^{-|z|}]$), or fixed precision ($Prot(z, fix) \in [z - \varepsilon, z + \varepsilon]$).

Algorithm 1: General communication protocol.

Data: Dyadic rational z (possibly padded with 0's)
Set cannon's position to $z' = Prot(z, mode)$;
Wait $T(|z|)$ units of time ;
if *The particle is detected in the right box* **then**
  | A transition to the state q_r occurs ;
end
if *The particle is detected in the left box* **then**
  | A transition to the state q_l occurs ;
end
if *The particle is not detected in any box* **then**
  | A transition to the state q_t occurs ;
end

Definition 1 Let $A \subseteq \{0, 1\}^\star$. We say that an error-free $SmSM$ $\mathcal{M}$, clocked with runtime $\tau : \mathbb{N} \to \mathbb{N}$, decides A if, for every $w \in \{0, 1\}^\star$, $\mathcal{M}$ accepts w in at most $\tau(|w|)$ steps if $w \in A$ and $\mathcal{M}$ rejects w in at most $\tau(|w|)$ steps if $w \notin A$.

Definition 2 Let $A \subseteq \{0, 1\}^\star$. We say that an error-prone $SmSM$ $\mathcal{M}$, with unbounded or fixed precision, clocked in runtime $\tau : \mathbb{N} \to \mathbb{N}$, decides A if there exists γ, $1/2 < \gamma \leq 1$, such that, for every $w \in \{0, 1\}^*$, $\mathcal{M}$ accepts w in at most $\tau(|w|)$ steps with probability γ if $w \in A$ and $\mathcal{M}$ rejects w in at most $\tau(|w|)$ steps with probability γ if $w \notin A$.

4.3.2 Measurement Algorithms

The measurement of the vertex position depends on the protocol, so that different protocols may originate different measurement algorithms. From now on we denote by $m|_\ell$ the first ℓ digits of m, if m has ℓ or more than ℓ digits, otherwise it represents m padded with k zeros, for some k such that $|m0^k| = \ell$. The pruning or the padding technique is used to control the time schedule during the measurement process. For the infinite precision we have the measurement Algorithm 2.

In Algorithm 2 we have no errors associated with the protocol since the cannon's position is always set to the desired position. Thus, doing this search around the vertex, we can approximate it up to any precision with no errors. The following result was already proved in [10, 13].

Algorithm 2: Measurement algorithm for infinite precision.

Data: Positive integer ℓ representing the desired precision
$x_0 = 0$;
$x_1 = 1$;
$z = 0$;
while $x_1 - x_0 > 2^{-\ell}$ **do**
  $z = (x_0 + x_1)/2$;
  $s = Prot\_IP(z|_\ell)$;
  if $s ==$ "q_r" **then**
    $x_1 = z$;
  end
  if $s ==$ "q_l" **then**
    $x_0 = z$;
  else
    $x_0 = z$;
    $x_1 = z$;
  end
end
return Dyadic rational denoted by x_0

Proposition 2 Let s be the result of $Prot\_IP(z|_\ell)$, y the vertex position of the $SmSE$, and T the time schedule. If $s =$ "q_l", then $z|_\ell < y$; if $s =$ "q_r", then $z|_\ell > y$; otherwise $|y - z|_\ell| < \frac{C}{T(\ell)}$, for some constant $C > 0$.

Proof The first two cases are obvious. In the third case we have

---

Algorithm 3: Measurement algorithm for unbounded precision.

Data: Positive integer ℓ representing the precision
$x_0 = 0$;
$x_1 = 1$;
$z = 0$;
while $x_1 - x_0 > 2^{-\ell}$ **do**
    $z = (x_0 + x_1)/2$;
    $s = Prot\_UP(z|_\ell)$;
    if $s == $ "q_r" **then**
        $x_1 = z$;
    end
    if $s == $ "q_l" **then**
        $x_0 = z$;
    else
        $x_0 = z$;
        $x_1 = z$;
    end
end
return *Dyadic rational denoted by* x_0

$$\frac{C}{|y - z|_\ell|} = t(z|_\ell) > T(\ell),$$

for some constant $C > 0$. Whence, $|y - z|_\ell| < \frac{C}{T(\ell)}$.   □

Unbounded precision requires the measurement Algorithm 3.

In the Algorithm 3 we have errors associated with the protocol since the cannon's position is set to a value uniformly chosen in the closed interval $[z - 2^{-|z|}, z + 2^{-|z|}]$. But, according to the definition of $Prot\_UP$ (see Protocol 2), the interval that is considered could be increasingly smaller by increasing the precision, and so the error will be increasingly smaller too.

We proved in [10, 14] the following result, where ℓ is such that $[z|_\ell - 2^{-\ell}, z|_\ell + 2^{-\ell}] \in]0, 1[$.

Proposition 3 *Let* s *be the result of* $Prot\_UP(z|_\ell)$, y *the vertex position of the SmSE, and* T *the time schedule. If* $s = $ "q_l", *then* $z|_\ell < y + 2^{-\ell}$, *if* $s = $ "q_r", *then* $z|_\ell > y - 2^{-\ell}$, *otherwise* $|y - z|_\ell| < \frac{C}{T(\ell)} + 2^{-\ell}$, *for some constant* $C > 0$.

Proof Let z' be the position uniformly chosen by the $Prot\_UP(z|_\ell)$. Thus $z' \in [z|_\ell - 2^{-\ell}, z|_\ell + 2^{-\ell}]$ and $|z|_\ell - z'| \leq 2^{-\ell}$. From this it follows that, if $s = $ "q_l", then $z' < y$ and therefore $z|_\ell < y + 2^{-\ell}$; if $s = $ "q_r", then $z' > y$ and therefore $z|_\ell > y - 2^{-\ell}$; otherwise, $|y - z'| < \frac{C}{T(\ell)}$, for some constant $C > 0$. Therefore $|y - z|_\ell| < \frac{C}{T(\ell)} + 2^{-\ell}$.   □

Proposition 4 *Given a SmSM with unbounded or infinite precision, vertex position at* y *and time schedule* $T(\ell) = C2^\ell$,

1. *The time complexity for the measurement algorithm for input ℓ is $\mathcal{O}(\ell T(\ell))$;*
2. *The output of the algorithm, for input ℓ, is a dyadic rational z such that $|y - z|$*
   $< 2^{-\ell+1}$.

Proof Statement (1) comes from the fact that the call to the protocol is repeated ℓ times, each taking $\mathcal{O}(T(\ell))$ steps. To prove statement (2) note that: (a) in the infinite precision case, the execution of the algorithm halts when $|y - z| < \frac{C}{T(\ell)}$; (b) in the worst case of the unbounded precision, the execution of the algorithm halts when $|y - z| < \frac{C}{T(\ell)} + 2^{-\ell}$, i.e., $|y - z| < 2^{-\ell+1}$; in both cases we have $|y - z| < 2^{-\ell+1}$. $\qquad\qquad\qquad\qquad\qquad\qquad\qquad\qquad\qquad\qquad\qquad\qquad\qquad\quad \square$

For the fixed precision protocol the measurement algorithm is different from the previous two.

Assume that initially we have a $SmSE$ with vertex at any real number $s \in {]}0, 1{[}$. We consider another $SmSE$ with vertex at position $y = 1/2 - \varepsilon + 2s\,\varepsilon$ and repeat ξ times the $Prot\_FP$ protocol with input $z' = 1$, i.e., with the vertex at position $1/2$. Consider a fixed precision $\varepsilon = 2^{-q}$, for some $q \in \mathbb{N}$, and physical time given by the Eq. 4.1. Consider also a Turing machine coupled with this $SmSE$ with a fixed schedule T. Then we will have three possible outcomes after running the experiment: q_r, q_l, or q_t. If η represents the position of the cannon where the time schedule exceeds the limit, for some precision ℓ, we have that the output q_r occurs for cannon positions in the interval $[y + \eta, 1/2 + \varepsilon]$, the output q_t occurs for cannon positions in the interval $[y - \eta, y + \eta]$ and the output q_l occurs for cannon positions in the interval $[1/2 - \varepsilon, y - \eta]$. Therefore we have that each outcome occurs with the following probability:

1. For the output q_r we have $p_r = (1/2 + \varepsilon - y - \eta)/(2\varepsilon)$.
2. For the output q_t we have $p_t = (y + \eta - y + \eta)/(2\varepsilon)$.
3. For the output q_l we have $p_l = (y - \eta - 1/2 + \varepsilon)/(2\varepsilon)$.

So, our experiment can be modeled as a multinomial distribution with three categories of success, each one with the stated probabilities. Let α, β, and γ be random variables used to count outcomes q_l, q_r, and q_t, respectively, and consider a new random variable $X = \frac{2\alpha + \delta}{2\xi}$. The expected value of X is

$$\bar{X} = \frac{2E[\alpha] + E[\delta]}{2\xi} = \frac{-1 + 2y + 2\varepsilon}{4\varepsilon} = \frac{4s\,\varepsilon}{4\varepsilon} = s.$$

The variance of X is

$$Var(X) = \left(\frac{1}{\xi}\right)^2 Var(\alpha) + \left(\frac{1}{2\xi}\right)^2 Var(\delta) + 2\left(\frac{1}{\xi}\right)\left(\frac{1}{2\xi}\right) Covar(\alpha, \delta).$$

Simplifying we get

$$Var(X) = -\frac{-4s\,\varepsilon + 4s^2\,\varepsilon + \eta}{4\varepsilon\xi} \le \frac{4s\,\varepsilon(1 - s)}{4\varepsilon\xi} \le \frac{1}{\xi}.$$

Using the Chebyshev's Inequality, for $\Delta > 0$, we get:

$$P(|X - \bar{X}| > \Delta) \leq \frac{Var(X)}{\Delta^2}$$

$$P(|X - s| > \Delta) \leq \frac{1}{\xi \Delta^2}$$

For precision $\Delta > 1/2^\ell$, we get

$$P(|X - s| > 1/2^\ell) \leq \frac{2^{2\ell}}{\xi} .$$

If we consider

$$\frac{2^{2\ell}}{\xi} < 2^{-h},$$

for $h \in \mathbb{N} - \{0\}$, then we get $\xi > 2^{2l+h}$.

Finally, we get the Algorithm 4 for the fixed precision case, where h is a positive integer chosen in order to get the error less than 2^{-h}.

Algorithm 4: Measurement algorithm for fixed precision.

Data: Integer ℓ representing the precision
$c = 0$;
$i = 0$;
$\xi = 2^{2\ell+h}$;
while $i < \xi$ **do**
$\quad$ $s = Prot\_FP(1|_\ell)$;
$\quad$ **if** $s == $ "q_l" **then**
$\quad\quad$ | $c = c + 2$;
$\quad$ **end**
$\quad$ **if** $s == $ "q_t" **then**
$\quad\quad$ | $c = c + 1$;
$\quad$ **end**
$\quad$ i++ ;
end
return $c/(2\xi)$

The following statements have been proved in [10].

Proposition 5 *For any real number $s \in]0, 1[$, fixed precision $\varepsilon = 2^{-q}$ for $q \in \mathbb{N}$, error probability 2^{-h} for $h \in \mathbb{N} - \{0\}$, vertex position at $y = 1/2 - \varepsilon + 2s\,\varepsilon$, and time schedule T,*

1. *The time complexity of the measurement algorithm for fixed precision with input ℓ is $\mathcal{O}(2^{2\ell}T(\ell))$;*
2. *The output of the algorithm is a dyadic rational m such that $|y - m| < 2^{-\ell}$.*

Proof Statement (1) derives from the fact that the procedure calls the oracle $2^{2\ell+h}$ times. Statement (2) is justified by observing that an approximation to $y \pm 2^{-\ell}$ with error probability less than 2^{-h} requires $2^{2\ell}/2^{-h}$ calls to the oracle. $\square$

We are now ready to investigate the computational power of analogue-digital machines clocked in polynomial time, using (without loss of generality) the smooth scatter machine. Note that, on input ℓ, all the measurement algorithms output a value with ℓ bits of precision, or a value with the maximum number of bits of precision $(< \ell)$ allowed by the schedule.

4.4 Computational Power of the Smooth Scatter Machine

The three types of $Sm\,SM$ obtained in the Sect. 4.3 express different computational powers. In this section, we first recall some concepts of nonuniform complexity classes (see [17, 18]).

4.4.1 Nonuniform Complexity Classes

An *advice function* is a total map $f : \mathbb{N} \to \{0, 1\}^*$ and a *prefix advice function* is an advice function with the extra condition that $f(n)$ is a prefix of $f(n + 1)$. These functions have an important role in nonuniform complexity as they provide external information to the machines that depend only in the input size. We recall the definition of nonuniform complexity class.

Definition 3 Let $\mathbb{C}$ be a class of sets and $\mathbb{F}$ be a class of advice functions. We define $\mathbb{C}/\mathbb{F}*$ as the class of sets B for which there exists a set $A \in \mathbb{C}$ and a prefix advice function $f \in \mathbb{F}$ such that, for every word $w \in \Sigma^*$ with size less or equal to n, $w \in B$ iff $\langle w, f(n)\rangle \in A$.

We consider deterministic Turing machines clocked in polynomial time and log-arithmic prefix advice functions (see [18]), obtaining the nonuniform complexity classes

$$P/\log\star \text{ and } P/\log^2\star.$$

We also consider probabilistic Turing machines clocked in polynomial time. Thus, based on the Definition 3, we get the following:

Definition 4 $BPP/\log\star$ is the class of sets B for which a probabilistic Turing machine $\mathcal{M}$, a prefix advice function $f \in \log$ and a constant $\gamma < 1/2$ exist such that, for every word w with size less or equal to n, $\mathcal{M}$ rejects $\langle w, f(|w|)\rangle$ with probability at most γ if $w \in B$ and accepts $\langle w, f(|w|)\rangle$ with probability at most γ if $w \notin B$.

The above definition is too restrictive, forcing the advice function to be chosen after the Turing machine (see Appendix A). Thus we choose to use a more relaxed definition where the Turing machine is chosen after the advice function.

Definition 5 $BPP//\log\star$ is the class of sets B for which, given a prefix advice function $f \in \log$, a probabilistic Turing machine $\mathcal{M}$ and a constant $\gamma < 1/2$ exist such that, for every word w with size less or equal to n, $\mathcal{M}$ rejects $\langle w, f(|w|)\rangle$ with probability at most γ if $w \in B$ and accepts $\langle w, f(|w|)\rangle$ with probability at most γ if $w \notin B$.

Similarly we define the nonuniform class $BPP//\log^2\star$, just changing the class of advice functions.

At this point we know that the Turing machine has an oracle that measures approximations of the vertex position. Since we are talking about nonuniform complexity classes, the information given by the advice is codified in the position of the vertex of the $SmSE$.

4.4.2   The Cantor Set C_3

The Cantor set C_3 is the set of ω-sequences x of the form

$$x = \sum_{k=1}^{+\infty} x_k 2^{-3k},$$

for $x_k \in \{1, 2, 4\}$. This means that C_3 corresponds to the set of elements composed by the triples 001, 010, or 100. This set is often used to codify real numbers. For example, in [19], the Cantor codification with base 4 and 9 is used to codify the real weights of neural nets.

This type of codification is required in order to be able to distinguish close values. For example, in order to describe the first bit of $011\cdots 1$ and $100\cdots 0$, we must read the whole number. To enforce gaps between close values, we encode the binary representations of the values in elements of the cantor set C_3.

We will work with prefix advice $f : \mathbb{N} \to \{0, 1\}^\star$, such that $f \in \log$. We denote by $c(w)$, with $w \in \{0, 1\}^\star$, the binary sequence obtained from the binary representation of w where each 0 is replaced by 100 and each 1 is replaced by 010. Given f, we denote its encoding as a real number by $y(f) = \lim y(f)(n)$ recursively defined as follows:

1. $y(f)(0) = 0.c(f(0))$;
2. If $f(n + 1) = f(n)s$, then $y(f)(n+1) = \begin{cases} y(f)(n)c(s) & \text{if } n + 1 \text{ is not power of 2} \\ y(f)(n)c(s)001 & \text{if } n + 1 \text{ is power of 2} \end{cases}$.

By means of this definition, $y(f)(n)$ is logarithmic in n and the encoding function returns a word of size $\mathcal{O}(f(n))$. Note that separators are added only at positions that are a power of 2. If we want to decode $f(|w|)$, such that $2^{m-1} < |w| \le 2^m$, we need to read $y(f)$ in triplets until we reach the $(m+1)$th separator. To reconstruct $f(2^m)$, we eliminate the separators and replace each triplet for the corresponding value. Since $|c(f(2^m))| = am + b$, for some constants a and b, the number of binary digits needed to reconstruct $f(2^m)$ is linear in m.

Note that we considered a prefix advice function instead of an advice function, otherwise the encoding would not be logarithmic in the size of the input.[4]

As first proved in [10, 14], these Cantor sets have the following property (see Appendix B):

Proposition 6 *For every* $y \in C_3$ *and for every dyadic rational* $z \in]0, 1[$, *such that* $|z| = m$, *if* $|y - z| \le 1/2^{i+5}$, *then the binary expansion of* y *and* z *coincide in the first* i *bits and* $|y - z| > 1/2^{-(m+10)}$.

4.4.3 Lower Bound for the Infinite Precision

We specify smooth scatter machines that decide the sets of some suitable nonuniform complexity class.

Theorem 2 *If* $B \in P/\log\star$, *then there exists an error-free* $SmSM$, *clocked in polynomial time, that decides* B.

Proof If $B \in P/\log\star$, then there exists a set $A \in P$ (i.e., A is decided by a deterministic Turing machine $\mathcal{M}_A$ clocked in polynomial time p_A) and a prefix advice function $f \in \log$, such that $w \in B$ iff $\langle w, f(|w|)\rangle \in A$. We can thus compute the pairing of w and $f(|w|)$ in polynomial time p, and check in polynomial time p_A, using Turing machine $\mathcal{M}_A$, if such a pair belongs to A.

To get $f(|w|)$, $\mathcal{M}$ reads some binary places of the vertex position $y(f)$ using the $SmSE$ (Algorithm 2) with an exponential schedule $T(\ell) = C2^\ell$. Since $|w|$ may not be a power of 2 the $SmSM$ reads $f(n)$, where $n \ge |w|$ and $n = 2^{\lceil \log|w|\rceil}$. We have that $|f(|w|)| \le a\log(|w|) + b$, for some constants $a, b \in \mathbb{N}$. Consequently $|f(n)| \le a\lceil\log(|w|)\rceil + b$, so that $\mathcal{M}$ reads

$$\ell = 3(a\lceil\log(|w|)\rceil + b) + 3(\lceil\log(|w|)\rceil + 1)$$

binary places of the vertex, where $3(\lceil\log(|w|)\rceil + 1)$ denotes the number of bits used in the separators.

Using Algorithm 2 on $\ell + 5 - 1$, by Proposition 4, the algorithm returns a dyadic rational m such that $|y(f) - m| < 2^{-\ell-5}$. Hence, by Proposition 6, $y(f)$ and m coincide in the first ℓ bits.

Again, by Proposition 4, we know that the time complexity of the Algorithm 2 is $\mathcal{O}(\ell T(\ell))$. Since ℓ is logarithmic in the size of the input word and T is exponential in the size of the query word we find that the Algorithm 2 takes polynomial time p_a in

---

[4]We would get $y(f) = 0.c(f(0))001c(f(1))001\cdots$ and since each $c(f(i))$ has a logarithmic size in i, the sequence $y(f)(n)$ would have a size $\mathcal{O}(n \times \log(n))$.

the size of the input word. We conclude that the total time for the whole computation is $\mathcal{O}(p_a + p + p_A)$, that is polynomial in the size of the input.   □

4.4.4 Smooth Scatter Machine as a Biased Coin

To prove the lower bounds for an error-prone $SmSM$ with unbounded or fixed precision we need some preliminary work. The first two statements—proved in [12, 13, 15]—explain how the $SmSE$ can be seen as a biased coin. The third statement—already proved in [14, 16]—state that, given a biased coin it is possible to simulate a sequence of fair coin tosses.

Proposition 7 *Given an error-prone smooth scatter machine with unbounded precision, vertex position at y, experimental time t, and time schedule T, there is a dyadic rational z and a real number $\delta \in]0, 1[$ such that the outcome of $Prot\_UP$ on z is a random variable that produces q_l with probability δ.*

Proof Consider an error-prone $SmSM$ with unbounded precision, vertex position y, experimental time t and time schedule T. Fix a positive integer ℓ such that $t(0) < T(\ell)$ and $t(1) < T(\ell)$, which means that if we run the experiment in the position $0|_\ell$ or if we run the experiment in the position $1|_\ell$, we will get an answer within $T(\ell)$ units of time.[5]

Given y and $T(\ell)$ we know that there exists $y' < y$ such that $t(y') = T(\ell)$. Consider now a dyadic rational z' such that $|z'| = \ell$ and $0 < z' - 2^{-\ell} < y' \le z'$. Observe that $0 < y' < y < 1$ and so $0 < z' < 1$ (see Fig. 4.8). The value of ℓ is fixed once and for all; however, it is supposed that ℓ can be fixed to a value large enough to get $z' - 2^{-\ell} > 0$. This restriction is easy to satisfy since we only require $t(0) < T(\ell)$ and $t(1) < T(\ell)$, which is obviously true for a large ℓ since the schedule is exponential.

If we run the $Prot\_UP$ on z', it will choose a value $z \in [z' - 2^{-\ell}, z' + 2^{-\ell}]$ uniformly. Thus, the probability that the protocol returns q_l is

$$\delta = \frac{y' - z' + 2^{-\ell}}{z' + 2^{-\ell} - (z' - 2^{-\ell})} = \frac{1}{2} - \frac{z' - y'}{2 \times 2^{-\ell}} .$$

Therefore $0 < \delta < 1$ and the probability that the protocol returns q_t or q_r is $1 - \delta$.   □

Proposition 8 *Given an error-prone smooth scatter machine with fixed precision, vertex position at y, experimental time t, and time schedule T, there is a dyadic rational z and a real number $\delta \in]0, 1[$ such that the outcome of $Prot\_FP$ on z is a random variable that produces q_l with probability δ.*

---

[5]In this case the position $0|_\ell$ or $1|_\ell$ means $0.\underbrace{0\cdots0}_{\ell}$ and $1.\underbrace{0\cdots0}_{\ell}$, respectively, not the dyadic position.

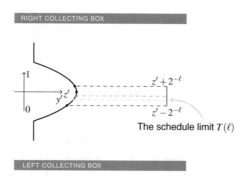

Fig. 4.8 The *SmSE* with unbounded precision as a coin

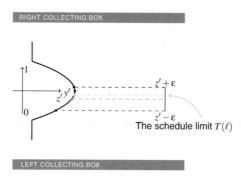

Fig. 4.9 The *SmSE* with fixed precision as a coin

Proof Consider an error-prone *SmSM* with fixed precision ε, vertex position at y, experimental time t and time schedule T. Fix a positive integer ℓ that verifies the following conditions: $2^{-\ell} \le \varepsilon$, $t(0) < T(\ell)$, and $t(1) < T(\ell)$. The last restriction means that if we run the experiment with the cannon either at $0|_\ell$ or at $1|_\ell$, we will get an answer within $T(\ell)$ units of time.[6]

Given y and T we know that there exists $y' < y$ such that $t(y') = T(\ell)$. Consider now a dyadic rational z' such that $|z'| = \ell$ and $0 < z' \le y' \le z' + 2^{-\ell} < 1$ (Figs. 4.9 and 4.10). The value of ε is fixed once and for all as is the value of y. However, it is supposed that ε can be fixed to a value that although fixed is very small, such that, $z' - \varepsilon > 0$.

If we run the *Prot_FP* on z' it will choose a value $z \in [z' - \varepsilon, z' + \varepsilon]$ uniformly. Thus the probability of the protocol return q_l is

$$\delta = \frac{y' - z' + \varepsilon}{z' + \varepsilon - (z' - \varepsilon)} = \frac{1}{2} - \frac{z' - y'}{2\varepsilon}.$$

Therefore, $0 < \delta < 1$ and the probability of outcome q_t or q_r is $1 - \delta$. ☐

---

[6]Id.

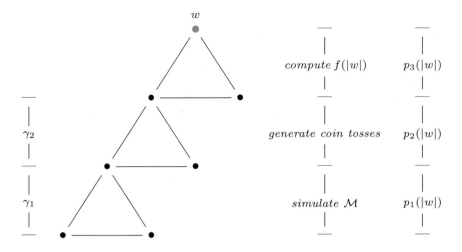

Fig. 4.10 Schematic description of the behavior of the *SmSM* with unbounded precision

Proposition 9 *Given a biased coin with probability of heads $q \in]\delta, 1 - \delta[$, for some $0 < \delta < 1/2$, and $\gamma \in]0, 1[$, we can simulate, up to probability $\geq \gamma$, a sequence of independent fair coin tosses of length n by doing a linear number of biased coin tosses.*

See the proof in Appendix C.

4.4.5   Lower Bound for the Unbounded and Fixed Precisions

Theorem 3 *If $B \in BPP//\log\star$, then there exists an error-prone smooth scatter machine with unbounded precision, clocked in polynomial time, that decides B.*

Proof If $B \in BPP//\log\star$, then there exists a prefix advice function $f \in \log$, a probabilistic Turing machine $\mathcal{M}$ working in polynomial time p_1, and a constant $\gamma_1 < 1/2$, such that, for every word w with size less or equal to $|w|$, $\mathcal{M}$ rejects $\langle w, f(|w|)\rangle$ with probability at most γ_1 if $w \in B$ and $\mathcal{M}$ accepts $\langle w, f(|w|)\rangle$ with probability at most γ_1 if $w \notin B$.

We compute $f(|w|)$ from $y(f)$ (the vertex of our *SmSM*) and use $\mathcal{M}$ to decide B. For this purpose consider γ_2 such that $\gamma_1 + \gamma_1\gamma_2 < 1/2$ and an exponential time schedule T.

By Proposition 7 there is a dyadic rational z, depending on y and T, that can be used to produce independent coin tosses with probability of heads $\delta \in]0, 1[$. By Proposition 9, we can use the biased coin to simulate a sequence of fair coin tosses of size $p_1(n)$, with probability of failure γ_2.

Similarly to the proof of the Theorem 2, if $|w| = n$, using Algorithm 3 on input $\ell + 5 - 1$, for $\ell = 3(a\lceil \log(n) \rceil + b) + 3(\lceil \log(n) \rceil + 1)$, we can extract, by Proposition 4, a dyadic rational m such that $|y(f) - m| < 2^{-\ell-5}$. Hence, by Proposition 6, $y(f)$ and m coincide in the first ℓ bits, so we can use $\mathcal{M}$ to compute $f(2^{\lceil \log(n) \rceil})$.

Thus, after using the $SmSE$ as oracle in order to compute $f(2^{\lceil \log(n) \rceil})$, it is used as a generator of a biased coin and thus as a generator of a sequence of size $p_1(n)$ of random coin tosses. Then we just need to simulate $\mathcal{M}$ on $\langle w, f(2^{\lceil \log(n) \rceil}) \rangle$.

The behavior of the smooth scatter machine is the following (Fig. 4.10):

Firstly, the $SmSM$ uses the $SmSE$ to compute $f(2^{\lceil \log(n) \rceil})$; then it uses the $SmSE$ as a generator of a fair sequence of coin tosses (with error probability γ_2); finally the $SmSM$ uses this sequence to guide its computation on $\langle w, f(2^{\lceil \log(n) \rceil}) \rangle$.

Therefore, if $w \in A$, then the $SmSM$ rejects w if $\mathcal{M}$, guided by the sequence of coin tosses, rejects $\langle w, f(2^{\lceil \log(n) \rceil}) \rangle$, a situation that happens with probability at most $\gamma_2 \gamma_1 + \gamma_1 < 1/2$. Similarly, the probability that $\mathcal{M}$ accepts w, for $w \notin A$, is less than $1/2$.

We can conclude that the $SmSM$ decides A with an error probability less than $1/2$. To see that it decides A in polynomial time note that Proposition 4 states that the running time of the measurement algorithm is $\mathcal{O}(\ell T(\ell))$. Since ℓ is logarithmic in input size and T is exponential in ℓ, the result is polynomial in the size of the input. Let such time be denoted by p_3.

Let p_4 denote the time needed to encode the pair w with $f(|w|)$, that is $\mathcal{O}(|w| + |f(|w|)|)$, and that p_2 denote the polynomial time needed to generate the sequence of coin tosses. Then we can conclude that the total time for the whole computation is $\mathcal{O}(p_1 + p_2 + p_3 + p_4)$, which is a polynomial in the size of the input. □

Theorem 4 *If $B \in BPP//\log\star$, then there exists an error-prone smooth scatter machine with fixed precision ε, clocked in polynomial time, that decides B.*

Proof The proof is similar to that of Theorem 3 but instead of Proposition 7 it uses Proposition 8; instead of using Proposition 4 it uses the Proposition 5; and instead of using measurement Algorithm 3 it uses Algorithm 4.

Note that in the fixed precision case the Algorithm 3 has an error γ_3. Thus, in this case, we consider γ_2 and γ_3 such that $\gamma_1 + \gamma_2 + \gamma_3 < 1/2$. In these conditions, if $w \in A$, then the $SmSM$ rejects w with probability at most $\gamma_3(\gamma_2 \gamma_1 + \gamma_1) + \gamma_2 \gamma_1 + \gamma_1 < 1/2$. Similarly, the probability that $w \notin A$ is accepted by $\mathcal{M}$ is less than $1/2$. □

4.4.6 Boundary Numbers

For the purpose of proving upper bounds, we introduce the *boundary numbers*.

Definition 6 Let $y \in]0, 1[$ be the vertex position and T the time schedule for a smooth scatter machine. For every $z \in \{0, 1\}^\star$, we define $l_{|z|}$ and $r_{|z|}$ as the two real numbers in $]0, 1[$ that satisfy the equation $t(l_{|z|}) = t(r_{|z|}) = T(|z|)$, with $l_{|z|} < y < r_{|z|}$.[7]

---

[7]There are always two boundary numbers satisfying this equation, as Fig. 4.11 shows.

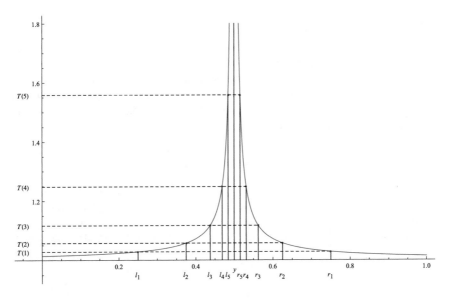

Fig. 4.11 The boundary numbers

Suppose that we want to query the oracle with z. If we query the $SmSE$, with vertex at y, we have three possible answers: q_l, if $z < y$ and $t(z) \leq T(|z|)$; q_r, if $z > y$ and $t(z) \leq T(|z|)$; q_t otherwise. We thus arrive at the Algorithm 5.

Therefore we may replace oracle consultation by a comparison of the query word (dyadic rational) with both $l_{|z|}$ and $r_{|z|}$. As we are going to see, in a sense to be precise later on, knowing enough bits of the boundary numbers is like querying the oracle.

Algorithm 5: Oracle simulation.

Data: Dyadic rational z representing the query, boundary numbers $l_{|z|}$ and $r_{|z|}$
if $z \leq l_{|z|}$ **then**
| A transition to the state q_l occurs ;
end
if $z \geq r_{|z|}$ **then**
| A transition to the state q_r occurs ;
end
if $l_{|z|} < z < r_{|z|}$ **then**
| A transition to the state q_t occurs ;
end

4.4.7 Upper Bound for the Infinite Precision

The precision we can get on the measurement of the vertex position determines the computational bounds of the smooth scatter machine. So far, the time schedule did not interfere in establishing the computational bounds. Now we apply the simulation technique, replacing the oracle by an advice function and prove upper bounds that, this time, seem to be sensitive to the time schedule.

Theorem 5 *If B is decidable by a smooth scatter machine with infinite precision and exponential protocol, clocked in polynomial time, then $B \in P/\log^2\star$.*

Proof Suppose that B is decidable by a $SmSM$ $\mathcal{M}$ with infinite precision and exponential time schedule T, clocked in polynomial time. Since T is exponential and $\mathcal{M}$ is clocked in polynomial time, we conclude that the size of the oracle queries grows at most logarithmically in the input size. This means that for any word w with size n, there exist constants a and b such that, during the computation, $\mathcal{M}$ only queries the oracle with words of size less or equal to $\ell = a\lceil \log(n) \rceil + b$. We consider a prefix advice function f such that $f(n)$ encodes the concatenation of boundary numbers needed to answer to all the queries of size ℓ:

$$l_1\rfloor_1 \# r_1 \rfloor_1 \# l_2 \rfloor_2 \# r_2 \rfloor_2 \# \cdots \# l_\ell \rfloor_\ell \# r_\ell \rfloor_\ell \#.$$

The prefix advice function f is such that $|f(n)| \in \mathcal{O}(2\ell + 2\sum_{i=1}^{\ell} i) = \mathcal{O}(\ell^2) = \mathcal{O}(\log^2(n))$. Therefore, to decide B in polynomial time, with prefix advice $f \in \log^2$, we simulate $\mathcal{M}$ on the input word but whenever a transition to the query state occurs and z is written in the query tape, we compare the query with the boundary numbers relative to $|z|$), i.e., with $l_{|z|}$ and $r_{|z|}$. The comparison, as explained in Sect. 4.4.6, gives us the same answer as $Prot\_IP(z)$, and thus the machine uses a similar measurement algorithm to approximate the vertex, replacing the call to the physical oracle by a comparison of the query word with $f(n)$. Since the comparison explained in Algorithm 5 can be done in polynomial time and $\mathcal{M}$ runs in polynomial time too, we can decide B in polynomial time given the advice in $P/\log^2\star$. $\square$

If we analyze better the binary expansion of each boundary number, we can change the advice function considered in the previous proof, which is quadratic in the size of the query word, in order to obtain an advice function linear in the size of the query word. The main idea of this result is based on the fact that the boundary number $l_{|z|+1}$ $(r_{|z|+1})$ can be obtained from $l_{|z|}$ $(r_{|z|}$, respectively), by adding a few more bits of information.

Proposition 10 *Given the boundary numbers for a smooth scatter machine with time schedule $T(k) \in \Omega(2^k)$ it is possible to define a prefix advice function f such that $f(n)$ encodes all the boundary numbers with size up to n and $|f(n)| \in \mathcal{O}(n)$.*

Proof Consider a $SmSM$ with vertex at y. If the time schedule associated with the $SmSM$ is $T(k) \in \Omega(2^k)$, then there exist α and k_0 in $\mathbb{N}$ such that for all $k \geq k_0$, $T(k) \geq \alpha 2^k$.

The value of the boundary number r_k is such that $y < r_k < y + 2^{-k+c}$, for some constant $c \in \mathbb{N}$ and for $k > k_0$. This means that, when we increase the size of k by one bit, we also increase the precision on y by one bit. Let us write the dyadic rational $r_k|_k$ as the concatenation of two strings, $r_k|_k = v_k \cdot w_k$, where w_k has size c and v_k has size $k - c$. Note that $r_k - 2^{-k+c} < v_k < r_k$, i.e., $y - 2^{-k+c} < r_k - 2^{-k+c} < v_k < r_k < y + 2^{-k+c}$, i.e., $|v_k - y| < 2^{-k+c}$.

The same reasoning applies to $l_k|_k = x_k \cdot y_k$, i.e., $|x_k - y| < 2^{-k+c}$, where y_k has size c and x_k has size $k - c$.

We show that we can obtain v_{k+1} from v_k with just two more bits. Suppose that v_k ends with the sequence $v_k = \cdots 10^\ell$. The only two possibilities for the first $k - c$ bits of y are $\cdots 10^\ell$ or $\cdots 01^\ell$. Thus, v_{k+1} must end in one of the following: $v_{k+1} = \cdots 10^\ell 1$ or $v_{k+1} = \cdots 10^\ell 0$ or $v_{k+1} = \cdots 01^\ell 1$ or $v_{k+1} = \cdots 01^\ell 0$. That is, *even though v_k is not necessarily a prefix of v_{k+1}*, the latter can be obtained from v_k by appending some information that determines which of the four possibilities is the case.[8] Suppose now that v_k ends with the sequence $v_k = \cdots 01^\ell$. The only two possibilities for the first $k - c$ bits of y are $\cdots 01^{\ell-1}1$ or $\cdots 01^{\ell-1}0$. Thus, v_{k+1} must end in one of the following: $v_{k+1} = \cdots 01^\ell 0$ or $v_{k+1} = \cdots 01^\ell 1$ or $v_{k+1} = \cdots 01^{\ell-1}00$ or $v_{k+1} = \cdots 01^{\ell-1}01$. In the same way, v_{k+1} can still be obtained from v_k by appending some information that determines which of the four possibilities is the case.

Similarly we obtain x_{k+1} from x_k.

We define the advice function inductively as follows: if $n < k_0$, then $f(n) = l_1|_1 \# r_1|_1 \# l_2|_2 \# r_2|_2 \# \cdots \# l_n|_n \# r_n|_n$; if $n = k_0$, then $f(k_0) = f(k_0 - 1) \# \# x_{k_0} \# y_{k_0} \# v_{k_0} \# w_{k_0}$; and if $n > k_0$, then $f(n) = f(n-1) \# b_{11}b_{12} \# y_n \# b_{21}b_{22} \# w_n$, where the b_{ij}'s denote the bits that distinguish between x_{n-1} and x_n and between v_{n-1} and v_n.

Considering $f(n)$, we can always recover $r_k|_k$ or $l_k|_k$, for $k \le n$, because, if $k \le n < k_0$, then the values of $r_k|_k$ and $l_k|_k$ are explicitly in $f(n)$; if $k = n = k_0$, then these values are explicitly in $f(n)$ and, moreover, the machine knows it is the last fully given boundary numbers $r_k|_k$ and $l_k|_k$ (with the two $\#\#$); and finally, if $n \ge k > k_0$, to obtain l_k and r_k after knowing l_{k-1} and r_{k-1} we only need to recalculate the two final bits of x_k and v_k and concatenate the result with either y_k or w_k, respectively.

To conclude, since y_n and w_n have constant size c, the value of $|f(n)|$ is asymptotically linear in n. □

With these different encodings we obtain a different upper bound for the infinite case.

Theorem 6 *If B is decidable by a smooth scatter machine with infinite precision and exponential protocol $T(k) \in \Omega(2^k)$, clocked in polynomial time, then $B \in P/\log\star$.*

---

[8]The following example helps to clarify the argument. Suppose that $y = 0.1100011000\ldots$ The sequence v_k can be taken as follows: $v_1 = 1$, $v_2 = 11$, $v_3 = 111$, $v_4 = 1101$, $v_5 = 11001$, $v_6 = 110010$, $v_7 = 1100100$, $v_8 = 11000111$, $v_9 = 110001100$, ...

Proof Suppose that B is decidable by a *SmSM* $\mathcal{M}$ with infinite precision and exponential time schedule $T(k) \in \Omega(2^k)$, clocked in polynomial time. Since T is exponential and $\mathcal{M}$ is clocked in polynomial time, we conclude that the size of the oracle queries can grow at most logarithmically in the size of the input. This means that for any word w with size n, there exist constants a and b such that, during the computation, $\mathcal{M}$ only queries the oracle with words of size less or equal to $\ell = a\lceil \log(n) \rceil + b$.

By Proposition 10 we can now define a prefix advice function f, encoding the boundary numbers, such that $|f(|z|)|$ is linear in size of the query $|z| \in \mathcal{O}(\ell)$, i.e., $|f(|z|)| \in \mathcal{O}(\log(n))$, where n is the size of the input.

Therefore, to decide B in polynomial time with prefix advice $f \in \log$, we can simulate $\mathcal{M}$ on the input word but whenever a transition to the query state occurs and z is written in the query tape we compare the query with the corresponding boundary numbers, i.e., with $l_{|z|}$ and $r_{|z|}$. The comparison, as explained in Sect. 4.4.6, provides the same answer as $Prot\_IP(z)$, and thus the machine uses a similar measurement algorithm to approximate the vertex, replacing the call to the physical oracle by a comparison with $f(n)$.

Since the comparison can be done in polynomial time and $\mathcal{M}$ runs in polynomial time too, we can decide B in polynomial time given the advice. □

The Theorems 2 and 6 allow us to prove the following corollary:

Corollary 1 *B is decidable by a smooth scatter machine with infinite precision and exponential protocol $T(k) \in \Omega(2^k)$, clocked in polynomial time, if and only if $B \in P/\log\star$.*

It is an open problem to know if the above corollary holds if we remove the schedule restriction.

4.4.8 Probabilistic Query Trees

The error-prone smooth scatter machine can obtain approximations to the vertex position through the measurement algorithm. After each run of the *SmSE*, the Turing machine is in one of the three possible states: q_r, q_t or q_l. The oracle consultations of a *SmSM* can then be seen as a ternary query tree since its (deterministic) computations are interspersed with calls to the oracle; after each call the machine is in one of the three above mentioned states.

Definition 7 A *query tree* is a rooted tree (V, E, ν) where each node in V is a configuration of the Turing machine in the query state (the root ν is the configuration of the first call to the oracle) or in a halting state, and each edge in E is a deterministic computation of the Turing machine between consecutive oracle calls or between oracle calls and halting configurations. The only nodes with zero children are the corresponding accepting and rejecting configurations.

Definition 8 A *m-ary query tree* is a query tree where each node except the leaves has m children.

Since we are not considering now the infinite precision case, we know that the behavior of the $SmSE$ is stochastic and thus, after each call to the oracle, the machine is in one of the three states with some probability. With this idea in mind, we can see all the oracle consultations by a $SmSM$ on an input w as a probabilistic ternary query tree, i.e., a ternary query tree where each edge is labeled by a probability. A single computation on w corresponds to a path in the tree, beginning in the root and ending in a leaf. The leaves of the tree are labeled with an A, if the computation on input w halts in an accepting configuration and are labeled with an R, if the computation on input w halts in a rejecting configuration (see Fig. 4.12).

Let $T_{n,m} = (V_{n,m}, E_{n,m}, \nu_{n,m})$ denotes a n-ary probabilistic query tree with depth m.

Definition 9 We define the set of all assignments of probabilities to the edges of $T_{n,m}$ as

$$\rho(T_{n,m}) = \{\sigma : E_{n,m} \to [0, 1] : \text{the sum of } \sigma\text{'s over the } n \text{ outcomes of every node is } 1\}.$$

Denote by $T_{n,m}^{\sigma}$ the n-ary probabilistic query tree with depth m and assignment σ.

Definition 10 The probability of one single path π of a n-ary probabilistic query tree $T_{n,m}^{\sigma}$ with depth m and assignment σ is

$$\prod_{i=1}^{m} \sigma(\pi[i]),$$

where $\pi[i]$ stands for the ith edge of the path from the root. The *acceptance probability* $P(T_{n,m}^{\sigma})$ is the sum of the probabilities of all *accepting paths*.

We define $D(\sigma_1, \sigma_2), \sigma_1, \sigma_2 \in \rho(T_{n,m})$ as the maximum distance between the two probabilistic query trees $T_{n,m}^{\sigma_1}$ and $T_{n,m}^{\sigma_2}$.

Definition 11 For every $\sigma_1, \sigma_2 \in \rho(T_{n,m})$, we define

$$D(\sigma_1, \sigma_2) = \max\{|\sigma_1(e) - \sigma_2(e)| : e \in E_{n,m}\}.$$

Now we define the largest possible difference in the acceptance probability for two different assignments, where the distance between the probabilities is less or equal to a value s.

Definition 12 For any $m \in \mathbb{N}$, $s \in [0, 1]$ and number of outcomes $out \in \mathbb{N}$, we define a function $\mathcal{A}_{out} : \mathbb{N} \times [0, 1] \to [0, 1]$ as

$$\mathcal{A}_{out}(m, s) = \sup\{|P(T_{out,m}^{\sigma'}) - P(T_{out,m}^{\sigma})| : \sigma, \sigma' \in \rho(T_{out,m}) \text{ and } D(\sigma, \sigma') \leq s\}.$$

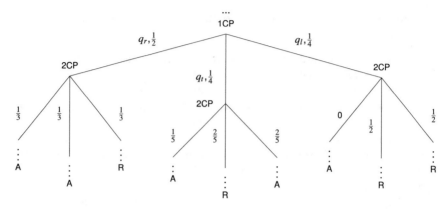

Fig. 4.12 The oracle calls by a smooth scatter machine as a ternary query tree, where iCP means the ith cannon position

This function satisfies the following relevant property:

Proposition 11 *For any $m \in \mathbb{N}$, $s \in [0, 1]$, and any number out $\in \mathbb{N}$ of children in the tree, $\mathcal{A}_{out}(m, s) \leq (out - 1)ms$.*

Proof The proof follows by induction in m. The result is straightforward for $m = 0$: we take $P(T_{out,0}^{\sigma}) = P(T_{out,0}^{\sigma'}) = 0$ for each rejecting leaf, and $P(T_{out,0}^{\sigma}) = P(T_{out,0}^{\sigma'}) = 1$ for each accepting leaf. Let the statement be true for m and consider the probabilistic query tree $T_{out,m+1}$ with out outgoing edges, $e_1, e_2, ..., e_{out}$, and depth $m + 1$. Each edge e_i is incident in a node $T_{out,m}(i)$, for $i = 1, \ldots, out$, respectively. Consider probability assignments $\sigma, \sigma' \in \rho(T_{out,m+1})$ such that $D(\sigma, \sigma') \leq s$. We have then

$$P(T_{out,m+1}^{\sigma}) = \sigma(e_1)P(T_{out,m}^{\sigma}(1)) + \sigma(e_2)P(T_{out,m}^{\sigma}(2)) + \cdots + \sigma(e_{out})P(T_{out,m}^{\sigma}(out))$$

$$P(T_{out,m+1}^{\sigma'}) = \sigma'(e_1)P(T_{out,m}^{\sigma'}(1)) + \sigma'(e_2)P(T_{out,m}^{\sigma'}(2)) + \cdots + \sigma'(e_{out})P(T_{out,m}^{\sigma'}(out)) .$$

As $\sigma(e_{out}) = 1 - \sigma(e_1) - \sigma(e_2) - \cdots - \sigma(e_{out-1})$ and $\sigma'(e_{out}) = 1 - \sigma'(e_1) - \sigma'(e_2) - \cdots - \sigma'(e_{out-1})$, we have that

$$\left| P(T_{out,m+1}^{\sigma}) - P(T_{out,m+1}^{\sigma'}) \right| = \left| (\sigma(e_1) - \sigma'(e_1))(P(T_{out,m}^{\sigma}(1)) - P(T_{out,m}^{\sigma}(out))) \right.$$
$$+ (\sigma(e_2) - \sigma'(e_2))(P(T_{out,m}^{\sigma}(2)) - P(T_{out,m}^{\sigma}(out)))$$
$$+ \cdots$$
$$+ (\sigma(e_{out-1}) - \sigma'(e_{out-1}))(P(T_{out,m}^{\sigma}(out - 1)) - P(T_{out,m}^{\sigma}(out - 1))$$
$$+ \sigma'(e_1)(P(T_{out,m}^{\sigma}(1)) - P(T_{out,m}^{\sigma'}(1)))$$
$$+ \sigma'(e_2)(P(T_{out,m}^{\sigma}(2)) - P(T_{out,m}^{\sigma'}(2)))$$
$$+ \cdots$$

$$+\sigma'(e_{out})(P(T^{\sigma}_{out,m}(out)) - P(T^{\sigma'}_{out,m}(out)))\Bigg| \ .$$

Since the difference of any two real numbers in $[0, 1]$ lies in $[-1, 1]$, we conclude that

$$\left| P(T^{\sigma}_{out,m+1}) - P(T^{\sigma'}_{out,m+1}) \right| \le |\sigma(e_1) - \sigma'(e_1)| + |\sigma(e_2) - \sigma'(e_2)| + \cdots + |\sigma(e_{out-1}) - \sigma'(e_{out-1})|$$
$$+ \sigma'(e_1)\mathcal{A}_{out}(m, s) + \sigma'(e_2)\mathcal{A}_{out}(m, s) + \cdots + \sigma'(e_{out})\mathcal{A}_{out}(m, s)$$
$$\le (out - 1)s + \mathcal{A}_{out}(m, s) \ .$$

Therefore, using the induction hypothesis,

$$\begin{aligned}
\mathcal{A}_{out}(m + 1, s) &\le \mathcal{A}_{out}(m, s) + (out - 1)s \\
&\le (out - 1)ms + (out - 1)s \\
&= (out - 1)(m + 1)s \ .
\end{aligned} \qquad \square$$

4.4.9   Upper Bound for the Unbounded Precision

Consider a $SmSM$ with vertex at position y and physical time t, and suppose that the $SmSM$ writes a query z with $|z| = k$, for $k \in \mathbb{N}$. Then consider the two boundary numbers l_k and r_k (see Sects. 4.4.6 and 4.4.7) for the schedule $T(k)$. Since the transition to one of the states q_r, q_t and q_l is probabilistic whenever the $SmSM$ uses the $SmSE$ with the protocol $Prot\_UP$, we conclude that approximations to the probabilities involved in these possible transitions are needed in order *to simulate the oracle calls*.

Protocol $Prot\_UP(z)$ chooses uniformly some position $z' \in [z - 2^{-k}, z + 2^{-k}]$ from where to shoot, originating eight possible shooting cases represented in Fig. 4.13. Assuming that we always have k large enough to obtain the shooting interval inside $]0, 1[$, we exclude the cases 6 and 8. Assuming that the protocol is exponential, we can discard also case 7, since the interval $]l_k, r_k[$ shrinks faster than the shooting interval. For each one of these cases, from 1 to 5, we will have the following probabilities:

Fig. 4.13 Shooting cases

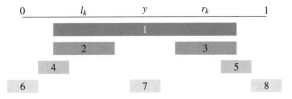

1. $0 < z - 2^{-k} < l_k < r_k < z + 2^{-k} < 1$

$$P(\text{``}q_l\text{''}) = \frac{l_k - z + 2^{-k}}{z + 2^{-k} - z + 2^{-k}} = \frac{1}{2} - \frac{z - l_k}{2 \times 2^{-k}}$$

$$P(\text{``}q_t\text{''}) = \frac{r_k - l_k}{z + 2^{-k} - z + 2^{-k}} = \frac{r_k - l_k}{2 \times 2^{-k}}$$

$$P(\text{``}q_r\text{''}) = \frac{z + 2^{-k} - r_k}{z + 2^{-k} - z + 2^{-k}} = \frac{1}{2} - \frac{r_k - z}{2 \times 2^{-k}} \ ;$$

2. $0 < z - 2^{-k} < l_k < z + 2^{-k} < r_k < 1$

$$P(\text{``}q_l\text{''}) = \frac{l_k - z + 2^{-k}}{z + 2^{-k} - z + 2^{-k}} = \frac{1}{2} - \frac{z - l_k}{2 \times 2^{-k}}$$

$$P(\text{``}q_t\text{''}) = \frac{z + 2^{-k} - l_k}{z + 2^{-k} - z + 2^{-k}} = \frac{1}{2} - \frac{l_k - z}{2 \times 2^{-k}}$$

$$P(\text{``}q_r\text{''}) = 0 \ ;$$

3. $0 < l_k < z - 2^{-k} < r_k < z + 2^{-k} < 1$

$$P(\text{``}q_l\text{''}) = 0$$

$$P(\text{``}q_t\text{''}) = \frac{r_k - z + 2^{-k}}{z + 2^{-k} - z + 2^{-k}} = \frac{1}{2} - \frac{z - r_k}{2 \times 2^{-k}}$$

$$P(\text{``}q_r\text{''}) = \frac{z + 2^{-k} - r_k}{z + 2^{-k} - z + 2^{-k}} = \frac{1}{2} - \frac{r_k - z}{2 \times 2^{-k}} \ ;$$

4. $0 < z - 2^{-k} < z + 2^{-k} < l_k$

$$P(\text{``}q_l\text{''}) = 1$$
$$P(\text{``}q_t\text{''}) = 0$$
$$P(\text{``}q_r\text{''}) = 0 \ ;$$

5. $r_k < z - 2^{-k} < z + 2^{-k} < 1$

$$P(\text{``}q_l\text{''}) = 0$$
$$P(\text{``}q_t\text{''}) = 0$$
$$P(\text{``}q_r\text{''}) = 1 \ .$$

Looking at the expressions, and considering error propagation, we can conclude that if we know $k + d$ bits of l_k and r_k we can approximate the probabilities within an error less than 2^{-d}.

Theorem 7 *If B is decidable by a smooth scatter machine with unbounded precision and exponential protocol T, clocked in polynomial time, then $B \in BPP//\log^2\star$.*

Proof Suppose that B is decidable by a *SmSM* $\mathcal{M}$ with unbounded precision and exponential protocol T, in polynomial time $\mathcal{O}(n^a)$. Since $\mathcal{M}$ on w runs in polynomial time in $n = |w|$ and it has an exponential time schedule $T(k)$, we can conclude that the size of the oracle queries must be at most logarithmic in the size of the input, that is, the size of the oracle queries must be less or equal to $b\lceil \log(n)\rceil + c$. The number of queries to the oracle cannot exceed the running time of the machine, so that the probabilistic query trees of the *SmSM* have a depth of at most αn^a, for some constant α.

Let γ be the error probability of machine $\mathcal{M}$ and $d \in \mathbb{N}$ such that $2^d > 2\alpha/(1/2 - \gamma)$. The probabilities of each outcome of all oracle queries will be truncated up to the precision $2^{-d-a\lceil \log(n)\rceil}$, according with Proposition 11, in order to obtain error probability of acceptance less than $1/2 - \gamma$:

$$\mathcal{A}(\alpha n^a, 2^{-d-a\lceil \log(n)\rceil}) \leq 2 \times \alpha n^a \times 2^{-d-a\lceil \log(n)\rceil}$$
$$= \frac{2\alpha n^a}{2^d \times 2^{a\lceil \log(n)\rceil}}$$
$$= \frac{2\alpha}{2^d} < (1/2 - \gamma).$$

Hence, as explained before the statement, to approximate the probabilities of all queries with precision $2^{-d-a\lceil \log(n)\rceil}$ we have to know $i + d + a\lceil \log(n)\rceil$ bits of l_i and r_i, for $1 \leq i \leq b\lceil \log(n)\rceil + c$. Consider now $\beta = max\{a, b\}$ and a prefix advice function f defined recursively as follows:

1. $f(0) = l_1\rfloor_{d+c} \# r_1 \rfloor_{d+c} \# l_2 \rfloor_{d+c} \# r_2 \rfloor_{d+c} \# \cdots \# l_c \rfloor_{d+c} \# r_c \rfloor_{d+c}$;
2. $f(x+1)$ is obtained by concatenating $f(x)$ with the bits $d + c + 2\beta x + 1$ to $d + c + 2\beta x + 2\beta$ of l_i and r_i, for $1 \leq i \leq \beta x + c$; and then by adding the first $d + c + 2\beta x + 2\beta$ bits of l_i and r_i for $\beta x + c + 1 \leq i \leq \beta x + c + \beta$. (All the blocks of bits separated by #.)

Advice f encodes approximations to the boundary numbers l_i and r_i, for $1 \leq i \leq \beta x + c$: a Turing machine can read $2(\beta x + c)$ nonsequential blocks of size 2β from $f(x)$, updating at the same time the approximations of l_i and r_i for $1 \leq i \leq \beta x + c$ and 2β nonsequential blocks of size $d + c + 2\beta x + 2\beta$, to get approximations of l_i and r_i, for $\beta x + c + 1 \leq i \leq \beta x + c + \beta$. Thus, a Turing machine can have access to approximations of l_i and r_i with $d + c + 2\beta x$ bits of precision. Analysing the advice function we can conclude that:

$$|f(x)| = 2 \times (d + c + 2\beta x) \times (c + \beta x) + \sum_{i=0}^{x} 2(\beta x + c) = \mathcal{O}(x^2).$$

Since we only consider x at most logarithmic in the input size, n, we have that $|f(\lceil \log(n)\rceil)| = \mathcal{O}(\log^2(n))$. Thus, the advice function $g(n) = f(\lceil \log(n)\rceil)$ provide approximations of l_i and r_i, for $1 \leq i \leq b\lceil \log(n)\rceil + c$, with at least $i + d + a\lceil \log(n)\rceil$ bits of precision, as desired.

Therefore, to decide B in polynomial time, using the prefix advice $f \in \log^2$, we construct a Turing machine $\mathcal{M}'$ that simulates $\mathcal{M}$ on the input word but, whenever $\mathcal{M}$ queries the oracle with z, $\mathcal{M}'$ compares the query z with the corresponding boundary numbers, i.e., with l_k and r_k, where $k = |z|$; checks the shooting cases; and computes the approximations to the probabilities with an error less than $2^{-d-a\lceil \log(n) \rceil}$. Then $\mathcal{M}'$ simulates a path in the probabilistic query tree, that represents the oracle consultation, by means of the computed probabilities, by tossing a coin $d + a\lceil \log(n) \rceil$ times. Note that this probabilistic tree has a depth of at most an^a and the edge difference is less than $2^{-d-a\lceil \log(n) \rceil}$. After simulating this path, the machine $\mathcal{M}'$ proceeds as $\mathcal{M}$. Since the difference in the probability of acceptance is bounded by $1/2 - \gamma$, $\mathcal{M}'$ gives a wrong answer with probability less than $\gamma + 1/2 - \gamma = 1/2$.

Recalling that the $SmSM$ $\mathcal{M}$ runs in polynomial time, and that comparing query words with boundary numbers and computing probabilities can also be done in polynomial time, we conclude that B is decidable in polynomial time with advice $g(n) = f(\lceil \log(n) \rceil)$. □

4.4.10 Upper Bound for the Fixed Precision

The error-prone $SmSM$ with fixed precision ε has also probabilistic computation trees. Thus, once again, we need approximations to the boundary numbers (see Sects. 4.4.6, 4.4.7 and 4.4.9) and to the probabilities in order to simulate the oracle whenever the $SmSM$ calls the $SmSE$ with the protocol $Prot\_FP$.

Consider a $SmSM$ with vertex at position y, fixed precision $\varepsilon = 2^{-q}$, for some positive integer q, and physical time t. Suppose that the $SmSM$ writes a query z with $|z| = k$ for $k \in \mathbb{N}$. Then consider the two boundary numbers l_k and r_k for the schedule $T(k)$.

Since $Prot\_FP(z)$ will choose uniformly some position $z' \in [z - \varepsilon, z + \varepsilon]$ from where to shoot, we have eight possible shooting cases represented also in Fig. 4.13. As previously discussed in Sect. 4.4.9, we assume that we always have k large enough to have the shooting interval inside $]0, 1[$, excluding the cases 6 and 8. We also know that the interval $]l_k, r_k[$ shortens, so that we consider that the case 7 does not occur also.

For each one of the remaining cases, from 1 to 5, we will have the following probabilities:

1. $0 < z - \varepsilon < l_k < r_k < z + \varepsilon < 1$

$$P(\text{``}q_l\text{''}) = \frac{l_k - z + \varepsilon}{z + \varepsilon - z + \varepsilon} = \frac{1}{2} - \frac{z - l_k}{2\varepsilon}$$

$$P(\text{``}q_t\text{''}) = \frac{r_k - l_k}{z + \varepsilon - z + \varepsilon} = \frac{r_k - l_k}{2\varepsilon}$$

$$P(\text{``}q_r\text{''}) = \frac{z + \varepsilon - r_k}{z + \varepsilon - z + \varepsilon} = \frac{1}{2} - \frac{r_k - z}{2\varepsilon} ;$$

2. $0 < z - \varepsilon < l_k < z + \varepsilon < r_k < 1$

$$P(\text{``}q_l\text{''}) = \frac{l_k - z + \varepsilon}{z + \varepsilon - z + \varepsilon} = \frac{1}{2} - \frac{z - l_k}{2\varepsilon}$$

$$P(\text{``}q_t\text{''}) = \frac{z + \varepsilon - l_k}{z + \varepsilon - z + \varepsilon} = \frac{1}{2} - \frac{l_k - z}{2\varepsilon}$$

$$P(\text{``}q_r\text{''}) = 0 ;$$

3. $0 < l_k < z - \varepsilon < r_k < z + \varepsilon < 1$

$$P(\text{``}q_l\text{''}) = 0$$

$$P(\text{``}q_t\text{''}) = \frac{r_k - z + \varepsilon}{z + \varepsilon - z + \varepsilon} = \frac{1}{2} - \frac{z - r_k}{2\varepsilon}$$

$$P(\text{``}q_r\text{''}) = \frac{z + \varepsilon - r_k}{z + \varepsilon - z + \varepsilon} = \frac{1}{2} - \frac{r_k - z}{2\varepsilon} ;$$

4. $0 < z - \varepsilon < z + \varepsilon < l_k$

$$P(\text{``}q_l\text{''}) = 1$$
$$P(\text{``}q_t\text{''}) = 0$$
$$P(\text{``}q_r\text{''}) = 0 ;$$

5. $r_k < z - \varepsilon < z + \varepsilon < 1$

$$P(\text{``}q_l\text{''}) = 0$$
$$P(\text{``}q_t\text{''}) = 0$$
$$P(\text{``}q_r\text{''}) = 1 .$$

Considering error propagation, we can conclude that if we know $q + d$ bits of l_k and r_k, then we can approximate these probabilities with an error at most 2^{-d}. (Note that, if we consider a real valued ε, we can also approximate the probabilities within the desired precision by providing the bits of ε too.)

We can now prove the upper bound for the fixed precision case.

Theorem 8 *If B is decidable by a smooth scatter machine with fixed precision $\varepsilon = 2^{-q}$, for some positive integer q, and exponential protocol T, clocked in polynomial time, then $B \in BPP//\log^2\star$.*

Proof Suppose that B is decidable by a $SmSM$ $\mathcal{M}$, with fixed precision $\varepsilon = 2^{-q}$, for some positive integer q, and exponential protocol T, in polynomial time $\mathcal{O}(n^a)$. Since $\mathcal{M}$ has an exponential time schedule $T(k)$, we can conclude that the size of the oracle queries must be at most logarithmic in the size of the input, i.e., less then or equal to $b\lceil \log(n) \rceil + c$. Moreover, there exists a constant α such that the number of queries does not exceed αn^a and, consequently, the probabilistic query trees of the $SmSM$ will have a depth of at most αn^a.

Let γ be the error probability of machine $\mathcal{M}$ and $d \in \mathbb{N}$ such that $2^d > 2\alpha/(1/2 - \gamma)$. The probabilities of each outcome of all oracle queries will be truncated up to the precision $2^{-d-a\lceil \log(n)\rceil}$, according with Proposition 11, in order to obtain error probability of acceptance less than $1/2 - \gamma$:

$$
\begin{aligned}
\mathcal{A}(\alpha n^a, 2^{-d-a\lceil \log(n)\rceil}) &\leq 2 \times \alpha n^a \times 2^{-d-a\lceil \log(n)\rceil} \\
&= \frac{2\alpha n^a}{2^d \times 2^{a\lceil \log(n)\rceil}} \\
&= \frac{2\alpha}{2^d} < (1/2 - \gamma) .
\end{aligned}
$$

Hence to approximate the probabilities of all queries with precision $2^{-d-a\lceil \log(n)\rceil}$ we have to know $q + d + a\lceil \log(n)\rceil$ bits of l_i and r_i, for $1 \leq i \leq b\lceil \log(n)\rceil + c$. Consider the prefix advice function f defined recursively as follows:

1. $f(0) = q\#l_1\rfloor_{q+d}\#r_1\rfloor_{q+d}\#l_2\rfloor_{q+d}\#r_2\rfloor_{q+d}\# \cdots \#l_c\rfloor_{q+d}\#r_c\rfloor_{q+d}$;
2. $f(x + 1)$ is obtained by concatenating $f(x)$ with the bits $q + d + ax + 1$ to $q + d + ax + a$ of l_i and r_i, for $1 \leq i \leq bx + c$; then by adding the first $q + d + ax + a$ bits of l_i and r_i for $bx + c + 1 \leq i \leq bx + c + b$. All the blocks of bits are separated by #.

From advice $f(x)$ a Turing machine has access to the value q (and then computes ε), and to the approximations of the boundary numbers l_i and r_i, for $1 \leq i \leq bx + c$ by doing the following: the machine reads $2(bx + c)$ nonsequential blocks of size a from $f(x)$, updating at the same time the approximations of l_i and r_i, for $1 \leq i \leq bx + c$, and $2b$ nonsequential blocks of size $q + d + ax + a$, to get approximations of l_i and r_i, for $bx + c + 1 \leq i \leq bx + c + b$. Thus, given advice $f(x)$, a Turing machine can approximate l_i and r_i with $q + d + ax$ bits of precision, for $1 \leq i \leq bx + c$.

Analyzing the advice function we can conclude that:

$$
|f(x)| \leq 2 \times (q + d + ax) \times (bx + c) + \sum_{i=0}^{x} 2(bx + c) = \mathcal{O}(x^2) .
$$

Since we only consider x at most logarithmic in the input size, n, we have that $|f(\lceil \log(n)\rceil)| = \mathcal{O}(\log^2(n))$. Thus, the advice function $g(n) = f(\lceil \log(n)\rceil)$ provides approximations of l_i and r_i, for $1 \leq i \leq b\lceil \log(n)\rceil + c$, with at least $q + d + a\lceil \log(n)\rceil$ bits of precision, as desired.

Therefore, to decide B in polynomial time with help by a prefix advice $f \in \log^2$, we construct a Turing machine $\mathcal{M}'$ that simulates $\mathcal{M}$ on the input word but whenever $\mathcal{M}$ queries the oracle with z, $\mathcal{M}'$ compares the query with the corresponding boundary numbers, i.e., with $l_{|z|}$ and $r_{|z|}$; checks the shooting cases; and computes the approximations to the probabilities with an error less than $2^{-d-a\lceil \log(n)\rceil}$. Then $\mathcal{M}'$ simulates a path in the probabilistic query tree, that represents the oracle consultation, by means of the computed probabilities, by tossing a coin $d + a\lceil \log(n)\rceil$ times. Note that this probabilistic query tree has a depth of at most αn^a and that the edge difference is less than $2^{-d-a\lceil \log(n)\rceil}$. After simulating this path the machine $\mathcal{M}'$ proceeds with the computation as $\mathcal{M}$.

Turing machine $\mathcal{M}'$ gives a wrong answer with probability less than $\gamma + 1/2 - \gamma = 1/2$. Recalling that the $SmSM$ $\mathcal{M}$ runs in polynomial time, that comparing query words with boundary numbers and computing probabilities can also be done in polynomial time, we conclude that B is decidable in polynomial time with advice $g(n) = f(\lceil \log(n)\rceil)$. □

4.5 Upper Bound with Explicit Time Technique

As we have seen in the previous sections, the use of boundary numbers raises the question of whether we really can achieve the upper bound of $BPP//\log^2\star$. We can equally ask, under what circumstances is the upper bound actually $BPP//\log\star$? Here we consider a special case where the upper bound does reduce to $BPP//\log\star$.

Using the physical time explicitly means that, given an exact expression for the experimental time for a $SmSM$, we take good use of it to compute the boundary numbers. We assume that we have an exact expression for the experimental time $t(z) = f(z - y)$ for cannon position z. As usual, the vertex position y is unknown, but we have the explicit form of the function f. Of course, this approach can cost a lot of computational resources since the function f may be computationally difficult to compute.

In order to understand better the explicit time idea note that, in the unbounded precision case, we use the advice function to encode approximations of the boundary numbers, allowing us to simulate the oracle answers and to compute approximations to the probabilities. The simplest formula for an explicit time consistent with our assumptions in (4.1) would be, for some constant $C > 0$,

$$t(z) = \frac{C}{|y - z|}. \tag{4.3}$$

To make use of this formula, we need C to be computable, but also we need bounds on how quickly we can compute approximations to C. To further simplify matters, we shall assume that $C = 1$.

Suppose that the $SmSM$ writes a query z with $|z| = k$ for $k \in \mathbb{N}$. Then consider the two boundary numbers l_k and r_k for the schedule $T(k)$. If we consider the explicit time, as the boundary numbers satisfy the property $t(l_k) = t(r_k) = T(k)$, we have

$$l_k = y - \frac{1}{T(k)}, \quad r_k = y + \frac{1}{T(k)}. \tag{4.4}$$

Thus, given approximations to y, we can obtain approximations of l_k and r_k. We consider that the schedule T is internal to the Turing machine, i.e. the Turing machine is capable of computing it up to any precision. Thus, by looking at the expressions and considering the error propagation rules, we conclude that if we have $d + 1$ bits of precision of y and $A = 1/T(k)$, we can compute the boundary numbers with an error less than 2^{-d}.

Theorem 9 *If B is decidable by a smooth scatter machine with unbounded precision and exponential protocol T, clocked in polynomial time, then, considering explicit time in the form (4.3) with $C = 1$, $B \in BPP//\log\star$.*

Proof We refer to the proof of Theorem 7, providing now the advice function that solves the problem in $BPP//\log\star$. All variables and constants are as in the proof of Theorem 7.

Consider a $SmSM$ $\mathcal{M}$ running in polynomial time and with schedule $T(k)$, exponential in k. By Theorem 7, we know that B can be decided by a probabilistic Turing machine, $\mathcal{M}'$ in polynomial time with access to an advice function f of size $\mathcal{O}(\log^2(n))$, which contains the first $d + c + 2\beta x$ bits of l_i and r_i for $1 \le i \le \beta x + c$.

We consider another function g, defined recursively as follows: $g(0)$ is the concatenation of the first $d + c$ bits of y; $g(x + 1)$ is the concatenation of $g(x)$ with the bits $d + c + 2\beta x + 1$ to $d + c + 2\beta x + 2\beta$ of y. We can use $g(x)$ to get the first $d + c + 2\beta x$ bits of y, therefore, by the previous reasoning, we can use the approximations of y in order to compute the approximation of all l_k and r_k, for $1 \le k \le bx + c$, with $d + c + 2\beta x$ bits of precision as the Turing machine can compute in polynomial time A with $d + c + 2\beta x$ bits of precision.

Analyzing $g(x)$ we conclude that $|g(x)| = (d + c + 2\beta x) + (x + 1) = \mathcal{O}(x)$. As in our case, as x will be at most logarithmic in the size of the input, we conclude that $|g(\lceil \log(n) \rceil)| = \mathcal{O}(\lceil \log(n) \rceil)$.

Thus, we define a probabilistic Turing machine $\mathcal{M}''$ that on input w simulates $\mathcal{M}'$ on w but instead of using the advice $f(|w|)$, uses the advice $g(|w|)$. Since we can recover the information of f from g in polynomial time and $\mathcal{M}'$ runs in polynomial time too, we conclude that our Turing machine runs in polynomial time and decides B. ☐

Using Theorems 3 and 9 we can trivially state the following corollary.

Corollary 2 *Considering explicit time in the form (4.3) with $C = 1$, B is decidable by a smooth scatter machine with unbounded precision and exponential protocol T, clocked in polynomial time, if and only if $B \in BPP//\log\star$.*

Theorem 10 *If B is decidable by a smooth scatter machine with fixed precision $\varepsilon = 2^{-q}$, for some positive integer q, and exponential protocol T, clocked in polynomial time, then, considering explicit time in the form (4.3) with $C = 1$, $B \in BPP//\log\star$.*

Proof We refer to the proof of Theorem 8, providing now the advice function that solves the problem in $BPP//\log\star$. All variables and constants are as in the proof of Theorem 8.

Consider a $SmSM$ $\mathcal{M}$ running in polynomial time, with fixed precision $\varepsilon = 2^{-q}$, for some positive integer q, and a schedule $T(k)$, exponential in k. By Theorem 8, we know that B can be decided by a probabilistic Turing machine $\mathcal{M}'$ in polynomial time with access to an advice function f of size $\mathcal{O}(\log^2(n))$, which contains the first $q + d + ax$ bits of l_i and r_i for $1 \leq i \leq bx + c$ and the the value q.

We consider another function g, defined recursively as follows: $g(0)$ is the concatenation of q with the first $q + d$ bits of y; $g(x + 1)$ is the concatenation of $g(x)$ with the bits $q + d + 1 + ax$ to $q + d + 1 + ax + a$ of y. We can use $g(x)$ to get ε and to get the first $q + d + ax$ bits of y. Therefore, by the previous reasoning, we can use the approximations of y to compute the approximation of all l_k and r_k for $1 \leq k \leq bx + c$ with $q + d + ax$ bits of precision, as the Turing machine can compute in polynomial time A with $q + d + ax$ bits of precision.

Analyzing $g(x)$ we conclude that $|g(x)| = (q + d + ax) + (x + 2) = \mathcal{O}(x)$. As in our case x will be at most logarithmic in the size of the input we conclude that $|g(\lceil \log(n) \rceil)| = \mathcal{O}(\log(n))$.

Thus, we define a probabilistic Turing machine $\mathcal{M}''$ that on input w simulates $\mathcal{M}'$ on w but instead of using the advice $f(|w|)$, uses the advice $g(|w|)$. Since we can recover the information of f from g in polynomial time and $\mathcal{M}'$ runs in polynomial time too we conclude that our Turing machine runs in polynomial time and decides B. □

Using Theorems 4 and 10 we can trivially state the following corollary:

Corollary 3 *Considering explicit time in the form (4.3) with $C = 1$, B is decidable by a smooth scatter machine with fixed precision $\varepsilon = 2^{-q}$, for some positive integer q, and exponential protocol T, clocked in polynomial time, if and only if $B \in BPP//\log\star$.*

4.6 Conclusion

In the past two decades there was a growing of interest in non-conventional models of computation inspired by the natural processes in biology, physics and chemistry. Some of these models explore parallel processing, some others see advantage in analogue components translated into real numbers, appearing as parameters in the systems. In this paper we explored an abstraction of the last category of models, studying an analogue-digital (hybrid) model of computation where the Turing machine is coupled with a physical oracle.

Abstracting from other models, the model we propose is introduced as a laboratory, where the computational power can be studied depending on the protocol between the Turing machine and the analogue component—being it infinite precision, unbounded precision and fixed precision, and still open to other forms of communication. The physical oracle itself is a measurement that the Turing machine performs in the physical world as an abstract scientist. The communicating between the Turing machine and the physical oracle is made through a query tape where the parameter needed to initialize the experiment is written. The consultation of the analogue oracle has a cost that is not just one time step of computation as a consequence of the unavoidable time costs inherent to the physical process.

We considered a particular physical oracle, the smooth scatter experiment or $SmSE$. This experiment is a symmetric two-sided measurement of distance and it is governed by elementary Newtonian mechanics. The $SmSE$ belongs to the class of physical oracles with exponential physical time t (the intrinsic time of the experiment) characterized by the following axioms:

1. Real values—The experiment is designed to find a physical unknown parameter $y \in]0, 1[$;
2. Queries—Each query is a binary string $z_1 z_2 \cdots z_k$ denoting a dyadic rational $z = 0.z_1 z_2 \cdots z_k$;
3. Finite output—The outcome is either $y < z$, $y > z$, with possible mistakes, or timeout;
4. Protocol timer—There is a time schedule $T : \mathbb{N} \to \mathbb{N}$, so that the time given to any query of length k is bounded by $T(k)$;
5. Sufficiency of the protocol—If $|y - z| > 2^{-|z|}$, then the result is not timeout;
6. Repeatability—Identical queries will result in identical results including identical timeouts.

This particular set of oracles led to a conjecture, already discussed in [5, 20], stating that for all "reasonable" physical theories and for all measurements based on them, the physical time of the experiment is at least exponential, i.e., the time needed to access the nth bit of the parameter being measured is at least exponential in n.

4.6.1   The Computational Power of the Analogue-Digital Machine

Using different protocols we get different ways of communicating between the Turing machine and the analogue device—the $SmSE$ in the present case (see Sect. 4.3). Different protocols relate with different measurement algorithms (see Sect. 4.3.2), defining three different types of smooth scatter machine or $SmSM$.

Codifying in the vertex of a $SmSE$ enough information, we were able to use the oracle to both generate sequences of non-biased coin tosses and, performing a measurement, solving decision problems of a suitable nonuniform complexity class. To measure the position of the vertex, we considered a bound for the consultation time—a time schedule—exponential in the precision (number of bits) of the query. Note that, although the lower bounds were proved for a specific analogue-digital machine—the $SmSM$—they are the same for every oracle we studied (see [7]).

Then we constructed advice functions encoding enough information to simulate $SmSE$ queries and established upper bounds, namely we proved that logarithmic squared size advice suffices to encode approximations to the so-called boundary numbers (see Sect. 4.4.6) as in [10]. Afterwards, again as in [10], we used an explicit time technique to reduce logarithmic squared advice just to logarithmic advice. Since boundary numbers exist, at least for the two-sided oracles with exponential physical time, we also conclude that the upper bounds are common to all two-sided physical oracles with such physical times.

Table 4.1 Main non-uniform complexity classes relative to the different protocols of the analogue-digital machine clocked in polynomial time

	Infinite	Unbounded	Fixed
Lower Bound	$P/\log\star$	$BPP//\log\star$	$BPP//\log\star$
Upper Bound	$P/\log\star$	$BPP//\log^2\star$ Exponential schedule	$BPP//\log^2\star$ Exponential schedule
Upper Bound Explicit Time	—	$BPP//\log\star$ Exponential schedule	$BPP//\log\star$ Exponential schedule

Our statements on the computational power of the analogue-digital machine clocked in polynomial time are summarized in the Table 4.1.

4.6.2 Open Problems

We have two non-trivial open problems related to oracles with exponential consultation time: (a) in the infinite precision case, to know if the lower and the upper bounds can be made to coincide without assumptions on the time schedule and (b) in the error-prone cases, to know if the lower and the upper bounds can be made to coincide without using the explicit time technique, namely, it is not known if there exists a set not belonging $BPP//\log\star$, decidable by a smooth scatter machine (or any other equivalent two-sided machine) in polynomial time.

Acknowledgments To Bill Tantau for the use of pgf/TikZ applications.

Appendix A: Nonuniform Complexity Classes

A nonuniform complexity class is a way of characterising families $\{C_n\}_{n\in\mathbb{N}}$ of finite machines, such as logic circuits, where the element C_n decides a restriction of some problem to inputs of size n. Nonuniformity arises because for $n \neq m$, C_n may be unrelated to C_m; eventually, there is a distinct algorithm for each input size (see [17]). The elements of a nonuniform class can be unified by means of a (possibly noncomputable) advice function, as introduced in Sect. 4.4, making up just one algorithm for all input sizes. The values of such a function provided with the inputs add the extra information needed to perform the computations (see [18]).

The nonuniform complexity classes have an important role in the *Complexity Theory*. The class P/poly contains the undecidable halting set $\{0^n: n$ is the encoding of a Turing machine that halts on input $0\}$, and it corresponds to the set of families decidable by a polynomial size circuit. The class $P/\log$ also contains the halting set defined as $\{0^{2^n}: n$ is the encoding of a Turing machine that halts on input $0\}$.

The definition of nonuniform complexity classes was given in Sect. 4.4. Generally we consider four cases: $\mathbb{C}/\mathbb{F}$, $\mathbb{C}/\mathbb{F}\star$, $\mathbb{C}//\mathbb{F}$, and $\mathbb{C}//\mathbb{F}\star$. The second and the fourth cases are small variations of the first and third, respectively. To understand the difference, note that $\mathbb{C}/\mathbb{F}$ is the class of sets B for which there exists a set $A \in \mathbb{C}$ and an advice function $f \in \mathbb{F}$ such that, for every $w \in \{0, 1\}^\star$, $w \in B$ iff $\langle w, f(|w|)\rangle \in A$. In this case, the advice function is fixed after choosing the Turing machine that decides the set A. As it is more intuitive to fix the Turing machine after choosing the suitable advice function, we considered a less restrictive definition, the type $\mathbb{C}//\mathbb{F}$: the class of sets B for which, given an advice function $f \in \mathbb{F}$, there exists a set $A \in \mathbb{C}$ such that, for every $w \in \{0, 1\}^\star$, $w \in B$ iff $\langle w, f(|w|)\rangle \in A$.

The following structural relations hold between the nonuniform classes used throughout this paper:

$$P/\log\star \subseteq BPP/\log\star \subseteq BPP//\log\star .$$

This result is trivial since we can just use the same Turing machine and the same advice function.

Appendix B: The Cantor Set

We prove Proposition 6 (see [10, 12] for further details). This proposition allows us to frame the distance between a dyadic rational and a real number. Recall that a dyadic rational is a number of the form $n/2^k$, where n is an integer and k is a positive integer. If such a number belongs to C_3 then it is composed by triplets of the form $001, 010$ or 100.

Proposition 12 *For every $x \in C_3$ and for every dyadic rational $z \in \,]0, 1[$ with size $|z| = m$, if $|x - z| \leq 1/2^{i+5}$, then the binary expansion of x and z coincide in the first i bits and $|y - z| > 1/2^{-(m+10)}$.*

Proof Suppose that x and z coincide in the first $i - 1$ bits and differ in the ith bit. We have two possible cases:

$z < x$: In this case $z_i = 0$ and $x_i = 1$ and the worst case for the difference occurs when the binary expansion for z after the ith position begins with a sequence of 1s and the binary expansion for x after ith position begins with a sequence of 0s.

$z > x$: In this case $z_i = 1$ and $x_i = 0$ and the worst case for the difference occurs when the binary expansion for z after ith position begins with a sequence of 0s and the binary expansion for x after the ith position begins with a sequence of 1s:

We can conclude that in any case $|x - z| > 2^{-(i+5)}$. Thus, if $|x - z| \leq 2^{-(i+5)}$, then x an z coincide in the first i bits.

The binary expansion of z after some position m is exclusively composed by 0s and since $x \in C_3$, it has at most 4 consecutive 0s after the mth bit. Thus, supposing that x and z coincide up to mth position, after this position they can coincide at most in the next 4 positions so they cannot coincide in $m + 5$ bits. Therefore, by the first part of the statement, $|x - z| > 2^{-(m+10)}$. $\square$

Appendix C: Random Sequences

Propositions 7 and 8 show how the $SmSE$ with unbounded or fixed precision could be seen as a biased coin. Given a biased coin, as stated by Proposition 9, we can simulate a fair sequence of coin tosses. Herein, we present the proof of such a statement.

Proposition 13 *Given a biased coin with probability of heads $\delta \in \,]0, 1[$ and a constant $\gamma \in \,]0, 1[$, we can simulate, up to probability $\geq \gamma$, a sequence of independent fair coin tosses of length n by performing a linear number of biased coin tosses.*

Proof Consider that we have a biased coin with probability of heads $\delta \in \,]0, 1[$. To simulate a fair coin toss we perform the following algorithm: Toss the biased coin twice and,

1. If the output is HT then output H;
2. If the output is TH then output T;
3. If the output is HH or TT then repeat algorithm.

As the probability of HT is equal to TH, we have the same probability of getting a H and a T and thus we simulate a fair coin. The probability that the algorithm halts in one run is $r = 2q(1 - q)$ and the probability of running it again is $s = 1 - \delta$. We want to run the algorithm until we get a sequence of fair coin tosses with size n. To get this sequence we may need to run the algorithm more than n times and thus we will study the total number of coin tosses required by considering the variable T_n denoting the number of runs until we get n fair coin tosses. The value T_n is a random variable that is given by the negative binomial distribution

$$T_n \overset{d}{=} NB(n, s) .$$

In this case we have the following mean and variance:

$$\mu = \frac{ns}{r} + n = \frac{n}{r}, \qquad \upsilon = \frac{ns}{r^2} .$$

Now, using the Chebyshev's inequality, we get

$$P(|\,T_n - \mu\,| \geq t) \leq \frac{\upsilon}{t^2} .$$

And thus, by considering $t = \alpha n$, for some α, we get

$$P(T_n \geq \mu + \alpha n) \leq \frac{ns}{r^2(\alpha n)^2} < \frac{1}{r^2\alpha^2 n} .$$

Since the worst case is for $n = 1$, in order to get the probability of failure less than $1 - \gamma$ we need

$$\alpha \geq \frac{1}{r\sqrt{(1 - \gamma)}} \,.$$

Noticing that $T_n \geq \mu + \alpha n$, we find that the total number of runs is

$$\frac{n}{r} + \frac{1}{r\sqrt{(1 - \gamma)}} \times n = \frac{n}{r}\left(1 + \frac{1}{\sqrt{1 - \gamma}}\right).$$

Since we toss a coin two times in each run, we get that the total number of coin tosses is linear in n

$$\frac{n}{\delta(1 - \delta)}\left(1 + \frac{1}{\sqrt{1 - \gamma}}\right).$$

□

References

1. Siegelmann, H.T., Sontag, E.D.: Analog computation via neural networks. Theor. Comput. Sci. **131**(2), 331–360 (1994)
2. Woods, D., Naughton, T.J.: An optical model of computation. Theor. Comput. Sci. **334**(1–3), 227–258 (2005)
3. Bournez, O., Cosnard, M.: On the computational power of dynamical systems and hybrid systems. Theor. Comput. Sci. **168**(2), 417–459 (1996)
4. Carnap, R.: Philosophical Foundations of Physics. Basic Books, New York (1966)
5. Beggs, E., Costa, J., Tucker, J.V.: Computational models of measurement and Hempel's axiomatization. In: Carsetti, A. (ed.), Causality, Meaningful Complexity and Knowledge Construction, vol. 46. Theory and Decision Library A, pp. 155–184. Springer, Berlin (2010)
6. Geroch, R., Hartle, J.B.: Computability and physical theories. Found. Phys. **16**(6), 533–550 (1986)
7. Beggs, E., Costa, J., Tucker, J.V.: Three forms of physical measurement and their computability. Rev. Symb. Log. **7**(4), 618–646 (2014)
8. Hempel, C.G.: Fundamentals of concept formation in empirical science. Int. Encycl. Unified Sci. **II**, 7 (1952)
9. Krantz, D.H., Suppes, P., Luce, R.D., Tversky, A.: Foundations of Measurement. Dover, New York (2009)
10. Beggs, E., Costa, J.F., Poças, D., Tucker, J.V.: Oracles that measure thresholds: the Turing machine and the broken balance. J. Log. Comput. **23**(6), 1155–1181 (2013)
11. Beggs, E., Costa, J.F., Poças, D., Tucker, J.V.: Computations with oracles that measure vanishing quantities. Math. Struct. Comput. Sci., p. 49 (in print)
12. Beggs, E.J., Costa, J.F., Tucker, J.V.: Limits to measurement in experiments governed by algorithms. Math. Struct. Comput. Sci. **20**(06), 1019–1050 (2010)
13. Beggs, E., Costa, J.F., Tucker, J.V.: The impact of models of a physical oracle on computational power. Math. Struct. Comput. Sci. **22**(5), 853–879 (2012)
14. Beggs, E., Costa, J.F., Loff, B., Tucker, J.V.: Computational complexity with experiments as oracles. Proceedings of the Royal Society, Series A (Mathematical,Physical and Engineering Sciences), **464**(2098), 2777–2801 (2008)

15. Beggs, E., Costa, J.F., Loff, B., Tucker, J.V.: Computational complexity with experiments as oracles. II. Upper bounds. Proceedings of the Royal Society, Series A (Mathematical,Physical and Engineering Sciences) **465**(2105), 1453–1465 (2009)
16. Beggs, E.J., Costa, J.F., Tucker, J.V.: Axiomatizing physical experiments as oracles to algorithms. Philosophical Transactions of the Royal Society, Series A(Mathematical, Physical and Engineering Sciences) **370**(12), 3359–3384 (2012)
17. Balcázar, J.L., Díaz, J., Gabarró, J.: Structural Complexity I, vol. 11. Theoretical Computer Science. Springer, Berlin (1990)
18. Balcázar, José L., Hermo, Montserrat: The structure of logarithmic advice complexity classes. Theor. Comput. Sci. **207**(1), 217–244 (1998)
19. Siegelmann, H.T.: Neural Networks and Analog Computation: Beyond the Turing limit. Birkhäuser, Boston (1999)
20. Beggs, E., Costa, J.F., Poças, D., Tucker, J.V.: An analogue-digital Church-Turing thesis. Int. J. Found. Comput. Sci. **25**(4), 373–389 (2014)

Chapter 5
Physical and Formal Aspects of Computation: Exploiting Physics for Computation and Exploiting Computation for Physical Purposes

Bruce J. MacLennan

Abstract Achieving greater speeds and densities in the post-Moore's Law era will require computation to be more like the physical processes by which it is realized. Therefore we explore the essence of computation, that is, what distinguishes computational processes from other physical processes. We consider such issues as the topology of information processing, programmability, and universality. We summarize general characteristics of analog computation, quantum computation, and field computation, in which data is spatially continuous. Computation is conventionally used for information processing, but since the computation governs physical processes, it can also be used as a way of moving matter and energy on a microscopic scale. This provides an approach to programmable matter and programmed assembly of physical structures. We discuss *artificial morphogenesis*, which uses the formal structure of embryological development to coordinate the behavior of a large number of agents to assemble complex hierarchical structures. We explain that this close correspondence between computational and physical processes is characteristic of *embodied computation*, in which computation directly exploits physical processes for computation, or for which the physical consequences of computation are the purpose of the computation.

5.1 Introduction

"Unconventional computation" is, of course, a negative term, and is defined by reference to "conventional computation," which is quite specific. Characteristics of conventional computation include digital (in fact binary) data and program representation, von Neumann architecture, (primarily) sequential program execution,

B.J. MacLennan (✉)
Department of Electrical Engineering and Computer Science,
University of Tennessee, Knoxville, TN, USA
e-mail: maclennan@utk.edu

© Springer International Publishing Switzerland 2017
A. Adamatzky (ed.), *Advances in Unconventional Computing*,
Emergence, Complexity and Computation 22,
DOI 10.1007/978-3-319-33924-5_5

addressable random-access memory, information processing implemented through sequential electronic binary logic, irreversible operations, classical (non-quantum) operation, etc. Unconventional computation may be defined, then, as computation that differs in one or more of these characteristics.

Given the success of conventional computation, it is reasonable to ask the reasons for studying unconventional computation. One motive is purely scientific: we would like to understand the full range of computational processes, in natural systems as well as in computers. Information processing is widespread in nature, but for the most part natural computation does not have the characteristics of conventional computation, and therefore we need to understand computation in a broader sense.

5.1.1   Post-Moore's Law Computation

The second motive for studying unconventional computation is technological, for it is apparent that Moore's Law must come to an end. First, the atomic structure of matter places limits on the smallness of electronic components and the density with which they can be assembled. Moreover, it is likely that economics will defeat Moore's Law even before it reaches these physical limits [60]. Therefore, in the post-Moore's Law world, progress in computation will depend on processing information in new ways, that is, on unconventional computation. The end of Moore's Law is on the horizon, and so it is important that we develop post-Moore's Law technologies to the point of practicality before the end is reached.

What might be the characteristics of post-Moore's Law computation? Conventional computer technology has benefitted from clearly separated hierarchical levels. Programming abstractions, such as data structures, are implemented in terms of primitive data elements, such as floating-point numbers and pointers, which are implemented in terms of many bits, each of which is represented by many electrons. Similarly, conceptually primitive operations, such as floating-point division, may be implemented by iterative algorithms, themselves implemented in sequential logic. In particular, the Boolean logic level is largely independent from those above and below it. That is, on one hand, Boolean logic can be used to implement various computer architectures, and on the other, Boolean logic can be implemented in many different technologies (e.g., relay, vacuum tube, transistor, VLSI).

Computing abstractions are implemented in terms of lower-level abstractions, and ultimately in the laws of physics, but post-Moore's Law computing technologies cannot afford these multiple hierarchical levels. To permit greater densities and speeds, computing abstractions and physical laws will need to be brought closer together, but we cannot change the laws of physics, so this assimilation of computation and physics will have to be accomplished by developing computing paradigms that are more like the laws of physics. Therefore post-Moore's Law computing will have more of the characteristics of the underlying physical processes.

For example, the laws of physics are fundamentally concurrent; individual particles respond in parallel to fields, forces, and other particles. Therefore, we expect parallel computation to be the norm in the post-Moore's Law era.

The laws of physics are expressed in differential equations (or partial differential equations), which describe continuous change in continuous quantities. Therefore, analog computation can be expected to increase in importance. Often operations can be implemented in a few analog components, which would require many digital components (see Sect. 5.3.2 below). Therefore analog representations are preferable for achieving higher densities.

It might be objected that quantum mechanics applies at the smallest scales, and therefore that digital computation is better matched to the physics at these scales. It is true that at very small scales certain quantities, such as charge, spin, and energy, are quantized. On the other hand, quantum wave functions are continuous functions of space and time, and the Schrödinger equation is a differential equation. Even qubits are continuous linear combinations of the basis states.

Conventional computation takes place in *discrete* or *sequential time* (see Sect. 5.2.3 below), in which operations take place in sequence at discrete times. (Parallel computation does not contradict the essentially sequential execution of digital computation.) At both the classical and the quantum levels, however, the laws of physics are expressed by differential equations. Therefore, as our computational processes become more like physical processes, we expect continuous-time processes to play an increasing role in post-Moore's Law computation.

As we approach very small scales, noise, uncertainty, defects, imperfections, and faults all become more likely, and ultimately unavoidable. Therefore, in the post-Moore's Law era we will have to abandon the idea that we are striving for systems that approximate ever more closely an ideal, which is perfect, noiseless, fault-free, etc. Rather, we will take these phenomena as a given, and design systems that exploit them rather than trying to avoid or mitigate them. Natural computation, which we find in living systems, has much to teach us about exploiting physical phenomena for robust and efficient information processing. For example, "noise" can be reconceptualized as a source of *free variability*, which can be used for escaping from local optima and for many other purposes [31, 37].

5.2   The Essence of Computation

Given a growing assimilation of computation to physics, one might wonder what distinguishes computational processes from other physical processes. Is every physical process a computation? What common properties distinguish information processing, computation, and control from other physical processes [40]?

All of these computational processes use physical arrangements (physical *form*) to represent something else, and use physical rearrangement (*transformation*) to represent some abstract process. What distinguishes a computation from other physical processes is that in principle the goals of the process could be achieved by any physical system that realized the same abstract transformation (a property called *multiple realizability*) [29, 35]. That is, while computation must be physically realized, it does not depend essentially on a particular physical realization.

5.2.1 The Four Whys of Computation

The preceding observation raises several issues, which are addressed best in terms of Aristotle's "four whys" (commonly known as his "four causes") (Aris., *Phys.* II 194b–195a, *Met.* 983a–b, 1013a–1014a). These are four sorts of answers to the question of *why* an object or process is what it is. In general, none of the four is sufficient on its own, and they differ in explanatory value depending on the subject matter and motivation of the question. His taxonomy is useful for classifying explanations not only in biological systems but also in artifacts, and therefore it can be applied to both natural and artificial information processing and control [40].

One of the *whys* answers the question: *What* is this? Traditionally it is called the *formal cause* because it accounts for an object or process in terms of its form or pattern: the formula that describes it. The second *why* answers: *From what* is this made? It is known traditionally as the *material cause*, since its explanation is in terms of the unformed stuff which gains its specific properties through the formal cause. These two *whys* are central to our topic, since information is realized by material forms, and information processing by physical rearrangement of these forms [24]. Although computation requires *some* material realization of its formal processes, it is independent (qua computation) of its *particular* material realization. Moreover, form and matter are relative terms, and the formed matter at one level of organization can provide the unformed matter for a higher level. For example, in conventional computers the addressable bytes and basic operations are the medium that software formally organizes, but the bytes and operations are themselves organized structures of lower-level objects and processes (logic gates).

The third *why* answers: *By what* is this object or process created and sustained? It is the familiar *efficient cause*. Computation must be powered, either by an initial supply of energy or by a continuing supply. Without its efficient cause, the computation is a potentiality that is not actualized.

The last *why* answers the question: *For the sake of what* does this object exist or this process take place? Traditionally it is called the *final cause* since it addresses the end, goal, or purpose of something. Artifacts, such as manufactured computers, are designed for some purpose, but biological systems also have functions that they fulfill. The heart circulates the blood, the immune system fights infections, the nervous system coordinates behavior and cognition, and so on.

The final cause is essential to the definition of computation, for what distinguishes computational and information processing systems from other physical systems is that their goal or function can be performed by any system with the same formal structure, independent of its material realization [31, 38, 40]. That is, their function is *information* processing as opposed to some other physical process. One test of whether a system is computational is to ask whether it could be replaced with another system with the same formal structure and still achieve its ends (i.e., whether it is multiply realizable). Of course, many natural systems serve multiple functions (e.g., circulation of the blood distributes oxygen, but also hormones, which transmit information), and so they might not be purely computational [40]. From this perspective, it is remarkable the extent to which the function of the nervous system is pure computation and control.

The relationship between a computational or information processing system and its physical realization can be expressed by the *realization homomorphism* [31], which says that a physical system realizes an abstract computation if there is a homomorphism from the physical system to the abstract system. The significance of the homomorphism is that it preserves some of the algebraic structure of the physical system, but not all of it. This captures our intuition that the physical system can have many properties that are irrelevant to its realization of the abstract system. However, we must also recognize that many realizations are only approximate. For example, the abstract computation might involve real numbers that are only approximately realized by floating-point numbers in a physical computer.

5.2.2 The Topology of Information

Rolf Landauer reminded us that information is physical; it must be represented in some physical medium [24]. But its essence—what makes it *this* information versus *that* information—lies in its form. Therefore differences of information are differences of abstract form, which can be described by topology, the science of abstract form and similarity.

The realization homomorphism $\mathcal{H}$ is a surjection mapping the physical state space S onto the abstract state space Σ, that is, $\mathcal{H} : S \twoheadrightarrow \Sigma$. In the physical state space we distinguish the *information-bearing degrees of freedom* (IBDF) from the *non-information-bearing degrees of freedom* (NIBDF) [4]. For example, the IBDF may be macrostates representing, for example, the bits 0 and 1, while the NIBDF may include the positions, momenta, etc. of individual particles, which do not represent information and which manifest as heat, noise, etc. The IBDF are managed by the computational process in order to realize the computation, but the NIBDF are not managed or are managed only in aggregate to keep them from interfering with the computation. Let $E \subset S \times S$ be the equivalence relation between physical states that are equivalent in their IBDF, and define the quotient space $Q = S/E$, which represents the IBDF. Then the realization homomorphism can be factored $\mathcal{H} = \mathcal{A} \circ I$, where $I : S \twoheadrightarrow Q$ is a surjection from the physical state space onto the IBDF, and $\mathcal{A} : Q \leftrightarrow \Sigma$ is a bijection between the IBDF and the abstract state space.

Conventional computation is digital; it uses discrete representations of information. Formal models of digital computation, such as finite-state machines and Turing machines, use finite, discrete alphabets of symbols or states, in which each element is identical to itself and completely different from every other element. This is a *discrete topology* (Σ, δ), in which the metric is defined $\delta(x, x) = 0$ for $x \in \Sigma$ and $\delta(x, y) = 1$ for all $y \neq x$. There are only two possible distances in a discrete metric space. A finite discrete space with 2^n elements is homeomorphic to the space $\{0, 1\}^n$ with the ∞-product metric[1]; this is of course the basis of binary representations on digital computers. Other, less trivial, topologies can be defined over these discrete spaces for the purposes of computation. For example, the discrete space $\{0, 1\}^n$ can represent the integers $\{0, 1, \ldots, 2^n - 1\}$ with their usual metric.

Traditionally, analog computers have operated on bounded real numbers represented by a physical quantity, but they are also capable of operating on other continuous quantities, such as complex numbers (represented, for example, by the phase and amplitude of a periodic signal). Moreover, quantum computers operate on complex linear superpositions of basis states (e.g., $z|0\rangle + z'|1\rangle$, with $z, z' \in \mathbb{C}$). Further, *analog field computers* can operate on *fields* [33], that is, spatially continuous distributions of continuous quantity (see Sect. 5.3.4). Images and continuous signals are examples of fields. All of these information spaces are *continua*, which may be defined formally as connected second-countable metric spaces. Second-countability means that they have a countable dense subset (as, for example, the rationals are a countable dense subset of the reals).

More generally, unconventional computers may be hybrid, that is, capable of operating on both discrete and continuous information spaces. These are products of spaces that are individually discrete or continuous spaces. The U-machine model encompasses both digital and analog computing (Sect. 5.3.5).

5.2.3 The Topology of Information Processing

Computation takes place in time. In conventional digital computation, operations are performed at discrete points in time. This is properly described as *sequential-time* computation since there is no implication that the operations be performed at regular time intervals, as in *discrete-time* computation [66]. More generally computation may be described by a partial order defining how later computations depend on earlier ones, thus permitting operations to be performed concurrently.

Analog computation can also be defined over either discrete or sequential time, as in the BSS model of computation over the reals [5]. Most artificial neural network models are sequential-time analog computations, since the neural operations do not happen at specific times, but require only that their inputs be available. Even most recurrent neural networks operate in sequential time, since the sequence of their outputs depends only on the sequence of their inputs. Neural networks can of course operate in discrete time, but only if the neural computations are clocked.

Traditionally, however, analog computers have operated in continuous time, integrating differential equations, which serve as programs. But there is also a continuous

---

[1]The ∞-product metric on the Cartesian product of two spaces (X, δ_1), (Y, δ_2) is defined $\delta_\infty[(x, y), (x', y')] = \max[\delta_1(x, x'), \delta_2(y, y')]$.

version of sequential computation, in which the computation is defined over a set of "instants" homeomorphic to a closed or half-open interval of the real numbers. That is, sequence is defined, but not duration or rate. More generally, both discrete and continuous concurrent computation can be defined by a partial order that defines the dependence of later computations on earlier ones. Operations are permitted to take place sequentially or concurrently, so long as this partial order is respected.

5.2.4 Programmability

Programmability is an important property of many computation systems. A system is *programmable* if it is capable of performing a wide range of functions depending on some finite systematic external specification (a *program*). Programmable systems are valuable because they can be used for many different purposes and their behavior can be adapted to changing circumstances, simply by changing the program.

Computational programs are usually described textually; for example sequential-time computations (digital or analog) are described by programs in programming languages, and continuous-time computations are described by ordinary or partial differential equations. Programs can also be described by diagrams, such as flow-charts for sequential-time computations and block diagrams for continuous-time processes.

The preceding are examples of discrete programs, but continuous programs are also possible. For example, some analog computers permit functions to be described by continuous graphs (Sect. 5.3.2.1). Moreover, many useful computations can be defined as *relaxation processes* in which the state descends a potential surface to approach an attractor, which is the solution to the problem (Sect. 5.3.2.3). Such processes may be continuous-, discrete-, or sequential-time depending on how the state changes as the system computes. In these cases the program, which governs the computation, is defined by the potential surface, which therefore defines a continuous program, which we might call a *guiding image* [29, 31]. The metaphors are different: instead of *writing a program*, we could say we are *drawing* or *sculpting a guiding image*. While it is certainly possible to create such a continuous program manually, more likely it will emerge from an adaptive or training process, as happens in artificial neural networks.

5.2.5 Universality

Programmability raises the issue of universality: Is a computer capable of computing anything, given an appropriate program? For example, we know the Universal Turing Machine (UTM) is capable of computing any Turing-computable function. That is, for any Turing machine there is a UTM plus program combination that is *equivalent* (computes the same function). However, we must be careful applying these familiar ideas to unconventional computation.

When comparing the power of different models of computation, it is important to remember that all models are idealizations of the things they are modeling, and these idealizations are intended to make the model more tractable than the original system *for some purpose*. Therefore, each model exists in a (usually implicit) *frame of relevance*, which delimits the sort of questions it is suited to answer [31, 35]. A model cannot be expected to give useful answers when applied outside of its frame of relevance; indeed, the answers are often misleading, more a reflection of the model than the system under investigation.

Therefore, while it is very tempting to compare various models of unconventional computation to the Church-Turing model, we must be cautious doing so. This model was developed to address questions of effective calculability in the foundations of mathematics, and they delimit its primary frame of relevance. It makes many idealizing assumptions, such as that tokens are discrete and can be perfectly discriminated from their background and classified as to type, etc. [31]. Two machines or programs are considered equivalent in power if they compute the same input–output function. Efficiency in analyzed in terms of asymptotic complexity, which ignores constant scale factors. And so forth.

While the conventional theory of Church-Turing computation has proved enormously fruitful, there are many important issues that are outside of its frame of relevance. For example, an important question in natural computation is how brains are able to process complex, noisy sensorimotor information in real time using relatively slow, low-precision computing devices (neurons). The conventional theory of computation is not equipped to deal with issues in real-time control. Further, asymptotic analysis is not very useful because (1) the constants matter, and (2) the size of the input is usually bounded. Therefore, in many of the contexts in which unconventional computation is relevant, such as natural computation and post-Moore's Law computation, the idealizing assumptions of Church-Turing computation are inappropriate, and different models, which make different assumptions, are more useful [31]. In these contexts, it is not usually appropriate to consider two computations equivalent solely because they compute the same function, and therefore it is not very useful to measure the power of a model of computation in terms of the class of functions it computes. This is only one of the criteria by which models of computation can be compared. In the context of unconventional computing there are many dimensions for comparing the capability of computational models.

5.3 Computation for Formal Ends

In order to understand the full range of unconventional computation, it is useful to explore the relation between the computational processes and their physical realizations. In this section we address computation in its usual sense, wherein the principal goal is an abstract information process, and the realization is a means to this end.

That is, the *material* processes are serving *formal* purposes. In Sect. 5.4 we consider the opposite situation, which is less familiar.

5.3.1 General Considerations

What are the requirements for unconventional realizations of abstract information and control processes or computations? In general, any reasonably controllable, mathematically describable physical process can be used for computation, including living systems, such as slime molds and bacterial mats [1]. We can outline some more specific considerations [35]. First we need a physical process that has at least the algebraic structure of the desired computation, so that the realization homomorphism holds. Therefore, we need to have sufficient control over the physical arrangements to implement the required structure. For general-purpose computation, we will want some flexibility in making these arrangements, so that any computation in a useful class can be implemented. In this case, we also may consider programmability, that is, whether there is some systematic way to set up the physical process in accord with an abstract description (the program).

Of course, the application may place additional restrictions on the class of admissible realizations. For example, some physical processes might be too slow or consume too much energy for the application. On the other hand, many potential applications do not require high speed, and a slower physical process, which is better matched to the application requirements, may have other advantages, such as energy efficiency, power source, stability, robustness, programmability, or precision. Moreover, many applications do not require high precision or faultless operation, and computation and control in nature provide many examples of how to tolerate and even exploit noise, errors, faults, imprecision, defects, indeterminacy, etc. For example, they can be used as sources of *free variability* for escaping from local optima, breaking deadlocks, driving exploration, etc. [41].

Useful computations require *transduction*, that is, the transfer of information from the environment into the computation, and the transfer of information and control from the computation back out into the environment [31, 35]. Both computation and transduction involve the formal and material aspects of physical processes. Computation, as we've seen, is *generically* realizable; that is, it can be realized by any physical process with the required formal structure. On the other hand, transduction provides the interface between the computational medium and *specific* physical systems (e.g., a photoreceptor or temperature sensor for input, an LED display or servomotor for output). In principle, a pure input transducer transfers the form from one material (the input medium) to another material (the computational medium), and a pure output transducer transfers the form from the computational medium to the output medium. In practice, pure transducers are rare, for there is usually some (intended or unintended) change in the form in addition to the intended material change; for example, the input might be filtered, digitized, limited, etc. Thus most transducers combine some information processing or computation with the change of medium.

5.3.2 Analog Computation

Analog computation is an important unconventional computing paradigm. Since the laws of physics are continuous, it is likely to become more important in the post-Moore's Law era, because it can be more directly realized [44]. In principle, any continuous physical quantities can be used as a medium for analog computing. Electronic analog computing, in which real numbers are represented typically by current, voltage, or charge, is most familiar, but there are many other possibilities. For example, mechanical analog computers have represented numbers by angular or linear displacement. Concentrations of substances that are continuous or approximately continuous can be used (as in reaction-diffusion computation [2]). Light is an attractive medium [3].

In choosing an analog computation medium, we must also consider the physical realization of the abstract operations required by the computation (e.g., addition, subtraction, multiplication, integration, filtering, various transcendental functions). The virtue of analog computation is that common, useful operations often have simple realizations. For example, addition can be performed by simply combining currents, charges, or light intensities; integration can be performed by charging a capacitor or by accumulation of a chemical reaction product.

One critical question in any analog computing technology is *precision*, which refers to the quality of a representation. Precision has two major components: *resolution*, which refers to the fineness of the representation, and *stability*, which refers to its ability to maintain its value over the duration of the computation. Precision can be expressed as a fraction of *full-scale variation* of a variable (the difference of its maximum and minimum values). Doubling the precision of an analog representation or computation can be very expensive compared to doubling digital precision (add one more bit), since it requires higher quality devices [32]. Fortunately, high precision is not required for many applications and for some approaches to analog computing, such as neural networks. In general, natural computation provides many examples of the utility of low-precision analog computing.

5.3.2.1 Programming Techniques

Certain basic operations are simple to implement in many analog computing technologies. As mentioned, direct combination of physical quantities can often be used to implement analog addition, $u(t) = v(t) + w(t)$. Some physical quantities are signed (e.g., voltage, current, charge) and can be used directly to represent signed quantities, others are not (e.g., intensities, concentrations of chemicals). In the latter case, signed quantities can be represented as differences of positive quantities. That is, instead of one signed variable $v(t)$, we use two non-negative variables, $v^+(t)$ and $v^-(t)$, that implicitly represent $v(t) = v^+(t) - v^-(t)$. The analog algorithm must be re-expressed in terms of the differential quantities. Given a signed representation, subtraction $[u(t) = v(t) - w(t)]$ and negation $[u(t) = -v(t)]$ are easy to implement.

Positive constant multiplication, $u(t) = cv(t)$ for $c > 0$, can be implemented by passive attenuation or active amplification. Signed constant multiplication can be implemented directly or in terms of the signed operations. The assumption here is that the scale factor c must be programmed, either externally (e.g., by adjusting a potentiometer) or internally (e.g., by programming a floating-gate transistor), and that this is a relatively slow process, which might not be under analog program control. Therefore, we contrast it with full variable multiplication, $u(t) = v(t) \times w(t)$, in which both factors can be the result of ongoing analog computation. Direct analog implementation can be more difficult than constant multiplication, but it can be accomplished. For example, a squaring operation can be used to implement multiplication by [56, p. 92]:

$$v \times w = 0.25[(v + w)^2 - (v - w)^2].$$

Squaring can be implemented directly without multiplication [56, chap. 3]. This illustrates an important principle of analog computing: we cannot transfer our digital intuitions about what is simple into the analog domain. In the analog domain, apparently complicated operations, such as square, square-root, logarithm, and exponential, can have more direct implementations than apparently simpler operations, such as multiplication. Certain nonlinear and transcendental functions can be built into an analog computer as basic operations.

Division, $u(t) = v(t)/w(t)$, has to be handled carefully, since a small divisor can saturate the quotient register. Similarly, although analog implementation of differentiation, $u(t) = \dot{v}(t)$, is generally simple, the operation is problematic since it is sensitive to high-frequency noise, which it amplifies. One solution is to apply a low-pass filter to the differentiator's input. Alternatively, analog computations involving differentiation can be recast as integrations.

Integration usually has a straightforward implementation as the accumulation of some quantity:

$$u(t) = u_0 + \int_0^t v(\tau)d\tau.$$

The integrator is initialized to the constant of integration u_0 at the beginning of the computation. This implements a differential equation $\dot{u}(t) = v(t)$ with an initial value $u(0) = u_0$.

For some applications (such as real-time control programs) the integration will be with respect to real time. In others, time in the analog computer will be independent of time in the abstract computation; it might integrate slower or faster. To ensure accurate results, the rate of analog integration has to be considered, since if it is too fast it may exceed the high-frequency response of the integrator, and if it is too slow, quantities will drift. Therefore analog integration often involves *time scaling*, in which time t in the computer is related to time τ in the abstract computation by $t = b\tau$ for some $b > 0$. To integrate the abstract differential equation $\dot{u}(\tau) = v(\tau)$, that is, $u(\tau) = \int_0^\tau v(\tau')d\tau'$, the analog computer uses the scaled integration $u(t) = b^{-1} \int_0^t v(t')dt'$. In electronic analog computers this can be accomplished by decreasing the integrator input gain by a factor of b.

Since analog computing represents abstract quantities directly by physical quantities, *magnitude scaling* is another important consideration. A variable x in the abstract computation, with a certain range of values, must be mapped into a physical quantity v, with a dynamic range and precision limited by the physical device. Exceeding the device's operating range can lead to inaccuracy through distortion. Magnitude scaling is accomplished by choosing a scale factor, $v = ax$, which is small enough to stay within the device's dynamic range, but not so small that important differences are less than the device's resolution. Therefore, the variables in the abstract computations have to be scaled, and differential equations (or integrations) need to be adjusted to incorporate the scale factors. Moreover, in addition to the explicit variables, there are implicit variables corresponding, for example, with derivatives $\dot{x}$, $\ddot{x}$, etc. These too need to be scaled with the equations adjusted accordingly.

Some analog computers provide tunable band-pass filters, which can be used to perform a discrete Fourier transform on a signal. Others provide analog matrix–vector multiplication in which the elements of the matrices and vectors are continuous quantities, and the multiplications and additions are implemented by analog computation. That is, $\mathbf{u}(t) = \mathbf{Mv}(t)$, where $u_j(t) = \sum_{k=1}^n M_{jk} v_k(t)$. This operation can be used to implement linear operators, such as filters. Another useful operation is a noise generator, which produces Gaussian white noise, which can be adjusted and filtered to have desired characteristics. Randomness is useful in some analog algorithms. Simple decision making can be implemented by sigmoid functions:

$$\sigma(s) = \frac{1}{1 + e^{-\beta s}}.$$

Then a differential equations such as $\dot{x} = \sigma(s - \theta)F(x, y, \ldots) + \sigma(\theta - s) G(x, y, \ldots)$ will be governed by $F(x, y, \ldots)$ if s is above the threshold, $s > \theta$, and by $G(x, y, \ldots)$ if $s < \theta$, with β controlling the sharpness of the transition.

Some analog computers provide means for computing arbitrary functions by means of a continuous version of table lookup. This mechanism allows the computation of functions for which there is no known closed-form description, or that would be too complicated to compute from their closed forms. To implement such a function, its graph $\{(x, f(x)) | x \in [x_{lwb}, x_{upb}]\}$ is represented in a suitable two-dimensional medium. When this medium is loaded in the computer, it can compute $u(t) = f[v(t)]$. Similarly, an arbitrary binary function g can be computed by representing its graph $(x, y, g(x, y))$ in a suitable three-dimensional medium. These are examples of guiding images, i.e., continuous representations of analog algorithms (Sect. 5.2.4 above).

5.3.2.2 General-Purpose and Universal Computation

Universality is an important question for any computing paradigm, for it tells us what are the minimal requirements for performing any computation in a large class of possible computations. Claude Shannon proved fundamental universality theorems for the differential analyzer, which were completed, corrected, and extended by Pour-El, Lipshitz, and Rubel [25, 57, 62, 63].

A related question is the power of analog computing relative to Turing computability, but it presents an immediate paradox. On the one hand, it is easy to show that the ability to operate on arbitrary real numbers confers super-Turing power (e.g., there is a real constant whose bits encode the solutions to the Halting Problem). On the other hand, analog computers are routinely simulated on ordinary digital computers, suggesting that analog computers have no more than Turing power. There are a variety of theorems in the literature, proving sub-Turing, Turing, or super-Turing power depending on the premises (representative citations can be found elsewhere [44]). The resolution of these apparently contradictory conclusions is that analog computation is not in the frame of relevance of Church-Turing computation (recall Sect. 5.2.5), and therefore the results are more a reflection of the idealizing assumptions of the various models than of the computational systems being modeled (more details are provided elsewhere [35]).

There are a number of ways to program analog computers. Sequential analog computations can be described in programming languages similar to those for digital computers, the principal difference being that the primitive operations are analog rather than digital. However, some caution is necessary. For example, exact equality and inequality tests, which are unproblematic in digital computation, may be infeasible in the analog domain, where infinite precision would be required. In the context of analog computing, it is more reasonable to test if the difference of two numbers is less than some ε.

Continuous-time analog computations are most often described by differential equations. They are also represented by block diagrams in which the differential equations are recast as explicit integrations (e.g., Fig. 5.1).

In principle, analog programs can contain constants that are not rational or even Turing-computable. Such constants cannot be represented finitely in discrete symbols, but they can be represented directly as continuous quantities. In a practical sense, however, due the limited precision of analog computing, constants can be represented digitally to the accuracy required. Nevertheless, it is important to broaden the notion of a program to include representations that are not textual, such as guiding images (Sect. 5.2.4 above).

5.3.2.3 Dynamical Systems

Dynamical systems are an attractive approach to analog computation; the system is defined so that the point attractors are solutions to the problem. Examples include analog solutions to traditional digital problems, such as sorting [8] and Boolean sat-

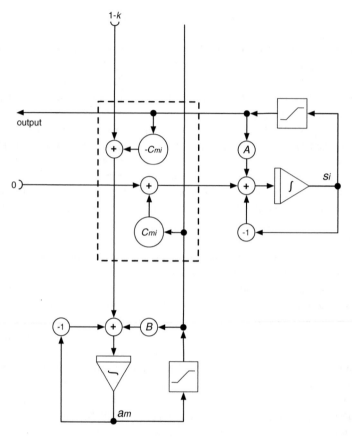

Fig. 5.1 Example analog algorithm implementing a dynamical system Boolean satisfiability [12]. The block enclosed in *dotted lines* is repeated for $m = 1, \ldots, M$ and $i = 1, \ldots, N$

isfiability [17, 48]. In the latter case, to solve a k-SAT Boolean satisfiability problem with M clauses and N variables, Ercsey-Ravasz and her colleagues define a dynamical system by the differential equations:

$$\dot{s}_i(t) = -s_i(t) + Af[s_i(t)] + \sum_{m=1}^{M} c_{mi} g[a_m(t)],$$

$$\dot{a}_m(t) = -a_m(t) + Bg[a_m(t)] - \sum_{i=1}^{N} c_{mi} f[s_i(t)] + 1 - k.$$

A particular problem instance is defined by the c_{mi} matrix elements: $c_{mi} = +1$ if variable i is positive in clause m, $c_{mi} = -1$ if variable i is negative in clause m, and $c_{mi} = 0$ if variable i is not in clause m. The f and g activation functions are linear squashing functions that map the s and a values into $[-1, 1]$ and $[0, 1]$, respectively. The s_i converge on a solution to the problem, if one exists.

Figure 5.1 displays an analog algorithm for implementing this dynamical system [12]. The overall structure is a cross-bar between the M integrators for the a_m and the N integrators for the s_i; thus $M + N$ integrators are required. A particular instance is programmed by setting the c_{mi} and $-c_{mi}$ connections as required for the problem. The integrators are initialized to small values to start the computation; non-zero offset or noise in the hardware integrators might have the same effect.

5.3.3 Quantum Computation

Quantum computation is another promising approach to post-Moore's Law computing. Because the units of information representation are *qubits* (quantum bits), it is often supposed that quantum computation is a species of digital computation, but in fact it is hybrid analog–digital computation. Quantum computation gets its power from being able to operate simultaneously on superpositions (complex linear combinations) of digital basis states. Quantum operations are unitary operators that operate on the continuous complex coefficients of the basis states. Fundamentally, "binary" quantum computation is computation over finite-dimensional complex vector spaces. One of the remarkable properties of quantum computation, which gives it an advantage over classical analog computation, is that it is possible to do error correction to eliminate noise in the complex coefficients [52, Sect. 10.6.4]. Some quantum computation takes place in continuous time, such as adiabatic quantum computing and quantum annealing [13, 61]. *Continuous-value quantum computation* is another approach to analog quantum computation [26].

5.3.4 Field Computation

While ordinary differential equations (ODEs) are adequate for describing some systems, spatially extended systems normally require partial differential equations (PDEs). Although most historical analog computers processed ODEs, already in the nineteenth century there were developments such as the "field analogy method" [33, 44]. Sometimes the state was represented in a continuous medium, such as a rubber sheet or an electrolytic tank; in other cases a sufficiently dense array of discrete components was used. Therefore, we may define a *field* as either a spatially continuous distribution of continuous quantity, or a discrete distribution that is sufficiently dense to be treated as continuous. (Physicists similarly distinguish *physical fields*, which are literally continuous, such as electromagnetic fields, from *phenomenological fields*, which can be treated as though continuous, such as fluids.) Thus we can have real- or complex-valued scalar fields or vector fields (more generally, fields over any continuous algebraic field).

Field computation, then, may be defined as computation in which the state is represented by one or more fields [27, 33]. It is also a natural way of describing image processing or other computation with spatially extended data, and field computers have operations, such as convolution, that operate in parallel on entire fields. The original motivation for the theory of field computation was to describe neural information processing in regions of cortex large enough to be considered fields (typically 0.1 mm^2 or larger) and in neurocomputers with comparable numbers of spatially organized neurons [27].

Mathematically, fields are treated as continuous functions over some spatial domain Ω. More precisely, they are elements of a Hilbert space of square-integrable functions on Ω, which we denote $\Phi(\Omega)$. Its metric is determined by the inner product; for $\phi, \psi \in \Phi(\Omega)$,

$$\langle \phi \mid \psi \rangle = \int_\Omega \overline{\phi(u)} \psi(u) \mathrm{d}u,$$

where $\overline{\phi(u)}$ denotes the complex conjugate (in case the fields are complex-valued). *Field transformations* are functions (linear or nonlinear) that map fields into fields; that is, they are operators on Hilbert spaces. One especially useful field transformation is the *field product* $\Psi\phi \in \Phi(\Omega')$, for $\Psi \in \Phi(\Omega' \times \Omega)$ and $\phi \in \Phi(\Omega)$, which is defined by the Hilbert–Schmidt integral, $(\Psi\phi)(u) = \int_\Omega \Psi(u, v)\phi(v)\mathrm{d}v$, for all $u \in \Omega'$. It is the field analog of a matrix–vector product. The outer product also has a field analogue: if $\phi \in \Phi(\Omega)$ and $\psi \in \Phi(\Omega')$, then $\phi \wedge \psi \in \Phi(\Omega \times \Omega')$ is defined $(\phi \wedge \psi)(u, v) = \phi(u)\psi(v)$. Other useful operations include the gradient, Laplacian, convolution, cross-correlation, and point-wise arithmetic operations between fields.

Two questions immediately arise: Are there universal field computers? And (more practically), what operations should be provided by a general-purpose field computer? These questions can be answered in the context of approximation theory for operators on Hilbert spaces. For example, there is a sort of field-polynomial approximation based on an analogue of Taylor's theorem for functional derivatives [27, 28, 30, 33]. Also, since a field can be considered a continuum of (infinitesimal) neurons, many neural network approximation theorems can be adapted to field computation [20, pp. 166–168, 219–220, 236–239, 323–326]. For example, one universal set of operations is the field product (Hilbert–Schmidt integral), pointwise addition, and scalar multiplication [33, 44].

5.3.5   The U-Machine

We commonly classify computation as digital or analog, or as "hybrid" if it combines both, but does digital (computation on discrete spaces) and analog (computation on continua) exhaust the possibilities of computation? What other topologies might there be for information and computation?

We have explored computation on second-countable metric spaces because they include both discrete spaces and continua, and have developed a corresponding machine model, the *U-machine* [38]. It gets its name from Urysohn's theorem, which states that any second-countable metric space is homeomorphic (topologically equivalent) to a subset of a Hilbert space [51, pp. 324–326]. Therefore, computations in second-countable spaces have realizations in Hilbert spaces, that is, they can be implemented by field computations. Indeed, the details of the Urysohn embedding imply that they can be approximated by computations over finite-dimensional vector spaces (and, in particular, neural networks).

Because the Urysohn embedding is a homeomorphism, any continuous computational process in a second-countable metric space has a continuous image in the subset of the Hilbert space. Further, for any continuous function on a second-countable space, there is a corresponding continuous function on the Hilbert space. Therefore, computations in second-countable spaces can implemented by computations in these Hilbert spaces, which can be implemented via the various universal approximation theorems on Hilbert spaces (Sect. 5.3.4). These provide the basic operations required for general-purposes computation on the U-machine.

Other sorts of physical media can be used to realize computational processes, for example, molecular computation. Next, however, I will address a different aspect of unconventional computation: how computation can be used directly to control physical processes.

5.4 Computation for Material Ends

In computation we have a relationship between a physical system and a formal system in which the formal system is a (typically incomplete) description of the physical system; this is the import of the realization homomorphism (Sect. 5.2.1). In conventional computation, as well as in the unconventional computation discussed in Sect. 5.3, the end (goal) is the abstract formal process, and the physical process is a means to that end. Furthermore, in a programmable computer, the program controls the physical processes so that they realize the abstract process described by the program. In particular, in the process of computation, matter and energy is reorganized in the computer, and this reorganization is under control of the program. Therefore we can look at the formal-material relationship from a different perspective in which the end is the physical process and the computation is the means to this end; that is, we have formal processes serving material purposes.

The tradeoffs are different. When the material processes are serving formal purposes, we usually try to minimize the energy and matter reorganized by computation, in order to decrease size, power requirements, and computation time. In contrast, when formal purposes are used to serve material ends, we might want to rearrange *more* matter or energy.

Computation for material purposes is different from a traditional control system, in which information processes (realized physically) govern a separate physical system via transducers. Here, we are describing a situation in which the physical realization of the computation *is* the physical process that is the goal. In particular, there are no transducers because there is no distinction between the information system and the controlled system; they are simply the formal and material aspects of the same process.

A simple example is provided by chemical reaction-diffusion (RD) systems [2]. On the one hand, an RD system can be viewed as a formal process and analyzed mathematically, as Turing did [65]. On the other hand, RD systems can be realized chemically so that the chemical reaction and diffusion processes are essential to both the computation and the physical patterns it creates. Such processes underlie patterns in animal skin colors and hair coats [46]. Algorithmic assembly by DNA is another example in which the molecules realize a process that computes a desired physical structure [58, 59, 64].

In general, programs are hierarchical structures that, when executed, generate complex dynamics, which is capable of generating complex structures. That is, complex hierarchical temporal patterns can generate complex hierarchical spatial patterns. When we look at the physical realization of a computation, we realize that these intricate data structures are realized in correspondingly intricate arrangements of matter and energy.

5.4.1 Programmable Materials

The value of this inverted perspective on computations and their realizations is that it is an approach to *programmable matter*, that is, to controlling systematically the properties and behavior of physical systems on a small scale [19, 45].

A step in this direction is provided by what can be called *programmable materials*, that is, materials whose physical properties vary widely and can be controlled systematically (i.e., programmed) [45]. Some of the many properties we might like to control are hardness, elasticity, flexibility, density, relative resistance, permittivity, photoconductivity, opacity, and refractive index. Moreover, we would like a combinatorially rich code for determining these physical properties; by analogy with biology, we may call the code the *genotype*, and the physical substance the *phenotype*.

It might seem unlikely that such a versatile material could exist, but nature provides an example: proteins. Proteins are coded by the four nucleotide bases of DNA and so, effectively, by strings over the alphabet {A, C, T, G}, a simple, but combinatorially rich code. Nevertheless, proteins, which are the primary elements of living things, have an enormous range of physical properties and have both active and passive functions. Proteins are the constituents of keratin (the material of horns, nails, and feathers), connective tissue (collagen and elastin), cellular skeletons (microtubules), enzymes, ion channels, signaling molecules, receptor and sensor molecules (such as rhodopsin), transporter and motor proteins, and so forth. The DNA code defines long sequences of a few different building blocks (amino acids), but the resulting polymers fold into complex three-dimensional shapes that give them a wide variety of physical properties. Some allosteric protein molecules even make simple decisions, responding to various combinations of regulators [7, pp. 63–65, 78–79]. One approach to programmable materials builds on proteins (natural and artificial), but once we understand the principle by which a simple, but combinatorially rich code can create structures with diverse physical properties, we can design new programmable materials based on different substrates.

5.4.1.1 Artificial Morphogenesis

Programmable materials may be very valuable, but much of the behavioral richness of living things comes from their complex hierarchical structure: from cells up to tissues, organs, and organisms, and from cells down to vesicles, membranes, and molecules (including proteins). There are many applications for which we would like to be able to build complex systems hierarchically structured from the microscale up to the macroscale. For example, we would like to be able to build robots with artificial nervous systems of comparable complexity and density to mammalian nervous systems, with similarly complex sensors and effectors to permit fluent, real-time behavior [42].

This raises the question of how to coordinate the self-assembly of vast numbers (millions or billions) of microscopic components into macroscopic complex systems. The problem might seem hopeless, but once again nature proves that it can be done. A human body has trillions of cells, yet during embryological development the cells self-organize into tissues, organs, and other structures. This suggests that embryological *morphogenesis*—the creation of physical form—can provide a model for the self-assembly of complex systems [6, 14, 15, 18, 23, 49, 50, 64]. Artificial systems may be very different from biological systems, but we can abstract the formal computation and control processes of morphogenesis from their biological realizations and apply them in artificial systems.

Our own approach to artificial morphogenesis is directed to the development of self-assembly processes that scale up to very large numbers of components (hundreds of thousands to millions or more) [34, 36, 37, 39, 41–43, 45]. To reach this goal, we describe morphogenetic processes by partial differential equations, effectively treating tissues as continuous media, and we use the mathematics of continuum mechanics. This is a reasonable approximation if the number of cells or agents is large, and is in fact commonly used in embryology and developmental biology. Using PDEs effects a useful separation of scales. The algorithms are developed and operate in terms of the dimensions of the object under assembly; this is the basis for determining parameters such as diffusion rates and agent velocities. These morphogenetic processes are independent of the scale of the "particles" (cells, agents, microrobots, etc.) constituting the medium, so long as it can be approximated as a continuum. Therefore, the algorithms do not depend on the size or number of agents; they scale.

To facilitate the expression of morphogenetic programs, we have developed a PDE-based programming language, which can be realized by computer simulation or, in principle, by microscopic physical agents [34]. The notation is designed to be interpretable in discrete or continuous time in order to facilitate a variety of realizations in simulation and physical agents.

5.5   Embodied Computation

Artificial morphogenesis is an example of *embodied computation*, which may be defined as "computation in which the physical realization of the computation or the physical effects of the computation are essential to the computation" [41]. The term is inspired by the theories of *embodied cognition* and *embodied artificial intelligence*, which call attention to the role that the body plays in control and information processing in humans and other animals [9–11, 16, 21, 22, 47, 53–55]. Formal structures emerge from the possibilities of physical interaction between a body and its environment, and these physical processes can substitute for information processes, thus decreasing the computational load on the nervous system.

In embodied computation, the formal and material aspects are not so separable as they are in conventional computation. On the one hand, information processing and control may depend for its correctness and effectiveness on realization in a specific kind of physical system. However, the specifics of the physical systems also limit the purpose of the computation, that is, the final cause, since the computation is not required to operate in other situations. The specifics may also provide material realizations of the computation that are available for the specific computational systems, but not necessarily for others. That is, a specific embodiment restricts the final, material, and efficient causes (e.g., possible energy sources), but these same restrictions may afford a wider range of formal causes (i.e., information and control processes) to accomplish its purpose. To take an example from nature, the specific embodiment of *E. coli* and the properties of its environment facilitate its use of chemical gradients to control its metabolically-powered movement toward more favorable locations. Indeed, all living systems use embodied computation, and they suggest ways of designing artificial embodied computation systems.

5.6   Conclusions

Computation is physical, but conventional computing technology has been built on a hierarchy of abstractions. In the post-Moore's Law era, computational processes will need to be more like the physical processes by which they are realized, which implies a greater role for analog, parallel, and stochastic models of computation. The increasing assimilation of computation to physics raises the question: What distinguishes computational processes from other physical processes? The answer is that the purpose of the system could be accomplished as well by other physical realizations with the same formal structure but different material realizations (multiple realizability). Therefore, any formal process can be considered computation (information processing, control), and it is apparent that there is a wide variety of possible unconventional computing paradigms. The computational state space can be discrete or continuous, and information processing can proceed in continuous, discrete, or sequential time, either serially or concurrently. As we journey out from the familiar domain of conventional computation, we must leave behind familiar notions of programming and universality, whose assumptions may be misleading outside of their frame of relevance. Promising unconventional computing paradigms include analog computation, quantum computation, field computation, and computation over second-countable metric spaces (which subsumes both analog and digital computation).

Traditionally, the purpose of a computation is a certain formal process, and the accompanying physical processes are merely a means to that end. However, we may turn the tables, and use computation for the sake of these physical processes, using the formal power and flexibility of computation to control the assembly and behavior of physical objects. This approach provides a path towards programmable matter and artificial morphogenesis. More generally, embodied computation takes advantage of a closer assimilation of computation to physics by exploiting physical processes more directly for computation, and by using computational techniques to govern physical processes.

References

1. Adamatzky, A.: Physarum Machines: Computers from Slime Mould. World Scientific Series on Nonlinear Science Series A, vol. 74. World Scientific, Singapore (2010)
2. Adamatzky, A., De Lacy Costello, B., Asai, T.: Reaction-Diffusion Computers. Elsevier, Amsterdam (2005)
3. Ambs, P.: Optical computing: A 60-year adventure. Adv. Opt. Technol. 2010, Article ID 372,652 (2010). doi:10.1155/2010/372652
4. Bennett, C.H.: Notes on Landauer's principle, reversible computation, and Maxwell's demon. Stud. Hist. Philos. Mod. Phys. **34**, 501–510 (2003)
5. Blum, L., Cucker, F., Shub, M., Smale, S.: Complexity and Real Computation. Springer, Berlin (1998)
6. Bourgine, P., Lesne, A. (eds.): Morphogenesis: Origins of Patterns and Shapes. Springer, Berlin (2011)
7. Bray, D.: Wetware: A Computer in Every Living Cell. Yale University Press, New Haven (2009)
8. Brockett, R.: Dynamical systems that sort lists, diagonalize matrices and solve linear programming problems. In: Proceedings of the 27th IEEE Conference Decision and Control, pp. 799–803. Austin, TX (1988)
9. Brooks, R.: Intelligence without representation. Artif. Intell. **47**, 139–159 (1991)
10. Clark, A.: Being There: Putting Brain, Body, and World Together Again. MIT Press, Cambridge (1997)
11. Clark, A., Chalmers, D.J.: The extended mind. Analysis **58**(7), 10–23 (1998)
12. Connor, R.J., Holleman, J., MacLennan, B.J., Smith, J.M.: Simulation of analog solution of Boolean satisfiability. Technical Report UT-EECS-15-735, University of Tennessee, Department of Electrical Engineering and Computer Science, Knoxville (2015)

13. Das, A., Chakrabarti, B.K.: Colloquium : quantum annealing and analog quantum computation. Rev. Mod. Phys. **80**, 1061–1081 (2008). http://link.aps.org/doi/10.1103/RevModPhys.80.1061

14. Doursat, R.: Organically grown architectures: creating decentralized, autonomous systems by embryomorphic engineering. In: Würtz, R.P. (ed.) Organic Computing, pp. 167–200. Springer, Heidelberg (2008)

15. Doursat, R., Sayama, H., Michel, O. (eds.): Morphogenetic Engineering: Toward Programmable Complex Systems. Springer, Heidelberg (2012)

16. Dreyfus, H.L.: What Computers Still Can't Do. MIT Press, New York (1992)

17. Ercsey-Ravasz, M., Toroczkai, Z.: Optimization hardness as transient chaos in an analog approach to constraint satisfaction. Nature Phys. **7**, 966–970 (2011)

18. Giavitto, J., Spicher, A.: Computer morphogenesis. In: Bourgine, P., Lesne, A. (eds.) Morphogenesis: Origins of Patterns and Shapes, pp. 315–340. Springer, Berlin (2011)

19. Goldstein, S.C., Campbell, J.D., Mowry, T.C.: Programmable matter. Computer **38**(6), 99–101 (2005)

20. Haykin, S.: Neural Networks and Learning Machines, 3rd edn. Pearson Education, New York (2008)

21. Iida, F., Pfeifer, R., Steels, L., Kuniyoshi, Y.: Embodied Artificial Intelligence. Springer, Berlin (2004)

22. Johnson, M., Rohrer, T.: We are live creatures: Embodiment, American pragmatism, and the cognitive organism. In: Zlatev, J., Ziemke, T., Frank, R., Dirven, R. (eds.) Body, Language, and Mind, vol. 1, pp. 17–54. Mouton de Gruyter, Berlin (2007)

23. Kitano, H.: Morphogenesis for evolvable systems. In: Sanchez, E., Tomassini, M. (eds.) Towards Evolvable Hardware: The Evolutionary Engineering Approach, pp. 99–117. Springer, Berlin (1996)

24. Landauer, R.: The physical nature of information. Phys. Lett. A **217**, 188 (1996)

25. Lipshitz, L., Rubel, L.A.: A differentially algebraic replacment theorem. Proc. Am. Math. Soc. **99**(2), 367–372 (1987)

26. Lloyd, S., Braunstein, S.L.: Quantum computation over continuous variables. Phys. Rev. Lett. **82**, 1784–1787 (1999). http://link.aps.org/doi/10.1103/PhysRevLett.82.1784

27. MacLennan, B.J.: Technology-independent design of neurocomputers: the universal field computer. In: Caudill, M., Butler, C. (eds.) In: Proceedings of the IEEE First International Conference on Neural Networks, vol. 3, pp. 39–49. IEEE Press (1987)

28. MacLennan, B.J.: Field computation in the brain. In: Pribram, K. (ed.) Rethinking Neural Networks: Quantum Fields and Biological Data, pp. 199–232. Lawrence Erlbaum, Hillsdale (1993). http://web.eecs.utk.edu/~mclennan

29. MacLennan, B.J.: Continuous formal systems: A unifying model in language and cognition. In: Proceedings of the IEEE Workshop on Architectures for Semiotic Modeling and Situation Analysis in Large Complex Systems, pp. 161–172. Monterey, CA (1995). http://web.eecs.utk.edu/+mclennan and http://cogprints.org/541

30. MacLennan, B.J.: Field computation in natural and artificial intelligence. Inf. Sci. **119**, 73–89 (1999). http://web.eecs.utk.edu/~mclennan

31. MacLennan, B.J.: Natural computation and non-Turing models of computation. Theor. Comput. Sci. **317**, 115–145 (2004)

32. MacLennan, B.J.: Analog computation (chap. 1, entry 19). In: Meyers, R. et al. (ed.) Encyclopedia of Complexity and System Science, pp. 271—294. Springer, Heidelberg (2009). doi:10.1007/978-0-387-30440-3_19. Reprinted in *Computational Complexity: Theory, Techniques, and Applications*, ed. by Meyers, R.A. et al., Springer, 2012, pp. 161–184

33. MacLennan, B.J.: Field computation in natural and artificial intelligence (chap. 6, entry 199). In: Meyers, R. et al. (ed.) Encyclopedia of Complexity and System Science, pp. 3334–3360. Springer, Heidelberg (2009). doi:10.1007/978-0-387-30440-3_199

34. MacLennan, B.J.: Preliminary development of a formalism for embodied computation and morphogenesis. Technical Report UT-CS-09-644, Department of Electrical Engineering and Computer Science, University of Tennessee, Knoxville, TN (2009)

35. MacLennan, B.J.: Super-Turing or non-Turing? Extending the concept of computation. Int. J. Unconv. Comput. **5**(3–4), 369–387 (2009)
36. MacLennan, B.J.: Models and mechanisms for artificial morphogenesis. In: Peper, F., Umeo, H., Matsui, N., Isokawa, T. (eds.) Natural Computing, Springer series, Proceedings in Information and Communications Technology (PICT) vol. 2, pp. 23–33. Springer, Tokyo (2010)
37. MacLennan, B.J.: Morphogenesis as a model for nano communication. Nano Commun. Netw. **1**(3), 199–208 (2010). doi:10.1016/j.nancom.2010.09.007
38. MacLennan, B.J.: The U-machine: a model of generalized computation. Int. J. Unconv. Comput. **6**(3–4), 265–283 (2010)
39. MacLennan, B.J.: Artificial morphogenesis as an example of embodied computation. Int. J. Unconv. Comput. **7**(1–2), 3–23 (2011)
40. MacLennan, B.J.: Bodies – both informed and transformed: Embodied computation and information processing. In: Dodig-Crnkovic, G., Burgin, M. (eds.) Information and Computation. World Scientific Series in Information Studies, vol. 2, pp. 225–253. World Scientific, Singapore (2011)
41. MacLennan, B.J.: Embodied computation: applying the physics of computation to artificial morphogenesis. Parallel Process. Lett. **22**(3) (2012)
42. MacLennan, B.J.: Molecular coordination of hierarchical self-assembly. Nano Commun. Netw. **3**(2), 116–128 (2012)
43. MacLennan, B.J.: Coordinating massive robot swarms. Int. J. Robot. Appl. Technol. **2**(2), 1–19 (2014). doi:10.4018/IJRAT.2014070101
44. MacLennan, B.J.: The promise of analog computation. Int. J. Gen.Syst. **43**(7), 682–696 (2014). doi:10.1080/03081079.2014.920997
45. MacLennan, B.J.: The morphogenetic path to programmable matter. Proc. IEEE **103**(7), 1226–1232 (2015)
46. Meinhardt, H.: Models of Biological Pattern Formation. Academic Press, London (1982)
47. Menary, R. (ed.): The Extended Mind. MIT Press, Cambridge (2010)
48. Molnár, B., Ercsey-Ravasz, M.: Asymmetric continuous-time neural networks without local traps for solving constraint satisfaction problems. PLoS ONE **8**(9), e73,400 (2013). doi:10.1371/journal.pone.0073400
49. Murata, S., Kurokawa, H.: Self-reconfigurable robots: shape-changing cellular robots can exceed conventional robot flexibility. IEEE Robot. Autom. Mag. pp. 71–78 (2007)
50. Nagpal, R., Kondacs, A., Chang, C.: Programming methodology for biologically-inspired self-assembling systems. In: AAAI Spring Symposium on Computational Synthesis: From Basic Building Blocks to High Level Functionality (2003). http://www.eecs.harvard.edu/ssr/papers/aaaiSS03-nagpal.pdf
51. Nemytskii, V.V., Stepanov, V.V.: Qualitative Differential Equations, Reprint of 1960 Princeton Univ, Press edn. Dover, New York, NY (1989)
52. Nielsen, M.A., Chuang, I.L.: Quantum Computation and Quantum Information, 10th anniversary edn. Cambridge University Press, Cambridge (2010)
53. Pfeifer, R., Bongard, J.: How the Body Shapes the Way We Think – A New View of Intelligence. MIT Press, Cambridge (2007)
54. Pfeifer, R., Lungarella, M., Iida, F.: Self-organization, embodiment, and biologically inspired robotics. Science **318**, 1088–93 (2007)
55. Pfeifer, R., Scheier, C.: Understanding Intelligence. MIT Press, Cambridge (1999)
56. Popa, C.R.: Synthesis of Computational Structures for Analog Signal Processing. Springer, New York (2011)
57. Pour-El, M.: Abstract computability and its relation to the general purpose analog computer (some connections between logic, differential equations and analog computers). Trans. Am. Math. Soc. **199**, 1–29 (1974)
58. Rothemund, P., Papadakis, N., Winfree, E.: Algorithmic self-assembly of DNA Sierpinski triangles. PLoS Biol. **2**(12), 2041–2053 (2004)
59. Rothemund, P., Winfree, E.: The program-size complexity of self-assembled squares. In: Symposium on Theory of Computing (STOC), pp. 459–68. Association for Computing Machinery, New York (2000)

60. Rupp, K., Selberherr, S.: The economic limit to Moore's law. IEEE Trans. Semicond. Manuf. **24**(1), 1–4 (2011). doi:10.1109/TSM.2010.2089811
61. Santoro, G.E., Tosatti, E.: Optimization using quantum mechanics: quantum annealing through adiabatic evolution. J. Phys. A: Math. Gen. **39**(36), R393 (2006). http://stacks.iop.org/0305-4470/39/i=36/a=R01
62. Shannon, C.E.: Mathematical theory of the differential analyzer. J. Math. Phys. Mass. Institute Technol. **20**, 337–354 (1941)
63. Shannon, C.E.: Mathematical theory of the differential analyzer. In: Sloane, N.J.A., Wyner, A.D. (eds.) Claude Elwood Shannon: Collected Papers, pp. 496–513. IEEE Press, New York (1993)
64. Spicher, A., Michel, O., Giavitto, J.: Algorithmic self-assembly by accretion and by carving in MGS. In: Proceedings of the 7th International Conference on Artificial Evolution (EA '05), no. 3871 in Lecture Notes in Computer Science, pp. 189–200. Springer, Berlin (2005)
65. Turing, A.: The chemical basis of morphogenesis. Philos. Trans. R. Soc. B **237**, 37–72 (1952)
66. van Gelder, T.: Dynamics and cognition (chap. 16). In: Haugeland, J. (ed.) Mind Design II: Philosophy, Psychology and Artificial Intelligence, revised & enlarged edn., pp. 421–450. MIT Press, Cambridge (1997)

Chapter 6
Computing in Perfect Euclidean Frameworks

Jérôme Durand-Lose

Abstract   This chapter presents what kind of computation can be carried out using an Euclidean space—as input, memory, output...—with dedicated primitives. Various understandings of computing are encountered in such a setting allowing classical (Turing, discrete) computations as well as, for some, hyper and analog computations thanks to the continuity of space. The encountered time scales are discrete or hybrid (continuous evolution between discrete transitions). The first half of the chapter presents three models of computation based on geometric concepts—namely: ruler and compass, local constrains and emergence of polyhedra and piece-wise constant derivative. The other half concentrates on signal machines: line segments are extended; when they meet, they are replaced by others. Not only are these machines capable of classical computation but moreover, using the continuous nature of space and time they can also perform hyper-computation and analog computation. It is possible to build fractals and to go one step further on to use their partial generation to solve, e.g., quantified SAT in "constant space and time".

6.1   Introduction

This chapter provides some insight on the following question: *What can be done with an Euclidean space with dedicated primitives and controls?* Space is not considered as the place to stack gates and wires for a computer (would it be sequential, parallel or distributed) but rather as the substrate of computation itself. The general framework is not machines or automata but some Euclidean space where information is displayed and evolved according to some dynamics.

J. Durand-Lose (✉)
Laboratoire d'Informatique Fondamentale d'Orléans, Université d'Orléans,
Batiment IIIA Rue Léonard de Vinci B.P. 6759 F-45067 Orléans Cedex 2,
Orléans, France
e-mail: jerome.durand-lose@univ-orleans.fr

© Springer International Publishing Switzerland 2017   141
A. Adamatzky (ed.), *Advances in Unconventional Computing*,
Emergence, Complexity and Computation 22,
DOI 10.1007/978-3-319-33924-5_6

The approaches considered here are: constructions with ruler and compass, polyhedra emerging from local constrains, extending a sequence of line segments crossing polyhedral regions, extending line segments until they intersect... In each case, localization, distance, carried information, available room, encounters... are elements where *space matters*.

Space is Euclidean, this means, on the one hand, that it is continuous and, on the other hand, that the underlying geometry is the one of points, lines and circles. This geometrical point of view is the prevalent as shown by the illustrations in this chapter.

This general framework has limitations: no differential equation that would not be trivial, no algebraic geometry... Outside of instantaneous "border" crossing or apparatus operation, all is straightforward and absolutely plain. The models presented here belongs to a more general framework: hybrid systems with continuous traits (related to the nature of space and possibly time) and discrete (phase transition, collision...).

Continuity opens the way to *Zenon effect*: an infinite number of discrete transitions during a finite (continuous) duration (in a finite space). Many models use this capability to *hyper-compute* (solving the Halting problem and even less "computable" problems).

These are idealized models: line have zero width, positions are exact... Physically, they are not very realistic: unbounded density of information, Euclidean at every scale space...

Each presented model is described, main results and references are provided. Of course it is not possible to give detailed proofs nor get into complex results; clues, sketches are provided as long as they remain intelligible.

This chapter has two parts. The first part presents three computing models in which primitives rely on the Euclidean nature of space. The first model, the *Geometric Computation Machines* of Huckenbeck [30, 31], uses an automaton to activate ruler and compass and generates points, lines and circles. The second one, the *Mondrian Automata* of Jacopini and Sontacchi [32], starts from uniform local constrains (on open balls from $\mathbb{R}^n$) on space-time diagrams ensuring causality; from these emerges polyhedra at the usual scale. The third one, the *Piecewise Constant Derivative* of Asarin and Maler [2, 3], partitions space into polyhedral regions corresponding to constant speeds; the orbit stating from a single point can be very complex with possibly infinitely many region changes during a finite portion of it.

The second part concentrates on one model: the signal machines of Durand-Lose [18]. After the definition of the model, a simulation of Turing machine is presented (thus asserting its computing capability). Using the continuity of both space and time, it is possible to dynamically scale down the computation and accelerate to implement a form of the Black Hole model of computation (and to hyper-compute). Fractal generation scheme can be used in order to dispatch sub-computations and to achieve *fractal computation* (allowing, e.g., to solve quantified SAT in constant space and time). This part ends by showing that the model is capable of analog computation (computing over real numbers).

This survey of computing models involving space is not comprehensive. Some models like cellular automata, tile assembling systems, tilings... have so much literature about that each would spread over a few chapters; it would be pointless to present them in a few pages. Models using higher level mathematics (differential equations, algebraic geometry...) would not fit here and neither would computational geometry algorithms. Many others (e.g. continuous counterparts of cellular automata like [28, 38]) are not addressed just because the purpose of this article is to show the variety and not to start a zoo recollection.

6.2 Three Models Operating on Euclidean Geometry

The first model uses ruler/straightedge and compass. The second one relies on local constraints. The third one concentrates on orbits when speed is constant on polyhedral regions.

6.2.1 Ruler and Compass

This section is devoted to the work of U. Huckenbeck on *Geometric Computation Machines* [30, 31]. The primitives of these machines are the usual geometric operations that can be carried out with ruler and compass. The purpose is not to do geometry algorithmic (would it be discrete, symbolic or algebraic) but to use these primitives to construct in a two dimensional Euclidean space.

Each machine is an automaton (or program) equipped with a finite number of registers. There are three kinds of register: for points, for lines and for circles. States are used to represent both the program counter and to record the state of the computation (i.e. Unfinished, Finished and Error, the last two ones are final).

The available operations are:

- output a value (point, line or circle),
- put in a register the intersection of two lines,
- put in a register one of the intersections of a line and a circle (optionally different from some point),
- put in a register one of the intersections of two circles (optionally different from some point),
- put in a register the line going through two points,
- put in a register the circle whose center is given (as a point) as well as its radius (as the distance between two points),
- copy a register, and
- Finished.

(a)

```
1: c₁ ← Circle ( center A, radius d(A,B) )
2: c₂ ← Circle ( center B, radius d(A,B) )
3: p₁ ← Intersection ( c₁, c₂ )
4: p₂ ← Intersection ( c₁, c₂ ) different from p₁
5: d₁ ← Line ( p₁, p₂ )
6: d₂ ← Line ( A, B )
7: p₃ ← Intersection ( d₁, d₂ )
8: Output p₃
9: Finished
```

(b)

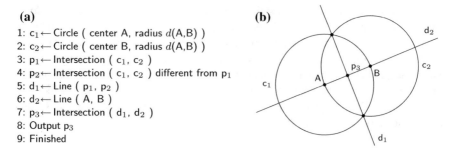

Fig. 6.1 Constructing the middle of the segment AB. **a** Program. **b** Construction

Intersections do not necessarily exist and neither are unique. This means that the execution of the automaton is non deterministic. Whenever an instruction cannot be carried out, the branch (of the tree of all possible executions) ends with Error.

Considering the whole tree of possible executions, if it is finite, has only Finished (i.e. no Error) leaf and all branches generate the same output, then the computation succeed and the output is the common output (it is generated by every branch). For example, the program of Fig. 6.1 computes the middle of a segment (whose extremities are A and B and are the only input). Please note that there are two possible executions (where are p_1 and p_2?), but their outputs are identical.

Conditional jump instructions are like "if $p_k \in E$ go to i: otherwise to j:" where p_k is a point-register and E is a predefined set used as an oracle.

A simple case is when E contains only the origin $(0, 0)$ and points $(1, 0)$ and $(0, 1)$ are provided as constants. The functions (computable in constant time) from an n-uplet of points to points are then exactly the ones where the coordinates of output points can be expressed from the one in the input using only piece-wise rational functions with integer coefficients.

This algebraic result is related to the possibility to implement the following primitives: on the one hand, projections $(x, y) \rightarrow (x, 0)$, $(x, y) \rightarrow (y, 0)$ and reconstruction $(x, 0) (y, 0) \rightarrow (x, y)$ and, on the other hand, addition, multiplication and division on the x axis. These constructions are left as exercises to the reader.

This corresponds to the classical construction of numbers computable with rule and compass [12, chap. 7]. Those are also closed by square rooting. Here, the condition that each branch should generate the same output makes it impossible for root to appear [31].

It should be noted that since the following operations can be performed: $(x, 0) \rightarrow (x + 1, 0)$, $(x, 0) \rightarrow (x - 1, 0)$, and test whether $(x, 0)$ is $(0, 0)$; an unbounded counter can be encoded with a point register. These machines can simulate any 2-counter automaton and are thus Turing-universal.

6.2.2   Mondrian Automata

The work of G. Jacopini and G. Sontacchi [32] starts from a space and time modelization of reality. Hypothesis are made from which follows local constraints that brought forth the emergence of polyhedra.

In a Euclidean space of any dimension, each point is associate to a state, a color. The hypothesis is made that color and neighborhood are linked: if two points have the same color, then there is a sufficiently small (non-zero) radius where neighborhoods match. This is depicted in Fig. 6.2.

As a consequence, if there is a ball of uniform color then any point of this color are only surrounded by this color. Topologically, this means that they form an open set.

Similarly, if there is a curve (of zero width) of a color then the curve must be a line segment (if neighborhoods are identical, so must be the derivatives). All the points of this color must be on parallel line segments and, following any direction, the surrounding colors should be the same. The extremities of the segments should have a color different from the line.

Each color corresponds to equal-dimension and parallel polyhedral regions. There frontiers of lesser dimensions should be colored differently (when restricted to their dimensions, they are open). Adjacent colors following another dimensions are always of the same color.

The other hypothesis made is that there is a finite number of colors. Hence, having a common neighborhood (up to re-scaling) for each color defines all the constraints. They provide all the information on the dimension and direction associated to each color as well as the color of the neighbors of higher dimensions.

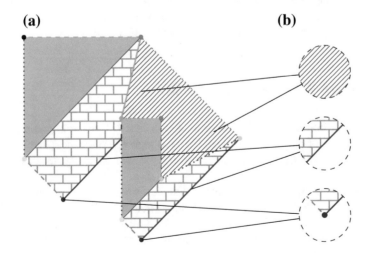

Fig. 6.2 Mondrian space. **a** Colored space. **b** Local constraints

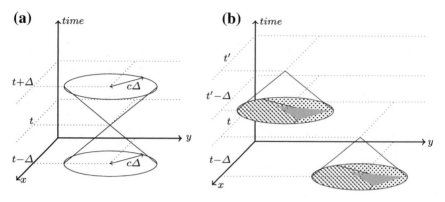

Fig. 6.3 Cones and causality. **a** Past and future cones. **b** Causality

Next step consists in adding one dimension for time and rules for causality. This is defined by a speed of light, c and the condition that the color of a point is uniquely defined by what is inside the past cone (delimited by the speed of light). Figure 6.3b shows two portions of space at different dates where colors are displayed similarly. The two cones based on this portions and delimited by the speed of light are thus identical.

Another argument from physical modelisation is that the system should be reversibility. This implies that the same constraint is also applied with time running in the opposite direction. This corresponds to exchanging the cones (pointing to past and future) in Fig. 6.3b.

Temporal constraints can also be read at the polyhedra scale. It is still possible to think in terms of intersections and collisions (this kind of approach is developed in the signal machine section). At this level, simulating a reversible Turing machine using rational positions for endpoints is not very difficult.

6.2.3 Piece-Wise Constant Derivative

In this model introduced by E. Asarin and O. Maler [2], space is partitioned into a finite number of polyhedral regions. On each region, a constant speed is defined. On Fig. 6.4, thick lines separate the regions and the arrows indicate the directions of speeds.

Starting from any point a trajectory is defined. When a region border is reached, movement just follows on the other side with the new speed. In Fig. 6.4, two trajectories are indicated. They both start on the left. The dashed trajectory changes direction twice and then goes away forever. The dotted one is wrapping itself infinitely around the intersection point of three regions.

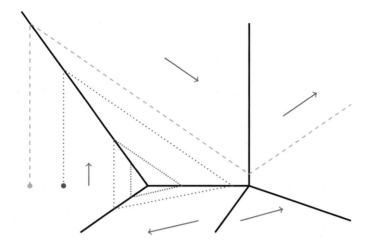

Fig. 6.4 Piece-wise constant derivative trajectories

This second trajectory is singular: it changes region infinitely often but neverthe-less reaches its limit in finite time (as a convergent geometrical sum) and stops there. There are two distinct time scales: a continuous time one where the limit is reached in finite time and an infinite discrete time one (of region change events). This is a *Zeno* phenomenon/effect.

The rest of the section is restricted to rational initial points and vertices (of poly-hedra). This allows to have exact manipulation on a computer and avoids having unbounded information (an oracle) encoded into a real number (for example the solution to the Halting problem as the Chaitin's omega number [11, Chap. 7]).

This system can compute considering that the input is the initial position (in a given zone) and that halt and result correspond to entering some other identified zone. In three dimensions, it is possible to encode the configuration of a Turing machine into a rational point and to design the PCD such that the trajectory loops and that each loop corresponds to a transition of the Turing machine. The Reachability problem—to decide whether a zone can be reached from a point—is thus undecidable.

Adding dimensions to the systems allows to add nested levels of Zeno effect and to climb hierarchies in the undecidable. With d dimensions, the level $d-2$ of the arithmetical hierarchy (viz. Σ_{d-2})[1] is decidable [2, 3]. The model is even more powerful: Reachability is complete on levels of the hyper-arithmetic hierarchy[2] [8–10].

---

[1] Σ_0 is the recursive sets, Σ_1 is recursively enumerable sets, e.g. the Halting problem.

[2] Extension of the arithmetical hierarchy to ordinal indices.

6.3   Signal Machines

This section is dedicated to one model that have been studied for over a decade: *signal machines* [16]. This model was born from reflections on the fabric of cellular automata [20]. Indeed, one of the key notions in cellular automata literature both for creating CA for special purposes or for understanding one is *signals*. They allow to store and transmit information, to start a process, to synchronize... Dynamics is often detailed as signals interacting in collisions resulting in the generation of new signals.

In the context of cellular automata, signals are discrete and often spread over more than one cell. They are often represented as line segments to stress on the global dynamics without blurring the picture with discretization details. The step forward is to get rid of discretization and to consider these ideal line segments as exact traces [20].

Signals are now dimensionless points on a one-dimensional Euclidean space. Typical properties of cellular automata (CA) are preserved: synchronicity and uniformity. Synchronicity means that signals move, at the same pace. Uniformity means that the dynamics of signals does not change, the speeds of signals only rely on their natures and their interactions only depend on their nature (like CA-patterns define the evolution of discrete signals). Signals have uniform movement. They "draw" line segments on space-time diagrams. The nature of a signal is called a *meta-signal*.

After defining signal machines, it is shown that they are able to compute (in the Turing understanding). Space and time continuity as well as malleability of computations are developed to implement the *Black Hole model of computation* and hyper-compute. Fractal generation is a natural feature of the model and unfinished fractal construction can be used to provide unbounded parallelism (and make *fractal computation*). Finally, the capacity of the model to do analog computation (over the reals) is considered.

6.3.1   Definitions

Definition 1 (*Signal machine*) A *signal machine* is defined by a triplet (M, S, R) where

- M is a finite set of *meta-signals*,
- S is function associating a *speed* to each meta-signal, and
- R is a set of collision rules. A *collision rule* associates to sets of at least two meta-signals of different speeds (*incoming*) a set of meta-signals of different speeds (*outgoing*). R is deterministic: a set appears at most once as the left (incoming) part of a rule.

In any configuration, there are finitely many signals and collisions. They are located in distinct places in space. Since a signal is completely defined by its associated meta-signal and a collision by a rule, a configuration is fully defined by associating to each point the real axes a meta-signal, a rule or nothing.

Definition 2 (*Configuration*) Let (M, S, R) be any signal machine. A *configuration* is totally defined by a function from space (i.e. $\mathbb{R}$) into meta-signals, collision rules and a special value *void* (from meta-signal and collision) such that the number of non-void positions is finite.

Let V be the set $M \cup R \cup \{\oslash\}$ where $\oslash \notin M \cup R$. A *configuration*, c, is a function from $\mathbb{R}$ into V such that: $|c^{-1}(M \cup R)| < \infty$.

In the following, *location* designates space-time coordinates while *position* is a spacial coordinate and *date* a temporal one. A location is denoted (x, t), space then time.

The dynamics from a signal machine is defined as time evolves: as long as signals do not meet, each one moves uniformly. Whereas as soon as two or more signals meet, they are replaced according to the corresponding collision rule.

Time is continuous but there is also a different discrete time scale: the one of collisions. Dynamics is defined using this scale considering dates with collision(s) and in-between times (simple propagation).

Definition 3 (*Dynamics and space-time diagram*) Let c be a configuration, let t be the date of the next collision if any, ∞ otherwise).

$$t = \min \left\{ d \in \mathbb{R}^{+*} \,\middle|\, \begin{array}{l} \exists x_1, x_2 \in \mathbb{R}, \; x_1 \neq x_2, \; \exists \mu_1, \mu_2 \in M, \\ \mu_1 \bowtie c(x_1) \wedge \mu_2 \bowtie c(x_2) \wedge (x_1 + S(\mu_1)d = x_2 + S(\mu_2)d) \end{array} \right\}$$

where the relation $\bowtie$ means "is equal (to the meta-signal) or belongs to the outgoing set of meta-signals (of the collision rule)".

The configurations between 0 and t have no collision and are defined by simple propagation of signals:

$$\forall d, 0 < d < t, c_d(x) = \left\{ \begin{array}{l} \mu \text{ if } \mu \bowtie c(x - S(\mu)d) \\ \oslash \text{ otherwise} \end{array} \right\}$$

where, by definition of t, μ is unique (if it exists).

If t is not ∞, the configuration at t is defined similarly with an extra case: if more than one signal arrives in x, then it is the corresponding collision rule.

Orbits are computed like that, one collision time after another. A *space-time diagram* gathers all the configurations between two dates.

This definition emphasizes the hybrid aspect of signal machines: continuous steps separated by discrete steps.

Example 1 The meta-signals and collision rules are defined in the left Fig. 6.5. On the right, is depicted a space-time diagram generated from a configuration where can be found (from left to right) signals of meta-signals μ_4, μ_1, μ_2 and μ_4.

(a)

Name	Speed
μ_1	1
μ_2	$-1/2$
μ_3	3
μ_4	0

(b)

$\{\,\mu_1, \mu_2\,\} \rightarrow \{\,\mu_2, \mu_1, \mu_3\,\}$
$\{\,\mu_3, \mu_4\,\} \rightarrow \{\,\mu_2\,\}$
$\{\,\mu_4, \mu_2\,\} \rightarrow \{\,\mu_2, \mu_4\,\}$

(c)

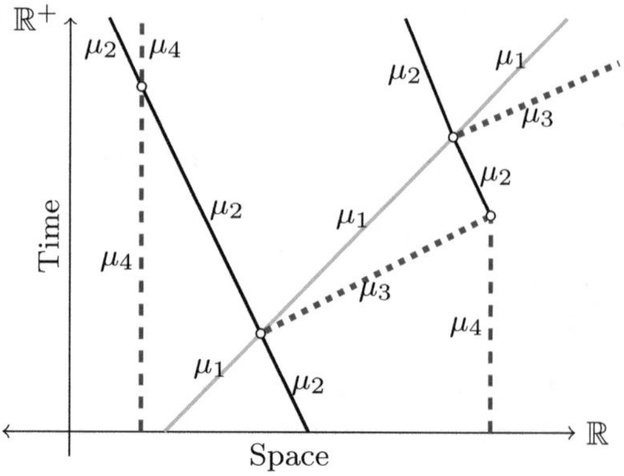

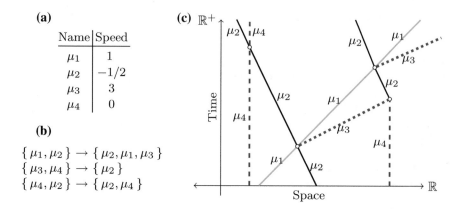

Fig. 6.5 Example of a signal machine and a generated space-time diagram. **a** Meta-signals. **b** Collision rules. **c** Space-time diagram

Used collision rules can also be read from the diagram. A rule is *blank* if it regenerates the same set of meta-signals (like the one for $\{\,\mu_4, \mu_2\,\}$ in the example). These rules are not indicated.

To find the location of a collision, a linear system of two equations in two variables has to be solved. Thus the location of any collision of signals whose speeds and initial locations are rational numbers, has to be rational (numbers). More generally, with rational speed and rational initial positions, all collisions happen at rational locations.

Definition 4 (*Rational signal machine*) It is a machine such that all speeds are rational numbers as well as any non-void positions in any initial configuration. In any generated space-time diagram, all collisions have rational locations and the positions of signals are rational at each collision time.

This is particularly useful to simulate signal machines: rational signal machines can be simulated exactly on a computer since rational numbers can be manipulated in exact precision. Such a simulation has indeed been implemented in Java[3] and was used to generate the figures.

Unless otherwise noted, all results are valid both on rational and unrestricted signal machines.

Example 2 (Finding the middle) It is possible to compute the middle of two signals, i.e. to position a signal exactly there. This is illustrated in Fig. 6.6 where a O signal is positioned exactly half-way between two W signals (bottom of Fig. 6.6c). This is started by the arrival of a Add signal on the left. When it encounters the left W, it is transformed into A and $\overrightarrow{\mathsf{R}}$. The latter is three times faster than the former and bounces on the right W; it becomes then $\overleftarrow{\mathsf{R}}$, still three times faster (but opposite direction). It encounters A exactly half-way between the two W.

[3]The author is not satisfied enough with its code to put it on the internet but send it on request.

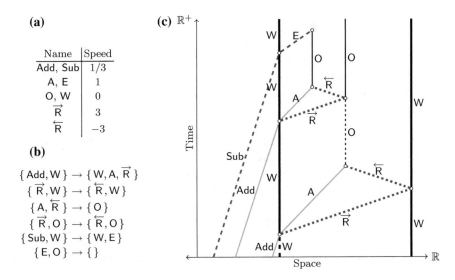

Fig. 6.6 Finding the middle and more. **a** Meta-signals. **b** Collision rules. **c** Space-time diagram

The correct positioning of this collision can be proved by computing the locations of all the intermediate collisions (each time a linear system of two equations in two variables). Almost all the *proofs of correction* of space-time diagrams (and hence the dynamics) are not more difficult.

Considering the rules of Fig. 6.6b, finding the middle only uses the three first ones. The fourth one allows to generate the middle between the left W and the first O on right of it. This is started by sending another Add from the left as illustrated in the middle of Fig. 6.6c.

It is also possible to suppress the first O on the right of the left W. To achieve this, a Sub *order* is sent from the left. It becomes E when *passing* over W. Signal E collides and destroys the first O it encounters. This corresponds to the last two rules and the top of Fig. 6.6c.

Adding another Sub on the right removes the other O. Adding another Sub on the right again makes it exit on the right. The reader is invited to add a rule to destroy this unwanted signal.

Finding the middle is a key primitive for designing signal machines. For example, as shown above, it is possible to use it repeatedly to record any natural number in unary (with O's as in Fig. 6.6c) in a bounded space.

6.3.2 Turing Computation

Signal machines are able to compute (in the Turing understanding) as shown by the simulation of a generic Turing machine. A Turing machine is composed of a finite automaton, an unbounded tape made of cells and a read/write head that allows the automaton to act and move on the tape. The tape is always finite but enlarged as needed.

Simulation is straightforward as shown in Fig. 6.7 (time always elapses upward). The evolution of the Turing machine in Fig. 6.7a can be seen in Fig. 6.7b. Vertical (null speed) signals encode each cell of the tape. Zigzagging signals indicate the position of the head and record the state of the automaton. The number of collisions is linear in the number of iterations of the Turing machine.

The only technical part is the dynamic enlargement of the tape. The speeds of head signals and $\overrightarrow{\#}$ are 1 and -1. Setting to -3 and 3 the speed of $\overleftarrow{\#}$ and $\overrightarrow{\#}$ ensures that each time the head goes further right, a new signal/cell is added with the same

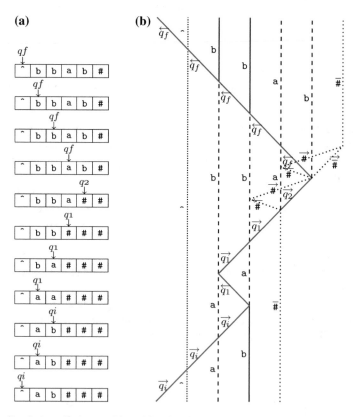

Fig. 6.7 Simulating a Turing machine with a signal machine. **a** Turing machine run. **b** Simulation by a signal machine

distance between vertical signals. This is the middle construction, but backwards! It is also possible to set these speeds such that the distance is halved each time. The width of the whole tape would then be bounded independently from the number of cells.

This construction works on rational signal machines that can be simulated exactly on any computer. Leaving open the definition of input, halt and output, rational signal machines have exactly the same computing power as Turing machines. Next section goes beyond this framework by considering accumulation points, then Sect. 3.5 considers non-rational signal machines and relates that to analog computation.

The simulation of Turing machines leads to the undecidability of the following problems for rational signal machines (expressible in classic context since everything is rational):

• decide whether the number of collisions is finite,
• decide whether a meta-signal appears,
• decide whether a signal participates in any collision,
• decide whether the computation remains spatially bounded...

The space-time diagram in Fig. 6.7b displays the trace of the Turing machine. Another interest of signal machines is also to provide graphical traces.

Using various meta-signals similar to O in Fig. 6.6, it is possible to encode sequences of letters functioning as a stack. These simulations can also be done with a bounded number of signals to encode the whole stack: positions are used to encode values instead of sequences of signals [18]. (With irrational positions, it is even possible to encode infinite stacks.) It is possible to simulate any Turing machine with a constant number of signals and collisions involving only two signals resulting in exactly two signals (conservation of the number of signals), but more over it remains true if rules should be injective: the rule is also defined by outgoing signals (reversibility) [25]. This simulation uses reversible universal Turing machines [5, 33, 35].

Signal machines can also be used to simulate the *Cyclic Tag Systems* introduced by Cook [13]. His work restarted the race to small universal machines, e.g. on Turing machines [40]. The smallest Turing-universal signal machine known simulates any CTS and has 13 meta-signals and 21 collision rules [22].

6.3.3 Malleability of Space-Time and the Black Hole Model

There is no scale nor origin in the figures and neither in the definition. The context is the continuum: scaling or translating the initial produces the same space-time diagram with all proportions preserved. In particular, if all distances are halved, then so are the durations.

It is possible to dynamically re-scale a configuration and then to restart it as shown in Fig. 6.8. On the left, there is the structure with the name of the meta-signals; on the middle, the dotted area at the bottom represents the initial active part of the space-time

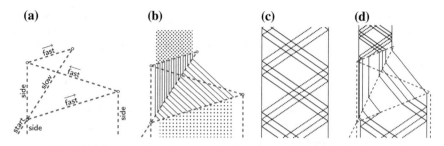

Fig. 6.8 Shrinking step. **a** Structure. **b** Modification. **c** Control. **d** Shrunk

diagram; it is frozen (parallel lines), scaled (change of direction) and then restarted (densely dotted area). On the right, an example is provided: control diagram alone (Fig. 6.8c) and with a shrinking (Fig. 6.8d).

To freeze a computation, a $\overrightarrow{\textsf{fast}}$ signal crosses the configuration and replaces anything it meets by a null-speed signal encoding what is encountered. The frozen system drifts as parallel signals. Being parallel, there is no collision and the (relative) distances are preserved. Is unfrozen by a signal of the same velocity as the freezing one ($\overrightarrow{\textsf{fast}}$ is reused here).

Scaling is done by changing the direction of parallel signals. It follows from Thales's theorem that each time proportion between distances is preserved. All distances are halved, so it the length of the $\overrightarrow{\textsf{fast}}$ signal.

On the continuum, there is no limit to scaling. It is possible to restart the shrinking process forever by sending a signal (**restart** in Fig. 6.9a) on the left to restart the process.

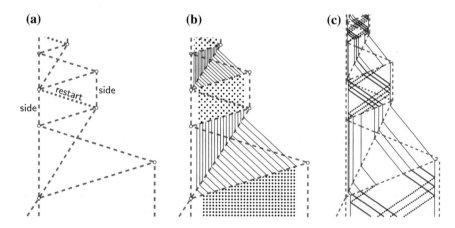

Fig. 6.9 Iterated shrinking. **a** Structure. **b** Modification. **c** Example

Each time the entangled original computation is activated with the same relative duration because although the activation duration is halved, since distances are halved, duration between collision is also halved. Altogether, in this finite portion of the space-time diagram, the whole infinite original space-time diagram is entangled.

With a bounded space simulation of a Turing machine entangled, if the computation stops, then it is in bounded time. With minor technical modification, it is possible, in case of halting to let some signal leave the iterated shrinking.

Outside of the structure, this signal (witness of the halt) could be collected by another signal. Recollection of the halting signal can only happen during a bounded delay. This bound on duration can be "implemented" in the space-time diagram by a collision with another signal. If the machine does not halt, then nothing is received before that collision. But, in case of halting, the witness is collected before. Outside of the shrinking structure, the halt is decided this way. Signal machines can hyper-compute by creating a local Zeno effect. This kind of construction was already presented in the first papers on the domain [17, 18].

The general principle behind this construction is to have two time-lines: one is infinitely accelerated and does the computation and possibly sends some signal while the other waits for it or timeout. This corresponds to the so-called *Black Hole model of computation* (the reader is refereed to the works of M. Hogarth [29] and I. Németi and colleagues [1, 27]). The accumulation on top of Fig. 6.9c corresponds to the Black Hole.

It is also possible to twist a space-time diagram by tapping the speed with a linear function locally without freezing the computation. A continuous shrinking is possible and can also be iterated [25] (in this article, the machine is also reversible and conservative).

6.3.4 Build and Use Fractals

Many fractals can be generated using signal machines in a straightforward way. For example, four meta-signals are enough to build the fractal accumulation in Fig. 6.10a. The space-time diagram is undefined at this accumulation *singularity*.

Recursively generating middles also generate a fractal as in Fig. 6.10b. By considering left and right thirds instead of halves, a classical construction of the Cantor set is generated as in Fig. 6.10c. By varying the speed and the proportion, it is possible to generate sets of any fractal dimension between 0 and 1 [36, Chap. 5].

The embedding structure in Fig. 6.9b is fractal. As seen previously, it can be used in depth to accelerate computations. In Fig. 6.10b spaces are sliced in half at each step. This one can be used breadth wisely to provide parallelism. For example, it is possible to carry out a case study on each side, e.g. for a boolean variable true on one side, false on the other.

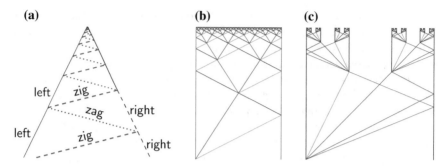

Fig. 6.10 Fractals. **a** Simple accumulation. **b** Half slicing. **c** Cantor set

Sub-cases (or other variables) can be treated by sub-slicing. For boolean formulas, it is possible to recursively slice until there is no more variable. All cases are thus considered. (Local results remains to be aggregated.)

If the variables appear in some boolean formula whose satisfiability is to be checked, then one gets a scheme to solve SAT: satisfaction of boolean formula (with $\neg$, $\vee$ and $\wedge$) over boolean variables $x_1, x_2, \ldots, x_n$.

This scheme can be further refined to deal with quantified variables. QSAT deals with quantified boolean formula, i.e. SAT formula prefixed by $Q_1 x_1 \, Q_2 x_2 \ldots Q_n x_n$ where Q_i belongs to $\{\forall, \exists\}$. The problem QSAT is PSPACE-complete [37, Sect. 8.3], i.e. complete for the polynomial reduction among problems solvable in polynomial space (deterministic or not by Savitch's theorem).

Going back to the scheme, if a boolean formula contains 10 variables, then 10 levels of slicing are done. What remains of the construction of the fractal in Fig. 6.10b is useless. Moreover since the diagram is not defined at the limit, the achievement of the fractal construction is unwanted.

There are many ways to stop the construction. The most basic one is to have a set of meta-signals for each level (unary counting in the machine definition). Another relies on a meta-signal used to initiate the construction of another layer (for ten layers, put ten copies of it). In the latter approach, the same machine can be used for any depth (genericity).

A QSAT formula, e.g. $\exists x_1 \forall x_2 \forall x_3 \; x_1 \wedge (\neg x_2 \vee x_3)$, is represented by a ray of signals encoding all its elements. Computation is organized following a complete binary tree (of depth 3 for the example, as can be seen in Fig. 6.11). At each node, a boolean variable is set to **true** on the right and to **false** on the left (Fig. 6.12a).

At the leaves, variables are all been replaced by values, the formula is evaluated (Fig. 6.12b). The value goes back in the tree which folds back: each node is a conjunction (resp. disjunction) when it is a universal (resp. existential) variable (Fig. 6.13). There are a lot of technical details; full presentation can be found in [14] and [36, Chap. 7].

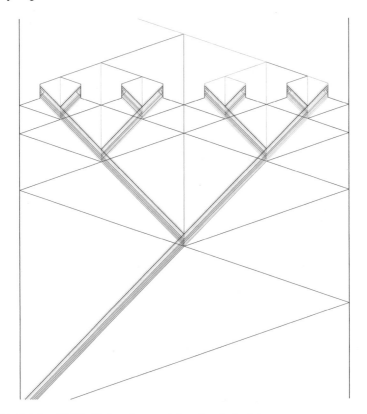

Fig. 6.11 Solving QSAT with fractal computation, the whole picture

A specific machine is generated for each formula. Using a more complex encoding of formulas, it is possible to use a unique machine. The formula is then totally encoded in the initial configuration [15] and [36, Chap. 8].

It is also possible to solve other problems on formulas: how many satisfying valuations (#SAT, #P-complete), what is the "smallest" satisfying valuation?… One "just" has to change the way the variable-free formulas are evaluated and the results are aggregated to generate an integer, a valuation… This is a modular parametrization of the construction.

If more levels are needed, one lets more levels of the fractal be generated. It does not need more time or space. Altogether, there is a signal machine able to decide any instance of QSAT in constant space and time.

Comparing to deciding the Halting problem in constant time in previous section, QSAT is trivial. Both rely on the continuity of space and time and the capability of unbounded acceleration. The difference is that, with QSAT, there is no accumulation, the key is to use the continuity to always be able to slice in two.

This technique of a controlled and unfinished fractal construction to display parallel computations and then aggregate the result is called *fractal computation*.

(a)   **(b)**

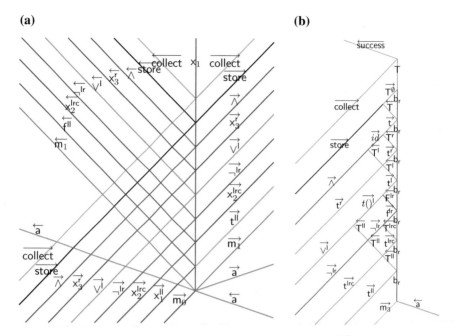

Fig. 6.12 Solving QSAT with fractal computation, details. **a** Setting $x_1(x_1{}^{\text{ll}})$ to false on the left $(\overleftarrow{f^{\text{ll}}})$ and true on the right $(\overleftarrow{t^{\text{ll}}})$. **b** Evaluation of the formula

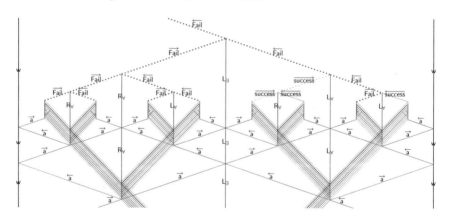

Fig. 6.13 Aggregating the results

Complexities.

Space and time as considered above are continuous measures of complexity. Yet, this is an hybrid model paced by discrete events: collisions. Discrete complexity measures can be defined by considering space-time diagrams as directed acyclic graphs. A *direct causal link* exists between two signals if the first ends in the collision where the second starts. A *causal link* is the transitive closure of direct causal links. *Time complexity* is then the size of the longest sequence of signals with direct causal link between each two consecutive signals (path in the DAG). *Space complexity* is the largest number of signals without any causal link.

With these definitions, complexity is quadratic in time (cubic for the generic case) but exponential in space.

Further readings.

Figure 6.10a shows that it is possible to build a fractal with only four different speeds. With two speeds or less, then number of collisions is finite and bounded. With three speeds, the situation is two-fold: with a rational signal machine, signals must travel on a regular mesh (without any accumulation). But accumlation might happen as soon as there is an irrational ratio between speeds or between initial positions. This can be understood by the presence of a mechanism computing the *gcd*, which only converges on rational [4].

What about the computing capability? In the rational case, the same mesh shows that usable memory is finite and bounded. Whereas in the other case, it is possible: a Turing machine is simulated and a fractal construction step is used to enlarge the tape [26].

6.3.5 Analog Computation

This section deals with computation on real numbers (not just rational, decimal, or algebraic). The only available real values in the model are distances between signals (if some unit is available) or the position of signals (if an origin is also provided) or the proportion between distances (orientation provides the sign).

Let there be a pair of (parallel) signals to be considered as unit. Then two (parallel) signals of distinct meta-signals (**base** or **value**)—or one encoding zero **zero**—can represent any real number. Dividing by two correspond to finding the middle. Multiplication by any constant can be done likewise.

Adding two numbers can be done as in Fig. 6.14. The presence of a parallelogram proves the equality of the distances.

Adding a negative value (or from the left) is done similarly. The sign of a value is deduced from the order the signals **base** and **value** are encountered. This provides a direct way to test the sign to make appropriate actions.

To add to a remote value, the number of values to pass has to counted, for example with a set of signals as the one of the bottom of the parallelogram ($bottom_k \ldots bottom_1 bottom_0$, side, top, side). This is used in Fig. 6.14 to skip values.

Starting from a sequence of real values (like the sequence of cells of the tape of a Turing machine), it is possible to multiply by constants and to add a value to another. Since values to skip are hard encoded into meta-signals, operations always happen on a bounded portion of the sequence, a window.

Fig. 6.14 Adding real 15 to
real −8

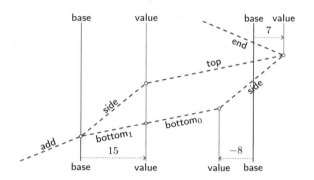

Increasing values could lead to have them overlap. The technical answer to this
is re-scaling all values as well as the unit pair (while preserving the distance from
a **base** to the next) thus preserving encoded values. This can be tested and realizes
dynamically.

Above primitive could be triggered by some deterministic finite automata (or
sequential program). The state of the automaton can be encoded in the meta-signals
used to carry out the operations (the number of meta-signals remains finite). Equally
it is possible to start the next operation from the same value (preserving the window).
It is also possible to move one value to the left or to the right, enlarging the sequence
when needed like for the Turing machine simulation (moving the window).

The automaton can be equipped with conditional transition: testing the sign of a
value can be used to branch. The automaton can have an initial and final state (no
more collision then). Constants may be provided in the initial configuration.

Altogether, starting from a finite sequence of real numbers (infinity extendable
on both side), it is possible to store in a cell the linear combination of values around
it, branch according to sign and move inside the sequence. This corresponds to the
Blum, Shub and Smale (BSS [6, 7]) model without inner multiplication. It is the
linear version of it: lin-BSS [10, 34] with an unbounded number of registers.

Signal machines are capable to implement lin-BSS. The converse is also true. The
configuration of a signal machine can be represented by a sequence of block each one
encoding the meta-signal, the distance to next plus various temporary registers. The
lin-BSS machine runs through the configuration and computes the minimal time to
a collision. Since speeds are constant of the signal machine, they become constants
of the lin-BSS machine, thus everything is linear.

Once this delay to the next collision time known, the automaton runs through
the configuration again. Distances are updated. When the distance is zero, then a
collision happens (it is tested whether more than two signals are involved). Involved
signals are replaced according to rules (there are hard encoded into the automaton).
If the number of signals is changed, all the values on the right are moved accordingly.
When room is needed (or should be removed) shifting the values in the cells is done
like for a Turing machine. The BSS automaton is like the one of a Turing machine:
it operates on a window (instead of a single cell) which he can moved.

Further readings.

This corresponds only to the *regular* operating of a signal machine. Taking accumulations and infinitely many signals during finite duration into account allows to go further on.

For example, it is possible to extract an infinite sequence of 0 and 1 representing the binary encoding of a real number. This flow can be used to make a multiplication: each time half and add or not depending on the received bit. Inner multiplication thus becomes possible and the whole classical BSS model can be implemented [19].

Accumulation can also be perceived as a convergent approximating process which is the foundation of recursive/computable analysis, type-2 Turing machine [39]. In this context, an input is an infinite stream of symbols representing a convergent approximation (approximation bound is known at each step) and the output is also such a stream (once something is output, it cannot be modified) with the same representation. It is possible to make a accumulation be located according to a process generating such a stream [21, 23] on a rational signal machine.

One last result about isolated accumulation on rational signal machine: not only they cannot happen everywhere (by a simple cardinality argument) but their possible locations are exactly characterized. They can only happen at dates that correspond to *computably enumerable (c.e.)* real numbers [11, Chap. 7], i.e. there is a Turing machine that produces an increasing and convergent infinite sequence (there is no hypothesis on the quality of the approximation). The positions of isolated accumulations are exactly the differences of two such numbers. Position and date can be handled independently. This is proved by a two scales construction: an embedded Turing machine is accelerated and stopped so that it provides the data on request in bounded time, the large scale directs the accumulation to the right spot according to the provided data [24].

6.4 Conclusion

Presented models operate inside continuous euclidean spaces. Their variety is huge as well as their computing capabilities. They bring forth a new kind of algorithmic where place, distance, relative positions… provide possibilities as well as constrains.

In most presented models, one key concept is signal: something continuous generating line segments.

Unsurprisingly, the capability to compute in the Turing understanding is common. As soon as it is possible to take advantage of continuity of space and time, analog computation and hyper-computation arise too (thus transcending the Church–Turing thesis). This remains true only in the ideal word of the model where there is no error no approximation nor limit to sub-division nor to density of information.

This is a limit to the realism of the models. Other arguments are the unbounded quantity of information that can be store and retrieve in a bounded space and the absence of Heisenberg's Uncertainty principle at any scale.

References

1. Andréka, H., Németi, I., Németi, P.: General relativistic hypercomputing and foundation of mathematics. Nat. Comput. **8**(3), 499–516 (2009)
2. Asarin, E., Maler, O.: Achilles and the Tortoise climbing up the arithmetical hierarchy. In: FSTTCS '95. LNCS, vol. 1026, pp. 471–483 (1995)
3. Asarin, E., Maler, O., Pnueli, A.: Reachability analysis of dynamical systems having piecewise-constant derivatives. Theor. Comp. Sci. **138**(1), 35–65 (1995)
4. Becker, F., Chapelle, M., Durand-Lose, J., Levorato, V., Senot, M.: Abstract geometrical computation 8: small machines, accumulations & rationality (2013, submitted)
5. Bennett, C.H.: Notes on the history of reversible computation. IBM J. Res. Dev. **32**(1), 16–23 (1988)
6. Blum, L., Cucker, F., Shub, M., Smale, S.: Complexity and Real Computation. Springer, New York (1998)
7. Blum, L., Shub, M., Smale, S.: On a theory of computation and complexity over the real numbers: NP-completeness, recursive functions and universal machines. Bull. Am. Math. Soc. **21**(1), 1–46 (1989)
8. Bournez, O.: Some bounds on the computational power of piecewise constant derivative systems (extended abstract). In: ICALP '97. LNCS, vol. 1256, pp. 143–153 (1997)
9. Bournez, O.: Achilles and the Tortoise climbing up the hyper-arithmetical hierarchy. Theor. Comp. Sci. **210**(1), 21–71 (1999)
10. Bournez, O.: Some bounds on the computational power of piecewise constant derivative systems. Theory Comput. Syst. **32**(1), 35–67 (1999)
11. Calude, C.S.: Information and Randomness: An Algorithmic Perspective. Texts in Theoretical Computer Science. An EATCS Series, 2nd edn. Springer, Heidelberg (2002)
12. Conway, J.H., Guy, R.L.: The Book of Numbers. Copernicus Series. Springer, Heidelberg (1996)
13. Cook, M.: Universality in elementary cellular automata. Complex Syst. **15**, 1–40 (2004)
14. Duchier, D., Durand-Lose, J., Senot, M.: Fractal parallelism: solving SAT in bounded space and time. In: Otfried, C., Chwa, K.-Y., Park, K. (eds.) International Symposium on Algorithms and Computation (ISAAC '10). LNCS, vol. 6506, pp. 279–290. Springer, Heidelberg (2010)
15. Duchier, D., Durand-Lose, J., Senot, M.: Computing in the fractal cloud: modular generic solvers for SAT and Q-SAT variants. In: Agrawal, M., Cooper, B.S., Li, A. (eds.) Theory and Applications of Models of Computations (TAMC '12). LNCS, vol. 7287, pp. 435–447. Springer, Heidelberg (2012)
16. Durand-Lose, J.: Calculer géométriquement sur le plan - machines à signaux. Habilitation à Diriger des Recherches, École Doctorale STIC, Université de Nice-Sophia Antipolis. In French (2003)
17. Durand-Lose, J.: Abstract geometrical computation for black hole computation (extended abstract). In: Margenstern, M. (ed.) Machines, Computations, and Universality (MCU '04). LNCS, vol. 3354, pp. 176–187. Springer, Heidelberg (2005)
18. Durand-Lose, J.: Abstract geometrical computation 1: embedding black hole computations with rational numbers. Fund. Inf. **74**(4), 491–510 (2006)
19. Durand-Lose, J.: Abstract geometrical computation and the linear Blum, Shub and Smale model. In: Cooper, B.S., Löwe, B., Sorbi, A. (eds.) Computation and Logic in the Real World, 3rd Conference on Computability in Europe (CiE '07). LNCS, vol. 4497, pp. 238–247. Springer, Heidelberg (2007)
20. Durand-Lose, J.: The signal point of view: from cellular automata to signal machines. In: Durand, B. (ed.) Journées Automates cellulaires (JAC '08), pp. 238–249 (2008)
21. Durand-Lose, J.: Abstract geometrical computation and computable analysis. In: Costa, J .F., Dershowitz, N. (eds.) International Conference on Unconventional Computation 2009 (UC '09). LNCS, vol. 5715, pp. 158–167. Springer, Heidelberg (2009)
22. Durand-Lose, J.: Abstract geometrical computation 4: small Turing universal signal machines. Theor. Comp. Sci. **412**, 57–67 (2011)

23. Durand-Lose, J.: Abstract geometrical computation 5: embedding computable analysis. Nat. Comput. **10**(4), 1261–1273 (2011). Special issue on Unconv. Comp. '09
24. Durand-Lose, J.: Geometrical accumulations and computably enumerable real numbers (extended abstract). In: Calude, C .S., Kari, J., Petre, I., Rozenberg, G. (eds.) International Conference on Unconventional Computation 2011 (UC '11). LNCS, vol. 6714, pp. 101–112. Springer, Heidelberg (2011)
25. Durand-Lose, J.: Abstract geometrical computation 6: a reversible, conservative and rational based model for black hole computation. Int. J. Unconv. Comput. **8**(1), 33–46 (2012)
26. Durand-Lose, J.: Irrationality is needed to compute with signal machines with only three speeds. In: Bonizzoni, P., Brattka, V., Löwe, B. (eds.) CiE '13, The Nature of Computation. LNCS, vol. 7921, pp. 108–119. Springer, Heidelberg (2013). Invited talk for special session Computation in nature
27. Etesi, G., Németi, I.: Non-Turing computations via Malament-Hogarth space-times. Int. J. Theor. Phys. **41**(2), 341–370 (2002)
28. Hagiya, M.: Discrete state transition systems on continuous space-time: a theoretical model for amorphous computing. In: Calude, C., Dinneen, M .J., Păun, G., Pérez-Jiménez, M.J., Rozenberg, G. (eds.) Proceedings of the 4th International Conference on Unconventional Computation, UC '05, Sevilla, Spain, October 3–7, 2005. LNCS, vol. 3699, pp. 117–129. Springer, Heidelberg (2005)
29. Hogarth, M.L.: Deciding arithmetic using SAD computers. Br. J. Philos. Sci. **55**, 681–691 (2004)
30. Huckenbeck, U.: Euclidian geometry in terms of automata theory. Theor. Comp. Sci. **68**(1), 71–87 (1989)
31. Huckenbeck, U.: A result about the power of geometric oracle machines. Theor. Comp. Sci. **88**(2), 231–251 (1991)
32. Jacopini, G., Sontacchi, G.: Reversible parallel computation: an evolving space-model. Theor. Comp. Sci. **73**(1), 1–46 (1990)
33. Lecerf, Y.: Machines de Turing réversibles. Récursive insolubilité en $n \in \mathbb{N}$ de l'équation $u = \theta^n u$, où θ est un isomorphisme de codes. Comptes rendus des séances de l'académie des sciences **257**, 2597–2600 (1963)
34. Meer, K., Michaux, C.: A survey on real structural complexity theory. Bull. Belg. Math. Soc. **4**, 113–148 (1997)
35. Morita, K., Shirasaki, A., Gono, Y.: A 1-tape 2-symbol reversible Turing machine. Trans. IEICE **E 72**(3), 223–228 (1989)
36. Senot, M.: Geometrical Model of Computation: Fractals and Complexity Gaps. Thèse de doctorat, Université d'Orléans (2013)
37. Sipser, M.: Introduction to the Theory of Computation. PWS Publishing Co, Boston (1997)
38. Takeuti, I.: Transition systems over continuous time-space. Electron. Notes Theor. Comput. Sci. **120**, 173–186 (2005)
39. Weihrauch, K.: Introduction to Computable Analysis. Texts in Theoretical Computer Science. Springer, Berlin (2000)
40. Woods, D., Neary, T.: The complexity of small universal Turing machines: a survey. Theor. Comp. Sci. **410**(4–5), 443–450 (2009)

Chapter 7
Unconventional Computers and Unconventional Complexity Measures

Ed Blakey

Abstract Unconventional computers—for example those exploiting chemical, analogue or quantum phenomena in order to compute, as opposed to those employing the standard, digital-computer approach of electronically implementing discrete-value logic gates—are widely studied both theoretically and experimentally. One notable motivation driving this study is the desire efficiently to solve classically difficult problems—we recall for example a chemical-computer approach to the NP-complete Travelling Salesperson Problem—, with computational complexity theory providing the criteria for judging this efficiency. However, care must be taken: *conventional* (Turing-machine-style) complexity analyses are in many cases inappropriate for assessing *unconventional* computers; new, non-standard computational resources, with correspondingly new complexity measures, may well be consumed during unconventional computation, and yet are overlooked by conventional analyses. Accordingly, we discuss in this chapter various resources beyond merely the conventional time and space, advocating such resources' consideration during analysis of the complexity of unconventional computers (and, more fundamentally, we discuss various interpretations of the term 'resource' itself). We hope that this acts as a useful starting point for practitioners of unconventional computing and computational complexity.

7.1 Background

We begin by outlining the prerequisite concepts of *computational complexity* and *unconventional computing*, and by recalling as a specific instance of the latter a certain *analogue integer-factorization device*.

E. Blakey (✉)
Oxford University Computing Laboratory, Oxford, UK
e-mail: ed.blakey@queens.oxon.org

© Springer International Publishing Switzerland 2017
A. Adamatzky (ed.), *Advances in Unconventional Computing*,
Emergence, Complexity and Computation 22,
DOI 10.1007/978-3-319-33924-5_7

165

7.1.1 Computational Complexity Theory

Computation—whether performed by a Turing machine (see [23]), a real-life digital computer, a quantum system, or any other device that has provision for accepting input and supplying corresponding output—must be *efficient* to be of practical use. The question of which aspects of a computation have a bearing on its efficiency or lack thereof is to some extent context-dependent (for example, portable but non-critical devices may favour space- over time-efficiency, whilst supercomputers processing meteorological data may not), though a widely accepted criterion for efficiency is offered by *computational complexity theory*.

Complexity theory suggests, and decades of collective practical experience corroborate, that a computer's efficiency corresponds to its using only a *polynomial* amount of computational resource (this polynomial function is of the *size* of the computation's input value). Since complexity theory has been developed primarily with the Turing machine and equivalent models/paradigms of computation in mind, these resources (that one hopes scale polynomially so as to be able to claim efficiency) are traditionally *run-time* and *memory space*; this limited view of resource is adequate when analysing the complexity of *standard* computers such as the (non-parallel, deterministic) Turing machine and closely related physical instantiations such as the real-life digital computer, but (as we see in Sect. 7.1.3) can lead to an unrealistically optimistic quantification of the complexity of *non-standard, unconventional* computers. The discrepancy arises since unconventional computers (for more about which see Sect. 7.1.2) may well consume unconventional resources (that is, those other than time and space), and the complexity measures corresponding to these resources should therefore be considered, though often are not.

7.1.2 Unconventional Computation

As we suggest above, traditional complexity theory is adequate for analysis of Turing-machine-style computers, but does not necessarily cater sufficiently for unconventional systems. Whilst it is unnecessary in the context of the present publication to describe or define unconventional computers (in their many forms: quantum, chemical/DNA, analogue, kinematic, optical, slime-mould, etc.), it is important to mention that *such computers consume resources other than time and space*, and that *these resources may contribute significantly to a system's complexity* in that they may impinge on the system's efficiency before availability of either time or space has become pressing. It is at least prima facie feasible, for instance, that a kinematic computer will need *energy* to compute, and that the time and space complexities of the system may be dwarfed by its 'energy complexity'; or that an analogue computer requires of the user a certain *precision* in order to function correctly, and that 'precision complexity' is more indicative of the system's efficiency than its time and space complexities. We consider now the latter (analogue/precision) example in more detail.

7.1.3 Motivating Example

We recall from [10] an analogue/optical system for factorizing natural numbers. Whilst we do not describe the system in detail here (a full account is available in [10], and an earlier, related system is described in [6] and covered by patent [11]), we note now some important features:

- the system can factorize numbers using time and space *polynomial* in the numbers' size (this is in contrast with the *exponentially* scaling time required by known conventional-computer solutions to the problem of factorization—see [14][1]), but
- the precision with which the user must manipulate/measure certain physical parameters (so as to effect input to/output from the system) increases as an *exponential* function of the size of the number being factorized.

For present purposes, then, the relevant point is that *the true complexity of the system, and the bars to its practical efficiency, arise because of issues relating to the resource of **required precision** rather than of **time** or **space***. This exemplifies the fact that insightful analysis of the complexity of unconventional computers entails consideration of accordingly unconventional resources. We consider now several such resources.

7.2 Commodity Resources

In this section, we detail certain resources consumed during unconventional computation.

7.2.1 What We Mean by 'Resource'

Note that, for present purposes, *resource* has a more specific meaning than its common usage[2] suggests. 'Resource' could validly be taken to mean

---

[1] Although [14] is some years old now, the stated exponential time requirement remains state-of-the-art.

[2] We find as a dictionary definition of 'resource' the following: "[a] means of supplying some want or deficiency; a stock or reserve upon which one can draw when necessary" [22].

- not only those *commodities* consumed/required during computation (i.e., "when necessary" as in the quotation in Footnote 2)—time, space, precision, etc.; but also
- the availability of non-determinism, oracles, etc. (that is, of features of the computational model/enveloping laws of physics, which features apparently augment computational power);
- the costs involved in manufacturing (rather than running) a computer (including, for example, distillation/manufacture of any entangled/cluster states from which respectively a general/measurement-based quantum computation proceeds; and including manufacturing costs viewed thermodynamically—see [18, 27]);
- etc.

One may, further, consider 'information-theoretic' resources and the inequalities (based essentially on simulability) between them—see Sect. 7.3.1.4; similar ideas arise in the context of trades-off between (computation-time, 'commodity') resources.

In the present chapter, we are concerned primarily with what we call *commodity resources*. These are computational resources consumed or required by a computer during execution of a computation (as opposed, for example, to resources used during a computer's manufacture, and to 'information-theoretic' resources such as communication channels; see Sect. 7.3 for more on non-commodity resources). Commodity resources include, but are far from limited to, time and space.

For the remainder of Sect. 7.2, by 'resource' we mean 'commodity resource'.

An important question here is:

($\star$) *what methods can be used to identify the relevant resources for a given computational model?*

This question may be posed either in generality—'given an *arbitrary* computational model, which resources should be considered?'—or for specific models; we consider both levels of generality (see Sects. 7.2.3 and 7.2.6 respectively).

Some clarification is needed on what it means for a resource to be 'relevant'. An intuitive understanding of this concept can be gleaned from the analogue factorization example of Sect. 7.1.3, which, we have commented, has time and space complexities (both polynomial) that fail to capture the system's actual, *exponential* complexity; the unconventional resource of *precision* captures the true complexity here, and, hence (unlike time and space), is *relevant*.

Note that, whereas the phrasing of our question ($\star$) suggests that resources' relevance is a function of the *computational model* under consideration, it may in fact vary between *instances* of the model: one computer of a certain model may have say time but not precision as a relevant resource, whilst, to another of the same model, precision but not time may be relevant. Nonetheless, when identifying *candidate* relevant resources, the choice of model seems more influential than the choice of specific computer. The complexity functions of these candidates (for the model(s) under consideration) can then be evaluated for the specific computer(s) of interest to see which are relevant (using for this evaluation notions such as *dominance*, which is defined and described in [9]).

Our focus is, of course, on *unconventional* models of computation. We claim as an aside that, in the conventional, Turing-machine model, the resources of *time* and *space* (and variations thereon) encompass all relevant complexity behaviour. We recall, for example, Păun's belief (see [20]) that "[t]he standard dimensions of computations are time and space"[3]; neither need one take Păun's word for it: note that many standard complexity classes are defined in terms of what is achievable by various computers within a certain *time* (loosely, these are P, NP (including the subclass of NP-complete problems), coNP, PH, EXP, NC, P/poly, BPP, BQP, PP) or using a certain amount of *space* (loosely, PSPACE, AC^0, NC, L).[4]

(Of course, should one deem parallel computation to be conventional, then to the conventional resources of *time* and *space* must be added *number of processors*. However, one may—and we do—view parallelism as an issue separate from our notions of (unconventional) complexity theory: susceptibility to a parallel approach is a property of a *problem*; of more interest here is how efficiently each parallel sub-computation can be performed (whether by Turing machine or unconventional computer); then account of parallelism is taken simply by summing respective sub-complexities. Similarly to parallelism, non-determinism may be—but here is not—viewed as a conventional computational practice, whence the resource *number of random bits consulted* (cf. Sect. 7.3.2) would be included in a list of conventional resources.)

7.2.2 Formalizing Commodity Resources

We model *commodity resources* (denoted *A*, *B*, *C*, etc.) as functions that depend upon the choice of *computer* and that map each *input value* to the corresponding amount[5] of *resource* used by the computer in processing the input value.

So, where Φ is a computer and *x* an input value for Φ, $A_\Phi(x)$ (or simply $A(x)$ if the choice of Φ is understood) is the amount—in fact, *number of units*: we stipulate that resource *A* have codomain $\mathbb{N} \cup \{\infty\}$ (the natural numbers[6] augmented with an infinity element)—of resource *A* consumed by Φ in processing *x*.

With each resource is associated a *complexity function*. For a given computer and resource, one may consider how the resource scales: one may be interested not in the

---

[3]Păun adds that "[t]his is true for computer science, not necessarily for the brain"; not wishing to deny the brain recognition as a computer, we must, and here do, question what resources are involved in unconventional computation.

[4]The 14 classes mentioned here are those in the 'Petting Zoo' section of Scott Aaronson's Complexity Zoo [1] (ignoring classes of function problems)—purportedly the most important (i.e., most referenced/fundamental/etc.) classes; those in the Petting Zoo but not amongst the 14 are MA, AM and SZK, which nonetheless are defined in terms of no other resources than time and space.

[5]This amount may, when no finite quantity of a resource is sufficient for a computation to complete satisfactorily, be infinite (e.g., no finite amount of the resource of *time* will suffice when a Turing machine has entered an infinite loop).

[6]Note that, in the present context, zero is deemed to be a natural number: $\mathbb{N} := \{0, 1, 2, \ldots\}$.

resource required by the computer in processing *some specific input value* (i.e., in some '*A*(*x*)'), but in the resource required *as a function of the input value's size*[7]—this is the *complexity function* corresponding to the resource. Specifically, we make the following definition.

Definition 1 (*Complexity function*) Let Φ be a computer with set X_Φ of possible input values and let A be a resource. The *complexity function* A_Φ^*, corresponding to resource A, is given by

$$A_\Phi^*(n) := \sup\{A_\Phi(x) \mid x \in X_\Phi \land |x| = n\} \ .$$

Whilst A, B, C, etc. denote types of *resource*, then, A^*, B^*, C^*, etc. denote types of *complexity*.

(Note that 'resource' has not been *defined*; we have described necessary, but not sufficient, properties of the notion, and consider further restrictions in Sect. 7.2.4, where the notion is summarized.)

7.2.3 Model-Independent Resources

We look now at some resources that apply to all computers, regardless of model.

7.2.3.1 Time and Space

Run-time and *memory space* are the standard resources considered in complexity analyses of Turing machines; indeed, up to variations (ink, head reversals, etc.), they are the only ones—recall the discussion of Sect. 7.2.1.

Whilst we see above that consideration of unconventional computation necessitates consideration of unconventional resources, this is not to say that time and space are not still relevant in unconventional contexts. Furthermore, it is clear for virtually all physical computing paradigms how to generalize these two resources from Turing machines to the wider class of physical computers. *Time* can be taken to be the number of units of physical time (measured in seconds, say[8]) that elapse during a computation (i.e., between input and output) performed by a physical system[9]; *space*

---

[7]We defer to [8, 12] and standard complexity theory further discussion of *size functions* (that map input values to their non-negative, real sizes), noting here by way of illustration only that the size of a natural number n expressed in place notation—with base b, say—can be taken to be the number of digits of n (or the often-convenient approximation $\log_b(n)$). We denote the size of an input value x by $|x|$.

[8]The use in complexity theory of $\mathcal{O}$-notation renders irrelevant this choice of the second as unit; it is made only for the sake of concreteness. Similarly the choice below of cubic metres.

[9]Here we implicitly assume that computations are not affected by relativistic effects, which is clearly not the case with so-called *relativistic computation*; for this model, then, our definition of time needs modification—see Sect. 7.2.6.3.

can be taken to be the physical volume (in cubic metres, say) occupied by the system (including any required storage space, electrical or not).

7.2.3.2 Material Cost

In addition to the material cost of *constructing* a computer (which may, from a commodity-resource point of view, be supposed to be a constant, one-off cost; or which can be treated as a non-commodity resource—see Sect. 7.3.1.1), there may be a cost in *running* it (over and above energy costs, which we discuss in Sect. 7.2.6 below). For example, if memory is implemented in such a way that each write (either to a fresh or used memory cell) costs a constant amount, then one may wish to consider the resource of 'ink'; this existing notion can be generalized to the resource of *material cost*, in what should in any given context be an obvious way.

7.2.3.3 Thermodynamic Cost

Strictly speaking, this resource is not applicable to computers from *all* paradigms (failing, in particular, for those from abstract, mathematical models), but is at least applicable to those that are *physically implemented*; Turing machines, then, are excluded, but digital computers that implement Turing machines are not.

The idea behind this resource is that computation typically *erases* information: evaluating a function (which will not in general be injective) in such a way that the input is destroyed and only the output is available after computation represents a loss of information, an increase in entropy and, hence, a thermodynamic cost; this concept was introduced by Landauer in [17] and developed by, amongst others, Bennett and Vitányi (who survey the notion in [4] and [25] respectively). Reference [27]— in which the corresponding complexity measure is explicitly introduced—notes that limits on the thermodynamic cost (a form of *computational* complexity) of a computation can be inferred by considering its *algorithmic* (that is, Kolmogorov) complexity. Note that, at least from the perspective of [27], this resource arises from information-theoretic (and, specifically, entropy-related) concerns; accordingly, (reversible) computation of an injective function is deemed to have negligible thermodynamic cost, regardless of (for example) the energy inefficiency of a physical instantiation of the computation.

In passing, we recall from [18] the notion of *thermodynamic depth*, which provides a measure of the amount of information lost in formation of an object (such as a computer) rather than in execution of a computational process. This is not, then, a commodity resource, but rather a manufacturing cost; see Sect. 7.3.1.1.

7.2.4   Model-Independent Features of Resources

Apart from consideration of individual *resources* (time, space, material/thermo-
dynamic costs, etc.) that are applicable in the context of arbitrary computational
models, one may consider broader *features* that (unspecified) such resources should
possess; we briefly discuss two such features now.

7.2.4.1   Blum's Axioms

These two axioms, introduced in [13], say of a resource that

1. the resource is defined if and only if computations during which the resource is
   consumed are themselves defined (this is the case, for example, with the Turing-
   machine resource of time: the number of time steps is a well-defined, finite natural
   number if and only if the computation is defined in the sense that it halts), and
   that
2. it is a decidable problem to check purported measurements of the resource's
   amount (again, time in the context of Turing machines satisfies this: given a
   Turing machine, an input value and a purported number n of time steps, one
   can check—simply by running the computation for n steps and checking for
   termination at that point and not before—whether the computation really does
   use n steps).

The axioms are, for our purposes, desirable,[10] and we stipulate that our commodity
resources satisfy them. However, necessary as we deem the axioms to be, they are
not *sufficient*; we note in [8] that a notion of commodity resource constrained only by
Blum's axioms leads to undesirable and deceptive complexity behaviour[11]—whereas
tools exist (see [8, 9]) to determine which of several resources are 'relevant' to a given
computation, these tools fail when our concept of resource is not restricted further
than by the axioms.

One restriction to the definition of resource that mitigates this deceptive complexity behaviour is to stipulate that resources be *normal*; we now briefly describe
the features of normalization of which we make use here, deferring to [8] a more
complete account.

---

[10]In particular, we do not consider here *non-deterministic* (commodity) resources; were we to, then
we should not want axiom 2 to take the form given here.

[11]Specifically, we demonstrate in [8] that the importance of certain resources can be artificially
exaggerated, as follows. One may, for example, count a Turing-machine computation's *time steps*
(let T be their number) and *tape cells* (S); then the measure T of time is more 'significant' than
that, S, of space in that $T \geq S$. However, one may (perversely but perfectly validly) measure space
instead as 2^S (with the mapping $k \mapsto 2^k$ establishing an order-isomorphism between the two spatial
measures, demonstrating their equivalence in some sense), whence it is for some Turing machines
the case that space appears more significant than time (in that $2^S > T$). It is this undesirable freedom
to engineer resources' apparent relative significance that motivates the below-described restriction
to *normal* resources.

7.2.4.2 Normalization

Roughly speaking, a *normal* resource is one that attains all natural-number values: a resource *A* is normal if and only if, for any natural number *n*, there exist a computer Φ and an input value *x* such that Φ, in processing *x*, consumes exactly *n* units of the resource (i.e., such that $A_\Phi(x) = n$). *Normalization* is a process whereby non-normal resources can be converted into normal resources that are order-isomorphic with the originals.[12]

If one were to allow as a resource an *arbitrary* function with codomain $\mathbb{N} \cup \{\infty\}$, then the resource would effectively return 'cardinals': the resource *counts* time steps, tape cells or similar, and one has an intrinsic *unit of measurement*. This seems resource-dependent and not conducive to comparison (for example, how many time steps, numerically speaking, should one deem of equivalent cost/value to one tape cell?). If, on the other hand, one allows only *normal* resources, then one has not 'cardinals' but 'ordinals', with 0 representing the least possible resource consumption, 1 the second-least, 2 the third-least, etc.; this is independent of the choice of resource, and of any unit of measurement suggested thereby, and so fairer, resource-heterogeneous comparison becomes possible.

If we stipulate that our commodity resources be as described in Sect. 7.2.2, satisfy Blum's axioms and be normal, then (as hinted at above and discussed in [8]) one finds that much of the deceptive complexity behaviour alluded to—notably, that of Footnote 11—is precluded.

7.2.4.3 Summary of 'Resource'

We summarize now our restrictions on the notion of resource. In the present chapter, a valid (commodity) resource

- is a function, dependent upon the choice of computer, that maps input values to natural numbers (or to ∞) (see Sect. 7.2.2);
- satisfies Blum's axioms (see Sect. 7.2.4.1); and
- is normal (see Sect. 7.2.4.2).

These are necessary conditions for a resource to be 'valid', though are still not between them sufficient[13]; a full *definition* of resource remains an open problem, to be investigated further in the wider project of which the present work is part.

---

[12]The resource *S* of Footnote 11, then, is normal, whereas 2^S—which normalizes to *S*—is not.

[13]An illustration of this insufficiency arises from the fact that, though normalization precludes *exaggeration* of a resource's importance via application of functions such as $k \mapsto 2^k$, it does nothing to preclude *understatement* of a resource's importance via application of functions such as $k \mapsto \lfloor \ln k \rfloor$.

7.2.5   Resource as a Lower Bound

We comment in passing that individual values of resources and of complexity functions are to be thought of as *lower bounds* on what is needed for a computation to succeed—recall (for example from Footnote 4 of [9]) the assumption that a computation can still proceed with more resource than is necessary (a desirable by-product of this is that our unconventional-complexity definitions are in some respects analogous to their traditional counterparts).[14] However, note that, in a quantum-mechanical (and, hence, quantum-computational) context, additional 'possibilities' (for example, potential routes taken by photons in a double-slit experiment) may interfere with *and cancel out* existing ones (see [21]); such phenomena should be considered, then, when selecting resources, so that provision of extra resource cannot, all else being equal, preclude a previously viable computation.

7.2.6   Specific, Model-Dependent Resources

We now consider specific (illustrative rather than exhaustive) models of computation, and ask which resources are likely to be relevant for instances thereof.

7.2.6.1   Actual, Physical Implementations of Turing Machines

We have already commented that, for the abstract, unimplemented Turing-machine model itself, the resources of time and space (and variants thereof) are sufficient for complexity-theoretic purposes. The resource of *precision*, then, is not a direct concern for Turing machines:

> "[w]hat is fundamental about the idea of a Turing Machine and digital computation in general, is that there is a perfect correspondence between the mathematical model and what happens in a reasonable working machine. Being definitely in one of two states is easily arranged in practice, and the operation of real digital computers can be (and usually is) made very reliable" [24].

However, both the *alphabet size* and *number of states* relate to precision in that distinguishing a greater number of distinct symbols/states entails there being smaller

---

[14]We must clarify this point: the sense in which resource and complexity offer *lower* bounds is that a computation *with fixed input value* may proceed with at least these bounds' allocation of resource; extra resource beyond that prescribed by a resource/complexity function is not problematic. This is in contrast with, and should not be confused with, the observation that a complexity function is the *maximum* (over input values of a certain size) amount of resource sufficient for a computation to succeed, in which sense complexity functions are *upper* bounds.

differences between them, resulting in smaller differences (in voltage or similar) between their respective real-world implementations.[15]

Nonetheless, the alphabet case is not problematic: 'meta-symbols' each comprising several symbols can be used instead of exponentially larger alphabets of individual symbols; similarly, the states may be encoded via such 'place notation'. Further, the Turing machine's unbounded tape seems unproblematic, as long as by 'computer' we mean not a fixed-memory machine, but the machine plus arbitrary additional memory (which must, of course, be accounted for in terms of the resource of space).

This suggests that consideration of time and space alone may still be enough, and (unsurprisingly) Turing [23] argues similarly (noting in particular the issue of alphabet size/symbol differentiation). However, from a *complexity* point of view, though real-world computers may offer a good, finite approximation of Turing machines, we are interested in *asymptotic* behaviour of resources, and so precision should be accounted for.

So, we consider, as we normally should, the resources of *time* and *space* (which latter accounts for the unboundedness of a Turing machine's tape: the resource quantifies how much space is needed given a certain input size, and may then determine adherence or lack thereof to financially/technologically/geographically imposed bounds on the space available to a physically implemented computer), and add to these *precision* so as to account for symbol and state numbers (and, hence, indirectly, for the size of the machine's transition table). These, it would seem, are the only significant differences between a Turing machine and a real-world implementation as far as complexity resources are concerned.

7.2.6.2 Analogue/Kinematic Computers

Time and space need, as always, to be considered when working with these models. However, they alone are not sufficient: recall the factorization system of Sect. 7.1.3. One additional relevant resource, as we have seen, is *precision* (this is also evident from, for example, the greatest-common-divisor system described in [7], the wedge-detection cannon system of [3] and the Differential Analyzer of [15]), and, we see below, there are others besides.

It seems intuitively clear that one should consider also the *energy* required to drive the computer. (Energy was not considered in the case of Sect. 7.2.6.1 (namely, real-world implementations of Turing machines) since, assuming 'ballistic' computation where the processor is used at capacity without, in particular, pauses for user interaction, energy consumption is linear in run-time and therefore redundant from a complexity-theoretic perspective (provided that one has not neglected to consider time).

---

[15]Strictly speaking, symbol and state numbers are a priori fixed, whereas we as complexity theorists should like to think in terms of functions of input size; accordingly, we may consider measures such as 'number of distinct symbols/states *used in the current computation*' as a function of input size.

In [24], the resource of precision is dealt with by acknowledging that an analogue computer has some 'ε' of imprecision, which value is fixed a priori and determines the maximum size of input that can be processed successfully, therefore not rendering precision a resource in our (commodity) sense. However, various standard and non-standard resources in our commodity sense—physical size, mass, initial stored energy, time interval of operation—are considered, though these happen to reduce (due to bounded density and similar) to the standard time and space.

In summary, then, relevant resources for analogue/kinematic computers include *time*, *space*, *precision* and *energy*.

7.2.6.3 Relativistic Computers

The broad idea of relativistic computation (of which the detailed physics is beyond the scope of the present chapter) is to exploit relativistic effects that allow a computer to experience time at a greater rate than its user; for then the user need wait less time (than if the user's and computer's clocks agreed) for an output value: more (computer) time steps are accommodated in each (user) second. Suggestions of how to achieve this effect include sending computers through wormholes, near black holes, etc.; the situation is sometimes contrived such that an *infinite* amount of computer time elapses in a finite amount of user time, whence hypercomputation becomes available.

When identifying resources for this model, there are two points to consider.

First, the resource of *time* is now ambiguous: one may consider seconds (say) counted by the *computer* or by the *user*. The tacit assumption is that the latter reference frame is the more telling and important, for else no computational speed-up is achieved. This assumption is perfectly reasonable: one views the computer, wormhole, etc. (but not the user) *together* as the computing system, and measures the time for which the user has to wait—this seems the more natural choice to resolve the ambiguity of the resource of time, and the most natural generalization of the resource as encountered in other models.

Secondly, note that, for present purposes, we adopt a fairly practical stance when considering models of computation: when identifying resources, for instance, we are careful to distinguish between (abstract) Turing machines and their (physical) implementations; and we consider practical issues such as achievable precision. Against this backdrop, then, it seems reasonable to exclude relativistic computers (at least as a form of hypercomputer) on the grounds of (for example) their energy consumption: although the user experiences only a finite amount of time, the computer needs to be *powered* for what the computer itself deems an eternity—we exploit relativity in an attempt to bypass *time* restrictions, but other resources' constraints (energy, durability of the physical machine, etc.) are still present. We see in another guise our original contention: time (and space) are not the only complexity-theoretic resources, and should not be treated as such. (There may, prima facie, still be an advantage offered by relativistic computers, in particular where computations are time-heavy; however, maintaining a computer's running appears to require other resources (energy or similar) in linear proportion to time, which suggests that a computation is never truly *uniquely* time-heavy.)

The power of this paradigm, then, comes primarily from neglect of the computer's time-frame in favour of the user's, though the availability (to the computer) of other resources such as energy is still a significant stumbling block.

(Commodity resources of user- and computer-time aside, there is clearly a significant *non*-commodity—specifically manufacturing—cost incurred during production (or at least discovery or similar) of the wormholes, black holes, etc. used by this computational paradigm; see Sect. 7.3.1.)

7.2.6.4 Optical Computers

Reference [26] introduces an optical system that computes via image manipulation; the resources considered are: time; number of grid images; spatial, amplitude and phase resolutions; dynamic range; and frequency of illumination. Without justification here, we state that these resources are akin to forms of *time*, *space* and *precision*, of which each is relevant for optical computers. We suggest also that *energy* is a relevant resource to the wider paradigm of optical computers, since some instances rely, for example, on the availability of electromagnetic waves of a prescribed *wavelength* (which may depend on the input size).

In summary, then, relevant resources for optical computers are *time*, *space*, *precision* and *energy* (and variants thereof).

7.2.6.5 Quantum Computers

We defer to future work a detailed account of the computational resources consumed by quantum computers, but make some general comments here.

In *circuit-model* quantum computation (wherein input is encoded via preparation of several quantum bits, processing takes the form of the application of unitary operations to subsets of these quantum bits, and output is via measurement of the system), we note that a commonly used complexity measure is the *number of invocations of unitary operations* [19]. This measure essentially captures the system's *run-time* (just as a Turing machine's run-time can be defined as the number of invocations of atomic, one-time-step operations whereby tape content, machine state and head position are updated), which is not, we suggest, a particularly insightful measure for quantum computers: the benefit enjoyed by quantum computers over their classical counterparts is gleaned in part from the use of superposition states, and the effective parallelism that this allows; a drawback is the strictly constrained way in which measurements can be taken of the system; this run-time measure, then, is a reflection of neither the 'amount of computation' performed (since 'parallelism' is not taken into account) nor the 'difficulty' in using the system (notably during measurement), of which two features at fewest one should arguably be captured by any useful complexity measure.

We consider now the *adiabatic* quantum model (where output values are encoded in the final ground state of an evolving system, with the evolution proceeding from an achievable initial ground state sufficiently slowly that no higher energy state is attained, so that our 'output-value' final ground state is indeed encountered). Standard expositions of the paradigm consider *time* as the only resource, as, indeed, is tacit in our "sufficiently slowly" above; however, determining this sufficient time makes use of trades-off with other resources. For example, the time sufficient for an evolution to remain in the ground state is a function of the minimum gap between the 0th (ground) and 1st energy states; *energy*, then, is an important (though commonly 'behind-the-scenes') resource for this computational model.

The *measurement-based* quantum model sees a computation take the form of several measurements (which can, in principle, be performed simultaneously), each of an individual quantum bit from an initial, large, entangled (cluster) state. As this description suggests, the complexity resources relevant to such a computation may be markedly different from those encountered in the circuit and other quantum models: enumerating invocations of unitary operations no longer applies in a straightforward manner; in fact, a relevant interpretation of '(non-commodity) complexity resource' in this context concerns the difficulty in producing the initial cluster state—see Sect. 7.3.1.1.

7.2.6.6   Chemical Computers

We mention here one specific resource that is particularly relevant to chemical computers, namely *mass*. We recall from [2] that DNA computers offer an approach to the (NP-complete) Travelling Salesperson Problem, and that the time (and, for that matter, energy) complexity of this method appears acceptable. However, as is pointed out in [16], the *mass* of DNA required by the method in processing non-trivial problem instances is greater than the mass of the Earth![16] We recall the long-held, de facto rule of thumb that tractability corresponds to polynomial resource consumption, and note that, in this case, the resource of *mass* imposes an exponential cost, and, therefore, intractability.[17]

(As an aside, note that, due to the bounded density of chemical-computing apparatus—including DNA itself—, mass is bounded by a constant multiple of space, and so the resource of mass tells us little new, provided that we already consider space. However, the distinction is illustrative of the unexpected ways in

---

[16]Hartmanis [16] writes of *weight*, but strictly means *mass*; the distinction is important since we are dealing with masses of the order of that of Earth, whence we may no longer assume negligible changes in gravitational strength from one part of the computational apparatus to another (nor, hence, negligible changes in the ratio between weight and mass).

[17]An alternative view of mass in this instance is as a measure of the number of parallel 'processors' at work during a chemical computation: the exponential speed-up observed with the TSP system and similar stems essentially from the presence of exponentially many DNA strands simultaneously testing one potential solution each. Recall, however, the discussion of parallelism in Sect. 7.2.1.

which complexity (in this example, space complexity) can be affected by strictly unconventional-computing concerns.)

7.3 Other Resources

We mention now in passing some (but by no means all) *non-commodity* interpretations of 'resource' (though justify briefly our focus[18] on commodity resources by citing (a) analogy with traditional complexity classes, where non-commodity resources such as the computational use of non-determinism are accounted for in the specifications of the classes themselves; and (b) the fact that many non-commodity resources can be viewed as features of the computational model, which is accommodated by our model-independent approach to complexity theory).

Hereafter, by 'resource', we do not necessarily mean 'commodity resource'.

7.3.1 Non-commodity Resource Types

We give now several non-commodity interpretations of 'resource'.

7.3.1.1 Manufacturing Costs

One may view as a resource the costs of *constructing* (rather than running) a computer. These may, for example, include the cost (whether this be financial, thermodynamic—see [18][19] and [27]—, or other) of manufacturing the system's physical structure (including the structure of wormholes, black holes, etc. in the case of relativistic computing—recall Sect. 7.2.6.3), or of producing an entangled state to be used in a quantum computation (e.g., a cluster state from which measurement-based quantum computation proceeds).

Viewing living brains as computers, one may also reasonably include under this type of resource such measures as time taken to evolve.

---

[18]Because of this focus, a more rigorous description of 'non-commodity resource' than is presented here is not necessary; furthermore, such description is rendered elusive by (amongst others) issues described in Sect. 7.3.2.

[19]We recall from [18] the complexity measure of *thermodynamic depth*, which gauges the loss of information during formation of an object (e.g., a computer). In relation to this measure, and in agreement with our general thesis that unconventional computers warrant consideration of unconventional resources, Lloyd and Pagels [18] write that, "if a definition of complexity is to be a useful measure for physical systems then it must be defined as a function of physical quantities, which in turn obey physical laws."

7.3.1.2 Features of Computational Models

Computation may be facilitated by taking advantage of 'permissions' granted by the computational model. For example, one may augment a Turing machine by allowing it to use *non-determinism*; then, although the same problems are *computable*, one does at least enjoy an apparent (and, if $P \neq NP$, an actual) increase in (time) *efficiency*. Similarly, a computer may be augmented by allowing consultation of *oracles*.

Such permissions are non-commodity resources on which one may draw to aid computation.

7.3.1.3 Features of Enveloping Physical Laws

Similarly to the features of the *computational model* (see above), one may exploit features of the *physics* describing the model. For example, a quantum computation may rely on the availability of *entanglement*, which entails one's adopting a non-classical physics, whereas, when using other computational models (such as centimetre-scale and larger kinematic computers), it may be convenient to assume that entanglement and other non-classical phenomena are *not* present, and that one may safely assume Newtonian dynamics.

Again, such features are non-commodity resources on which one may draw to aid computation (and to aid our describing and reasoning about computation).

(We mention in passing Heisenberg's uncertainty principle, closely related to which are trades-off between precision and other (commodity) resources such as energy. Further discussion of this idea is beyond the scope of the present chapter.)

7.3.1.4 Information-Theoretic Resources

Another class of resources that are non-commodity, but on which one can nonetheless draw for computational gain, is the class of 'information-theoretic' resources. These include (classical and quantum) *communication channels, entangled states*, etc.

We may, for example, summarize the information-theoretic content of the quantum teleportation protocol [5] with the inequality,

$$2cl\text{-}bit + e\text{-}bit \geq qubit \ ,$$

where '*cl-bit*' stands for the resource of being able to transfer a classical bit, '*qubit*' stands for the resource of being able to transfer a quantum bit, '*e-bit*' stands for the resource of having available an (entangled, two-party) Bell state, and '$X \geq Y$' indicates that resources X are at least as powerful as (i.e., they can simulate) resources Y.

7.3.2 Recasting as Different Resource Types

Note that the boundaries between 'types' of resource (commodity, model-feature, physics-feature, information-theoretic, etc.) are not always clear: there is not always a unique 'type' to which a given computational resource belongs. For example, the resource of *non-determinism* is discussed above in the context of being a model-feature resource; however, its presence or otherwise could equally be considered a physics-feature resource (where the presence or otherwise in physics of pre-ordainment is the determining factor), or even a commodity resource (where one may count, say, the number of random bits consulted during a computation).

However, the crucial point is not that resources fall neatly into these categories, but that practitioners of unconventional computing consider all resources (from whatever categories they may be) relevant to their systems.

7.4 Summary

We reiterate our main claim: that successful analysis of the complexity of *unconventional* computers requires consideration of correspondingly *unconventional* resources. This claim is justified by the motivating example of the factorization system (Sect. 7.1.3), the true complexity of which conventional resources alone fail to capture. We discuss above various non-standard (commodity) resources, which are relevant to various computational models; we discuss also different, non-commodity interpretations of the term 'resource'.

We hope that the issues raised here provoke thought amongst the unconventional-computing and computational-complexity communities, and lead to more rigorous and complete analyses of unconventional systems' computational complexity.

Acknowledgments We thank Andy Adamatzky for his kind invitation to contribute this chapter, as well as conference participants, collaborators, colleagues and reviewers who have helped to shape and direct this and related research.

References

1. Aaronson, S.: *Complexity Zoo*. (http://qwiki.stanford.edu/wiki/Complexity_Zoo, 2005 onwards)
2. Adleman, L.: Molecular computation of solutions to combinatorial problems. Science, vol. 266 (1994)
3. Beggs, E., Costa, J., Loff, B., Tucker, J.: The complexity of measurement in classical physics. TAMC, LNCS **4978** (2008)
4. Bennett, C.: The thermodynamics of computation—a review. Int. J. Theor. Phys. **21**(12) (1982)
5. Bennett, C., Brassard, G., Crépeau, C., Jozsa, R., Peres, A., Wootters, W.: Teleporting an unknown quantum state via dual classical and Einstein–Podolsky–Rosen channels. Phys. Rev. Lett, vol. 70 (1993)

6. Blakey, E.: An Analogue Solution to the Problem of Factorization. Oxford University Computing Laboratory Research Reports series, RR-07-04 (2007)
7. Blakey, E.: On the computational complexity of physical computing systems. Unconv. Comput. 2007 (2007)
8. Blakey, E.: Beyond Blum: what is a resource? Int. J. Unconv. Comput. (2008)
9. Blakey, E.: Dominance: Consistently Comparing Computational Complexity. Oxford University Computing Laboratory Research Reports series, RR-08-09 (2008)
10. Blakey, E.: Factorizing RSA keys, an improved analogue solution. New Gener. Comput. **27**(2), 159–176 (2008). doi:10.1007/s00354-008-0059-3
11. Blakey, E., co-prepared by patent attorneys: System and Method for Finding Integer Solutions. United States patent 7453574 (2008)
12. Blum, L., Cucker, F., Shub, M., Smale, S.: Complexity and Real Computation. Springer, Berlin (1997)
13. Blum, M.: A machine-independent theory of the complexity of recursive functions. J. Assoc. Comput. Mach. **14**(2) (1967)
14. Brent, R.: Recent progress and prospects for integer factorisation algorithms, LNCS **1858**. (2000)
15. Bush, V.: The differential analyzer: a new machine for solving differential equations. J. Franklin Inst, vol. 212 (1931)
16. Hartmanis, J.: On the weight of computations. Bull. EATCS **55** (1995)
17. Landauer, R.: Irreversibility and heat generation in the computing process. IBM J. Res. Dev. 5(3), (1961)
18. Lloyd, S., Pagels, H.: Complexity as thermodynamic depth. Ann. Phys., vol. 188 (1988)
19. Nielsen, M., Chuang, L.: Quantum Computation and Quantum Information. Cambridge University Press, Cambridge (2000)
20. Păun, G.: From cells to (silicon) computers, and back, New Computational Paradigms; Changing Conceptions of What is Computable. Springer, New York (2008)
21. Penrose, R.: The Emperor's New Mind: Concerning Computers, Minds, and the Laws of Physics. Oxford University Press, Oxford (1999)
22. *Resource* dictionary entry. Oxford English Dictionary. Oxford University Press, Oxford (1989)
23. Turing, A.: On computable numbers, with an application to the Entscheidungsproblem. Proc. Lond. Math. Soc. **2**(42) (1936)
24. Vergis, A., Steiglitz, K., Dickinson, B.: The complexity of analog computation. Math. Comput. Simul. **28**(2) (1986)
25. Vitányi, P.: Time, space, and energy in reversible computing. Proc. 2nd Conf. Comput. Front. (2005)
26. Woods, D.: Computational complexity of an optical model of computation. Ph.D. thesis, NUI Maynooth (2005)
27. Zurek, W.: Thermodynamic cost of computation, algorithmic complexity and the information metric. Nature, vol. 341 (1989)

Chapter 8
Decreasing Complexity in Inductive Computations

Mark Burgin

Abstract Kolmogorov or algorithmic complexity has found applications in many areas including medicine, biology, neurophysiology, physics, economics, hardware and software engineering. Conventional Kolmogorov/algorithmic complexity and its modifications are based on application of conventional, i.e., recursive, algorithms, such as Turing machines. Inductive complexity studied in this paper is based on application of unconventional algorithms such as inductive Turing machines, which are super-recursive as they can compute much more than recursive algorithm can. It is possible to apply inductive complexity in all cases where Kolmogorov complexity is used. In particular, inductive complexity has been used in the study of mathematical problem complexity. The main goal of this work is to show how inductive algorithms can reduce complexity of programs and problems. In Sect. 8.2, we build the constructive hierarchy of inductive Turing machines and study the corresponding hierarchy of inductively computable functions. Inductive Turing machines from the constructive hierarchy are very powerful because they can build (compute) the whole arithmetical hierarchy. In Sect. 8.3, it is proved that inductive algorithms from the constructive hierarchy can essentially reduce complexity of programs and problems and the more powerful inductive algorithms are utilized the larger reduction of complexity is achievable.

8.1 Introduction

Being a scientific reflection of efficiency, complexity has become a buzzword in contemporary science. There are different kinds and types of complexity with a diversity of different complexity measures. Kolmogorov, also called algorithmic, complexity has turned into an important and popular tool in many areas such as computer science, software development, probability theory, statistics, and information theory.

M. Burgin (✉)
University of California, Los Angeles, 405 Hilgard Ave., Los Angeles, CA 90095, USA
e-mail: mburgin@math.ucla.edu

© Springer International Publishing Switzerland 2017   183
A. Adamatzky (ed.), *Advances in Unconventional Computing*,
Emergence, Complexity and Computation 22,
DOI 10.1007/978-3-319-33924-5_8

Algorithmic complexity has found applications in medicine, biology, neurophysiology, physics, economics, hardware and software engineering. In biology, algorithmic complexity is used for estimation of protein identification [18, 19]. In physics, problems of quantum gravity are analyzed based on the algorithmic complexity of a given object. In particular, the algorithmic complexity of the Schwarzschild black hole is estimated [20, 21]. Benci et al. [1] apply algorithmic complexity to chaotic dynamics. Zurek elaborates a formulation of thermodynamics by inclusion of algorithmic complexity and randomness in the definition of physical entropy [32, 33]. Kreinovich and Kunin [24] apply Kolmogorov complexity to problems in mechanics. Tegmark [29] discusses what can be the algorithmic complexity of the whole universe. The main problem with this discussion is that the author identifies physical universe with physical models of this universe. To get valid results on this issue, it is necessary to define algorithmic complexity for physical systems because conventional algorithmic complexity is defined only for such symbolic objects as words and texts [26]. In addition, it is necessary to show that there is a good correlation between algorithmic complexity of the universe and algorithmic complexity of its model used by Tegmark [29].

In economics, a new approach to understanding of the complex behavior of financial markets using algorithmic complexity is developed [27]. In neurophysiology, algorithmic complexity is used to measure characteristics of brain functions [28]. Algorithmic complexity has been useful in the development of software metrics and other problems of software engineering [14, 17, 25]. Crosby and Wallach [16] use algorithmic complexity to study low-bandwidth denial of service attacks that exploit algorithmic deficiencies in many common applications' data structures.

Thus, we see that Kolmogorov/algorithmic complexity is a frequent word in the present days' scientific literature, in various fields and with diverse meanings, appearing in some contexts as a precise concept of algorithmic complexity, while being a vague idea of complexity in general in other texts. The reason for this is that people study and create more and more complex systems.

Algorithmic complexity in its classical form gives an estimate of how many bits of information we need to build or restore a given text by algorithms from a given class. Conventional Kolmogorov/algorithmic complexity and its modifications, such as uniform complexity, prefix complexity, monotone complexity, process complexity, conditional Kolmogorov complexity, time-bounded Kolmogorov complexity, space-bounded Kolmogorov complexity, conditional resource-bounded Kolmogorov complexity, time-bounded prefix complexity, and resource-bounded Kolmogorov complexity, are based on application of conventional, i.e., recursive, algorithms, such as Turing machines. Besides, researchers investigated quantum Kolmogorov complexity based on quantum algorithms [30]. Inductive complexity studied in this paper is based on application of unconventional algorithms such as inductive Turing machines, which are super-recursive as they can compute much more than recursive algorithm can. It is possible to apply inductive complexity in all cases where Kolmogorov complexity is used. In particular, inductive complexity has been used in the study of mathematical problem complexity [13, 15, 22].

The main goal of this work is to show how inductive algorithms can reduce the complexity of programs and problems. In Sect. 8.2, we build the constructive hierarchy of inductive Turing machines and study the corresponding hierarchy of inductively computable functions. The constructive hierarchy of inductive Turing machines also generates a hierarchy of algorithmic problems [9]. Inductive Turing machines from the constructive hierarchy are very powerful because they can build (compute) the whole arithmetical hierarchy [7]. In addition, as it is proved in Sect. 8.3, inductive algorithms from the constructive hierarchy can essentially reduce the complexity of programs and problems and the more powerful inductive algorithms are utilized the larger reduction of complexity is achievable.

Denotations

If X is an alphabet, then X^* is the set of all words (finite sequences) in the alphabet X.

If x is a word in an alphabet, then $l(x)$ is the length of x.

N is the set of all natural numbers.

A partial function $f : X^* \to N$ tends to infinity (we denote it by $f(x) \to \infty$) if for any number m from N, there is a number k such that $f(x) > m$ when $l(x) > k$.

A partial function $f : N \to N$ tends to infinity (we denote it by $f(n) \to \infty$) if for any number m from N, there is a number k such that $f(n) > m$ when $n > k$.

$IC_n(x)$ is the inductive complexity of an object (word) x on the level n.

$C(x)$ is the Kolmogorov complexity of an object (word) x.

8.2   Constructive Hierarchy of Inductive Turing Machines and Inductively Computable Functions

An inductive Turing machine has hardware, software and infware as any computer.

The *infware* of a computer is the system of all data that can be processed by the computer. The *infware* of inductive Turing machines, as in the case of the majority of other abstract automata, such as finite automata or Turing machines, consists of words in some alphabet. Here we consider only finite words in a finite alphabet. However, it is possible to think about infinite words and/or infinite alphabets, e.g., the alphabet of all real numbers.

The *hardware* of an inductive Turing machine M consists of three abstract devices: a *control device A*, which is a finite automaton and controls the performance of M; a *processor* or *operating device H*, which corresponds to one or several *heads* of a conventional Turing machine; and the *memory E*, which corresponds to the *tape* or tapes of a conventional Turing machine. The memory E of the simplest inductive Turing machine consists of three linear tapes, and the operating device consists of three heads, each of which is the same as the head of a Turing machine and works with the corresponding tape. Such machines are called *simple inductive Turing machines* [8].

The *control device A* of *M* is a finite automaton. It controls and regulates processes and parameters of the machine *M*: the state of the whole machine *M*, the processing of information by *H*, and the storage of information in the memory *E*.

The *memory E* of a general inductive Turing machine is divided into different but, as a rule, uniform cells. It is structured by a system of relations that organize memory as a well-structured system and provide connections or ties between cells. In particular, *input* registers, the *working* memory, and *output* registers of *M* are discerned in the structure of *E*. Connections between cells form an additional structure *K* of *E*. Each cell can contain a symbol from the alphabet of the language of the machine *M* or it can be empty. In what follows, we consider inductive Turing machine with a structured memory. Note that adding additional structure to a linear tape, it is possible to build many (even an infinite number of) tapes of different dimensions.

In a general case, cells may be of different types. Different types of cells may be used for storing different kinds of data. For example, binary cells, which have type B, store bits of information represented by symbols 1 and 0. Byte cells (type BT) store information represented by strings of eight binary digits. Symbol cells (type SB) store symbols of the alphabet(s) of the machine *M*. Cells in conventional Turing machines have SB type. Natural number cells, which have type NN, are used in random access machines (RAM). Cells in the memory of quantum computers (type QB) store qubits or quantum bits. Cells of the tape(s) of real-number Turing machines [8] have type RN and store real numbers. Cells in finite-dimensional machines of Blum, Shub and Smale have type RN and store real numbers [2]. When different kinds of devices are combined into one, this new complex device may have several types of memory cells. This feature makes inductive Turing machines an efficient model for heterogeneous computing systems and computations. In addition, different types of cells facilitate modeling the brain neuron structure by inductive Turing machines.

The *processor H* performs information processing in *M*. However, in comparison to computers, *H* performs very simple operations. When *H* consists of one unit, it can change a symbol in the cell that is observed by *H*, and go from this cell to another using a connection from *K*. It is possible that the processor *H* consists of several processing units similar to heads of a multihead Turing machine. This allows one to model various real and abstract computing systems: multiprocessor computers; Turing machines with several tapes; networks, grids and clusters of computers; cellular automata; neural networks; and systolic arrays.

The *software R* of the inductive Turing machine *M* is also a program that consists of simple rules similar to the rules of conventional Turing machines:

$$q_h a_i \rightarrow a_j q_k c \qquad (8.1)$$

In this formula, symbols q_h and q_k denote states of *A*, symbols a_i and a_j denote symbols from the alphabet of *M*, and *c* is a type of connections (links) in the memory *E*. The rule (8.1) means that if the state of the control device *A* of *M* is q_h and the processor *H* observes the symbol a_i in the observed cell, then the state of *A* becomes q_k, while the processor *H* writes the symbol a_j in the cell where it is situated and moves to the next cell by a connection of the type *c*. Each rule directs one step of computation of the inductive Turing machine M. Rules of the inductive Turing machine M define the transition function of M and describe changes of A, H, and E. Consequently, these rules also determine the transition functions of A, H, and E. Note that application of a rule (8.1) depends not on two variables q_h and a_i as in the case of conventional Turing machines but on three variables c, q_h and a_i, i.e. the rule is applicable only if there is a connection (link) of the type c from the observed cell. Thus, it is also possible to write this rule in the following form:

$$q_h a_i c \rightarrow a_j q_k c \qquad (8.2)$$

The described rules (8.1) and (8.2) represent only synchronous parallelism of computation. At the same time, it is possible to consider inductive Turing machines of any order in which their processor can perform computations in the concurrent mode. For instance, different heads can work in the asynchronous mode. Besides, it is also possible to consider inductive Turing machines with several processors. These more sophisticated computational models are studied elsewhere.

A general step of the machine M has the following form. At the beginning, the processor H observes some cell with a symbol a_i (it may be Λ as the symbol of an empty cell) and the control device A is in some state q_h. Then the control device A (and/or the processor H) chooses from the system R of rules a rule r with the left part equal to $q_h a_i$ and performs the operation prescribed by this rule. If there is no rule in R with such a left part, the machine M stops functioning. If there are several rules with the same left part, M works as a nondeterministic Turing machine, performing all possible operations. When A comes to one of the final states from F, the machine M also stops functioning. In all other cases, it continues operation without stopping.

In the output stabilizing mode, M gives the result when M halts and its control device A is in a final state from F, or when M never stops but at some step of the computation the content of the output register becomes fixed and does not change. The computed result of M is the word that is written in the output register of M. In all other cases, M does not give the result.

This means that an inductive Turing machine can do what a Turing machine can do but in addition, it produces its results without stopping. It is possible that in the sequence of computations after some step, the word (say, w) on the output tape (in the output register) is not changing, while the inductive Turing machine continues working. Then this word w is the output of the inductive Turing machine. Note that if an inductive Turing machine gives some output, it is produced after a finite number of steps (in finite time). So, contrary to confusing claims of some researchers, an inductive Turing machine does not need infinite time to produce a result.

Now let us build the *constructive hierarchy* of inductive Turing machines.

The memory E is called *recursive* if all relations that define its structure are recursive. Here recursive means that there are Turing machines that decide or build the structured memory E [6]. There are different techniques to organize this process. The simplest approach assumes that given some data, e.g., a description of the structure of E, a Turing machine T called the *construction machine* of the memory E builds all connections in the memory E before the machine M starts its computation. According to another methodology, memory construction by the machine T and computations of the machine M go concurrently, i.e., while the machine M computes, the machine T constructs connections in the memory E. It is also possible to consider a situation when some connections in the memory E are assembled before the machine M starts its computation, while other connections are formed parallel to the computing process of the machine M.

In real computers, all three techniques are employed although it is done in a very simple form. Indeed, when the hardware of a computer is manufactured, the connections in its memory are put together before the machine starts working. In contrast to this, utilization of external memory such as floppy discs, CD-ROM and flash drives involves formation of memory connections when computer is functioning.

Besides, it is possible to consider a schema when the construction machine T is separate from the machine M, while another structural design adopts the construction machine T as a part of the machine M. In some sense, T performs preprocessing of information for the machine M and works as an agent for M in the sense of agent technology. In [5], it is demonstrated how such preprocessing can essentially improve device performance.

Note that preprocessing is a process often utilized in natural information processing systems. For instance, visual information is at first preprocessed in eyes of animals and humans and only then it is transmitted to the brain. Psychologists call this preprocessing sensation, which is considered as an early stage of perception when neurons in a receptor create a pattern of nerve impulses that are transmitted to the brain for further processing [31]. This shows that inductive Turing machines perform biologically inspired computations.

Inductive Turing machines with recursive memory are called *inductive Turing machines of the first order*.

While in inductive Turing machines of the first order, the memory is constructed by Turing machines, it is also possible to use inductive Turing machines for memory construction for other inductive Turing machines. This brings us to the concept of inductive Turing machines of higher orders. For instance, in inductive Turing machines of the second order, the memory is constructed by Turing machines of the first order.

In general, we have the following definition.

The memory E is called *n-inductive* if its structure is constructed by an inductive Turing machine of the order n. Inductive Turing machines with n-inductive memory are called *inductive Turing machines of the order $n + 1$*. Namely, the memory of an inductive Turing machine of order n is constructed by Turing machines of order $n - 1$.

Note that any inductive Turing machine of order n is also an inductive Turing machine of order $n + 1$ because inductive Turing machines of order n can simulate any inductive Turing machine of order $n - 1$ [8].

Inductive algorithmic level n consists of all inductive Turing machines of order n. We denote the class of all inductive Turing machines of the order n by $\mathbf{IT}_n$ and define the *constructive hierarchy* of inductive Turing machines

$$\mathbf{IT} = \bigcup_{n=1}^{\infty} \mathbf{IT}_n$$

Machines from **IT** are called *constructively inductive Turing machines*.

The constructive hierarchy of inductive Turing machines induces hierarchic inductive computability for functions.

Definition 8.2.1 A function f is *inductively computable on the level n* if there is an inductive Turing machine of order n that computes f.

Note that an inductively computable on the level 0 function f is a recursively computable function and vice versa.

Results from [6, 8] show that each level of inductively computable functions has essentially more functions than the previous level. For instance, Kolmogorov complexity is computable on the first level but is not computable on the zero level being recursively non-computable [3]. The difference between the levels of inductive computability brings us to the following definition.

Definition 8.2.2 A function f is *inductively computable on the exact level n* if there is an inductive Turing machine of order n that computes f, while there is no inductive Turing machine of order $n - 1$ that computes f.

Note that an inductively computable on the exact level 0 function f is a recursively computable function and vice versa, i.e., the inductive algorithmic exact level 0 is the same as the inductive algorithmic level 0.

Proposition 8.2.1 *For any $n \geq 0$, function f is inductively computable on the level n, then it is inductively computable on the level $n + 1$.*

Indeed, any inductive Turing machine of order n is also an inductive Turing machine of order $n + 1$ that computes the same function and so what is computable on the level n is also computable on the level $n + 1$.

8.3 Inductive Complexity of Higher Orders

The constructive hierarchy of inductive Turing machines induces hierarchic inductive complexity for finite objects such as natural numbers or words in a finite alphabet.

Definition 8.3.1 For any $n \geq 0$, the *inductive complexity* $IC_n(x)$ of an object (word) x on the level n is defined as

$$IC_n(x) = \begin{cases} \min\{l(p); U(p) = x\} & \text{when there is } p \text{ such that } U(p) = x \\ \text{undefined} & \text{when there is no } p \text{ such that } U(p) = x \end{cases}$$

Here $l(p)$ is the length of the word p and U is a universal inductive Turing machine of order n.

Note that for $n = 0$, inductive complexity $IC_0(x)$ coincides with Kolmogorov complexity $C(x)$, while for $n = 1$, inductive complexity on the first level $IC_1(x)$ coincides with inductive complexity $IC(x)$ studied in [7, 8].

Definition 8.3.2 The *inductive complexity* $IC_T(x)$ of an object (word) x with respect to an inductive Turing machine T is defined as

$$IC_T(x) = \begin{cases} \min\{l(p); T(p) = x\} \text{ when there is } p \text{ such that } T(p) = x \\ \text{undefined when there is no } p \text{ such that } T(p) = x \end{cases}$$

Note that if T is a Turing machine, then inductive complexity $IC_T(x)$ with respect to T coincides with Kolmogorov complexity $C_T(x)$ with respect to T.

Relation $\preccurlyeq$ is basic for the theory of inductive complexity [4, 8]. Namely, if $f(n)$ and $g(n)$ are functions that take values in natural numbers, then

$$f(n) \preccurlyeq g(n) \text{ if there is a real number } c \text{ such that } f(n) \leq g(n) + c \text{ for almost all } n \in N$$

Let us consider a class **H** of functions that take values in natural numbers. Then a function $f(n)$ is called *optimal* for **H** if $f(n) \preccurlyeq g(n)$ for any function $g(n)$ from **H**.

Theorem 8.3.1 ([8]) *For any $n \geq 0$, the function $IC_n(x)$ is optimal in the class of all inductive complexities $IC_T(x)$ with respect to some inductive Turing machine T of order n.*

In what follows, it is possible to assume, without loss of generality (cf., for example, [8]), that all considered inductive Turing machines have only one-dimensional (linear) tapes and work with words in the alphabet $\{1, 0\}$.

As for any n, there is an inductive Turing machine M of order n such that $M(x) = x$ for all words x in the alphabet $\{1, 0\}$, we have the following result.

Proposition 8.3.1 *For any $n \geq 0$, $IC_n(x)$ is a total function on the set of all words in the alphabet $\{1, 0\}$.*

As it is demonstrated in [10], inductive complexity is intrinsically related to inductive decidability and undecidability of algorithmic problems.

For a given class of algorithms **K**, there are various algorithmic problems [11]. Here is one of them, which is closely related to inductive complexity. For simplicity, we consider automata that work with words in some alphabet A.

The Definability Problem R_D. Find an automaton W that for an arbitrary automaton D from **K** and an arbitrary word x in the alphabet A, informs whether $D(x)$ is defined, e.g., gives 1 when $D(x)$ is defined and 0 when $D(x)$ is not defined.

Finding when an algorithm gives a result and when it does not is an important task for many practical problems. Note that for Turing machines, the Definability Problem is equivalent to the Halting Problem.

Theorem 8.3.2   ([11]) *The Definability Problem* R_D *for the class* $\mathbf{IT}_n$ *of all inductive Turing machines of order n is undecidable in the class* $\mathbf{IT}_n$ *of all inductive Turing machines of order n.*

However, if we consider a larger class of inductive Turing machines, this algorithmic problems become decidable. Namely, we have the following result.

Theorem 8.3.3   *For any* $n \geq 0$, *the Definability Problem* R_D *for the class* $\mathbf{IT}_n$ *of all inductive Turing machines of order n is decidable in the class* $\mathbf{IT}_{n+1}$ *of all inductive Turing machines of order n + 1.*

Proof We need to find an inductive Turing machine W of order $n + 1$ that solves the Definability Problem R_D for the class of all inductive Turing machines of order n. However at first, we describe the inductive Turing machine H of order n that builds the structured memory of the machine W utilizing a binary codification $c : \mathbf{IT}_n \rightarrow N$ of all Turing machines of order n, for example, the codification constructed in [8].

The machine H has the following components (subroutines): a universal inductive Turing machine U of order n with a tester T, a connector L and a generator G of all words in the alphabet $\{1, 0\}$. A standard technique to include subroutines in the structure of Turing machines and inductive Turing machines is described, for example, in [8].

The machine H works in the following way. At first, the generator G generates all words in the alphabet $\{1, 0\}$ and writes them on the *word tape* of the machine H. Then the head of the machine H comes to the first cell of the *indicating tape* of the machine W and the connector L connects the first cell of the indicating tape with the second cell of the indicating tape by two links - one of the type a and another of the type d. After this, the tester T checks if the first word in the word tape has the form $c(K)*w$ where $c(K)$ is the code of some inductive Turing machine K of order n and w is a word in the alphabet $\{0, 1\}$. When this is not true, the head of the machine H comes to the second cell of the indicating tape of the machine W.

When the first word in the word tape has the form $c(K)*w$, the component U starts working with the word $c(K)*w$. If U gives a result, the connector L eliminates the link of the type a between the first and second cells of the indicating tape. Then L connects the first cell of the indicating tape with the second cell of the indicating tape by a link of the type b. After this, the head of the machine H comes to the second cell of the indicating tape of the machine W. If U does not give a result, the head of the machine H simply goes to the second cell of the indicating tape of the machine W.

After this, while the head of the machine H is in the second cell of the indicating tape of the machine W, the connector L connects the second cell of the indicating tape with the third cell of the indicating tape with two links - one of the type a and another of the type d. After this, the tester T checks if the second word in the word tape has the form $c(K)*w$ where $c(K)$ is the code of some inductive Turing machine K of order n and w is a word in the alphabet $\{0, 1\}$. When this is not true, the head of the machine H comes to the third cell of the indicating tape of the machine W.

When the second word in the word tape has the form $c(K)*w$, the component U starts working with the word $c(K)*w$. If U gives a result, the head of the machine H comes to the third cell of the indicating tape of the machine W, while the connector L eliminates a link of the type a between the second and third cells of the indicating tape and connects the second cell of the indicating tape with the third cell of the indicating tape by a link of the type b. If U does not give a result, the head of the machine H simply goes to the third cell of the indicating tape of the machine W.

The machine H continues to perform this procedure until all subsequent cells of the indicating tape of the machine W are connected by a link of the type a or b. In addition, the connector L connects all cells of other tapes in the conventional way where a cell is connected with its left and right neighbors. Then the inductive Turing machine W of order $n + 1$, which has a counter C and the same generator G as the machine H has as components (subroutines) of W, is ready to start working.

Given a word $u = c(K)*w$, the machine W uses the generator G for generating all words in the alphabet $\{1, 0\}$ writing them one by one on a word tape of the machine W. Note that generators in H and W are the same. So, all words in the alphabet $\{0, 1\}$ are written in the word tapes of the machines H and W in the same order. While the generator G is writing, the machine W is comparing each generated word with the word u and the counter C is counting the number of comparisons made.

When the comparison gives the positive result, i.e., when the number (say m) of the word u is found, this cycle finishes and a new cycle begins. In it, the head of the machine W that works in the indicating tape of the machine W uses three rules:

$$q_1 \Lambda b \rightarrow 1 q_{11} b \qquad (8.3)$$

$$q_1 \Lambda a \rightarrow 0 q_{10} a \qquad (8.4)$$

$$q_0 \Lambda d \rightarrow q_0 \Lambda d \qquad (8.5)$$

Using rule (8.5), the head of the machine W goes from a cell to the next cell, while on each step, the counter subtracts 1 from its content. When the head comes to the cell with number m, the register of the counter is empty and the machine W changes its state from q_0 to q_1. After this either rule (8.3) or rule (8.4) is used. Thus, the machine W changes its state. When the state becomes q_{11}, the machine W writes 1 into the output tape and stops. When the state becomes q_{10}, the machine W writes 0 into the output tape and stops.

We see that the machine W writes 1 when there is the link b from the cell with number m and writes 0 when there is the link a from the cell with number m. By construction, this means that the machine W writes 1 when the inductive Turing machine K of order n gives the result being applied to the word w and writes 0 when the machine K does not give the result being applied to the word w. It means that inductive Turing machine W of order $n + 1$ solves the Definability Problem R_D for the class of all inductive Turing machines of order n.

Theorem is proved.

Corollary 8.3.1 ([8]) *Inductive Turing machines of the first order can solve the Halting Problem for Turing machines.*

As it is possible to simulate any recursive algorithm by a Turing machine, we have the following result.

Corollary 8.3.2 ([12]) *Inductive Turing machines of the first order can solve the Definability Problem* R_D *for any class of recursive algorithms.*

Now let us consider computability of inductive complexities $IC_n(x)$.

Theorem 8.3.4 *For any* $n \geq 0$, *the function* $IC_n(x)$ *is inductively computable on the level* $n + 1$.

Proof The definition of the function $IC_n(x)$ depends on the choice of a universal inductive Turing machine U of order n. However, if we can compute this function defined by one universal inductive Turing machine of order n, then we can we can compute this function defined by any universal inductive Turing machine of order n due to the property of universality. Thus, we assume that the function $IC_n(x)$ is determined by a universal inductive Turing machine U of order n.

So, we need to find an inductive Turing machine W of order $n + 1$ that computes this function $IC_n(x)$. To do this, we describe the inductive Turing machine H of order n that builds the structured memory of the machine W utilizing a binary codification $c : \mathbf{IT}_n \to N$ of all inductive Turing machines of order n, for example, the codification constructed in [8].

Note that it is possible to assume that an inductive Turing machine with a structured memory has any (even a countable) number of tapes and works in the alphabet $\{1, 0\}$ [8].

The machine H has the following components (subroutines): the chosen universal inductive Turing machine U of order n; a connector L, which connects cells in the tapes of machine W; a forward counter F, which given a number m, bring the head of H to the cell with the number m; a generator G, which generates all words in the alphabet $\{1, 0\}$; and an enumerator N, which enumerates words generated by G so that if a word w is longer than a word u, then w has the larger number than u. A standard technique to include subroutines in the structure of Turing machines and inductive Turing machines is described, for example, in [8].

The machine H builds connections between cells from three linearly ordered tapes of the machine W: the *upper indicating tape*, the *lower indicating tape* and the *length tape*. The machine H works in the following way. At first, the generator G generates all words in the alphabet $\{1, 0\}$ and writes them on the *word tape* of the machine H. Then the enumerator N enumerates all generated by H words and writes their numbers on the *number tape* of the machine H.

At the beginning, the connector L connects the first cell of the lower indicating tape of the machine W with the first cell of the upper indicating tape and with the first cell of the length tape of the machine W by a link of the type a.

After this, the first cycle of the machine H starts. The head of the machine H comes to the first cell of the lower indicating tape of the machine W and the universal inductive Turing machine U begins processing the first word u_1 written in the word tape. If the machine U gives a result w, then the connector L connects the first cell of the lower indicating tape by a link of the type b with the cell of the upper indicating tape that has the number equal to the number of the word w. The cell with the necessary number is determined by the forward counter F of the machine H. In addition, the machine H determines the length of the word u_1 and the connector L connects the first cell of the lower indicating tape by a link of the type l with the cell of the length tape that has the number equal to the length of the word u_1. The cell with the necessary number is also determined by the forward counter F of the machine H. After this, the head of the machine H comes to the second cell of the lower indicating tape of the machine W. If U does not give a result being applied to u_1, the head of the machine H simply goes to the second cell of the lower indicating tape of the machine W.

Then the machine H performs the second cycle, the third cycle and so on. Beginning the nth cycle, the head of the machine H comes to the nth cell of the *lower indicating tape* of the machine W and the universal inductive Turing machine U begins processing the nth word u_n written in the word tape. If the machine U gives a result v, then the connector L connects the nth cell of the lower indicating tape by a link of the type b with the cell of the upper indicating tape that has the number equal to the number of the word v. The cell with the necessary number is determined by the forward counter F of the machine H. In addition, the machine H determines the length of the word u_n and the connector L connects the nth cell of the lower indicating tape by a link of the type l with the cell of the length tape that has the number equal to the length of the word u_n. The cell with the necessary number is also determined by the forward counter F of the machine H. After this, the head of the machine H comes to the $(n + 1)$th cell of the lower indicating tape of the machine W. If U does not give a result being applied to u_n, the head of the machine H simply goes to the $(n + 1)$th cell of the lower indicating tape of the machine W. Note that any cell from the lower indicating tape has a link of the type l if and only if it a link of the type b.

The machine W has the following components (subroutines): a backward counter B, which determines the number of the cell where the head of the machine W is situated m; a forward counter F, which given a number m, bring the head of W to the cell with the number m; and the enumerator N, from the machine H.

The machine W works with the input x according to the following algorithm. At the beginning, determines the number of the word x using the enumerator N and stores this number in the number tape of the machine W.

After this the first cycle of W starts. The head of the machine W comes to the first cell of the lower indicating tape. If this cell has a link of the type l, then the head of the machine W uses this link to go to the length tape of the machine W. Note that any cell of the lower indicating tape has at most one link of the type l. Therefore, this move is uniquely determined.

After the head of the machine W comes to the length tape using the link of the type l, the backward counter B finds the number of this cell and writes this number in the output register (tape). Then the head of the machine W comes to the first cell of the length tape and after that, it goes to the first cell of the lower indicating tape using the link of the type l. Performing the next step of the first cycle, the head of the machine W uses the link of the type b to go to the upper indicating tape of the machine W.

Then using backward counter B, the machine W finds the number of the cell where its head is situated and compares this number with the number of the word x stored in the memory. When the numbers coincide, the machine W stops, and the length k of the first word is its output. In this case, the value of $IC_n(x)$ is equal to k because $U(u_1) = x$ and u_1 has the minimal length with this property.

When the compared numbers do not coincide or the first cell of the lower indicating tape does not have a link of the type l, the machine W goes to the second cell of the lower indicating tape.

In a general case, the nth cycle of the machine W when its head is in the nth cell of the lower indicating tape. At first, the machine W stores the number n in its memory. Then if the nth cell has a link of the type l, then the head of the machine W uses this link to go to the length tape of the machine W. Note that any cell of the lower indicating tape has at most one link of the type l. Therefore, this move is uniquely determined.

After the head of the machine W comes to the length tape using the link of the type l, the backward counter B finds the number of this cell and writes this number in the output register (tape). Then the head of the machine W comes to the first cell of the length tape and after that, it goes to the first cell of the lower indicating tape using the link of the type l.

Performing the next step of the nth cycle, the head of the machine W uses the a forward counter F to bring the head of W back to the cell with the number n. Then the machine W uses the link of the type b to go to the upper indicating tape of the machine W.

Then using backward counter B, the machine W finds the number of the cell where its head is situated and compares this number with the number of the word x stored in the memory. When the numbers coincide, the machine W stops, and the length t of u_n is its output. n this case, the value of $IC_n(x)$ is equal to t because $U(u_n) = x$ and u_n has the minimal length with this property.

When the compared numbers do not coincide or the nth cell of the lower indicating tape does not have a link of the type l, the machine W goes to the $(n+1)$th cell of the lower indicating tape.

After a finite number of steps, the machine W computes the value of $IC_n(x)$ because there is an inductive Turing machine E that computes the identity function $e(x) = x$, i.e., $E(x) = x$ for all words x in the alphabet $\{1, 0\}$, and thus, $IC_n(x)$ is always defined and is not larger than $l(c(E)) + l(x)$ where $c(E)$ is the code $c(E)$ of E.

Theorem is proved.

Corollary 8.3.3 ([3, 8]) *Kolmogorov complexity $C(x)$ is inductively computable on the first level.*

We remind that $l(x)$ is the number of symbols in the word x in the alphabet $\{1, 0\}$.

Theorem 8.3.5 *For any* $n \geq 0$, $IC_n(x) \to \infty$ *when* $l(x) \to \infty$.

Proof The number of elements x for which $IC_n(x)$ is less than or equal to a given number m is finite because there are finitely many inductive Turing machines of order n the program of which has length less than or equal to the number m, so as m tends to infinity, the function $IC_n(x)$ does the same. Note that the program of an inductive Turing machine T of order n with a structured memory contains the program of the inductive Turing machine of order $n - 1$ that builds the memory of the machine T.

   Theorem is proved.
   This theorem implies the corresponding results for Kolmogorov complexity proved in [23] and inductive complexity on the level 1 proved in [8].

Corollary 8.3.4 ([23]). *Kolmogorov complexity* $C_n(x)$ *tends to* ∞ *when* $l(x) \to \infty$.

Corollary 8.3.5 ([8]) *Inductive complexity* $IC(x)$ *tends to* ∞ *when* $l(x) \to \infty$.

   Let us consider two partially ordered by relation $<$ sets X and Y. As always, $x \leq z$ means either $x < z$ or $x = z$.
   We remind that a function $f : X \to Y$ is called *increasing* if for any x and z from X, the inequality $x \leq z$ implies the inequality $f(x) \leq f(z)$ and is called *strictly increasing* if for any x and z from X, the inequality $x < z$ implies the inequality $f(x) < f(z)$.
   Although $IC_n(x)$ tends to infinity, on an infinite set of words, it grows slower than any total increasing inductively computable on the level n function.

Theorem 8.3.6 *For any* $n \geq 0$, *if* f *is a total strictly increasing inductively computable on the level n function with natural numbers as its values, then for infinitely many elements (words in the alphabet $\{1, 0\}$) w, we have* $f(w) > IC_n(w)$.

Proof Let us assume that all words in the alphabet $\{1, 0\}$ form an ordered sequence $u_1, u_2, u_3, \ldots, u_i, \ldots$ and there is some word u such that for all elements y that are larger than u, we have $f(w) \leq IC_n(w)$. Because $f(w)$ is an inductively computable on the level n function, there is an inductive Turing machine T of order n that computes $f(w)$. It is done in the following way. Given a word w, the machine T makes the first step, producing $f_1(w)$ on its output tape. Making the second step, the machine T produces $f_2(w)$ on its output tape and so on. After n steps, T has $f_n(w)$ on its output tape. Since the function is inductively computable on the level n, this process stabilizes on some value $f_n(w) = f(w)$, which is the result of computation with the input w.

   Taking the function $h(m) = \min\{x; f(x) \geq m\}$, we construct an inductive Turing machine M of order n that computes the function $h(x)$.
   The inductive Turing machine M contains a copy of the machine T. Utilizing this copy, M finds one after another the values $f_1(u_1), f_1(u_2), \ldots, f_1(u_{m+1})$ and compares these values to m. Then M writes into the output tape the least u for which the value $f_1(u)$ is larger than or equal to m. Then M finds one after another

the values $f_2(u_1), f_2(u_2), \ldots, f_2(u_{m+1})$ and compares these values to m. Then M writes the least u for which the value $f_2(u)$ is larger than or equal to m into the output tape. This process continues until the output value of M stabilizes. It happens for any number m due to the following reasons. First, $f(x)$ is a total function, so all values $f_i(u_1), f_i(u_2), \ldots, f_i(u_{m+1})$ after some step $i = t$ become equal to $f(u_1), f(u_2), \ldots, f(u_{m+1})$. Second, $f(x)$ is a strictly increasing function and thus, $f_i(u_{m+1}) > m$. In such a way, the machine M computes $h(m)$. Since m is an arbitrary number, the machine M computes the function $h(x)$.

Since for all elements y that are larger than w, we have $f(y) \leq \mathrm{IC}_n(y)$. Thus, there is an element m such that $\mathrm{IC}_n(h(m)) \geq f(h(m))$ and $f(h(m)) \geq m$ as $f(x)$ is a strictly increasing function and $h(m) = \min\{x; f(x) \geq m\}$. By definition, $\mathrm{IC}_T(h(m)) = \min\{l(x) ; T(x) = h(m)\}$. As $T(m) = h(m)$, we have $\mathrm{IC}_T(h(m)) \leq l(m)$. Thus, $l(m) \geq \mathrm{IC}_T(h(m)) \succcurlyeq \mathrm{IC}_n(h(m)) \geq m$. However, it is impossible that $l(m) \succcurlyeq m$. This contradiction concludes the proof of the theorem.

Corollary 8.3.6 ([8]) *If f is a total strictly increasing inductively computable on the first level function with natural numbers as its values, then for infinitely many elements (words in the alphabet $\{1, 0\}$) w, we have $f(w) > \mathrm{IC}_1(w)$.*

We can prove a stronger statement than Theorem 8.3.6. To do this, we assume for simplicity that inductive Turing machines are working with words in some finite alphabet and that all these words are well ordered, that is, any set of words contains the least element. It is possible to find such orderings, for example, in [26].

Theorem 8.3.7 *For any $n \geq 0$, if h is an increasing inductively computable on the level n function that has natural numbers as its values, is defined in an infinite inductively decidable on the level n set V and tends to infinity when $l(x)$ tends to infinity, then for infinitely many elements x from V, we have $h(x) > \mathrm{IC}_n(x)$.*

Proof Let us assume that there is some element z such that for all elements x that are larger than z, we have $h(x) \leq \mathrm{IC}(x)$. Because $h(x)$ an inductively computable on the level n function, there is an inductive Turing machine T of order n that computes $h(x)$. Taking the function $g(m) = \min\{x; h(x) \geq m$ and $x \in V\}$, we construct an inductive Turing machine M of order n that computes the function $g(x)$.

As V is an inductively decidable on the level n set, there is an inductive Turing machine H of order n that given an input x, produces 1 when $x \in V$, and produces 0 when $x \notin V$. It means that H computes the characteristic function $c_V(x)$ of the set V.

The inductive Turing machine M, which is constructed, contains a copy of the machine H and a copy of the machine T. These copies are constructed as subroutines of M with separate memories. As each of these memories is $(n - 1)$-inductive, the inductive Turing machine M has order n. Utilizing this copy of T, the machine M computes the value $h_1(1)$ and compares it to m. Utilizing this copy of H, the machine M computes the value $c_{V1}(1)$. If $h_1(1)$ is larger than m and $c_{V1}(1) = 1$, then M writes 1 into the output tape. Otherwise, M writes nothing into the output tape. After this, M finds the values $h_2(1)$ and $h_2(2)$ and compares these values to m. Concurrently, M finds the values $c_{V2}(1)$ and $c_{V2}(2)$. Then M writes into the output tape the least x for which the value $h_1(x)$ is larger than or equal to m and at the same time, $c_{V2}(x) = 1$. This process continues. Making cycle i of the computation, M computes the values $h_i(1), h_i(2), \ldots, h_i(i)$ and compares these values to m. We remind here that $h_i(j)$ is the result of i steps of computation of T with the input j. Concurrently, M computes the values $c_{Vi}(1), c_{Vi}(2), \ldots, c_{Vi}(i)$. Then M writes into the output tape the least x for which the value $h_i(x)$ is larger than or equal to m and at the same time, $c_{Vi}(x) = 1$. Such cycle is repeated until the output value of M stabilizes.

Each value $c_{Vi}(x)$ stabilizes at some step t because $c_V(x)$ is a total inductively computable function. In a similar way, each value $h_i(x)$ stabilizes at some step q because $h(x)$ is an inductively computable function defined for all $x \in V$. Thus, after this step $p = \max\{q, t\}$, the value $h_i(x)$ becomes equal to the value $h(x)$. In addition, there is such a step t when a number n is found for which $h(n) \geq m$. After this step, only such numbers x can go to the output tape of M that belong to V and are less than or equal to n.

This happens for any given number m due to the following reasons. First, $h(x)$ is defined for all elements from V, so those values $h_i(1), h_i(2), \ldots, h_i(m + 1)$, for which the argument of h_i belongs to V, after some step $i = r$ become equal to $h(1), h(2), \ldots, h(m)$. Second, $h(x)$ is an increasing function that tends to infinity.

This shows that the whole process stabilizes and by the definition of inductive computability, the machine M computes $g(m)$. Since m is an arbitrary number, the machine M computes the function $g(x)$.

To conclude the proof, we repeat the reasoning from the proof of Theorem 8.3.6. Since for all elements x that are larger than z, we have $h(x) \leq IC(x)$, there is an element m such that $IC(g(m)) \geq h(g(m))$ and $h(g(m)) \geq m$ as $h(x)$ is an increasing function and $g(m) = \min\{x; h(x) \geq m \text{ and } x \in V\}$. By definition, $IC_T(g(m)) = \min\{l(x) \; ; \; T(x) = g(m)\}$. As $T(m) = g(m)$, we have $IC_T(g(m)) \leq l(m)$. Thus, $l(m) \geq IC_T(h(m)) \succcurlyeq IC(h(m)) \geq m$. However, it is impossible that $l(m) \succcurlyeq m$. This contradiction concludes the proof of the theorem.

Corollary 8.3.7 ([8]) *If h is an increasing inductively computable on the first level function that has natural numbers as its values, is defined in an infinite recursively decidable set V and tends to infinity when $l(x)$ tends to infinity, then for infinitely many elements x from V, we have $h(x) > IC_1(w)$.*

Remark 8.3.1 Although Theorem 8.3.6 can be deduced from Theorem 8.3.7, we give an independent proof of Theorem 8.3.6 because it demonstrates another technique, which displays essential features of inductive Turing machines.

Corollary 8.3.8 *If h is a total increasing inductively computable on the level n function that tends to infinity when $l(x) \to \infty$, then for infinitely many elements x, we have $h(x) > IC_n(x)$.*

Since the composition of two increasing functions is an increasing function and the composition of a recursive function and an inductively computable on the level n function is an inductively computable on the level n function, we have the following result.

Corollary 8.3.9 *If $h(x)$ and $g(x)$ are increasing functions, $h(x)$ is inductively computable on the level n and defined in an inductively decidable on the level n set V, $g(x)$ is a recursive function, and they both tend to infinity when $l(x) \to \infty$, then for infinitely many elements x from V, we have $g(h(x)) > \mathrm{IC}(x)$.*

In addition to the function $\mathrm{IC}_n(x)$ with $i = 0, 1, 2, 3, \ldots$, we also introduce the function

$$\mathrm{mIC}_n(x) = \min\{\mathrm{IC}_n(y); l(y) \geq l(x)\}$$

It has the following properties.

Theorem 8.3.8 *(a) For any n, $\mathrm{mIC}_n(x)$ is a total increasing function.*
(b) For any n, $\mathrm{mIC}_n(x)$ is not inductively computable on the level n.
(c) For any n, $\mathrm{mIC}_n(x) \to \infty$ when $l(x) \to \infty$.
(d) For any n, $\mathrm{mIC}_n(x)$ is inductively computable on the exact level $n + 1$.

Proof (a) Since $\mathrm{IC}_n(x)$ is a total function, $\mathrm{mIC}_n(x)$ is also a total function. By definition, $\mathrm{mC}_n(x)$ is increasing.

(b) If $\mathrm{mIC}_n(x)$ is an inductively computable function on the level n, then by Theorem 8.3.6, for infinitely many elements x, we have $\mathrm{mIC}_n(x) > \mathrm{IC}_n(x)$. However, by the definition of $\mathrm{mIC}_n(x)$, we have $\mathrm{mIC}_n(x) \leq \mathrm{IC}_n(x)$ everywhere. This contradiction completes the proof of the part (b).

Part (c) follows from Theorem 8.3.3.

(d) As by Theorem 8.3.4, the function $\mathrm{IC}_n(x)$ is inductively computable on the level $n + 1$, there is an inductive Turing machine T of order $n + 1$ that computes $\mathrm{IC}_n(x)$. Then we can change the machine T so that the new inductive Turing machine M of order $n + 1$ will compute $\mathrm{mIC}_n(x)$. To achieve this, we add a new output tape to T making its output tape intermediate and include T as a subroutine of M.

Then with the input x, the machine M starts computing $\mathrm{IC}_n(y)$ for all words y such that $l(y) \geq l(x)$. When the value of $\mathrm{IC}_n(z)$ is computed (cf. Theorem 8.3.4), it is compared to all previously computed values $\mathrm{IC}_n(y)$ and the least is written into the output tape of the machine M. Because $\mathrm{IC}_n(x) \to \infty$ when $l(x) \to \infty$, after some number of steps the output stabilizes and by the definition of inductive computation, this will be the value $\mathrm{mIC}_n(x)$.

Theorem is proved.

Theorem 8.3.9 *For any $n \geq 0$, $\mathrm{IC}_n(x)$ is not inductively computable on the level n.*

Proof At first, we show that if $\mathrm{IC}_n(x)$ is inductively computable on the level n, then $\mathrm{mIC}_n(x)$ is also inductively computable on the level n. Let us assume that the function $\mathrm{IC}_n(x)$ is inductively computable on the level n. Then there is an inductive

Turing machine T of order n that computes $IC_n(x)$. Then we can change the machine T so that the new inductive Turing machine M of order n will compute $mIC_n(x)$. To achieve this, we add a new output tape to T making its output tape intermediate and include T as a subroutine of M.

Then with the input x, the machine M starts computing $IC_n(y)$ for all words y such that $l(y) \geq l(x)$. When the value of $IC_n(z)$ is computed, it is compared to all previously computed values $IC_n(y)$ and the least is written into the output tape of the machine M. Because $IC_n(x) \to \infty$ when $l(x) \to \infty$, after some number of steps the output stabilizes and by the definition of inductive computation, this will be the value $mIC_n(x)$.

At the same time, by Theorem 8.3.8, the function $mIC_n(x)$ tends to infinity when $l(x) \to \infty$ and is increasing but not inductively computable on the level n. This gives us a contradiction. Thus, the statement of Theorem 8.3.9 is true.

Theorem is proved.

Corollary 8.3.10 ([8]) *The inductive complexity $IC_1(x)$ is not inductively computable by inductive Turing machines with recursive memory.*

Theorems 8.3.4 and 8.3.9 imply the following result.

Corollary 8.3.11 *For any $n \geq 0$, the function $IC_n(x)$ is inductively computable on the exact level $n + 1$.*

Corollary 8.3.12 *There is no inductively computable on the exact level n function $f(x)$ defined for an infinite inductively decidable set that coincides with $IC(x)$ in the whole of the domain of definition of $f(x)$.*

Theorems 8.3.4, 8.3.7 and 8.3.8 allow us to prove the following result.

Theorem 8.3.10 *There are infinitely many elements x for which $IC_{n+1}(x) < IC_n(x)$.*

Proof By Theorems 8.3.4 and 8.3.8, the function $mIC_n(x)$ is inductively computable on the level $n + 1$. In addition, the function $mIC_n(x)$ tends to infinity when $l(x) \to \infty$ and is increasing but not inductively computable on the level n. Thus, by Theorem 8.3.7, there are infinitely many elements x for which $IC_{n+1}(x) < mIC_n(x)$. As $mIC_n(x) \leq IC_n(x)$, there are infinitely many elements (words) x for which $IC_{n+1}(x) < IC_n(x)$.

Theorem is proved.

Theorem 8.3.10 shows that for infinitely many elements x, their complexity on the level $n+1$ is less than their complexity on the level n.

Corollary 8.3.13 ([8]) *There are infinitely many elements x for which $IC_1(x) < C(x)$.*

Corollary 8.3.13 shows that for infinitely many elements x, their inductive complexity on the first level is less than their Kolmogorov complexity.

Corollary 8.3.14 *For any $n > 0$, there are infinitely many elements x for which $IC_n(x) < C(x)$.*

Corollary 8.3.14 shows that for any $n > 0$ and infinitely many elements x, their inductive complexity on the level n is less than their Kolmogorov complexity.

As composition of increasing functions is an increasing function, Theorems 8.3.7 and 8.3.8 imply the following result.

Theorem 8.3.11 *For any increasing recursive function $h(x)$ that tends to infinity when $l(x) \rightarrow \infty$ and any inductively decidable on the level n set V, there are infinitely many elements x from V for which $IC_{n+1}(x) < h(C_n(x))$.*

Corollary 8.3.15 ([8]) *In any inductively decidable on the first level set V, there are infinitely many elements x for which $C(x) > IC(x)$.*

Corollary 8.3.16 ([8]) *In any recursive set V, there are infinitely many elements x for which $IC_1(x) < C(x)$.*

Corollary 8.3.17 *For any n and any inductively decidable on the first level set V, there are infinitely many elements x in V for which $IC_{n+1}(x) < C_n(x)$.*

Corollary 8.3.18 *For any n and any recursive set V, there are infinitely many elements x in V for which $IC_{n+1}(x) < C_n(x)$.*

Corollary 8.3.19 *In any inductively decidable (recursive) set V, there are infinitely many elements x for which $\ln_2(C(x)) > IC(x)$.*

If $\ln_2(C(x)) > IC(x)$, then $C(x) > 2^{IC(x)}$. At the same time, for any natural number k, the inequality $2^n > k \cdot n$ is true almost everywhere. This and Corollary 8.3.19 imply the following result.

Corollary 8.3.20 *For any natural number k and in any inductively decidable (recursive) set V, there are infinitely many elements x for which $C(x) > k \cdot IC(x)$.*

Corollary 8.3.21 *For any natural number a, there are infinitely many elements x for which $\ln_a(C(x)) > IC(x)$.*

8.4 Conclusion

We have demonstrated that, with respect to a natural extension of the Kolmogorov or algorithmic complexity, inductive Turing machines of order $n + 1$ are much more efficient than inductive Turing machines of order n. In particular, inductive Turing machines are much more efficient than any kind of recursive algorithms. Informally, this means that in comparison with recursive algorithms, super-recursive programs for solving the same problem are shorter, have lower branching (i.e., less instructions of the form IF *A* THEN *B* ELSE *C*), make fewer reversions and unrestricted transitions (i.e., fewer instructions of the form GO TO *X*) for infinitely many problems solvable by recursive algorithms. In addition, the higher level of inductive algorithms is utilized the larger reduction of complexity is achievable.

Researchers introduced and studied different modifications of Kolmogorov/ algorithmic complexity, such as uniform complexity, prefix complexity, monotone complexity, process complexity, conditional Kolmogorov complexity, time-bounded Kolmogorov complexity, space-bounded Kolmogorov complexity, conditional resource-bounded Kolmogorov complexity, time-bounded prefix complexity, and resource-bounded Kolmogorov complexity (cf. [8, 26]). It would be interesting to define and study corresponding modifications of inductive complexity.

References

1. Benci, V., Bonanno, C., Galatolo, S., Menconi, G., Virgilio, M. Dynamical systems and computable information, Preprint in Physics cond-mat/0210654, 2002 (electronic edition: http://arXiv.org)
2. Blum, L., Shub, M., Smale, S.: On a theory of computation and complexity over the real numbers: NP-completeness, recursive functions and universal machines. Bull. Am. Math. Soc. **21**(1), 1–46 (1989)
3. Burgin, M.S.: Inductive Turing machines. Not. Acad. Sci. USSR **27**(3), 1289–1293 (1983)
4. Burgin, M.S.: Generalized Kolmogorov complexity and other dual complexity measures. Cybern. Syst. Anal. **26**(4), 481–490 (1990)
5. Burgin, M.: Super-recursive algorithms as a tool for high performance computing. In: Proceedings of the High Performance Computing Symposium, pp. 224–228. San Diego (1999)
6. Burgin, M.: Nonlinear phenomena in spaces of algorithms. Int. J. Comput. Math. **80**(12), 1449–1476 (2003)
7. Burgin, M.: Algorithmic complexity of recursive and inductive algorithms. Theor. Comput. Sci. **317**(1/3), 31–60 (2004)
8. Burgin, M.: Superrecursive Algorithms. Springer, New York (2005)
9. Burgin, M. Superrecursive hierarchies of algorithmic problems. In: Proceedings of the 2005 International Conference on Foundations of Computer Science, pp. 31–37. CSREA Press, Las Vegas (2005)
10. Burgin, M.: Algorithmic complexity as a criterion of insolvability. Theor. Comput. Sci. **383**(2/3), 244–259 (2007)
11. Burgin, M.: Algorithmic complexity of computational problems. Int. J. Comput. Inf. Technol. **2**(1), 149–187 (2010)
12. Burgin, M.: Measuring Power of Algorithms, Computer Programs, and Information Automata. Nova Science Publishers, New York (2010)
13. Burgin, M., Calude, C.S., Calude, E.: Inductive Complexity Measures for Mathematical Problems. CDMTCS Research, Report 416 (2011)
14. Burgin, M., Debnath, N.C.: Complexity of algorithms and software metrics. In: Proceedings of the ISCA 18th International Conference "Computers and their Applications, pp. 259–262. International Society for Computers and their Applications, Honolulu, Hawaii (2003)
15. Calude, C.S., Calude, E., Queen, M.S.: Inductive complexity of P versus NP Problem. Unconv. Comput. Nat. Comput. Lect. Notes Comput. Sci. **7445**, 2–9 (2012)

16. Crosby, S.A., Wallach, D.S.: Denial of Service via Algorithmic Complexity Attacks, Technical Report TR-03-416. Department of Computer Science, Rice University (2003)
17. Debnath, N.C., Burgin, M.: Software metrics from the algorithmic perspective. In: Proceedings of the ISCA 18th International Conference "Computers and their Applications", pp. 279–282. Honolulu, Hawaii (2003)
18. Dewey, T.G.: The algorithmic complexity of a protein. Phys. Rev. E **54**, R39–R41 (1996)
19. Dewey, T.G.: Algorithmic complexity and thermodynamics of sequence: structure relationships in proteins. Phys. Rev. E **56**, 4545–4552 (1997)
20. Dzhunushaliev, V.D.: Kolmogorov's algorithmic complexity and its probability interpretation in quantum gravity. Class. Quantum Gravity **15**, 603–612 (1998)
21. Dzhunushaliev, V.D., Singleton, D.: Algorithmic Complexity in Cosmology and Quantum Gravity, Preprint in Physics gr-qc/0108038, 2001 (electronic edition: http://arXiv.org)
22. Hertel, J.: Inductive complexity of Goodstein's theorem. Unconv. Comput. Nat. Comput. Lect. Notes Comput. Sci. **7445**, 141–151 (2012)
23. Kolmogorov, A.N.: Three approaches to the definition of the quantity of information. Probl. Inf. Trans. **1**, 3–11 (1965)
24. Kreinovich, V., Kunin, I.A.: Application of Kolmogorov Complexity to Advanced Problems in Mechanics, University of Texas at El Paso, Computer Science Department Reports, UTEP-CS-04-14 (2004)
25. Lewis, J.P.: Limits to software estimation. Softw. Eng. Notes **26**, 54–59 (2001)
26. Li, M., Vitanyi, P.: An Introduction to Kolmogorov Complexity and its Applications. Springer, New York (1997)
27. Mansilla, R.: Algorithmic Complexity in Real Financial Markets, Preprint in Physics cond-mat/0104472 (2001) (electronic edition: http://arXiv.org)
28. Shaw, F.Z., Chen, R.F., Tsao, H.W., Yen, C.T.: Algorithmic complexity as an index of cortical function in awake and pentobarbital-anesthetized rats. J. Neurosci. Methods **93**(2), 101–110 (1999)
29. Tegmark, M.: Does the universe in fact contain almost no information? Found. Phys. Lett. **9**, 25–42 (1996)
30. Vitanyi, P.M.B.: Quantum Kolmogorov complexity based on classical descriptions. IEEE Trans. Inf. Theory **47**(6), 2464–2479 (2001)
31. Zimbardo, P.G., Weber, A.L., Jonson, R.L.: Psychology. Allyn and Bacon, San Francisco (2003)
32. Zurek, W.H.: Algorithmic randomness and physical entropy. Phys. Rev. A (3) **40**(8), 4731–4751 (1989)
33. Zurek, W.H.: Algorithmic information content, Church-Turing thesis, physical entropy, and Maxwell's demon. Information dynamics (Irsee, 1990): NATO Adv. Sci. Inst. Ser. B Phys., 256. Plenum, New York, 245–259 (1991)

Chapter 9
Asymptotic Intrinsic Universality and Natural Reprogrammability by Behavioural Emulation

Hector Zenil and Jürgen Riedel

Abstract We advance a Bayesian concept of *intrinsic asymptotic universality*, taking to its final conclusions previous conceptual and numerical work based upon a concept of a *reprogrammability* test and an investigation of the complex qualitative behaviour of computer programs. Our method may quantify the trust and confidence of the computing capabilities of natural and classical systems, and quantify computers by their degree of reprogrammability. We test the method to provide evidence in favour of a conjecture concerning the computing capabilities of Busy Beaver Turing machines as candidates for Turing universality. The method has recently been used to quantify the number of *intrinsically universal* cellular automata, with results that point towards the pervasiveness of universality due to a widespread capacity for emulation. Our method represents an unconventional approach to the classical and seminal concept of Turing universality, and it may be extended and applied in a broader context to natural computation, by (in something like the spirit of the Turing test) observing the behaviour of a system under circumstances where formal proofs of universality are difficult, if not impossible to come by.

H. Zenil (✉)
Department of Computer Science, University of Oxford, Oxford, UK
e-mail: hector.zenil@algorithmicnaturelab.org

H. Zenil
SciLifeLab, Unit of Computational Medicine, Department of Medicine
Solna, Karolinska Institute, Stockholm, Sweden

H. Zenil
Algorithmic Nature Group, LABORES, Paris, France

J. Riedel
Institut Für Physik, Universität Oldenburg, Oldenburg, Germany

J. Riedel
Algorithmic Nature Group, Laboratoire de Recherche Scientifique
LABORES, Paris, France

© Springer International Publishing Switzerland 2017
A. Adamatzky (ed.), *Advances in Unconventional Computing*,
Emergence, Complexity and Computation 22,
DOI 10.1007/978-3-319-33924-5_9

205

9.1   Introduction

Attempts to answer even the simplest questions about the behaviour of computer programs are bedevilled by uncomputability. The concept of asymptotic intrinsic universality introduced here is based upon a Bayesian approach to emulation by computer programs of other computer programs. The method provides a means to quantify their reprogramming capabilities, associating them with a deciding procedure that asymptotically recognizes computation with a confidence value and sets forth a hierarchy of reprogrammability (see [21]) based upon the likelihood of a system being, in one degree or another, close to (or removed from) Turing universality.

In [18], a related conjecture concerning other kinds of simply defined programs was presented, suggesting that all Busy Beaver Turing machines may be capable of universal computation, as they seem to share some of the informational and complex properties of systems known to be capable of universal computational behaviour.

We have recently found that most computer programs can be reprogrammed to emulate an increasing number of other (different) computer programs of the same size [8] under a similar *block emulation* transformation or set of compilers of increasing size. We also previously advanced a conceptual framework for reprogrammability based upon the display of different qualitative output behaviours [20] and modelled as a type of Turing test to determine computational capabilities [21]. This has been used in connection with an instance of natural computation–in an *in-silico* simulation of Porphyrin molecules [12] in the context of spatial computing.

Here we advance a Bayesian method, namely *asymptotic intrinsic universality*, that draws everything together and translates the seminal concept of computation universality to degrees of belief and confidence based upon emulation and reprogrammability capabilities applicable to natural computation. We test the method with a case study of the set of Turing machines defined by the Busy Beaver functions.

9.2   Methods

9.2.1   The Classical Turing Machine Model

A Turing machine consists of a finite alphabet set with symbols $\sum = \{0, 1, \ldots, k\}$ and states $\{1, 2, \ldots n\} \bigcup \{0\}$, with 0 the "halting state". The Turing machine "runs" on an one-way unbounded tape and for each pair:

- the machine's current "state" n'; and
- the tape symbol k' the machine's head is "reading".

For each pair (n', k') there is a corresponding instruction (n'', k'', d):

- a state n'' to transition into (which may be the same as the one it was in). If 0 the machine halts;

- a unique symbol k'' to write on the tape (the machine can overwrite a 1 on a 0, a 0 on a 1, a 1 on a 1, and a 0 on a 0), and
- a direction to move in d: -1 (left), 1 (right) or 0 (none, when halting).

For $k = 2$, there are $(4n + 2)^{2n}$ Turing machines with n states according to this formalism. The output string is taken from the number of contiguous cells on the tape the head has gone through.

Definition 9.2.1 We denote by $(n, 2)$ the set (or space) of all n-state 2-symbol Turing machines (with the halting state not included among the n states) and by $T(n, k)$ a specific Turing machine with n states and k symbols.

9.2.2 The Busy Beaver Functions

A *Busy Beaver Turing machine* [11] is a Turing machine that, when provided with a blank tape, does a lot of work. Formally, it is an n-state k-symbol Turing machine started on an initially blank tape that writes a maximum number of 1s or moves the head a maximum number of times upon halting. An online computer program showing the behaviour of these computer programs can be found in [16].

Most Turing machines never halt, yet Busy Beavers do halt (by definition over the empty tape). We know from algorithmic information theory that among those Turing machines that do halt, most will halt quickly or will perform very little work, yet by definition Busy Beavers are those that perform the greatest amount of work. In a recent investigation [8] focused on cellular automata (CA), we have also shown that most computer programs are candidates for intrinsic universality, and thus for Turing universality.

There are known values for all 2-symbol Busy Beavers up to 4-state Turing machines, and explicit constructions give exact or lower bounds for other state and symbol pairs.

Definition 9.2.2 If σ_T is the number of 1s on the tape of a Turing machine T upon halting, then: $\sum(n) = \max\{\sigma_T : T \in (n, 2)\ T(n)\ halts\}$.

Definition 9.2.3 If t_T is the number of steps that a machine T takes upon halting, then $S(n) = \max\{t_T : T \in (n, 2)\ T(n)\ halts\}$.

$\sum(n)$ and $S(n)$ are noncomputable functions by reduction to the halting problem. Yet values are known for $(n, 2)$ with $n \leq 4$.

Busy beavers are the Turing machines that perform more computation among the machines if their same size (by number of states but more appropriately by program length in bits) needed. This follows from Rado's definitions and it means that Busy Beavers have also the greatest Logical Depth, as defined by Bennett [1]. Yet a Busy beaver is required to halt. When running for the longest time or writing the largest number of non-blank symbols, $bb(n)$ has to be clever enough to make wise use

H. Zenil and J. Riedel

of its resources and an instruction away to halt at the end. There is thus evidence that these machines are far from trivial and that for several important measures of complexity they are among most complex, if not the most, yet their computational power is unknown and its investigation would represent a way to connect complexity to computational power. Here we undertake first steps with interesting results.

9.2.3 Block Emulation and Intrinsic Universality

The notion of *intrinsic* computational universality used for cellular automata was an adaptation of classical Turing-universality [6]. *Intrinsic universality* is stronger than Turing-universality [9, 10] and the concept can be extended and adapted to other computing systems, including computer programs in general.

Definition 9.2.4 A computer program of a given size is intrinsically universal if it is able to emulate the output behaviour of any other computer program under a coarse-graining compiler [9].

The so-called *Game of Life* is an example of 2-dimensional cellular automaton that is not only Turing-universal but also intrinsic universal [5]. This means that the Game of Life does not only compute any computable function but can also emulate the behavior of any other 2D-dimensional cellular automaton (under rescaling).

Definition 9.2.5 (*emulation/simulation by rescaling/coarse-graining*) Let A and B be two computer programs. Then A emulates/simulates B if there exists a rescaling/projection P of A such that $f_A^P = f_B$, where f_A and f_B are the computed functions of A and B. We consider P a compiler to translate A into B (see Fig. 9.1).

The exploration of the computing capabilities of computer programs can then proceed by *block emulation*, whereby the scale of space-time diagrams of a computation are found and rescaled/coarse-grained.

The emulations here explored are related to an even stronger form of *intrinsic universality*, namely linear-time intrinsic universality [9], which implies that all emulations carry only a linear overhead as a result of our brute force exploration of the compiler and rule space. This is because the coarse-graining emulation is of a block of fixed length and therefore what one can consider a compiler (another computer program of fixed size).

Following these ideas, one can try out different possible compilers and see what type of computer programs a specific computer program is able to emulate. The *linear* block transformation was suggested in [13, 14].

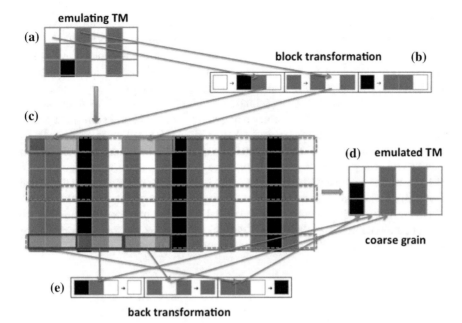

Fig. 9.1 Illustration of the process of one Turing machine emulating another via a block transformation. In this case **a** shows a $bb(2, 3)$ with initial tape ☐☐■☐■☐ after 2 steps. By performing the block transformation of length 3 **b** on the initial condition of **a**, after 6 steps using the same $bb(2, 3)$ rule one gets **c**. If the output of every 3rd step is taken and the back transformation **d** performed on these outputs, one gets the output **e**. This is identical with the output of $TM(2, 3)$ with rule number 2 797 435 run on the same initial condition as in **a** for 2 steps. In other words, **e** is the coarse-grained version of the block transformed $bb(4, 2)$ **a** which in turn produces the same output as $TM(4, 2)$ of rule number 2 797 435. In this picture one cannot see the compiler directly as it is encoded within the internal states of the $bb(4, 2)$. **a** emulating TM, **b** block transformation, **c** coarse grain, **d** emulated TM, **e** back transformation

9.2.4 Turing Machine Emulation

The exploration of the emulating space of Turing machines (TM) is more complicated than for Cellular Automata because the space-time diagram does not contain the head configuration state of the Turing machine.

We ran the random TMs and the Busy Beaver Turing machines for the number of steps given by $S(n)$. For example, for $n = 4$ states, $S(n) = 107$, given by the Busy Beaver $bb(4)$. We looked for all transformations which allow a back transformation for block sizes 2 to 4 and only considered (2-symbol, 4-state) and (3-symbol, 2-state) Busy Beaver Turing machines and a sample of random Turing machines of the same size.

To ascertain which TM from the same rule space corresponded to the emulated Busy Beaver or TM, we adopted the following algorithm:

For a n-state and k-symbol Turing machine (TM), we enumerated all possible block transformations $P(n, k)$ of given block size n (n-tuples), e.g. $P(2, 3) = \Box \rightarrow$ ■■, ■ $\rightarrow$ ■□, ■ $\rightarrow$ ■■ for a 3-symbol, 2 state TM. We found a total of k^{2n} possible transformations. We applied each member of the set of possible transformations to a TM of the corresponding rule space, in this paper that of a Busy Beaver (bb) or a randomly selected TM given a randomly initialized tape. We then let the TM evolve for n steps. Then we took every n output line of the TM and performed a back transformation on the output, e.g. $P(2, 3)^{-1} =$ ■■ $\rightarrow \Box$, ■□ $\rightarrow$ ■, ■■ $\rightarrow$ ■. At the same time we drew a TM of the same rule space out of a random sample and let it evolve for n steps using the same initial tape. If the output was a valid output of a TM, we tried to match it with the output of the Busy Beaver or random TM described above. In order to exclude trivial emulations, we filtered out all those emulated TMs which are just a n-time repetition of the initial tape. It is important to note that we are not taking the initial states of the TMs into account. We are just focusing on the output of TMs when performing the block transformations (Figs. 9.1 and 9.2).

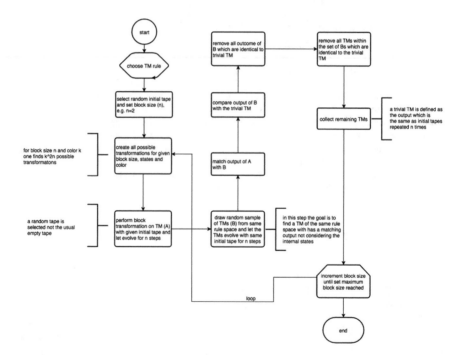

Fig. 9.2 Flow diagram of emulation of TMs

9.2.4.1 Busy Beaver Conjectures

These facts suggest the following conjectures, which are also relevant to the dynamic behaviour of a set of simply-described machines characterized by universal behaviour.

In previous work we explored these conjectures relating to Busy Beavers with numerical approximations of their sensitivity to initial conditions [18] and the qualitative behaviour that initial conditions induce over space-time diagrams [17]. Which was similar to work we did on the Game of Life [19].

Definition 9.2.6 A *weak* universal Turing machine is a machine that allows its initial tape to be in a non all-'blank' configuration.

If $bb(n)$ is weak universal, then it is allowed to start either from a periodic tape configuration or an infinite sequence produced by e.g. a finite automaton. In other words, these are machines that are Turing universal not necessarily running on non-empty tapes.

Conjecture 9.2.1 The Busy Beaver conjecture(s) as advanced in [18] establish(es) that:

- (strong version): For all $n > 2$, $bb(n)$ is Turing universal.
- (sparse version): For some $n > 2$, $bb(n)$ is Turing universal.

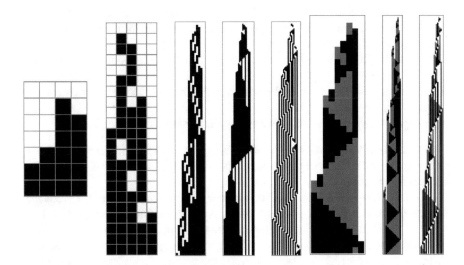

Fig. 9.3 Typical space-time evolution/behaviour of *Busy Beaver Turing machines*. The first 6 figures from *left* to *right* correspond to Busy Beaver machines with 2-symbols and 2 to 6 states (for illustration purposes only those <4 were plotted with a background mesh) for which the first 3 have exact ($S(n)$) runtime values (6, 21, 107). For the rest a cutoff value was arbitrarily chosen, so as to provide an optimally effective illustration. The behaviour of a Busy Beaver cannot be a trivial repetition because it does have to avoid getting into an infinite cycle in order to halt

- (weak version): For all $n > 2$, $bb(n)$ is weak Turing universal.
- (weakest version): For some $n > 2$, $bb(n)$ is (weak) Turing universal.

Here we provide evidence in favour of all conjectures in the form of an increasingly monotonic asymptotic intrinsic universal behaviour.

It is known that among all 2-state 2-symbol Turing machines, none can be universal. $bb(n)$, as defined by Rado [11], is a Turing machine with n states plus the halting state. $bb(2)$ is thus actually $bb(2, 3)$, a 3-state 2-symbol machine in which one state is specially reserved for halting only. If bb is unary, then it will be assumed to be a 2-symbol Turing machine, otherwise it will be denoted by $bb(n, k)$ (Fig. 9.3).

9.3   Results

9.3.1   A Bayesian Approach to Turing Universality

We looked into the number of compilers up to a certain size for which a computer program can emulate other computer programs of the same size (e.g. in number of states for TMs, number neighbours for CAs, or description bits in general). Given all the unknown priors and the uncertainty in the degree of belief, we need a basic function that:

- Is increasingly monotonic. Normalizing by total number of explored compilers should provide a measure for comparison, but the function itself should only count the number of emulations.
- $f(x) > 0$ when $x > 0$. Evidently any emulation should amount to a non-zero value.
- Nonlinearly converges to 1. We want a function that "slowly" converges to a positive value and
- Incorporates a degree of belief weighting the number of emulations found.

Because intrinsic universality implies Turing universality [9], this approach is of relevance in finding the reprogramming capabilities of classical and unconventional computing systems.

The exact shape of the function has no essential meaning as long as it is concave and complies with the above requirements. A canonical function is $ax/(ax + 1)$, where $x \in \mathbb{N}^+$ is the number of different non-trivial emulations of a system under evaluation and $a \in (0, 1]$ the degree of belief modifying the rate of convergence, in this case $a = 1$ (see Fig. 9.4). We then define the asymptotic intrinsic universality of a computing system s as:

Definition 9.3.1 (*asymptotic intrinsic universality*) Let s be a computing machine of fixed size and x the number of non-equivalent (e.g. under coarse-graining) emulations of other computing systems of the same size that s can emulate, then

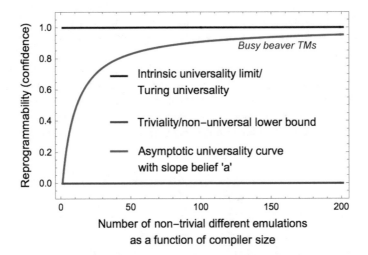

Fig. 9.4 Asymptotic intrinsic universality curve ($ax/(ax + 1)$, made continuous for illustration purposes) is the Bayesian approach to the otherwise seminal but abstract concept of computation universality applicable to both abstract and natural/unconventional computation. For example, we found evidence in favour of a conjecture postulating that Busy Beaver Turing machines are somewhere on the asymptotic universality curve, highly so if the degree of belief according to *a* assigns it a higher confidence every time that such a machine in question is able to emulate some other

$\Delta(s) = ax/(ax + 1)$ is the function that retrieves a confidence value of reprogrammability based upon the intrinsic universality of *s* according to belief *a*.

9.3.2 Case Study: Busy Beaver Functions

Here we provide evidence in favour of the Busy Beaver conjectures by way of the different qualitative behavioural properties they display and their intrinsic universality capabilities.

9.3.2.1 Qualitative Behaviour Analysis

Among the intuitions suggesting the truth of one of these conjectures, is that it is easier to find a machine capable of halting and performing unbounded computations for a Turing machine if the machine already halts after performing a sophisticated calculation, than it is to find a machine showing sophisticated behaviour whose previous characteristic was simply to halt. This claim can actually be quantified, given that the number of Turing machines that halt after $t = n$ for increasing values of *n* decreases exponentially [2, 4, 22]. In other words, if a machine capable of halting is chosen by chance, there is an exponentially increasing chance of finding that it will halt sooner rather than later, meaning that most of these machines will behave trivially because they will not have enough time to do anything interesting before halting.

Figure 9.5 provides a summation of the behavioural investigation of Busy Beaver machines. Histograms show the different qualitative behaviour in bimodal and multimodal discrete distributions. The multimodality is not an effect of the size of the initial condition that grows smoothly by $\log(n)$, nor of the stepwise behaviour of the lossless compression algorithm (Compress based upon Deflate). If it were an effect

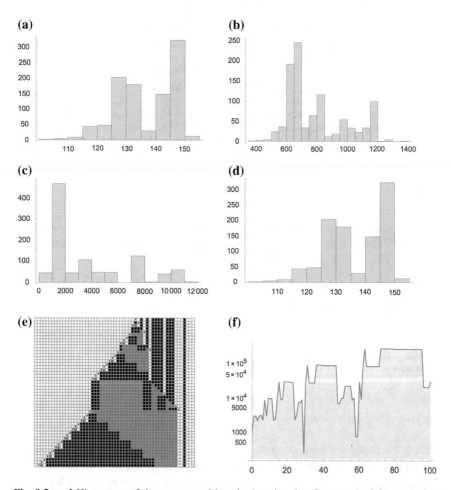

Fig. 9.5 **a–d** Histograms of the compressed lengths (x axis using Compress) of the space-time diagrams of $bb(n)$ for $n = 3$ to 6 for 1×10^3 steps each, showing accumulation of different qualitative behaviours. **e** A right-left compressed behaviour of a Busy Beaver runnning for 1.5×10^3. Only rows for which the head has moved further to the right or left than ever before are kept, a method suggested in [15]. **f** Function computed by the Busy Beaver $b(5, 2)$ for consecutive initial conditions 1 to 100 in binary

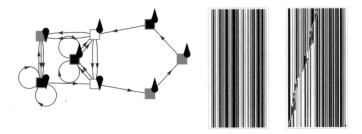

Fig. 9.6 *Left* State diagram after 20 steps (state 1 is a down-tick, state 2 is an up-tick). *Right* Two runs from different "random" initial conditions of length 100 bits showing (*left*) a quick halting (computation of the identity) and (*right*) an apparently random movement of the head for another initial condition running on the same 4-state 3-symbol Busy Beaver Turing machine

of the length of the initial condition, then Fig. 9.5b–d would look like Fig. 9.5a, which is not the case. They display genuinely different behaviours (see Fig. 9.6(right)).

The state diagram in Fig. 9.6(left) suggests how to choose an initial configuration for the machine to enter into an infinite loop (e.g. connected cycle on the left), and therefore how to enter into a never-halting computation, a requirement for (weak) Turing universality. Fig. 9.6(right) shows the behaviour of $bb(4, 3)$ for 2 different initial conditions, one for which it halts (or "computes" the identity) and another for which the computation goes on in a rather complex head movement fashion.

9.3.2.2 Reprogrammability of Busy Beavers by Block Emulation

Figure 9.7 shows that Busy Beavers are much more capable of emulating the behaviour of other (non-trivial) Turing machines than the control case, a sample of random Turing machines from the same rule space size (i.e. all machines are of the same size). This is consonant with theoretical expectations [2].

9.3.2.3 Busy Beavers Are Candidates for Turing Universality

The capacity for universal behaviour implies that a system is capable of being reprogrammed and is therefore reactive to external input. It is no surprise that universal systems should be capable of responding to their input and doing so succinctly, if the systems in question are efficient universal systems. If the system is incapable of reacting to any input or if the output is predictable (decidable) for any input, the system cannot be universal.

We have here provided evidence that Busy Beavers comply with all the requirements for Turing universality and must therefore be considered a very interesting non-trivial set of Turing machines that are candidates for Turing universality.

Evidence in favour of the conjectures is based upon the following observations:

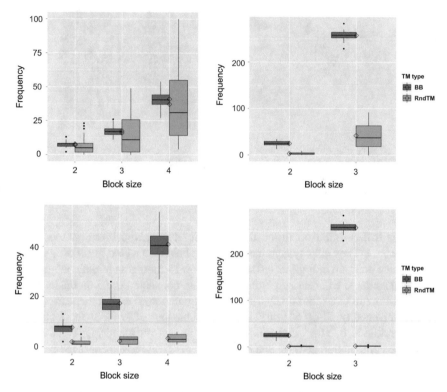

Fig. 9.7 *Top* Boxplots showing the differences in the emulation power of (*left*) $bb(4, 2)$ versus a set of randomly selected TMs in $(4, 2)$, and (*right*) $bb(2, 3)$ versus a set of randomly selected TMs in $(2, 3)$. The data show how many emulations on average a set of Busy Beavers of a given rule space and a set of random TMs selected from the same rule space can produce for given block sizes. The data shows a variance for both TM types, since the output of valid block transformations is compared with the output of a TM sample taken from the same rule space. Trivial TMs (c.f. flow chart in Fig. 9.2) are excluded. Each emulation is counted, even if it corresponds to the same TM. The *diamond shapes* represent the mean of the data points. *Bottom* Same plots, but only TM evolutions with different hash values (from their evolution) are counted, i.e. only distinct TMs are counted, rendering the difference between Busy Beavers and random Turing machines even more prominent

- Busy beavers produce space-time diagrams of the highest complexity compared to the space-time diagrams of other rules in the same rule space.
- Busy beavers show qualitatively different behaviour for different initial conditions; they can halt and it is not difficult to devise ways to perform non-halting computations based upon infinite loops, especially for non-empty inputs.
- The small set of Busy Beavers investigated emulate a larger number of other (non-trivial) Turing machines on average compared to random Turing machines of the same size. In other words, we found evidence indicating that $\Delta(bb(n, k)) >$

$\Delta(RndTM(n, k))$, where $RndTM(n, k) \in bb^c$ is a random Turing machine in the complement set of the Busy Beavers bb^c, and the confidence level a is fixed.

- Thus the measure of asymptotic intrinsic universality that we defined $\Delta(bb(n, k))$ converges to 1 much faster than $\Delta(RndTM(n, k))$.

Asymptotic intrinsic universality is strictly stronger than Turing universality. Fig. 9.7: Random TM statistics serve as a control because we know that the set of machines that either quickly halt or never halt are of density measure 1, and will therefore end up dominating the average emulation with $\Delta(RndTM(n, k))/nk \sim 0$ for $n, k \to \infty$. So if we find, as we in fact did, that $\Delta(bb(n, k))/nk$ grows faster than $\Delta(RndTM(n, k))/nk$, we would be demonstrating with a high level of confidence that Busy Beaver Turing machines have greater reprogramming capabilities and are candidates for intrinsic universality, and therefore Turing universality.

9.4 Discussion

9.4.1 Universality Versus Reprogrammability in Natural Computation

We have brought together several concepts that are relevant and applicable to natural computation where, e.g., resources are often scarce and computation occurs independently of the substrate, making for concepts that are disembodied, independent not only of specific hardware but of models and formalisms (e.g. whether one can define a halting configuration).

On the one hand, there is the use of the concept of intrinsic universality, which our definition of asymptotic universality relies upon. Intrinsic universality as originally formulated for cellular automata does not require a halting configuration. This makes it applicable to natural computation, because a halting state is an arbitrary choice–the option to design a state as a halting one–which is meaningless in natural computation. Furthermore, the coarse-graining only takes into consideration the output configuration rather than the state configuration, which is consistent with recent extensions for *membrane computing* or P systems [7], where a computation with only one possible output can be reached through many different paths, regardless of the internal states transited through en route. Indeed, since we are not looking 'inside' the TM (its internal states), we are treating it as a black box (see Fig. 9.1) on which we perform an external observer test. The compiler used to look at the internal states is a behaviourally shallow one. Interestingly, the transformed TM (in Fig. 9.1, a Busy Beaver) does lock immediately into the same pattern. Again, one would need to visualize the internal states to see a difference between other emulations producing the same output.

On the other hand, in a world where "emptiness" or simple/completely regular initial conditions cannot be guaranteed, weak universality is more realistic. The concept of asymptotic universality is based upon and adapted to deal with these situations in the context of natural computation where a system may be a black box but its behaviour can be reinterpreted (by emulation) and exploited. Of course one difficulty is to identify different behaviours in order to undertake a behavioural comparison, and this is why we have also introduced complexity indices that can serve as tools to quantify the space-time evolutions of systems or their representations.

A limitation of any empirical approach is that of a non-universal system able to emulate an increasing number of non-universal systems may not be universal but an open question is whether there is a threshold N above which a system is universal after simulating N number of other systems. This is also why the emulation results should be complemented by an analysis of the complexity of the emulations themselves. While we do know that, for example, finite automata are bounded by the kind of complexity they can produce by the complexity of the set of regular languages, the connection between these qualitative differences and the computational power of the systems in this context is a future subject of further investigation.

The chief advantage of this approach is the amenability to non-formal evidence of the reprogrammability of less conventional systems, where formal proofs of universality are difficult, if not impossible to come by. In other words, while the method does disclose universal systems at the limit, it does not rule out non-universal ones, thus producing possible false positives. However limited, any false positive is still a reprogrammable system, thereby providing a more natural/pragmatic definition of *natural universality*.

9.4.2 The Busy Beaver Conjectures

It would not have been possible to anticipate that the behaviour displayed would have been that of Busy Beavers, despite their complexity for empty inputs. Nor could the low emulation capabilities of all other trivial and non-trivial machines in the complement set of the Busy Beavers bb^c have been anticipated, because they are no longer being tested and quantified over the full set of possible initial conditions but over the subset that allow the emulation of other computer programs (Turing machines) of the same (growing) size. In other words, what we are exploring is the Cartesian product $P \times C$ of the pairs (p, c), where $p \in P$ is a computer program (e.g. a Turing machine) and $c \in C$ a compiler that maps p onto $p' \in P$ of size $|p'| = |p|$ (in this case the number of states, but in the general the number of bits, i.e. its Kolmogorov complexity [3]).

Here we explored the reprogrammable space, a subset of the the space of all computer programs for either a specific input or, equivalently (per Turing universality), for all inputs. This also means that most of the machines that either halt almost immediately and therefore do nothing interesting, or else never halt, can actually be effectively reprogrammed, and the results obtained here and in [8, 17] strongly suggest that they may even be candidates for intrinsic universality (i.e. the ability to emulate any other computer program under a coarse-graining compiler), a stronger concept than that of Turing universality.

9.5   Conclusion

The set of Busy Beaver machines describes an (enumerable) infinite set of Turing machines characterized by a particular specific behaviour. If the conjectures are true according to the evidence we have provided, the result is more surprising, because a describable property determines the computational power of this non-trivial infinite set of Turing machines. Here we have taken these ideas a step further in the direction of an empirical proposal for considering statistical computational evidence of computational universality. Because of the undecidability of the halting problem we may never obtain stronger evidence of the computational capabilities of these computer programs.

We have introduced a novel experimental and methodological Bayesian approach to theoretical computing challenges that circumvents traditional limitations imposed by classical definitions, in particular related to undecidability, unreachability and universality and deals with pragmatic unconventional reprogramming by behavioural emulation rather than through attempting producing formal analytical proofs, which are not only difficult, but impossible in general, specially in the realm of natural computation where we think these new concepts and methods are more relevant.

References

1. Bennett, C.H.: Logical depth and physical complexity. The Universal Turing Machine A Half-Century Survey, pp. 207–235. Springer, Heidelberg (1995)
2. Calude, C.S., Stay, M.A.: Most programs stop quickly or never halt. Adv. Appl. Math. **40**(3), 295–308 (2008)
3. Chaitin, G.J.: Computing the Busy Beaver Function. Open Problems in Communication and Computation, pp. 108–112. Springer, New York (1987)
4. Delahaye, J.-P., Zenil, H.: Numerical Evaluation of the Complexity of Short Strings: a Glance Into the Innermost Structure of Algorithmic Randomness. Appl. Math. Comput. **219**, 63–77 (2012)
5. Durand, B., Zsuzsanna R.: The game of life: universality revisited. Cellular Automata, pp. 51–74. Springer, Heidelberg (1999)
6. von Neumann J.: Theory of Self-reproducing Automata, Burks A.W. (ed.), University of Illinois Press, Urbana, Ill., (1966)
7. Păun, G.: Introduction to Membrane Computing. Applications of Membrane Computing. pp. 1–42. Springer, Berlin (2006)
8. Riedel J., Zenil H.: Cross-boundary Behavioural Reprogrammability Reveals Evidence of Pervasive Turing Universality, (ArXiv preprint). arXiv:1510.01671
9. Ollinger N.: The quest for small universal cellular automata. In: Widmayer P., Triguero F., Morales R, Hennessy M., Eidenbenz S. and Conejo R. (eds.) Proceedings of International Colloquium on Automata, languages and programming (ICALP'2002), LNCS, vol. 2380, pp. 318–329, (2002)

10. Ollinger N., The intrinsic universality problem of one-dimensional cellular automata, *Symposium on Theoretical Aspects of Computer Science (STACS'2003)*, Alt H. and Habib M. (eds.), *LNCS* 2607, pp. 632–641, 2003
11. Rado, T.: On Noncomputable Functions. Bell Syst. Tech. J. **41**(3), 877–884 (1962)
12. Terrazas, G., Zenil, H., Krasnogor, N.: Exploring Programmable Self-Assembly in Non DNA-based Computing. Nat. Comput. **12**(4), 499–515 (2013)
13. Wolfram, S.: Table of Cellular Automaton Properties. Theory and Applications of Cellular Automata, pp. 485–557. World Scientific, Singapore (1986)
14. Wolfram, S.: Statistical mechanics of cellular automata. Rev. Mod. Phys. **55**, 601–644 (1983)
15. Wolfram, S.: A New Kind of Science. Wolfram Science. Il, Chicago (2002)
16. Zenil, H., Beaver, B.: Wolfram Demonstrations Project. http://demonstrations.wolfram.com/BusyBeaver/
17. Zenil, H., Villarreal-Zapata, E.: Asymptotic behaviour and ratios of complexity in cellular automata rule spaces. Int. J. Bifurc. Chaos, **13**(9) (2013)
18. Zenil, H.: On the Dynamic Qualitative Behaviour of Universal Computation. Complex Systems **20**(3), 265–277 (2012)
19. Zenil, H.: Algorithmicity and programmability in natural computing with the game of life as an In silico case study. J. Exp. Theor. Artif. Intell. **27**(1), 109–121 (2015)
20. Zenil, H.: A behavioural foundation for natural computing and a programmability test. In: Dodig-Crnkovic, G., Giovagnoli, R. (eds.) Computing Nature: Turing Centenary Perspective, SAPERE Series, vol. 7, pp. 87–113. Springer, Heidelberg (2013)
21. Zenil, H.: What is nature-like computation? A behavioural approach and a notion of programmability. Philos. Tech. **27**(3), 399–421 (2014) (online: 2012)
22. Zenil, H.: From computer runtimes to the length of proofs: With an algorithmic probabilistic application to waiting times in automatic theorem proving. In: Dinneen, M.J., Khousainov, B., Nies, A. (eds.) Computation, Physics and Beyond International Workshop on Theoretical Computer Science, WTCS 2012, LNCS, vol. 7160, pp. 223–240. Springer, Heidelberg (2012)

Chapter 10
Two Small Universal Reversible Turing Machines

Kenichi Morita

Abstract We study the problem of constructing small universal Turing machines (UTMs) under the constraint of *reversibility*, which is a property closely related to physical reversibility. Let URTM(m,n) denote an m-state n-symbol universal reversible Turing machine (URTM). Then, the problem is to find URTM(m,n) with small m and n. So far, several kinds of small URTMs have been given. Here, we newly construct two small URTMs. They are URTM(13,7) and URTM(10,8) that can simulate cyclic tag systems, a kind of universal string rewriting systems proposed by Cook. We show how these URTMs can be designed, and compare them with other existing URTMs.

10.1 Introduction

A universal Turing machine (UTM) is a TM that can simulate any TM. Turing himself showed it is possible to construct such a machine [21]. Since then, there have been many researches on UTMs, in particular, on finding small UTMs. If we write an m-state n-symbol UTM by UTM(m,n), then the problem is to find UTM(m,n) with small values of m and n. This problem attracted many researchers, and various small UTMs have been presented till now (see e.g., a survey paper [23]). In the early stage of this study, a direct simulation method of TMs was employed. Later, an indirect method of simulating universal systems that are much simpler than TMs was proposed to construct very small UTMs. Minsky [7] presented a method of simulating 2-tag systems, which are universal string rewriting systems, and gave a UTM that has seven states and four symbols. After that, such an indirect simulation method has mainly been used to give small UTMs. Rogozhin [20] designed small UTMs for many pairs of m and n. They are UTM(24,2), UTM(10,3), UTM(7,4), UTM(5,5), UTM(4,6), UTM(3,10), and UTM(2,18) that also simulate 2-tag systems. Some of these results were improved later. Kudlek and Rogozhin [6] gave UTM(3,9) that simulates 2-tag

K. Morita (✉)
Hiroshima University, Higashi-hiroshima 739-8527, Japan
e-mail: km@hiroshima-u.ac.jp

© Springer International Publishing Switzerland 2017   221
A. Adamatzky (ed.), *Advances in Unconventional Computing*,
Emergence, Complexity and Computation 22,
DOI 10.1007/978-3-319-33924-5_10

systems, and Neary and Woods [18] constructed UTM(15,2), UTM(9,3), UTM(6,4), and UTM(5,5) that simulate bi-tag systems.

Here, we study the problem of constructing small universal reversible Turing machines (URTMs). *Reversible computing* is a paradigm of computing that reflects physical reversibility, one of the fundamental microscopic properties of physical systems. It is thus related to quantum computing, since the evolution of a quantum system is reversible. A reversible Turing machine (RTM) is a standard model in the theory of reversible computing. In fact, it was shown by Bennett [2] that for any (irreversible) TM, there is an RTM that simulates the former.

Roughly speaking, an RTM is a "backward deterministic" TM, where each computational configuration has at most one predecessor (a precise definition will be given in the next section). Although its definition is simple, it has a close relation to reversible physical systems. It has been shown that any RTM can be implemented as a circuit composed of reversible logic element with 1-bit memory very simply [8, 11, 14]. It is also known that a reversible logic element with 1-bit memory can be realized in the billiard ball model (BBM) [9, 17]. BBM is an idealized mechanical model of a reversible physical system proposed by Fredkin and Toffoli [5], where computing is carried out by collisions of balls and reflectors. Hence, the whole system of an RTM can be embedded in such a reversible physical model.

So far, there have been several researches on small URTMs. Let URTM(m,n) denote an m-state n-symbol URTM. Morita and Yamaguchi [15] first constructed URTM(17,5) that simulates cyclic tag systems, which are another kind of universal string rewriting systems proposed by Cook [4]. Axelsen and Glück [1] studied a different type of URTM that computes all computable injective functions, but their objective was not finding a small URTM. Later, Morita [9, 13] constructed URTM(15,6), URTM(24,4), and URTM(32,3), which also simulate cyclic tag systems. On the other hand, it is in general difficult to design simple URTMs with only two symbols, or with a very small number of states. As for a 2-symbol URTM, we can use a general procedure for converting a many-symbol RTM into a 2-symbol RTM [16]. By this, we obtain URTM(138,2) [13]. In [10], methods for converting an m-state n-symbol RTM into 4-state $(2mn + n)$-symbol RTM, and 3-state $O(m^2 n^3)$-state RTM were given. Applying these methods to URTM(17,5) and URTM(32,3), we obtain URTM(4,175) and URTM(3,36654), respectively. Note that, in Sect. 10.3, we newly give URTM(10,8). Applying the conversion method to it, we have URTM(4,168) that is slightly simpler than URTM(4,175).

In this chapter, we give two new URTMs, and explain how we can design small URTMs. They are URTM(13,7), and URTM(10,8), which again simulate cyclic tag systems. In Sect. 10.2, we give basic definitions on RTMs, 2-tag systems (2-TSs), and cyclic tag systems with halting condition (CTSHs). We also show how a CTSH can simulate a 2-TS. In Sect. 10.3, we construct URTM(13,7) and URTM(10,8). In Sect. 10.4, we compare these two URTMs with other small URTMs, and summarize the results.

10.2   Reversible Turing Machines and Tag Systems

In this section, we give definitions and basic properties on reversible Turing machines, m-tag systems, and cyclic tag systems.

10.2.1   Reversible Turing Machines

There are two kinds of formulations for reversible Turing machines. They are the quadruple formulation [2], and the quintuple formulation [9]. They can be easily converted each other keeping reversibility. Here, we use the quintuple formulation, because the number of states of a Turing machine of this form can be about a half of that in the quadruple form. Also, most classical universal Turing machines are given in the quintuple form.

Definition 1 A *one-tape Turing machine* (TM) in the quintuple form is defined by

$$T = (Q, S, q_0, s_0, \delta),$$

where Q is a non-empty finite set of states, S is a non-empty finite set of symbols, q_0 is an initial state ($q_0 \in Q$), s_0 is a special blank symbol ($s_0 \in S$). δ is a move relation, which is a subset of $(Q \times S \times S \times \{-, +\} \times Q)$. The symbols "$-$", and "$+$" are shift directions of the head, which stand for "left-shift", and "right-shift", respectively. Each element of δ is a quintuple of the form $[p, s, s', d, q]$. It means if T reads the symbol s in the state p, then writes s', shifts the head to the direction d, and goes to the state q.

In the above definition, final states that halt for any symbol are not specified according to the designing convention of universal Turing machines (in other words, final states are not counted in the number of states of a universal Turing machine). Also note that δ is defined as a "relation" rather than a "function", This is because determinism and reversibility will be defined almost symmetrically by this definition (but, they are slightly asymmetric since the head-shift operation is performed after the read/write operation).

Let $w \in S^*, q \in Q$, and $h \in \{0, 1, \ldots, |w| - 1\}$. A triplet $[w, q, h]$ is called a *computational configuration* (or simply a *configuration*) of $T = (Q, S, q_0, s_0, \delta)$. The configuration $[w, q, h]$ means that the tape contains w (all the other squares of the tape have the blank symbol s_0), the state is q, and the head position is at the hth symbol of w (the position of the leftmost symbol of w is the 0-th). In the following, we use such an expression to write a configuration of T.

Determinism and reversibility of TM is defined as follows. T is called a *deterministic TM* iff the following holds for any pair of distinct quintuples $[p_1, s_1, s_1', d_1, q_1]$ and $[p_2, s_2, s_2', d_2, q_2]$ in δ.

$$\text{If } p_1 = p_2, \text{ then } s_1 \neq s_2$$

T is called a *reversible TM* iff the following holds for any pair of distinct quintuples $[p_1, s_1, s'_1, d_1, q_1]$ and $[p_2, s_2, s'_2, d_2, q_2]$ in δ.

$$\text{If } q_1 = q_2, \text{ then } s'_1 \neq s'_2 \wedge d_1 = d_2$$

The above is called the *reversibility condition*. It is easy to see that if T is reversible, then there is at most one reversely applicable quintuple to each configuration, and thus every configuration of T has at most one predecessor.

In the following, we consider only deterministic (irreversible or reversible) TMs, and thus the word "deterministic" is omitted. Hence, by a "reversible TM" (RTM), we mean a deterministic reversible TM.

10.2.2 *m-Tag Systems*

A tag system is a string rewriting system originally proposed by Post [19], and an *m*-tag system ($m = 1, 2, \ldots$) is a variant of it. In the *m*-tag system, rewriting of strings is performed in the following way. Let $\alpha = a_1 \ldots a_n$ be a string over an alphabet A. If the system has a production rule $a_1 \rightarrow b_1 \ldots b_k$ and $n \geq m$, then we can obtain a new string $a_{m+1} \ldots a_n b_1 \ldots b_k$. Namely, if the first symbol of α is a_1 and $|\alpha| \geq m$, then remove the leftmost m symbols, and append the string $b_1 \ldots b_k$ at the right end of it as shown in Fig. 10.1. Repeating this procedure, we can obtain new strings successively. If we reach a string β to which there is no applicable production rule, or $|\beta| < m$, then the rewriting process terminates.

We now define *m*-tag systems based on the definition by Rogozhin [20].

Definition 2 An *m-tag system* (*m*-TS) is defined by $T = (m, A, P)$, where m is a positive integer, A is a finite alphabet, and $P : A \rightarrow A^* \cup \{halt\}$ is a mapping that gives a set of production rules (we assume $halt \notin A$). Let $a \in A$. If $P(a) = b_1 \ldots b_k \in A^*$, we write it by $a \rightarrow b_1 \ldots b_k$, and call it a *production rule* of T. If $P(a) = halt$, then a is called a *halting symbol*. We usually write P as the set of production rules: $\{a \rightarrow P(a) \mid a \in A \wedge P(a) \neq halt\}$.

The *transition relation* $\underset{T}{\Rightarrow}$ on A^* is defined as follows. For any $a_1, \ldots, a_m, a_{m+1}, \ldots, a_n, b_1, \ldots, b_k \in A$ such that $n \geq m$,

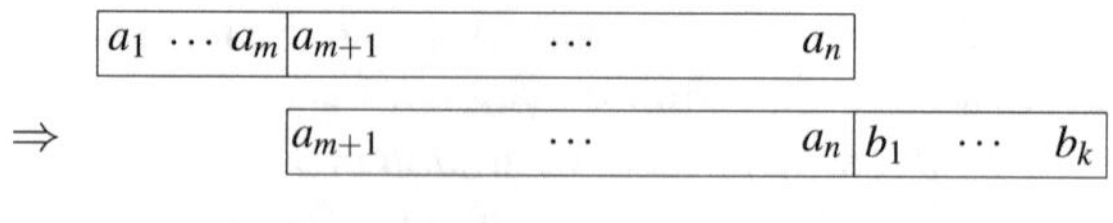

Fig. 10.1 Rewriting in an *m*-TS. If there is a production rule $a_1 \rightarrow b_1 \ldots b_k$ and $n \geq m$, then the first m symbols are removed, and the string $b_1 \ldots b_k$ is appended at the *right end*. If a_1 is a halting symbol or $n < m$, then the rewriting process terminates

$$a_1 \ldots a_m a_{m+1} \ldots a_n \underset{T}{\Rightarrow} a_{m+1} \ldots a_n b_1 \ldots b_k \quad \text{iff} \quad a_1 \to b_1 \ldots b_k \in P.$$

When there is no ambiguity, we use $\Rightarrow$ instead of $\underset{T}{\Rightarrow}$. Let $\alpha \in A^*$. By the above definition of $\Rightarrow$, if the first symbol of α is a halting symbol, or $|\alpha| < m$, then there is no $\alpha' \in A^*$ that satisfy $\alpha \Rightarrow \alpha'$. Such α is called a *halting string* or a *final string*. The reflexive and transitive closure of $\Rightarrow$ is denoted by $\overset{*}{\Rightarrow}$. Let $\alpha_i \in A^*$ ($i \in \{0, 1, \ldots, n\}$, $n \in \mathbb{N}$). We say $\alpha_0 \Rightarrow \alpha_1 \Rightarrow \cdots \Rightarrow \alpha_n$ is a *complete computing process* of T starting from α_0 if α_n is a halting string.

In [3, 7], it is shown that for any TM there is a 2-TS that simulates the TM. Hence, the class of 2-TS is computationally universal.

Theorem 1 ([3, 7]) *For any one-tape two-symbol TM, there is a 2-TS that simulates the TM.*

10.2.3 Cyclic Tag Systems

A cyclic tag system (CTS) is a variant of a tag system proposed by Cook [4]. He used CTS to prove computational universality of the elementary cellular automaton of rule 110. Since CTS has two kinds of symbols, we fix its alphabet as $\{Y, N\}$. In CTS, there are k ($= 1, 2, \ldots$) production rules $Y \to w_0, Y \to w_1, \ldots, Y \to w_{k-1}$, which are used one by one cyclically in this order. More precisely, the pth production rule $Y \to w_p$ is applicable at time t, if $p = t \bmod k$. If the first symbol of the string at time t is Y, then it is removed, and w_p is appended at the end of the string. On the other hand, if the first symbol is N, then it is removed, and nothing is appended. Hence, we assume that the production rule $N \to \lambda$ is always applicable. Figure 10.2 shows this process. In the following, we write the set of production rules as $(w_0, \ldots, w_{k-1})$, since the left-hand side of each production rule is always Y. CTSs are simpler than m-TSs because of the following reasons: they have only two kinds of symbols Y and N, production rules are used one by one in the specified order (hence there is no need of table lookup), and a string is appended only when the first symbol is Y.

In the original definition of a CTS in [4], the notion of halting was not defined explicitly. In fact, it halts only if the string becomes the empty string λ. Hence, the final configuration of simulated TM cannot be retrieved from the halting string (i.e., λ) of a CTS. Therefore, when we use CTS as an intermediate system for making a UTM, then some halting mechanism should be incorporated. Though there will be several ways of defining the notion of halting, we use the method employed in [15], which is given in the following definition.

Definition 3 A *cyclic tag system with halting condition* (CTSH) is a system defined by

$$C = (k, (\mathrm{halt}, w_1, \ldots, w_{k-1})),$$

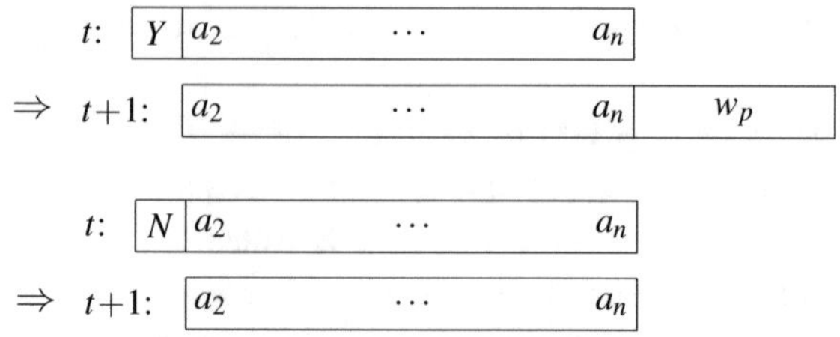

$$
\begin{array}{ll}
t: & \boxed{Y \mid a_2 \quad\cdots\quad a_n} \\
\Rightarrow\; t+1: & \boxed{a_2 \quad\cdots\quad a_n \mid w_p}
\end{array}
$$

$$
\begin{array}{ll}
t: & \boxed{N \mid a_2 \quad\cdots\quad a_n} \\
\Rightarrow\; t+1: & \boxed{a_2 \quad\cdots\quad a_n}
\end{array}
$$

Fig. 10.2 Rewriting in a cyclic tag system at time t. Here, we assume its cycle length is k, and the pth production rule is $y \to w_p$, where $p = t \bmod k$. If the first symbol of the string at time t is Y, then it is removed, and w_p is appended at the *right end*. If the first symbol is N, then it is removed, and nothing is appended

where $k \in \mathbb{Z}_+$ is the length of a cycle, and $(w_1, \ldots, w_{k-1}) \in (\{Y, N\}^*)^{k-1}$ is a $(k-1)$-tuple of *production rules*. A pair (v, m) is an *instantaneous description* (ID) of C, where $v \in \{Y, N\}^*$ and $m \in \{0, \ldots, k-1\}$. m is called the *phase* of the ID. The transition relation $\underset{C}{\Rightarrow}$ is defined below. For any $v \in \{Y, N\}^*$, $m, m' \in \{0, \ldots, k-1\}$,

$$(Yv, m) \underset{C}{\Rightarrow} (vw_m, m') \text{ iff } (m \neq 0) \wedge (m' = m+1 \bmod k),$$
$$(Nv, m) \underset{C}{\Rightarrow} (v, m') \quad \text{ iff } m' = m+1 \bmod k.$$

If there is no ambiguity, we use $\Rightarrow$ instead of $\underset{C}{\Rightarrow}$. By the definition of $\Rightarrow$, we can see that, for any $v \in \{Y, N\}^*$ and $m \in \{0, \ldots, k-1\}$, IDs $(Yv, 0)$ and (λ, m) have no successor ID. Hence, an ID of the form $(Yv, 0)$ or (λ, m) is called a *halting ID*. Let $v_i \in \{Y, N\}^*$, $m_i \in \{0, \ldots, k-1\}$ ($i \in \{0, 1, \ldots, n\}, n \in \mathbb{N}$). We say $(v_0, m_0) \Rightarrow (v_1, m_1) \Rightarrow \cdots \Rightarrow (v_n, m_n)$ is a *complete computing process* of C starting from an *initial string* v if $(v_0, m_0) = (v, 0)$ and (v_n, m_n) is a halting ID. Here, v_n is called a *final string*. The reflexive and transitive closure of $\Rightarrow$ is denoted by $\overset{*}{\Rightarrow}$. An n-step transition is denoted by $\overset{n}{\Rightarrow}$.

We give a simple example of a CTSH $\hat{C}$ in Example 1. In Sect. 10.3 it will be used to explain how constructed URTM simulate CTSHs.

Example 1 Consider a CTSH $\hat{C} = (3, (\text{halt}, NY, NNY))$. A complete computing process of $\hat{C}$ starting from the initial string NYY is as follows.

$$
\begin{aligned}
(NYY, 0) \quad &\Rightarrow (YY, 1) \quad &&\Rightarrow (Y\,NY, 2) \\
\Rightarrow (NY\,NNY, 0) \quad &\Rightarrow (Y\,NNY, 1) &&\Rightarrow (NNY\,NY, 2) \\
\Rightarrow (NY\,NY, 0) \quad &\Rightarrow (Y\,NY, 1) &&\Rightarrow (NY\,NY, 2) \\
\Rightarrow (Y\,NY, 0) \quad &&&
\end{aligned}
$$

The last ID $(YNY, 0)$ is a halting ID, and YNY is the final string. □

We now show that any 2-TS can be simulated by a CTSH. Thus, from Theorem 1, the class of CTSHs is computationally universal. The proof method is due to Cook [4] except that halting of CTSH is properly managed here.

Theorem 2 *For any 2-TS T, we can construct a CTSH C that simulates T.*

Proof Let $T = (2, A, P)$. We define A_N and A_H as follows: $A_N = \{a \mid P(a) \neq \text{halt}\}$ and $A_H = \{a \mid P(a) = \text{halt}\}$. They are the sets of non-halting symbols, and halting symbols, respectively. Thus, $A = A_N \cup A_H$. We denote $A_N = \{a_1, \ldots, a_n\}$ and $A_H = \{b_0, \ldots, b_{h-1}\}$. Let $k = \max\{n, \lceil \log_2 h \rceil\}$. Let $\text{bin}_k : \{0, \ldots, 2^k - 1\} \to \{N, Y\}^k$ be the function that maps an integer j ($0 \leq j \leq 2^k - 1$) to the k-bit binary number represented by N and Y, where N and Y stand for 0 and 1, respectively. For example, $\text{bin}_4(12) = YYNN$. Now, we define a coding function $\varphi : A^* \to \{N, Y\}^*$. It is a string homomorphism that satisfies the following.

$$\varphi(a_i) = N^i Y N^{k-i} \ (1 \leq i \leq n)$$
$$\varphi(b_i) = Y \text{bin}_k(i) \ (0 \leq i \leq h - 1)$$

Namely, each symbol in A is coded into a string of length $k + 1$ over $\{N, Y\}$. Now, the CTSH C that simulates T is given as follows.

$$C = (2k + 2, (\text{halt}, w_1, \ldots, w_{2k+1}))$$
$$w_i = \begin{cases} \varphi(P(a_i)) & (1 \leq i \leq n) \\ \lambda & (n + 1 \leq i \leq 2k + 1) \end{cases}$$

Let $s_1 \cdots s_m$ be a string over A, where $s_j \in A$ ($j \in \{1, \ldots, m\}$). In C, it is represented by $(\varphi(s_1 \cdots s_m), 0)$. First, consider the case $s_1 = a_i$ for some $a_i \in A_N$. Thus, in T, $s_1 \cdots s_m \underset{T}{\Rightarrow} s_3 \cdots s_m P(a_i)$ holds. Since $\varphi(s_1 \cdots s_m) = N^i Y N^{k-i} \varphi(s_2) \varphi(s_3 \cdots s_m)$, this transition is simulated by C in $2k + 2$ steps as below.

$$(N^i Y N^{k-i} \varphi(s_2) \varphi(s_3 \cdots s_m), 0)$$
$$\underset{C}{\overset{i}{\Rightarrow}} (Y N^{k-i} \varphi(s_2) \varphi(s_3 \cdots s_m), i)$$
$$\underset{C}{\Rightarrow} (N^{k-i} \varphi(s_2) \varphi(s_3 \cdots s_m) \varphi(P(a_i)), i + 1)$$
$$\underset{C}{\overset{k-i}{\Rightarrow}} (\varphi(s_2) \varphi(s_3 \cdots s_m) \varphi(P(a_i)), k + 1)$$
$$\underset{C}{\overset{k+1}{\Rightarrow}} (\varphi(s_3 \cdots s_m P(a_i)), 0)$$

Second, consider the case $s_1 = b_i$ for some $b_i \in A_H$. In this case, $s_1 \cdots s_m$ is a halting string in T. Since $\varphi(s_1 \cdots s_m) = Y \text{bin}_k(i) \varphi(s_2 \cdots s_m)$, the ID $(Y \text{bin}_k(i) \varphi(s_2 \cdots s_m), 0)$ is also a halting ID in C.

By above, if

$$\alpha_0 \underset{T}{\Rightarrow} \alpha_1 \underset{T}{\Rightarrow} \cdots \underset{T}{\Rightarrow} \alpha_{l-1} \underset{T}{\Rightarrow} \alpha_l$$

is a complete computing process of T, then it is simulated by

$$(\varphi(\alpha_0), 0) \overset{2k+2}{\underset{C}{\Rightarrow}} (\varphi(\alpha_1), 0) \overset{2k+2}{\underset{C}{\Rightarrow}} \cdots \overset{2k+2}{\underset{C}{\Rightarrow}} (\varphi(\alpha_{l-1}), 0) \overset{2k+2}{\underset{C}{\Rightarrow}} (\varphi(\alpha_l), 0),$$

which is a complete computing process of C.   □

10.3 Constructing Small Universal Reversible Turing Machines

In this section, we give URTM(13,7) and URTM(10,8). If codes (descriptions) of a CTSH C and an initial string $\alpha_0 \in \{Y, N\}^*$ are given, each of these URTMs simulates the rewriting process of C from the initial ID $(\alpha_0, 0)$ step by step until C halts. The URTM U has a one-way infinite tape, and keeps the codes of C and an ID of C as shown in Fig. 10.3. The production rules of C are stored in the left-side segment of the tape. Initially, the segment of "removed symbols" on the tape is empty, and the initial string α_0 is kept in the segment of "current string". To indicate the border between the removed symbols and the current string, different kinds of symbols are used for the removed ones, and for the leftmost one of the current string (or, temporarily pointed by the head). Each time the leftmost symbol of the current string is removed by a rewriting in C, this border is shifted to the right by one square. Thus, if the ID of C is (α, m), then α is stored in the segment of the current string. The phase m of the ID is recorded by putting a "phase marker", which is also a specified symbol of U, at the mth production rule of C on the tape. If the first symbol of the current string is Y and $m > 0$, then the right-hand side w_m of the mth production rule $Y \to w_m$ is appended at the right end of the current string. If the first symbol is N, then nothing is appended. In both cases, the phase marker is moved to the position of the next production rule. If C enters a halting ID, then U halts.

10.3.1 13-State 7-Symbol URTM

We first give URTM(13,7) U_{13_7}. It is defined as follows.

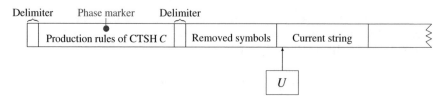

Fig. 10.3 A configuration of a URTM U that simulates a CTSH C

$$U_{13\_7} = (Q_{13\_7}, \{b, y, n, Y, N, *, \$\}, q_{begin}, b, \delta_{13\_7})$$

$$Q_{13\_7} = \{q_{begin}, q_{case\_y\_1}, q_{case\_y\_2}, q_{case\_y\_n}, q_{copy\_start}, q_{copy\_y\_1}, q_{copy\_y\_2},$$
$$q_{copy\_y\_3}, q_{copy\_n\_1}, q_{copy\_n\_2}, q_{copy\_n\_3}, q_{copy\_end}, q_{cycle\_end}\}$$

The move relation δ_{13_7} is described in Table 10.1. It contains 57 quintuples. In this table, "halt" means that the simulated CTSH halts with an ID $(Yv, 0)$ for some $v \in \{Y, N\}^*$, while "null" means that it halts with an ID (λ, m) for some $m \in \mathbb{N}$. We can verify that U_{13_7} satisfies the reversibility condition by a careful inspection of δ_{13_7}. It was also verified by a computer program. Note that, if reversibility is not required, then, for example, the states $q_{case\_y\_2}$ and $q_{copy\_y\_2}$ could be merged to reduce the number of states. However, since there are quintuples $[q_{case\_y\_2}, y, y, +, q_{case\_y\_2}]$, and

Table 10.1 The move relation δ_{13_7} of U_{13_7}

	b	y	n	
q_{begin}	(null)	$Y, -, q_{case\_y\_1}$	$N, -, q_{case\_y\_n}$	
$q_{case\_y\_1}$	(halt)	$y, -, q_{case\_y\_1}$	$n, -, q_{case\_y\_1}$	
$q_{case\_y\_2}$		$y, +, q_{case\_y\_2}$	$n, +, q_{case\_y\_2}$	
$q_{case\_y\_n}$	$*, -, q_{copy\_start}$	$y, -, q_{case\_y\_n}$	$n, -, q_{case\_y\_n}$	
q_{copy_start}	$b, +, q_{cycle\_end}$	$b, +, q_{copy\_y\_1}$	$b, +, q_{copy\_n\_1}$	
$q_{copy\_y\_1}$	$y, +, q_{copy\_y\_2}$	$y, +, q_{copy\_y\_1}$	$n, +, q_{copy\_y\_1}$	
$q_{copy\_y\_2}$	$b, -, q_{copy\_y\_3}$			
$q_{copy\_y\_3}$	$y, -, q_{copy\_start}$	$y, -, q_{copy\_y\_3}$	$n, -, q_{copy\_y\_3}$	
$q_{copy\_n\_1}$	$n, +, q_{copy\_n\_2}$	$y, +, q_{copy\_n\_1}$	$n, +, q_{copy\_n\_1}$	
$q_{copy\_n\_2}$	$b, -, q_{copy\_n\_3}$			
$q_{copy\_n\_3}$	$n, -, q_{copy\_start}$	$y, -, q_{copy\_n\_3}$	$n, -, q_{copy\_n\_3}$	
q_{copy_end}		$y, +, q_{copy\_end}$	$n, +, q_{copy\_end}$	
q_{cycle_end}		$y, +, q_{cycle\_end}$	$n, +, q_{cycle\_end}$	
	Y	N	$*$	$\$$
q_{begin}				
$q_{case\_y\_1}$			$*, +, q_{case\_y\_2}$	$\$, -, q_{case\_y\_1}$
$q_{case\_y\_2}$	$Y, -, q_{case\_y\_n}$			$\$, +, q_{case\_y\_2}$
$q_{case\_y\_n}$			$*, -, q_{case\_y\_n}$	$\$, -, q_{case\_y\_n}$
q_{copy_start}			$b, +, q_{copy\_end}$	
$q_{copy\_y\_1}$	$Y, +, q_{copy\_y\_1}$	$N, -, q_{copy\_y\_3}$	$*, +, q_{copy\_y\_1}$	$\$, +, q_{copy\_y\_1}$
$q_{copy\_y\_2}$				
$q_{copy\_y\_3}$	$Y, -, q_{copy\_y\_3}$		$*, -, q_{copy\_y\_3}$	$\$, -, q_{copy\_y\_3}$
$q_{copy\_n\_1}$	$Y, +, q_{copy\_n\_1}$	$N, -, q_{copy\_n\_3}$	$*, +, q_{copy\_n\_1}$	$\$, +, q_{copy\_n\_1}$
$q_{copy\_n\_2}$				
$q_{copy\_n\_3}$	$Y, -, q_{copy\_n\_3}$		$*, -, q_{copy\_n\_3}$	$\$, -, q_{copy\_n\_3}$
q_{copy_end}	$y, +, q_{begin}$	$n, +, q_{begin}$	$*, +, q_{copy\_end}$	$\$, +, q_{copy\_end}$
q_{cycle_end}			$*, +, q_{cycle\_end}$	$\$, -, q_{copy\_start}$

$[q_{copy\_y\_1}, b, y, +, q_{copy\_y\_2}]$, they cannot be merged without violating the reversibility condition.

We now give a string homomorphism $\varphi_1 : \{Y, N\}^* \to \{y, n\}^*$ as follows: $\varphi_1(Y) = y$, $\varphi_1(N) = n$. Note that φ_1 simply converts the uppercase Y and N into lower case y and n. Let $C = (k, (halt, w_1, \ldots, w_{k-1}))$ be an arbitrary CTSH, and $v_0 \in \{Y, N\}^*$ be an initial string. Then the initial tape for U_{13_7} is as follows, where \$ and the leftmost b are used as delimiters (see Fig. 10.3). Here, w^R denotes the reversal of the string w.

$$b\, \varphi_1(w_{k-1}^R) * \cdots * \varphi_1(w_2^R) * \varphi_1(w_1^R) * b\, \$ \, \varphi_1(v_0)\, b$$

In the case of CTSH $\hat{C}$ with $v_0 = NYY$ in Example 1, the initial tape for it is $b\, ynn * yn * b\, \$\, nyy\, b$. Snapshots of computation of U_{13_7} is as below. It simulates the complete computing process of $\hat{C}$: $(NYY, 0) \Rightarrow (YY, 1) \Rightarrow (YNY, 2) \Rightarrow (NYNNY, 0) \Rightarrow (YNNY, 1) \Rightarrow (NNYNY, 2) \Rightarrow (NYNY, 0) \Rightarrow (YNY, 1) \Rightarrow (NYNY, 2) \Rightarrow (YNY, 0)$. In each computational configuration of U_{13_7}, the head position is also indicated by the underline.

$$
\begin{aligned}
t = 0 &: [\, b\, y\, n\, n * y\, n * b\, \$\, \underline{n}\, y\, y\, b,\ q_{begin},\ 10\,] \\
7 &: [\, b\, y\, n\, n * y\, n\, b * \$\, n\, y\, y\, b,\ q_{begin},\ 11\,] \\
8 &: [\, b\, y\, n\, n * y\, n\, b * \$\, \underline{n}\, Yy\, b,\ q_{case\_y\_1},\ 10\,] \\
18 &: [\, b\, y\, n\, n * y\, \underline{n} * * \$\, n\, Yy\, b,\ q_{copy\_start},\ 6\,] \\
19 &: [\, b\, y\, n\, n * y\, b\, \underline{*} * \$\, n\, Yy\, b,\ q_{copy\_n\_1},\ 7\,] \\
25 &: [\, b\, y\, n\, n * y\, b * * \$\, n\, Yy\, \underline{b},\ q_{copy\_n\_1},\ 13\,] \\
26 &: [\, b\, y\, n\, n * y\, b * * \$\, n\, Yy\, n\, \underline{b},\ q_{copy\_n\_2},\ 14\,] \\
27 &: [\, b\, y\, n\, n * y\, b * * \$\, n\, Yy\, \underline{n}\, b,\ q_{copy\_n\_3},\ 13\,] \\
35 &: [\, b\, y\, n\, n * \underline{y}\, n * * \$\, n\, Yy\, n\, b,\ q_{copy\_start},\ 5\,] \\
36 &: [\, b\, y\, n\, n * b\, \underline{n} * * \$\, n\, Yy\, n\, b,\ q_{copy\_y\_1},\ 6\,] \\
44 &: [\, b\, y\, n\, n * b\, n * * \$\, n\, Yy\, n\, \underline{b},\ q_{copy\_y\_1},\ 14\,] \\
45 &: [\, b\, y\, n\, n * b\, n * * \$\, n\, Yy\, n\, y\, \underline{b},\ q_{copy\_y\_2},\ 15\,] \\
46 &: [\, b\, y\, n\, n * b\, n * * \$\, n\, Yy\, n\, \underline{y}\, b,\ q_{copy\_y\_3},\ 14\,] \\
56 &: [\, b\, y\, n\, n\, \underline{*}\, y\, n * * \$\, n\, Yy\, n\, y\, b,\ q_{copy\_start},\ 4\,] \\
57 &: [\, b\, y\, n\, n\, b\, \underline{y}\, n * * \$\, n\, Yy\, n\, y\, b,\ q_{copy\_end},\ 5\,] \\
64 &: [\, b\, y\, n\, n\, b\, \underline{y}\, n * * \$\, n\, y\, \underline{y}\, n\, y\, b,\ q_{begin},\ 12\,] \\
174 &: [\, \underline{b}\, y\, n\, n * y\, n * * \$\, n\, y\, Yn\, y\, n\, n\, y\, b,\ q_{copy\_start},\ 0\,] \\
175 &: [\, \underline{b}\, y\, n\, n * y\, n * * \$\, n\, y\, Yn\, y\, n\, n\, y\, b,\ q_{cycle\_end},\ 1\,] \\
184 &: [\, b\, \underline{y}\, n\, n * y\, n * \underline{*} \$\, n\, y\, Yn\, y\, n\, n\, y\, b,\ q_{copy\_start},\ 8\,] \\
185 &: [\, b\, y\, n\, n * y\, n * b\, \underline{\$}\, n\, y\, Yn\, y\, n\, n\, y\, b,\ q_{copy\_end},\ 9\,] \\
291 &: [\, b\, y\, n\, n\, b\, y\, n * * \$\, n\, y\, y\, n\, y\, \underline{n}\, n\, y\, n\, y\, b,\ q_{begin},\ 15\,] \\
292 &: [\, b\, y\, n\, n\, b\, y\, n * * \$\, n\, y\, y\, n\, y\, \underline{N}\, n\, y\, n\, y\, b,\ q_{case\_y\_n},\ 14\,] \\
303 &: [\, b\, y\, n\, \underline{n} * y\, n * * \$\, n\, y\, y\, n\, y\, N\, n\, y\, n\, y\, b,\ q_{copy\_start},\ 3\,] \\
315 &: [\, b\, y\, n\, b * y\, n * * \$\, n\, y\, y\, n\, y\, \underline{N}\, n\, y\, n\, y\, b,\ q_{copy\_n\_1},\ 15\,] \\
316 &: [\, b\, y\, n\, b * y\, n * * \$\, n\, y\, y\, n\, \underline{y}\, N\, n\, y\, n\, y\, b,\ q_{copy\_n\_3},\ 14\,] \\
665 &: [\, b\, y\, n\, n * y\, n * b\, \$\, n\, y\, y\, n\, y\, n\, n\, y\, n\, y\, \underline{n}\, y\, n\, y\, b,\ q_{begin},\ 19\,] \\
676 &: [\, b\, y\, n\, n * y\, n * \underline{b}\, \$\, n\, y\, y\, n\, y\, n\, n\, y\, n\, Yn\, y\, b,\ q_{case\_y\_1},\ 8\,]
\end{aligned}
$$

We explain how U_{13_7} simulates CTSH by this example. Production rules of $\hat{C}$ is basically expressed by the string $ynn * yn **$. However, to indicate the phase m of an ID (v, m), the mth $*$ from the right is altered into b (where the rightmost $*$ is the 0-th). This b is used as a "phase marker". Namely, $ynn * yn * b$, $ynn * yn\, b *$, and $ynn\, b\, yn* *$ indicate the phase is 0, 1, and 2, respectively. Hence, in the configuration at $t = 0$, the string $ynn * yn * b$ is given on the tape. To the right of the production rules the initial string nyy is given. Between them, there is a delimiter \$ that is not rewritten into another symbol throughout the computation. In the state q_{begin}, the leftmost symbol of the current string is pointed by the head. Then, it is changed to the uppercase letter Y or N to indicate the leftmost position of the current string.

The state q_{begin} (appearing at time $t = 0, 7, 64, 291$, and 665) reads the first symbol y or n of the current string, and temporarily changes it into Y or N, respectively. Depending on the read symbol y or n, U_{13_7} goes to either $q_{case\_y\_1}$ ($t = 8$), or $q_{case\_y\_n}$ ($t = 292$). If the read symbol is b, U_{13_7} halts, because it means the string is null. If the symbol is y, U_{13_7} performs the following operations (the case n is explained in the next paragraph). By the state $q_{case\_y\_1}$ ($t = 8$), the URTM moves leftward to find the delimiter \$, and then visits the left-neighboring square by $q_{case\_y\_1}$. If it reads b, then it halts ($t = 676$), because the phase is 0. If otherwise, U_{13_7} returns to the delimiter \$. Then, using $q_{case\_y\_2}$, U_{13_7} goes to the state $q_{case\_y\_n}$, and moves leftward to find the phase marker b that indicates the position of the next production rule. By the state q_{copy_start} ($t = 18$, and 35) U_{13_7} starts to copy each symbol of the production rule. If U_{13_7} reads a symbol y ($t = 18$) (or n ($t = 35$), respectively), then it shifts the marker b to this position, and goes to $q_{copy\_y\_1}$ ($t = 36$) (or $q_{copy\_n\_1}$ ($t = 19$)) to attach the symbol at the end of the string to be rewritten. On the other hand, if it reads symbol $*$ in q_{copy_start} ($t = 56$), then it goes to q_{copy_end} ($t = 57$), which mean the end of execution of a production rule, and thus it starts to read the next symbol in the rewritten string ($t = 64$). Likewise, if it reads symbol b in q_{copy_start} ($t = 174$), then it goes to q_{cycle_end} ($t = 175$), which mean the end of one cycle, and thus the phase is set to 0 ($t = 185$). The state $q_{copy\_y\_1}$ is for moving rightward to find the first b that is to the right of the current string ($t = 44$), and rewrites it into y ($t = 45$). The states $q_{copy\_y\_2}$ and $q_{copy\_y\_3}$ ($t = 46$) are for returning to the marker position and for repeating the copying procedure. $q_{copy\_n\_1}, \ldots, q_{copy\_n\_3}$ are for copying the symbol n, which are similar to the case of y ($t = 19, 25, 26, 27$).

On the other hand, if U_{13_7} reads a symbol n in the state q_{begin} ($t = 291$), then it enters the state $q_{case\_y\_n}$ ($t = 292$), and tries to copy symbols as in the case of y. At $t = 303$ it starts to copy a symbol n in q_{copy_start}. However, since it finds a symbol N in the state $q_{copy\_n\_1}$ ($t = 315$), it enters the state $q_{copy\_n\_3}$ ($t = 316$) without attaching the symbol n at the right end. By above, the phase marker is finally shifted to the next production rule without copying the symbols of the current production rule.

Repeating the above procedure, U_{13_7} simulates a given CTSH step by step, and halts in the state $q_{case\_y\_1}$ reading the symbol b if the CTSH halts in an ID with phase 0. If the string of the CTSH becomes null, U_{13_7} halts in the state q_{begin} reading the symbol b. In the above example, U_{13_7} halts at $t = 676$, and the final string YNY of $\hat{C}$ is obtained at as a suffix of the string (excluding the last blank symbol b) starting from the symbol Y.

10.3.2 10-State 8-Symbol URTM

Next, we give URTM(10,8) U_{10_8}. It is defined as below.

$$U_{10\_8} = (Q_{10\_8}, \{b, y, n, n', Y, N, *, \$\}, q_{begin}, b, \delta_{10\_8})$$
$$Q_{10\_8} = \{q_{begin}, q_{case\_y\_1}, q_{case\_y\_2}, q_{case\_y\_n}, q_{copy\_start}, q_{copy\_y\_1}, q_{copy\_y\_2},$$
$$q_{copy\_n\_1}, q_{copy\_n\_2}, q_{copy\_end}\}$$

Table 10.2 shows the move relation δ_{10_8}. It contains 61 quintuples. Reversibility of U_{10_8} is verified by a careful checking of δ_{10_8}. It was also checked by a computer program. The URTM U_{10_8} is constructed by modifying U_{13_7} in the previous subsection. Thus, the initial tape of U_{10_8} is just the same as that of U_{13_7}. Furthermore, the simulation time for a given CTSH is also the same.

The difference between U_{13_7} and U_{10_8} is as follows. First, the removed symbols from the string are indicated by the uppercase letters Y and N. Second, if the leftmost symbol of the current string is n, then it is temporarily changed to n', which is a newly added symbol in U_{10_8}. Third, if the current phase is m, then the symbols of the

Table 10.2 The move relation δ_{10_8} of U_{10_8}

	b	y	n	n'
q_{begin}	(null)	$y, -, q_{case\_y\_1}$	$n', -, q_{case\_y\_n}$	
$q_{case\_y\_1}$	(halt)			
$q_{case\_y\_2}$	$b, -, q_{copy\_y\_2}$	$y, -, q_{case\_y\_n}$		
$q_{case\_y\_n}$	$*, -, q_{copy\_start}$			
q_{copy_start}	$b, +, q_{begin}$	$b, +, q_{copy\_y\_1}$	$b, +, q_{copy\_n\_1}$	
$q_{copy\_y\_1}$	$y, +, q_{case\_y\_2}$	$y, +, q_{copy\_y\_1}$	$n, +, q_{copy\_y\_1}$	$n', -, q_{copy\_y\_2}$
$q_{copy\_y\_2}$	$Y, -, q_{copy\_start}$	$y, -, q_{copy\_y\_2}$	$n, -, q_{copy\_y\_2}$	
$q_{copy\_n\_1}$	$n, +, q_{copy\_end}$	$y, +, q_{copy\_n\_1}$	$n, +, q_{copy\_n\_1}$	$n', -, q_{copy\_n\_2}$
$q_{copy\_n\_2}$	$N, -, q_{copy\_start}$	$y, -, q_{copy\_n\_2}$	$n, -, q_{copy\_n\_2}$	
q_{copy_end}	$b, -, q_{copy\_n\_2}$	$Y, +, q_{begin}$		$N, +, q_{begin}$
	Y	N	$*$	$\$$
q_{begin}	$y, +, q_{begin}$	$n, +, q_{begin}$	$*, +, q_{begin}$	$\$, -, q_{copy\_start}$
$q_{case\_y\_1}$	$Y, -, q_{case\_y\_1}$	$N, -, q_{case\_y\_1}$	$*, +, q_{case\_y\_2}$	$\$, -, q_{case\_y\_1}$
$q_{case\_y\_2}$	$Y, +, q_{case\_y\_2}$	$N, +, q_{case\_y\_2}$		$\$, +, q_{case\_y\_2}$
$q_{case\_y\_n}$	$Y, -, q_{case\_y\_n}$	$N, -, q_{case\_y\_n}$	$*, -, q_{case\_y\_n}$	$\$, -, q_{case\_y\_n}$
q_{copy_start}			$b, +, q_{copy\_end}$	
$q_{copy\_y\_1}$	$Y, +, q_{copy\_y\_1}$	$N, +, q_{copy\_y\_1}$	$*, +, q_{copy\_y\_1}$	$\$, +, q_{copy\_y\_1}$
$q_{copy\_y\_2}$	$Y, -, q_{copy\_y\_2}$	$N, -, q_{copy\_y\_2}$	$*, -, q_{copy\_y\_2}$	$\$, -, q_{copy\_y\_2}$
$q_{copy\_n\_1}$	$Y, +, q_{copy\_n\_1}$	$N, +, q_{copy\_n\_1}$	$*, +, q_{copy\_n\_1}$	$\$, +, q_{copy\_n\_1}$
$q_{copy\_n\_2}$	$Y, -, q_{copy\_n\_2}$	$N, -, q_{copy\_n\_2}$	$*, -, q_{copy\_n\_2}$	$\$, -, q_{copy\_n\_2}$
q_{copy_end}	$Y, +, q_{copy\_end}$	$N, +, q_{copy\_end}$	$*, +, q_{copy\_end}$	$\$, +, q_{copy\_end}$

production rules $w_1, \ldots, w_{m-1}$ are changed into the uppercase letters. By above, the states $q_{copy\_y\_2}$ and $q_{case\_y\_2}$ in U_{13_7} can be merged into one state without violating the reversibility condition. Likewise, the states $q_{copy\_n\_2}$ and q_{copy_end} in U_{13_7} can be merged into one state. Hence, in U_{10_8}, the old state $q_{copy\_y\_2}$ ($q_{copy\_n\_2}$, respectively) is removed, and the old state $q_{copy\_y\_3}$ ($q_{copy\_n\_3}$) is renamed to $q_{copy\_y\_2}$ ($q_{copy\_n\_2}$). Furthermore, the states q_{begin} and q_{cycle_end} in U_{13_7} can be merged into one state. Therefore, in U_{13_7}, q_{cycle_end} is removed. By above, the number of states of U_{10_8} is reduced to 10.

Snapshots of computation process of U_{10_8} for the CTSH $\hat{C}$ with the initial string NYY is as below.

$$
\begin{aligned}
t = 0 : &\; [\, b\,y\,n\,n * y\,n * b \,\$\, \underline{n}\,y\,y\,b, \; q_{begin}, \; 10 \,] \\
7 : &\; [\, b\,y\,n\,n * y\,n\,b * \,\$\, N\underline{y}\,y\,b, \; q_{begin}, \; 11 \,] \\
8 : &\; [\, b\,y\,n\,n * y\,n\,b * \,\$\, \underline{N}\,y\,y\,b, \; q_{case\_y\_1}, \; 10 \,] \\
18 : &\; [\, b\,y\,n\,n * y\,\underline{n} * * \,\$\, N\,y\,y\,b, \; q_{copy\_start}, \; 6 \,] \\
19 : &\; [\, b\,y\,n\,n * y\,b\,\underline{*} * \,\$\, N\,y\,y\,b, \; q_{copy\_n\_1}, \; 7 \,] \\
25 : &\; [\, b\,y\,n\,n * y\,b * * \,\$\, N\,y\,y\,\underline{b}, \; q_{copy\_n\_1}, \; 13 \,] \\
26 : &\; [\, b\,y\,n\,n * y\,b * * \,\$\, N\,y\,y\,n\,\underline{b}, \; q_{copy\_end}, \; 14 \,] \\
27 : &\; [\, b\,y\,n\,n * y\,b * * \,\$\, N\,y\,y\,\underline{n}\,b, \; q_{copy\_n\_2}, \; 13 \,] \\
35 : &\; [\, b\,y\,n\,n * y\,N * * \,\$\, N\,y\,y\,n\,b, \; q_{copy\_start}, \; 5 \,] \\
36 : &\; [\, b\,y\,n\,n * b\,\underline{N} * * \,\$\, N\,y\,y\,n\,b, \; q_{copy\_y\_1}, \; 6 \,] \\
44 : &\; [\, b\,y\,n\,n * b\,N * * \,\$\, N\,y\,y\,n\,\underline{b}, \; q_{copy\_y\_1}, \; 14 \,] \\
45 : &\; [\, b\,y\,n\,n * b\,N * * \,\$\, N\,y\,y\,n\,y\,\underline{b}, \; q_{case\_y\_2}, \; 15 \,] \\
46 : &\; [\, b\,y\,n\,n * b\,N * * \,\$\, N\,y\,y\,n\,\underline{y}\,b, \; q_{copy\_y\_2}, \; 14 \,] \\
56 : &\; [\, b\,y\,n\,n\,\underline{*}\,Y\,N * * \,\$\, N\,y\,y\,n\,y\,b, \; q_{copy\_start}, \; 4 \,] \\
57 : &\; [\, b\,y\,n\,n\,b\,\underline{Y}\,N * * \,\$\, N\,y\,y\,n\,y\,b, \; q_{copy\_end}, \; 5 \,] \\
64 : &\; [\, b\,y\,n\,n\,b\,Y\,N * * \,\$\, N\,Y\underline{y}\,n\,y\,b, \; q_{begin}, \; 12 \,] \\
174 : &\; [\, \underline{b}\,Y\,N\,N * Y\,N * * \,\$\, N\,Y\underline{y}\,n\,y\,n\,n\,y\,b, \; q_{copy\_start}, \; 0 \,] \\
175 : &\; [\, b\,\underline{Y}\,N\,N * Y\,N * * \,\$\, N\,Y\,y\,n\,y\,n\,n\,y\,b, \; q_{begin}, \; 1 \,] \\
184 : &\; [\, b\,y\,n\,n * y\,n * \,\underline{*} \,\$\, N\,Y\,y\,n\,y\,n\,n\,y\,b, \; q_{copy\_start}, \; 8 \,] \\
185 : &\; [\, b\,y\,n\,n * y\,n * b\,\underline{\$}\, N\,Y\,y\,n\,y\,n\,n\,y\,b, \; q_{copy\_end}, \; 9 \,] \\
291 : &\; [\, b\,y\,n\,n\,b\,Y\,N * * \,\$\, N\,Y\,Y\,N\,\underline{n}\,n\,y\,n\,y\,b, \; q_{begin}, \; 15 \,] \\
292 : &\; [\, b\,y\,n\,n\,b\,Y\,N * * \,\$\, N\,Y\,Y\,N\,\underline{Y}\,n'\,n\,y\,n\,y\,b, \; q_{case\_y\_n}, \; 14 \,] \\
303 : &\; [\, b\,y\,n\,\underline{n} * Y\,N * * \,\$\, N\,Y\,Y\,N\,Y\,n'\,n\,y\,n\,y\,b, \; q_{copy\_start}, \; 3 \,] \\
315 : &\; [\, b\,y\,n\,b * Y\,N * * \,\$\, N\,Y\,Y\,N\,Y\,\underline{n}'\,n\,y\,n\,y\,b, \; q_{copy\_n\_1}, \; 15 \,] \\
316 : &\; [\, b\,y\,n\,b * Y\,N * * \,\$\, N\,Y\,Y\,N\,\underline{Y}\,n'\,n\,y\,n\,y\,b, \; q_{copy\_n\_2}, \; 14 \,] \\
665 : &\; [\, b\,y\,n\,n * y\,n * b\,\$\, N\,Y\,Y\,N\,Y\,N\,N\,Y\,N\,\underline{y}\,n\,y\,b, \; q_{begin}, \; 19 \,] \\
676 : &\; [\, b\,y\,n\,n * y\,n * \,\underline{b}\,\$\, N\,Y\,Y\,N\,Y\,N\,N\,Y\,N\underline{y}\,n\,y\,b, \; q_{case\_y\_1}, \; 8 \,]
\end{aligned}
$$

Comparing the above computational configurations with the ones of U_{13_7}, we can see that, e.g., at time $t = 45$ the state $q_{case\_y\_2}$ is used instead of $q_{copy\_y\_2}$, and at time $t = 175$ the state q_{begin} is used instead of q_{cycle_end}. However, the essentially the same operation as in U_{13_7} is performed at each step, and thus at time $t = 676$ the final string YNY is obtained.

10.4   Comparison with Other Small URTMs

Besides URTM(13,7) and URTM(10,8), which are constructed here, several URTMs that simulates CTSH have been given in [9, 13, 15]. They are URTM(15,6), URTM(17,5), URTM(24,4), and URTM(32,3).

On the other hand, it is generally difficult to design an RTM that has only two symbols, or a very small number of states. To obtain an RTM with a small number of states, general procedures for converting a given many-state RTM into a 4-state and 3-state RTMs are given in [10], though the number of symbols of the resulting RTMs becomes very large.

Theorem 3 ([10]) *For any one-tape m-state n-symbol RTM T, we can construct a one-tape 4-state $(2mn + n)$-symbol RTM $\tilde{T}$ that simulates T.*

Theorem 4 ([10]) *For any one-tape m-state n-symbol RTM T, we can construct a one-tape 3-state RTM $\hat{T}$ with $O(m^2n^3)$-symbols that simulates T.*

Applying the method of Theorem 3 to URTM(10,8), we obtain URTM(4,168). Likewise, by the method of Theorem 4, we obtain URTM$(3, 36654)$ from URTM(32,3).

To construct a 2-symbol URTM, we can use a method of converting a many-symbol RTM into a 2-symbol RTM shown in [16]. In particular, the following lemma is shown in [13] to convert a 4-symbol RTM to a 2-symbol RTM.

Lemma 1 ([13]) *For any one-tape m-state 4-symbol RTM T, we can construct a one-tape m'-state 2-symbol RTM $T^{\dagger}$ that simulates T such that $m' \leq 6m$.*

By this method, we can obtain URTM(138,4) from URTM(24,4) [13].

These results are summarized as follows.

- URTM$(3,36654)$ with 37936 quintuples [10]
- URTM(4,168) with 381 quintuples [10]
- URTM(10,8) with 61 quintuples
- URTM(13,7) with 57 quintuples
- URTM(15,6) with 62 quintuples [9]
- URTM(17,5) with 67 quintuples [15]
- URTM(24,4) with 82 quintuples [13]
- URTM(32,3) with 82 quintuples [13]
- URTM(138,2) with 220 quintuples [13]

We can see URTM(10,8) has the minimum value of $m \times n$ among the above URTM(m, n)'s. On the other hand, URTM(13,7) has the smallest number of quintuples among them. The pairs of numbers of states and symbols of these URTMs, as well as the smallest UTMs so far known, are plotted in Fig. 10.4.

The products of the numbers of states and symbols of URTM(10,8), URTM(13,7), URTM(15,6), URTM(17,5), URTM(24,4), and URTM(32,3) are all less than 100, and thus relatively small. However, those of URTM$(3,36654)$, URTM(4,168), and URTM(138,2), which are converted from the above ones, are very large, and it is not

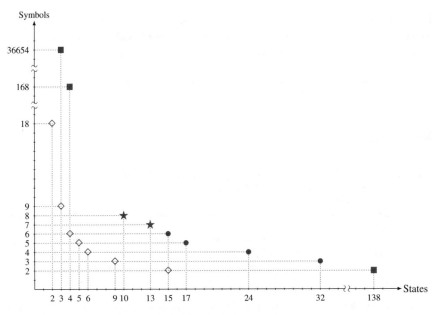

Fig. 10.4 State-symbol plotting of small URTMs and UTMs. ★ shows URTMs newly given in this paper that simulate cyclic tag systems. ● indicates URTMs given in [9, 13, 15] that simulate cyclic tag systems. ■ indicates URTMs converted from other URTMs. ◇ shows UTMs given in [6, 18, 20] that simulate 2-tag systems or bi-tag systems

known whether there are much smaller URTMs. Also, small URTM(m,n)'s such that $5 \leq m \leq 9$ have not yet been constructed till now.

Examples of computing processes of the nine URTMs listed above were simulated by a computer program. Animation-like figures of the computer simulation results, as well as description files of the URTMs, are available in [12].

10.5 Concluding Remarks

We studied the problem of constructing small URTMs, which are universal TMs that satisfy the reversibility condition. For this, we used a method of simulating cyclic tag systems with halting condition. In this way, we newly obtained URTM(13,8), and URTM(10,8).

Woods and Neary [22] proved that both cyclic tag systems, and 2-tag systems can simulate TMs in polynomial time, and thus the small UTMs of Minsky [7], Rogozhin [20], Kudlek and Rogozhin [6], Neary and Woods [18], and others can simulate TMs efficiently. Although we did not discuss time complexity of the URTMs in detail, it is easy to see that the URTMs given here simulate cyclic tag systems in polynomial time. Therefore, these URTMs also simulate TMs in polynomial time.

In this study, we used a method of simulating cyclic tag systems with halting condition to construct small URTMs. However, it is not known whether there are better methods other than it. Also, it is not known whether a 2-state URTM exists. Since there have been only several researches on small URTMs so far, there seems much room for improvement, and thus they are left for the future study.

Acknowledgments This work was supported by JSPS KAKENHI Grant Number 15K00019.

References

1. Axelsen, H.B., Glück, R.: A simple and efficient universal reversible Turing machines. In: Proceedings of the LATA 2011, LNCS, vol. 6638, pp. 117–128 (2011). doi:10.1007/978-3-642-21254-3_8
2. Bennett, C.H.: Logical reversibility of computation. IBM J. Res. Dev. **17**, 525–532 (1973). doi:10.1147/rd.176.0525
3. Cocke, J., Minsky, M.: Universality of tag systems with P = 2. J. Assoc. Comput. Mach. **11**, 15–20 (1964). doi:10.1145/321203.321206
4. Cook, M.: Universality in elementary cellular automata. Complex Syst. **15**, 1–40 (2004)
5. Fredkin, E., Toffoli, T.: Conserv. Log. Int. J. Theoret. Phys. **21**, 219–253 (1982). doi:10.1007/BF01857727
6. Kudlek, M., Rogozhin, Y.: A universal Turing machine with 3 states and 9 symbols. In: Proceedings of the DLT 2001, LNCS, vol. 2295, pp. 311–318 (2002). doi:10.1007/3-540-46011-X_27
7. Minsky, M.L.: Computation: Finite and Infinite Machines. Prentice-Hall, Englewood Cliffs (1967)
8. Morita, K.: A simple reversible logic element and cellular automata for reversible computing. In: Proceedings of the MCU 2001, LNCS 2055, pp. 102–113 (2001). doi:10.1007/3-540-45132-3_6
9. Morita, K.: Reversible computing and cellular automata – A survey. Theoret. Comput. Sci. **395**, 101–131 (2008). doi:10.1016/j.tcs.2008.01.041
10. Morita, K.: Reversible Turing machines with a small number of states. In: Proceedings of the NCMA 2014, pp. 179–190 (2014). Slides with figures of computer simulation: Hiroshima University Institutional Repository, http://ir.lib.hiroshima-u.ac.jp/00036075
11. Morita, K.: Constructing reversible Turing machines by reversible logic element with memory. In: Adamatzky, A. (ed.) Automata, Computation, Universality, pp. 127–138. Springer-Verlag (2015). doi:10.1007/978-3-319-09039-9_6. Slides with figures of computer simulation: Hiroshima University Institutional Repository, http://ir.lib.hiroshima-u.ac.jp/00029224
12. Morita, K.: Constructing small universal reversible Turing machines (slides with figures of computer simulation). Hiroshima University Institutional Repository (2015). http://ir.lib.hiroshima-u.ac.jp/00036736
13. Morita, K.: Universal reversible Turing machines with a small number of tape symbols. Fundam. Inform. **138**, 17–29 (2015). doi:10.3233/FI-2015-1195
14. Morita, K., Suyama, R.: Compact realization of reversible Turing machines by 2-state reversible logic elements. In: Proceedings of the UCNC 2014, LNCS, vol. 8553, pp. 280–292 (2014). doi:10.1007/978-3-319-08123-6_23. Slides with figures of computer simulation: Hiroshima University Institutional Repository, http://ir.lib.hiroshima-u.ac.jp/00036076
15. Morita, K., Yamaguchi, Y.: A universal reversible Turing machine. In: Proceedings of the MCU 2007, LNCS, vol. 4664, pp. 90–98 (2007). doi:10.1007/978-3-540-74593-8_8
16. Morita, K., Shirasaki, A., Gono, Y.: A 1-tape 2-symbol reversible Turing machine. Trans. IEICE Japan **E-72**, 223–228 (1989)
17. Mukai, Y., Morita, K.: Realizing reversible logic elements with memory in the billiard ball model. Int. J. Unconv. Comput. **8**, 47–59 (2012)

18. Neary, T., Woods, D.: Four small universal Turing machines. Fundamenta Informaticae **91**, 123–144 (2009). doi:10.3233/FI-2009-0036
19. Post, E.L.: Formal reductions of the general combinatorial decision problem. Am. J. Math. **65**, 197–215 (1943). doi:10.2307/2371809
20. Rogozhin, Y.: Small universal Turing machines. Theoret. Comput. Sci. **168**, 215–240 (1996). doi:10.1016/S0304-3975(96)00077-1
21. Turing, A.M.: On computable numbers, with an application to the Entscheidungsproblem. Proc. Lond. Math. Soc. Ser. 2 **42**, 230–265 (1936)
22. Woods, D., Neary, T.: On the time complexity of 2-tag systems and small universal Turing machines. In: Proceedings of the 47th Symposium on Foundations of Computer Science, pp. 439–446 (2006). doi:10.1109/FOCS.2006.58
23. Woods, D., Neary, T.: The complexity of small universal Turing machines: a survey. Theoret. Comput. Sci. **410**, 443–450 (2009). doi:10.1016/j.tcs.2008.09.051

Chapter 11
Percolation Transition and Related Phenomena in Terms of Grossone Infinity Computations

Dmitry I. Iudin and Yaroslav D. Sergeyev

Abstract In this chapter, a number of traditional models related to the percolation theory is taken into consideration: site percolation, gradient percolation, and forest-fire model. They are studied by means of a new computational methodology that gives a possibility to work with finite, infinite, and infinitesimal quantities *numerically* by using a new kind of a computer—the Infinity Computer—introduced recently. It is established that in light of the new arithmetic using grossone-based numerals the phase transition point in site percolation and gradient percolation appears as a critical interval, rather than a critical point. Depending on the 'microscope' we use, this interval could be regarded as finite, infinite, or infinitesimal interval. By applying the new approach we show that in vicinity of the percolation threshold we have *many* different *infinite clusters* instead of *one infinite cluster* that appears in traditional considerations. With respect to the cellular automaton forest-fire model, two traditional versions of the model are studied: a real forest-fire model where fire catches adjacent trees in the forest in the step by step manner and a simplified version with instantaneous combustion. By applying the new approach there is observed that in both situations we deal with the model but with different time resolutions. We show that depending on 'microscope' we use, the same cellular automaton forest-fire model reveals either the instantaneous forest combustion or the step by step firing.

D.I. Iudin (✉)
Institute of Applied Physics of Russian Academy of Science, Nizhni Novgorod, Russia,
Lobachevsky State University, Nizhni Novgorod, Russia
e-mail: iudin_di@nirfi.sci-nnov.ru

Y.D. Sergeyev
University of Calabria, Rende, Italy
e-mail: yaro@dimes.unical.it

Y.D. Sergeyev
Lobachevsky State University of Nizhni Novgorod, Nizhni Novgorod, Russia

© Springer International Publishing Switzerland 2017
A. Adamatzky (ed.), *Advances in Unconventional Computing*,
Emergence, Complexity and Computation 22,
DOI 10.1007/978-3-319-33924-5_11

239

11.1   Introduction

There exist several important difference in the usage of, on the one hand, finite quantities and, on the other hand, infinities and infinitesimals in science. The notions of infinite and infinitesimal are usually used in pure mathematics. In their turn, applied mathematics, physics, and engineering work either with finite quantities or use limits trying in any case to obtain finite answers in order to know, for example, the behavior of some processes at infinity. One of the main differences consists of the fact that the Common Notion 5 of Euclid '*The whole is larger than a part*' observed in the world around us does not hold true for symbols traditionally used to work with infinity. For instance, we have $\infty - 1 = \infty$ and infinite numbers introduced by Cantor it also follows $x - 1 = x$, if x is an infinite cardinal, although for any finite x we have $x - 1 < x$.

Due to the enormous importance of the concepts of infinite and infinitesimal in science, people try to introduce them in their work with computers, too (see, e.g. the IEEE Standard for Binary Floating-Point Arithmetic). However, the work of people with infinities and infinitesimals and, in particular, with non-standard Analysis remains a very theoretical field because various arithmetics (see, e.g., [6, 9, 26, 32, 35, 62]) developed for infinite and infinitesimal numbers are quite different with respect to the finite arithmetic we are used to deal with. It happens that certain operations with infinite numbers can be undeterminate (e.g., $\infty - \infty$, $\frac{\infty}{\infty}$, sum of infinitely many items, etc.). Very often representations of infinite numbers are based on infinite sequences of finite numbers and it is not clear how to store infinite quantities in a finite computer memory. These crucial difficulties did not allow people to construct computers that would be able to work with infinite and infinitesimal numbers *in the same manner* as we are used to do with finite numbers and to study infinite and infinitesimal objects numerically.

Numerous trials having as a goal an evolvement of existing counting systems in such a way that they could include in the process of computing infinite and infinitesimal numbers were done during the centuries (see, e.g., [6, 9, 26, 32, 35, 62] and references given therein). In spite of these numerous efforts, the work with infinities and infinitesimals remained symbolic (i.e., non numeric) until a new applied point of view on infinite and infinitesimal numbers has been introduced recently in [36, 41, 47]. The new approach does not use Cantor's ideas and describes infinite and infinitesimal numbers that are in accordance with Euclid's Common Notion 5 mentioned above. It gives a possibility to work with finite, infinite, and infinitesimal quantities *numerically* by using: (i) a new kind of computers—the Infinity Computer—introduced in [37, 42]; (ii) a new numeral system[1] with an infinite radix.

---

[1] We are reminded that a *numeral* is a symbol or a group of symbols that represents a *number*. The difference between numerals and numbers is the same as the difference between words and the things they refer to. A *number* is a concept that a *numeral* expresses. The same number can be represented by different numerals. For example, the symbols '8', 'eight', and 'IIIIIIII' are different numerals, but they all represent the same number..

It is worthwhile noticing that the new approach does not contradict Cantor. In contrast, it can be viewed as an evolution of his deep ideas regarding the existence of different infinite numbers in a more applied way. For instance, Cantor has shown that there exist infinite sets having different cardinalities $\aleph_0$ and $\aleph_1$. In its turn, the new approach specifies this result showing that in certain cases within each of these classes it is possible to distinguish sets with the number of elements being different infinite numbers. We emphasize that the new approach has been introduced as an evolution of standard and non-standard Analysis and not as a contraposition to them. One or another version of Analysis can be chosen by the working mathematician in dependence on the problem he deals with.

In order to see the place of the new approach in the historical panorama of ideas dealing with infinite and infinitesimal, see [27, 45, 46, 54, 56]. In particular, connections of the new approach with bijections is studied in [28] and metamathematical investigations on the new theory and its non-contradictory can be found in [27]. The new methodology has been successfully applied for studying percolation and biological processes (see [22, 23, 49, 61]), infinite series (see [45, 63]), hyperbolic geometry (see [29, 30]), fractals (see [22, 23, 38, 43, 49]), numerical differentiation and optimization (see [13, 51]), the first Hilbert problem, Turing machines, and lexicographic ordering (see [46, 53, 56–58]), cellular automata (see [11, 12]), ordinary differential equations (see [51, 52]), etc.

In this chapter, we consider a number of applications related to the theory of percolation and study them using the new approach. On the one hand, percolation represents the simplest model of a disordered system. Disordered structures and random processes that are self-similar on certain length and time scales are very common in nature. They can be found on the largest and the smallest scales: in galaxies and landscapes, in earthquakes and fractures, in aggregates and colloids, in rough surfaces and interfaces, in glasses and polymers, in proteins and other large molecules. Disorder plays a fundamental role in many processes of industrial and scientific interest. On the other hand, percolation reveals a concept of self-similarity and demonstrates numerous fractal features. Owing to the wide occurrence of self-similarity in nature, the scientific community interested in this phenomenon is very broad, ranging from astronomers and geoscientists to material scientists and life scientists. From the mathematical point of view, self-similarity implies a recursive process and, consequently, is tightly connected with the concept of infinity. This turns us to an idea that percolation is very suitable to demonstrate advantages of the new computational approach proposed in [36, 41].

The outline of the chapter is as follows. In Sect. 11.2 we briefly describe the new approach that allows one to write down different finite, infinite, and infinitesimal numbers by a finite number of symbols as particular cases of a unique framework and to execute numerical computations with all of them. Then in Sect. 11.3 we apply the new methodology to the percolation phase transition. Generalized percolation problem known as gradient percolation are analyzed in terms of infinity computations in Sect. 11.4 while forest fires models are discussed in Sect. 11.5. In the final section, the applications are summarized and discussed briefly.

11.2   Expressing Infinities and Infinitesimals with a High Accuracy

In order to understand how it is possible to study percolation and fractals at infinity with an accuracy that is higher than the traditional one, let us remind the important difference that exists between numbers and numerals (see footnote 1): a *numeral* is a symbol or a group of symbols used to represent a *number*. A *numeral system* consists of a set of rules used for writing down numerals and algorithms for executing arithmetical operations with these numerals. It should be stressed that the algorithms can vary significantly in different numeral systems and their complexity can be also dissimilar. For instance, division in Roman numerals is extremely laborious and in the positional numeral system it is much easier.

Notice that numeral systems strongly bound the possibilities to express numbers and to execute mathematical operations with them. For instance, the Roman numeral system lacks a numeral expressing zero. As a consequence, such expressions as X–X and III–XVI in this numeral system are indeterminate forms. The introduction of the positional numeral system has allowed people to avoid indeterminate forms of this type and to execute the required operations easily.

One of the simplest existing numeral systems that allows its users to express very few numbers is the system used by Warlpiri people, aborigines living in the Northern Territory of Australia (see [5]) and by Pirahã people living in Amazonia (see [18]). Both peoples use the same very poor numeral system for counting consisting just of three numerals—one, two, and 'many'—where 'many' is used for all quantities larger than two.

As a result, this poor numeral system does not allow Warlpiri and Pirahã to distinguish numbers larger than 2, to execute arithmetical operations with them, and, in general, to say a word about these quantities because in their languages there are neither words nor concepts for them. In particular, results of operations $2 + 1$ and $2 + 2$ are not 3 and 4 but just 'many' since they do not know about the existence of 3 and 4. It is worthy to emphasize thereupon that the result 'many' is not wrong, it is correct but *its accuracy is low*. Analogously, when we look at a mob, then both phrases 'There are 3405 persons in the mob' and 'There are many persons in the mob' are correct but the accuracy of the former phrase is higher than the accuracy of the latter one.

Our interest to the numeral system of Warlpiri and Pirahã is explained by the fact that the poorness of this numeral system leads to such results as

$$\text{'many'} + 1 = \text{'many'}, \quad \text{'many'} + 2 = \text{'many'}, \tag{11.1}$$

$$\text{'many'} - 1 = \text{'many'}, \quad \text{'many'} - 2 = \text{'many'}, \tag{11.2}$$

$$\text{'many'} + \text{'many'} = \text{'many'} \tag{11.3}$$

that are crucial for changing our outlook on infinity. In fact, by changing in these relations 'many' with ∞ we get relations that are used for working with infinity in the traditional calculus:

$$\infty + 1 = \infty, \quad \infty + 2 = \infty, \quad \infty - 1 = \infty, \quad \infty - 2 = \infty, \quad \infty + \infty = \infty.$$
$$(11.4)$$

We can see that numerals 'many' and ∞ are used in the same way and we know that in the case of 'many' expressions in (11.1)–(11.3) are nothing else but the result of the lack of appropriate numerals for working with finite quantities. This analogy allows us to conclude that expressions in (11.4) used to work with infinity are also just the result of the lack of appropriate numerals, in this case for working with infinite quantities. As the numeral 'many' is not able to represent the existing richness of finite numbers, the numeral ∞ is not able to represent the richness of the infinite ones.

As was mentioned above, in order to give people the possibility to write down more infinite and infinitesimal numbers, a new numeral system has been introduced recently in [36, 41, 54]. It allows people to express a variety of different infinities and infinitesimals, to perform numerical computations with them, and to avoid both expressions of the type (11.1)–(11.4) and indeterminate forms such as $\infty - \infty$, $\frac{\infty}{\infty}$, $0 \cdot \infty$, etc. that can occur in the traditional calculus and related to limits. Notice that even though the new methodology works with infinite and infinitesimal quantities, it is not related to symbolic computations practiced in non-standard analysis (see [35]) and has an applied, computational character.

The new numeral system and the related computational methodology are based on the introduction in the process of computations of a new numeral, ①, called *grossone*. It is defined as the infinite integer being the number of elements of the set, $\mathbb{N}$, of natural numbers.[2] Thanks to the introduction of the new numeral the set $\mathbb{N}$ can be written in the form

$$\mathbb{N} = \{1, 2, 3, 4, 5, 6, \quad \ldots \quad ① - 2, \quad ① - 1, \quad ①\},$$

whereas positive integers larger than grossone are called *extended natural numbers*. Notice that symbols used traditionally to deal with infinite and infinitesimal quantities (e.g., ∞, Cantor's ω, $\aleph_0$, $\aleph_1$, ..., etc.) are not used together with ①. Similarly, when the positional numeral system and the numeral 0 expressing zero had been introduced, symbols I, IV, VI, XIII, and other symbols from the Roman numeral system had been substituted by the respective Arabic symbols.

The numeral ① allows one to express a variety of numerals representing different infinities and infinitesimals, to order them, and to execute numerical computations with all of them in a handy way. For example, for ① and $①^{4.1}$ (that are examples of infinities) and $①^{-1}$ and $①^{-4.1}$ (that are examples of infinitesimals) it follows

---

[2]Nowadays not only positive integers but also zero is frequently included in $\mathbb{N}$. However, since zero has been invented significantly later than positive integers used for counting objects, zero is not include in $\mathbb{N}$ in this text.

$$0 \cdot ① = ① \cdot 0 = 0, \quad ① - ① = 0, \quad \frac{①}{①} = 1, \quad ①^0 = 1, \quad 1^① = 1, \quad 0^① = 0,$$

$$\text{(11.5)}$$

$$0 \cdot ①^{-1} = ①^{-1} \cdot 0 = 0, \quad ①^{-1} > 0, \quad ①^{-4.1} > 0, \quad ①^{-1} - ①^{-1} = 0,$$

$$\frac{①^{-1}}{①^{-1}} = 1, \quad \frac{5 + ①^{-4.1}}{①^{-4.1}} = 5①^{4.1} + 1, \quad (①^{-1})^0 = 1, \quad ① \cdot ①^{-1} = 1,$$

$$① \cdot ①^{-4.1} = ①^{-3.1}, \quad \frac{①^{4.1} + 6①}{①} = ①^{3.1} + 6, \quad \frac{①^{4.1}}{①^{-4.1}} = ①^{8.2},$$

$$(①^{4.1})^0 = 1, \quad ①^{4.1} \cdot ①^{-1} = ①^{3.1}, \quad ①^{4.1} \cdot ①^{-4.1} = 1.$$

It follows from (11.5) that a finite number b can be represented in this numeral system simply as $b①^0 = b$, since $①^0 = 1$, where the numeral b itself can be written down by any convenient numeral system used to express finite numbers. The simplest infinitesimal numbers are represented by numerals having only negative finite powers of $①$ (e.g., the number $70.12①^{-10.23} + 5.84①^{-80.37}$ consists of two infinitesimal parts, see also examples above). Notice that all infinitesimals are not equal to zero. For instance, $①^{-4.1} = \frac{1}{①^{4.1}}$ is positive because it is the result of division between two positive numbers.

In the context of the present chapter it is important that in comparison to the traditional mathematical tools used to work with infinity the new numeral system allows one to obtain more precise answers in certain cases. For instance, Table 11.1 compares results obtained by the traditional Cantor's cardinals and the new numeral system with respect to the measure of some infinite sets (for a detailed discussion regarding the results presented in Table 11.1 and for more examples dealing with infinite sets see [27, 28, 46, 47, 56]). Notice, that in $\mathbb{Q}$ and $\mathbb{Q}'$ we calculate different numerals and not numbers. For instance, numerals $\frac{3}{1}$ and $\frac{6}{2}$ have been counted two times even though they represent the same number 3.

Then, four sets of numerals having the cardinality of continuum are shown in Table 11.1. Among them we denote by A_2 the set of numbers $x \in [0, 1)$ expressed in the binary positional numeral system, by A_2' the set being the same as A_2 but with x belonging to the closed interval $[0, 1]$, by A_{10} the set of numbers $x \in [0, 1)$ expressed in the decimal positional numeral system, and finally we have the set $C_{10} = A_{10} \cup B_{10}$, where B_{10} is the set of numbers $x \in [1, 3)$ expressed in the decimal positional numeral system. It is worthwhile to notice also that $①$-based numbers present in Table 11.1 can be ordered as follows

$$\lfloor \sqrt{①} \rfloor < \frac{①}{2} < \frac{①}{2} + 2 < ① - 3 < ① - 4 < ① < 2① < 2① + 1 <$$

$$①^2 < 2①^2 + 1 < 2^① < 2^① + 1 < 10^① < 3 \cdot 10^①.$$

Table 11.1 Measuring infinite sets using ①-based numerals allows one in certain cases to obtain more precise answers in comparison with the traditional cardinalities, $\aleph_0$ and C, of Cantor

Description of sets	Cardinality	Number of elements
The set of natural numbers $\mathbb{N}$	Countable, $\aleph_0$	①
$\mathbb{N} \setminus \{3, 5, 10\}$	Countable, $\aleph_0$	①-3
$\mathbb{N} \setminus \{3, 5, 10, 23\}$	Countable, $\aleph_0$	①-4
The set of even numbers $\mathbb{E}$	Countable, $\aleph_0$	$\frac{①}{2}$
The set of odd numbers $\mathbb{O}$	Countable, $\aleph_0$	$\frac{①}{2}$
$\mathbb{O} \bigcup \{-2, 10, 23\}$	Countable, $\aleph_0$	$\frac{①}{2} + 2$
The set of square natural numbers $\mathbb{G} = \{x : x = n^2, n \in \mathbb{N}, x \in \mathbb{N}\}$	Countable, $\aleph_0$	$\lfloor \sqrt{①} \rfloor$
The set of integer numbers $\mathbb{Z}$	Countable, $\aleph_0$	$2①+1$
The set of pairs of natural numbers $\mathbb{P} = \{(p, q) : p \in \mathbb{N}, q \in \mathbb{N}\}$	Countable, $\aleph_0$	$①^2$
The set of numerals $\mathbb{Q}' = \{-\frac{p}{q}, \frac{p}{q} : p \in \mathbb{N}, q \in \mathbb{N}\}$	Countable, $\aleph_0$	$2①^2$
$\mathbb{Q}' \bigcup \{0\}$	Countable, $\aleph_0$	$2①^2 + 1$
The set of numerals A_2	Continuum, C	$2^①$
The set of numerals A_2'	Continuum, C	$2^① + 1$
The set of numerals A_{10}	Continuum, C	$10^①$
The set of numerals C_{10}	Continuum, C	$3 \cdot 10^①$

It can be seen from Table 11.1 that Cantor's cardinalities, $\aleph_0$ and C, say only whether a set is countable or uncountable while the ①-based numerals allow us to express the exact number of elements of the infinite sets. Notice that both numeral systems—the new one and the numeral system of infinite cardinals—do not contradict one another. Both numeral systems provide correct answers, but their answers have different accuracies related to the numeral systems the respective numerals belong to. By using an analogy from physics we can say that the lens of our new 'telescope' used to observe infinite sets is stronger and where Cantor's 'telescope' allows one to distinguish just two dots (countable sets and the continuum) we are able to see instead of these two dots many different dots (infinite sets having different number of elements).

It is worthwhile to emphasize also that the new methodology does not contradict ideas of the one-to-one correspondence. For instance, in the new fashion, the set, $\mathbb{E}$, of even natural numbers can be written in the form

$$\mathbb{E} = \{2, 4, 6, \quad \dots \quad ① - 4, \quad ① - 2, \quad ①\}. \tag{11.6}$$

It follows from [36, 41] that the number of elements of the set of even numbers is equal to $\frac{①}{2}$ and the set of odd numbers has the same number of elements (see

Table 11.1). Thus, ① is even. Note that the next even number is ①+2 but it is not natural because ①+2 > ①, it is extended natural (see [36, 41] for a detailed discussion). Thus, we can write down not only the initial (as it is done traditionally) part of the record

$$
\begin{array}{cccccc}
2, & 4, & 6, & 8, & 10, & 12, \ldots \\
\updownarrow & \updownarrow & \updownarrow & \updownarrow & \updownarrow & \updownarrow \\
1, & 2, & 3, & 4 & 5, & 6, \ldots
\end{array} \tag{11.7}
$$

but also the final part of (11.7)

$$
\begin{array}{ccccccccc}
2, & 4, & 6, & 8, & 10, & 12, \ldots & ①-4, & ①-2, & ① \\
\updownarrow & \updownarrow & \updownarrow & \updownarrow & \updownarrow & \updownarrow & \updownarrow & \updownarrow & \updownarrow \\
1, & 2, & 3, & 4 & 5, & 6, \ldots & \frac{①}{2}-2, & \frac{①}{2}-1, & \frac{①}{2}
\end{array}
$$

concluding so (11.7) in a complete accordance with the Common Notion 5 of Euclid. We end this brief acquaintance with the new computational methodology by noticing that the new numeral system allows us to solve (it is better to say 'to avoid') many other paradoxes related to infinite and infinitesimal quantities (see [36, 41, 44]).

11.3 Geometric Phase Transition and Square Lattice Percolation

In 1957, two mathematicians, S.R. Broadbent and J.M. Hammersley, have published an article [4] where they have shared with readers an idea of probabilistic formalizations of water infiltration in electric coffee maker. Their description, named later *percolation theory*, represents one of the simplest models of a disordered system.

Consider a square lattice, where each site is occupied randomly with probability p or empty with probability $1 - p$. Occupied and empty sites may stand for very different physical properties [20, 21, 60]. For simplicity, let us assume that the occupied sites are electrical conductors (represented by warm colored pixels in Fig. 11.1), the empty sites (shown by black pixels in Fig. 11.1) represent insulators, and that electrical current can flow only between nearest neighbor conductor sites.

At a low concentration p, the conductor sites are either isolated or form small clusters of nearest neighbor sites (see Fig. 11.1). We suppose that two conductor sites belong to the same cluster if they are connected by a path of nearest neighbor conductor sites, and a current can flow between them. At low p values, the mixture is an insulators, since a conducting path connecting opposite edges of our lattice does not exist. At large p values, on the other hand, many conducting paths between opposite edges exist, where electrical current can flow, and the mixture is a conductor (see Fig. 11.2).

At some concentration in between, therefore, a threshold concentration p_c must exist where for the first time electrical current can percolate from one edge to the

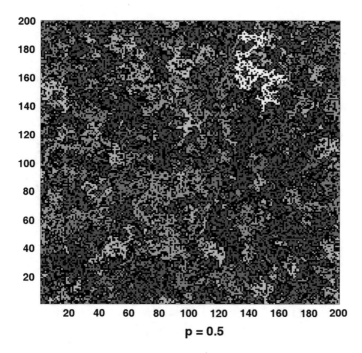

Fig. 11.1 Site percolation on a square lattice. Warm *colored cells* of a square lattice correspond to conducting pixels, *black* stand for non-conducting, *white* cells belong to maximal conducting cluster. Concentration of conducting pixels equals to $p = 0.5$

other (see Fig. 11.3). Thus, for the values $p < p_c$ we have an insulator, and for $p \geq p_c$ we have a conductor. The threshold concentration is called the *percolation threshold*, or, since it separates two different phases, the *critical concentration*. For a site problem on a square lattice the percolation threshold is approximately equal to 0.59, i.e., $p \approx 0.59$. A situation for a value p close to the threshold is displayed in Fig. 11.3.

If the occupied sites are superconductors and the empty sites are conductors, then p_c separates a normal-conducting phase for values $p < p_c$ transition from a superconducting phase where $p \geq p_c$. Another example is a mixture of magnets and paramagnets, where the system changes at p_c from a paramagnet to a magnet.

In contrast to the more common thermal phase transitions, where the transition between two phases occurs at a critical temperature, the percolation transition described here is a geometrical phase transition, which is characterized by the geometric features of large clusters in the neighborhood of p_c. At low values of p only small clusters of occupied sites exist. When the concentration p increases, the average size of the clusters increases, as well. At the critical concentration p_c, a large cluster appears which connects opposite edges of the lattice. This cluster commonly named *spanning cluster* or *percolating cluster*. In the thermodynamic limit, i.e. in the infinite system limit spanning cluster named *infinite cluster*, since its size diverges

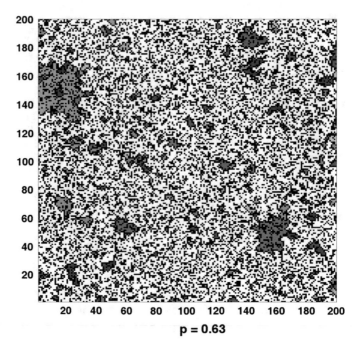

Fig. 11.2 Site percolation on a square lattice. Concentration of conducting pixels is equal to $p = 0.63$. Warm *colored cells* of a square lattice correspond to the conducting pixels isolated from maximal (*white*) cluster

when the size of the lattice increases to infinity. It should be emphasized here that from traditional standpoint there exist unique *infinite* cluster and this *infinite* cluster always coincides with *spanning* cluster.

When p increases further, the density of the infinite cluster also increases, since more and more sites start to be a part of the infinite cluster. Simultaneously, the average size of the finite clusters, which do not belong to the infinite cluster, decreases. At $p = 1$, trivially, all sites belong to the infinite cluster.

In percolation, the concentration p of occupied sites plays the same role as the temperature in thermal phase transitions. Similar to thermal transitions, long range correlations control the percolation transition and the relevant quantities near p_c are described by power laws and critical exponents.

The percolation transition is characterized by the geometrical properties of clusters for values of p that are close to p_c. One of important characteristics describing these properties is the probability, P_∞, that a site belongs to the infinite cluster. For $p < p_c$, only finite clusters exist, and, therefore, it follows $P_\infty = 0$. For values $p > p_c$, P_∞ behaves similarly to the magnetization below critical temperature, and increases with p by a power law

$$P_\infty \sim (p - p_c)^\beta ,$$ (11.8)

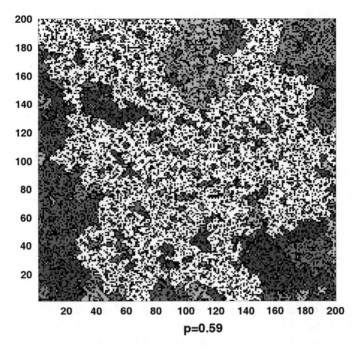

Fig. 11.3 Site percolation on a square lattice. Concentration of conducting pixels is equal to $p = 0.588$

where $\beta = 5/36$ is critical exponent in 2D case.

The linear size of the finite clusters, below and above percolation transition, is characterized by the *correlation length* ξ. The correlation length is defined as the mean distance between two sites on the same finite cluster. When p approaches p_c, ξ increases as

$$\xi \simeq a \cdot |p - p_c|^{-\nu}, \tag{11.9}$$

with the same exponent $\nu = 4/3$ below and above the threshold.

To obtain ξ averages over all finite clusters in the lattice are required.

There is Whelsy to note that all quantities described above are defined in the thermodynamic limit of large systems. In a finite system, P_∞, for example, is not strictly zero below p_c.

The structure of percolation cluster can be well described in the framework of the fractal theory. We begin by considering the percolation cluster at the critical concentration p_c. A representative example of the *spanning* clusters shown in Fig. 11.3. As seen in the figure, the *infinite* cluster contains holes of all sizes. The cluster is self-similar on all length scales (larger than the unit size and smaller than the lattice size), and can be regarded as a fractal. The *fractal dimension*, d_f, describes how, on the average, the mass, M, of the cluster within a sphere of radius r scales with the r,

$$M\left(r\right) \sim r^{d_f}.\tag{11.10}$$

In random fractals, $M\left(r\right)$ represents an average over many different cluster configurations or, equivalently, over many different centers of spheres on the same *infinite*
cluster. Below and above p_c, the mean size of the finite clusters in the system is
described by the correlation length ξ. At p_c, ξ diverges and holes occur in the *infinite*
cluster on all length scales. Above p_c, ξ also represents the linear size of the holes in
the *infinite* cluster. Since ξ is finite above p_c, the *infinite* cluster can be self-similar
only on length scales smaller than ξ. We can interpret $\xi\left(p\right)$ as a typical length up to
which the cluster is self-similar and can be regarded as a fractal. For length scales
larger than ξ, the structure is not self-similar and can be regarded as homogeneous.
If our length scales is smaller than ξ, we see a fractal structure. On length scales
larger than ξ, we see a homogeneous system which is composed of many unit cells
of size ξ. Mathematically, this can be summarized as

$$M\left(r\right) \sim \begin{cases} r^{d_f}, & r \ll \xi, \\ r^d, & r \gg \xi. \end{cases}\tag{11.11}$$

One can relate the fractal dimension d_f of percolation cluster to the exponents β
and ν. The probability that an arbitrary site within a circle of radius r smaller than ξ
belongs to the *infinite* cluster, is the ratio between the number of sites on the *infinite*
cluster and the total number of sites,

$$P_\infty \sim \frac{r^{d_f}}{r^2}, \qquad r < \xi.\tag{11.12}$$

This equation is certainly correct for $r = \lambda\xi$, where λ is an arbitrary constant smaller
than 1. Substituting $r = \lambda\,\xi$ in (11.12) yields

$$P_\infty \sim \lambda^{d_f-2} \cdot \frac{\xi^{d_f}}{\xi^2} \sim \frac{\xi^{d_f}}{\xi^2}.\tag{11.13}$$

Both sides are powers of $p - p_c$. By substituting (11.8) and (11.9) into (11.13) we
obtain,

$$d_f = 2 - \frac{\beta}{\nu}.\tag{11.14}$$

Thus the fractal dimension of the *infinite* cluster at p_c is not a new independent
exponent but depends on β and ν. Since β and ν are universal exponents, d_f is also
universal. It can be shown [60] that (11.14) also represents the fractal dimension of
the finite clusters at p_c and below p_c, as long as their linear size is smaller than ξ.

The exponents β, ν, and γ describe the critical behavior of typical quantities
associated with the percolation transition, and are called the *critical exponents*.
The exponents are universal and depend neither on the structural details of the lattice

(e.g., square or triangular) nor on the type of percolation (site, bond, or continuum), but only on the dimension d of the lattice ($d = 2$ in our present consideration).

This universality is a general feature of phase transitions, where the order parameter vanishes continuously at the critical point (second order phase transition). In below table, the values of the critical exponents β, ν, and γ in percolation are listed for 2D case [20].

Percolation	$d = 2$
Order parameter P_∞ : β	5/36
Correlation length ξ : ν	4/3
Mean cluster size S : γ	43/18
Fractal dimension	91/48

The fractal dimension, however, is not sufficient to fully characterize a percolation cluster. For a further intrinsic characterization of a fractal we consider the shortest path between two sites on the cluster. We denote the length of this path, which is called the 'chemical distance', by l. The *graph* dimension d_l, which is also called the 'chemical' or 'topological' dimension, describes how the cluster mass M within the chemical distance l from a given site scales with l,

$$M\,(l) \sim l^{d_l}. \tag{11.15}$$

While the fractal dimension d_f characterizes how the mass of the cluster scales with the Euclidean distance r, the graph dimension d_l characterizes how the mass scales with the chemical distance l.

The concept of the chemical distance also plays an important role in the description of spreading phenomena such as epidemics and forest fires, which propagate along the shortest path from the seed.

Let us investigate the percolation problem from positions of the new arithmetics of infinite and infinitesimal numbers (see [36, 38, 41]). Consider a 2D square lattice with period a and the linear size $L = a \cdot \textcircled{1}$. The full number of cells of such a lattice is, therefore, infinite and is equal to $V = \textcircled{1}^2$. Since the critical parameter is defined as the attitude of the occupied sites number N to their full number $p = N/V = N/\textcircled{1}^2$ then the smallest change in concentration $\delta p = \textcircled{1}^{-2}$ is equivalent to adding or subtracting only one occupied site. The infinitesimal small value δp is the maximum precision level we can distinguish by considering the critical parameter p on the $\textcircled{1} \times \textcircled{1}$ lattice. In order to obtain a higher precision level we should increase our lattice linear size. For example, if we use a lattice with period a and linear size $L = a \cdot \textcircled{1}^{1+\vartheta/2}$, where $\vartheta > 0$, the maximum precision level we can distinguish by considering the critical parameter p is $\delta p = 1/V = \textcircled{1}^{-(2+\vartheta)}$.

When we investigate the percolation problem we increase or decrease the critical parameter p using an appropriate precision level δp starting from an arbitrary point in between $p = 0$ and $p = 1$. According to the **Postulate 1** we are able to execute only

Fig. 11.4 Correlation length
versus p

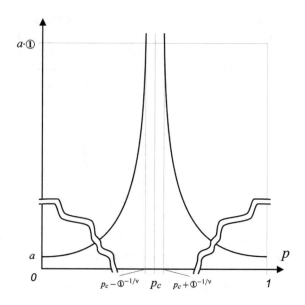

a finite number of steps with length δp. Therefore, the length of critical parameter p interval that we can investigate is determined by the precision level we chose.

Consider the behavior of correlation radius. In the vicinity of percolation threshold the correlation radius diverges according to (11.9). On the other hand, the radius of correlation cannot exceed the system linear size $\xi \lesssim \xi_{max} = L = a \cdot ①$, where $\xi_{max} = a \cdot ①$ is the maximal correlation length. The situation is depicted in Fig. 11.4.

We see that in the range $[p_c - ①^{-1/\nu}, \ p_c + ①^{-1/\nu}]$ the radius of correlation in our $① \times ①$ lattice does not change and keeps the value $\xi_{max} = a \cdot ①$.

Now we should decide which step we shall use to express different points on p axis. Infinitely many variants can be chosen dependent on the precision level we want to obtain. All these variants form three groups. The first group appears when in order to change p we use a small but still finite step $\delta p \ll 1$. In the case the phase transition is infinitely sharp because $\delta p \gg ①^{-1/\nu}$. The second group appears when $\delta p = c \cdot ①^{-1/\nu}$, where c is a finite grossdigit that is less than one. In the case the phase transition occupies the finite interval $[p_c - ①^{-1/\nu}, \ p_c + ①^{-1/\nu}]$. The third group appears when $\delta p = ①^{-\varsigma}$, where $\dfrac{1 + \nu}{\nu} \leq \varsigma \leq 2.$[3] In the case phase transition interval contains more than $①$ different points and if we execute a finite number of steps with length δp along this infinite transition area there exist three possibilities: (1) system contains a lot of finite and *infinite* clusters that coagulate but *spanning* cluster is still absent; (2) *spanning* cluster already exists and absorbs finite and infinite clusters; (3) at the beginning of our execution *spanning* cluster is absent but it appears after finite number of steps. This appearance is due to adding only one occupied site

---

[3]For example, if we add only one occupied site in our greed, then p increases by $\delta p = ①^{-2}$, and that is the smallest step along p we can distinguish in our $① \times ①$ lattice.

Fig. 11.5 Embedded infinite clusters of different scales

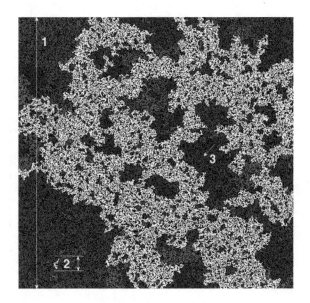

in our grid that produce confluence of either two infinite clusters or one finite and one infinite clusters.

Figure 11.5 shows that when the positive tune-out of critical parameter from the threshold value is infinitesimal number the *spanning* cluster envelops a set of embedded infinite clusters of different scales. The image linear size on Fig. 11.5 is denoted by 1 and makes ① pixels Some of the embedded infinite clusters are comparable with the *spanning* cluster (e.g., small white cluster on Fig. 11.5 denoted by 2) and have linear sizes R that could be expressed by following:

$$R = \frac{a}{K}①, \tag{11.16}$$

where $K > 1$ is a finite number. Remainder of the embedded infinite clusters has linear sizes that are indefinitely small as compared with ① (e.g., point feature on Fig. 11.5 denoted by 3) and could be expressed as $R = ①^{\varepsilon}$, where $0 < \varepsilon < 1$.

On the step when *spanning* cluster appears the order parameter jumps from zero value up to the infinitesimal value as seen in Fig. 11.6

$$P_{\infty min} \sim C_1 ①^{d_f - 2} = C_1 ①^{-\beta/\nu} = C_1 ①^{-\frac{5}{48}}, \tag{11.17}$$

where C_1 is a finite number. The first equality in (11.17) defines $P_{\infty min}$ as a measure of the relative size of *spanning* cluster expressible as a proportion of elements number of spanning cluster $C_1 ①^{d_f}$ to total number of grid elements $①^2$. The second equality in (11.17) appears as a consequence of Eq. (11.14).

Fig. 11.6 P_∞ versus p

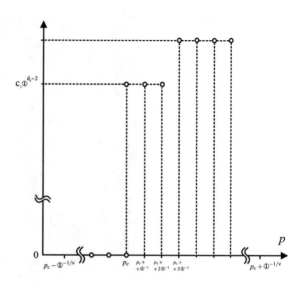

We can see that application of the new arithmetic of infinite and infinitesimal numbers gives us a unique opportunity to consider a point of phase transition in more detail (viewed just like a point with respect of traditional approach).

11.4 Gradient Percolation

An important site-percolation problem generalization appears when the concentration p of occupied sites varies with the vertical distance z in our square grid. In literature (see [19]), this generalization is commonly named as the *gradient percolation*. It can be conveniently pictured in a geographical description in which the set of sites connected to the area $p \lesssim 1$ is called the 'land'. In Fig. 11.7, it is shown by white pixels. In this geographical language the set of connected empty sites not surrounded by land is called the 'sea', in Fig. 11.7, it is shown by black pixels. Then, there naturally exist groups of occupied sites that are not connected with the land called 'islands'. They are shown by orange pixels in Fig. 11.7. Analogously, there exist also connected empty sites surrounded by the land. They are called 'lakes', which are shown also in black in Fig. 11.7. In this geographical description, the part of the land in contact with the sea is called the 'seashore'. In [19] this line is attributed as the *diffusion front*.

The diffusion front is conveniently described (see [19]) by its average width h_f, that can be related easily to the concentration gradient dp/dz at the position of the front. We see in Fig. 11.7 that, far from the front, islands or lakes are very small, whereas, near the front, their size becomes comparable to the width of the front. The islands correspond to the finite clusters in a percolation system, and the lakes

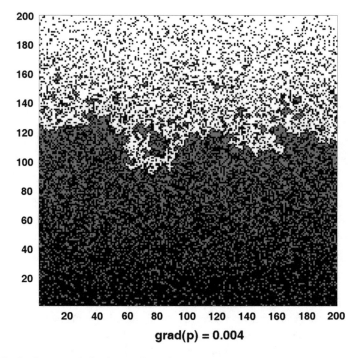

Fig. 11.7 Gradient percolation in two dimensions

correspond to the finite holes. The typical linear size of both quantities scales as ξ. Relation (11.9) tells that the size of the islands or lakes should increase when approaching the mean position of the front. But this size, even at z_f, is bounded due to the finite gradient of $p(z)$. The maximum typical size of islands and lakes is then given by the width of the front, which represents the only characteristic length scale in the problem, and we can assume

$$h_f \simeq \xi(z_c \pm h_f). \tag{11.18}$$

This assumption expresses our observation that islands or lakes near the front have the size comparable to the width of the front. Using (11.18) and expanding $p(z)$ around $z = z_c$ we obtain

$$h_f \simeq a|p(z_c \pm h_f) - p_c|^{-\nu} \simeq a\left|h_f \frac{dp}{dz}(z_c)\right|^{-\nu},$$

which gives

$$h_f \simeq a^{\frac{\beta_f}{\nu}}\left|\frac{dp}{dz}(z_c)\right|^{-\beta_f}, \tag{11.19}$$

where

$$\beta_f = \frac{v}{1+v}. \tag{11.20}$$

As percolation is a critical phenomenon, the exponent β_f depends only on the dimensionality of the system (for $d = 2$ it follows $\beta_f = 4/7$), and not on the particular lattice structure (square, triangular, etc.).

Let us assume now, that we examine the gradient percolation phenomenon on a square lattice N^2 where $N = ①$, and the critical parameter p changes linearly, accepting infinitesimal value $p(z = a) = \frac{1}{a} \cdot ①^{-1}$ (value equal to zero) in the first line of lattice cells and value equal to unit $p(z = a \cdot ①) = 1$ in the last, $①$-s. In other words,

$$p(z) = A \cdot z,$$

where $A = \frac{1}{a} \cdot ①^{-1}$ and z changes discretely. Then the value of the derivative in (11.19) is $\frac{1}{a} \cdot ①^{-1}$, and the diffusion front width makes

$$h_f \simeq a \cdot ①^{\beta_f} = a \cdot ①^{4/7}, \tag{11.21}$$

Thus, on scales of the observation commensurable with the size of the entire system, the diffusion front width is viewed as infinitesimally small and it is represented by the sharp border of two contrast phases—'sea' and 'land'. On the contrary, length scales commensurable with the finite number of the lattice periods a are completely absorbed by huge fluctuations of the front. At last, on scales proportional with h_f the width of front appearers to the observer as a finite value.

11.5 Forest Fires Model

In this section, we are going to apply the new arithmetic to a self-organized critical forest-fire model (see [15, 16]) which is tightly connected with the percolation methodology and in some sense combines the dynamic and the static percolation problems. Let us assume, that we examine the forest fire model on a d-dimensional hypercubic lattice with the lattice spacing a and on the linear scale $L = a \cdot ①$. Then, the lattice contains the infinite number of sites $N = ①^d$. A lattice site can be in one of the following three states: empty, a tree, or a burning tree. The forest-fire model is a stochastic cellular automaton in which a configuration at every time step evolves according to the following rules [2]:

1. A tree grows in an empty site with a probability p;
2. A site with a burning tree becomes an empty site;
3. A tree becomes a burning tree if at least on of its neighbors is a burning tree;
4. A tree without burning neighbors catches fire spontaneously with a probability ε.

Standing with arbitrary initial conditions, the system approaches after a transition period a steady state the properties of which depend only on the parameter values.

Let ρ_e, ρ_t, and ρ_f be the mean overall density of empty sites, of trees, and of burning trees in the system in the steady state, respectively. These densities are related by the equations

$$\rho_e + \rho_t + \rho_f = 1 \tag{11.22}$$

and

$$p\rho_e = \rho_f. \tag{11.23}$$

The last relation says that the mean number of growing trees equals the mean number of burning trees in the steady state [8].

During one time step, there are

$$\varepsilon \rho_t N \tag{11.24}$$

lightning strokes in the system and

$$p\rho_e N \tag{11.25}$$

growing trees. Therefore in the steady state the mean number $\bar{s}$ of trees that are destroyed by a lightning stroke is

$$\bar{s} = p\rho_e N (\varepsilon \rho_t N)^{-1} = \frac{p}{\varepsilon} \frac{1 - \rho_t - \rho_f}{\rho_t} \tag{11.26}$$

When the fire density is large, trees cannot live long enough to become a part of large forest clusters. So, large-scale structures we are interested in can only occur when the fire density ρ_f is small. Equation (11.23) shows that the fire density becomes small when the tree growth rate p approaches zero. Therefore value p should be represented as an infinitesimal small number, say $p = \textcircled{1}^{-\phi}$, where $\phi > 0$. When $p \ll 1$ and consequently in compliance with (11.23) $\rho_f \ll \rho_e$ one could rewrite the last equation in the following approximate form [8, 15]

$$\bar{s} = p\rho_e N (f\rho_t N)^{-1} = \frac{p}{\varepsilon} \frac{1 - \rho_t}{\rho_t}. \tag{11.27}$$

As it was already mentioned above, in the model discussed the forest fire can also be considered as a percolation process [25]. This description is 'mean field' in the sense that spatial correlations are neglected and trees are considered as uniformly randomly distributed over the lattice at a density equal to ρ_t. A cluster is defined as a set of trees that are all connected through nearest neighbor links. If one tree in the cluster catches fire then the whole cluster will eventually burn down, because the fire will be able to spread through nearest neighbor links to all trees in the cluster. This way of considering the forest fire model allows one to make use of several results from the percolation theory. First of all, in the mean field description $\rho_t \lesssim x_c$, where $x_c \leqslant 1$ is the d-dimensional percolation threshold, for example, $x_c \simeq 0.59$ and $x_c \simeq 0.35$ in 2D and 3D site percolation problem correspondingly. Therefore, the second factor of the right-hand side of Eq. (11.27) is of the order of one [8, 15], and Eq. (11.27) then represents a power law [8, 15]

$$\bar{s} \sim \left(\frac{\varepsilon}{p}\right)^{-1}. \tag{11.28}$$

Moreover, we have to choose the tree growth rate p so small that even the largest forest cluster burns down rapidly, before new trees grow up at its edge. The last statement implies that

$$p \ll T_L^{-1}, \tag{11.29}$$

where T_L is the time the fire needs to burn down. In addition, the lightning probability ε must satisfy

$$\varepsilon \ll p \tag{11.30}$$

Otherwise, a tree is destroyed by lightning before its neighbors grow up, and no large-scale structures can be formed. The inequalities (11.29) and (11.30) represent a double separation of time scales

$$T_L \ll p^{-1} \ll \varepsilon^{-1}, \tag{11.31}$$

which is the condition for self-similar behavior in the forest fire model. In this case, the dynamics of the system depends only on the ratio ε/p, but not on ε and p separately. The values ε, and ε/p also could be represented as infinitesimal small numbers. The most essential is the ratio ε/p, and for the beginning we choose $\varepsilon/p \simeq ①^{-\theta}$, where $\theta > 0$ is a finite number. So, we can rewrite expression (11.28) as following

$$\bar{s} \sim ①^{\theta}. \tag{11.32}$$

The mean number $\bar{s}$ of trees that are destroyed by a lightning stroke is obviously less than $N = ①^d$. Therefore in the steady state it follows that

$$\theta < d. \tag{11.33}$$

Lightning will strike the system every $T_f = (\varepsilon \rho_t N)^{-1}$ time steps on average. In order to obtain nontrivial dynamics we should keep the lightning waiting time T_f short enough in comparison with time interval p^{-1} that system requests to be overgrown with forest trees. Otherwise the model discussed will demonstrate saw-tooth overall forest density oscillations: firstly, the model grid is overgrown with forest trees, then one lightning stroke sets fire to the bush, completely destroying the forest and prepare a place for the new generation. Moreover, increasing the system size in order to prevent the fire from dying out we observe the fire fronts in the form of more or less smooth and regular spirals. These spirals represent self-sustained dissipative structures or combustion autowaves. The characteristic spatial scale of the autowaves, is of the order of $a \cdot p^{-1}$ [31]. Thus, in order to obtain self similar forest fires dynamics we have to fulfil the following condition

$$T_f = (\varepsilon \rho_t N)^{-1} \ll p^{-1} \tag{11.34}$$

and again so long as $N = ①^d$ and $\varepsilon / p \simeq ①^{-\theta}$ we have to satisfy the following inequality

$$\theta < d. \tag{11.35}$$

Mean field consideration of the forest fire model allows us to make use of a couple more results from percolation theory. The first is the result that the number of trees in a forest cluster is simply $(R/a)^{d_f}$ where d_f is the percolation cluster fractal dimension and R is the cluster gyration radius (see (11.10)). Then the largest forest cluster with gyration radius of the order of $R \sim L = a \cdot ①$ contains approximately $(L/a)^{d_f} = ①^{d_f}$ elements. The number of the burnt out trees (the size of a fire) is simply the number of destroyed cluster elements. For the largest fire, this number under the order of size will make $(L/a)^{d_f} = ①^{d_f}$. So, instead of (11.35) we have to satisfy the following inequality

$$\theta < d_f. \tag{11.36}$$

The second is the result that the time T_R the fire needs to burn down a forest cluster of size R is determined by the shortest pass dimension d_{min} on percolation cluster

$$T_R = \left(\frac{R}{a}\right)^{d_{min}}. \tag{11.37}$$

The scale of the largest fire which can arise on our lattice is comparable to lattice size $R \sim L = a \cdot ①$. Thus, the maximal forest fire duration will make

$$T_L \simeq \left(\frac{L}{a}\right)^{d_{min}} \simeq ①^{d_{min}}. \tag{11.38}$$

Let us introduce the observer of our system, that uses a new time scale with step equal to the maximal forest fire duration, i.e. $①^{d_{min}}$ steps of initial modelling time. For such an observer even the largest forest cluster is burned down instantaneously, i.e., during one time step when one of its trees is struck by lightning. A typical example of the forest fire model time evolution that our observer could observe is shown in Fig. 11.8. The top panel of the figure represents burning tree number time evolution. Time scales with step equals to T_L. The ordinate axis tick labels should be multiplied by $\frac{L}{a}$ factor. One may choose an extremely huge value of $\frac{L}{a}$, even $\frac{L}{a} = ①$. The bottom panel of the figure represents time evolution of the overall tree density. From the point of view of our observer the overall tree density slumps instantly during forest fire sparks. The tree growth rate p and lightning probability ε for such an observer will be changed by the following values

$$\grave{p} = p \cdot T_L, \quad \grave{\varepsilon} = \varepsilon \cdot T_L, \tag{11.39}$$

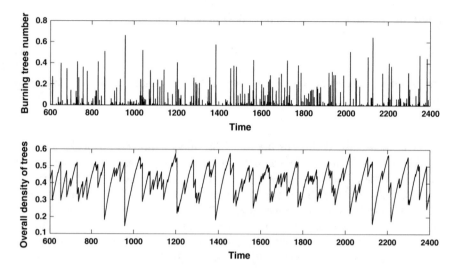

Fig. 11.8 The forest fire model time evolution in two dimensions. Time scales with steps equal to T_L

and can be represented by the infinitesimal numbers

$$\grave{p} \simeq ①^{d_{min}-\phi}, \quad \grave{\varepsilon} \simeq ①^{d_{min}-\phi-\theta}. \tag{11.40}$$

The first expression in (11.40) represents an infinitesimal number when

$$\phi > d_{min}, \tag{11.41}$$

in compliance with inequality (11.29) and estimation (11.38).

From the point of view of our observer the system time evolution looks like a set of sparks of different intensity (see Fig. 11.9 that just zooms up an image patch in Fig. 11.8). One separate spark could be represented as a product of discrete delta-functions (one step—one tree in fire) and the number of trees in a forest cluster that catches the fire. But when we are going to investigate the forest fire internal structure, its inherent dynamic, our observer could not help us. In the case we have to use more powerful 'microscope'. The situation is depicted in Fig. 11.10 that in its turn zooms up an image patch in Fig. 11.9. In the figure time makes only a couple of steps each equals to $①^{d_{min}}$. To launch a narrow analysis of fire spreading we should choose the time resolution that is considerably less than T_L. The model with instantaneous combustion is referred to as a simplified version of the real forest-fire model [8, 15]. Now one can see that we deal with the same model but with different time resolution.

Large-scale structures and therefore criticality can only occur when different forest fires or different sparks are well separated from each other. Otherwise the system will undergo exposure to several fires simultaneously. This separation is well reflected in Fig. 11.9. Different forest fires in our system are well separated in time when

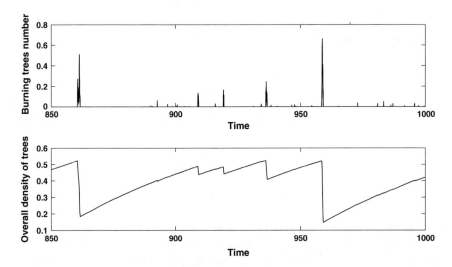

Fig. 11.9 The forest fire model time evolution in two dimensions. Zoom of image patch in Fig. 11.8. Time scales with step equals to T_L

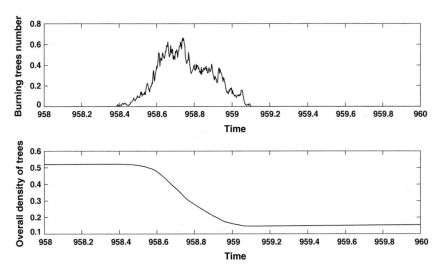

Fig. 11.10 The forest fire model time evolution in two dimensions. Zoom of image patch in Fig. 11.9. Time scales in T_L units

following condition is fulfilled

$$\grave{\varepsilon} \cdot N \rho_t \sim ①^{d+d_{min}-\theta-\phi} \ll 1. \tag{11.42}$$

It means that

$$\theta + \phi > d + d_{min}. \tag{11.43}$$

The last inequality combined with inequality (11.36) provide us with more rigid condition in comparison with inequality (11.41)

$$\phi > d_{min} + (d - d_f) \tag{11.44}$$

as far as in any dimension $d > d_f$.

Thus, to observe criticality and large-scale structures we have to fulfil the following system of inequalities

$$\begin{cases} \phi > d_{min} + (d - d_f) \\ \theta < d_f \\ \theta + \phi > d + d_{min}. \end{cases} \tag{11.45}$$

Infinite and infinitesimal numbers introduced in [36, 41, 47] allow us to decide on the order of priorities in our model space. When we determine the system linear size $L = a \cdot ①^1$ the following choice of the model parameters waits on the results of the choice of the exponents ϕ and θ

$$p = ①^{-\phi}; \quad \varepsilon = ①^{-(\phi+\theta)}. \tag{11.46}$$

In the 2D case, for example, we can use the following parameter values in concordance with (11.45)

$$\phi = 2; \quad \theta = \frac{3}{2}, \tag{11.47}$$

and consequently

$$p = ①^{-2}; \quad \varepsilon = ①^{-3.5}. \tag{11.48}$$

11.6 Conclusion

In this chapter, it has been shown that infinite and infinitesimal numbers introduced in [36, 41, 47] allow us to obtain exact numerical results at different points at infinity instead of traditional asymptotic forms. We consider a number of traditional models related to the percolation theory using the new computational methodology. It has been shown that the new computational tools allow one to create new, more precise models of percolation and to study the existing models more in detail. The introduction in these models of new, computationally manageable notions of the infinity and infinitesimals gives a possibility to pass from the traditional qualitative analysis of the situations related to these values to the quantitative one. Naturally, such a transition is very important from both theoretical and practical viewpoints.

The point of view on computations presented in this chapter uses strongly two methodological ideas borrowed from Physics: relativity and interrelations holding between the object of an observation and the tool used for this observation. Site percolation and gradient percolation were studied by applying the new computational tools. It has been established that in infinite the system phase transition point is not really a point as it seems if one uses the traditional approach. In light of the new arithmetic it appears as a critical interval, rather than a critical point. Depending on 'microscope' we use this interval could be regarded as finite, infinite, and infinitesimal interval. Using the new approach we observed that in vicinity of percolation threshold we have *many* different *infinite clusters* instead of *one infinite cluster* that appears in the traditional consideration. Moreover, we have now a tool to distinguish those infinite clusters. In particular, we can distinguish *spanning infinite* clusters from *embedded* infinite clusters.

Then we consider gradient percolation phenomenon on an infinite square lattice with an infinitesimal gradient of the critical parameter p that changes linearly, accepting infinitesimal value $p(z = a) = \frac{1}{a} \cdot ①^{-1}$ (value equal to zero) in the first line of lattice cells and value equal to unit $p(z = a \cdot ①) = 1$ in the last, $①$-s line of lattice cells. We observe that diffusion front width in this case stretches for an infinite number of lattice spacing: $h_f \simeq a \cdot ①^{\beta_f} = a \cdot ①^{4/7}$. And again this value could be regarded as finite, infinite and infinitesimal short depending on 'microscope' we use.

Scientists that deal with forest-fire model and its applications, distinguish between two versions of the model: real forest-fire model where fire catches adjacent trees in the forest in the step by step manner and simplified version with instantaneous combustion [8, 15]. Using the new approach we show that in both situations we deal with the same model but with different time resolution. We observe that depending on the 'microscope' we use the same cellular automaton forest-fire model reveals either instantaneous forest combustion or step by step firing. By means of the new approach it was also observed that the scaling properties of the system to be very sensitive to the trees growing rate and to the ratio between the ignition probability and the growth probability. As far as we choose infinitesimal values of the two parameters we immediately determine the measure or extent of the system size infinity that provides the criticality of the system dynamics. Correspondent inequalities for grosspowers are derived.

Acknowledgments This work was supported in part by grant from the Government of the Russian Federation (contract No. 14.B25.31.0023) and by the Russian Foundation for Basic Research (projects No. 13-05-12102 ofi_m, No. 13-05-01100 A, No. 15-01-06612 A).

References

1. Abbott, L., Rohrkemper, R.: Prog. Brain Res. **165**, 13 (2007)
2. Bak, P., Chen, K., Tang, C.: A forest-fire model and some thoughts on turbulence. Phys. Lett. A **147**, 297 (1990)
3. Benayoun, M., Cowan, J.D., van Drongelen, W., Wallace, E.: PLOS Comput. Biol. **6**, e1000846 (2010)
4. Broadbent, S.R., Hammersley, J.M.: Percolation processes i. crystals and mazes. Proc. Camb. Phil. Soc. **53**, 629–641 (1957)

5. Butterworth, B., Reeve, R., Reynolds, F., Lloyd, D.: Numerical thought with and without words: evidence from indigenous Australian children. Proc. Natl. Acad. Sci U. S. A. **105**(35), 13179–13184 (2008)
6. Cantor, G.: Contributions to the Founding of the Theory of Transfinite Numbers. Dover Publications, New York (1955)
7. Cauchy, A.L.: Le Calcul infinitésimal. Paris (1823)
8. Clar, S., Drossel, B., Schwabl, F.: Scaling laws and simulation results for the self-organized critical forest-fire model. Phys. Rev. E **50**(2), 1009–1018 (1994)
9. Conway, J.H., Guy, R.K.: The Book of Numbers. Springer, New York (1996)
10. d'Alembert, J.: Encyclopédie, ou dictionnaire raisonné des sciences, des arts et des métiers. Différentiel **4** (1754)
11. D'Alotto, L.: Cellular automata using infinite computations. Appl. Math. Comput. **218**(16), 8077–8082 (2012)
12. D'Alotto, L.: A classification of two-dimensional cellular automata using infinite computations. Indian J. Math. **55**, 143–158 (2013)
13. De Cosmis, S., De Leone, R.: The use of Grossone in mathematical programming and operations research. Appl. Math. Comput. **218**(16), 8029–8038 (2012)
14. de Pablo, J.J.: Coarse-grained simulations of macromolecules: from DNA to nanocomposites. Annu. Rev. Phys. Chem. **62**, 555–574 (2011). doi:10.1146/annurev-physchem-032210-103458
15. Drossel, B., Schwabl, F.: Self-organized critical forest-fire model. Phys. Rev. Lett. **69**(11), 1629–1632 (1992)
16. Drossel, B., Schwabl, F.: Self-organized criticality in a forest-fire model. Physica A **191**, 47–50 (1992)
17. Feder, J.: Fractals. Plenum, New York (1988)
18. Gordon, P.: Numerical cognition without words: evidence from Amazonia. Science **306**, 496–499 (2004)
19. Gouyet, J.F.: Dynamics of diffusion and invasion fronts: on the disconnection-reconnection exponents of percolation clusters. In: Rabin, Y., Bruinsma, R. (eds.) Soft Order in Physical Systems, pp. 163–166. Springer, New York (1994)
20. Halvin, S., Bunde, A.: Fractals and Disordered Systems. Springer, Berlin (1995)
21. Halvin, S., Bunde, A.: Fractals in Science. Springer, Berlin (1995)
22. Iudin, D.I., Sergeyev, Ya.D., Hayakawa, M.: Interpretation of percolation in terms of infinity computations. Appl. Math. Comput. **218**, 8099–8111 (2012)
23. Iudin, D.I., Sergeyev, Ya.D., Hayakawa, M.: Infinity computations in cellular automaton forest-fire model. Commun. Nonlinear Sci. Numer. Simul. **20**(3), 861–870 (2015)
24. Izhikevich, E.M., Gally, J.A., Edelman G.M.: Cereb Cortex **14**, 933 (2004); Izhikevich, E.M.: Neural Comput. **18**, 245 (2006)
25. Jensen, H.J.: Self-Organized Criticality. Cambridge university press, Cambridge (1998)
26. Leibniz, G.W., Child, J.M.: The Early Mathematical Manuscripts of Leibniz. Dover Publications, New York (2005)
27. Lolli, G.: Metamathematical investigations on the theory of grossone. Appl. Math. Comput. **255**, 3–14 (2015)
28. Margenstern, M.: Using Grossone to count the number of elements of infinite sets and the connection with bijections, p-adic numbers. Ultrametr. Anal. Appl. **3**(3), 196–204 (2011)
29. Margenstern, M.: An application of Grossone to the study of a family of tilings of the hyperbolic plane. Appl. Math. Comput. **218**(16), 8005–8018 (2012)
30. Margenstern, M.: Fibonacci words, hyperbolic tilings and grossone. Commun. Nonlinear Sci. Numer. Simul. **21**(1–3), 3–11 (2015)
31. Moβner, W.K., Drossel, D., Schwabl, F.: Computer simulations of the forest-fire model. Physica A **190**, 205–217 (1992)
32. Newton, I.: Method of Fluxions (1671)
33. Pica, P., Lemer, C., Izard, V., Dehaene, S.: Exact and approximate arithmetic in an amazonian indigene group. Science **306**, 499–503 (2004)

34. Quintanilla, J.A., Ziff R.M.: Near symmetry of percolation thresholds of fully penetrable disks with two different radii. Phys. Rev. E **76**(5), 051115 [6 pages]. doi:10.1103/PhysRevE.76. 051115 (2007)
35. Robinson, A.: Non-standard Analysis. Princeton Univ. Press, Princeton (1996)
36. Sergeyev, Ya.D.: Arithmetic of Infinity, 2nd edn. Edizioni Orizzonti Meridionali CS (2003)
37. Sergeyev, Ya.D.: http://www.theinfinitycomputer.com (2004)
38. Sergeyev, Ya.D.: Blinking fractals and their quantitative analysis using infinite and infinitesimal numbers. Chaos, Solitons Fractals **33**(1), 50–75 (2007)
39. Sergeyev, Ya.D.: Infinity computer and calculus. In: Simos, T.E., Psihoyios, G., Tsitouras, Ch. (eds.) AIP Proceedings of the 5th International Conference on Numerical Analysis and Applied Mathematics, vol. 936, pp. 23–26. Melville, New York (2007)
40. Sergeyev, Ya.D.: Measuring fractals by infinite and infinitesimal numbers. Math. Methods, Phys. Methods Simul. Sci. Technol. **1**(1), 217–237 (2008)
41. Sergeyev. Ya.D.: A new applied approach for executing computations with infinite and infinitesimal quantities. Informatica **19**(4), 567–596 (2008)
42. Sergeyev, Ya.D. : Computer system for storing infinite, infinitesimal, and finite quantities and executing arithmetical operations with them. EU patent 1728149, 03.06.2009
43. Sergeyev, Ya.D.: Evaluating the exact infinitesimal values of area of Sierpinski's carpet and volume of Menger's sponge. Chaos, Solitons Fractals **42**, 3042–3046 (2009)
44. Sergeyev, Ya.D.: Numerical computations and mathematical modelling with infinite and infinitesimal numbers. J. Appl. Math. Comput. **29**, 177–195 (2009)
45. Sergeyev, Ya.D.: Numerical point of view on Calculus for functions assuming finite, infinite, and infinitesimal values over finite, infinite, and infinitesimal domains. Nonlinear Anal. Ser. A: Theory, Methods Appl. **71**(12), 1688–1707 (2009)
46. Sergeyev, Ya.D.: Counting systems and the first Hilbert problem. Nonlinear Anal. Ser. A: Theory, Methods Appl. **72**(3-4), 1701–1708 (2010)
47. Sergeyev, Ya.D.: Lagrange lecture: methodology of numerical computations with infinities and infinitesimals. Rendiconti del Seminario Matematico dell'Università e del Politecnico di Torino **68**(2), 95–113 (2010)
48. Sergeyev, Ya.D.: Higher order numerical differentiation on the infinity computer. Opt. Lett. **5**(4), 575–585 (2011)
49. Sergeyev, Ya.D.: Using blinking fractals for mathematical modelling of processes of growth in biological systems. Informatica **22**(4), 559–576 (2011)
50. Sergeyev, Ya.D.: Numerical computations with infinite and infinitesimal numbers: theory and applications. In: Sorokin, A., Pardalos, P.M. (eds.) Dynamics of Information Systems: Algorithmic Approaches, pp. 1–66. Springer, New York (2013)
51. Sergeyev, Ya.D.: Solving ordinary differential equations by working with infinitesimals numerically on the infinity computer. Appl. Math. Comput. **219**(22), 10668–10681 (2013)
52. Sergeyev, Ya.D.: Numerical infinitesimals for solving ODEs given as a black-box. In: Simos, T.E., Tsitouras, Ch. (eds.) AIP Proceedings of the International Conference on Numerical Analysis and Applied Mathematics (ICNAAM-2014), vol. 1648. Melville, New York, 150018 (2015)
53. Sergeyev, Ya.D.: The Olympic medals ranks, lexicographic ordering and numerical infinities. Math. Intell. **37**(2), 4–8 (2015)
54. Sergeyev, Ya.D.: Un semplice modo per trattare le grandezze infinite ed infinitesime, Matematica nella Società e nella Cultura: Rivista della Unione Matematica Italiana **8**(1), 111–147 (2015)
55. Sergeyev, Ya.D.: The exact (up to infinitesimals) infinite perimeter of the Koch snowflake and its finite area, Commun. Nonlinear Sci. Numer. Simul. **20**(3), 861–870 (2016)
56. Sergeyev, Ya.D., Garro, A.: Observability of turing machines: a refinement of the theory of computation. *Informatica* **21**(3), 425–454 (2010)
57. Sergeyev, Ya.D., Garro, A.: Single-tape and multi-tape turing machines through the lens of the Grossone methodology. J. Supercomput. **65**(2), 645–663 (2013)

58. Sergeyev, Ya.D., Garro, A.: The Grossone methodology perspective on turing machines. In: Adamatzky, A. (ed.) Automata, Universality, Computation. Springer Series "Emergence, Complexity and Computation", vol. 12, pp. 139–169 (2015)
59. Shante, K.S., Kirkpatrick, S.: An introduction to percolation theory. Adv. Phys. **20**(85), 325–357 (1971)
60. Stauffer, D.: Introduction to Percolation Theory. Taylor & Francis, Berlin (1985)
61. Vita, M.C., De Bartolo, S., Fallico, C., Veltri, M.: Usage of infinitesimals in the Menger's Sponge model of porosity. Appl. Math. Comput. **218**(16), 8187–8196 (2012)
62. Wallis, J.: Arithmetica Infinitorum (1656)
63. Zhigljavsky, A.A.: Computing sums of conditionally convergent and divergent series using the concept of grossone. Appl. Math. Comput. **218**(16), 8064–8076 (2012)

Chapter 12
Spacetime Computing: Towards Algorithmic Causal Sets with Special-Relativistic Properties

Tommaso Bolognesi

Abstract *Spacetime computing* is undoubtedly one of the most ambitious and less explored forms of unconventional computing. Totally unconventional is the medium on which the computation is expected to take place—the elusive texture of physical spacetime—and unprecedentedly wide its scope, since the emergent properties of these computations are expected to ultimately reproduce *everything* we observe in nature. First we discuss the distinguishing features of this peculiar form of unconventional computing, and survey a few pioneering approaches. Then we illustrate some novel ideas and experiments that attempt to establish stronger connections with advances in quantum gravity and the physics of spacetime. We discuss techniques for building *algorithmic causal sets*—our proposed deterministic counterpart of the *stochastic* structures adopted in the Causal Set programme for discrete spacetime modeling—and investigate, in particular, the extent to which they can reflect an essential feature of continuous spacetime: Lorentz invariance.

12.1 Introduction

Most approaches in the broad field of unconventional computing are tightly related to structures and functions that can be observed in the natural world (natural computing).

On one hand, smart solutions that have emerged during the multi-billion-year evolution of life on Earth provide valuable inspiration for developing novel algorithms meant to run on traditional computers, or novel computing paradigms and architectures to be implemented by ad-hoc, human-designed electronic hardware (bio-inspired computing).

On the other hand, in the last two decades interest has grown for experiments in which the 'hardware' itself is provided by nature. For example, Rubel's Extended Analog Computer (1993) takes advantage of materials that are unfit for conventional computation, but still contribute to the machine functionality, just based

T. Bolognesi (✉)
CNR/ISTI, 1, Via Moruzzi, Pisa, Italy
e-mail: t.bolognesi@isti.cnr.it

© Springer International Publishing Switzerland 2017
A. Adamatzky (ed.), *Advances in Unconventional Computing*,
Emergence, Complexity and Computation 22,
DOI 10.1007/978-3-319-33924-5_12

on the laws of nature that they follow [22, 26]. As another example, in 1994 Adleman successfully used DNA molecules to solve a graph theoretic, combinatorial problem [2], thus starting the field of biomolecular computing. In [1], computing in 'reaction-diffusion' excitable media is shown to involve new computational paradigms, advanced non-standard architectures and novel materials. 'Natural hardware' may indeed take several forms, including chemical soups, cellular systems, bacteria, ant colonies, or various other biological substrata, as documented elsewhere in this volume. One of the main challenges that these unconventional, often massively parallel systems pose is how to harness their computing capabilities by adequate programming paradigms and techniques.

How about conceiving, as 'natural hardware', the elusive, ultimate fabric of the universe, namely spacetime? This bold question immediately raises two problems.

First, spacetime, the mathematical structure defined by the Einstein equations, is classically conceived as a *continuous* entity (a pseudo-riemannian manifold), while we usually associate the concept of computation to *discrete* entities, such as the state or the tape of a Turing machine. This objection is easily answered. On one hand, some of the above mentioned examples of natural computing prove that computation with continuous media is indeed definitely feasible. On the other hand, several recent theories of quantum gravity (e.g. Loop Quantum Gravity [29], Causal Dynamical Triangulations [3], Causal Sets [7]) adopt discrete models of spacetime that appear as perfectly adequate for supporting computation, as we shall soon illustrate.

The second difficulty is severe. While the biosphere offers several examples of brilliant information processing activities whose operation and purpose we now understand well, from the processing of genetic information as encoded in DNA to that of sensory data by various receptors and organs (say, echolocation in bats), we currently have no *direct* clues about information processing activities and functions that can be attributed to the discrete texture of spacetime, and no idea of what type of algorithm, if any, might be working at those ultra-low scales.

Thus, it would be inappropriate, at least as of today, to talk about *spacetime-inspired computing* in the same way as we talk about bio-inspired computing. Similarly, it would be extremely hazardous to imagine that physical spacetime might one day become the ultimate 'natural hardware' for human-controlled computations at the Plank scale (10^{-35} m); and not just because current experiments in molecular electronics and DNA-based computer circuits still take place at a much higher scale (10^{-9} m, the nano-scale) [16], but because, under the conjecture that human actions are themselves ultimately emergent from computations at ultra-low scales, the idea that we could manipulate those levels appears as highly questionable, if not a loopy logical impossibility.

Then, does it make sense to talk about *spacetime computing*?

The main purposes of this chapter are to provide some arguments in favour of a positive answer, and to illustrate a few past and present research and experimentation activities which, in our opinion, can be reasonably collected under this bold name.

Two warnings are in order. First, the field is still fuzzily defined and might still take advantage from several research tracks, across several disciplines. But it is not our aim to be exhaustive in this respect; rather, we shall mainly focus on our own approach to the matter, presenting some of our recent results as well as original material. Second, we insist on the highly speculative nature of this research and on the fact that we still lack experimental evidence for punctual connections between computational processes and spacetime features.

In Sect. 12.2 we mention some solid, general reasons for believing that spacetime computes, although we are still very far from understanding how it does it.

In Sect. 12.3 we mention some early steps in this research area, mainly centered on the model of cellular automata. In Sect. 12.4 we consider other models of computation and their possible use for modeling a discrete, algorithmic, evolving space.

In Sect. 12.5 we move from a Newtonian view at space and time, intended as absolute, independent entities, to a relativistic, integrated view at spacetime. This leads us to deal with *causal sets*, or *causets*. After recalling a standard technique for building these discrete, *stochastic* models of spacetime, we discuss two general methods for obtaining discrete, algorithmic, *deterministic* versions of them, and introduce EH-causets ('Event-History') and PA-causets ('Permutation-Ant'). We also introduce an automaton, that we call 'Ring Ant', which produces both types of causet.

In Sect. 12.6 we identify a key feature of continuous spacetime that any discrete model of a computational spacetime must cope with: Lorentz invariance. We then introduce a relatively rough but practical indicator meant to assess the 'Lorentzianity' of the investigated spacetime models.

In Sect. 12.7 we illustrate a number of concrete examples of EH-causets derived from computations of the Ring Ant automaton, and show how they perform in terms of our Lorentzianity indicator. Section 12.8 is devoted to the illustration of PA-causets from computations of the same Ring Ant automaton. We find that these causets can perform better than EH-causets in terms of Lorentzianity.

In Sect. 12.9 we summarise our viewpoints about spacetime computing and its possible developments.

12.2 An Algorithmic Bottom Layer

A quick argument in support of the idea that nature is fundamentally algorithmic is offered by the 'typing monkeys' metaphor.

In its original version, used, among others, by Borel and Eddington in the context of statistical mechanics, the metaphor suggests that an *infinite* sequence of random characters must include with probability 1, say, all sonnets by Shakespeare. However, if the sequence is created progressively, one can easily calculate that the expected time for the first complete sonnet to appear is incomparably longer than the age of our universe.

A modern version of the metaphor combines the use of old-fashioned typewriters with that of computer terminals. Assume, for example, that eight monkeys are typing at random on eight special typewriters with only three characters. The characters have the same shape—a square—but three different colors: white, grey, black.

Fig. 12.1 *Upper* random typing on eight paper sheets, using a three-character (three-color) typing machine. *Lower* results of 80,000-step computations of eight randomly chosen 3-state, 3-color Turing machines, each running with the corresponding sheet above as input

Each monkey fills a 30-line sheet, where each line contains 30-characters. The eight completed sheets are shown in the upper row of Fig. 12.1; they appear statistically indistinguishable from one another, and no clue of order or structure can be detected in any of them: no shakespearean sonnet in sight. If the origin and unfolding of our universe were based on this kind of mechanism, we would get a totally random, structureless world, unable to support particles, atoms, not to mention stars and life.

When randomness and computation are combined, the picture changes. Each of the lower diagrams in Fig. 12.1 shows the result of feeding the corresponding upper sheet to a different Turing machine selected at random from the set of $(4 \times 3 \times 3)^{3 \times 3} = 101, 559, 956, 668, 416$ two-dimensional, 3-state, 3-color Turing machines, running for 80,000 steps. The potential 'universe' picture is now different: in spite of the randomness in the inputs and in the choice of the machines, some order emerges, manifested as an unbalance among the three colors, a tendency to distinguish between background and foreground 'objects', some alignment, and repeated patterns.

A more formal treatment of the typing monkey metaphor is possible via the notion of *algorithmic probability* of strings. Consider a string s of n bits. In the absence of any information on the origin of s, we usually assume that it was picked at random from the set of all 2^n bit strings of length n, thus we assign to it a probability 2^{-n}. If, however, we have reasons to believe that s was produced algorithmically, we can use the universal a priori probability, or Solomonoff-Levin algorithmic probability [14]:

$$m(s) = \sum_{p:U[p]=s} 1/2^{|p|}$$

where the summation involves all programs p of length $|p|$ such that a universal, prefix-free Turing machine terminates with output string s when running p. The use of this probability is legitimate regardless of the details of the algorithmic process producing s, as guaranteed by Levin's coding theorem (see [14] for details and further references).

Consider the set S of all bit strings of length 5. In Fig. 12.2-left we plot the individual algorithmic probabilities of the strings of S as estimated by the Online

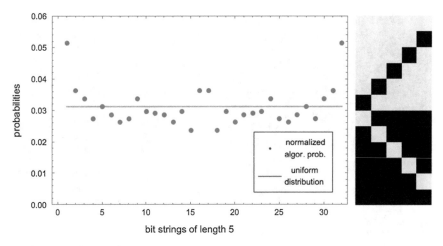

Fig. 12.2 *Left* The algorithmic probabilities of the 32 bit strings of length 5. *Right* The 12 strings with algorithmic probability greater than or equal to 1/32

Algorithmic Complexity Calculator tool [30].[1] This distribution is normalized—for each string s the plotted value is indeed $m(s)/\sum_{x \in S} m(x)$—and compared with the uniform distribution of the 32 strings, each occurring with probability 1/32. Figure 12.2-right shows the 12 strings for which the (normalized) algorithmic probability is greater than or equal to 1/32.

This simple example shows that algorithmic probability favours regular strings—in this case, those with all bits equal, with at most one exception. In an algorithmic universe order and disorder still coexist, but the former is given more chances to emerge.

12.3   Cellular Automata: From Zuse to Wolfram

The idea of a physical space that computes is attributed to the German engineer Konrad Zuse (1910–1995), one of the fathers of the modern computer, although a related concept of a universe made of interacting elementary automaton-like entities—the *monads*—had been formulated much earlier by Gottfried Wilhelm von Leibnitz (1646–1716).

In his 1969 technical report *Rechnender Raum* (Calculating Space [40], re-edited in [39]) Zuse considered discretised versions of continuous fields in 1-D and 2-D space, describing, in particular, molecule velocity and pressure in a gas-filled cylinder, and characterised their discrete dynamics in terms of higher order cellular automata rules directly derived from the differential equations of the continuous

---

[1]http://www.complexitycalculator.com/.

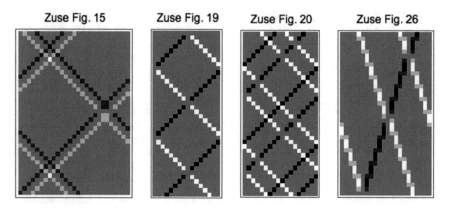

Fig. 12.3 Zuse's particles in higher-order cellular automata from discretised velocity-pressure 1-D fields. The diagrams show expansions of the computations illustrated in Figures 15, 19, 20, 26 of reference [40], using a circular topology for the CA cells. The integer values represented by the *grey* levels of the cells only refer to the *velocity* field; plots for the *pressure* field are similar

dynamics. Zuse's interest focused on what he baptized 'digital particles' (see [40], Sect. 3.1)—localised structures that emerge and possibly interact as the dynamics of the discretised field evolves. These 'particles' are abstract patterns that move in a spacetime diagram in which space extends horizontally and time flows downward; they should not be confused with the actual gas molecules, in the same way as an ocean wave is well distinguished from individual water molecules.

We have implemented some of the 1-D CA discussed in Sect. 3 of [40]; Fig. 12.3 shows the associated emergent particles, that correspond to the computations carried out by Zuse—only manually, and to a rather limited depth—in Figs. 15, 19, 20 and 26 of his paper.

Zuse observed that, as a consequence of a collision, particles may undergo a slight displacement of their initial trajectories, depending on their relative phase, and took this as a sign that '*a certain reaction process in particle interaction*' is possible. The phenomenon can be observed in the rightmost diagram of Fig. 12.3.

The discovery of particles in CA computations and the intuition that their interactions resemble those of particle physics led Zuse to conjecture that CA might not just be useful discrete approximations of a supposedly continuous physical reality, but a perfect reflection of what reality ultimately is, and of how it operates: the physical universe as a giant Cellular Automaton.

Some limited experiments with particle emergence and interaction in 2D CA are also discussed by Zuse in [40], but it is only with Conway's Game of Life, divulged by Martin Gardner in 1970 [13] (the same year of the English translation of Zuse's *Rechnender Raum*) that the spectacular potentialities of these automata became known to the wider public. *Blinkers, gliders, spaceships, pulsars* are just a few examples of the various localised structures that emerge in the Game of Life, and that are just elaborated instances of Zuse's digital particles. However, the large majority of the scientific community has constantly refused to attribute any deep theoretical meaning to these emergent phenomena, relegating them to the whimsical arena of recreational mathematics.

Very important contributions in support to the conjecture of a discrete, algorithmic spacetime have been given, after Zuse, by Ed Fredkin, with his work on Digital Philosophy.[2] Fredkin's Finite Nature assumption states that space and time are ultimately discrete, and that the number of possible states of any volume of spacetime is also finite [11]. Furthermore, information and computation assume, in Fredkin's work, a more fundamental status than matter, energy, and their transformations. As a consequence, Fredkin's interest, like Zuse's, focuses on cellular automata, in particular on second-order, Reversible Universal Cellular Automata (RUCA) and a model called SALT [21], in which two distinct components are arranged in a regular 3D lattice resembling *NaCl* crystals. According to Fredkin, RUCA reflect exactly and efficiently CPT symmetry, a fundamental property of physical laws, and can be regarded as the fundamental, information-processing mechanism at the roots of the physical universe.

Important theoretical and experimental developments in the field of cellular automata, reversible computation and their applications to the modelling of physical processes are due, among others, to Toffoli and Margolous [18, 32–35] (see [8] for a survey).

Perhaps the strongest impulse to the investigation and divulgation of the computational universe conjecture is due, in more recent years, to Stephen Wolfram. With his monumental and controversial volume 'A New Kind of Science' [38], appeared in 2002, Wolfram has somehow reversed the approach by Zuse and Fredkin: rather than deriving a specific model of computation from the consideration of specific physical systems, he has undertaken a rather systematic, abstract exploration of the wide space of models of computation—from cellular automata to Turing machines, from register machines to string and graph rewrite systems, and more—with the objective to classify their emergent behaviours.

When a model of computation is sufficiently simple, it is possible to exhaustively explore all its instances: this is the case for Elementary Cellular Automata (ECA). An ECA is a linear arrangement of potentially infinite binary cells, typically represented as black (for '1') and white (for '0') squares, representing discrete, 1-D space. Time is also discrete, and, for any given ECA, the binary value $c_i(t + 1)$ of cell c_i at time $t + 1$ depends on the binary values at time t of c_i itself and of its immediate neighbors, namely $c_{i-1}(t), c_i(t), c_{i+1}(t)$. This dependency is expressed by a boolean function of three boolean arguments; since there are 256 such functions, we have 256 distinct ECAs, whose behaviours have been thoroughly studied by Wolfram, starting both from simple and from random initial configurations of the cells.

ECAs are the simplest form of CA, considerably simpler than the higher-order CAs investigated by Zuse and Fredkin, and yet they turn out to be very effective in illustrating the creative power of 'spontaneous' computations. Emergent features observed in CA include self-similarity, pseudo-randomness (deterministic chaos), and digital particles, as illustrated in Fig. 12.4.

---

[2]http://www.digitalphilosophy.org/.

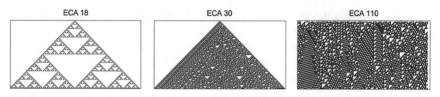

Fig. 12.4 Three ECAs illustrating self-similarity, pseudo-randomness, digital particles

Fig. 12.5 A 3-color cellular
automaton with elementary
initial condition, exhibiting
pseudo-randomness, particle
trajectories and selfsimilarity

Interestingly, it is enough to move to a slightly more complex class of CAs—those that operate on ternary rather than binary cells—for finding a single CA in which the three mentioned features coexist, even when starting from an elementary initial condition. The automaton, discovered by Remko Siemerink[3], is illustrated in Fig. 12.5.

The properties illustrated in Figs. 12.4 and 12.5 appear to reflect some of the most fundamental, recurring patterns of Nature. Another crucial ingredient for the existence of what we regard as the most complex layer of our universe—the biosphere—is of course self-replication. In fact, CAs have been originally conceived by John von Neumann, after a suggestion by Stanislaw Ulam, exactly for studying and simulating this phenomenon. The pervasive presence of these features both in Nature and in the abstract space of CAs provides further evidence that the universe is fundamentally algorithmic.

Beside this ability to replicate key aspects of Nature, two of the principles behind the operation of CAs appear, apriori, quite attractive to physicists: *uniformity*—the same boolean function is used for all cells, and *locality*—the cell transition function only involves neighbouring cells. Their relevance for applications to physics is rather obvious: they reflect the views that physical laws should not change with space or time, and that effects are transmitted by contact. However, a third principle behind CAs, namely *parallelism*—the synchronous operation of the unbounded set of cells—is much less appealing, since it hints at the idea of a global clock.

This reserve on parallelism, the fact that all Turing-universal models are equivalent, at least from the point of view of their computing power, and the observation that interesting emergent properties are found, more or less evident, also outside the realm of CAs, have provided momentum for the investigation of the wider space of non parallel, Turing-universal models.

---

[3] http://www.wolframscience.com/summerschool/2009/alumni/siemerink.html.

12.4 Beyond CA: Ant-Based Models of a Dynamic Space

The diagram of an elementary cellular automaton is a two-dimensional array of bits in which space extends horizontally and time flows vertically. Each row is a snapshot of space at a given time. Beside its 0/1 state, each cell in the array is characterised by the unique, absolute values of its space and time coordinates. Indeed, ECAs reflect exactly a Newtonian concept of space and time.

But there are several other ways to conceive an algorithmic, netwtonian, evolving space. By dropping the requirement of parallel operation, one is led to consider models in which the data structure manipulated by the computation, for example a 1-D array of binary cells, is modified locally rather than globally, for example one cell at each step.

12.4.1 Turing Machines

Elementary Turing machines (TM) are the most obvious example of this computational paradigm, that we shall call *ant-based*. The 'ant', in the case of a TM, is the *control head*—the finite state, read/write unit that reads the current cell and reacts, depending also on its own state, by writing a new a bit in the cell and moving one step left or right, as established by the state transition table.

A TM binary tape is analogous to the row of an ECA array, and can be interpreted as a snapshot of space. By packing the successive tape configurations of a TM we obtain a 2-D array conceptually equivalent to an ECA diagram—a discrete spacetime. However, in the TM case the change from one row to the next is confined to one location, all the rest of the tape being unaffected, so that in general one can trace the ant motion across spacetime: a TM-based toy universe evolves much more slowly than an ECA universe.

12.4.2 Turmites

Similar to CAs, TMs admit higher-dimensional variants. For example, in two-dimensional Turing machines, called 'turmites' [15], the control head moves on a two dimensional square array. The most famous turmite is *Langton ant* [37], a machine that gets trapped in a rather complex periodic behavior (a 'highway') only after over a thousand, pseudo-random initial steps, but many more behaviors are possible in this model, as illustrated in [15].

12.4.3   Network Mobile Automata

The cell array of a Turing machine is a rigid support similar to an immutable, infinite newtonian spatial background expected to exist before starting the computation. As an alternative, and being inspired by the Big-Bang concept, it is attractive to investigate algorithmic models in which space is not a predefined infinite rigid structure, but an emergent product itself of the computation.

Wolfram [38] has widely explored the idea of a graph-based computational Big-Bang, one in which space is modeled by a graph $G(N, E)$—a set N of nodes interconnected, two-by-two, by a set E of edges. A *Network Mobile Automaton* is somewhat analogous to a Turing Machine, except that the ant does not move on a tape but on a graph, and modifies the latter locally, step by step. Space starts as a tiny graph, and evolves into a gigantic network of nodes due to the graph-rewrite rules applied at each step. These rules change the local topology of the graph and, most importantly, may introduce new nodes and edges: space evolves and grows with the computation. Trivalent (or 'cubic') graphs —ones in which each node has exactly three neighbours—are sufficient for 'implementing' spaces of any dimensionality, including 3D space ([38], Chap. 9).

In [4] we have explored variants of Network Mobile Automata for creating planar trivalent networks by using only two simple rewrite rules, namely the 2D Pachner rules, sometimes called *Expand-Contract* and *Exchange*. These rules have found application also in Loop Quantum Gravity [29], where they are used for the dynamics of spin networks. In spite of the planarity restriction, our experiments have yielded a wide variety of interesting regular 1-D ('polymer-like') and 2D networks, as well as oscillating rings, semi-regular hexagonal grids, up to totally irregular patterns.

12.4.4   A Multi-threaded Universe

A typical objection against the ant-based computational universe conjecture is that we perceive the world as a concurrent, multi-threaded system, not as the single-threaded one that the ant-based view seems to imply.

Wolfram [38] has an appealing argument to dismiss this objection. In essence, viewing a sentient being itself as a bounded region R of the toy universe, and associating R's perception to a change of its internal configuration/state, an act of perception will only occur when the ant visits R. But during the inter-visit intervals the ant has the opportunity to modify many of the other regions, thus creating R's subjective illusion of multiple parallel changes. The sequential ant behaviour is detected by the external observer, not by the internal one.

12.5 From Absolute Space to Relativistic Spacetime: Algorithmic Causets

Although with graph oriented models we have gotten rid of the cumbersome, regular and rigid spatial background of CAs and TMs, we have still been reasoning in newtonian terms, dealing with a sequence, in *absolute time*, of snapshots of *absolute space*. But Minkowski and Einstein have taught us that space and time, taken separately, have no absolute value, since different inertial observers, say Bob and Alice, register different spatial distances and different time intervals between the same two events: in particular, Bob may perceive them as simultaneous when Alice does not, and vice versa. The only absolute distance between events—one on which all inertial observers agree—is *spacetime distance*, i.e. Lorentz distance.

To most physicists, no spacetime model should ignore the lesson of Special Relativity. Thus, let us briefly summarize some basic features of this integrated view at space and time, and the associated notion of causality.

12.5.1 Lorentz Distance and Lightcones

Let us consider Minkowski spacetime $M^{(1,3)}$, which describes the simplest form of a matter-free, flat universe. $M^{(1,3)}$ can be understood as Euclidean 4-D space E^4, with spatial dimensions w, x, y, z, in which one of the dimensions, say w, is interpreted as a time dimension t, and where the Euclidean distance is replaced by the Lorentz distance. While the Euclidean distance between two points $p(w, x, y, z)$ and $p'(w', x', y', z')$ is given by $d^2(p, p') = (w - w')^2 + (x - x')^2 + (y - y')^2 + (z - z')^2$, their Lorentz distance is expressed by:

$$L^2(p, p') = +(t - t')^2 - (x - x')^2 - (y - y')^2 - (z - z')^2.$$

Due to its $+ - - -$ signature, the squared Lorentz distance can be *positive, null,* or *negative*: correspondingly, the two points are said to be in *time-like, light-like,* or *space-like* relation.

The *lightcone* of point p is the set of all points q in light-like relation with p, including p itself. The *future* (resp. *past*) *lightcone* of p is the subset of the lightcone whose points have time coordinate larger (resp. smaller) than that of p.[4] A physical process taking place at p can only influence the processes taking place at points on or inside the future lightcone of p: causality is limited by the speed of light.

---

[4]Note that the difference between the time coordinates of two points p and q that are in time-like or light-like relation depends on the frame of reference, and is affected by the Lorentz transformation between inertial frames, but only in its absolute value, not in its sign. If the points are in space-like relation, on the contrary, the sign itself may change, so that different observers may disagree on the time ordering of the events.

The Lorentz distance immediately induces a *partial order* '$\prec$' among spacetime points: $p \prec q$ whenever $L^2(p, q) \geq 0$ and q is on or inside the future lightcone of p. Given a set of points S with partial order '$\prec$', we can define intervals between points: the *order interval* $I[s, t]$ between points s and t is the set of points $\{p \in S | s \prec p \wedge p \prec t\}$, which includes s and t.

12.5.2 Stochastic Causal Sets from 'sprinkling'

If the fundamental structure of a *continuous* spacetime manifold is its causal, light-cone structure, it seems natural to conceive *discrete* spacetime as a partially ordered set of events, or a *causal set* ('*causet*') [7].

A *causet* is a set with a partial order relation. As such, it can be represented as a *directed acyclic graph* (DAG) $C(N, E)$, where N is the set of nodes and E is the set of edges that define the partial order among nodes. We shall assume causets to be *transitively reduced*, in which case their edges are called *links*. (The transitive reduction of a DAG G is the unique smallest graph that has the same transitive closure as G.)

The *sprinkling technique* is a stochastic method that allows one to directly derive such a DAG from a continuous, Lorentzian spacetime—one in which the Lorentz metric is defined. Consider 2-D Minkowski space $M^{(1,1)}$, the simplest toy model of a Lorentzian spacetime, with one time dimension (vertical) and one space dimension (horizontal).

In Fig. 12.6-left we show an interval of $M^{(1,1)}$, between the points labelled 0 and 9—*source* and *sink*—and a set S of 8 points uniformly sprinkled in it. We also show all directed edges that connect point-pairs that are in time-like relation (the probability of finding two points in light-like relation is zero). In Fig. 12.6-right we show, upside-down, the corresponding causet $C(S, E)$, obtained by taking the transitive reduction of the 'raw' graph on the left and disregarding node coordinate information.

In our opinion, one of the attractive efforts in the field of spacetime computing, that we begin to illustrate in the next subsection, is to reverse the above logic: under the assumption of a fundamentally discrete and algorithmic universe, the plan is to directly build a discrete, algorithmic model—say, an *n*-node causet—from scratch, without resorting to an underlying continuum, while expecting the familiar properties manifested by continuum models—e.g. dimension, curvature, Lorentz invariance—to emerge as $n \to \infty$.[5]

Note that this asymptotic perspective implies that those familiar properties might emerge only after some coarse-graining of the causet, e.g. by focusing only on a fraction of the available points. This view would leave room for 'wild' behaviours

---

[5]Most physicists would favour the inclusion of a quantum-mechanical perspective to this effort, trying to handle collections of causets rather than individual instances, in the spirit of 'sum over histories', or 'path integrals'. We do not cover this aspect here, except for a few short comments in the conclusive section.

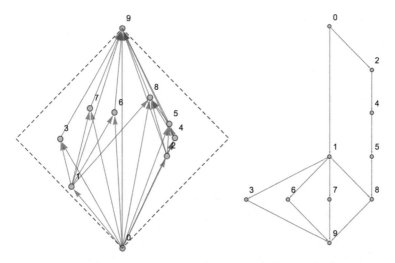

Fig. 12.6 *Left* Sprinkling 8 points in an interval (*red dotted lines*) between two fixed points, labelled 0 and 9, of 2-D Minkowski space $M^{(1,1)}$. *Right* Deriving a transitively reduced causet from the points, based on their mutual Lorentz distances; causal links here flow downward

of the causet at its smallest scales; for example, the causet *as is* might turn out not to be faithfully embeddable in any manifold: "physics near the Plank scale need not be continuum-like" [25].

12.5.3 Algorithmic EH-causets ('Event-History')

In light of the variety of interesting emergent properties offered by the models of computation mentioned in the previous sections, and of the importance that we have come to attribute to causal sets for correct, post-newtonian spacetime modeling, we are interested in the possibility to *directly derive causal sets from the computations of those simple models*. The ultimate, ambitious goal of this approach would be not only to obtain discrete spacetime models that exhibit the right properties of dimensionality, curvature, Lorentz invariance, but that also emergent properties such as fractals, or periodic localised structures analogous to CA 'particles', an effect that we certainly cannot expect from a purely stochastic approach! In essence, the plan is to merge two well distinct research efforts that are referred to as the 'Computational Universe Conjecture' and the 'Causal Set Programme'.

Can we recast the computations of, say, Turing Machines or Network Mobile Automata in terms of causal sets? Bizarre as the question may sound, the answer is definitely positive.

The idea of conceiving the steps of a computation as a set of causally related, partially ordered events was first explored in [12], but the purpose there was to characterise computable functions. It was Wolfram [38] who first proposed to view these graphs as instances of spacetime.

A general method for deriving a DAG from a sequential computation is easily defined [5], as long as we can represent the computation C as a sequence of steps that create, destroy, write and read state variables:

$$C = ((-, W_0), (R_1, W_1), \ldots, (R_n, W_n), \ldots)$$

Each event (R_i, W_i) in the sequence reads the elements of some set R_i of state variables, and writes those of some set W_i. We conceive state variables, and associated read and write operations, in a rather broad sense: a state variable is not only a slot in some memory support, a cell on a tape, the state of a Turing Machine control unit; it can also be a node or an edge in a trivalent graph or, generally, any atomic component of some complex data structure. Then, read, write, creation or elimination operations are just manipulations of these items. The initial configuration of the system is created—written—by event 0, which does not read anything.

Once the above sequence C of computation steps is provided, a causet $C(N, E)$ is readily obtained: nodes $N = \{1, 2, \ldots n, \ldots\}$ are in one-to-one correspondence with the events, and an edge $i \rightarrow j$ is created in E whenever $W_i \cap R_j \neq \emptyset$: this means that some variable has been written (or created) by event i and read (or destroyed) by event j. State variables play the role of *causality mediators* between events, and organise events in a partial order which describes the history of the computation. For this reason we shall sometimes refer to these DAGs as *EH-causets*, for 'Event History'. (Note that in this model the actual values assumed by the state variables play no role.)

Before showing to the reader some examples of application of this general technique, we introduce a second approach for building algorithmic causets, more directly related to the stochastic, sprinkling procedure of Sect. 12.5.2. For doing this, it is convenient to represent sprinklings by permutations.

12.5.4 Correspondence Between Sprinklings and Permutations

There exists a tight correspondence between a k-point sprinkling S in 2-D Minkowski space and a particular permutation π of the first k positive integers. All the information necessary for deriving a causet from S, via the Lorentz distance, can be compactly recorded in a permutation π.

To see this, we must view S under a different angle, literally. Without loss of generality, assume that the sprinkling—e.g. the one of Fig. 12.6-left —has taken place in the interval of Minkowski space $M^{(1,1)}$ between points $s(0, 0)$ and $t(0, \sqrt{2})$, so that the interval, identified by the red dotted lines, is indeed a square with sides of length 1.

Let us now rotate by $-\pi/4$ this 'diamond' and its content around the origin $(0, 0)$, so that the points fall in the unit box between $(0,0)$ and $(1,1)$. Let now S_x be the list of the points, after rotation, sorted by ascending x-coordinate, and S_y be the list of the same points sorted by ascending y-coordinate. Finally, let $\pi = (\pi_1, \pi_2, \ldots, \pi_k)$ be the list of integers where π_i indicates the rank in S_y of the i-th point in S_x. Clearly π is a permutation of the first k positive integers. The patient reader may check that the permutation derived from the set S of 8 sprinkled points in Fig. 12.6-left is (4, 8, 7, 6, 1, 2, 3, 5): the first element is 4 because the first point in S_x, labeled '1' in Fig. 12.6-left, is the 4th point in S_y; the second element is 8 because the second point in S_x, labeled '3' in Fig. 12.6-left, is the 8th point in S_y; and so on.

Consider now the following procedure for deriving a causet from a *generic* permutation π.

Permutation-based causet construction procedure

A causet $C_\pi(\Pi, F)$ is derived from a permutation π of the first k positive integers as follows:

- **Nodes**
$$\Pi = \{(i, \pi_i)|i = 1, 2 \ldots k\} \cup \{(0, 0), (k + 1, k + 1)\}$$

  The $k + 2$ nodes are points with integer-valued coordinates. Of course these coordinates are not part of the causet structure: they are only used for defining the causet links.
- **Links**

  A link $(i, h) \rightarrow (j, k)$ from node (i, h) to node (j, k) is created in F if and only if $i < j, h < k$, and the rectangle identified by the two points (as lower left and upper right vertices, respectively) is empty, i.e. no other node is found inside it. $\diamond$

Nodes labeled $(0, 0)$ and $(k + 1, k + 1)$ are the source and the sink of the causet. Note that graph $C_\pi(\Pi, F)$ is acyclic and transitively reduced by construction.

Going back to our original sprinkling S and to the permutation π derived from it, we can now establish (without proof) the following simple fact.

Fact 1 *The causet $C(S, E)$ obtained directly from sprinkling S and the causet $C_\pi(\Pi, F)$ obtained from permutation π (in turn derived from S) are isomorphic. More precisely, there is a link $p_i \rightarrow p_j$ in E if and only if there is a link $(i, \pi_i) \rightarrow (j, \pi_j)$ in F.*

Additionally, it is easy to see that the process of obtaining causets from k-point random sprinklings in $M^{(1,1)}$ intervals, via the Lorentz distance, is *statistically equivalent* to that of obtaining causets from random permutations of the first k positive integers, via the permutation-based causet construction procedure above. The ultimate reason is that the x and y coordinates of the rotated sprinkled points are uniform and independent random variables.

In conclusion, one can safely and conveniently build sprinkled, 2-D interval causets by just using random permutations.

Let us now consider algorithmic techniques for generating and manipulating permutations, and therefore causets.

12.5.5 Algorithmic PA-causets ('Permutation Ant')

In a *Permutation Ant Automaton* [6] the 'ant' moves up and down a finite, one-dimensional array of cells $A(c_1, c_2, \ldots, c_n)$ by short steps or jumps, while performing operations such as reading, writing, swapping, creating or deleting cells. No matter how the array evolves, at any stage it stores a permutation $\pi = (\pi_1, \pi_2, \ldots, \pi_n)$ of the first n integers, one integer in each cell. Based on this permutation, at any step we can derive a *PA-causet*, for 'Permutation Ant', by the permutation-based construction procedure of the previous subsection.

The model can be enriched in various ways (see [6]). For example, the ant may be stateless, or follow a finite-state behaviour, like in Turing Machines. Furthermore, array cells may store bits, beside the elements of the permutation. The ant may read and write the bits in its neighborhood, as in CAs, and may react depending on these bits and on its current state (if any); it may move by unit steps, or may jump to other locations of array A, as addressed by the π_i of the current cell.

In the sequel we describe an algorithm that combines some of these features in a way that allows us to derive from the same computation *two distinct causets*: an EH-causet and a PA-causet. The advantage is to use a single model for illustrating the two causet construction mechanisms and for exploring two causet spaces.

12.5.6 Ring Ant

This model borrows and combines ideas from automata introduced in [6] and in [5]. The support of the computation is a circular tape with a distinguished first cell c_1. Each cell c_i stores a pair (b_i, π_i), where b_i is a bit and π_i is a positive integer representing an element of the stored permutation π, which must be read starting from c_1.

The ant has 4 possible states $\{s_1, s_2, s_3, s_4\}$; its behavior depends on the current detected *situation* and is manifested by a set of possible *reactions*.

Situation. This is coded by 5 bits: 3 bits (b_{i-1}, b_i, b_{i+1}) are those found in the cell c_i where the ant is currently positioned, and in the two neighboring cells, with indices treated circularly. The two remaining bits code the state of the ant (i.e. $s_1 \leftrightarrow (0, 0)$, $s_2 \leftrightarrow (0, 1)$, $s_3 \leftrightarrow (1, 0)$, $s_4 \leftrightarrow (1, 1)$). Thus, there are 32 possible situations.

Reaction. This is also expressed by 5 bits $(b_1 \ldots b_5)$ whose interpretation is as follows.

- b_1 – insert a new cell $(b_1, n + 1)$ at the current location, where n is the current tape length, so that $(n + 1)$ is a fresh new element of the permutation.

- (b_2, b_3) identify 4 cases:
  - $(0, 0)$: ant moves left one step;
  - $(0, 1)$: ant moves right one step;
  - $(1, 0)$: ant moves left by π_i steps, where c_i is the current cell;
  - $(1, 1)$: ant moves right by π_i steps, where c_i is the current cell;
- (b_4, b_5) identify the new state of the ant.

The last two of the four ant moves are reminiscent of the GOTO command of various programming languages. Since for each of the 32 situations we can associate 1 out of 32 reactions, we can conceive 32^{32} distinct instances of the automaton—a huge space that we can only explore by random samplings or by a genetic algorithm approach.[6]

In our experiments we have started each computation with a two-cell circular tape storing permutation $\pi = (1, 2)$, with bits set to 0, and with the ant in state s_1, and we have run the automaton for an arbitrary number of steps—typically a few thousands. Note that at each step the circular tape and the stored permutation grow by one unit.

The derivation of an EH-causet from a computation of this automaton, along the lines described in Sect. 12.5.3, needs some clarification. We let the bits stored in the tape cells play the role of causality mediators among the events that write and read them; the permutation elements can't be used for this purpose, because they are never read. Since each event reads the bits of three cells, we obtain a 'raw' causet in which nodes—representing events—have in-degree 3. The out-degree of an event equals the number of times subsequent events have read the cell created by that event.

On the other hand, deriving a PA-causet from the final permutation is straightforward using the procedure described in Sect. 12.5.4. Note that the final permutation, read from cell c_1, always starts with '1'. This means that the derived PA-causet has always the node with integer coordinates $(1, 1)$ as the root; this is also the only source node of the graph.

We ask again the reader to be patient: before providing examples of causets obtained by the above techniques we need to discuss an important criterion that we shall use for their assessment.

12.6 Causal Sets and Lorentz Invariance

We have mentioned earlier the requirement for discrete, algorithmic spacetime models to reflect as much as possible the features of continuous, physical spacetime. For example, we might attempt to reproduce the standard 4 dimensions of relativistic spacetime, or even aim at 10, 11 or 26 dimensions, as suggested by more recent theories such as String, Superstring and M-theory. In fact, one may expect spacetime dimensionality to depend on the observation scale. For example, two recent quantum gravity theories—Causal Dynamical Triangulations [3] and Quantum Einstein Gravity [24]—agree in picturing a four-dimensional universe which turns

---

[6]Some of the computations presented here have been indeed selected by a genetic algorithm, using appropriate fitness functions. These aspects are not discussed here.

two-dimensional when observed at ultra low scales. A few techniques are available for estimating causet dimensionality [20, 23], and some applications to algorithmic causets have been investigated in [5].

However, the fundamental property of continuous spacetime on which we want to focus now is *Lorentz invariance*.

A physical entity, e.g. the distance between two spacetime events, or a physical law, e.g. the Maxwell equations, is Lorentz invariant if it does not change its value, or its form, under the Lorentz transformation, which describes the *change of spacetime coordinates* in passing from one inertial (non accelerating) frame of reference to another. According to the principle of relativity, the laws of physics must be invariant for all inertial frames of reference. It was the consideration of the Maxwell equations that led Einstein to abandon the Galilean principle of relativity and to adopt the one based on the Lorentz transformation, since only the latter can account for the constant speed of light that comes with the Maxwell equations.

The issue of Lorentz invariance for causets is delicate, and we shall try to illustrate its essence without introducing excessive technicalities.

To begin with, a causet $C(N, E)$ is an abstract graph structure *without coordinates*, thus, applying a Lorentz transformation to it appears totally meaningless. For the idea to make sense we must still refer to an embedding of C in a manifold M, where each node has its own coordinates. The embedding must be consistent with the partial order expressed by the links E, that is, for any edge $p \to q$ in the transitive closure of E, p and q must be causally related also in M, via the Lorentz metric, and vice versa.

Embeddability is one of the hard problems studied in the field of causal sets. A generic DAG $C(N, E)$ with a sufficient number of nodes is very unlikely to be embeddable in a manifold. However, once C is embedded in some manifold M, it can be embedded in any other manifold M' obtained by applying a Lorentz transformation to M: this is because the transformation drags, with M, the nodes N embedded in it, and does so while preserving their mutual Lorentz distances, so that consistency between the partial orders—in the discrete and in the continuum—is preserved. From the above remarks one might be tempted to conclude that as soon as a causet is embeddable in a Lorentzian manifold, it is also Lorentz invariant. But there is a complication.

A key point of Lorentz invariance is that all reference frames must appear equivalent—none must be singled out as preferred. If we decide to look at the pure causal structure of the graph, without coordinates, then we have no means to discriminate among Lorentz-interrelated reference frames: they all correspond to one and the same graph. But if we allow to access the coordinate information coming with the embeddings, then it turns out that for some graphs all embeddings are equivalent, while for other graphs they are not. In the latter case, a preferred frame may emerge.

The typical examples used for illustrating these two cases are a sprinkled 2-D causet and a regular square grid. A remarkable feature of a uniform Poisson distribution of points in a region of 2-D Minkowski space $M^{(1,1)}$ is that a Lorentz transformation will change the overall shape of the cloud of points, but will leave the local picture unchanged: an observer sitting on one of the points will notice no change in its neighbourhood—in statistical sense. No frame is special.

This is not the case for a regular, directed square grid embedded in $M^{(1,1)}$, one where each node has two incoming and two outgoing links at $+45$ or -45 degrees from the vertical, time axis—links that partition the plane into square tiles. Here a Lorentz transformation does induce local changes: as the frame speed increases, points get packed with increasing density along lines that get increasingly separated, breaking the symmetry of the original grid, and allowing us to single out the latter as the preferred, rest frame.

The notion of 'causet Lorentz invariance' or 'causet Lorentzianity' that uses embeddings and coordinates has, in our opinion, strong and weak aspects. It is strong because the original notion of Lorentz invariance does rest, crucially, upon that of reference frame—an embedding manifold. It is weak because causets were not conceived to inhabit a continuous background spacetime, but to be themselves the only existing, discrete spacetime—a concept known as *background independence*. Their 'Lorentzianity' should be directly manifested by their features—their nodes and edges—without need to refer to manifolds and their coordinate systems. Of course, when following this latter path we should be ready to give up the rich tool set that comes with continuous manifolds. For example, we immediately get into trouble when trying to define 'reference frame' purely in terms of DAGs.

What we show next is an alternative and simplified approach to causet Lorentzianity that avoids embeddings and reference frames, and only looks at DAG properties. This will be applied to algorithmic causets in the next section.

12.6.1 The Interplay of Longest and Shortest Paths

A rather counterintuitive feature of Minkowski space and its Lorentz distance L is the *reversed triangular inequality*—the fact that, given three points p, q and x, with q in the future lightcone of p and x in the order interval $I[p, q]$ (see Sect. 12.5.1), we have:

$$L(p, q) \geq L(p, x) + L(x, q).$$

The Lorentz distance measures the time elapsed along a trajectory between two points, and the above inequality reflects the well known twin paradox of Special Relativity, by which the travelling twin, going through point x, ages more slowly than his sedentary brother. In fact, beside the longest path—a straight line from p to q—there is an infinite number of alternative paths that register shorter, or even *much* shorter time delays between those points, up to the limit case of two segments of light rays forming a $\pi/2$ angle at x, corresponding to a null time delay.

In [6] we have proposed a notion of causet Lorentzianity that takes into account, in a graph-oriented setting, this peculiar, wide range of path lengths that characterises Lorentzian manifolds. The technique consists in collecting and aggregating statistical data about path lengths in the causet C under investigation. Given interval causet

$C[s, t]$, we compute the lengths $lpl_s(x)$ and $spl_s(x)$ of, respectively, the longest and shortest paths from s to any given element x of the set of causet *Nodes*. We then aggregate the data into function $msp(l)$, which provides the *mean shortest path length* associated to each possible longest path length l:

$$msp(l) := Mean|\{spl_s(x)|x \in Nodes \wedge lpl_s(x) = l\}|$$

A very slow growth of this function reveals the presence of a wide gap between the lengths of the longest and shortest path from s to the other nodes. This is indeed what we observe in the *longest/shortest path plot*—the plots of the above *msp* function— for interval causets obtained from sprinkling in $M^{(1,1)}$. Instances of this plot will be included in many of the forthcoming figures (see, for example, the lowest function plots in the bottom row of Fig. 12.7) as a benchmark for analogous plots of other causets.

Having introduced our loose but practical indicator of 'causet Lorentzianity', we are ready to examine some empirical results. In the next two sections we explore the two classes of causets that we can derive from the computations of the Ring Ant automata of Sect. 12.5.6, namely the EH-causets (Event-History), in which computation steps are partially ordered through the mediation of write and read operations, and PA-causets (Permutation-Ant), in which a partial order is directly derived from the final permutation computed by the ant.

12.7   EH-causets from Ring Ant Automata

By randomly sampling the huge space of EH-causets we could establish the following facts.

- All the EH-causets that we have examined—in the order of a few thousand—are planar. Recall that these causets are obtained by transitively reducing raw graphs: the latter in general turn out not to be planar.
- We find that around 70–80% of the causets are linear paths—totally ordered sequences of nodes—or other slightly more elaborate periodic patterns that still grow, essentially, one-dimensional. Often the periodic phase is reached after an initial random-like transient phase. Emergent periodic patterns, called 'highways', are commonly observed in many other models of computation, e.g. in two-dimensional Turing machines [15, 37].
- In the remaining cases we find either random-like planar graphs, or regular planar graphs that we call *tiling causets*, for reasons to be clarified later, or intermediate cases in which randomness and regularity coexist.

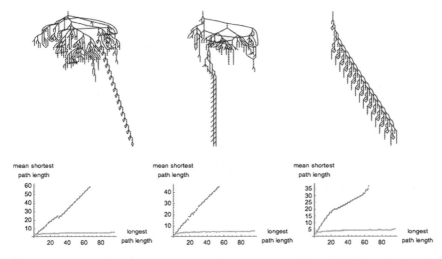

Fig. 12.7 *Upper row* EH-causets from Ring Ant computations with emergent highways. *Lower row* corresponding longest/shortest path plots, compared with the longest/shortest path plot of a 2500-node, sprinkled, 2-D Minkowski interval causet (lower function plots)

12.7.1 Highways

In Fig. 12.7 we show three examples of emergent highways in EH-causets from Ring Ant computations. Causets are shown in the upper row. In the lower row we present the corresponding longest/shortest path plots, each compared with the analogous plot for an interval causet obtained from sprinkling in $M^{(1,1)}$, which grows much more slowly. Eventually the functions for these causets grow linear: due to the periodic highway, the longest and (mean) shortest path lengths get coupled by a constant proportionality factor.

Under a spacetime perspective, these cases are not very interesting: establishing an analogy with the 'digital particles' that emerge in some cellular automata seems inappropriate, since in that case the localised structures move on a background structure, possibly interpreted as empty spacetime, which is missing here.

12.7.2 Random-Like Causets

The two rows in Fig. 12.8 show two different, random-like EH-causets from Ring Ant computations. Each graph is shown in two alternative renderings (left and center). In each of the two diagrams on the r.h.s., the upper and lower functions represent, respectively, the longest/shortest path plots for the corresponding causet and for a sprinkled 2-D Minkowski causet (not shown in figure). The large gap between the two plots, in both cases, indicates that, in spite of their random-like character, these causets perform poorly w.r.t. our Lorentzianity criterion.

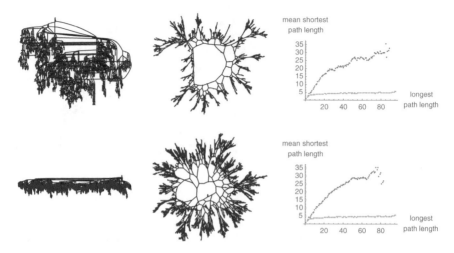

Fig. 12.8 Two random-like EH-causets from Ring Ant computations. At each row the same graph is shown with two different graph drawing algorithms (*left* and *center*). *Right* the longest/shortest path plot for the causet at the *left* (upper function) is compared with the analogous plot for a 2500-node, sprinkled 2-D Minkowski causet, not shown (lower function)

We may wonder whether better, i.e. lower longest/shortest path plots can be obtained by a randomised version of the EH-causet construction technique, one in which each time a new node n is added to the raw causet, three edges $p \to n, q \to n$, $r \to n$ are added to the graph, with p, q, r *chosen at random* among the existing nodes. The experiment is illustrated in Fig. 12.9, where the longest/shortest path plot for a causet obtained by such a randomised procedure is compared, as usual, to that of a sprinkled causet. The randomisation yields an improvement over the random-like cases of Fig. 12.8, but not enough to achieve the performance of sprinkled causets.[7] We note, incidentally, that these randomised causets are not planar.

12.7.3 Regular Tiling Causets

Figure 12.10 shows two regular EH-causets from Ring Ant computations, and their longest/shortest path plots. The graph on the left repeats the basic pattern of the graph on the right, and should not be likened to the periodic patterns—highways— of Fig. 12.7. Its corresponding longest/shortest path plot exhibits a slight improvement w.r.t. the linear growth of those periodic causets, but is still very far from the

---

[7]Due to computational bottlenecks, in Fig. 12.9 and in subsequent analogous figures we tolerate possible differences between the maximum longest path length (about 100 links) achieved by the 2500-node, 2-D Minkowski sprinkled causet constantly used as a benchmark, and the maximum longest path lengths obtained for the various causets under scrutiny, as long as the growth trends for these functions are sufficiently clear.

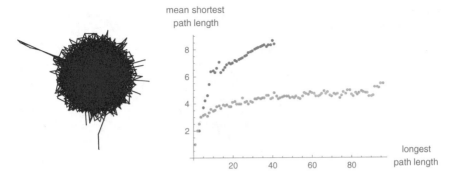

Fig. 12.9 *Left* causet from randomised version of the Ring Ant EH-causet construction technique. *Right* corresponding longest/shortest path plot (upper function), compared with the plot for a sprinkled causet (lower function)

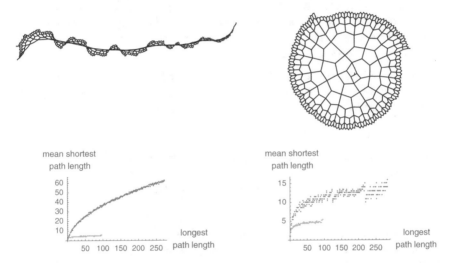

Fig. 12.10 Two regular EH-causets from Ring Ant computations (*upper row*), and their corresponding longest/shortest path plots (*lower row*). As in Fig. 12.8, each plot is compared with the analogous plot for a sprinkled 2-D Minkowski causet, which appears, in both cases, as the lower function

performance of random, sprinkled causets. The longest/shortest path plot for the graph on the right of Fig. 12.10 performs better, with a considerably slower growth of the mean shortest path length.

These regular, 'tiling causets' are reminiscent of the tessellations of the hyperbolic plane, whose patterns are often represented on the Poincaré disc [9].

We recall that regular tessellations of the sphere, of the Euclidean plane and of the hyperbolic plane, can be represented by the Schläfli symbol $\{p, q\}$ which indicates that q regular p-gons meet at each vertex. Based on the value of $1/p + 1/q$ the following can be established:

Fig. 12.11 Pentagonal
tessellations {5, 3} and {5, 4}

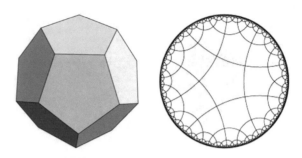

- $1/p + 1/q > 1/2$: the integer solutions are $\{\{3, 3\}, \{3, 4\}, \{3, 5\}, \{4, 3\}, \{5, 3\}\}$,
  which are the Schläfli symbols for the five Platonic solids—tetrahedron, octahe-
  dron, icosahedron, cube, dodecahedron.
- $1/p + 1/q = 1/2$: the integer solutions are $\{\{3, 6\}, \{4, 4\}, \{6, 3\}\}$, corresponding
  to the familiar tilings of the plane by equilateral triangles, squares, hexagons.
- $1/p + 1/q < 1/2$: there are infinite integer solutions, and as many regular tessellations of the hyperbolic plane.

Figure 12.11 shows two tessellations with regular pentagons meeting at vertices
in groups of three (left) and four (right) yielding, respectively, positive and negative
curvature. Curvature can indeed be defined also for planar graphs.

Definition 1 *(Combinatorial curvature)*

For a planar graph $G(N, E)$, the combinatorial curvature cc of a node $x \in N$ is
defined as:

$$cc(x) := 1 - degree(x)/2 + \sum_{f \sim x}(1/size(f)),$$

where summation is over all faces f incident with x. ⋄

Based on this definition, the dodecahedron {5, 3} has constant positive curvature
1/10, while hyperbolic tessellation {5, 4} has constant negative curvature –1/5.

We have mentioned regular tessellations and their curvature for comparisons with
the causet in Fig. 12.10-right. This planar graph has less symmetries than those of
the tessellations in Fig. 12.11, and is essentially formed by two concentric spirals of
pentagonal faces; these faces meet at vertices in groups of three *or* four, with the
exception of a pair of adjacent faces—a heptagon and an exagon—found at each
round of one of the spirals. Following the spiral paths we find that nodes with degree
3 and degree 4 alternate, with the degree understood now as the sum of the in-degree
and out-degree. Thus, roughly half of the nodes have positive curvature, and half have
negative curvature. Link orientation is not shown in figure: both the radial links and
those on the spiralling paths point outwards. Note that a hypothetical arrangement of
the pentagons into a single spiral would yield a totally ordered, totally uninteresting
causet: the presence of at least two spirals is essential for avoiding this collapse.

An interesting effect of the spiral arrangement is that a spiral path provides the
longest path from the root, at the center of the graph, to any given node x, while an

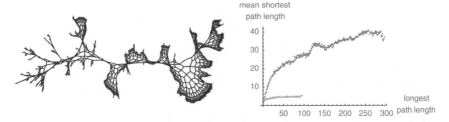

Fig. 12.12 *Left* EH-causet from a Ring Ant computation, in which the pattern observed in the causets of Fig. 12.10 appears mixed with random-like elements. *Right* longest/shortest path plot for this causet (upper function) compared with the analogous plot of a sprinkled causet (lower function)

essentially radial path will provide a substantially shorter, alternative path to x. This 'trick' implemented by the graph explains the relatively good performance of the longest/shortest path plot. Note that a similar pair of paths—a long, spiralling route and a short, mainly radial one—can be found for any pair of nodes.

12.7.4   Mixed Cases

Several cases were found in which elements of order—e.g. the tiling structure—are mixed with random-like components. One example is the EH-causet shown in Fig. 12.12.

The mix of regularity and pseudo-randomness is one of our key motivations for investigating algorithmic causets. However, the performance of this causet in terms of our rough Lorentzianity indicator is quite poor, as revealed by the plot in the r.h.s. of the figure.

12.7.5   Causet from the Fractal Sequence

When discussing random-like EH-causets, we have introduced a randomised causet construction procedure in which each new node n is connected to previously created from-nodes p, q, r, chosen at random.

Now that we have seen examples of regular causets it is useful to explore the opposite solution and directly select those from-nodes by a completely deterministic procedure.

The fractal sequence [36] is a sequence of natural numbers defined as:

$$a(n) = k \quad \text{if} \quad n = (2k - 1) * 2^m,$$

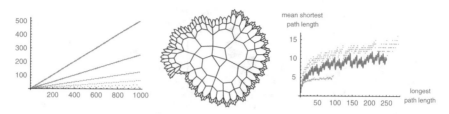

Fig. 12.13 *Left* Fractal sequence. *Center* Causet obtained from the sequence of pairs of elements of the fractal sequence. *Right* longest/shortest path plot of the causet (*intermediate solid line*), compared with analogous plots for sprinkled causet (*lower*) and causet in Fig. 12.10-right (*upper*)

where $m = 0, 1 \ldots$ and $k = 1, 2 \ldots$ Its first 12 values are (1, 1, 2, 1, 3, 2, 4, 1, 5, 3, 6, 2); in Fig. 12.13-left we plot the first 1000 values. The center of the figure shows the 502-node causet derived by splitting that 1000-element sequence into 500 consecutive pairs and using each pair as the from-nodes of each new node.[8] More precisely: we start with two nodes, labeled 1 and 2; the from-nodes of new node 3 are (1, 1), since this is the first pair of the fractal sequence, thus parallel edges $1 \rightarrow 3$ and $1 \rightarrow 3$ are added; then node 4 is added, with edges $2 \rightarrow 4$ and $1 \rightarrow 4$, since (2, 1) is the second pair of the sequence; then edges $3 \rightarrow 5$ and $2 \rightarrow 5$ are added, and so on. The longest/shortest path plot of the causet is shown at the right: it is the solid line that appears in the middle, between the lower reference plot for a sprinkled causet and the upper longest/shortest path plot for the example of Fig. 12.10-right, reproduced here for comparison.

The resulting causet is remarkably similar to the causet in Fig. 12.10-right. The procedures for their construction are quite different, and the raw causets appear rather different too. But when transitively reduced, the two graphs reveal the same basic structure, formed by two concentric spirals of planar faces, although with the fractal sequence one of the spirals is formed by heptagons, not pentagons. Again the relatively good longest/shortest path plot is due to the simultaneous presence of long spiral and short radial paths between nodes. This appears to be a recurrent 'trick' in algorithmic causets, for keeping the growth rate of the longest/shortest path plots under control.

12.8 PA-causets from Ring Ant Automata

As explained in Sect. 12.5.6, the deterministic Ring Ant automaton whose computations can be represented as partially ordered sets of events—yielding the EH-causets just discussed—also keeps a permutation π of the first n naturals, with n growing by one unit at each step. At any time the permutation can be readily turned into a PA-causet ('Permutation Ant'), as described in Sect. 12.5.4.

---

[8]It turns out that segmenting the fractal sequence into triples, quadruples, etc., in place of pairs, does not yield interesting causets.

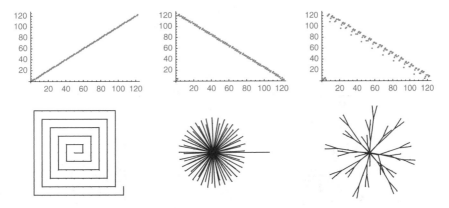

Fig. 12.14 *Upper row* Final permutations computed by three runs of the Ring Ant automaton. *Lower row* PA-causets derived from the permutations

Analogous to the case of EH-causets, about 60 % of these PA-causets are uninteresting, 1-D linear graphs. The remaining graphs split between regular structures, such as trees, and random-like structures.

12.8.1 Regular Causets

In Fig. 12.14 we show three very simple cases, meant primarily to further clarify the process of deriving the PA-causet (bottom row) from the final permutation (upper row). The elements of the permutation, intended as nodes of the causet, 'see' in their future lightcone only the elements/nodes that appear up-right to them in the permutation plot. Hence, with the permutation in Fig. 12.14-left, that moves upward, all nodes are causally related with one another, thus yielding a linear path structure. With the two remaining permutations, that move downward, one or a few nodes near the origin of the plot 'see' almost all the remaining nodes in their future lightcone, while all these nodes are totally or largely causally unrelated, thus yielding a tree structure.

Two further regular cases are illustrated in Fig. 12.15. The graphs look similar, but differ in link orientation. In the graph on the right all radial edges point outward, while in the graph on the left the outer radial edges point inward. The reader may easily deduce the impact of this difference on the two longest/shortest path plots. In particular, the graph on the right is the first we have found whose longest/shortest path plot outperforms that of the sprinkled causet. The 'trick' is trivial, even more than that of the spiralling graphs: all nodes that are reached from the root—the central node—by a longest path longer than 2 can be reached by a shortest path of length 2, thus the longest/shortest path plot is constant.

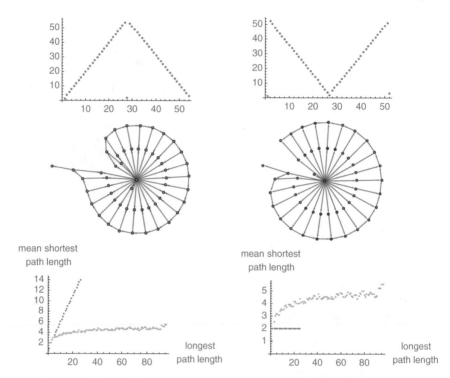

Fig. 12.15 *Upper row* final permutations computed by two runs of the Ring Ant automaton. *Central row* PA-causets derived from the permutations. *Lower row* longest/shortest path plots. In both plots we compare the longest/shortest path plot of the corresponding causet—a *straight line*—with that for a sprinkled causet

12.8.2 Random-Like Causets

Let us now consider some random-like cases. Two of these PA-causets are shown in the central row of Fig. 12.16; the permutation from which each is derived appears in the upper row, and the corresponding longest/shortest path plots (dotted lines), compared with the analogous plot for sprinkled causets (solid lines), is shown in the lower row.

At a simple visual inspection, these deterministic causets appear indistinguishable from the stochastic causets obtained by sprinkling. Most importantly, their longest/shortest path plots are equivalent to those from sprinkled causets. Thus, we have eventually found algorithmic causets that satisfy our test for Lorentzianity.

The significance of this success should be correctly assessed, and perhaps demystified: the result ultimately confirms that permutations are indeed equivalent to sprinklings, as discussed in Sect. 12.5.4, and that some instances of our Ring Ant automaton behave as 'good' pseudo-random number generators—an ability that, according to a conjecture proposed in [38], would indicate computational universality.

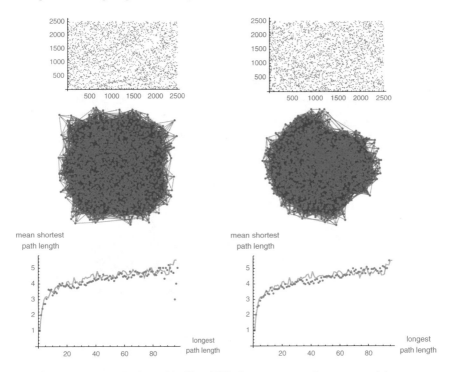

Fig. 12.16 *Upper row* final, random-like, 2500-element permutations computed by two runs of the Ring Ant automaton. *Central row* PA-causets derived from the permutations. *Lower row* longest/shortest path plots. The *dotted line* is the longest/shortest path plot for the PA-causet, while the *solid line* is that of a sprinkled causet of the same size (2500 nodes)

On the other hand, the crucial challenge—one that justifies our insistence on determinism—would be to find cases in which passing our Lorentzianity test combines with the presence of order, or of some mix of order and disorder in the causet, where 'order' is simply understood as a regularity that can be easily detected by visual inspection. Would this be possible?

Note that with PA-causets we can visually inspect two types of diagram—the permutation and the graph—with the idea that possible emergent order might be more apparent in one than in the other. By exploiting this advantageous circumstance, we have identified some additional interesting computations that seem to match, at least to some extent, our objective.

12.8.3   Mixed Cases

In the case illustrated in Fig. 12.17, the final permutation, shown on the left, is somewhat similar to the fractal sequence of Fig. 12.13-left. A peculiarity of the

Fig. 12.17 A peculiar, regular Ring Ant computation. *Left* final permutation. *Center* PA-causet derived from the permutation. *Right* longest/shortest path plot (the short horizontal segment), compared with the analogous plot for a sprinkled causet

Fig. 12.18 The distilled structure of the causet in Fig. 12.17

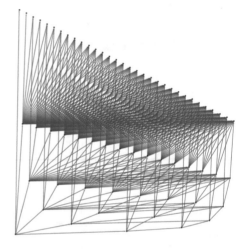

corresponding causet, shown at the center of the figure, is that the shortest path from the root to any node is never longer than 2. For the 1000-node causet shown, the longest path has length at most 10. The resulting longest/shortest path plot is shown in Fig. 12.17-right, where it is compared, as usual, with the analogous plot for a 2500-node sprinkled causet.

Given the high regularity detected in the permutation diagram—a regularity that goes unnoticed in the plot of the graph—we have reconstructed a version of the permutation by an ad-hoc algorithm, and plotted the graph with integer node coordinates as defined in Sect. 12.5.4, in order to better expose its structure. This is shown in Fig. 12.18.

The reader may easily check that the shortest path from the root—the leftmost node at the bottom—to any other node is never longer than 2, while the longest path does grow, but quite slowly, presumably as the logarithm of the number of nodes. This graph implements another brute-force 'trick' for keeping the growth of the longest/shortest path plot under control, alternative to the double spiral of Fig. 12.10-right or the circular pattern of Fig. 12.15-right.

Of course a flat longest/shortest path plot—one in which the mean shortest path length is constant, and independent from the longest path length—is as bad as one that grows linearly; our aim is to approximate the longest/shortest path plots of sprinkled causets.

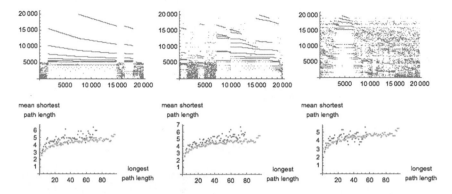

Fig. 12.19 *Upper* permutations from 20,000-step computations of the Ring Ant automaton. *Lower* longest/shortest path plots for the same computations, but limited to 2500 steps. These are compared, as usual, with the plot for a 2500-node sprinkled causet

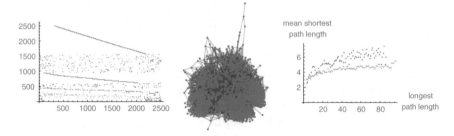

Fig. 12.20 A Ring Ant computation producing a mixed, ordered and random-like pattern. *Left* final permutation. *Center* PA-causet derived from the permutation. *Right* longest/shortest path plot

It turns out that several permutations can be obtained in which aspects of regularity and of pseudo-randomness are mixed to varying degrees, with a beneficial effect on longest/shortest path plots. This phenomenon is illustrated in Fig. 12.19, where three permutations of this kind and their corresponding longest/shortest path plots are presented.

A rather peculiar case of mix between order and randomness, in which the two components are sharply separated, is illustrated in Fig. 12.20. As in the previous example of Fig. 12.17, the plot of the final permutation is much more informative than the (default rendering of the) graph. This permutation appears as a mix of the descending 'lines' already seen in Fig. 12.17, and a series of random-like slabs, and its structure appears to grow indefinitely. In Fig. 12.21 we show the final permutation after 21,000 steps of the automaton. The appearance of these slabs plays an essential role in keeping the longest/shortest path plot close to that of sprinkled causets.

Do the 'lines' that appear in the permutation of Fig. 12.17, or the remarkable mix of order and disorder in the permutation of Fig. 12.21, correspond to properties of physical relevance for the associated causet/spacetime?

Fig. 12.21 Final
permutation after 21,000
steps of the automaton of
Figure 12.20

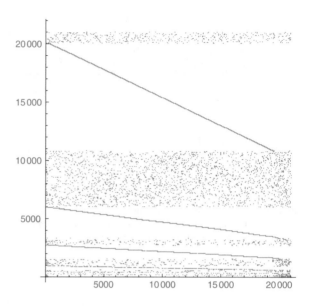

A first simple remark is that in both cases the 'lines' are formed by points that are in space-like relation with one another. However, these sets are not maximal space-like regions, since each line has points that are space-like related also to additional, external points. Furthermore, it would be wrong to take these lines as separators between past and future, since links may well cross them, as documented in Fig. 12.18. In conclusion, whether or not these lines might represent some meaningful 2-D spacetime pattern is still unclear to us.

On the other hand, in search for properties of physical relevance, we may wonder how these two cases perform under the Lorentz transformation.

In the left column of Fig. 12.22 we show the sets of points of our two permutations after a 45 degree rotation, thus going back to their interpretation as events in Minkowski space $M^{(1,1)}$. In the column at the right we show their Lorentz transformations, relative to a reference system that moves at 1/3 of the speed of light c with respect to the system at rest.

Let us focus on the upper case. Consider the descending lines that form the original permutation π (Figs. 12.17-left and 12.18), and assign them indices $i = 1, 2,...$, starting from the top. Line i is formed by points whose y coordinates decrease by unit steps while x-coordinates are evenly spaced by steps of length 2^i. As a consequence, the angular coefficient of line i is $m(i) = -2^{-i}$. After the rotation by $\pi/4$, line i has angular coefficient $Tan(\theta(i) + \pi/4)$, where $\theta(i) = ArcTan(-2^{-i})$. After some manipulation, using:

$$Sin(ArcTan(-2^{-i})) = -2^{-i}/\sqrt{1 + 4^{-i}},$$
$$Cos(ArcTan(-2^{-i})) = 1/\sqrt{1 + 4^{-i}},$$

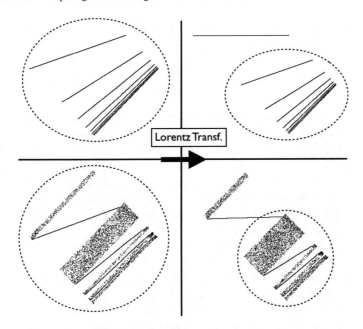

Fig. 12.22 Lorentz transformation of the two permutations of Figs. 12.17 (32,767 nodes) and 12.21 (21,000 nodes). The points are plotted as seen by a system at rest (*left column*) and by a system moving at 1/3 of the speed of light *c* (*right column*). Selfsimilarity is manifested in the *upper* case, and approximated in the *lower* case, as highlighted by the *dotted* regions

we obtain the values $mr(i)$ of the angular coefficients of the rotated lines in the upper-left diagram of Fig. 12.22:

$$mr(i) = \frac{2^i - 1}{2^i + 1}.$$

We wish now to apply the 2-D Lorentz transformation LT_v, for a frame moving at constant speed v, to the points (x, t) of the rotated lines:

$$LT_v(x, t) = (\gamma_v(x - vt), \gamma_v(t - xv)),$$

where γ_v is the Lorentz factor $1/\sqrt{1 - v^2}$.

We can now compute the slope $mr\,LT_v(i)$ of the Lorentz-transformation LT_v of the rotated line i—another line—based on v and on the slope $mr(i)$ of the latter. Letting $LT_v(1, mr(i)) = (x_v(i), t_v(i))$, we have:

$$mr\,LT_v(i) = t_v(i)/x_v(i).$$

In general, a Lorentz transformation of the set of rotated lines, relative to a generic speed, will yield slopes that do not compare with those of the original line set. But if, for any integer j, we select speed $v(j) = (2^j - 1)/(2^j + 1)$, intended as a fraction of the speed of light c, then:

$$mr\,LT_{v(j)}(i) = \frac{t_{v(j)}(i)}{x_{v(j)}(i)} = \frac{mr(i) - v(j)}{1 - v(j)mr(i)} = mr(i - j).$$

The last equality indicates that the Lorentz transformation, for these specific $v(j)$ speeds, shifts the original lines so that they overlap with themselves.

For example, the set of lines with angular coefficients (1/3, 3/5, 7/9, 15/17, 31/33) after a Lorentz transformation $LT_{1/3}$ (i.e. for $j = 1$) become lines with coefficients (0, 1/3, 3/5, 7/9, 15/17). This is exactly reflected in the upper row of Fig. 12.22, where the upper line on the r.h.s. is flat ($mr\,LT_{v(1)}(1) = 0$), while the lines below it repeat the slopes of the lines at the l.h.s.. After transformation $LT_{3/5}$ ($j = 2$), the new slopes are (−1/3, 0, 1/3, 3/5, 7/9), and so on.

The fact that this peculiar form of invariance is achieved only for a discrete set of inertial observer speeds is strongly reminiscent of the so called *Lorentzian lattices* [27], which are invariant only under a discrete subgroup of the Lorentz group.

In the case illustrated in the lower row of Fig. 12.22, the overall structure of the cloud of spacetime points as seen from the system at rest (l.h.s. diagram) is only qualitatively repeated in a subset of the Lorentz-transformed set (r.h.s. diagram), due to the fact that the thickness values for the successive random-like slabs do not seem to follow a regular progression. Recall, however, that the Lorentz transformation leaves unaffected, in statistical sense, a cloud of points uniformly distributed in a region of $M^{(1,1)}$, such as these slabs.

We believe that the examples illustrated in this subsection represent promising preliminary steps in the search for algorithmic causets that mix regular and pseudo-random features while attempting to match the requirement of Lorentz invariance.

12.9   Conclusions

Does it make sense to talk about *spacetime computing*? In this chapter we hope we have identified a few attractive research items that can be legitimately grouped under this name, and be seen as dealing with a rather extreme form of natural computing.

As anticipated in the introduction, our presentation has mainly focused on some specific issues related to the modelling of discrete, algorithmic spacetime by causal sets, while several other relevant aspects have been left uncovered. A much wider treatment of the relations between nature and computation can be found, for example, in [39].

One of the aspects we have largely ignored here is quantum mechanics.

Although causal sets reflect in their basic structure the quantisation of spacetime, each node of the graph corresponding to a 'quantum' (an 'atom') of the latter, much more would be needed, e.g. in terms of dynamical laws, in order to set up a fully blown quantum-mechanical, algorithmic, causet-based theory of the natural universe, one involving Lagrangians, amplitudes, path integrals or sums over histories and all the conceptual tools that make quantum mechanics and quantum field theory so powerful (and so difficult). It is fair to say that very little progress has been done so far by theoretical physicists in this direction: for our purposes here, any detailed discussion on these aspects would be inappropriate.

It is worth mentioning, however, that the related question of whether and how we can use computers to *fully* and *exactly* simulate Physics, *and quantum mechanical features in particular*, was already addressed by Richard Feynman in 1981, in a famous keynote speech at a conference on the 'Physics of Computation' [10]. In that speech, also due to previous interactions with Fredkin, Feynman suggested, for this simulation, a visionary computer architecture based on cellular automata enriched with quantum mechanical capabilities, thus promoting the development of quantum computing.

The influence of these ideas is particularly evident in the work of Seth Lloyd, where all the different physical phenomena of the quantum world—e.g. all sub-atomic particle interactions—are interpreted as different quantum information processing activities, and the universe is seen as a huge network of programs that collectively determine the evolution ... of the universe itself [17].

If this multiplicity of different quantum computing processes is a correct picture of our world, then we might hope to be eventually able to 'crack' the code of some of these programs and profit from their imitation, in the same way as we do with bio-inspired computing.

However, the boldest conjecture about spacetime computing hints at the existence of a *single* algorithm at the root of everything, with that multiplicity of computing processes only emerging from this unique source, and Schmidhuber [28] goes as far as suggesting that it is cheaper for a Turing machine to compute *all* possible computable universes rather than just one (ours), thus outlining an ultra-concise, computational theory of the multiverse.

Whether *this* is instead the correct picture—whether there is indeed a single, possibly elementary, possibly immutable, perhaps even deterministic and non-quantistic algorithm at the bottom of the universe or multiverse[9]—is still completely unknown. What is certain is that its discovery would deeply revolutionise the landscape of theoretical physics, catapulting spacetime computing to its forefront.

In light of the above conjecture, a sensible research track in the spacetime computing agenda is, in our opinion, the exploration of *abstract* algorithms for building discrete models of spacetime - models that, in the present chapter, we have identified with causal sets. By 'abstract' we mean that we abstain from coding into these algorithms any knowledge from theoretical physics—e.g. constants such as the speed of light—since everything should emerge from the algorithm itself, a posteriori.

One may object that the space of algorithms is potentially infinite, and that searching it blindly is unreasonable. We believe that these difficulties can be in part mitigated. First, the notion of computational universality (or Turing-completeness) acts

---

[9]The possible existence of non-quantum mechanical laws at the roots of reality, below the layer of quantum mechanics, has been recently envisaged by G.'t Hooft [31].

as a sort of unifying factor for all models above a certain threshold of sophistication. Second, the space of qualitative behaviours that characterise the computations of potentially all conceivable algorithms is much smaller than the space of those algorithms, as widely shown by Wolfram [38].

Furthermore, the search for 'promising' algorithms is not totally blind, but should be guided by the early appearance of interesting emergent patterns—patterns that we should be able to recognise as useful for setting up, in the long, or *very long* run, the features of our familiar physical world. Among the valuable clues we include periodicity, self-similarity, pseudo-randomness, and 'digital particles'.

All of these features have been already found in the artificial, computational universe, but mainly in models such as cellular automata and 2-D Turing Machines that, due to other intrinsic limitations, lack physical realism. In this chapter we have therefore devoted much attention to the alternative model of (algorithmic) causal sets, and have extended the list of desirable emergent features to one of great physical significance: Lorentz invariance. We have discussed an indicator—the longest/shortest path plot—meant to reveal the closeness of an algorithmic causet to the ideal (2-D) Lorentzian causet, and have shown the extent to which regular, random-like, or mixed causets perform against this benchmark. Finally, we have identified causets that satisfy a form of discretised Lorentz invariance while offering a remarkable mix of regularity and pseudo-randomness.

Causal sets represent a more promising and physically realistic model than cellular automata or Turing machines. Of course, several of the regular, often planar causets that we have introduced appear naively simple, and remote from the complexity of the 4-dimensional causets that we might expect to represent spacetime, at least at sufficiently high scales of observation. Still, they have been helpful for elucidating the variety of emergent properties that the model can offer.

Cracking the code that animates the elusive, discrete texture of the physical universe is the most ambitious goal of spacetime computing. As pointed out in [19], this goal would greatly benefit from the cooperation of various research areas, most notably quantum gravity and cosmology, complex networks and the theory of computing. Algorithmic causal sets seem to represent an ideal choice also in this respect: causets have acquired enduring attention and esteem from various schools of thought in quantum gravity, and, given their high abstraction level, they also offer a relatively easy access point for investigations and simulations by the curious computer scientist!

Acknowledgments I express my gratitude to Alex Lamb and Marco Tarini for useful technical discussions, and to Stephen Wolfram for having initially stimulated my interest in this area of research.

References

1. Adamatzky, A.: Computing in Nonlinear Media and Automata Collectives. IoP Institute of Physics Publishing, London (2001)
2. Adleman, L.M.: Molecular computation of solutions to combinatorial problems. Science **266**(11), 1021–1024 (1994)
3. Ambjørn, J., Jurkiewicz, J., Loll, R.: The self-organizing quantum universe. Sci. Am. **299**, 42–49 (2008)
4. Bolognesi, T.: Planar trinet dynamics with two rewrite rules. Complex Syst. **18**(1), 1–41 (2008)
5. Bolognesi, T.: Algorithmic causal sets for a computational spacetime. In: Zenil., H. (ed.) A Computable Universe. World Scientific, Singapore (2013)
6. Bolognesi, T., Lamb, A.: Simple Indicators for Lorentzian Causets (2014). http://arxiv.org/abs/1407.1649v2 [gr-qc]
7. Bombelli, L., Lee, J., Meyer, D., Sorkin, R.D.: Space-time as a causal set. Phys. Rev. Lett. **59**(5), 521–524 (1987)
8. Brown, J.: Minds, Machines, and the Multiverse—The Quest for the Quantum Computer. Simon & Schuster, New York (2000)
9. Conway, J.H., Burgiel, H., Goodman-Strauss, C.: The Symmetries of Things. A K Peters CRC Press, Massachusetts (2008)
10. Feynman, R.: Simulating physics with computers. Int. J. Theor. Phys. **21**(6–7), 467–488 (1982)
11. Fredkin, E.: Five big questions with pretty simple answers. IBM J. Res. Dev. **48**(1), 31–45 (2004)
12. Gacs, P., Levin, L.A.: Causal nets or what is a deterministic computation? Inf. Control **51**, 1–19 (1981)
13. Gardner, M.: Mathematical games: the fantastic combinations of John Conway's new solitaire game 'Life'. Sci. Am. **223**(4), 120–123 (1970)
14. Hutter, M., Legg, S., Vitanyi, P.M.B.: Algorithmic probability. Scholarpedia, **2**(8), 2572 (2007). (Revision 151509)
15. Pegg, E., jr.: Turmite, From MathWorld–A Wolfram Web Resource, created by Eric W. Weisstein. http://mathworld.wolfram.com/Turmite.html, Jan. 21 (2011)
16. Livshits, G.I., et al.: Long-range charge transport in single G-quadruples DNA-molecules. Nat. Nanotechnol. **9**, 1040–1046 (2014)
17. Lloyd, S.: Universe as quantum computer. Complexity **3**(1), 32–35 (1997)
18. Margolus, N.: Crystalline computation. In: Hey, A.J.G. (ed.) Feynman and Computation, pp. 267–305. Perseus Books, Cambridge (1999)
19. Markopoulou, F.: The computing spacetime. In: Cooper, S.B., Dawar, A., Löwe, B. (eds.) How the World Computes—Turing Centenary Conference and 8th Conference on Computability in Europe, CiE 2012, Cambridge, UK, June 18-23, 2012. Proceedings, Lecture Notes in Computer Science, vol. 7318, pp. 472–484. Springer, Heidelberg (2012)
20. Meyer, D.A.: The dimension of causal sets, Ph.D. Thesis, MIT (1989)
21. Miller, D.B., Fredkin, E.: Two-state, reversible, universal cellular automata in three dimensions. In: CF '05: Proceedings of the 2nd Conference on Computing Frontiers, pp. 45–51. ACM, New York (2005)
22. Mills, J.W.: The nature of the extended analog computer. Physica D Nonlinear Phenomena **237**, 1235–1256 (2008)
23. Nowotny, T., Requardt, M.: Dimension theory of graphs and networks. J. Phys. Math. Gen. **31**(10), 2447 (1998)
24. Reuter, M., Saueressig, F.: Quantum Einstein gravity. New J. Phys. **14**(5), 055022 (2012)
25. Rideout, D., Wallden, P.: Emergence of spatial structure from causal sets. Proceedings DICE 2008, J. Phys. Conf. Ser. 174:012017, April 30 2009. doi:10.1088/1742-6596/174/1/012017
26. Rubel, L.A.: The extended analog computer. ADVAM Adv. Appl. Math. **14**, 39–50 (1993)
27. Saravani, M., Aslanbeigi, S.: On the causal set-continuum correspondence. Class. Quantum Gravity **31**(20), 205013 (2014)

28. Schmidhuber, J.: A computer scientist's view of life, the universe, and everything. In: Freksa, Christian, Jantzen, Matthias, Valk, Rüdiger (eds.) Foundations of Computer Science. Lecture Notes in Computer Science, vol. 1337, pp. 201–208. Springer, Heidelberg (1997)

29. Smolin, L.: Atoms of space and time. Sci. Am. 66–75 (2004)

30. Soler-Toscano, F., Zenil, H., Delahaye, J.-P., Gauvrit, N.: Calculating Kolmogorov complexity from the output frequency distributions of small Turing machines. PLOS one 9(5), e96223 (2014)

31. Gerard 't, H.: The Cellular Automaton Interpretation of Quantum Mechanics, June (2014). http://arxiv.org/abs/1405.1548 [quant-ph]

32. Toffoli, T.: Non-conventional computers. In: Webster, J. (ed.) Encyclopedia of Electrical and Electronics Engineering, pp. 455–471. Wiley, New York (1998)

33. Toffoli, T., Margolus, N.: Cellular automata machines. Complex Syst. 1 (1987)

34. Toffoli, T., Margolus, N.: Programmable matter: concepts and realization. Int. J. High Speed Comput. 5, 155–170 (1993)

35. Toffoli, T., Margolus, N.: Invertible cellular automata: a review. In: International Symposium on Physical Design (1994)

36. Weisstein, E.W.: Fractal sequence. From MathWorld–A Wolfram Web Resource, created by Eric W. Weisstein. http://mathworld.wolfram.com/FractalSequence.html, valid on Nov. 25, 2015

37. Weisstein, E.W.: Langton's ant. From MathWorld–A Wolfram Web Resource, created by Eric W. Weisstein. http://mathworld.wolfram.com/LangtonsAnt.html, valid on Nov. 26, 2015

38. Wolfram, S.: A New Kind of Science. Wolfram Media, Inc., Champaign (2002)

39. Zenil, H. (ed.) A Computable Universe. World Scientific, Singapore (2013)

40. Zuse, K.: Calculating space. Technical report, Proj, MAC, MIT, Cambridge, Mass., 1970. Technical Translation AZT-70-164-GEMIT. Original title: "Rechnender Raum"

Chapter 13
Interaction-Based Programming in MGS

Antoine Spicher and Jean-Louis Giavitto

Abstract The modeling and simulation of morphogenetic phenomena require to take into account the coupling between the processes that take place in a space and the modification of that space due to those processes, leading to a chicken-and-egg problem. To cope with this issue, we propose to consider a growing structure as the byproduct of a multitude of *interactions* between its constitutive elements. An interaction-based model of computation relying on spatial relationships is then developed leading to an original style of programming implemented in the MGS programming language. While MGS seems to be at first glance a domain specific programming language, its underlying interaction-based paradigm is also relevant to support the development of generic programming mechanisms. We show how the specification of space independent computations achieves *polytypism* and we develop a direct interpretation of well-known differential operators in term of data movements.

13.1 Introduction

The development of the MGS unconventional programming language was driven by a motto: *computations are made of local interactions*. This approach was motivated by difficulties encountered in the modeling and simulation of morphogenetic processes [13, 16] where the construction of an organism in the course of time, from a germ cell to a complete organism, is achieved through a multitude of local interactions between its constitutive elements. In this kind of systems, the spatial structure varies over time and must be calculated in conjunction with the state of the system.

A. Spicher (✉)
LACL, Université Paris-Est Créteil, 61 rue du général de Gaulle, 94010 Créteil, France
e-mail: antoine.spicher@u-pec.fr

J.-L. Giavitto
UMR 9912 STMS, IRCAM – CNRS – Paris Sorbonne University,
UPMC – INRIA, 1 Place Igor-Stravinsky, 75004 Paris, France
e-mail: jean-louis.giavitto@ircam.fr

© Springer International Publishing Switzerland 2017
A. Adamatzky (ed.), *Advances in Unconventional Computing*,
Emergence, Complexity and Computation 22,
DOI 10.1007/978-3-319-33924-5_13

As an example, consider the diffusion in a growing medium of a morphogen that controls the speed of growth of this medium. The coupling between the processes that take place in a space and the modification of the space due to the processes leads to a chicken-and-egg problem: what comes first?

Recently, a new approach has emerged in theoretical physics to overcome the same difficulty in general relativity where the mass moves along a space-time geodesic while space-time geometry is defined by the distribution of mass. In this approach, space is not seen as a primitive structure but rather as a byproduct of the causal relationships induced by the interaction between the entities of the system.

Contributions

In this chapter, we propose to reconstruct the MGS mechanisms of computation from this point of view. This abstract approach is illustrated in Sect. 13.2 by a down-to-earth example, the bubble sort, and we show how the basic interactions at work (the swap of two adjacent elements) build an explicit spatial structure (a linear space). Our motivation in the development of this example, is to make explicit the basic ingredient of a generic notion of interaction: interactions define a neighborhood and, from neighbor to neighbor, a global shape emerges. This leads naturally to use topological tools to describe computations in MGS.

Section 13.3 presents the notion of topological rewriting and illustrates this notion on various paradigmatic examples of morphogenesis. These examples are only sketched to show the expected relevance of the MGS computational mechanisms in the simulation of physical or biological systems.

In the Sect. 13.4, we show that these computational mechanisms are also relevant to support the development of generic programming mechanisms, not necessarily related to natural computations. Here we address genericity from the point of view of *polytypism*, a notion initially developed in functional programming.

The MGS approach of polytypism boils down to the specification of patterns of data traversals. In Sect. 13.5, we show these patterns of data movements at work in the interpretation of differential operators. This interpretation gives a computational content to well-known differential operators subsuming discrete and continuous computations. As an illustration, we provide a generic formulation that encompasses discrete and continuous equations of diffusion that can be used in hybrid diffusion. The section ends by sketching how sorting can be achieved by a set of differential equations.

13.2   From Physics to Computation: Interactions

The simulation of morphogenetic systems leads to a chicken-and-egg situation between the processes taking place in a space and the modification of the space due to these processes. This problem has been pointed out by A. Turing in his seminal study of morphogenesis [54].

In general relativity, a similar issue occurs where the mass moves along geodesics while spacial geometry is defined by the distribution of the mass. Physicists achieved to deal with the interdependence between mass and space through the concept of *causality* in a space-time structure. In this section, we propose to transpose this idea into computations.

13.2.1   Spatial Structure, Causality and Interaction

Space-Time and Causality

To address the interdependence problem, a classical solution consists in considering a space-time as a manifold M endowed with a differentiable structure with respect to which a metric g is defined. Then a causal order is derived from the light cones of g. However, it has been known for some time that one can also go the other way [26, 29]: considering only the events of space-time M and an order relation $\prec$ such that $x \prec y$ if event x may influence what happens at event y, it is possible to recover from $\prec$ the topology of M, its differentiable structure and the metric g up to a scalar factor [5]. Moreover, it can be done in a purely order theoretic manner [30].

The causal relation $\prec$ is regarded as the fundamental ingredient in the description of the system evolution, the topology and geometry being secondary in the description of the dynamics. As advocated by the *causal set* program developed by Sorkin et al. [40, 41], this approach is compatible with the idea of "becoming", making possible to see a system more naturally as a "growing being" than as a "static thing", a mandatory characteristic of genuine morphogenetic processes.

Causality and Interaction

A similar path can be followed for the development of a framework suitable for the computer modeling of morphogenesis. The idea is to describe the evolution of the system as a set of interactions (read: computation). These interactions entail a causal relation. In this way, starting from the set of potential entities in the system and from a set of interactions acting on these entities, one may reconstruct incrementally the spatial structure of the whole system as a byproduct of the causal structure of the system's interactions (computations).

Focusing on interactions rather than on the spatial background in which the evolution takes place is a solution to the problem of describing morphogenetic processes as dynamical systems. In this view interactions are local by definition. In the manner of the aforementioned causal relation $\prec$, interactions are regarded as the fundamental ingredient in the description of the system evolution, the topology and geometry being secondary.

The study of causality in computations can be traced back at least to the sixties with the development of Petri nets, where an event (i.e., firing of an enabled transition) is a local action and where there is a clear notion of event independence. Since, the subject has been extensively studied, for example with the notion of *event structure* [56] whose intersection with causal sets in physics has been noticed [35].

13.2.2 Computing with Interactions

To illustrate the previous idea in an algorithmic context, let us revisit a classical algorithm, the *bubble sort*, by considering first the interactions at work, and then by trying to reconstruct a data structure from them.

Mathematical Notations

The set of functions (*resp.* total functions) from a domain D to a range R is written $D \to R$ (*resp.* $D \hookrightarrow R$). If $s \in D \to R$ and $D' \subseteq D$, then $s|_{D'}$ is the restriction of s to D'. The expression $[u_1 \to a_1, \ldots, u_n \to a_n]$ is an element s of $\{u_1, \ldots, u_n\} \hookrightarrow \{a_1, \ldots, a_n\}$ such that $s(u_i) = a_i$. The expression $s \cdot s'$ denotes a function s'' of domain $\mathrm{dom}(s) \cup \mathrm{dom}(s')$ such that $s''(u) = s'(u)$ if $u \in \mathrm{dom}(s')$ else $s''(u) = s(u)$. Given a set $D = \{u_1, u_2, \ldots\}$ we form the set of formal elements $\widehat{D} = \{\widehat{u}_1, \widehat{u}_2, \ldots\}$. Let e be an expression where the elements of $\widehat{D}$ appear as variables and let s be some function of $D \hookrightarrow R$, $e[s]$ denotes the evaluation of expression e where all occurrences of $\widehat{u}$ are replaced by $s(u)$ for all $u \in D$.

The "memory" where the data are stored is specified as a set of places named *positions*. A position plays the role of the spatial part of an event in physics. By denoting $\mathcal{V}$ the set of values that can be stored, a state of a computation is represented by a total function s from a set of positions to $\mathcal{V}$. Let $\mathcal{P} = \mathrm{dom}(s)$ be the set of positions in some state s, elements of $\widehat{\mathcal{P}}$ can be used in some expression e as placeholders replaced by their associated values in $e[s]$.

Interactions as Rules

An interaction is defined by a reciprocal action between positions. The effect of the interaction is to change the values associated with the involved positions. Because the spatial structure can be dynamic, for a given interaction I we consider a set l_I of input positions and a set r_I of output positions. We do not require $l_I \subseteq r_I$ (meaning that some positions may appear during the interaction), nor $r_I \subseteq l_I$ (meaning that some positions may disappear during the interaction). Finally, an interaction is guarded by some condition, that is a Boolean expression C_I, which controls the occurrence of the interaction. We will write an interaction I as a rule

$$l_I \ / \ C_I \longrightarrow R_I$$

where R_I is an expression to be evaluated to a function of $r_I \hookrightarrow \mathcal{V}$.

The semantics of an interaction is as follows: an interaction I may occur in a state s if and only if $C_I[s]$ evaluates to true. Then, the result of the interaction is

$$s|_{\mathrm{dom}(s)\setminus l_I} \cdot R_I[s]$$

The expression $s|_{\mathrm{dom}(s)\setminus l_I}$ restricts s to the positions that do not take part into the interaction. This partial state is then augmented by the local result $R_I[s]$ of the interaction. After the interaction the set of positions is given by $(\mathrm{dom}(s)\setminus l_i) \cup \mathrm{dom}(R_I)$.

Bubble Sort As a Set of Interactions

For the sake of simplicity, we focus on sorting sequences of three numbers taken in $\mathcal{V} = \{1, 2, 3\}$. Initially, the sequence is encoded using three symbolic positions p_1, p_2 and p_3, so that for instance the initial sequence $[3, 1, 1]$ is represented by the function:

$$s_0 = [p_1 \rightarrow 3, p_2 \rightarrow 1, p_3 \rightarrow 1]$$

The elementary instruction at work in the bubble sort consists in swapping two neighbor elements that are not well ordered. The algorithm can then be described by the two following interactions:

$$I_1 : \{p_1, p_2\} / (\widehat{p_2} < \widehat{p_1}) \longrightarrow [p_1 \rightarrow \widehat{p_2}, p_2 \rightarrow \widehat{p_1}]$$
$$I_2 : \{p_2, p_3\} / (\widehat{p_3} < \widehat{p_2}) \longrightarrow [p_2 \rightarrow \widehat{p_3}, p_3 \rightarrow \widehat{p_2}]$$

In state s_0, only I_1 can occur since numbers 3 and 1 are not well ordered on positions p_1 and p_2. The result of the interaction gives the sequence $[1, 3, 1]$:

$$
\begin{aligned}
s_1 &= s_0|_{\mathrm{dom}(s_0) \backslash l_{I_1}} \cdot R_{I_1}[s_0] \\
&= [p_3 \rightarrow 1] \cdot [p_1 \rightarrow 1, p_2 \rightarrow 3] \\
&= [p_1 \rightarrow 1, p_2 \rightarrow 3, p_3 \rightarrow 1]
\end{aligned}
$$

Figure 13.1 gives the state space generated by these interactions where a computation is a trajectory going from bottom to top.

Interactions I_1 and I_2 alone do not modify the underlying set of positions. To illustrate such a modification, let us consider the removal of duplicate values with some additional interactions which merge neighbor positions sharing the same value:

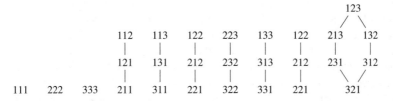

Fig. 13.1 The state space of interactions $\{I_1, I_2\}$ with $\mathcal{V} = \{1, 2, 3\}$ and $\mathcal{P} = \{p_1, p_2, p_3\}$ as set of positions. We write *abc* for the function $[p_1 \rightarrow a, p_2 \rightarrow b, p_3 \rightarrow c]$. Starting from an initial state $s = a_1 a_2 a_3$, the final state of the computation must be $s' = a_i a_j a_k$ such that $\{i, j, k\} = \{1, 2, 3\}$ and $a_i \leq a_j \leq a_k$. There are $|\mathcal{V}^{\mathcal{P}}| = 3^3 = 27$ states. One goes from one state to another by the application of one interaction (edges are oriented from *bottom* to *top*). The branching from 321 shows the possibility to apply either I_1 or I_2 leading to a non-deterministic behavior

$$I_3 : \{p_1, p_2\} / (\widehat{p_2} = \widehat{p_1}) \longrightarrow [p_4 \rightarrow \widehat{p_1}]$$
$$I_4 : \{p_2, p_3\} / (\widehat{p_3} = \widehat{p_2}) \longrightarrow [p_5 \rightarrow \widehat{p_2}]$$
$$I_5 : \{p_1, p_5\} / (\widehat{p_5} < \widehat{p_1}) \longrightarrow [p_1 \rightarrow \widehat{p_5}, p_5 \rightarrow \widehat{p_1}]$$
$$I_6 : \{p_4, p_3\} / (\widehat{p_3} < \widehat{p_4}) \longrightarrow [p_4 \rightarrow \widehat{p_3}, p_3 \rightarrow \widehat{p_4}]$$
$$I_7 : \{p_1, p_5\} / (\widehat{p_5} = \widehat{p_1}) \longrightarrow [p_6 \rightarrow \widehat{p_1}]$$
$$I_8 : \{p_4, p_3\} / (\widehat{p_3} = \widehat{p_4}) \longrightarrow [p_6 \rightarrow \widehat{p_3}]$$

Interaction I_3 (*resp.* I_4) expresses the merge of positions p_1 and p_2 into p_4 (*resp.* p_2 and p_3 into p_5) when they are labeled with the same value. Interactions I_5 and I_6 specify the sorting of values when p_4 and p_5 are labeled. Finally, I_7 and I_8 describe the merge of positions into a unique position p_6.

With these additional interactions, another outcome is possible from state s_0 by applying interaction I_4:

$$s_1' = s_0|_{\text{dom}(s_0)\backslash l_{I_4}} \cdot R_{I_4}[s_0] = [p_1 \rightarrow 3, p_5 \rightarrow 1]$$

As expected, the set of positions is modified to $\{p_1, p_5\}$. The corresponding state space is of course bigger than the previous one and exhibits more branching (i.e., non-determinism) but remains confluent.

The Spatial Organization of Positions

Although the set of positions comes without any structure, the interactions make a specific use of it so that in general any position is not involved with all the others. As a consequence, the interactions induce a notion of locality leading to a spatial organization of the set of positions. This space can be made explicit by building the minimal structure such that, for any interaction I, the elements of l_I are "neighbors". Noticing that for an interaction I, the interaction also involves any subset of l_I, the neighborhood must be *closed by inclusion*. It turns out that this property defines a combinatorial spatial structure called an *abstract simplicial complex* (ASC) [22].

An ASC is a collection $\mathcal{K}$ of non-empty finite sets such that $\sigma \in \mathcal{K}$ and $\tau \subseteq \sigma$ implies $\tau \in \mathcal{K}$. The elements of $\mathcal{K}$ are called *simplices*. The *dimension* of a simplex $\sigma \in \mathcal{K}$ is $\dim(\sigma) = |\sigma| - 1$ and the dimension of a complex is the maximum dimension of any of its simplices when it exists. A simplex of dimension n is called a n-simplex. A vertex is a 0-simplex. The *vertex set* of an ASC $\mathcal{K}$ is the union of all its simplices, $\text{Vert}(\mathcal{K}) = \cup_{\sigma \in \mathcal{K}} \sigma$. An edge is a 1-simplex whose border consists of two vertices. A graph is an ASC of dimension 1 built with vertices and edges. An ASC of dimension 2 also contains triangular surfaces bounded by 3 edges. ASCs of dimension 3 contain tetrahedrons bounded by four 2-simplices. And so on and so forth.

Let $\mathscr{I}$ be a set of interactions. We call the *interaction complex* of $\mathscr{I}$, the smallest ASC $\mathcal{K}_{\mathscr{I}}$ containing all the l_I as simplices for $I \in \mathscr{I}$. For the set of interactions $\{I_1, I_2\}$, the associated interaction complex is:

$$\mathcal{K}_{\{I_1, I_2\}} = \big\{\{p_1\}, \{p_2\}, \{p_3\}, \{p_1, p_2\}, \{p_2, p_3\}\big\}$$

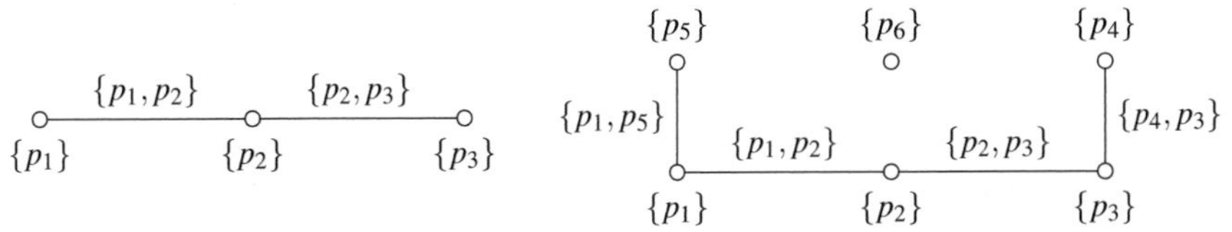

Fig. 13.2 Examples of interaction complex: $\mathcal{K}_{\{I_1,I_2\}}$ on the *left* and $\mathcal{K}_{\{I_1,...,I_8\}}$ on the *right*

The complex is pictured on the left of Fig. 13.2. As expected for a bubble sort, the induced spatial organization is a sequential 1-dimensional structure, here $[p_1, p_2, p_3]$. The bubble sort without duplicate values gives raise to the interaction complex $\mathcal{K}_{\{I_1,...,I_8\}}$ pictured on the right of Fig. 13.2. The complex exhibits the four sequential organizations that the system can take over time: $[p_1, p_2, p_3]$, $[p_4, p_3]$ (after the merge of p_1 and p_2), $[p_1, p_5]$ (after the merge of p_2 and p_3), and $[p_6]$ (if the three values were initially the same). Notice that $\mathcal{K}_{\{I_1,I_2\}}$ is a sub-complex of $\mathcal{K}_{\{I_1,...,I_8\}}$ since $\{I_1, \ldots, I_8\}$ contains I_1 and I_2.

Asymmetry of Interaction

The interactions involved in the bubble sort are asymmetric. For example, the roles played by p_1 and p_2 in I_1 are not interchangeable so that I_1 differs from:

$$I'_1 : \{p_2, p_1\} / (\widehat{p_1} < \widehat{p_2}) \longrightarrow [p_2 \to \widehat{p_1}, p_1 \to \widehat{p_2}]$$

although $l_{I_1} = l_{I'_1}$. In fact, using a set to track the input positions l_I does not catch all the information contained in the interaction. As a consequence, some different sets of interactions may have the same interaction complex. For example, $\mathcal{K}_{\{I'_1,I_2\}}$ is exactly the same as $\mathcal{K}_{\{I_1,I_2\}}$.

We can get round this issue by considering a *directed* spatial structure rather than an ordinary ASC. The notion of directed graph exists, as well as the notion of directed ASC [21] (direction in ASC differs from the notion of orientation for dimension greater than 2, e.g., there are two orientations for a 2-simplex but three directions). For the sake of simplicity, let us put aside the asymmetry issue and restrict the formal descriptions to undirected structures.

13.3 An Interaction-Based Programming Language

Following the approach given above, sorting n elements requires n positions and $(n-1)$ interactions, and sorting n elements without duplicate values requires $n(n+1)/2$ positions and $\sum_{i=0}^{n} 2i(n-i)$ interactions. However, these rules are pretty similar and can be captured by some "meta-rules". For example, the previous interactions can be represented by the following generic rules:

$$\rho_{x,y} : \{x, y\} \,/\, \widehat{y} < \widehat{x} \;\longrightarrow\; [x \to \widehat{y}, y \to \widehat{x}]$$
$$\rho'_{x,y,z} : \{x, y\} \,/\, \widehat{y} = \widehat{x} \;\longrightarrow\; [z \to \widehat{x}]$$

These rules mean to denote a whole family of interactions that are obtained by substituting positions for x, y and z. In fact, x, y and z are *position variables* instead of actual positions. For example, interaction I_1 is got by applying substitution $[x \to p_1, y \to p_2]$, that is, $I_1 = \rho_{p_1,p_2}$; in the same way, $I_3 = \rho'_{p_1,p_2,p_4}$.

However, while some substitutions are desired, a lot of them are not. For example, ρ_{p_1,p_6} is definitively not part of the original specification of the bubble sort. Indeed, the authorized substitutions have to respect some knowledge that was built implicitly in the original interactions. In our example, x and y should stand for two positions so that x is "before" y. As a matter of fact, this knowledge is the one captured by the interaction complex. Thus, a necessary condition for a substitution of $\rho_{x,y}$ (*resp.* $\rho'_{x,y,z}$) to be accepted is that set $l_{\rho_{x,y}}$ (*resp.* $l_{\rho'_{x,y,z}}$) corresponds to a simplex of $\mathcal{K}_{\{I_1,\ldots,I_8\}}$.

This approach has been used to design the interaction-based programming language MGS. In MGS, a computation is specified through sets of "meta-rules" called *transformations*. Such a transformation is to be applied on a *topological collection*, that is, a set of labeled positions equipped with an interaction complex. The application is done by matching some positions in the collection which respect the inputs and conditions of some rule of the transformation. The instance of the rule gives raise to an interaction which modifies locally the state of the collection and possibly its structure. In this section, the syntax of the language is briefly described and its use is illustrated with a light survey of examples involving dynamic organizations (where the set of positions is not fixed once and for all) and higher dimensional structures (beyond graphs).

13.3.1 A Brief Description of the MGS Language

MGS provides *topological collections*, an original data structure for representing the state of a system based on the topology of interactions, and *transformations*, a rule-based definition of functions on collections for specifying the interaction laws of the system.

Topological Collections

Topological collections are the unique data structure available in MGS. They define the interaction structure of a dynamic system. They can also be seen as a *field* associating a value with each element of a combinatorial structure modeling the topology of a space.

In the previous section, we focused on ASC to model the spatial structure. This structure is the natural choice for the spatial constraint arising from the interactions. However, other combinatorial structures extending the notion of ASC can be used to get more concision and flexibility in the representation. In the MGS language, *cell spaces* [53] are used to subsume ASC and other kinds of spatial organization, so that a topological collection is a labeled cell space.

Cell Spaces

Formally, cell spaces are made of an assembly of elementary objects called *topological cells* (*cells* for short). For the sake of simplicity, let us assume the existence of a set of topological cells $\mathscr{P}$ together with a function $\dim : \mathscr{P} \hookrightarrow \mathbb{N}$ associating a *dimension* with each cell. Cells $\sigma \in \mathscr{P}$ such that $\dim(\sigma) = n$ are called *n-cells*. Cells represent elementary pieces of space: 0-cells are vertices, 1-cells are edges, 2-cells are surfaces, 3-cells are volumes, etc.

A *cell space* $\mathscr{K}$ is a partially ordered subset of $\mathscr{P}$, that is a couple $\mathscr{K} = \langle S_{\mathscr{K}}, \prec_{\mathscr{K}} \rangle$ such that $S_{\mathscr{K}} \subset \mathscr{P}$ and $\prec_{\mathscr{K}}$ is a strict partial order[1] over S such that the restriction of the dimension function on $S_{\mathscr{K}}$ is strictly monotonic: for all $\sigma, \tau \in S_{\mathscr{K}}, \sigma \prec_{\mathscr{K}} \sigma' \Rightarrow \dim(\sigma) < \dim(\sigma')$. If it exists, the dimension of a cell space is the maximal dimension of its cells.

The relation $\prec_{\mathscr{K}}$ is called the *incidence relationship* of the cell space $\mathscr{K}$, and if $\sigma \prec_{\mathscr{K}} \sigma'$, σ and σ' are said *incident*. Contrary to ASC, the number of cells in the boundary of a cell is not constrained.

We call *closure* (resp. *star*) of a cell σ the set $\mathrm{Cl}\,\sigma = \{\sigma' \mid \sigma' \preceq \sigma\}$ (resp. $\mathrm{St}\,\sigma = \{\sigma' \mid \sigma' \succeq \sigma\}$). Operator Cl is a closure operator that can be used to equip the set of cells $S_{\mathscr{K}}$ with a topology. Numerous operators can be defined to exploit the induced space. The notions of face and (p, q)-neighborhood are especially used in MGS. The *faces* of a cell σ are the cells σ' that are immediately incident: $\sigma' \prec_{\mathscr{K}} \sigma$ and $\dim(\sigma') = \dim(\sigma) - 1$; σ is called a *coface* of σ' and we write $\sigma' < \sigma$. Two cells are *q-neighbor* if they are incident to a common q-cell. If the two cells are of dimension p, we say that they are (p, q)-*neighbor*. A (p, q)-*path* is then a sequence where any two consecutive cells are (p, q)-neighbor.

Cell spaces are very general objects allowing sometimes unexpected constructions. For example, an edge with three vertices in its border is a regular cell space. Additional properties are often considered leading to particular classes of cell spaces, such as the *abstract cell complexes* of A. Tucker [52], the *CW-complexes* of J. H. Whitehead, the *combinatorial manifolds* of V. Kovalevsky [25], to cite a few. ASCs also form a class of cell spaces. Figure 13.3 shows an example of cell space.

Topological Collections

A *topological collection* C is a function that associates values from an arbitrary set $\mathscr{V}$ with cells of some cell space (see Fig. 13.3). Thus the notation $C(\sigma)$ refers to the value of cell σ in collection C. We call *support* of C and write $|C|$ for the set of cells for which C is defined. Set $\mathscr{V}$ is left arbitrary to allow the association of any kind of information with the topological cells: for instance geometric properties ($\mathscr{V} = \{-1, 0, 1\}$ for representing orientation or $\mathscr{V} = \mathbb{R}^n$ for Euclidean positions) or

---

[1]i.e. a irreflexive, transitive and antisymmetric binary relation on $S_{\mathscr{K}}$: for x, y and z in $S_{\mathscr{K}}$, we have $x \prec_{\mathscr{K}} y \prec_{\mathscr{K}} z \Rightarrow x \prec_{\mathscr{K}} z$ and we never have $x \prec_{\mathscr{K}} y \prec_{\mathscr{K}} x$.

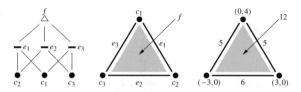

Fig. 13.3 On the *left*, the Hasse diagram of the incidence relationship of the cell space given in the middle: it is composed of three 0-cells (c_1, c_2, c_3), of three 1-cells (e_1, e_2, e_3) and of a single 2-cell (f). The *closure* of cell e_1 is composed of e_1, c_1 and c_2. The *faces* of cell f are e_1, e_2 and e_3. The *cofaces* of cell c_1 are e_1 and e_3. On the *right*, a topological collection associates data with the cells: positions with vertices, lengths with edges and area with f

arbitrary state of a subsystem (a mass, a concentration of chemicals, a force acting on certain cells, etc).

The collection C can be written as a formal sum

$$\sum_{\sigma \in |C|} v_\sigma \cdot \sigma \qquad \text{where} \quad v_\sigma \overset{\text{def.}}{=} C(\sigma)$$

With this notation, the underlying cell space is left implicit but can usually be recovered from the context. By convention, when we write a collection C as a sum

$$C = v_1 \cdot \sigma_1 + \cdots + v_p \cdot \sigma_p$$

we insist that all c_i are distinct.[2] Notice that this addition is associative and commutative: the specific order of operations used to build a topological collection is irrelevant. Using this notation, a *subcollection* S of a collection C is defined as a collection forming a subpart of the sum: $C = S + S'$; subcollection S' is then called the *complement* of S in C and we write $S' = C - S$.

The current implementation of MGS provides the programmer with different types of collections, namely seq, array, set, bag, etc. Actually, the topological collection approach makes possible to unify various data structures as sketched in Sect. 13.4.2.

Transformations

Transformations of topological collections embody the concepts of interaction and interaction complex introduced in Sect. 13.2 with the notion of *topological rewriting*. A transformation T is a function specified by a set $\{r_1, \ldots, r_n\}$ of rewriting rules of the form $p \Rightarrow e$ where the left hand side (l.h.s.) p is a *pattern* and the right hand side (r.h.s.) e is an MGS expression. For example, the bubble sort algorithm is defined in MGS by:

---

[2]The formal sum notation is borrowed from algebraic topology where set $\mathcal{V}$ is taken with a commutative group structure which gives an abelian group structure to topological chains and cochains [33]. See the elaboration in Sect. 13.5.

```
trans bubble_sort = {
   x, y / y < x => y, x;
   x, y / y == x => x;
}
```

An application of a transformation rule on a collection C selects a subcollection S of C matching with the pattern p that is then substituted by the subcollection resulting from the evaluation of the expression e.

Patterns

Patterns are used to specify subcollections where interactions may occur. They play the exact same role as l_I and C_I in the interaction notation $l_I \, / \, C_I \longrightarrow R_I$ of Sect. 13.2.2. However, the cell space setting used in the definition of topological collections requires a more elaborate tool for this specification instead of a simple set l_I of interacting positions (which is restricted to the simplices of an ASC). Transformation rule patterns allow the programmer to describe the local spatial organizations (and states) leading to some interactions.

Let us describe the core part of the pattern syntax which is based on three constructions summarized in the following grammar:

$$\Pi \ ::= \ id \ | \ \Pi \, \Omega \, \Pi \ | \ \Pi \, / \, \Lambda$$
$$\Omega \ ::= \ \varepsilon \ | \ < \ | \ > \ | \ ,$$

In this grammar, Λ represents an MGS expression, id corresponds to an identifier, and ε denotes the empty string. The grammar can be described as follows:

Pattern Variable: An identifier x of id, called in this context a *pattern variable*, matches a n-cell σ labeled by some value $C(\sigma)$ in the collection C to be transformed.

The same pattern variable can be used many times in a pattern; it then always refers to the same matched cell. Moreover, patterns are *linear*: two distinct variables always refer to two distinct cells.

Incidence: A pattern $p_1 \oplus p_2$ of $\Pi \, \Omega \, \Pi$ specifies a constraint $\oplus \in \Omega$ on the incidence between the last element matched by pattern p_1 and the first element matched by pattern p_2. The pattern $x \ < \ y$ (*resp.* $x \ > \ y$) matches two cells σ_x and σ_y such that σ_x is a face (*resp.* coface) of σ_y. The lack of operator (ε) denotes the independence of the cells (there is no constraint between them).

Binary interactions between two elements, say x and z, are frequent in models and can be specified with the pattern $x \ < \ y \ > \ z$. Variable pattern y stands for some k-cell making x and z $(k-1, k)$-neighbors, e.g., two vertices linked by an edge in a graph. Often, the naming of y does not matter and the syntactic sugar $x, \ z$ is used instead.

Although directed structures are out of the scope of this chapter, one may mention that it exists a directed extension of cell spaces that is actually used in MGS. The direction is encoded in the reading so that $x \ > \ y$ is directed from x to y and will not match the same subcollections than $y \ < \ x$ in directed collections.

Guard: Assuming a pattern p of Π and an MGS expression e of Λ, p / e matches
a subcollection complying with p such that the expression e evaluates to true.

We do not detail here the syntax of MGS expressions which is not of main importance but two elements have to be clarified. Firstly, in this chapter,[3] function
application is expressed using currying as in functional programming: $f\ e_1\ e_2$
means that function f is applied with two arguments e_1 and e_2. The benefit of
this syntax lies in the left associativity of the application so that feeding a binary
function with the first argument builds a unary function waiting for the second,
like in $(f\ e_1)\ e_2$. This principle extends to any n-ary functions. Secondly, any
pattern variable, say x, can be used in guard expressions (as well as in the r.h.s.
expression of a rule) where it denotes the label of the matched k-cell. Its faces
(*resp.* cofaces, *resp.* $(k, k+1)$-neighbor cells) are accessed by `faces x` (*resp.*
`cofaces x`, *resp.* `neighbors x`).

Since MGS is a dynamically typed language (that is, types are checked at run time),
types can be seen as predicates checking if their argument is of the right type.
Some syntactic sugar has been introduced to ease the reading of type constraints
in patterns so that the pattern `x / int(x)`, matching a cell labeled by some
integer, can be written `x:int`. On the contrary, `x:!int` matches a cell labeled by
something but an integer.

For example, the pattern

```
v1 < e12 > v2 < e23 > v3 < e31 > v1
f > e12 f > e23 f > e31 / (e12 == e31)
```

matches the entire collection of Fig. 13.3 with, for instance, the following association:

$$v1 \mapsto (0, 4) \cdot c_1 \qquad\qquad e12 \mapsto 5 \cdot e_1 \qquad\qquad f \mapsto 12 \cdot f$$
$$v2 \mapsto (3, 0) \cdot c_2 \qquad\qquad e23 \mapsto 6 \cdot e_2$$
$$v3 \mapsto (-3, 0) \cdot c_3 \qquad\quad e31 \mapsto 5 \cdot e_3$$

Rule Application

Let $T = \{r_1, \ldots, r_n\}$ be an MGS transformation. Following the interaction-based
computation model described in Sect. 13.2.2, the application of a rule p => e of T
on a collection C consists in finding some subcollection S of C matching pattern p
then replacing S by the evaluation S' of e in C. One point not discussed earlier is the
choice of the rule(s) to be applied when a number of instances exist. This choice is
called *rule application strategy*.

The default rule application strategy in MGS is qualified *maximal-parallel*. It
consists in choosing a maximal set $\{S_1, S_2, \ldots\}$ of *non-intersecting* subcollections
of C each matched by some pattern of T. In this context, *non-intersecting* means that

---

[3]Since the expression syntax is secondary, the authors made the choice to use in papers an ideal
syntax that may differ from the syntax currently implemented (which may by the way change from
a major release to another).

for any two subcollections S_i and S_j, their supports check that $|S_i| \cap |S_j| = \emptyset$. The application is then done in parallel as represented by the following diagram:

$$
\begin{array}{ccccccccc}
C & = & S_1 & + & S_2 & + & \cdots & + & R \\
\downarrow{\scriptstyle T} & & \downarrow{\scriptstyle r_{i_1}} & & \downarrow{\scriptstyle r_{i_2}} & & & & \downarrow \\
T(C) & = & S'_1 & + & S'_2 & + & \cdots & + & R
\end{array}
$$

where each S'_k results from the evaluation of e_{i_k} and $R = C - (S_1 + S_2 + \cdots)$ consists of the untouched part of C. Since it may exist different ways to decompose collection C w.r.t. transformation T, only one of the possible outcomes (randomly chosen) is returned by the transformation. The formal semantics is given in [47].

On the contrary, the strategy informally used in Sect. 13.2.2 is qualified *asynchronous* since only one rule application is considered at a time. Its diagram is as follows:

$$
\begin{array}{ccccc}
C & = & S & + & (C - S) \\
\downarrow{\scriptstyle T} & & \downarrow{\scriptstyle r} & & \downarrow \\
T(C) & = & S' & + & (C - S)
\end{array}
$$

The asynchronism means that two events cannot take place simultaneously. Between the synchronous and the asynchronous strategy, there is considerable room for alternative rule application strategies.

For instance, asynchronism is often assumed for stochastic processes on populations where simultaneous events are unlikely (e.g., in Poisson processes). These kinds of processes are often used for stochastic simulation, for example of chemical or biochemical systems of reactions. In this context, pure asynchronism is not enough: a *stochastic constant* is attached to each reaction (that is, rule) and expresses "how fast" it is. A continuous-time Markov chain can then be derived from the trajectories generated by the iterations of the transformation. We name the continuous-time extension of the asynchronous rule application strategy with stochastic constants, the *Gillespie rule application strategy* after the name of D.T. Gillespie. Gillespie proposed in [20] an algorithm for the exact stochastic simulation of well-stirred reaction systems which is implemented in MGS. The MGS Gillespie strategy has been used in different applications in integrative and synthetic biology, see for example [46, 49].

13.3.2 Reviews of Some Applications to Complex Systems

We advocate that MGS is adequate for the modeling and simulation of dynamical systems. In this section, we show various examples that support this assertion. These examples are only sketched to support our claim and the interested reader may refer to the references given for the technical details.

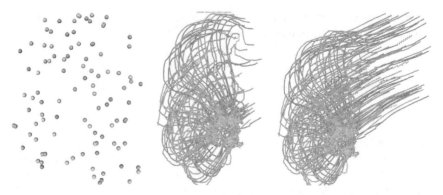

Fig. 13.4 Trajectory of a flock of 50 birds. *Left plot* the initial state where each bird has a randomly chosen direction. *Center plot* the configuration after 300 iterations. *Right plot* after 900 iterations of the transition function

Flocking Birds

In [18] the classical example of a simulation of a flock of birds has been considered. The simulation is the direct implementation of a model of flocking birds proposed by U. Wilensky and by the development of steering behaviors of boids (generic simulated flocking creatures) invented by C. Reynolds [39].

Here, no creation nor destruction of birds happen, but the neighborhood structure changes in time with the movements of the birds. This example uses the Delaunay topological collection [32, 42] where the neighborhood structure is not built explicitly by the programmer but is computed implicitly at run time using the positions of the birds in a Euclidean space represented as labels of 0-cells in the collection. Figure 13.4 illustrates three iteration steps of the simulation. The transformation corresponding to the dynamics specifies three rules corresponding to the three behaviors described by Wilensky: *separation* (when a bird is at close range of a neighbor, it changes direction), *cohesion* (when a bird is too far from all its neighbors, it tries to join the group quickly) and *alignment* (when the neighbors of a bird are neither too far nor too close, the bird adjusts its direction following the average directions of its neighbors).

Diffusion Limited Aggregation

Diffusion Limited Aggregation, or DLA, is a fractal growth model studied by two physicists, T.A. Witten and L.M. Sander, in the 80s [57]. The principle of the model is simple: a set of particles diffuses randomly on a given spatial domain. Initially one particle, the seed, is fixed. When a mobile particle collides a fixed one, they stick together and stay fixed. For the sake of simplicity, we suppose that they stick together forever and that there is no aggregate formation between two mobile particles. This process leads to a simple CA with an asynchronous update function or a lattice gas automaton with a slightly more elaborated set of rules.

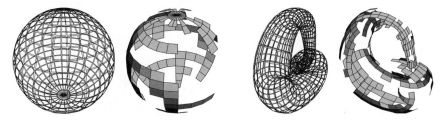

Fig. 13.5 DLA on complex objects (topology and final state). On the *left*: a sphere with 18 parallels and 24 meridians. On the *right*: a Klein's bottle

The MGS approach enables a generic specification of such a DLA process which works on various kinds of space [45]. Figure 13.5 shows applications of the *same* DLA transformation on two different topologies: it is an example of the *polytypic* [24] capabilities of MGS.

Declarative Mesh Subdivision

Mesh subdivision algorithms are usually specified informally with the help of graphical schemes defining local mesh refinements. For example, the Loop subdivision scheme [28], working on triangular meshes, is described with the *local rule*

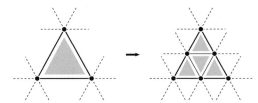

These specifications are then implemented efficiently in an imperative framework. These implementations are often cumbersome and imply some tricky indices management. Smith et al. [38] asked the question of the declarative programming of such algorithms in an index-free way that has been positively answered in [47] with the MGS specification of some classical subdivision algorithms in terms of transformations (see Fig. 13.6).

Coupling Mechanics and Topological Surgery

Developmental biology investigates highly organized complex systems. One of the main difficulties raised by the modeling of these systems is the handling of their dynamical spatial organization: they are examples of dynamical systems with a dynamical structure.

In [43] a model of the shape transformation of an epithelial sheet requiring the coupling of a mechanical model with an operation of topological surgery has been considered. This model represents a first step towards the declarative modeling of neurulation. Neurulation is the topological modification of the back region of the embryo when the neural plate folds; then, this folding curves the neural plate until the two borders touch each other and make the plate becomes a neural tube (see Fig. 13.7).

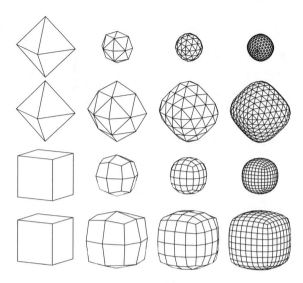

Fig. 13.6 Results of the application of subdivision algorithms. From *top* to *bottom*: the Loop's algorithm, the Butterfly algorithm, the Catmull-Clark's algorithm and the Kobbelt's algorithm. From the *left* to the *right*, the initial state then 3 iteration steps. These pictures have been generated in MGS

Fig. 13.7 Simulation of a neurulation-like process in MGS: from the *left* to the *right*, a sheet of epithelial cells is curving until the hems sew together to form a tube

Modeling the Growth of the Shoot Apical Meristem

Understanding the growth of the shoot apical meristem at a cellular level is a fundamental problem in botany. The protein PIN1 has been recognized to play an important role in facilitating the transport of auxin. Auxin maxima give the localization of organ formation. In 2006, Barbier de Reuille et al. investigated in [4] a computational model to study auxin distribution and its relation to organ formation. The model has been implemented in the MGS language using a Delaunay topological collection. Figure 13.8 shows the results of a simulation done in MGS of a model of meristem growth.

Integrative Modeling

Systems biology aims at integrating processes at various time and spatial scales into a single and coherent formal description to allow analysis and computer simulation. Rule-based modeling is well fitted to model biological processes at various levels of

Fig. 13.8 Results, at time steps 3, 18, 27, 35, 41 and 59, for the simulation of a virtual meristem done by Barbier de Reuille in [3]. *Red dots* correspond to auxin and each primordium cell is shown in a different color

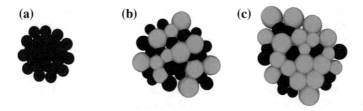

Fig. 13.9 Results of an integrative model. Germinal cells are in *dark gray* and somatic cells in *light gray*. **a–c** correspond to an initial population and its evolutions at logical time 43 and 62. Refer to [48]

description. This approach has been validated through the description of various models of a synthetic bacterium designed in the context of the iGEM competition [48], from a very simple biochemical description of the process to an individual-based model (see Fig. 13.9).

This model, as well as the previous one, aims at the modeling of an entire population (of cells, of bacteria) through an explicit representation of the individuals with mechanical, chemical and biological (i.e., gene expression) behaviors, integrated with the specification of entity/entity interaction and dynamic neighborhood computation.

Algorithmic Problems

The use of MGS is not restricted to the modeling and simulation of complex systems. Many other applications not given here have been developed. For example, purely algorithmic applications include the Needham–Schroeder public-key protocol [31], the computation of prime numbers using Eratosthene's sieve, the normalization of Boolean formulas, the computation of various algorithms on graphs like the computation of the shortest distance between two nodes or the maximal flow, to cite a few. Moreover, any computation described in an unconventional framework can be programmed in MGS since the language unifies many models of computation as we will see in Sect. 13.4.2. Detailed examples can be found on the MGS web page.[4]

---

[4]http://www.spatial-computing.org/mgs.

13.4   Generic Programming in Interaction-Oriented Programming

While MGS was initially designed as a domain specific programming language dedicated to the modeling of dynamical systems with a dynamical structure, it allows an elegant and concise formulation of classical algorithms (as illustrated with the bubble sort in Sect. 13.2.2). The main difference with other general-purpose programming languages lies in its interaction-based style of programming. In this section, we outline the new perspectives on *genericity* opened by the interaction-oriented approach.

We advocate that the design of unconventional models of computation has an impact not only in the study of alternative models of computation (with alternative calculability and complexity classes) but may also impact questions raised in "classical programming languages". This development offers also a link, investigated in the next section, between the notion of data structure/traversal and the notion of differential operators.

13.4.1   Polymorphism, Polytypism and Generic Programming

Genericity in a programming language is the crucial ability to abstract away irrelevant requirement on data types to produce, once and for all, a piece of code that can be reused in many different situations. It is then a central question in software engineering.

Several mechanisms have been proposed to support genericity. In Musser and Stepanov's approach [34], the fundamental requirements on data structures are formalized as *concepts*, a notion more general than a type, with generic functions implemented in terms of these concepts. The best known examples are the STL in C++ or the Java Collection interface.

Let us take a look at the Java Collection interface. This interface is implemented by all data containers of the Java standard API. Besides the usual container methods (size, emptiness, membership, etc.), any Java Collection has to implement an iterator over its elements to be compatible with the enhanced for loop syntax. Iterators decouple data structure traversals from data types, which enables the definition of the same algorithm operating on different data type.

This property has been formalized in type theory as *polymorphism* available in many programming languages. For example, *Parametric polymorphism* [8] allows the definition of functions working uniformly (i.e., the same code) on different types. The uniformity comes from the use of a type variable representing a type *to-be-specified-later* and instantiated usually at function application. For example, the OCaml type declaration for "lists of something" is as follows:

```
type α list = Empty | Cons of α * α list
```

In this definition (specifying that a list is either empty or built from an element prepended to a list), the type variable α can be instantiated by any types: $\alpha = $ int for a list of integers, α = string for a list of strings, α = β list for a list of lists of something-abstracted-by-type-variable-β, etc. Any function acting on the structure of lists independently of the content type has a polymorphic type. For example, the classical map function applying some function f on each element of a list is defined once and for all by:

```
let rec map f = function
  | Empty -> Empty
  | Cons (h, t) -> Cons (f h, map f t)
```

and is of type $(\alpha \rightarrow \beta) \rightarrow \alpha$ list $\rightarrow \beta$ list. This sole definition works for any instance of type variables α and β.

We can go further by considering *polytypism* where the type of the container is also abstracted. For example, consider the following type definition of binary trees:

```
type α tree = Leaf of α | Node of α tree * α * α tree
```

As for list, a map function can be defined on trees:

```
let rec map f = function
  | Leaf e -> Leaf (f e)
  | Node (l, e, r) -> Node (map f l, f e, map f r)
```

of types $(\alpha \rightarrow \beta) \rightarrow \alpha$ tree $\rightarrow \beta$ tree. Both map functions actually work the same way: they transform all elements of type α in the structure by traversing it recursively. This traversal can be specified by associating a combinator with each constructors of the data type (Empty, Leaf, etc).

Generic programming as proposed in [23, 24] generalizes this idea by allowing to program only one map algorithm (in our example, the map function is defined twice). This approach is applicable for a class of types called *algebraic data types* (ADT). Roughly speaking, ADT are specified inductively using the unit type, type variables, disjoint union of types (operator | in the previous definitions), product of types (operator *), composition of types, and a fixpoint operator. Generic programming uses this uniformity to provide a way to express inductively polytypic algorithms for ADT by associating a combinator with each of these data type constructions. Refer to [23] for a detailed presentation.

13.4.2 From Data Structures to Topological Collections

In MGS, topological collections can be polymorphic in the sense mentioned above since there is no constraint on the set of labels: the same cell space can be labeled by integers, strings, or anything else [9]. In addition, a transformation relies on the (local) notion of neighborhood which is constitutive of the notion of space without constraining a particular (global) shape of space. In other words, the same transformation can be used on any collection providing the programmer with a form of polytypism.

However, the spatial point of view goes beyond polymorphism and polytypism as found in "classical programming languages". The previous discussion makes apparent that generic programming relies on abstracting away some information on data type to keep the information relevant to data traversal. This can be explicit through the notion of iterators in Java, or implicit through the notion of *natural homomorphism* (combinator associated to a constructor) in ADT.

This information is alternatively expressed in term of space, where the notion of (data-)movement find a natural setting. The spatial point of view provides both a richer set of data structures than those described by the algebraic approach[5] and more expressive mechanisms to express elementary movements (e.g., iterators impose an ordering which can be detrimental).

In the following we investigate the idea that a data structure corresponds to a topological collection with a specific kind of topology. The polytypism of MGS transformations is presented together with some examples. In the next section, we introduce a family of generic operators which are finally related with differential operators considered in continuous computation.

Data Structures

In computer science, the notion of *data structure* is used to organize a collection of data in a well organized manner. The need of structuring data is twofold: firstly, programmers are interested in capturing in a data structure the logical relationship between the data it contains. This is for example one of the primary idea behind database tools like *entity–relationship models*, UML diagrams, or XML *document type definitions* when used to specify how the represented objects are ontologically related to each other (e.g., a *book* has an *ISBN*, a *title*, some *authors*, etc.). The data structure becomes a model of some organization in the world.

Secondly, from a computational perspective, data structures are designed to efficiently access the data. (In this respect, the art of database design consists in modeling consistently reality while being queried efficiently.) Algorithms are often expressed as nested traversals of some structures. The choice of a data structure is then highly coupled to the algorithm to be implemented. For example, the `list` structure defined above allows a linear traversal of the data in order of appearance in the list, while the `tree` structure allows the so-called prefix, infix and postfix traversals. In the object-oriented programming paradigm, the *iterator* design pattern (mandatory for any Java class implementing the `Collection` interface for example) can be understood as an attempt to catch this notion of traversals.

MGS Collection Types

By confronting the concept of traversal with the concept of topological collection, it is possible to retrieve the conventional notion of data structure in MGS. The main idea consists in extracting from each data structure a specific topology capturing the graph

---

[5]Incidentally ADT are restricted to tree-shaped structures while topological collections are able to deal with a wider class of data structures; for instance, the generic handling of arrays, circular buffers or graphs cannot be adequately done in the ADT framework.

of its traversals: the nodes represent positions for storing the data and the edges are defined so that two elements are neighbor in the collection (i.e., may interact) if they are consecutive in some traversal of the data structure. Let us review some classical data structures and their MGS interpretations. Each of them is associated with a dedicated collection type corresponding to a specific (constrained) topology. Each collection type comes with some syntactic facilities in the current implementation of the language.

Sequences

As already said, such structures are linear, meaning that elements are accessed one after the other. The associated topological structure is then a linear graph as the one obtained in Sect. 13.2. These structures exhibit two natural traversals that we can refer to *left-to-right* and *right-to-left*. Considering only one of these two traversals leads to a directed structure similar to a linked list while considering both leads to doubly linked list structures.

In its current implementation, MGS provides the programmer with two linear left-to-right directed collection types: seq and array. They differ by the nature of the underlying space; the former is a *Leibnizian* collection type where the structure is generated by a specific relationship between the data (here the order of insertion) while the latter is *Newtonian*[6] where the structure is firstly specified (or allocated) and then inhabited. The following example of MGS program illustrates this difference:

```
trans rem = { x / even (right x) => <undef> }

rem (1, 2, 3, 4, 5)       ↝ (2, 4, 5)
rem [| 1, 2, 3, 4, 5 |]  ↝ [| <undef>, 2, <undef>, 4, 5 |]
```

Transformation rem removes all labels of positions having an even number on its right. The predicate right can be expressed as a straightforward expression involving the generic comma operator ",”; it takes its specific meaning on sequence from the underlying topology. In the case of a seq collection, the removal modifies the space from a 5-node graph to a 3-node graph. In the case of an array collection, the underlying space remains unchanged but the affected positions are left empty (i.e., without any label).

(Multi-)Sets

The main difference with linear structures is that sets are unordered collections, so that when iterated, no order gets the priority. As a consequence, for any pair of elements there exists at least one iteration making them neighbors. The obtained topological structure is a complete graph.

---

[6]The qualifiers *Newtonian* and *Leibnizian* have been chosen after the names of I. Newton and G.W. Leibniz who had completely different understanding of the concept of space: the former thought space as a container, that is, an *absolute space* pre-existing to the bodies it contains; the latter understood space as the expression of a *relationship* between bodies.

In the current implementation of MGS, collection types set and bag are available: bag collections allow multiple occurrences of the same element while set collections do not.

Records

A record corresponds to a collection of data aggregated together in a single object. Each data can be referred by a different field name. This is equivalent to a struct construction in the C programming language. The elements of a record are not generally accessed one from the other (they are accessed from the whole record) leading to consider an empty neighborhood between the data. The topological structure of records is then a graph with no edge: each field is represented by an isolated vertex labeled by its associated data.

Group Based Fields (GBF)

Algorithms on matrices often work by traversing data regularly by row and/or by column. Forgetting what happens on the boundary, the induced topological structure exhibits a very regular pattern where each element has four neighbors: its immediate successors and predecessors on its row and column. Of course this reasoning can be extended to any multidimensional arrays.

GBF are the generalization of multidimensional arrays where the regular structure corresponds to the Cayley graph of an abelian group presentation. The abelian group is finitely generated by a set of directions from an element to its neighbors (e.g., north, south, east and west for matrices) together with a set of relations between these directions (e.g., north and south are opposite directions). GBF are a powerful tool that allows the specification in few lines of complicated regular structures. Figure 13.10 illustrates the specification of an infinite hexagonal grid using GBF.

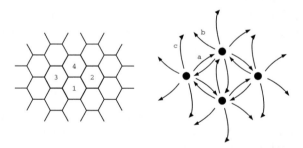

Fig. 13.10 Infinite hexagonal grid generated by 3 directions a, b and c related by $a + b = c$. On the *right*, a local view of the associated Cayley graph focused on the 4 positions marked in the grid on the *left*. Each *arrow* represents the displacement from a position to another following one of the directions or their inverses

Unifying Natural Models of Computation

The previous paragraph shows that the notion of topological collection subsumes a large variety of data structures. As a consequence, it appears that various natural models of computation can be adequately described in the MGS framework simply by choosing the right collection type and the right rule application strategy:

- *Artificial Chemistry* [2] corresponds to a space where any two entities may interact. We have seen that this corresponds to a complete graph, hence to the set or bag collection types. The application of transformation rules on these structures achieves the same effect as multiset rewriting [10] (rewriting on associative-commutative terms).
- *Membrane Computing* [14, 36] extends the idea of multiset programming by considering nested multisets (membranes) and transport between them. The corresponding space can be represented as a multiset containing either ordinary values and nested multisets. This example shows the interest to reflect the spatial structure also in the labels, see [19]. The rule application strategy is usually the maximal-parallel one.
- *Lindenmayer Grammars* [27] correspond to parallel string rewriting and hence to a linear space and a maximal-parallel application strategy.
- *Cellular Automata* (CA) [55] and *Lattice Gas Automata* [50] correspond to maximal-parallel rewriting in a regular lattice. Such a lattice can be easily specified as a GBF collection type. For example the classical square grids with von Neuman or Moore neighborhoods are defined in MGS by:

```
gbf NEWS = < N, E, W, S; W + E = S + N = 0 >
gbf Moore = < N, E, NE, SE; N + E = NE, E - N = SE >
```

13.4.3 Polytypism in MGS

Transformations allow the programmer to iterate over the neighbors of an element (e.g., with the neighbors, faces and cofaces primitives) or over the pairs of neighbor elements in a topological collection.

Since transformation patterns express interactions in terms of cells incidence in full generality, the same pattern can be applied on any collection of any collection type, so that a transformation can be viewed as a polytypic computation in an interaction-based style.

The following paragraphs present some examples of MGS polytypic transformations.

Map-Reduce in MGS

The MGS counterpart of the aforementioned polytypic map function can be implemented as follows:

```
trans map f = { x => f x }
```

In this declaration an extra-parameter f is expected so that map f is a transformation applying f on each element of a collection. For example:

```
map succ (1,2,3,4,5) ⤳ (2,3,4,5,6)
```

In this example, f is set to the successor function succ so that the elements of the collection are incremented by one.

In functional programming, the map function appears in conjunction with another polytypic function, reduce. This function iterates over the elements of some structure to build up a new value in an accumulator. It is parameterized by the accumulation function. For example, reduce can be used to compute the sum of the elements of a list of integers. The MGS counterpart of this function can be implemented as follows:

```
trans reduce op = { x, y => op x y }
```

This rule collapses two neighbor elements into one unique value computed from the combination function op. Obviously this rule has to be iterated until a fixpoint is reached to finally get the reduction of the whole structure. For example:

```
(reduce add) [fixpoint] (1,2,3,4,5) ⤳ (15)
```

sums up all the elements of a sequence. The application of the transformation (reduce add) is annotated by the qualifier fixpoint.

Bio-Inspired Algorithms

Numerous distributed algorithms are inspired by the behavior of living organisms. For instance *ant colony optimization algorithms* use ants ability to seek the shortest path between the nest and a source of food [12]. Such algorithms are often specified as reactive mutli-agent systems where the individual behavior is defined independently from the spatial organization of the underlying structure. In this respect, these algorithms are polytypic and can be easily specified in MGS. We focus on two toy but representative examples of such computations.

Random Walk

One of the key behaviors of agent-based distributed algorithms is a random walk allowing agents to scatter everywhere in the space so that each place is visited at least once by an agent with high probability. The expression of such a walk in MGS is as follows:

```
trans walk agent empty = { p:agent, q:empty => q, p }
```

Transformation walk is composed of a simple rule specifying that if an agent has an empty place in its neighborhood, it moves to that place leaving its previous location empty. To get even more genericity the predicates for being an agent or an empty place are given as parameters agent and empty. This rule is not deterministic since a choice as to be done when several empty places surround the agent; in such a case, one of these places is randomly chosen uniformly, ensuring no bias in the walk. This transformation operates on any type of collection.

Propagation

Another fundamental procedure consists in broadcasting some information in space. This mechanism is at work in propagated outbreak of epidemics, in graph flooding in networks or in the spread of fire in a burning forest. A simple MGS transformation can implement this last example as follows:

```
trans fire on_fire to_fire to_ash = {
  p:!on_fire / (neighbors_exists on_fire p) => to_fire p;
  p:on_fire => to_ash p;
}
```

Two rules are given to (1) set into fire an unburnt place which is neighbor with a burning one (primitive `neighbors_exists` checks for some neighbor of `p` where predicate `on_fire` holds), and to (2) extinguish places in fire. Once again, the genericity of this specification (especially of primitive `neighbors_exists` which visits all the neighbors of an element whatever the collection type is) makes this transformation polytypic.

The genericity of these specifications allows the development of tools and techniques operable in many different situations. For example, we have developed in [37] the tracking of the *spatial activity* in such algorithms, leading to a generic optimized simulation procedure usable with any collection type.

13.5   From Computation to Physics: Differential Calculus in MGS

Cell spaces are defined in algebraic topology for a discrete (and algebraic) description of spaces. Therefore, one can expect to find on these spaces the operators mathematicians have imagined on more usual spaces, like differential operators. This is indeed the case and the interested reader may refer to [11] for an introduction.

Topological collections are directly inspired by the notions of topological chain and cochain built upon cell spaces, but with a weaker structure. In the previous sections we have established a direct link between the notion of data structure and the notion of topological collection, it is then tempting to investigate in the context of data structures what a differential operator is. All the mathematical background required to understand this relation is detailed in the Appendix.

In this section, we show how a set of operators can be derived as computation patterns of MGS programs. These operators, which can be seen as movements of data on data structures [15], inherit algebraic properties from differential calculus, providing the programmer with the ability to express a program as a set of differential equations.

13.5.1 Transport of Data

The main motivation of this section is to relate the formal definition of topological collections to the discrete counterpart of differential forms as developed in [11].

As defined in Sect. 13.3.1, a topological collection C is a function associating values from a set $\mathcal{V}$ with the cells of a cell space $\mathcal{K}$. This structure is almost a *discrete form*: a discrete form further constrains the set $\mathcal{V}$ to be equipped with a structure of abelian group (i.e., with an addition operator $+_\mathcal{V}$) and the cell space $\mathcal{K}$ to be equipped with a structure of chain complex (i.e., with a boundary operator ∂). Topological chains are the basic ingredient to define a boundary operator. As a matter of fact, the boundary of a $(n+1)$-cell is a n-chain, that is a function associating integers with the n-cells of a cell space. These integers are used to represent the multiplicity and the orientation of a cell in the complex. This operator is linearly extended from cells to chains.

From now on, we consider that the set $\mathcal{V}$ of values of a topological collection is an abelian group[7] written additively $+_\mathcal{V}$, and that a boundary operator ∂ is defined on its support cell space. Collections C are then forms. As functions on cells, they can be linearly extended to n-chains of $(\mathcal{K}, \partial)$, so that:

$$C(n_1.\sigma_1 + n_2.\sigma_2 + \cdots) = n_1 C(\sigma_1) +_\mathcal{V} n_2 C(\sigma_2) +_\mathcal{V} \cdots$$

where the σ_i are cells of $\mathcal{K}$.

The application of a collection C on a chain c, usually written $[C, c]$, leads to consider a *derivative operator* **d** as the "adjoint operator" of ∂ defined by:

$$[\mathbf{d}C, c] = [C, \partial c]$$

By duality, the same can be done to get a dual derivative operator $\star\mathbf{d}$, called here *coderivative*, from a dual boundary operator $\star\partial$. Both operators are then related by a *correspondence operator* $\ast$ analogous to the Hodge dual. All technical details are given in Appendix.

The point is that the three operators **d**, $\star\mathbf{d}$ and $\ast$ act on topological collections (forms), and can be defined as transformations parameterized by the group operator $+_\mathcal{V}$ and the boundary operators ∂ and $\star\partial$:

- the correspondence operator $\ast$ behaves like a map function and transforms the value associated with a cell,
- the derivative operator **d** transfers values from cells to their cofaces, and
- the coderivative operator $\star\mathbf{d}$ transfers values from cells to their faces.

---

[7]In MGS, the set $\mathcal{V}$ of values is usually arbitrary with no meaningful addition. In such a case, instead of working in $\mathcal{V}$, one may consider the free abelian group $\langle\mathcal{V}\rangle$ finitely generated by the elements of $\mathcal{V}$. This group is in a way universal since any group on $\mathcal{V}$ can be recovered from an homomorphism $h : \langle\mathcal{V}\rangle \to \mathcal{V}$ such that $h(1.v) = v$ and $h(g_1 +_{\langle\mathcal{V}\rangle} g_2) = h(g_1) +_\mathcal{V} h(g_2)$.

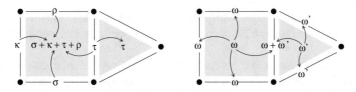

Fig. 13.11 Transport of values under the actions of differential operators: on the *left*, the derivative operator where only some 1-cells are labeled; on the *right*, the dual derivative operator where only the 2-cells are labeled (the action is represented on the primal mesh)

Figure 13.11 illustrates these data movements. From the point of view of data movements, the group operations can be interpreted as follows:

- *Identity*: the zero of $\mathcal{V}$ coincides with the absence of label. It can then be used to deal with partially defined collection.
- *Addition*: is used to combine multiple labels moving to the same cell due to the action of an operator. Examples of such collisions are pictured on Fig. 13.11. The commutativity means that the order of combinations does not matter.
- *Negation*: this operator is essential to program the relative orientation[8] between cells and to get the nilpotence of the boundary operator ($\partial \circ \partial = 0$).

The three differential operators can straightforwardly be translated in MGS following the definitions given in Appendix. As an example, the derivative is defined in MGS as follows:

```
trans derivative add mul zero inc = {
    x => faces_fold (fun y -> add (mul (inc x y) y)) zero x
}
```

The transformation is parameterized by four arguments. The three first arguments specify the considered abelian group structure[9] over $\mathcal{V}$, and function `inc` gives the incidence number of a pair of cells (which defines the boundary operator). Transformation `derivative` works on each cell σ of a collection by summing up (`add`) all the labels associated with the faces of σ (`faces_fold`) with respect to the relative orientation between incident cells (`inc`).

13.5.2 Programs and Differential Equations

The previous operators are generic: they are polymorphic because they apply irrespectively of $\mathcal{V}$ (as soon as it is an abelian group) and polytypic because they apply

---

[8]Orientation is only a matter of convention. It is always possible to consider the opposite orientation to change the sign of a value: for example, a negative flow labeling an edge represents a positive flow going against the chosen orientation of the edge.

[9]Expression add v_1 v_2 evaluates the addition $v_1 +_{\mathcal{V}} v_2$ of two values, expression `mul` n v evaluates the multiplication $\underbrace{v +_{\mathcal{V}} \cdots +_{\mathcal{V}} v}_{n \text{ times}}$ of a value by an integer, and `zero` gives the identity element.

on cell spaces of any shape. We qualify these operators as "discrete differential operators" because they exhibit the same formal properties as their continuous counterparts.

These basic operators can be composed to get more complex data transports. By mimicking their continuous counterparts, they can be used as building blocs to define more elaborated data circulation in the data structures, like gradient, curl, divergence, etc. For example, we have proposed in [15] an MGS implementation of the *Laplace–Beltrami* operator Δ, a general Laplacian operator, defined in our notations by:

$$\Delta = * \star \mathbf{d} * \mathbf{d} + \mathbf{d} * \star \mathbf{d} *$$

We have also shown that the straightforward translation of differential equations results in effective MGS programs.

In the rest of this paragraph, we illustrate the approach on the modeling of two classical physical phenomena: diffusion and wave propagation. However, the challenge at stake is not restricted to the generic coding of (the numeric simulation of) physical models. We believe that the abstract spatial approach of computation, based on the notion of interaction, opens the way to a new comprehension of algorithms through the physical modeling of data circulation. To illustrate this idea, we present in the last paragraph of this chapter an analysis of a system of differential equations that exhibits a sorting behavior.

Programming of Differential Equations

Let us consider the modeling of two classical physical systems, the diffusion and the wave propagation, given by the following equations:

$$\dot{U} = D\Delta U \qquad \ddot{V} = C^2 \Delta V$$

where D, C are respectively the diffusion coefficient and the wave propagation speed, U, V stand for the collections to be transformed, and $\dot{U}, \ddot{V}$ are their first and second temporal derivatives respectively. The behavior of the corresponding MGS programs depends on the parameterization chosen for the differential operators. Figure 13.12 shows simulations where the differential operators are parameterized so that the abelian group of labels corresponds to the real numbers under the usual addition, and the boundary, coboundary and correspondence operators encode a classical square grid with step dx. The temporal derivatives are interpreted using the forward difference (e.g., $\dot{U}(t) = \frac{U(t+dt)-U(t)}{dt}$ for some time t and duration dt).

With these parameters, the MGS programs coincide with the numerical resolution of the equations using a finite difference method. Some applications of Sect. 13.3.2 (neurulation, meristem growth and the integrative model) can also be derived from a model originally described in terms of differential equations.

More complex numerical schemes (equivalent or improved compared to the usual finite element method in terms of error control, convergence and stability) can be expressed in this framework as shown in [11, 51]. All these works rely on the under-

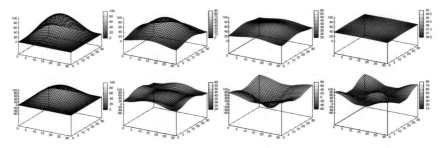

Fig. 13.12 Simulations of a diffusion and a wave propagation in MGS from the straightforward implementation of the differential equations: diffusion with coefficients $D = 10^{-2}$ and $dt = dx = 1$ on *top line*, from *left* to *right* at time $t = 0, 1000, 3500$ and 10000; wave propagation with coefficients $C = 5.10^{-3}$ and $dt = dx = 1$ on *bottom line*, from *left* to *right* at time $t = 0, 1500, 3000$ and 4500

standing of physical laws with a discrete interpretation of the differential calculus; MGS lends itself to the implementation of these theories.

Furthermore, the genericity of the spatial point of view makes possible to encompass, in the same computational formulation, different physical models of the same phenomena. For example, we can change the nature of the labels in $\mathcal{V}$ from real numbers to multisets of symbols. While the former corresponds to concentrations in the previous setting, the later can be interpreted as individual particles. In this setting, the Laplacian operator Δ expresses the jump of particles from a position to some neighbor in the collection. A multiplication between two collections with same support specifies reactions between particles. We find here the basic ingredients of membrane computing that we have mentioned earlier in Sect. 13.4.2: diffusing between membranes and reacting.

Subsuming different kinds of models is fruitful to get refined models of complex systems. Another application consists in coupling different models relying syntactically on the same differential description. For example, we have been able to simulate a 1D hybrid diffusion system partitioned into subsystems each governed by its own diffusion model, either the Fick's second law or a random walk of particles both specified by the same generic equation but parameterized by a specific $(\mathcal{V}, +_{\mathcal{V}})$. The interface between two models is simply driven by the conversion laws between the involved parameters (here particles and concentrations). See Fig. 13.13 for an illustration.

Programming with Differential Equations

Continuous formalisms are sometimes used to describe the asymptotic behavior of a discrete computation. As an example, we have been able in [44] to provide a differential specification of *population protocols*, a distributed computing model [1], allowing us to study the asymptotic behavior of such programs.

We further believe that the formulation of classical (combinatorial, discrete) computations through differential operators acting on a data structure, is able to bring new understanding on old problems and to make a bridge with the field of *analog com-*

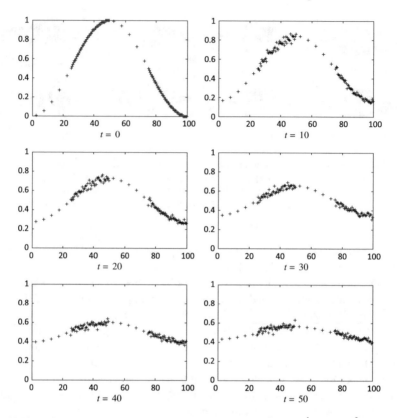

Fig. 13.13 Hybrid diffusion in 1D. The initial distribution is given by $\frac{1}{2}(1 - \cos(\frac{2\pi x}{L}))$ for $x \in [0, L]$ with $L = 100$. The system is divided into 4 parts: on intervals $[0, 25]$ and $[50, 75]$ the diffusion is governed by the Fick's second law (FL) solved by a finite difference method with space step $\Delta x = 5$ and time step $\Delta t = 1$; on intervals $[25, 50]$ and $[75, 100]$ the diffusion is governed by a discrete uniform random walk (RW) with space step $\delta x = 0.5$ and time step $\delta t = 0.1$. The correspondence between the two models is given by a unit of matter in FL for 10^4 particles in RW. The figures give the states taken by the system at times $t = 0, 10, 20, 30, 40$ and 50 for a diffusion coefficient $D = 10$

putation. In this perspective, a computation is seen as a dynamical system (see [17] for an application in the field of autonomic computing).

To illustrate this point, we elaborate on an example introduced by R.W. Brockett about the computational content of the systems of ordinary differential equations of the form $\dot{H} = [H, [H, N]]$ where H and N are symmetric matrices and $[\cdot, \cdot]$ is the commutator operator [6, 7]. It has been shown that by choosing appropriately N and the initial value of H, the system is able to perform many combinatorial algorithms, like sorting sequences, emulating finite automata or clustering sets of data.[10]

In the case of sorting, the system corresponds to a non-periodic finite Toda lattice, a simple model for a one-dimensional crystal given by a chain of particles with

---

[10]http://hrl.harvard.edu/analog/.

nearest neighbor interaction. The equations of motion of a particle are given by:

$$\begin{cases} \dot{p}_i = e^{(q_{i-1}-q_i)} - e^{(q_i-q_{i+1})} \\ \dot{q}_i = p_i \end{cases} \qquad i \in [1..n]$$

where q_i is the displacement of the ith particle from its equilibrium position, and p_i is its momentum. To integrate the system, one uses the following change of variables:

$$u_i = -\frac{1}{2}p_i \qquad v_i = \frac{1}{2}e^{(q_i-q_{i+1})/2}$$

giving the following system:

$$\begin{cases} \dot{u}_i = v_i^2 - v_{i-1}^2 \\ \dot{v}_i = 2(u_{i+1} - u_i)v_i \end{cases} \qquad i \in [1..n]$$

The behavior of this system is as follows: starting from an initial sequence of values $u_i(0)$ (and small non-zero values for the $v_i(0)$), the $u_i(t)$ asymptotically converge to s_i where s_i is the ith value in the *sorted* sequence of the $u_i(0)$.

This system can be specified in the differential setting of MGS by:

$$\begin{cases} \dot{U} = \nabla(V^2) \\ \dot{V} = 2(\nabla U)V \end{cases}$$

where U and V are coupled topological collections defined on the same one-dimensional cell space, and ∇ is a gradient-like operator inducing the direction followed by the sort.

From this specification, many applications can be derived. Obviously the original formulation can be retrieved when the underlying cell space is a sequence and labels are real numbers. By understanding literally the specification as a usual system of differential equations, the implementation computes the sorting of a continuous field. In fact, the formulation can be interpreted in n dimensions giving raise to a fully polytypic specification of sorting in several dimensions. By relying on genericity and by switching the labels from numbers to sets of symbols, the system turns to be a distributed sorting algorithm in the artificial chemistry style. See Fig. 13.14 for illustrations.

13.5.3 Future Research Directions

The development presented here only scratch the surface of the subject and many works remain to be done to investigate and to understand the contribution of the differential formulation in MGS. For example, while the development of MGS has put emphasis on the spatial structure induced by the interactions, the temporal aspects

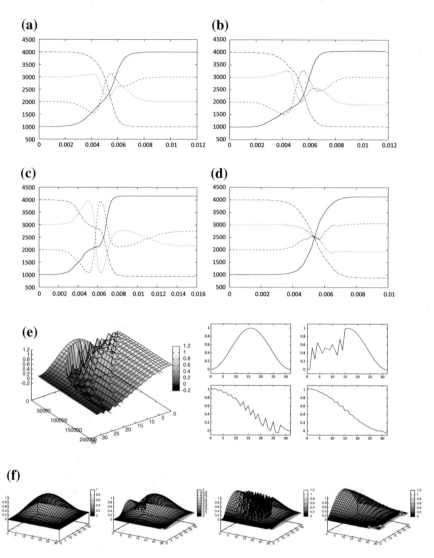

Fig. 13.14 Different applications of the Brockett's analog sort in MGS. **a** presents the results of an MGS simulation of the original system sorting the sequence [1000, 2000, 3000, 4000] into [4000, 3000, 2000, 1000]. Time flows from *left* to *right*; each curve in a plot shows the evolution of an element of the sequence. **b–d** show 3 runs of the same system in the artificial chemistry interpretation. While the original system is deterministic, the chemical version is stochastic (relying on the Gillespie rule application strategy) and exhibits a wide variety of trajectories; however almost all runs converge to the correct sorted sequence (with more or less accuracy). **e–f** illustrate the sort of continuous fields respectively in 1D (on the *left*, the space-time diagram of the sort; on the *right*, states of the system after 0, 100, 1000 and 10000 steps of simulation) and 2D following the x-axis (from *left* to *right*, states of the system after 0, 100, 200 and 3000 steps of simulation)

have been relegated to the choice of the rule application strategy. In differential calculus, time is considered homogeneously with space with the use of a temporal derivative (written $\dot{X}$ in the previous differential equations) revealing the complex and algebraic nature of time in computation. A future work must relate the dot operator with the causal structure mentioned in Sect. 13.2 and revisit the concept of rule application strategy in consequence. In the previous example, the transport of data seems more effective to express patterning rather than structural evolution. The handling of dynamical structures with the sole use of the MGS differential operators remains an open question.

Appendix

This section introduces some elements of algebraic topology and discrete differential calculus used in Sect. 13.5. Algebraic topology (and more especially homology) extends the notion of cell space with an algebraic structure. The key ingredients of this extension are the so-called *topological chains* and *boundary operators*.

Topological Chains

Given a cell space $\mathcal{K}$ and a non-negative integer n, *topological chains of dimension n* (or *n-chains*) are the elements of the free abelian group $C_n(\mathcal{K})$ finitely generated by the n-cells of $\mathcal{K}$. A chain $c \in C_n(\mathcal{K})$ can be understood as a function of $S_{\mathcal{K}} \hookrightarrow \mathbb{Z}$ null everywhere but on a finite set of n-cells of $\mathcal{K}$. Consequently they can be represented by finite formal sum of the form:

$$c = c(\sigma_1).\sigma_1 + \cdots + c(\sigma_p).\sigma_p = \sum_{\sigma \in S_{\mathcal{K}}} c(\sigma).\sigma$$

where $\{\sigma_1, \ldots, \sigma_p\}$ is the set of cells of $\mathcal{K}$ where c is not null.

Topological chains can be interpreted in various ways. They provide a mean to count the cells of a cell space; the possibility to count cells negatively allows to consider orientation of cells. Topological chains are sometimes defined with values in an arbitrary group.[11] Here we restrict ourselves to the group $\mathbb{Z}$.

Boundary Operators

By definition, cell spaces cannot take into account multi-incidence, that is the number of times a cell is incident to another. A solution [53] consists in considering the *incidence numbers* i_σ^τ for any pair of cells σ and τ, so that:

---

[11]The group of n-chains with values in an abelian group G is denoted $C_n(\mathcal{K}, G)$. One can show that $C_n(\mathcal{K}, G) \cong C_n(\mathcal{K}) \otimes G$ where $\otimes$ denotes the tensor product of groups.

$$\forall \sigma \quad \partial \sigma = \sum_{\tau < \sigma} i_\sigma^\tau . \tau \quad \Rightarrow \quad \partial c = \sum_\sigma c(\sigma) \partial \sigma$$

Of course, $i_\sigma^\tau = 0$ if σ and τ are not incident, and i_σ^τ can be negative to take orientation into account. The operator ∂ is linearly extended to any chain of $C_n(\mathcal{K})$.

Homology uses the operator ∂ to study holes in a cell space. In such a case, the operator is called a *boundary operator* and has to respect the nil-potent property: $\partial \circ \partial = 0$, which can be interpreted as "the boundary of a boundary is empty" or "a boundary has no boundary". Such a boundary operator gives raise to a mathematical structure called a *chain complex*:

$$C_0(\mathcal{K}) \quad \xleftarrow{\partial} \quad C_1(\mathcal{K}) \quad \xleftarrow{\partial} \quad \ldots$$

Discrete Forms and Derivative

The discrete counterpart of differential forms coincides with the notion of *cochain* [11]. The set of *discrete forms of dimension n* (or *n*-forms[12]) over a cell-space $\mathcal{K}$ with values in an abelian group G consists of the group homomorphisms of $C^n(\mathcal{K}, G) = \mathrm{Hom}(C_n(\mathcal{K}), G)$ from *n*-chains to the group G. This set inherits naturally the group structure of $C_n(\mathcal{K})$ and its elements can be uniquely specified by the value of G they associate with each cell of $\mathcal{K}$. Like chains, forms can be represented by formal sums with a slight difference, the sum can be infinite. For example, the action $[F, c]$ of a *n*-form F on a *n*-chain c works as follows:

$$[g_1 \cdot \sigma_1 + g_2 . \sigma_2 + g_3 . \sigma_3, 2 . \sigma_1 - 4 . \sigma_3] = 2g_1 - 4g_3$$

where the g_i are elements of G. This application is the discrete analogue of integration of forms on some domain represented by a chain.

The *derivative* $\mathbf{d}F$ of a *n*-form F is then defined to implement a discrete Stokes' theorem, that is, $\mathbf{d}F$ is the $(n+1)$-form adjoint of the boundary operator with respect to application:

$$\forall c \, [\mathbf{d}F, c] = [F, \partial c] \quad \Rightarrow \quad \mathbf{d}F = \sum_\sigma \left(\sum_{\tau < \sigma} i_\sigma^\tau F(\tau) \right) \cdot \sigma$$

Informally the derivative $\mathbf{d}F$ associates with a cell σ the sum of the values associated with the incident cells of σ in F with respect to the incident numbers. One can show easily that $\mathbf{d} \circ \mathbf{d} = 0$ leading to the mathematical structure of *cochain complex* used in cohomology:

---

[12]We choose the term "form" instead of "cochain" in reference to the work of Desbrun et al. [11] about a discrete counterpart of differential calculus. However our concern is more symbolic compared to the numerical issues investigated in discrete differential calculus.

$$C^0(\mathcal{K}, G) \xrightarrow{\ \mathbf{d}\ } C^1(\mathcal{K}, G) \xrightarrow{\ \mathbf{d}\ } \ \cdots$$

Duality

Since a cell space is a partially ordered set, one may consider its inverse order. With this respect, one associates with any cell space $\mathcal{K}$ a cell space $\star\mathcal{K}$, called the *dual of* $\mathcal{K}$, as a formal copy of $\mathcal{K}$ where the incidence relationship is reversed. By referring by $\star\sigma$ to the copy of σ in $\star\mathcal{K}$, we get

$$\forall \sigma, \tau \quad \star\sigma \prec \star\tau \Leftrightarrow \tau \prec \sigma$$

When $\mathcal{K}$ is of dimension n, so does $\star\mathcal{K}$ and $\dim(\star\sigma) = n - \dim(\sigma)$.

Like any cell complex, $\star\mathcal{K}$ can be equipped with a boundary operator $\star\partial$ and its dual derivative operator $\star\mathbf{d}$, so that, considering a well chosen correspondence operator $\ast$ between the primal and dual forms, we get the following diagram:

$$
\begin{array}{ccccccc}
\cdots & \xleftarrow{\ \partial\ } & C_p(\mathcal{K}) & \xleftarrow{\ \partial\ } & C_{p+1}(\mathcal{K}) & \xleftarrow{\ \partial\ } & \cdots \\
\cdots & \xrightarrow{\ \mathbf{d}\ } & C^p(\mathcal{K}, G) & \xrightarrow{\ \mathbf{d}\ } & C^{p+1}(\mathcal{K}, G) & \xrightarrow{\ \mathbf{d}\ } & \cdots \\
 & & \ast\updownarrow\ast & & \ast\updownarrow\ast & & \\
\cdots & \xleftarrow{\ \star\mathbf{d}\ } & C^{n-p}(\star\mathcal{K}, G) & \xleftarrow{\ \star\mathbf{d}\ } & C^{n-p-1}(\star\mathcal{K}, G) & \xleftarrow{\ \star\mathbf{d}\ } & \cdots \\
\cdots & \xrightarrow{\ \star\partial\ } & C_{n-p}(\star\mathcal{K}) & \xrightarrow{\ \star\partial\ } & C_{n-p-1}(\star\mathcal{K}) & \xrightarrow{\ \star\partial\ } & \cdots
\end{array}
$$

In this presentation, the choice of $\star\partial$ (i.e., of the incidence numbers $i^{\star\sigma}_{\star\tau}$ of $\star\mathcal{K}$) and the correspondence operator $\ast$ (as well as the group G) are left as parameters since it depends on the application. For example, while [53] chooses $i^{\star\sigma}_{\star\tau} = i^{\tau}_{\sigma}$, [11] uses $i^{\star\sigma}_{\star\tau} = -1^{\dim(\sigma)} i^{\tau}_{\sigma}$. Moreover the correspondence operator takes an important place in the discrete calculus of [11] as it corresponds to the discrete counterpart of the Hodge operator.

References

1. Aspnes, J., Ruppert, E.: An introduction to population protocols. In: Garbinato, B., Miranda, H., Rodrigues, L. (eds.) Middleware for Network Eccentric and Mobile Applications, pp. 97–120. Springer, Heidelberg (2009)
2. Banâtre, J.P., Le Métayer, D.: Programming by multiset transformation. Commun. ACM **36**(1), 98–111 (1993)
3. Barbier de Reuille, P.: Vers un modèle dynamique du méristème apical caulinaire d'Arabidopsis thaliana. Theses, Université Montpellier II - Sciences et Techniques du Languedoc (2005)
4. Barbier de Reuille, P., Bohn-Courseau, I., Ljung, K., Morin, H., Carraro, N., Godin, C., Traas, J.: Computer simulations reveal properties of the cell-cell signaling network at the shoot apex in Arabidopsis. PNAS **103**(5), 1627–1632 (2006)

5. Bombelli, L., Lee, J., Meyer, D., Sorkin, R.: Space-time as a causal set. Phys. Rev. Lett. **59**(5), 521 (1987)
6. Brockett, R.W.: Dynamical systems that sort lists, diagonalize matrices, and solve linear programming problems. Linear Algebra Appl. **146**, 79–91 (1991)
7. Brockett, R.W.: Differential geometry and the design of gradient algorithms. In: R. Green, e. S.T. Yau (eds.) Symposia in Pure Mathematics, vol. 54, pp. 69–92 (1993)
8. Cardelli, L., Wegner, P.: On understanding types, data abstraction, and polymorphism. ACM Comput. Surv. **17**(4), 471–523 (1985)
9. Cohen, J.: Typing rule-based transformations over topological collections. Electron. Notes Theor. Comput. Sci. **86**(2), 1–16 (2003). Elsevier
10. Dershowitz, N., Jouannaud, J.P.: Rewrite systems. In: Handbook of Theoretical Computer Science, vol. B, pp. 243–320. MIT Press, Cambridge (1991)
11. Desbrun, M., Kanso, E., Tong, Y.: Discrete differential forms for computational modeling. In: ACM SIGGRAPH 2006 Courses, SIGGRAPH '06, pp. 39–54. ACM, New York, NY, USA (2006)
12. Dorigo, M., Stützle, T.: Ant Colony Optimization. Bradford Company, Scituate (2004)
13. Giavitto, J.L., Michel, O.: MGS: a programming language for the transformations of topological collections. Technical Report 61-2001, LaMI – Université d'Évry Val d'Essonne (2001)
14. Giavitto, J.L., Michel, O.: The topological structures of membrane computing. Fundamenta Informaticae **49**, 107–129 (2002)
15. Giavitto, J.L., Spicher, A.: Topological rewriting and the geometrization of programming. Physica D **237**(9), 1302–1314 (2008)
16. Giavitto, J.L., Spicher, A.: Morphogenesis: Origins of Patterns and Shapes. Computer Morphogeneis, pp. 315–340. Springer, Berlin (2011)
17. Giavitto, J.L., Michel, O., Spicher, A.: Spatial organization of the chemical paradigm and the specification of autonomic systems. Software-Intensive Systems and New Computing Paradigms. Lecture Notes in Computer Science, vol. 5380, pp. 235–254. Springer, Berlin (2008)
18. Giavitto, J.L., Michel, O., Spicher, A.: Interaction based simulation of dynamical system with a dynamical structure (ds)2 in mgs. In: Summer Computer Simulation Conference, pp. 99–106 (2011)
19. Giavitto, J.L., Michel, O., Spicher, A.: Unconventional and nested computations in spatial computing. Int. J. Unconv. Comput. (IJUC) **9**(1–2), 71–95 (2013)
20. Gillespie, D.T.: Exact stochastic simulation of coupled chemical reactions. J. Phys. Chem. **81**(25), 2340–2361 (1977)
21. Grandis, M.: Ordinary and directed combinatorial homotopy, applied to image analysic and concurrency. Homol. Homotopy Appl. **5**(2), 211–231 (2003)
22. Henle, M.: A Combinatorial Introduction to Topology. Dover publications, New York (1994)
23. Jansson, P., Jeuring, J., Meertens, L.: Generic programming: an introduction. In: 3rd International Summer School on Advanced Functional Programming, pp. 28–115. Springer, Heidelberg (1999)
24. Jeuring, J., Jansson, P.: Polytypic programming. Advanced Functional Programming, pp. 68–114. Springer, Berlin (1996)
25. Kovalevsky, V.A.: Geometry of Locally Finite Spaces. Editing House Dr. Baerbel Kovalevski, Berlin (2008)
26. Kronheimer, E., Penrose, R.: On the structure of causal spaces. In: Mathematical Proceedings of the Cambridge Philosophical Society, vol. 63, pp. 481–501. Cambridge University Press (1967)
27. Lindenmayer, A.: Mathematical models for cellular interaction in development, Parts I and II. J. Theor. Biol. **18**, 280–315 (1968)
28. Loop, T.L.: Smooth subdivision surfaces based on triangle. Master's thesis, University of Utah (1987)
29. Malament, D.B.: The class of continuous timelike curves determines the topology of spacetime. J. Math. Phys. **18**(7), 1399–1404 (1977)

30. Martin, K., Panangaden, P.: A domain of spacetime intervals in general relativity. Commun. Math. Phys. **267**(3), 563–586 (2006)
31. Michel, O., Jacquemard, F.: An analysis of the Needham–Schroeder public-key protocol with MGS. In: Mauri, G., Paun, G., Zandron, C. (eds.) In: Preproceedings of the Fifth workshop on Membrane Computing (WMC5), pp. 295–315. EC MolConNet - Universita di Milano-Bicocca (2004)
32. Michel, O., Spicher, A., Giavitto, J.L.: Rule-based programming for integrative biological modeling - application to the modeling of the lambda phage genetic switch. Natural Comput. **8**(4), 865–889 (2009)
33. Munkres, J.: Elements of Algebraic Topology. Addison-Wesley, Boston (1984)
34. Musser, D.R., Stepanov, A.A.: Generic programming. In: Gianni, P. (ed.) Symbolic and Algebraic Computation. Lecture Notes in Computer Science, vol. 358, pp. 13–25. Springer, Berlin (1989)
35. Panangaden, P.: Causality in physics and computation. Theor. Comput. Sci. **546**, 10–16 (2014)
36. Paun, G.: Computing with membranes: An introduction. Bull. Eur. Assoc. Theor. Comput. Sci. **67**, 139–152 (1999)
37. Potier, M., Spicher, A., Michel, O.: Topological computation of activity regions. In: Proceedings of the 2013 ACM SIGSIM Conference on Principles of Advanced Discrete Simulation. SIGSIM-PADS '13, pp. 337–342. ACM, New York, NY, USA (2013)
38. Prusinkiewicz, P., Samavati, F.F., Smith, C., Karwowski, R.: L-system description of subdivision curves. Int. J. Shape Model. **9**(1), 41–59 (2003)
39. Reynolds, C.W.: Flocks, herds, and schools: A distributed behavioral model. In: Stone,M.C. (ed.) Computer Graphics. In: Proceedings of the SIGGRAPH '87. vol. 21, pp. 25–34 (1987)
40. Sorkin, R.: Geometry from order: causal sets. Einstein Online **2**, 1007 (2006)
41. Sorkin, R.D.: Relativity theory does not imply that the future already exists: a counter example. In: Relativity and the Dimensionality of the World, pp. 153–161. Springer, Heidelberg (2007)
42. Spicher, A., Michel, O.: Using rewriting techniques in the simulation of dynamical systems: Application to the modeling of sperm crawling. In: Proceedings of the Fifth International Conference on Computational Science (ICCS'05), vol. I, pp. 820–827 (2005)
43. Spicher, A., Michel, O.: Declarative modeling of a neurulation-like process. BioSystems **87**, 281–288 (2006)
44. Spicher, A., Verlan, S.: Generalized communicating p systems working in fair sequential mode. Sci. Ann. Comput. Sci. **21**(2), 227–247 (2011)
45. Spicher, A., Michel, O., Giavitto, J.L.: A topological framework for the specification and the simulation of discrete dynamical systems. In: Proceedings of the Sixth International conference on Cellular Automata for Research and Industry (ACRI'04), Lecture Notes in Computer Science, vol. 3305. Springer, Amsterdam (2004)
46. Spicher, A., Michel, O., Cieslak, M., Giavitto, J.L., Prusinkiewicz, P.: Stochastic p systems and the simulation of biochemical processes with dynamic compartments. BioSystems **91**(3), 458–472 (2008)
47. Spicher, A., Michel, O., Giavitto, J.L.: Declarative mesh subdivision using topological rewriting in mgs. In: Proceedings of the International Conference on Graph Transformations (ICGT) 2010, Lecture Notes in Computer Science, vol. 6372, pp. 298–313 (2010)
48. Spicher, A., Michel, O., Giavitto, J.L.: Understanding the dynamics of biological systems, chap. Interaction-based simulations for Integrative Spatial Systems Biology, pp. 195–231. Springer, Heidelberg (2011)
49. Spicher, A., Michel, O., Giavitto, J.L.: Interaction-based simulations for integrative spatial systems biology. In: Understanding the Dynamics of Biological Systems: Lessons Learned from Integrative Systems Biology. Springer, New York (2011)
50. Toffoli, T., Margolus, N.: Cellular Automata Machines: A New Environment for Modeling. MIT Press, Cambridge (1987)
51. Tonti, E.: A direct discrete formulation of field laws: the cell method. CMES - Comput. Model. Eng. Sci. **2**(2), 237–258 (2001)
52. Tucker, A.: An abstract approach to manifolds. Ann. Math. **34**(2), 191–243 (1933)

53. Tucker, A.W.: Cell spaces. Ann. Math. **37**(1), 92–100 (1936)
54. Turing, A.M.: The chemical basis of morphogenesis. Philoso. Trans. Royal Soc. Lond. B: Biol. Sci. **237**(641), 37–72 (1952)
55. Von Neumann, J.: Theory of Self-Reproducing Automata. University of Illinois Press, Champaign (1966)
56. Winskel, G.: Event structures. Springer, Heidelberg (1987)
57. Witten, T., Sander, L.: Diffusion-limited aggregation. Phys. Rev. B **27**(9), 5686 (1983)

Chapter 14
Cellular Automata in Hyperbolic Spaces

Maurice Margenstern

Abstract The chapter presents a bit more than fifteen years of research on cellular automata in hyperbolic spaces. After a short historical section, we remind the reader what is needed to know from hyperbolic geometry. Then we sum up the results which where obtained during the considered period. We focus on results about universal cellular automata, giving the main ideas which were used in the quest for universal hyperbolic cellular automata with a number of states as small as possible.

14.1 Introduction

The first cellular automaton in the hyperbolic plane was devised in 1999, in a join paper with Kenichi Morita, see [42]. This paper strikingly showed the possibility to solve NP-problems in polynomial time, a property which was anticipated by [48], a paper the authors of [42] were not aware of. It is important to remark that in contrast with [48], the paper [42] provided an explicit implementation of the process: namely, the authors constructed a cellular automaton based on a tessellation of the hyperbolic plane based on the regular convex pentagon with right angle: the pentagrid, see Sect. 14.2.

In 2001, the author found a new, very easy way to implement cellular automata in the hyperbolic plane. He found a coordinate system of the tiles of the pentagrid which allows to find a path from the tile to a specific one fixed in advance and once for all, the computation of the path being in a linear time in the size of the coordinate.

The results which where obtained from this point can be divided into four parts: complexity results about computations performed with hyperbolic cellular automata, see Sect. 14.3.2, universality results for this model of computation, see Sect. 14.4, theoretical results about tilings in the hyperbolic plane and cellular automata in

M. Margenstern (✉)
LITA, EA 3097, Université de Lorraine, UFR MIM, Campus du Saulcy,
57045 METZ Cédex 1, France
e-mail: maurice.margenstern@univ-lorraine.fr

© Springer International Publishing Switzerland 2017
A. Adamatzky (ed.), *Advances in Unconventional Computing*,
Emergence, Complexity and Computation 22,
DOI 10.1007/978-3-319-33924-5_14

this setting, see Sect. 14.3.3, and a few possible applications in three directions, see Sect. 14.3.4.

14.2 Hyperbolic Geometry: What to Know

In this section, we start with a very short look on the history of hyperbolic geometry, see Sect. 14.2.1. Then, we present it through one of its most popular model, Poincaré's disc, see Sect. 14.2.2. Then we more precisely look at tilings in hyperbolic spaces, see Sect. 14.2.3 and them at three of them, two ones in the plane, see Sect. 14.2.4 and the last one in the space, see Sect. 14.2.5.

14.2.1 A Remarkable History

The history of hyperbolic geometry goes back to Euclid himself. It can be the concern of a whole book. By the way, the author wrote a popularization book on the topic, see [41], in French.

The cause of the history lies in the parallel axiom of Euclidean geometry. In his *Elements*, see [4], Euclid remarks that some objects cannot be defined otherwise than naming them in natural language. On the same line, he states several axioms in order to fix obvious properties which cannot be proved without some circularity. The first properties in Euclid's line are simple properties which do not raise objection. His last axiom says that in the plane, from a point P out of a line ℓ, there is precisely one parallel to ℓ which passes through P. Euclid's formulation is equivalent to the just here formulated one but it is not that. It is a more complex statement, see for instance [4, 41]. Later on, we shall call it **Euclid's axiom**, *EA* for short. Around two centuries after the writing of the *Elements*, Posidonius stated another property which is equivalent to Euclid's axiom. Still five centuries later, Proclus thought to have proved Euclid's axiom. In fact, he found a new property which is equivalent to *EA*. This started a more than one thousand year search in order to prove *EA* from the other axioms of the *Elements*. We have no room here to give the details of this history. It is a very interesting one, both for the personalities of the discoverers of the new geometry, also for the difficulties met by the discovery itself in order to be accepted by the mathematical community. It should never be forgotten that science is made by men and women and that the scientific adventure involves all human passions alike any other adventure: the best ones as well as the worst ones. The history of hyperbolic geometry gives a full illustration of that.

What can be concluded from this history? In short, the first thousand year gave people the feeling that they were running along a circle. This was not the case. As we know the end of the story, we can say that they found a lot of properties which are equivalent in the Euclidean geometry but not in the new one. In the 17th century, somebody tried a new direction: assume a property which negates *EA*, and try to find out a contradiction. If this try works, it provides us with a proof of *EA*. Still two centuries were needed to reach the end of this new track. Around the end of the first third of the 19th century, two men, independently of each other and almost at the same time found out the end of the new path. It was not a proof of *EA*: it was a new geometry, the hyperbolic geometry, as it was called much later. For reasons we cannot tell here, the discovery became known to the world more than thirty years later, around the beginning of the last third of the 19th century. And that time, people understood what was the meaning of this new geometry: it was a proof that there is no proof for *EA*. Because models of this geometry were found in the Euclidean one. Accordingly, you can either accept *EA* or you can reject it. In fact, this splits geometry into two parts: the Euclidean one, and the new one. Soon, a third geometry appeared, again motivated by some aspects of the Euclidean geometry, the so called elliptic geometry. A bit later, generalizations of these geometries and new geometries appeared, transforming the initial book of geometry into a big collection of books. The biggest part still belongs to Euclidean geometry. A smaller part is devoted to hyperbolic geometry.

Before leaving history, note that it was the first time in the history of mathematics that people met with something which cannot be proved. As at that time logic started to leave its Middle-Age beginnings and to reach its mathematical basis, the discovery of hyperbolic geometry, appeared to be a natural example of what is an independent axiom. The way to prove *EA* gave rise to a whole method, then a realm of methods gathered into what is now called model theory: the models of hyperbolic geometry found in the second half of the 19th century were the first models of a geometry where *EA* is false. The discovery of hyperbolic geometry was an actual revolution in the scientific world. It had echoes in philosophy too. The wave risen by the discovery is not over: the birth of computer science can easily be traced to it through the logical path we just described, were an intermediate stage was the foundation crisis which is presently dormant but not at all solved.[1]

14.2.2 Poincarés Disc

Poincaré's models appeared in 1872. There are two of them which are image of each other through what mathematicians call a conformal transformation. One model is a half-plane of the Euclidean plane. The other model is in an open disc of the Euclidean plane. We take this latter model for the following reason. Euclidean models of hyperbolic geometry necessarily introduce distortions. In our familiar Euclidean world we can only see a very small part of the hyperbolic plane. The disc model gives a more clear image of this property. It is also the case because Poincaré's models keep angles, which means that the Euclidean angles between images of secant hyperbolic lines are the same as the hyperbolic angles. Accordingly, Poincaré's disc can be viewed as a window over the hyperbolic plane, like a window in a plane flying in the

---

[1] Not on fashion problems are not necessarily uninteresting problems.

Fig. 14.1 Poincaré's disc.
The lines s, p, q and m pass
through A which lies out of ℓ.
The line s cuts ℓ, the lines p
and q are parallel to ℓ. The
line m is non-secant with ℓ

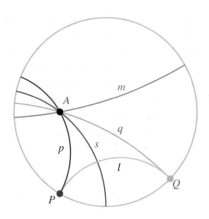

air, which focuses on the point at which we look. The rest of the plane vanishes at the
border of the disc: the closer we are to this border, the further we are from the point
which we look at. Figure 14.1 illustrates a particular situation in this plane. In order
to understand it, let us say that the points of the hyperbolic plane are represented by
the points of the open disc whose border is the green circle on the picture of Fig. 14.1.
Lines of the plane are either traces in the disc of diametral lines or of circles which
are orthogonal to the border of the disc.

The figure illustrates a point A which lies outside a line ℓ of the plane. The line s
passes through A and cuts ℓ in the hyperbolic plane. The lines p and q have a common
point with ℓ which is on the border of the disc. From our definition, such a point
does not belong to the hyperbolic plane. We say that a point of the border of the disc
is a **point at infinity**. Note that a line has two points at infinity and that two distinct
points at infinity define a single line. We say that two lines which pass through the
same point at infinity are **parallel**. As shown by Fig. 14.1, from a point A out of
a line ℓ there are always two distinct lines which are parallel to ℓ and which pass
through A. They are distinct: p passes through a point at infinity of ℓ and q passes
through the other point at infinity of ℓ.

For our sequel, we just need an additional property.

Theorem 1 *Two lines of the hyperbolic plane are non-secant if and only if they have
a common perpendicular.*

Also, we need to define a particular transformation: the **reflection in a line**.
Consider a line ℓ. If ℓ is represented by a diameter, the reflection in ℓ is the Euclidean
one. If ℓ is represented by a circle S which is orthogonal to the border of the disc, the
image of a point P by reflection in ℓ is the inverse P' of P with respect to the circle S.
More precisely, if C is the centre of S and r is its radius, then P, P' and C are on the
same Euclidean line d, P and P' are on the same side of d and $CP \cdot CP' = r^2$. The
line ℓ is called the **axis** of the reflection. The inversion keeps angles in the Euclidean
plane. It is the reason why Poincaré's model also keeps the Euclidean angles. In the
hyperbolic plane, a finite product of reflections in lines is called an **isometry**. In

fact, the isometries of the hyperbolic plane, alike those of the Euclidean plane, are products of at most three reflections in lines.

14.2.3 Tilings

We specialize our look at hyperbolic geometry by focusing on tilings, especially on **tessellations**. Here, we define a tessellation by the following process:

Algorithm 1 *Construction process of a tessellation:*
1. Fix a regular convex polygon P: its sides are equal, its interior angles are also equal.
2. Replicate P by refection in its sides.
3. Recursively replicate the images in their sides.

Call **copy** of P, any image obtained in this process as well as P itself. We say that the process constructed by Algorithm 1 defines a **tessellation based on** P if and only if all the copies of P cover the hyperbolic plane and any two copies either coincide or their interiors do not meet. We also say that P **generates** a tessellation. Note that this definition holds in any geometry where the notion of isometry is defined.

In the Euclidean plane, up to similarities, there are three tessellations only: they are based on a square, on an equilateral triangle or on a regular convex hexagon. In the hyperbolic plane, the situation is very different. First, there is no similarity: triangles with equal angles are equal. This fourth case of equality of triangles holds in hyperbolic geometry, of course not in the Euclidean one. Second, there are infinitely many tessellations based on a regular convex polygon. From a theorem of Poincaré concerning triangles, tessellations based on a regular convex polygon can easily be reduced to triangles, one easily proves the following property:

Theorem 2 (Poincaré 1882) *Let P be a regular convex polygon of the hyperbolic plane. Let p be the number of its sides and let $\dfrac{2\pi}{q}$ be its interior angle. Then the tessellation based on P exists if and only if*

$$\frac{1}{p} + \frac{1}{q} < \frac{1}{2}. \tag{14.1}$$

The theorem says that **any** regular convex polygon of the hyperbolic plane generates a tessellation: indeed, for any such polygon P with p sides and with $\dfrac{2\pi}{q}$ as interior angle, the inequality (1) holds. Usually, the tessellation generated by P is denoted by $\{p, q\}$. Note that q is also the number of copies of P around a vertex of a copy in the tessellation. Note that the inequality in (1) entails that there are infinitely many different tessellations based on a regular polygon in the hyperbolic plane: they are all different in shape. Remark that the tessellations of the Euclidean plane are obtained if we replace the inequality sign in (1) by the equality sign. Then we easily check that there are only three solutions for the corresponding equation in p and q.

14.2.4 The Pentagrid and the Heptagrid

Now, we focus on two particular tessellations of the hyperbolic plane: the tessellations $\{5, 4\}$ and $\{7, 3\}$ which we call **pentagrid** and **heptagrid** respectively (Fig. 14.2).

Note that in (1), $q \geq 3$. When $q = 3$, the smallest value for p is 7. Note that $p = 6$ is the case of the Euclidean regular hexagon. When $q = 4$, the smallest value for p is 4. Here too, $p = 4$ corresponds to a Euclidean figure: the square. Accordingly, for the considered angles, the values of p we have are the smallest ones. In some sense, they are the first cases to be investigated. Another interest lies in the following. From a theorem of Coxeter and Moses, see [3], the tessellation $\{p, q\}$ is a Cayley graph of a group of displacements of the hyperbolic plane when p is even. When p is odd this may not be the case: the pentagrid is an example as shown in [38]. When both p and q are odd, the heptagrid is an example that, again, this may be not the case as shown in [38] too.

14.2.5 The Dodecagrid

In higher dimension, in the Euclidean plane, the cube and hypercube always generate a tessellation. In the hyperbolic 3D-space, there are four tessellations generated by a regular convex polyhedron, and there are five of them in the hyperbolic 4D-space generated by a regular convex polytope. In the hyperbolic nD-space, for $n \geq 5$, there is no tessellation generated in a regular convex polytope.

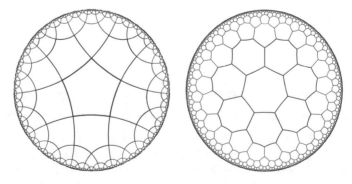

Fig. 14.2 *To left* the pentagrid. *To right* the heptagrid. Both tessellations are represented in Poincaré's disc

Fig. 14.3 Schlegel representation of a dodecahedron. In Sect. 14.3.1, we shall see how the same projection can be used to represent the dodecagrid itself

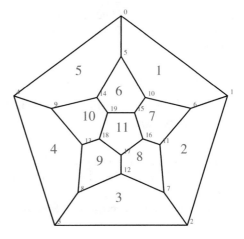

We shall consider a tessellation of the hyperbolic 3D-space: we call it the **dodecagrid**, defined by the signature {5, 3, 4} of the generating dodecahedron: 5 means that its faces are pentagon, 3 is the number of pentagons meeting at a vertex of the dodecahedron, 4 is the number of dodecahedra which share an edge in the tessellation. It is shown that the dodecahedron is the natural generalization to the hyperbolic 3D-space from the regular convex pentagon in the same way as it is in the Euclidean 3D-space. However, in the Euclidean plane there is a single regular convex pentagon, up to similarities. In the hyperbolic plane, there are infinitely many regular convex pentagons. The dodecagrid is based on the dodecahedron whose faces are pentagons of the pentagrid. Such a dodecahedron is called **Poincaré's dodecahedron**. To represent the dodecagrid we use a central projection on a face of a dodecahedron called **Schlegel** projection, as illustrated by Fig. 14.3.

Note that there is another tessellation of the hyperbolic 3D-space based on a regular convex dodecahedron. The signature of this tessellation is {5, 3, 5}. Its faces are also regular convex pentagons, three of them meet at a point on the dodecahedron, but this time, five dodecahedra share an edge in the tessellation. Of course, the angles involved in this dodecahedron are different from the angles of Poincaré's dodecahedron. They are smaller. Accordingly, the new dodecahedron is bigger than Poincaré's one. For this reason, it is called the **big** dodecahedron. A study of the big dodecahedron can be found in [51].

14.3 Results on Hyperbolic Cellular Automata

In this section, we look at the implementation of cellular automata and at the results which were established in this new frame. The implementation itself rests on an important result, see Sect. 14.3.1. Very striking properties appeared in the field of complexity theory in this context, see Sect. 14.3.2. Also striking results were obtained in the field explored in Sect. 14.3.3 about tilings in the hyperbolic plane and cellular automata in general. The section is concluded by Sect. 14.3.4 listing three possible applications.

14.3.1 Implementation of Cellular Automata in Hyperbolic Spaces

In this sub-section, we shall investigate the implementation of cellular automata in the hyperbolic plane, see Sect. 14.3.1.1, and then the implementation in the hyperbolic 3D-space, see Sect. 14.4.1. In Sect. 14.3.1.1 we investigate this implementation in the pentagrid and in the heptagrid. In Sect. 14.4.1, we investigate it in the dodecagrid.

14.3.1.1 Implementation in the Pentagrid and in the Heptagrid

The definition of cellular automata is based on three principles of homogeneity: homogeneity of space, of time, of programs. This is the reason why, most often, the cells of a cellular automaton are tiles of a tessellation and, again for homogeneity reason, of a tessellation based on a regular convex polygon. The homogeneity in time together with the computational conditions involves a discrete time represented by non negative integers. The homogeneity of programs says that each cell is dotted with a copy of the same finite automaton. As in cellular automata in the Euclidean plane, at each tick of the clock, the cell updates its own state depending on its sates and on the states of its neighbours. This is also why cellular automata are based on tessellations: the neighbourhood of a cell must be everywhere the same.

Once the tessellation is chosen, the first question is to locate the cells. This problem is particularly easy to solve in the Euclidean tessellations of the plane. They have a very easy system of coordinates. Just a look at it. Most often, when working on the Euclidean plane, people use cellular automata on the square grid. In most papers, the tiling is identified by $\mathbb{Z}^2$. You will probably never find the explicit identification of the square grid with $\mathbb{Z}^2$. In fact, the square grid has no specially marked tile. However, one of them must be identified with (0, 0), which is called **origin**. Then, one of its neighbours must be identified with (1, 0). If you have an analogous watch, which is not a very mathematically axiomatic object, you can fix which second neighbour of the origin is identified with (0, 1). If you have no analogous watch, you choose[2] the neighbour of the origin which is identified with (0, 1): this is also a way to fix the orientation.

Now, let us look at the same problem in the hyperbolic plane, first, in the pentagrid.

In [43], a first attempt to fix coordinates in the pentagrid was proposed together with an important property: the bijection of a quarter of the pentagrid with a finitely

---

[2]Don't worry: nobody will ask you how you proceed for this identification. Such a question has no place in a serious, objective paper.

Fig. 14.4 The bijection
between the Fibonacci tree
and a quarter of the pentagrid

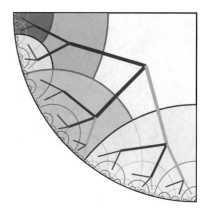

generated tree. The bijection was not clearly proved in that paper. A better proof is
given in [17]. A definitely clear one, I hope so, is given in [40]. A new system of
coordinates is given in [12], and I systematically used it afterwards. This system is
based on the fact that the tree which is in bijection with the tessellation is connected
with the Fibonacci sequence, see Fig. 14.4.

The connection with the Fibonacci tree can be seen on the figure. We have two
kinds of nodes, white ones which have three sons, black ones which have two of them
only. Now, number the nodes from the root, level by level and, on each level, from
left to right, and call **coordinate** of the node the representation of its number in the
Fibonacci basis. Remember that each positive number is a sum of distinct Fibonacci
numbers and that this representation can be made unique by requiring it should be
the longest one. Then, it appears that for each node, if v is its coordinate, exactly
one of its sons has $v00$ as its coordinate. We cannot prove this property here, see for
instance [17]. We also cannot give details. We just mention that from this property,
there is an important corollary.

Theorem 3 (Margenstern 2001, see [12]) *There is an algorithm which computes the
path in the Fibonacci tree from the root to a node from the coordinate of the node,
which is linear time with respect to the length of the coordinate.*

Now, from this it is easy to extend this system of coordinate to the whole pentagrid.
Note that around a central tile, fixed once and for all, it is possible to place five quarters
exactly so that the union of the tiles exactly gives the pentagrid, with no overlapping.

The same can be done for the heptagrid for a central tile and seven sectors, see
Fig. 14.5. A striking property is that the bijection between the Fibonacci tree and a
quarter of the pentagrid also holds for the same tree and an angular sector of the
heptagrid, see [17]. It is not possible here to give more details than Figs. 14.5 and
14.6.

We can now go back to the location problem in the pentagrid, or in the heptagrid.
Fix a tile which will be the **origin**. Its neighbours are identified with the root of
the Fibonacci tree. One neighbour of the root is identified with the leftmost son of

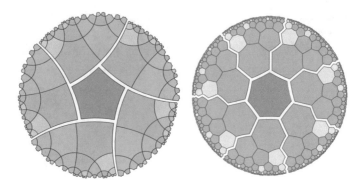

Fig. 14.5 *To left* splitting the pentagrid into a central tile and five quarters. *To right* splitting the heptagrid into a central tile and seven sectors

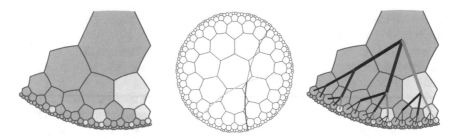

Fig. 14.6 *To left* a sector of the heptagrid. *Middle* how to delimit a sector of the heptagrid. Mid-points of edges of heptagons of the heptagrid lie on a line, the **mid-point line**. A sector is delimited by two rays supported by mid-point lines. By definition, the heptagons which have at least five vertices in the sector belong to the sector. *To right* the bijection between the Fibonacci tree and the sector illustrated by the left-hand side picture

the root. A common neighbour to the root and to the leftmost son is identified as the middle son of the root or, if we make use of an analogous watch, the middle son comes just after the leftmost one while counter-clockwise turning around the root. As the Fibonacci tree is used both for the pentagrid and for heptagrid, the just mentioned sequence of three choices works for both the pentagrid and for the heptagrid. This allows us to identify these tessellations with a big tree whose root is the central tile and whose sub-trees rooted at the sons of the root are roots of copies of the Fibonacci tree: five, seven of them for the pentagrid, heptagrid respectively. Now, we can see that this identification with the Fibonacci tree is very close to the identification of the Euclidean square grid with $\mathbb{Z}^2$.

The method we used to study the heptagrid and the heptagrid can be generalized to all tilings $\{p, q\}$ of the hyperbolic plane. This was performed in [17] where references to papers can be found. It is interesting to note that this method can be generalized to more general situations. In particular, it was applied to a tessellation of the hyperbolic 3*D* space and to another one of the hyperbolic 4*D* space, again see [17]. Before turning to the study of the dodecagrid in the hyperbolic 3*D* space, let us mention

Fig. 14.7   Splitting of the dodecagrid: the three types of regions

another interesting property. We noticed that both the pentagrid and the heptagrid can be made in bijection with a few copies of the Fibonacci tree. This property can be generalized to all tilings $\{p, 4\}$ and $\{p+2, 3\}$ with $p \geq 5$. For each p, these tilings are in bijection with a few copies of the same tree which is associated with another recurrence relation attached to some polynomial with integer coefficients, see [17].

14.3.1.2   Localization of Tiles in the Dodecagrid

Now, we turn to the implementation in the hyperbolic 3D-space. Basically, we shall perform the same method as what we did in the pentagrid. However, the 3D-context will strongly change many things.

Let us see how we can manage things.

Remember that in order to represent a dodecahedron D, we used the central projection from a point of D onto the plane supporting one face F of D. We can imagine that the centre C of the projection lies on the line which is perpendicular to F and that it passes through the centre of the pentagon constituting F: we call such a line the **axis** of F. Also, we can imagine that the face which is opposite to F lies in between F and C.

We shall use this representation to give a *dynamic* representation of the dodecagrid itself which is illustrated by Fig. 14.7.

Let us explain this representation. Look at the leftmost picture of the figure. Faces 1 and 5 of the dodecahedron D are coloured in green: we say that they are **shadowed**. The meaning is that we forbid the reflection of D in these faces. We also forbid the reflection of D in face 0 whose supporting plane is the plane onto which the projection is performed, see Fig. 14.3, so that face 0 is also shadowed. On the other faces, we allow the reflection of D in its faces. This is illustrated by drawing the projection on the dodecahedron reflected in the face i, say D_i, on the plane supporting this face. Now, on D_i, we have to shadow the faces which are on the same plane than those which are already shadowed and we also must shadow a new one. Take, as an example, faces 6 and 10. Face 6 has a face on the same plane as face 5 and another one on the same plane as face 1. Moreover, faces 6 and 10 share a side s. In D_6 and in D_{10}, each dodecahedron has a single face which shares s, say G_6 and G_{10} respectively. Now, by definition of the dodecagrid, s is shared by four dodecahedra in the tessellation. We have already identified three of them: D itself, D_6 and D_{10}. The fourth one is either the reflection of D_6 in G_6 or the reflection of D_{10} in G_{10}. In order to obtain a bijection, we forbid one reflection by shadowing the corresponding face and we force the other. On the illustration, we can see that we shadowed G_{10}. Now, we successively perform the construction of the D_i's in this order: 2, 3, 4, 10, 9, 8, 7, 6, 11. On faces 2 and 3, we can put a dodecahedron with three shadowed faces as D. The, on faces 4, 10, 9, 8, 7 and 6, we can put a dodecahedron with four shadowed faces as we have seen for D_{10}. At last, on face 11, we can see that D_{11} must have six shadowed faces. This gives us three types of regions of the hyperbolic 3D-space if we consider the region which is the intersection of the half-spaces defined by the planes supporting the shadowed faces, taking the half-space which contains the considered dodecahedron. These regions are denoted by O, H, T illustrated by the pictures of Fig. 14.7 in that order. The figure shows us that the decomposition of the regions represented by D_{10} and D_{11} do not involve new regions. From this, we deduce the possibility to construct a finitely branched tree which is in bijection with an eighth of the hyperbolic 3D-space illustrated by the leftmost picture of Fig. 14.7. From this figure, we easily derive the following rules for constructing the tree:

$$O \to O^2 H^6 T \qquad\qquad H \to OH^6 T \qquad\qquad T \to H^5 T$$

This splitting improves the one which is given in [17]. There, four regions are involved. However, both splittings are very similar: they are connected with the same algebraic number β, the greatest positive zero of $X^2 - 8X + 1$. Representing a suitable numbering of the tree, the same basis for representing the numbers of the nodes allows us to define coordinates. These coordinates, as in the pentagrid allow us to compute a path from the node to the root. However, this computation is cubic in the dodecagrid while it is linear in the α-grids.

In fact, for implementing cellular automata in the hyperbolic 3D, the model which will be used allow us to devise a much simpler implementation based on the pentagrid. We postpone this representation to Sect. 14.4.

14.3.2 Complexity Results

We can define the hyperbolic **P**-class as the set of cellular automata in the pentagrid or in the heptagrid, which work in polynomial time. We denote it by $\mathbf{P}_h$. Define the hyperbolic **NP**-class as the set of cellular automatic which, given the solution of the problem, checks the solution in polynomial time. Denote it by $\mathbf{NP}_h$.

The main complexity result is the following property.

Theorem 4 (Iwamoto-Margenstern-Morita-Worsch 2002) [7] *For cellular automata either on the pentagrid or in the heptagrid, we have* $\mathbf{P}_h = \mathbf{NP}_h = \mathbf{PSPACE}$.

This result was anticipated by [48] who announced the possibility to solve **NP**-problems in polynomial time in the hyperbolic plane. In [43], an explicit cellular

Fig. 14.8 Simulation of
non-deterministic hyperbolic
cellular automaton by a
deterministic one

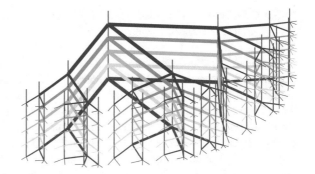

automaton in the pentagrid was given to solve **3-SAT** in polynomial time in the hyperbolic space. Note that the authors of [43] were not aware of [48] when their paper appeared. The results contains a proof that a non-deterministic hyperbolic cellular automaton can be simulated by a deterministic one in polynomial time, in fact in cubic time, see [7]. The proof is illustrated by Fig. 14.8.

Theorem 4 makes us conjecture that the complexity hierarchy for cellular automata in hyperbolic spaces is much reduced compared to the classical one. It is confirmed by the following result:

Theorem 5 (Iwamoto-Margenstern 2003 [8]) *For complexity classes of cellular automata in hyperbolic spaces, we have the following inclusions:*

$$DLOG_h = NLOG_h = P_h = NP_h = PSPACE$$
$$\nsubseteq PSPACE_h = EXPTIME_h = NEXPTIME_h = EXPSPACE$$

Note that since this time, no new result was obtained on this line. I just mention [10] which makes more precise what was written in [48]. Also, a few people considered necessary to include space considerations in the definition of the complexity classes like P_h or $DLOG_h$. There is no reason why this condition of space would be included in the hyperbolic case and not in the Euclidean case. If we accepted such a distinction, it would ruin a possibility to prove that $\mathbf{P} \neq \mathbf{NP}$. Indeed, due to Theorem 4, a path for a proof of $\mathbf{P} \neq \mathbf{NP}$ could be to prove, on the assumption of $\mathbf{P} = \mathbf{NP}$ that we can construct an isometric embedding of a part of the hyperbolic plane in the Euclidean one, which is not possible.

14.3.3   Application to Undecidability Results

The localization technique mentioned in Sect. 14.3.1 allows us to tackle general theorems about cellular automata as well as problems about tilings in general. The first direction is dealt with in Sect. 14.3.3.1 while the second one is dealt with in Sect. 14.3.3.3.

14.3.3.1   General Properties About Cellular Automata
in Hyperbolic Spaces

In this sub-subsection, we mention the theorems which are the hyperbolic analogs
of famous theorems on cellular automata in the Euclidean square grid. The first
theorem concerns the general characterization of cellular automata themselves. For
this purpose, people consider the transformation induced by a cellular automaton on
the object which is called the space of configurations. We remind the reader that a
configuration is the set of states on all the cells of the cellular automaton at a given
instant. In the Euclidean square grid identified with $\mathbb{Z}^2$, the space of configurations
is $Q^{\mathbb{Z}^2}$ where Q is the finite set of states of the automaton. Now, if $c \in Q^{\mathbb{Z}^2}$, denote
by $c(x, t)$ the state of the cell at x at time t. When we do not consider the time, simply
the state of the cell at x, we denote it $c(x)$. Denote by f the local transition function:
it indicates how the cell at x changes its state depending on $c(x + V)$ which gives the
states of all cells in $x + V$, where V is a neighbour of the origin. This allows us to
define the **global transition function** F defined by $F(c)(x) = f(c(x + V))$. Clearly,
F is a mapping from $c \in Q^{\mathbb{Z}^2}$ into itself. Hedlund's theorem says that a mapping
from $c \in Q^{\mathbb{Z}^2}$ into itself is the global function of a cellular automaton if and only if it
is continuous and it commutes with shifts. Remember that a shift is a mapping from
$\mathbb{Z}^2$ into itself of the form $x \mapsto x + v$, where v is a fixed element of $\mathbb{Z}^2$. More-Myhill
theorem characterizes the surjectivity of the global function of a cellular automaton
by saying that such a function is surjective if and only if it is injective on finite
configurations, where a finite configuration is the restriction of a configuration to a
finite subset of $\mathbb{Z}^2$.

What can be said for cellular automata on the pentagrid or on the heptagrid?
Call α-grid the pentagrid, heptagrid, depending on whether α is 5, 7 respectively.
With the localization performed in Sect. 14.3.1, we the α-grid with $\mathcal{F}_\alpha = 1 + \overset{\alpha}{\underset{i=1}{\cup}} \mathcal{F}$,
where $\mathcal{F}$ is the Fibonacci tree and where $\overset{\alpha}{\underset{i=1}{\cup}} \mathcal{F}$ denotes the union of α copies of
the Fibonacci tree. It is easy to endow $\mathcal{F}_\alpha$ with a metric which ignores what is at
large from the origin. We call this metric the discrete metric. Then we can define the
global function of a hyperbolic cellular on the α-grid as the function F defined on
$Q^{\mathcal{F}_\alpha}$ by $F(c)(x) = f(c(V(x)))$, where $V(x)$ is the neighbourhood of x. We say that
a cellular automaton on the α-grid is **rotation invariant** if and only if $f(c(V(x)))$
is not change if we replace V by V^ρ where V^ρ is obtained from V by a circular
permutation. Indeed, the neighbours of x different of x share an edge with x so that
they can be numbered from 1 to α by counter-clockwise turning around x, the side 1
being the side to the father of x. We say that the origin is the father of the root and
we fix once and for all the side 1 of the origin. Then a circular permutation of V
boils down to a circular permutation on $[1 \dots \alpha]$. At last, a shift on $\mathcal{F}_\alpha$ is induced on
this space by a shift in the α-grrid which leaves the grid globally invariant. Then we
have:

Theorem 6 (Margenstern 2008 [18]) *A mapping from $Q^{\mathcal{F}\alpha}$ into itself is the global function of a rotation invariant cellular automaton if and only if it is continuous on $Q^{\mathcal{F}\alpha}$ endowed with the discrete metrics and it commutes with the shifts.*

Note the additional hypothesis of rotation invariance which is not required for the Euclidean square grid.

It is interesting to remark that if the analog of Hedlund's theorem almost holds in the hyperbolic case, it is no more true for the More-Myhill theorem as stated by the following theorems. In the statements, if X is a cellular automaton, F_X is its global function.

Theorem 7 (Margenstern-Kari 2009 [24]) *There is a cellular automaton A on the α-grid such that F_A is injective but F_A is not surjective. There is also a cellular automaton B on the α-grid such that F_B is surjective but F_B is not injective.*

Theorem 8 (Margenstern-Kari 2009 [24]) *There is a rotation invariant cellular automaton on the α-grid which is surjective but not injective even on finite configurations.*

Note that in Theorem 8 we assume the stronger hypothesis of a rotation invariant cellular automaton. The existence of a rotation invariant cellular automaton C on the α-grid such that F_C which would be injective but F_C would be not surjective is still open.

14.3.3.2   Undecidable Problems About Cellular Automata

It is know that the injectivity of the global function of a cellular automaton on $\mathbb{Z}^2$ is an algorithmically unsolvable problem. This was proved by Jarkko Kari in 1994. The same question may be raised for cellular automata on an α-grid. We have the following result:

Theorem 9 *There is no algorithm to decide whether the global function of a cellular automaton on the heptagrid is injective or not.*

As in the Euclidean case, the proof is very complex. It relies on the possibility to build a plane-filling path for the hyperbolic plane. Figure 14.9 shows us an origin-constrained plane-filling path. The path illustrated by the figure starts from a fixed in advance origin.

For the proof of Theorem 9, we need a path which fills infinitely many larger and larger parts of the hyperbolic plane, crossing each tile exactly once: it is not needed that the path fills the whole plane with this property, see [23]. An origin-constrained path is not enough for that purpose, but the path constructed in [23] is enough for proving Theorem 9.

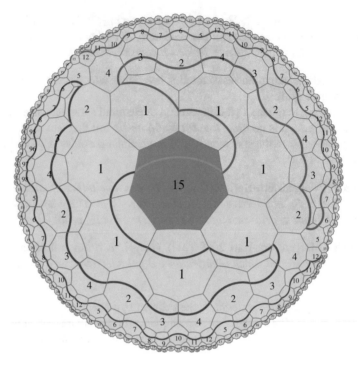

Fig. 14.9 An origin-constrained plane-filling path in the heptagrid

14.3.3.3   About Tiling Problems

An important application of the localization technique is the proof that the tiling problem is undecidable for the hyperbolic plane. The tiling problem consists in the following. We are given a finite set of tiles called the **prototiles** such that by copies of the prototiles we can tile the plane. By copies we define which isometries are allowed for producing a copy of a prototile. By tiling we also mean that if the sides bear decorations, the decorations match when the supporting edges belonging to different tiles coincide. We can solve the tiling problem if, for any set of prototiles we can say whether or not it tiles the plane. By a theorem of Berger, see [1], the tiling problem of the Euclidean plane is undecidable. In 1971, Robinson raised the same question for the hyperbolic plane, see [49], giving a solution for the origin-constrained problem in 1978: in the origin-constrained problem, the initial tile is given. This problem was solved in [19], in 2008. In fact, the proof was already published in [15, 16] in 2007:

Theorem 10   (Margenstern 2007 [15, 16, 19]) *The tiling problem is algorithmically unsolvable for the hyperbolic plane.*

Figure 14.10 illustrates the tiling which was used as a basis ingredient for the proof of Theorem 10. Note that the proof we gave in [15, 16, 19] proves a little bit stronger result: this algorithmic unsolvability result is established for the heptagrid. If it holds

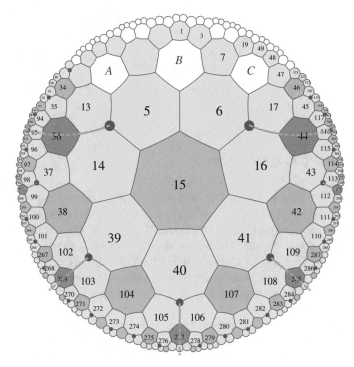

Fig. 14.10 The tiling which allowed us to prove the algorithmic unsolvability of the tiling problem in the hyperbolic plane

for the heptagrid, a fortiori, it cannot hold for any tessellation of the hyperbolic plane and, still a fortiori for any tiling of the hyperbolic plane.

The technique used to prove Theorem 10 allowed me to solve connected theorems:

Theorem 11 (Margenstern 2008 [20]) *The finite tiling problem is algorithmically unsolvable for the hyperbolic plane.*

Theorem 12 (Margenstern 2009 [22]) *The periodic tiling problem is algorithmically unsolvable for the hyperbolic plane, also in its domino version.*

Consider a finite set of prototiles T together with an additional prototile b with $b \notin T$ which is called the **blank**. A tiling of the plane with copies of tiles in $T \cap \{b\}$ is said **finite** if and only if there are only finitely many copies of tiles in T. Theorem 11 says that no algorithm can tell us for a given finite set of prototiles different from b whether this set of prototiles provides us with a finite tiling of the hyperbolic plane or not. The Euclidean case was formulated in [9] where it was proved algorithmically unsolvable.

The **periodic tiling problem** is a bit different question. Consider a finite set of prototiles T. A tiling of the Euclidean plane with copies of tiles in T is said **periodic** if it is globally invariant under two independent shifts. It is not clear how to define the notion of periodicity in the hyperbolic plane. As many authors do, we shall consider that a tiling of the hyperbolic plane is periodic if it is globally invariant under a non trivial shift. Accordingly, Theorem 12 says that no algorithm can tell us for a given finite set of prototiles whether this set of prototiles provides us with a periodic tiling of the hyperbolic plane or not. The Euclidean case was proved algorithmically unsolvable in [5].

14.3.4   Possible Applications

This section is a kind of tribute that nowadays scientists have to pay according to the scientific policy of too many countries, especially those of the EU. It may be a consolation for European scientists to know that the US policy is not very much wiser than the European one. The author considers that this tribute has a braking effect on the development of science. On the long run, this effect may kill science itself. However, as people usually do when they have to pay tribute, they try to counteract especially by humour. It is said that scientists like humour, even in a scientific publication...

This is why we propose here three applications of the previous sections. Please note that the content of this subsection is a consequence of the studies described in the previous sections. It is not at all the opposite situation. And so, let us give place to fantasy and humour! We shall successively see a hyperbolic colour chooser, see Sect. 14.3.4.1, then how to communicate between tiles in a hyperbolic tessellation, see Sect. 14.3.4.2 and we conclude by the presentation of three keyboards in Sect. 14.3.4.3: a French, a Japanese and a Chinese ones.

14.3.4.1   A Hyperbolic Colour Chooser

The chooser we present here, allows the user to select a colour in a display where the colours are dispatched according to their hue and then to their intensity in the hue, see [2].

Figure 14.11 illustrates the display presented by the colour chooser. The hues are distributed among the seven sectors which are displayed around the central cell in the right-hand side picture of Fig. 14.5. Inside a sector, the colours are displayed according to their intensity which decreases as we go towards the border of the Poincaré's disc. Figure 14.12 illustrates the working of the chooser. Thanks to a few keys of the keyboard, the user may choose to travel in the heptagrid. He/she may decide to go towards the border, going from a tile to one of its sons or, it may go to one of the two neighbours of the tile on the same level of the tree or it may decide to go back, to the father of the tile. As we indicated in the very beginning, we take advantage of the Poincaré's model in which the centre represents the point of the hyperbolic plane at which we look at.

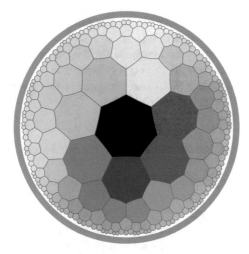

Fig. 14.11   The colour chooser

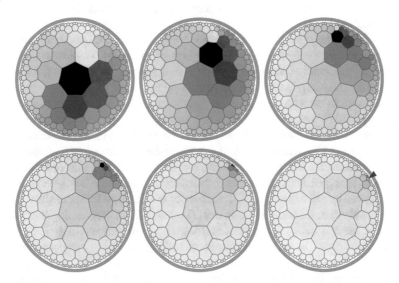

Fig. 14.12   How to use the colour chooser: when the *black tile* disappears, an indicator allows the user not to be lost. Pictures from [33]

Consequently, in the chooser colour, the chosen colour is always the one which is in the central cell.

In Fig. 14.12, we can see that the selected hue is blue and that we select lighter and lighter blue hues. Note that, symmetrically, the black tile moves towards the border. At some point, this black tile may be not observable in our representation of the Poincaré's disc and then, the user might be lost. But, fortunately, when it disappears, an arrow appears on the border in order to allow the user to go back to this cell by 'pulling' it back to the central position.

Note that we have chosen the heptagrid for the colour chooser. There are two reasons for that. Seven sectors allow a bigger choice than five one. Moreover, for the eye, heptagons look like very much like hexagons so that the grid itself looks like closer to a honeycomb which is more pleasant.

14.3.4.2 Communication Between Cells in the Hyperbolic Plane

Now, let us turn to the communication between cells of the α-grid. We shall find there two well known characters in many scientific papers: Alice and Bob. We find them during a visit in the hyperbolic plane. They are on the heptagrid. Alice is a very absent-minded person. She lost the coordinate of her friend. What to do? Fortunately, the heptagrid was recently endowed with a message system, see [14] with an account on a computer experiment, see [30]. Accordingly, she sends a message to all tiles. When the message will reach Bob, he will answer giving its coordinate together with a message.

The idea of the system is simple and it works both in the pentagrid and in the heptagrid, so that we shall use the terminology of an α-grid. The hardware system is based on the location tools we have described in Sect. 14.3.1.1. At the software level, the system is equipped a system of coordinates based on the numbering of the sides of a tile. In each tile, the sides are numbered from 1 to α, starting from 1 and increasing while counter-clockwise turning around a tile. Accordingly, for each tile, it is enough to fix which side is the side 1. For the central tile it is fixed once and for all. For the other tiles, it is the side to the father in the Fibonacci tree associated to the sector containing the tile, the father of the root being the central tile. To facilitate the message system, each side receives two numbers: the number it has in each tile, as a side is shared by exactly two tiles. The order of the numbers is that of the message. Such a couple of numbers is called an **arc**. The sequence of arcs leading from a tile A to another tile B is called the **absolute address** of B with respect to A.

The diffusion of the message proceeds as follows. The sender considers itself as the central cell and dispatch the message to its neighbours. A tile which receives the message also receives its absolute address with respect to the sender. It also receives the information of its status in the tree associated to the sender which is called the **relative tree**. Remember that the status of a node is black, white if it has two, three sons respectively. This protocol allows the message to reach all tiles exactly once.

When a tile T wishes to reply to the message, it has its absolute address with respect to the sender. By reversing each arc and the order of the arcs, T gets the absolute address of the sender with respect to itself. He/she sends the replying message together with the absolute address which is used to reach the sender and which is transmitted to the sender as the absolute address of T with respect to the sender. Figure 14.13 illustrates this process.

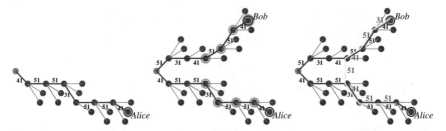

Fig. 14.13 *To left* the position of Alice in the heptagrid with respect to the centre. *Middle* the message sent by Alice reached Bob. *To right* Bob answered to Alice

Fig. 14.14 The French keyboard. The letters are displayed around the central cell

14.3.4.3   Hyperbolic Keyboards

In this sub-subsection, we shall see the application of a common idea. The idea is to use the pentagrid in order to display the symbols of an alphabet which is used by the user to create his/her messages. The pentagrid is used for several reasons. The first one is that the tiles of the pentagrid are bigger than those of the heptagrid. This spares the reader to some efforts in order to have precise movement of the hand on the screen of a tablet or of a smart phone. The second reason is the specificity of the language, this is particularly useful for the Japanese and for the Chinese keyboards.

A French Keyboard

The French keyboard is interesting in this regard that the performance of such a keyboard are a bit higher than the traditional keyboard of a cell phone. It is illustrated by Fig. 14.14.

On the figure, we can see that the basic five vowels are the immediate neighbours of the central tile. The other letters are alphabetically displayed sector by sector, counter-clockwise around the tile and, in each sector from left to right, and level by level. If the motion of the central cell is performed by keys, the farthest letters are reached in three strokes. Figure 14.15 illustrates the selection of the letter q which appears in the central tile.

A Japanese Keyboard

The Japanese keyboard is based on the use of *hiraganas* and *katakanas* which are syllabic alphabets used for the phonetic transcription of the Japanese language. More-

Fig. 14.15 *To left* the keyboard in its idle position. *Middle* the target letter is in the sector headed by **o**. *To right* the letter **q** is displayed

Fig. 14.16 The Japanese keyboard: the hiragana syllabic alphabet

over, traditionally, the display of these syllabic alphabet is based on the five vowels of the Japanese language. A similar keyboard was realized with katakana signs.

In Fig. 14.16, we can see the display of the hiragana syllabic alphabet. We can see that each sector of the pentagrid around the central tile is devoted to a vowel and to the hiragana signs based on this vowel. Note that in each sector, the order of the signs is the same and follows the traditional order.

Figure 14.17 illustrates the working of the keyboard. By pressing appropriate keys, the right syllable is pulled onto the central tile, here **ha**. First, the vowel **a** is selected, then the syllable **ka** and then, the syllable **ha**.

It was planned to also implement plates with kanji characters which would be called from the hiragana or katakana keyboard, but this was never realized, see [47].

A Chinese Keyboard

With the Chinese keyboard, we use both α-grids. The situation is more complex than in the Japanese keyboard as the phonetic representation is different and that we have to produce Chinese characters. Note that in the figures we shall consider simplified Chinese characters only. The syllables of the Chinese language are written with the latin alphabet through a system called **pin-yin**. The heptagrid is used to display the syllables, see [46], and Fig. 14.18.

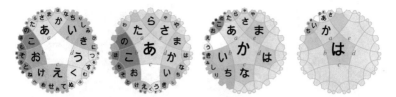

Fig. 14.17 Working of the Japanese keyboard. *Leftmost picture* the keyboard. *Next pictures* selecting the vowel **a**, then the syllable **ka** and, at last, the syllable **ha**

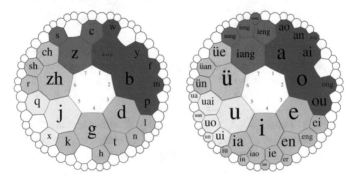

Fig. 14.18 The display of the consonants and the vowels for the Chinese keyboard. Note the cell with a, o and e in the display of the consonants

The left-hand side picture of the figure represents the syllables introduced by a consonant, the right-hand side one represents the syllables introduced with a vowel. Ounce a syllable is selected through the two plates of Fig. 14.18, a plate with the Chinese characters corresponding to these syllables is displayed. Figure 14.19 displays two plates, one with the characters associated with the pin-yin **men** and the other with the pin-yin **shi**.

On both plates, we observe the same convention. The Chinese language uses tones for each syllable. There are four of them, in fact there is a five one when a syllable is pronounced in a neutral tone. Accordingly, characters are grouped according to the tone with which they are pronounced.

[b] The working of these keyboards is the same as for the Japanese keyboard.

In many cases, the distribution of tones for a given syllable is not uniform. As can be seen on the right-hand side picture, there are more characters which are pronounced **shì**, fourth tone, than with another tone.

14.4   Universal Hyperbolic Cellular Automata

In this last section, we look at a problem to which I devoted much time: looking after universal cellular automata in hyperbolic spaces with a number of states as small as possible. Now, in this search, we have to carefully look at the terms we use as the

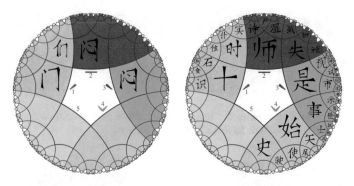

Fig. 14.19 *Left-hand side* the characters associated with the pin-yin **men**. *Right-hand side* the characters associated with the pin-yin **shi**. On both pictures, note the representation of the tone attached to each sector in the central tile

word *universality* is understood in sometime very different meanings. In this study, inspired by the definition of universality for Turing machines, we shall distinguish two types of universality: **weak universality** and **strong universality**. By strong universality, we mean that the cellular automaton starts from a finite configuration: all cells outside a large enough disc are in the **quiescent** state: as long as all the neighbours of the cell and the cell itself are in this state, they remain in this state. The quiescent states plays for cellular automata the role of the blank square for Turing machines. Also, in strong universality, when the computation is completed, the cellular automaton simply replicate the same configuration at each time after the end of the computation: we say that the cellular automaton **halts**. In strong universality, the simulating cellular automaton itself starts from a finite configuration, and if the simulated computation halts on the considered input, the simulating cellular automaton also halts when the simulated device halted. Weak universality keeps the idea of simulating the computation of some computing device, but it relax the constraint of finite initial configuration as well as that of halting when the simulated computation halts. In particular, when the simulating cellular automaton halted, the configuration is finite. However, the relaxation on finite initial configuration does not mean that any initial configuration is accepted. Usually, the restriction means that outside a large disc, or a large ball in nD-space with $n \geq 3$, the configuration is periodic. It is also accepted that the configuration of the simulating cellular automaton goes on changing after the simulated computation halted. IN that case, it is simply required that the halting of the simulated computation is signalized in some way. Section 14.4.2 deals with strong universality, Sect. 14.4.3. deals with the weak one. Section 14.4.2 deals with the pentagrid, the heptagrid and the dodecagrid at the same time for a reason which will be clear in the sub-section. Section 14.4.3 deals with weak universality. All the corresponding results are based on the implementation of the same model of computation, see Sect. 14.4.3.1. Now, as the implementation of cellular automata in the dodecagrid is not based on the study we schematically described in Sect. 14.3.1.2, we devote a sub-section for that purpose, see Sect. 14.4.1.

14.4.1 Implementing Cellular Automata in the Dodecagrid

The model we use in Sect. 14.4.3 is basically a planar model, see Sect. 14.4.3.1. Its transfer into the dodecagrid will use the advantage of the addition of the third dimension, but this will not induce deep transformation of the model which, in some sense, remains basically planar. We shall discuss this point in Sect. 14.4.3.4.

Consequently, we shall use the localization tool for the pentagrid in order to locate cells of the dodecagrid as, most of them will involve tiles for which a face lies on a fixed plane Π. Of course, the trace on Π of the dodecagrid is a copy of the pentagrid. Figure 14.20 illustrates how we proceed.

We fix a dodecahedron D_0 and a face F_0 of D_0 and we call Π the plane which supports F_0. Consider the set $\mathcal{P}$ of all dodecahedra which have a face on Π and which are on the same side of Π as D_0. Then we apply the central projection on each dodecahedron of $\mathcal{P}$, projecting the dodecahedron on its face on Π. In this way, we obtain the pictures of Fig. 14.20. The point is how to represent the tiles. The first idea is to colour the tiles inside the face which is on Π: this is illustrated by the left-hand side picture of the figure, say Fig. 14.20$_L$. But in this case, no additional information can be given. The right-hand side picture of Fig. 14.20, say Fig. 14.20$_R$ gives another way. We remark that two neighbouring dodecahedra of $\mathcal{D}$ share a face. Consider for instance the central tile and its neighbour which is in the orange colour in the Fig. 14.20$_L$. The face is numbered 4 in the central tile of Fig. 14.20$_L$. The same face appears in the orange dodecahedron as the orange face which shares and edge with the face 4 of the central tile. In Fig. 14.20$_R$, we take advantage of this situation in

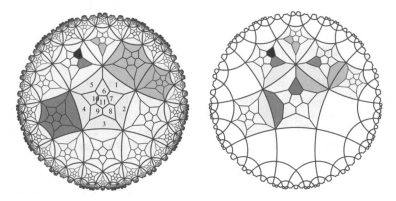

Fig. 14.20 Two different ways for representing a pseudo-projection on the pentagrid. On the *left-hand side* the tiles have their colour. On the *right-hand side* the colour of a tile is reflected by its neighbours only

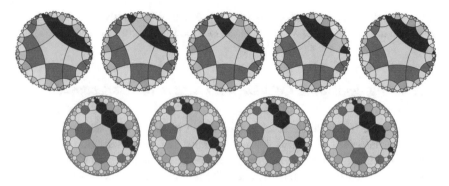

Fig. 14.21 Propagation of the halting signal in the α-grids. *Above* the pentagrid. *Below* the heptagrid

order to represent the colour of a tile T by the faces of neighbouring dodecahedra in contact with T. Accordingly, the orange colour appears on the face 4 of the central tile of Fig. 14.20_R while the yellow colour appears in the face which shares an edge with face 4. This allows us to mention other neighbours as in Fig. 14.20_R.

This representation allows us to use the very convenient coordinate system of the pentagrid for the simulation of local parts of a cellular automaton in the dodecagrid. We shall this more precisely in Sect. 14.4.3.4.

14.4.2   Strong Universality Results

The first ingredient of the result is the cellular automaton of [11] which works on a line. However, the cellular automaton with 7 states indicated in that paper is not strictly strongly universal. The point is that the initial configuration is not strictly finite. With four additional states, as proved in [32], we obtain a strongly universal cellular automaton $\mathcal{L}_{11}$ which starts from a finite configuration and which halts when the simulated computation halts.

The idea of [32] is to implement $\mathcal{L}_{11}$ in the grids we consider in hyperbolic spaces. This implementation is split into two parts. As $\mathcal{L}_{11}$ is a cellular automaton on a line, we first construct a segment of line which grows along a line as the computation proceeds. This provides us with a cellular automaton $\mathcal{P}$ which propagates this linear structure. Then, we merge $\mathcal{P}$ with $\mathcal{L}_{11}$ and we obtain what is proved in the following statement. See the illustrations in the plane, Fig. 14.21, and in the 3D-space, Fig. 14.22.

Theorem 13 (Margenstern 2013 [32]) *In each of the following tilings: the pentagrid and the heptagrid of the hyperbolic plane and also the dodecagrid of the hyperbolic 3D-space, there is a deterministic, rotation invariant cellular automaton with radius 1 which has* 10 *states and which is strongly universal.*

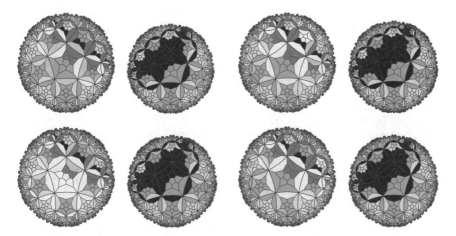

Fig. 14.22 Propagation of the halting signal in the dodecagrid. In the dodecagrid, we need tow additional lines under the plane Π in order to construct the line. In each picture, the small pentagrid represents the projections onto Π of the dodecahedra which hang below that plane

The figures represent the moment when the simulated computation halts. A signal passes to a part of the line where it reaches the end of the linear structure: this end continues the construction until it is stopped by the green signal.

14.4.3   Weak Universality Results

As mentioned in the introduction to Sect. 14.4, all the computations we present in this sub-section implement the same model which is sketchily described in Sect. 14.4.3.1. Section 14.4.3.2 gives the general properties used to implement the railway model in hyperbolic tessellations. Then Sects. 14.4.3.3–14.4.3.6 show the implementation of the best result so far in the pentagrid, heptagrid, dodecagrid and the grid {11, 3}.

14.4.3.1   The Railway Model

The railway model was created by Ian Stewart, see [50] in the Euclidean plane. There, the model simulates a Turing machine. In the papers I devoted to universality of cellular automata in hyperbolic spaces, the model is a bit changed in order to simulate a register machine. A full explanation can be found in [34].

The idea of the model is to mimic a railway circuit on which a single locomotive is running. The railway consists of tracks which are segments and quarters of a circle which are put together. The system also allows crossings and switches. The crossings allow two tracks to cross each other. For the switches, there are three kinds of them. Figure 14.23 illustrates them in a symbolic representation.

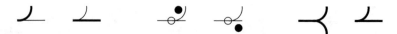

Fig. 14.23 The three types of switches for the railway circuit, in symbolic notations. From *left* to *right* fixed switch, flip-flop and memory switch

Fig. 14.24 The basic element: the flip-flop is at W, the memory switch is at R

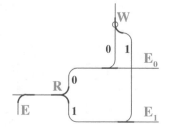

Let us explain the working of the switches. They all have a common structure: three tracks abut a point, the centre of the switch and call them a, b and c. There are two ways for crossing a switch. During an **active passage**, we also say **active crossing**, the locomotive arrives to the centre through a and it leaves the switch either through b or through c. The track though which the locomotive leaves the switch is called the **selected track**. The **passive passage**, we also say **passive crossing**, occurs in the opposite direction: the locomotive arrives from b or from c and it leaves the switch through a. The fixed and the memory switches accept both crossings. The flip-flop can be crossed actively only. In Fig. 14.23 there are two pictures for each switch: each one corresponds to a different selected track. In a fixed switch, as its name suggests it, the selected track is fixed once and for all. And so there are two kinds of them. However, note that the combination of a fixed switch in one direction with a crossing allows us to mimic the fixed switch in the other direction. In the flip-flop, the selected track switches to the other direction after each passage of the locomotive. In the memory switch, the selected track is that of the last passive crossing by the locomotive.

Figure 14.24 shows us the basic element, see [50], which allows to construct a circuit which simulates a Turing machine or, as we did, a register machine. It associates a flip-flop with a memory switch. The element contains one bit of information which is represented by a certain position of its flip-flop and of its memory switch. The other position of both switch defines the other value. Figure 14.25 illustrates the working of the element. The element can be crossed in four ways: depending on its content and on the operation to be performed. Two operations can be done: reading and rewriting. The reading is illustrated by the first row: as shown by the pictures, the locomotive enters through the memory switch. As can be seen, the way followed by the locomotive depends on which track is the selected one. Then, the locomotive leaves the element without changing the switches.

The rewriting is illustrated by the second row of Fig. 14.25. As shown by the picture, the locomotive enters through the flip-flop. Through a possible crossing and

Fig. 14.25 The working of the element. *Upper row* reading the content of the element. *To left* reading 0; *to right* reading 1. *Lower row* rewriting the content of the element. *To left* rewriting 0 into 1; *to right* rewriting 1 to 0

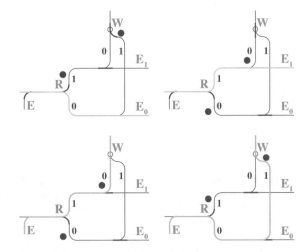

a fixed switch, the locomotive is led to the memory switch which it passively crosses, taking the way which corresponds to value which is opposite to that of the way on which it was sent by the flip-flop. From the working of the flip-flop and of the memory switch, we can see that in this way, when the locomotive has left the memory switch, the content of the element was changed to the opposite value. At last, just after the memory switch, the locomotive is sent on the exit way by a suitable fixed switch: **E** on the pictures of Fig. 14.25.

Combining such elements in appropriate structures, one can simulate a register machine, for instance, see [34].

14.4.3.2   Hyperbolic Railway: General Features

In this sub-subsection, we provide the reader with some properties about the implementation of the railway circuit in hyperbolic spaces. Here, we give the mainlines of these features which were used in all the implementations given in the next sub-subsections. In each of them, we shall see further refinements about the general features described in the present sub-subsection.

The first point is the implementation of the basic element described in Sect. 14.4.3.1. In the Euclidean plane, the construction relies on the existence of a grid which is the actual support of the structure. The circle arcs are there for both aesthetic and technical results: real-life railways do not offer curves with sharp angles. If we can at least locally transpose the structure of a grid, it will be possible to implement the basic element. For this, the Fibonacci tree offers a simple solution: the role of the verticals can be played by the branches of the tree while the role of the horizontals can be played by its levels. The connection from verticals to horizontals and conversely requires that the tracks follow paths which makes the sharpest angle on the tiling. This constraint has to be respected: otherwise the tracks will never

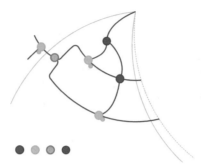

Fig. 14.26 Implementation of the basic element in the α-grid. The *blue disc* represents the crossing, the grid one, a fixed switch, the *orange one*, a memory switch and the *red one*, a flip-flop. The *small green disc* indicates the selected track of the fixed witch

go on the level, they will go further and further from the level and no cycle can be constructed.

Once this constraint is observed, it is easy to implement the crossings and the switches and to dispatch them as required in order to follow the pattern given in Fig. 14.24. This is illustrated by Fig. 14.26 which can held either in the pentagrid or in the heptagrid.

Another point has to be solved: the necessity to have parallel tracks. Depending on the purpose of a parallel track, another branch of the tree can be used. The use of another branch can also be used on a very little scale, the returning close the initial track and then repeat the process as long as needed. The advantage is that if another parallel is needed, the trick can again be repeated. We can also do a similar thing for horizontals: a parallel track can be obtained by using a further level. In any case, a parallel track is usually much longer than the initial track. Figure 14.27 presents to left a toy register machine implemented as a railway system in the Euclidean plane and, to right, its hyperbolic translation in the heptagrid. We can see the need of parallel tracks and how they can be implemented in the heptagrid. This implementation also works for the pentagrid and, hence for the dodecagrid as will be seen in Sects. 14.4.3.3–14.4.3.5. It also works for the cellular automaton constructed in Sect. 14.4.3.6

14.4.3.3 In the Pentagrid

We start the implementation of the railway circuit by the pentagrid. We shall present two implementations. One with many states, and then the presently best result. The first result in the pentagrid presented a weakly universal cellular automaton with 22 states, see [6]. In that implementation, the locomotive was represented by two contiguous cells travelling on the tracks. The front of the locomotive was green, its rear was red, the track was blue. Each centre of a crossing or of the switches had a specific colour.

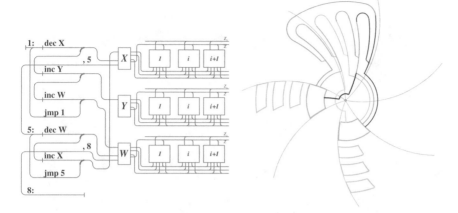

Fig. 14.27 *To left* a toy example of the railway implementation of a register machine in the Euclidean plane. *To right* a sketchy implementation of the hyperbolic railway model in the heptagrid

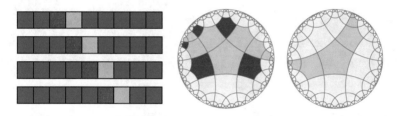

Fig. 14.28   The basic motion of the locomotive

Here, we present a solution with less states, in fact 9 of them, see [45]. In this implementation, the tracks are still in blue, and the locomotive is still represented by two contiguous cells, a green one, the front, and a red one, the rear. A big difference with the automaton of [6] is that the centre of a crossing or of a switch has no special colour: it is blue as any cell of the tracks. The difference is made by the neighbourhood of the centre. In Fig. 14.28, the left-hand side picture represents the motion of the locomotive on the tracks. The right-hand side picture represents the two types of tracks: first, the horizontal one and then, the vertical one.

Figures 14.29 and 14.30 represent the motion of the locomotive on these tracks. On the horizontal tracks, the locomotive goes from right to left, see Fig. 14.29. On the vertical ones, it goes from up to down, see Fig. 14.30. Of course, the motions in opposite directions have been tested in both cases. However, we have no room to display them here. The reader is referred to [34, 45]. At last, Fig. 14.31 shows us the idle configuration of the crossing, the fixed switch, the flip-flop and the memory switch in this order, from left to right. By idle configuration, we mean a local configuration where the locomotive is not present. From lack of room, we cannot show the motion of the locomotive when it crosses the corresponding configurations.

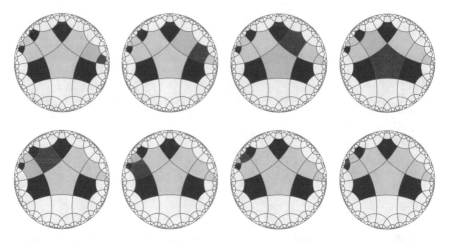

Fig. 14.29 Implementation of horizontal tracks in the pentagrid: from *right* to *left*

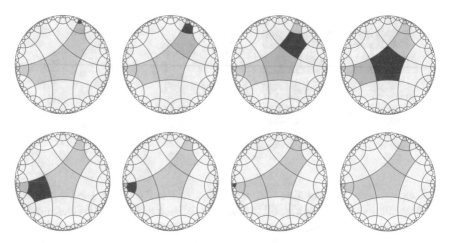

Fig. 14.30 Implementation of vertical tracks in the pentagrid: from *up* to *down*

Note that, the crossing needs four tests, depending from which track arriving to the centre of the crossing the locomotive is arriving. The fixed switch needs three tests. As already noticed, we may assume that the fixed switch always selects the left-hand side track as it is the case in Fig. 14.31. As the flip-flop can be crossed actively only, and as both idle configurations are required, we need two tests. At last, the memory switch may be crossed both actively and passively, and there are two possible idle configuration depending on which track is the selected one. This represents six tests. We refer the reader to [45] for details, in particular for the rules of the automaton which were checked by a computer program.

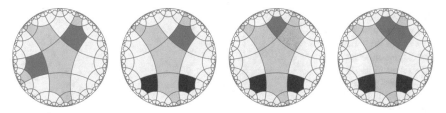

Fig. 14.31   Implementation of vertical tracks in the pentagrid: from *up* to *down*

Fig. 14.32   The elements of the tracks. *Leftmost picture* the standard element. Second and third pictures from *left* the element which allows to perform sharp turns. Fourth and fifth pictures, illustration for a sharp turn

The best result for the pentagrid is a weakly universal cellular automaton with five states, see [35, 36]. This cellular automaton takes advantage of the new ideas introduced since the previous one with nine states.

Theorem 14 (Margenstern 2014 [35, 36]) *There is a weakly universal cellular automaton on the pentagrid with five states which is planar and rotation invariant.*

The cellular automaton is **planar** as it always has infinitely many cycles of cells which are not in the quiescent state. This means that the circuit can never be reduced to a line. Without this restriction, it would be possible to construct a weakly universal cellular automaton on the pentagrid with two states which also would be rotation invariant, see [26]. The implementation of the railway circuit for the proof of Theorem 14 makes use of three new ingredients. Using what was used in a cellular automaton of Sect. 14.4.3.4, the cells of the tracks are in a quiescent state. The tracks are indicated by the neighbourhood of its cells. Also, the tracks of the circuit are now one-way. This simplifies something for the fixed switch, but this entails some complexification for the memory switch which is now split into two pars: the passive, active switch, concerned by a passive, active crossing respectively of the switch by the locomotive. The second ingredient is used by an automaton we quote in Sect. 14.4.3.6. Our railway circuit borrows an element which belongs to motorway traffic. Indeed, we replace crossings by **round-abouts**. The one-way condition is connected with the reduction of the locomotive to a single cell. As there is no indication of the direction, this feature must be indicated by the neighbours of a cell. The cells of the tracks are **white** as they are in the quiescent state. Their neighbours which are not white are the **milestones** which indicate the track and its direction (Fig. 14.32).

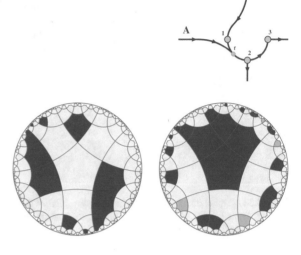

Fig. 14.33 The new crossing: the one-way tracks from **A** and **B** intersect. We have a three-quarters round-about. The small disc at **f** represents a fixed switch. Discs **1**, **2** and **3** represent the check-points

Fig. 14.34 The tracks. *To left*, a vertical one. *To right*, a horizontal one

The structure of the round-about is indicated by Fig. 14.33. The idea is that in a round about, the continuation of the track which arrive at it is on the second track which is met along the round about. As shown on Fig. 14.33, the round-about has three **check-points** and a fixed switch which, in a passive way, allows the arriving locomotive from one direction to enter the round-about and the locomotive arrives in this way to the second check-point. If the locomotive arrives from the other direction, it is directly sent on the first-check-point. Accordingly, it is needed to count up to two in some way. Here, we perform this task by the colour. The locomotive is usually green. When it arrives at a check-point for the first time, it is still green. At the check-point, it is changed to red. When the locomotive arrives at the next check-point, as it is red, it is both changed back to green and it is sent on the right track.

Figure 14.34 shows us how vertical and horizontal tracks look. From lack of room, we cannot display the motion of the locomotive on such tracks. On each of one there may be either a green or a red locomotive. Similarly, on the next figures, we do not show the motion of the locomotive. It would require much pictures to test each possibility.

Figure 14.35 shows the idle configuration of a check-point on a round-about, the fixed switch and the flip-flop. Note that in a passive crossing, the same fixed switch works for both tracks arriving at the centre.

Figure 14.36 illustrates the idle configurations of the memory switches. Remember that there are two memory switches: the active and the passive ones. Also remember that each one has two positions: one for selecting the left-hand side track, the other for selecting the right-hand side one. We can remark that the active memory switch looks very much like the flip-flop. Note that the active memory switch is *passive*: it is not changed after the passage of the locomotive. The passive switch is *active*. It

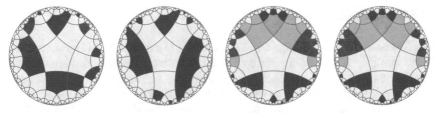

Fig. 14.35   *To left* the check-point of a round about. *Second picture* the fixed switch. *Last pictures* the flip-flop. First, the selected track is to *left*, then it is to *right*

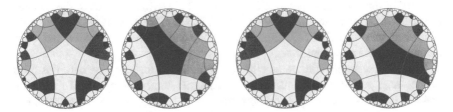

Fig. 14.36   The stable configuration of the active and passive memory switches. *To left*, the switches selected the *right-hand side track*. *To right*, they selected the *left-hand side track*

Fig. 14.37   The organisation of the memory switch

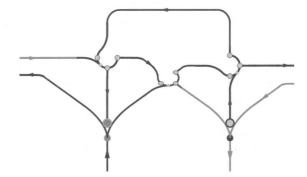

is changed after being crossed by the locomotive. But this change also triggers the change of the active switch. Figure 14.37 illustrates the organisation of the connection between the passive part of a memory switch with its active one. The figure shows that the change of selection in a memory switch entails a delay. Now, we may assume that the switches are distant enough so that the change is performed when the locomotive reaches the next switch on its way.

The figures corresponding to the test of all the configurations illustrated by Figs. 14.34, 14.35 and 14.36 are displayed in [36]. The rules for the automaton are also all given there. They were checked by a computer program, which completely proves Theorem 14.

14.4.3.4   In the Dodecagrid

The first result for the dodecagrid was obtained in [13]: the automaton has five states. This automaton is directly inspired by the one which works in [6] with this difference that the centre of the crossings and the switches has the same colour as a standard cell of the tracks. So it should be compared with the nine sates of [45] in the pentagrid. The reason of this reduction is simple: in the $3D$-space, either Euclidean or hyperbolic, it is possible to replace the crossings by bridges. Accordingly, the crossings are simply replaced by tracks thanks to the third dimension. The implementation of bridges is not very difficult in the hyperbolic $3D$ space, see [13]. Somehow later, the number of states was reduced to three of them, see [25]. Here, we present the best possible result with two states, see [27, 29]. The states are called **black** and **white**, that latter colour being the quiescent state.

The cellular automaton works with a locomotive reduced to one cell, this time of the same colour as the milestones, necessarily the black colour. This entails that the tracks are one-way. We know from Sect. 14.4.3.3 what it induces on the memory switch and what it simplifies for the fixed switch. Of course, the active memory switch and the passive one are implemented in a very different way from that of Sect. 14.4.3.3. We shall use the conventions and the representation given in Sect. 14.4.1 in the illustrations of this sub-subsection and in the further explanations.

Remember that most cells of the track have a face which is on the same plane Π_0 on which we perform the projection of each dodecahedron in the figures of this sub-subsection.

Figure 14.38 illustrates the implementation of a vertical while Fig. 14.39 illustrates the implementation of a horizontal. In both cases, a two-way track is implemented: one track over Π_0 and one track below it. As shown in Fig. 14.38, the milestones of the tracks mimic catenaries. Of course, if the way back is not needed, we keep the way which is over Π_0. For this figure and for all the others of this section, we do not show the motion of the locomotive which would require too much room.

Figure 14.40 illustrates the structure of a bridge. In the first row of the figure, the leftmost picture gives us a global view of the bridge. The next two pictures show the implementation of the pillars of the bridge. The second row of the figure illustrates the look of both ends of the bridge. The bridge is built along the plane Π_1 which is orthogonal to Π_0 and which supports faces of dodecahedra of the tessellation.

Figure 14.41 illustrates the implementation of the flip-flop and of the active memory switch. They are both active structures with a small different action. The left-hand side picture represents the flip-flop. The active memory switch has globally the same structure. The difference between these switches lies in their controller. In the left-hand side picture, we can see a purple face on the central cell. It is the place of the controller. On the right-hand half of the figure, we can see the controllers: first, that of the flip-flop and then that of the active memory switch.

Figure 14.42 illustrates the passive switches. First, it gives an illustration of the fixed switches: left- and right-hand side ones. As can be seen, they have the same passive structure and the difference lies in the return path which is under the plane Π_0. The other two pictures represent the passive memory switch. It looks like the fixed

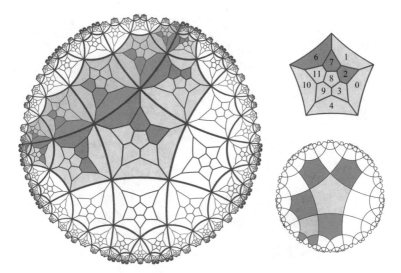

Fig. 14.38 A vertical track in the dodecagrid. *To right* below, a vertical cut of a dodecahedron; higher, the catenary structure

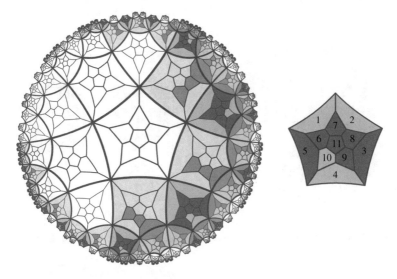

Fig. 14.39 A horizontal track in the dodecagrid

switches but, contrarily to them, it possesses a controller. The rightmost picture gives us a global view of the memory switch. We can see the two parts, below the active switch and above the passive one and, in orange, the path which goes from the passive switch to the active one. In [27], the reader may find all the details concerning this cellular automaton. In particular, the rules of the automaton are given. They have been controlled by a computer program, allowing us to prove the following result.

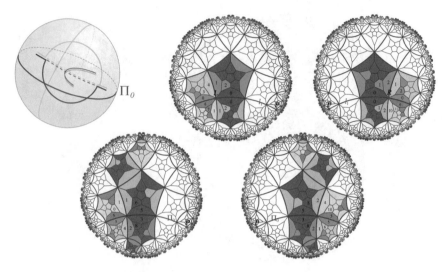

Fig. 14.40 Structure of a bridge in the dodecagrid. *First row. To left*, the global view. Then: the pillars of the bridge, *left-hand side* and *right-hand side*, projection on Π_0. *Second row* the ends of the bridge, both sides, projection on Π_1

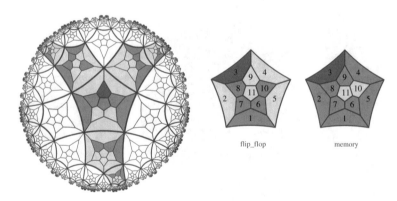

Fig. 14.41 Active switches. *To left*, the common structure of the switches. *To right* their controllers

Theorem 15 (Margenstern 2010 [27, 29]) *There is a weakly universal cellular automaton in the dodecagrid which is weakly universal and which has two states exactly, one state being the quiescent state. Moreover, the cellular automaton is rotation invariant and the set of its cells changing their state has infinitely many cycles both in two perpendicular planes of the dodecagrid.*

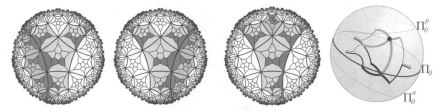

Fig. 14.42 Passive switches. *To left*, the fixed switch. *Middle*, the passive memory switch. *To right*, the connection between the two parts of the memory switch

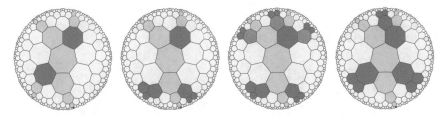

Fig. 14.43 Stable configurations for the weakly universal cellular automaton on the heptagrid with six states

14.4.3.5   In the Heptagrid

The first result in the heptagrid was [44] with a weakly universal cellular automaton with six states:

Theorem 16 (Margenstern-Song [44]) *There is a weakly universal cellular automaton in the heptagrid which has six states. The automaton is rotation invariant and the set of the cells which change their state has infinitely many cycles.*

Figure 14.43 illustrates the stable configurations of the crossing and of the switches for the cellular automaton of Theorem 16. This result was improved in [28] to a cellular automaton with four states. But, recently, this result was improved in [37]:

Theorem 17 (Margenstern 2014 [37]) *There is a weakly universal cellular automaton in the heptagrid which has three states. The automaton is rotation invariant and the set of the cells which change their state has infinitely many cycles.*

The automaton which allows us to prove Theorem 17 makes use of new ingredients which were found in the search of weakly universal cellular automata with two states, we shall this in Sect. 14.4.3.6.

We have already met such an ingredient in Sect. 14.4.3.3. Now, we shall meet a new one. The idea is to unify the working of the active switches. For this, we apply the same idea as the one which allowed to split the active memory switch from the passive one: this introduces a delay in a process which is performed within five steps by the automata which have much more states. Here too, the idea is to count up to

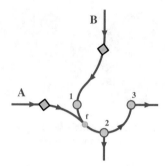

Fig. 14.44 The new round-about: a new control is appended to the already introduced check-points. See the structure of the doubler in Fig. 14.47

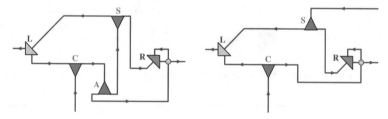

Fig. 14.45 *To left* the flip-flop. *To right* the active memory switch

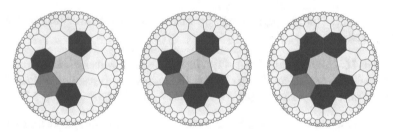

Fig. 14.46 The cells of the tracks

two to perform the working of the round-about. As we have three states, this cannot be performed by a change of colour. We proceed as follows: before reaching the first check-point on its way, the locomotive which arrives at a round-about meets a **doubler** which duplicates the locomotive, see [37]. After the doubler, two contiguous locomotives travel on the round-about. On the first check-point they meet the second locomotive is killed. When a check-point is crossed by a single locomotive, it sends it on the right way, performing the crossing (Fig. 14.44).

Figure 14.45 displays the active switches in a schematic way, showing the main ingredients which are detailed in Fig. 14.47. Figure 14.46 shows us the elements of the tracks. At last, Fig. 14.48 illustrates the passive switches: the fixed one and then the passive one. Figures 14.46, 14.47 and 14.48 make use of the same symbolism

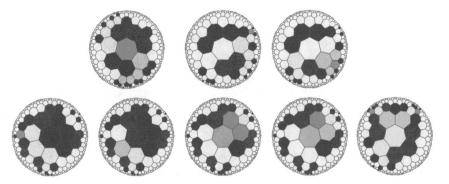

Fig. 14.47   *First row*, from *left* to *right* the doubler and then the check-point. *Second row*, from *left* to *right* the controller, with different visits and, *rightmost picture*, the fork

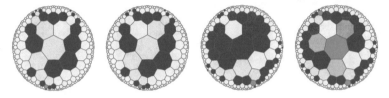

Fig. 14.48   *Left-hand side* the fixed switches, from the *left* and then from the *right*. It also works with two locomotives. *Right-hand side* the passive memory switch. First: passage through the selected track and then, passage through the non-selected track

to show the motion of the locomotive and its impact on the neighbouring cells. In Fig. 14.46, we can see the three steps from the entrance into the central cell of the element to its exit. The locomotive is about to enter the cell, it is in the neighbour coloured with yellow, then the locomotive at the centre which is orange in the figure and then, the locomotive is leaving the element: it is on the neighbour which is in rose. In Fig. 14.47, we again apply this convention. However, in the doubler, the colour switch from orange to mauve when the locomotive is doubled. In the second check-point, the mauve colour indicates that two locomotives enter. The green cell means that the considered cell flashed: it turned to white and then took back its colour. This allowed the check point to kill the cells and to create a new locomotive which appears in yellow at the centre of the check-point. The second row of Fig. 14.47, we can see the working of the controller: when its centre is blue, it let the locomotive cross the cell and go on its way, when it is read, it stops it. The next two pictures indicate that when the appropriate signal arrives, the centre change its colour to the opposite. The orange colour symbolizes the change from blue to red, the green one, the change from red to blue. The convention in the fork is clear. The same conventions are used in Fig. 14.48 for the passive memory switch when the locomotive arrives through the non-selected track, see the rightmost picture. The connection from the passive memory switch the active one is organized as in the cellular automaton of Theorem 14, see Fig. 14.37.

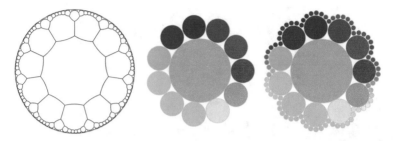

Fig. 14.49 *To left* the tiling {11, 3} in the Poincaré's model. *To right* another representation of a cell in the tiling {11, 3} and the same one together with its neighbours

Fig. 14.50 Organization of the passive memory switch with forks and sensors. Note that the sensors are not represented with the same symbol as the controllers in Fig. 14.45

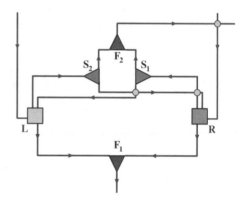

14.4.3.6   In the Grid {11, 3}

In this section, we give an account on the proof of the following result:

Theorem 18 (Margenstern 2015 [39]) *There is a weakly universal cellular automaton on the tessellation* {11, 3} *with two states which is rotation invariant and such that the set of its cells we change during the computation has infinitely many cycles.*

This result was anticipated by a similar statement in the tessellation {13, 3}, see [31]. This latter paper introduced the round-about structure with the doubler and the check-points. However, the fork and the controller where devised in [37]. Accordingly, here we just mention the construction of the tracks and sketchily indicate the working of the structures adapted to this specific tessellation. First, we have to represent, at least locally, a few cells of the tessellation. In Fig. 14.49, the leftmost picture represents the tiling in Poincaré's disc. The next picture of the figure shows us a symbolic representation which we shall use in this sub-subsection. The rightmost picture indicates the neighbours of the central cell which belong to the second generation (Fig. 14.50).

In this new setting, we also change the passive memory switch: we introduce a new device, the **sensor**: it is different from the controller of the active switch which stops

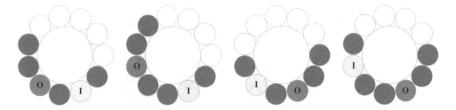

Fig. 14.51   The four possible elementary elements of the tracks. Note that on the pictures, *I* indicates the entry gate of the element and *O* indicates its exit gate

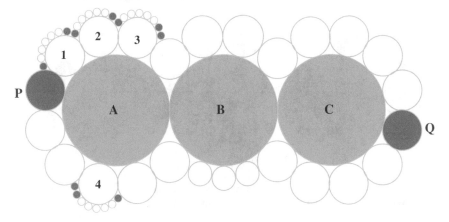

Fig. 14.52   The use of elements of the tracks in order to define a track going from the cell P to the cell Q

the locomotive when its centre is red. This is the reason while the sensor is illustrated by a symbol which is different from that used for the controller in Fig. 14.45.

First, we indicate the implementation of the tracks, a mandatory step. Figure 14.51 illustrates the elements of the tracks. This figure, combined with Fig. 14.52, shows how we can join any tile of the tessellation with another one by a tracks consisting of the elements of Fig. 14.51. This rely on the fact that we can join any tile of the tessellation to another one, see, for instance [17].

Next, we present the elements which implement fixed switch, the doubler, the check-point, the fork, the controller and the sensor.

Figure 14.53 illustrates the fixed switch and the basic structures needed to implement the round-about: the doubler and the check-point. On this figure and on the next ones of this sub-subsection, we can see colours which indicate the point where the locomotive enters the configuration and the point from where it leaves it. The conventions are the same as in Sect. 14.4.3.5.

Figure 14.54 makes two zooms on the check-point whose structure is a bit complex. The first picture of Fig. 14.54 illustrates the situation when a single locomotive arrives at the check-point. It is then sent to the cells which are around *D* on the figure.

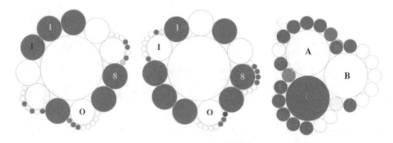

Fig. 14.53 Idle configurations. *To left* the fixed switch. *Middle* the doubler. *To right* the check-point

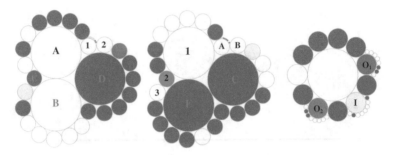

Fig. 14.54 *Leftmost* and *middle pictures* the idle configuration of the checkpoint of the round-about, zoom on D and then on E. *Rightmost picture* the idle configuration of the fork

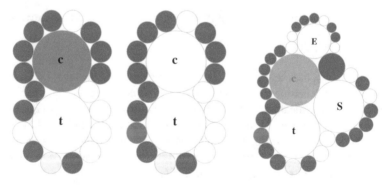

Fig. 14.55 *To left* the idle configurations of the controller in the active switches. *To right* the idle configuration of the sensor for the passive memory switch

The second picture illustrates the situation when two contiguous locomotives arrive at the check-point. Then the locomotive are killed and a new one is sent on the cells around C, going to the next check-point on the round about.

The last picture of Fig. 14.54 illustrates the fork. We can see that two locomotive are produced but they are sent in different directions.

In Fig. 14.55, we can see the controller used in the implementation of the flip-flop and of the active memory switch and the sensor of the passive memory switch. The first picture of the figure shows us that when the cell C of the controller is black, it stops the locomotive which enters the controller. The second picture shows us that when C is white, the controller let the locomotive cross it. The last picture of the figure is the sensor of the passive memory switch. The light colour of the cell C indicates that whether C is black or white, the locomotive crosses the sensor. Of course, when it is black, the cells S is activated. The flash of S triggers a locomotive which is sent to the active memory switch in order to change the signal of its controllers and this locomotive is further duplicated by a fork in order to change the signal of the other sensor of the passive memory switch. This second locomotive reaches the other sensor at E. The flash of E makes C change to black.

Temporary Conclusion

As can be seen from the sections of the chapter, there is still much work ahead, both for theoretical work and for applications.

References

1. Berger, R.: The undecidability of the domino problem. Mem. Am. Math. Soc. **66**, 1–72 (1966)
2. Chelghoum, K., Margenstern, M., Martin, B., Pecci, I.: Palette hyperbolique : un outil pour interagir avec des ensembles de données, IHM'2004, Namur (2004)
3. Coxeter, H.S.M., Moser, W.O.J.: Generators and Relations for Discrete Groups, II edn. Springer, Berlin (1965)
4. Euclid's Elements, Books I-XIII Complete and Unabridged, Translated by Sir Thomas Heath, Barnes & Noble (2006)
5. Gurevich, Yu., Koriakov, I.: A remark on Berger's paper on the domino problem. Sib. Math. J. **13**, 459–463 (1972)
6. Herrmann, F., Margenstern, M.: A universal cellular automaton in the hyperbolic plane. Theor. Comput. Sci. **296**, 327–364 (2003)
7. Iwamoto, Ch., Margenstern, M., Morita, K., Worsch, Th.: Polynomial-time cellular automata in the hyperbolic plane accept accept exactly the PSPACE languages, SCI'2002, Orlando, pp. 411–416 (2002)
8. Iwamoto, Ch., Margenstern, M.: A survey on the complexity classes in hyperbolic cellular automata. In: Proceedings of SCI'2003, vol. V, pp. 31–35 (2003)
9. Kari, J.: Reversibility and surjectivity problems of cellular automata. J. Comput. Syst. Sci. **48**, 149–182 (1994)
10. Kreinovich, V., Margenstern, M.: Some curved spaces, one can solve NP-hard problems in polynomial time. Zapiski nauchnyh seminarov POMI **358**, 224–250 (2008)
11. Lindgren, K., Nordahl, M.G.: Universal computations in simple one-dimensional cellular automata. Complex Syst. **4**, 299–318 (1990)
12. Margenstern, M.: New tools for cellular automata of the hyperbolic plane. J. Univers. Comput. Sci. **6**(12), 1226–1252 (2000)
13. Margenstern, M.: A universal cellular automaton with five states in the 3D hyperbolic space. J. Cell. Autom. **1**(4), 315–351 (2006)
14. Margenstern, M.: On the communication between cells of a cellular automaton on the penta- and heptagrids of the hyperbolic plane. J. Cell. Autom. **1**(3), 213–232 (2006)

15. Margenstern, M.: About the domino problem in the hyperbolic plane, a new solution (2007), 60p. arXiv:cs/0701096
16. Margenstern, M.: The domino problem of the hyperbolic plane is undecidable (2007), 18p. arXiv:0706.4161
17. Margenstern, M.: Cellular Automata in Hyperbolic Spaces, Vol. 1, Theory. Old City Publishing, Philadelphia (2007), 422p
18. Margenstern, M.: On a characteriztion of cellular automata in tilings of the hyperbolic plane. Int. J. Found. Comput. Sci. **19**(5), 1235–1257 (2008)
19. Margenstern, M.: The domino problem of the hyperbolic plane is undecidable. Theor. Comput. Sci. **407**, 29–84 (2008)
20. Margenstern, M.: The finite tiling problem is undecidable in the hyperbolic plane. Int. J. Found. Comput. Sci. **19**(4), 971–982 (2008)
21. Margenstern, M.: Cellular Automata in Hyperbolic Spaces, Vol. 2, Implementation and Computations. Old City Publishing, Philadelphia (2008), 360p
22. Margenstern, M.: The periodic domino problem is undecidable in the hyperbolic plane. Lect. Notes Comput. Sci. **5797**, 154–165 (2009)
23. Margenstern, M.: The injectivity of the global function of a cellular automaton in the hyperbolic plane is undecidable. Fundamenta Informaticae **94**(1), 63–99 (2009)
24. Margenstern, M.: About the Garden of Eden theorems for cellular automata in the hyperbolic plane. Electron. Notes Theor. Comput. Sci. **252**, 93–102 (2009)
25. Margenstern, M.: A weakly universal cellular automaton in the hyperbolic 3D space with three states, Discrete Mathematics and Theoretical Computer Science. In: Proceedings of AUTOMATA'2010, pp. 91–110 (2010)
26. Margenstern, M.: Towards the Frontier between decidability and undecidability for hyperbolic cellular automata. Lect. Notes Comput. Sci. **6227**, 120–132 (2010)
27. Margenstern, M.: A new weakly universal cellular automaton in the 3D hyperbolic space with two states (2010), 38p. arXiv:1005.4826v1
28. Margenstern, M.: A universal cellular automaton on the heptagrid of the hyperbolic plane with four states. Theor. Comput. Sci. **412**, 33–56 (2011)
29. Margenstern, M.: A new weakly universal cellular automaton in the 3D hyperbolic space with two states. Lect. Notes Comput. Sci. **6945**, 205–2017 (2011)
30. Margenstern, M.: A protocol for a message system for the tiles of the heptagrid, in the hyperbolic plane. Int. J. Satell. Commun. Policy Manag. **1**(2–3), 206–219 (2012)
31. Margenstern, M.: A family of weakly universal cellular automata in the hyperbolic plane with two states. (2012), 83p. arXiv:1202.1709
32. Margenstern, M.: About strongly universal cellular automata. Electron. Proc. Theor. Comput. Sci. **128**(17), 93–125 (2013)
33. Margenstern, M.: The challenge of hyperbolic cellular automata, from NP-problems to bacteria, WSEAS-CIMACCS'13, Nanjing, China, 17–19 November 2013
34. Margenstern, M.: Small Universal Cellular Automata in Hyperbolic Spaces: A Collection of Jewels. Springer, Berlin (2013), 331p
35. Margenstern, M.: A weakly universal cellular automaton in the pentagrid with five states. In: Calude, C.S., et al. (eds.) Lecture Notes in Computer Science, Gruska Festchrift, vol. 8808, pp. 99–113 (2014). arXiv:1403.2373
36. Margenstern, M.: A weakly universal cellular automaton in the pentagrid with five states (2014), 23p. arXiv:1403.2373
37. Margenstern, M.: A weakly universal cellular automaton in the heptagrid with three states (2014), 27p. arXiv:1410.1864v1
38. Margenstern, M.: Pentagrid and heptagrid: the Fibonacci technique and group theory. J. Autom. Lang. Comb. **19**(1–4), 201–212 (2014)
39. Margenstern, M.: A weakly universal cellular automaton with 2 states on the tiling {11, 3} (2015), 31p. arXiv:1501.06328
40. Margenstern, M.: About embedded quarters and points at infinity in the hyperbolic plane (2015), 17p. arXiv:1507.08495

41. Margenstern, M.: *Le rêve d'Euclide, Promenades en géométrie hyperbolique*, Éditions Le Pommier, Collection [Impromptus Le Pommier] (2015), 221p
42. Margenstern, M., Morita, K.: A polynomial solution for 3-SAT in the space of cellular automata in the hyperbolic plane. J. Univers. Comput. Sci. **5**(9), 563–573 (1999)
43. Margenstern, M., Morita, K.: NP problems are tractable in the space of cellular automata in the hyperbolic plane. Theor. Comput. Sci. **259**, 99–128 (2001)
44. Margenstern, M., Song, Y.: A universal cellular automaton on the ternary heptagrid. Electron. Notes Theor. Comput. Sci. **223**, 167–185 (2008)
45. Margenstern, M., Song, Y.: A new universal cellular automaton on the pentagrid. Parallel Process. Lett. **19**(2), 227–246 (2009)
46. Margenstern, M., Wu, L.: A propposal for a Chinese keyboard for cellphones, smartphones, ipads and tablets (2013), 28p. arXiv:1308.4965v1
47. Margenstern, M., Martin, B., Umeo, H., Yamano, S., Nishioka, K.: A Proposal for a Japanese Keyboard on Cellular Phones. Lecture Notes in Computer Science. In: Proceedings of ACRI'2008, Yokohama, Japan, vol. 5191, pp. 299-306, 24–26 September 2008
48. Morgenstein, D., Kreinovich, V.: Which algorithms are feasible and which are not depends on the geometry of space-time. Geocombinatorics **4**(3), 80–97 (1995)
49. Robinson, R.M.: Undecidability and nonperiodicity for tilings of the plane. Inventiones Mathematicae **12**, 177–209 (1971)
50. Stewart, I.: A subway named turing, mathematical recreations. Sci. Am. **271**, 90–92 (1994)
51. Weber, C., Seifert, H.: Die beiden Dodekaederräume. Math. Z. **37**, 237–253 (1933)

Chapter 15
A Computation in a Cellular Automaton Collider Rule 110

Genaro J. Martínez, Andrew Adamatzky and Harold V. McIntosh

Abstract A cellular automaton collider is a finite state machine build of rings of one-dimensional cellular automata. We show how a computation can be performed on the collider by exploiting interactions between gliders (particles, localisations). The constructions proposed are based on universality of elementary cellular automaton rule 110, cyclic tag systems, supercolliders, and computing on rings.

15.1 Introduction: Rule 110

Elementary cellular automaton (CA) rule 110 is the binary cell state automaton with a local transition function φ of a one-dimensional (1D) CA order ($k = 2, r = 1$) in Wolfram's nomenclature [57], where k is the number of cell states and r the number of neighbours of a cell. We consider periodic boundaries, i.e. first and last cells of a 1D array are neighbours. The local transition function for rule 110 is defined in Table 15.1, the string 01101110 is the number 110 in decimal notation:

G.J. Martínez (✉) · A. Adamatzky
Unconventional Computing Centre, University of the West of England,
Coldharbour Lane, Bristol BS16 1QY, UK
e-mail: genaro.martinez@uwe.ac.uk

A. Adamatzky
e-mail: andrew.adamatzky@uwe.ac.uk

H.V. McIntosh
Departamento de Aplicación de Microcomputadoras,
Universidad Autónoma de Puebla, 49 Poniente 1102, 72000 Puebla, Mexico
e-mail: mcintosh@unam.mx

© Springer International Publishing Switzerland 2017
A. Adamatzky (ed.), *Advances in Unconventional Computing*,
Emergence, Complexity and Computation 22,
DOI 10.1007/978-3-319-33924-5_15

Fig. 15.1 An example of CA rule 110 evolving for 384 time steps from a random configuration, where each cell assigned state '1' with uniformly distributed probability 0.5. The particles are filtered. Time goes down

$$
\begin{aligned}
\varphi(1, 1, 1) &\to 0 & \varphi(0, 1, 1) &\to 1 \\
\varphi(1, 1, 0) &\to 1 & \varphi(0, 1, 0) &\to 1 \\
\varphi(1, 0, 1) &\to 1 & \varphi(0, 0, 1) &\to 1 \\
\varphi(1, 0, 0) &\to 0 & \varphi(0, 0, 0) &\to 0
\end{aligned}
\tag{15.1}
$$

A cell in state '0' takes state '1' if both its neighbours are in state '1' or left neighbour is '0' and right neighbour is '1'; otherwise, the call remains in the state '0'. A cell in state '1' takes state '0' if both its neighbours are in state '1', or both its neighbours are in state '0' or it left neighbour is '1' and its right neighbour is '0'. Figure 15.1 shows an evolution of rule 110 from a random initial condition. We can see there travelling localisation: particles or gliders, and some stationary localisations: breathers, oscillators or stationary structures.

15.1.1 System of Particles

A detailed description of particles/gliders discovered in evolutions of CA rule 110 is provided in [32, 36].[1] Further, we refers to a train of n copies of particle A as A^n.

Figure 15.2 shows all known particles, and generators of particles, or glider guns. Each particle has its unique features, e.g. slopes, velocities, periods, contact points, collisions, and phases [33, 35, 37]. A set of particles in rule 110 is defined as:

---

[1] See also, http://uncomp.uwe.ac.uk/genaro/rule110/glidersRule110.html.

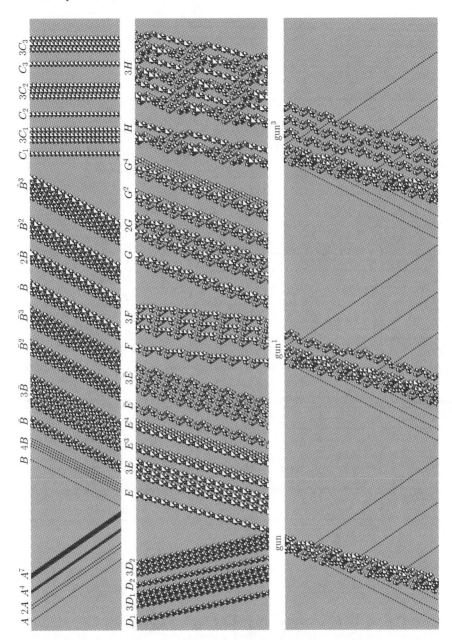

Fig. 15.2 Types of particles discovered in rule 110

Table 15.1 Properties of particles in rule 110

Structure	Margins				Velocity	Lineal volume
	Left – Right					
	ems	oms	ems	oms		
e_r	–	1	–	1	$2/3 \approx 0.666666$	14
e_l	1	–	1	–	$-1/2 = -0.5$	14
A	–	1	–	1	$2/3 \approx 0.666666$	6
B	1	–	1	–	$-2/4 = -0.5$	8
$\bar{B}^n$	3	–	3	–	$-6/12 = -0.5$	22
$\hat{B}^n$	3	–	3	–	$-6/12 = -0.5$	39
C_1	1	1	1	1	$0/7 = 0$	9–23
C_2	1	1	1	1	$0/7 = 0$	17
C_3	1	1	1	1	$0/7 = 0$	11
D_1	1	2	1	2	$2/10 = 0.2$	11–25
D_2	1	2	1	2	$2/10 = 0.2$	19
E^n	3	1	3	1	$-4/15 \approx -0.266666$	19
$\bar{E}$	6	2	6	2	$-8/30 \approx -0.266666$	21
F	6	4	6	4	$-4/36 \approx -0.111111$	15–29
G^n	9	2	9	2	$-14/42 \approx -0.333333$	24–38
H	17	8	17	8	$-18/92 \approx -0.195652$	39–53
Glider gun	15	5	15	5	$-20/77 \approx -0.259740$	27–55

$$\mathcal{G} = \{A, B, \bar{B}^n, \hat{B}^n, C_1, C_2, C_3, D_1, D_2, E^n, \bar{E}, F, G^n, H, gun^n\}.$$

where n means that a structure of the particle can be extendible infinitely, the rest of symbols denote types of particles as shown in Fig. 15.2. Table 15.1 summarizes key features of the particles: column *structure* gives the name of each particle including two more structures: e_r and e_l which represent the slopes of ether pattern (periodic background). The next four columns labeled *margins* indicate the number of periodic margins in each particle: they are useful to recognize contact points for collisions. The margins are partitioned in two types with even values *ems* and odd values *oms* which are distributed also in two groups: left and right margins. Column v_g indicates a velocity of a particle g, where g belongs to a particle of the set of particles $\mathcal{G}$. A relative velocity is calculated during the particle's displacement on d cells during period p. We indicate three types of a particle propagation via sign of its velocity. A particle travelling to the right has *positive velocity*, a particle travelling to the left has *negative velocity*. Stationary particle has zero velocity. Different velocities of particles allow us to control distances between the particle to obtained programmable reactions between the particles. Typically, larger particles has lower velocity values. No particle can move faster than v_{e_r} or v_{e_l}. Column *lineal volume* shows the minimum and maximum number of necessary cells occupied by the particle.

15.1.2 Particles as Regular Expressions

We represent CA particles as strings. These strings can be calculated using de Bruin diagrams [31, 32, 34, 37, 55] or with the tiles theory [16, 33, 35, 37].[2]

A regular language L_{R110} is based on a set of regular expressions Ψ_{R110} uniquely describing every particle of $\mathcal{G}$. A subset of the set of regular expressions

$$\Psi_{R110} = \bigcup_{i=1}^{p} w_{i,g} \ \forall \ (w_i \in \Sigma^* \wedge g \in \mathcal{G}) \qquad (15.2)$$

where $p \geq 3$ is a period, determines the language

$$L_{R110} = \{w|w = w_iw_j \vee w_i + w_j \vee w_i^* \text{ and } w_i, w_j \in \Psi_{R110}\}. \qquad (15.3)$$

From these set of strings we can code initial configurations to program collisions between particles [27, 36, 39].

To deriver the regular expressions we use the de Bruijn diagrams [31, 34, 55] as follows. Assume the particle A moves two cells to the right in three time steps (see Table 15.1). The corresponding extended de Bruijn diagram (2-shift, 3-gen) is shown in Fig. 15.3. Cycles in the diagram are periodic sequences uniquely representing each phase of the particle. Diagram in Fig. 15.3 has two cycles: a cycle formed by just a vertex 0 and another large cycle of 26 vertices composed by other nine internal cycles. The sequences or regular expressions determining the phases of the particle A are obtained by following paths through the edges of the diagram. There regular expressions and corresponding paths in Bruijn diagram are shown below.

  I. The expression $(1110)^*$: vertices 29, 59, 55, 46 determining A^n particles.
 II. The expression $(111110)^*$: vertices 61, 59, 55, 47, 31, 62 defining nA particles with a T_3 tile among each particle.

---

[2]See a complete set of regular expressions for every particle in rule 110 in http://uncomp.uwe.ac.uk/genaro/rule110/listPhasesR110.txt.

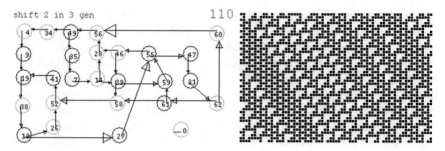

Fig. 15.3 De Bruijn diagram calculating A particles (*left*) and space-time configuration of automaton showing locations of periodic sequences produced (*right*)

Table 15.2 Four sets of phases Ph_i in rule 110

Phases level one $(Ph_1) \to \{f_1\_1, f_2\_1, f_3\_1, f_4\_1\}$
Phases level two $(Ph_2) \to \{f_1\_2, f_2\_2, f_3\_2, f_4\_2\}$
Phases level three $(Ph_3) \to \{f_1\_3, f_2\_3, f_3\_3, f_4\_3\}$
Phases level four $(Ph_4) \to \{f_1\_4, f_2\_4, f_3\_4, f_4\_4\}$

III. The expression $(11111000100110)^*$: vertices 13, 27, 55, 47, 31, 62, 60, 56, 49, 34, 4, 9, 19, 38 describing the periodic background configurations in a specific phase.

Cycle with period 1 (vertex 0) yields a homogeneous evolution in state 0. The evolution space in Fig. 15.3 shows different trains of A particles. The initial condition is constructed following some of the seven possible cycles of the de Bruijn diagram or a combination of them. In this way, the number of particles A or the number of intermediate tiles T_3 can be selected by moving from one cycle to another.

The alignment of the f_i_1 phases is analysed to determine the whole set of strings for every particle. We describe the form and limits of each particle by tiles. Then a phase is fixed (in our case the phase f_i_1) and a horizontal line is placed in the evolution space bounded by two aligned T_3 tiles. The sequence between both tiles aligned in each of the four levels determines a periodic sequence representing a particular structure in the evolution space of rule 110. All periodic sequences in a specific phase are calculated, enumerating the phases for each particle or non-periodic structure.

Table 15.2 represents disjoint subset of phases, each level contains four phases. Variable f_i indicates the phase of a particle, and the subscript j (in the notation f_i_j) indicates the selected set Ph_j of regular expressions. Finally, we use the next notation to codify initial conditions by phases as follows:

$$\#_1(\#_2, f_i\_1) \tag{15.4}$$

where $\#_1$ represents a particle according to Cook's classification (Table 15.1) and $\#_2$ is a phase of the particle with period greater than four.

15.2 Universal elementary CA

A concept of universality and self-reproduction in CA was proposed by von Neumann in [54] in his design of a universal constructor in a 2D CA with 29 cell-states. Architectures of universal CA have been simplified by Codd in 1968 [10], Banks in 1971 [7], Smith in 1971 [51], Conway in 1982 [8], Lindgren and Nordahl in 1990 [22], and Cook in 1998 [11].[3] Cook simulated a cyclic tag system, equivalent to a minimal Turing machine, in CA rule 110. In general, computation capacities are explores in complex CA and chaotic CA [40].

15.3 Cyclic Tag Systems

Cyclic tag systems are used by Cook in [11] as a tool to implement computations in rule 110. Cyclic tag systems are modified from tag systems by allowing the system to have the same action of reading a tape in the front and adding characters at its end:

1. Cyclic tag systems have at least two letters in their alphabet ($\mu > 1$).
2. Only the first character is deleted ($\nu = 1$) and its respective sequence is added.
3. In all cases if the machine reads a character zero then the production rule is always null ($0 \to \varepsilon$, where ε represents the empty word).
4. There are k sequences from μ^* which are periodically accessed to specify the current production rule when a nonzero character is taken by the system. Therefore the period of each cycle is determinate by k.

Such cycle determines a partial computation over the tape, although a halt condition is not specified. Let us see some samples of a cyclic tag system working with $\mu = 2$, $k = 3$ and the following production rules: $1 \to 11$, $1 \to 10$ and $1 \to \varepsilon$. To avoid writing a chain when there is no need to add characters, the $\vdash_k$ relation is just indicated. For example, the $00001 \vdash_1\vdash_2\vdash_3\vdash_1\vdash_2 10$ represents the relations $00001 \vdash_1 0001 \vdash_2 001 \vdash_3 01 \vdash_1 1 \vdash_2 10$. Each relation indicates which exactly sequence μ is selected.

Cyclic tag systems tend to growth quickly which makes it difficult to analyse their behaviour. Morita in [43, 44] demonstrated how to implement a particular halt condition in cyclic tag systems given an output string when the system is halting, and how a partitioned CA can simulate any cyclic tag system, consequently computing all the recursive functions.

Similar to Post's developments with tag systems, Cook determined that for a cyclic tag system with $\mu = 2$, $k = 2$, the productions $1 \to 11$ and $1 \to 10$, and starting evolution with the state 1 on the tape, it is impossible to decide if the process is terminal.

---

[3] A range of universal CA is listed here http://uncomp.uwe.ac.uk/genaro/Complex_CA_repository.html.

15.4   Cyclic Tag System Working in Rule 110

Let us see how a cyclic tag system operates in rule 110 [58]. We use a cyclic tag system with $\mu = 2$, $k = 2$ and the productions $1 \rightarrow 11$ and $1 \rightarrow 10$, starting its evolution in state 1 on the tape. A fragments of the systems' behaviour is shown below:

$1 \vdash_1 11 \vdash_2 110 \vdash_1 1011 \vdash_2 01110 \vdash_1\vdash_2 11010 \vdash_1 101011 \vdash_2 0101110 \vdash_1\vdash_2 0111010$
$\vdash_1\vdash_2 1101010 \vdash_1 10101011 \vdash_2 010101110 \vdash_1\vdash_2 010111010 \vdash_1\vdash_2 011101010 \vdash_1\vdash_2$
$110101010 \vdash_1 1010101011 \vdash_2 01010101110 \vdash_1\vdash_2 01010111010 \vdash_1\vdash_2 01011101010$
$\vdash_1\vdash_2 01110101010 \vdash_1\vdash_2 11010101010 \vdash_1 101010101011 \vdash_2 0101010101110 \vdash_1\vdash_2$
$0101010111010 \vdash_1\vdash_2 01010111010 10 \vdash_1\vdash_2 0101110101010 \vdash_1\vdash_2 0111010101010 \vdash_1\vdash_2$
$1101010101010 \vdash_1 10101010101011 \vdash_2 010101010101110 \vdash_1\vdash_2 010101010111010 \vdash_1\vdash_2$
$01010 1011101010 \vdash_1\vdash_2 010101110101010 \vdash_1\vdash_2 010111010101010 \ldots$

We start with the expression $1(10)^*$. The cyclic tag systems moves (from the right to the left) and adds a pair of bits. As soon as the expression $1(10)^*$ appears again, a number of relations selected in each interval in such a manner that the expressions grow lineally in order of $f_1 = 2(n + 1)$.

If we take consecutive copies of $1(10)^*$ with their respective intervals determined by the number of j productions (represented as $\vdash_i^j$), we obtain the following sequence: $1 \vdash_i^2 110 \vdash_i^4 11010 \vdash_i^6 1101010 \vdash_i^8 1101010 \vdash_i^{10} 110101010 \vdash_i^{12} 11010101010 \vdash_i^{14} 1101010101010 \vdash_i^{16}$ .... There are no states where to '0' appear together.

Further, we show how to interpret particles and their collisions to emulate a cyclic tag system in rule 110. We must use trains of particles to represent data and operators, their reactions, transform and deletion of data on the tape. A schematic diagram, where trains of particles are represented by lines, is shown in Fig. 15.4. The diagram is explained with details in the next sections.

15.4.1   Components Based on Sets of Particles

A construction of the cyclic tag system in rule 110 can be subdivided into three parts (Fig. 15.4). First part is the left periodic part controlled by trains of 4_A^4 particles. This part is static. It controls the production of 0's and 1's. The second part is the center determining the initial value on the tape. The third part is the right, cyclic, part which contains the data to process. It adds or removes data on the tape.

Set of particles 4_A^4

The four trains of A^4 particles are static but their phases change periodically. A key point is to implement these components by defining both distances and phases, because some choices of phases or distances might induce an undesirable reactions between the trains of particles.

Packages defined by particles A^4 have three different phases: f_1_1, f_2_1 and f_3_1. To construct the first train 4_A^4 we must establish the phase of each A^4. Let us assign

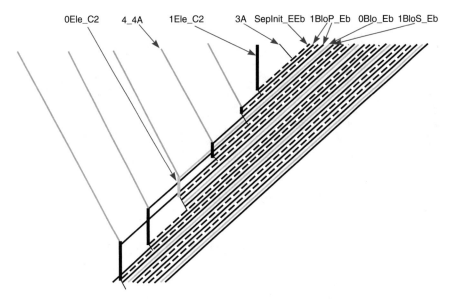

OEle_C2   4_4A   1Ele_C2   3A SepInit_EEb 1BloP_Eb 0Blo_Eb 1BloS_Eb

Fig. 15.4 Schematic diagram of a cyclic tag system working in rule 110

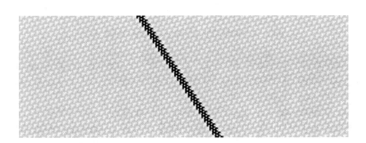

Fig. 15.5 Set of particles 4_A^4

phases as follows:

$$A^4(f_3\_1)\text{-}27e\text{-}A^4(f_2\_1)\text{-}23e\text{-}A^4(f_1\_1)\text{-}25e\text{-}A^4(f_3\_1),$$

see Fig. 15.5. Spaces between each train 4_A^4 are fixed but the phases change. The soliton-like collisions between the particles $\bar{E}$ occur:

$$\{649e\text{-}A^4(f_2\_1)\text{-}27e\text{-}A^4(f_1\_1)\text{-}23e\text{-}A^4(f_3\_1)\text{-}25e\text{-}A^4(f_2\_1)\text{-}649e\text{-}A^4(f_1\_1)\text{-}$$
$$27e\text{-}A^4(f_3\_1)\text{-}23e\text{-}A^4(f_2\_1)\text{-}25e\text{-}A^4(f_1\_1)]\text{-}649e\text{-}A^4(f_3\_1)\text{-}27e\text{-}A^4(f_2\_1)\text{-}23e\text{-}$$
$$A^4(f_1\_1)\text{-}25e\text{-}A^4(f_3\_1)\}*$$

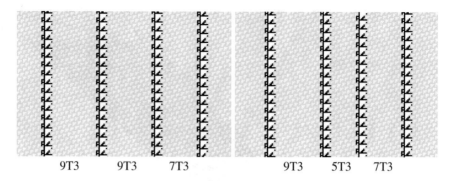

Fig. 15.6 Set of particles 1Ele_C_2 (*left*) and 0Ele_C_2 (*right*)

If for every 4_A^4 we take a phase representing the complete train, we can rename it as:

$$\{649e - 4\_A^4(F_2) - 649e - 4\_A^4(F_1) - 649e - 4\_A^4(F_3)\}*$$

this phase change is important to preserve good reactions coming to the left side of the system.

Set of particles 1Ele_C_2 and 0Ele_C_2

The central part is made of the state '1' written on the tape represented by a train of four C_2 particles. A set of particles 1Ele_C_2 represents '1' and a set of particles 0Ele_C_2 represents '0' on the tape.

The left configurations in Fig. 15.6 shows the set of particles 1Ele_C_2. We should reproduce each set of particles by the phases f$_i$_1. The phases are coded as follows: C_2(A,f$_1$_1)-2e-C_2(A,f$_1$_1)-2e-C_2(A,f$_1$_1)-e-C_2(B,f$_2$_1). The first three particles C_2 are in phase (A,f$_1$_1) and the fourth particle C_2 is in phase (B,f$_2$_1). The distances between the particles are 9T_3-9T_3-7T_3. To determine the distances, we count the number of tiles T_3 between particles. Similarly, we obtain the distances 9T_3-5T_3-7T_3 for the particles 0Ele_C_2.

Set of particles 0Blo_$\bar{E}$

The left part stores blocks of data without transformations in trains of E and the particles $\bar{E}$.

The set of particles 0Blo_$\bar{E}$ is formed by 12$\bar{E}$ particles as we can see in Fig. 15.7. There must be an exact phase and distance between each one of the particles, otherwise the whole system will be disturbed.

Set of particles 1BloP_$\bar{E}$ and 1BloS_$\bar{E}$

To write '1's we must use two set of particles—*primary* and *standard*.

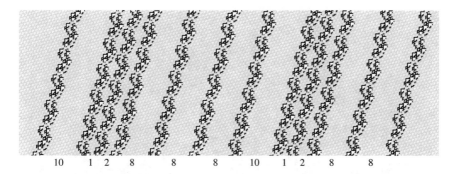

Fig. 15.7   Set of particles 0Blo_$\bar{E}$

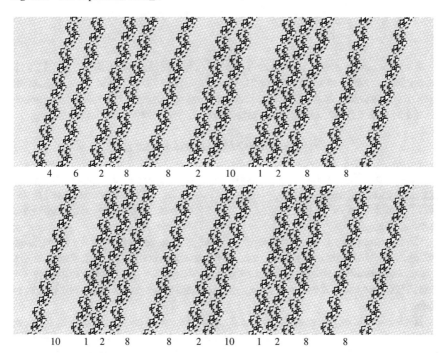

Fig. 15.8   Set of particles 1BloP_$\bar{E}$ (*up*) and 1BloS_$\bar{E}$ (*down*)

They are differences in distance between first two particles $\bar{E}$, as shown in Fig. 15.8. Both blocks produce the same set of particles 1Add_$\bar{E}$. The main reason to use both set of particles is because the CA rule 110 evolves asymmetrically and therefore we need a double set of particles to produce values 1 correctly.

Set of particles SepInit_$E\bar{E}$

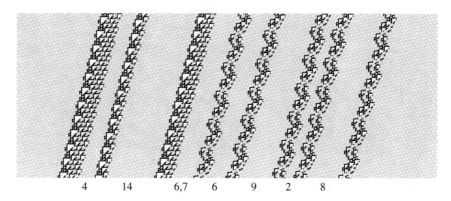

Fig. 15.9 Set of particles SepInit_$E\bar{E}$

A leader component renamed as the set of particles SepInit_$E\bar{E}$ (see Fig. 15.9) is essential to separate trains of data and to determine the incorporation of the data on the tape. Its has a small but detailed code determining which data without transformation would be added or erased from the tape, depending on the value that is coming.

Set of particles 1Add_$\bar{E}$ and 0Add_$\bar{E}$

Figure 15.10 illustrates the set of particles 1Add_$\bar{E}$ and 0Add_$\bar{E}$ produced by two previous different trains of data. A set of particles 1Add_$\bar{E}$ must be generated by the set of particles 1BloP_$\bar{E}$ or 1BloS_$\bar{E}$. This way, both set of particles can produce the same element.

On the other hand, a set of particles 0Add_$\bar{E}$ is generated by a set of particles 0Blo_$\bar{E}$. Nevertheless, we could produce $\bar{E}$ particles modifying their first two distances and preserving them without changing others particles to get a reliable reaction. This is possible if we want to experiment with other combinations of blocks of data.

If a leader set of particles SepInit_$E\bar{E}$ reaches a set of particles 1Ele_$\bar{E}$, it erases this value from the tape and adds a new data that shall be transformed. In other case, if it finds a set of particles 0Ele_$\bar{E}$, then it erases this set of particles from the tape and also erases a set of unchanged data which comes from the right until finding a new leader set of particles. This operation represents the addition of new values from periodic trains of particles coming from the right. Thus a set of particles 1Add_$\bar{E}$ is transformed into 1Ele_$\bar{E}$ colliding against a train of 4_A^4 particles representing a value 1 in the tape, and the set of particles 0Add_$\bar{E}$ is transformed into 0Ele_$\bar{E}$ colliding against a train of 4_A^4 particles representing a value 0 in the tape.

Table 15.3 shows all distances (in numbers of T_3 tiles) for every. We can code the construction of this cyclic tag system across phase representations in three main big sub systems:

left: ...-217e-4_A^4(F2)-649e-4_A^4(F1)-649e-4_A^4(F3)-649e-4_A^4(F2)-

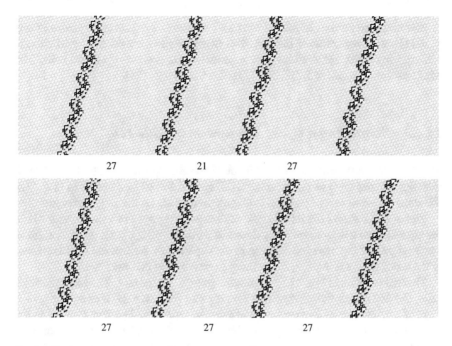

Fig. 15.10 Set of particles 1Add_Eb (*up*) and 0Add_$\bar{E}$ (*down*)

Table 15.3 Distances between sets of particles

Set of particles	Distance
1Ele_C_2	9-9-7
0Ele_C_2	9-5-7
1BloP_$\bar{E}$	4-6-2-8-8-2-10-1-2-8-8
1BloS_$\bar{E}$	10-1-2-8-8-2-10-1-2-8-8
0Blo_$\bar{E}$	10-1-2-8-8-8-10-1-2-8-8
SepInit_$E\bar{E}$	4-14-(6 or 7)-6-9-2-8
1Add_$\bar{E}$	27-21-27
0Add_$\bar{E}$	27-27-27

649e-4_A^4(F1)-649e-4_A^4(F3)-216e-

center: 1Ele_C_2(A,f$_1$_1)-e-A^3(f$_1$_1)-

right: SepInit_$E\bar{E}$(C,f$_3$_1)-1BloP_$\bar{E}$(C,f$_4$_1)-SepInit_$E\bar{E}$(C,f$_3$_1)-
       1BloP_$\bar{E}$(C,f$_4$_1)-0Blo_$\bar{E}$(C,f$_4$_1)-1BloS_$\bar{E}$(A,f$_4$_1)-
       SepInit_$E\bar{E}$(A,f$_2$_1)(2)-1BloP_$\bar{E}$(F,f$_1$_1)-SepInit_$E\bar{E}$(A,f$_3$_1)(2)-
       1BloP_$\bar{E}$(F,f$_1$_1)-0Blo_$\bar{E}$(E,f$_4$_1)-1BloS_$\bar{E}$(C,f$_4$_1)-e-
       SepInit_$E\bar{E}$(B,f$_1$_1)(2)-1BloP_$\bar{E}$(F,f$_3$_1)-e-
       SepInit_$E\bar{E}$(B,f$_1$_1)(2)-217e-….

The initial conditions in rule 110 are able to generate the serial sequence of bits 1110111 and a separator at the end with two particles. A desired construction is achieved in 57,400 generations and an initial configuration of 56,240 cells. The whole evolution space is 3,228,176,000 cells. See details [38].

15.4.2 Simulating a Cyclic Tag System in Rule 110

The cyclic tag system starts with the value '1' on the tape, see Fig. 15.4. We show a selection of snapshots of the machine working in rule 110 (see details in [38, 47]). We show different sets of particles with coloured labels on the snapshot below.

Figure 15.11 shows the initial stage of the cyclic tag system with the state '1' in the tape. This data is represented by the set of particles $1Ele\_C_2$. The snapshot shows a central part of the machine and a train of A^3 particles. We can see the first leader in the set of particles $SepInit\_E\bar{E}$ coming from the right periodic side.

The first reaction in Fig. 15.11 deletes the state '1' on the tape. The set of particles $1Ele\_C_2$) and the particles' separator are prepared for next data to be aggregated. If a set of particles $0Ele\_C_2$ is encountered on the tape then data is not added to the tape until another separator appears. The particles $\bar{E}$ left after the first production are invisible to the system, they do not affect any operations because they cross as solitons, without state modifications, the subsequent set of particles 4_A^4.

In Fig. 15.12 we see a set of particles $1Ele\_C_2$ constructed from a train of particles 4_A^4. These particles have a very short life because quickly a separator set of particles arrives. This separator erases the particles and prepares new data that would be aggregated to the tape.

Figure 15.13 presents the construction of a set of particles $1Ele\_C_2$. In this stage of the evolution, we can see how data is aggregated, based on their values, before they cross the tape. Similar reactions can be observed with the set of particles $0Ele\_C_2$.

Figure 15.14 shows a set of particles $0Ele\_C_2$ and its roles in the system. At the top, a set of particles $1Add\_\bar{E}$, previously produced by a standard component $1BloS\_\bar{E}$, crosses a set of particles $0Ele\_C_2$. A leader set of the particles deletes '0' from the tape and all the subsequent incoming data. There are $1BloP\_\bar{E}$, $0Blo\_\bar{E}$ and $1BloS\_\bar{E}$ set of particles in the illustrated sequence. The tile T_{14} is generated in the process. This differences in distances between the particles determine a change of phases which will lead to erasure of particles $\bar{E}$, instead of production of particles C. The reaction $A^3 \rightarrow \bar{E}$ is used to delete the particles.

Production rules in cyclic tag system specify that for the state '0' the first element of the chain must be erased and the other elements are conserved and no data are written on the tape. If the state is '1' the first element of the chain is deleted and 10 or 11 are aggregated depending of the k value. This behaviour is particularly visible when a separator finds 0 or 1 and deletes it from the tape. If the deleted data is '0', a separator does not allow the production of new data. If the deleted data is '1' the separator aggregates new elements 11 or 10, which are modified at later stages of the

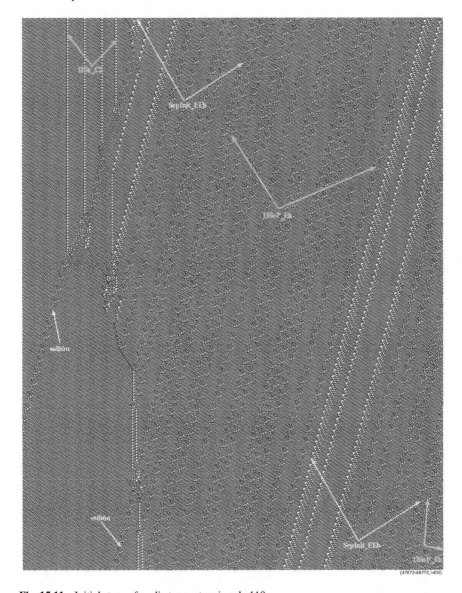

Fig. 15.11   Initial stage of cyclic tag system in rule 110

system's development. Using this procedure, we can calculate up to the sixth '1' of the sequence 011<1>0 produced by the cyclic tag system.

In terms of periodic phases, this cyclic tag system working in rule 110 can be simplified as follows:

left: $\{649e\text{-}4\_A^4(F\_i)\}^*$, for $1 \leq i \leq 3$ in sequential order

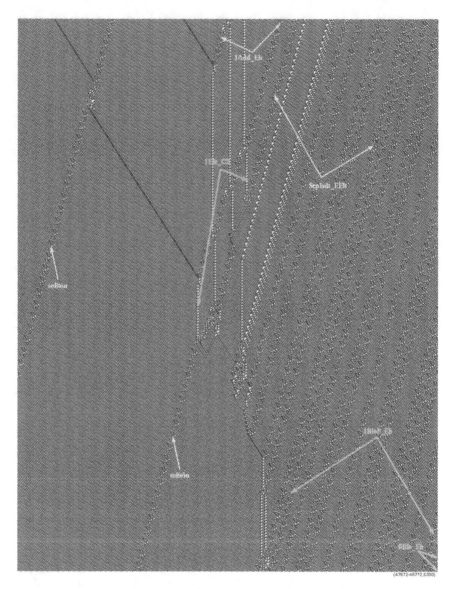

Fig. 15.12   Constructing an element 1Ele_C_2

center: $246e$-1Ele_C2(A,f_1_1)-e-A^3(f_1_1)

right: {SepInit_$E\bar{E}$(#,f_i_1)-1BloP_$\bar{E}$(#,f_i_1)-SepInit_$E\bar{E}$(#,f_i_1)-
        1BloP_$\bar{E}$(#,f_i_1)-0Blo_$\bar{E}$(#,f_i_1)-1BloS_$\bar{E}$(#,f_i_1)}* (where
        $1 \leq i \leq 4$ and # represents a particular phase).

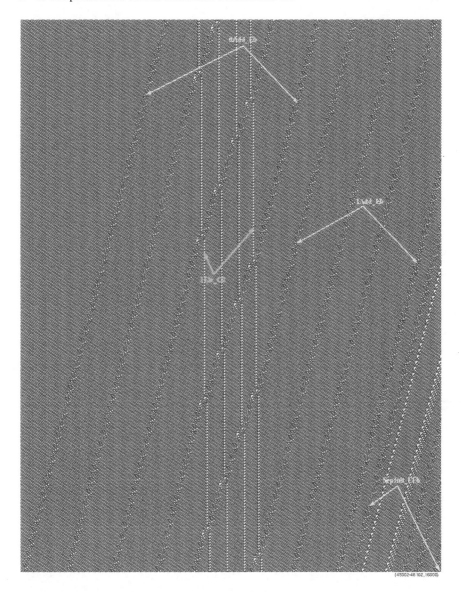

Fig. 15.13   Transformed data crossing the tape of values

These periodic coding will be very useful to design and synchronise three inter-linked rings of 1D CA (cyclotrons) to make a 'supercollider'.

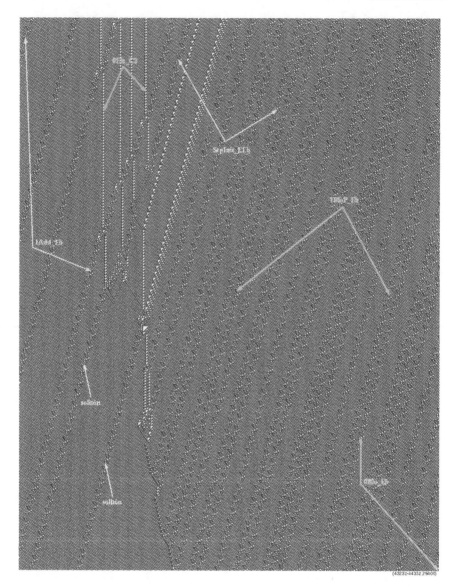

Fig. 15.14 Deleting a set of particles 0Ele_C2

15.5 Cellular Automata Supercollider

In the late 1970s Fredkin and Toffoli proposed a concept of computation based on ballistic interactions between quanta of information that are represented by abstract particles [53]. The Boolean states of logical variables are represented by balls or

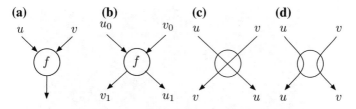

Fig. 15.15 Schemes of ballistic collision between localizations representing logical values of the Boolean variables u and v

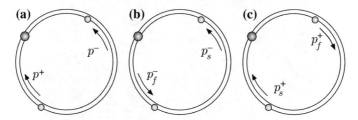

Fig. 15.16 Representation of abstract particles in a 1D CA ring

atoms, which preserve their identity when they collide with each other. Fredkin, Toffoli and Margolus developed a billiard-ball model of computation, with underpinning mechanics of elastically colliding balls and mirrors reflecting the balls' trajectories. Margolus proposed a special class of CA which implements the billiard-ball model [24]. Margolus' partitioned CA exhibited computational universality because they simulated Fredkin gates via collision of soft spheres [25, 26]. Also, we consider previous results about circular machines designed by Arbib, Kudlek, and Rogozhin in [5, 20, 21]. Initial reports about CA collider were published in [28–30].

The following functions with two input arguments u and v can be realised in collisions between two localizations:

- $f(u, v) = c$, fusion (Fig. 15.15a)
- $f(u, v) = u + v$, interaction and subsequent change of state (Fig. 15.15b)
- $f_i(u, v) \mapsto (u, v)$ identity, solitonic collision (Fig. 15.15c);
- $f_r(u, v) \mapsto (v, u)$ reflection, elastic collision (Fig. 15.15d);

To represent Toffoli's supercollider [53] in 1D CA we use the notion of an idealised particle $p \in \mathcal{G}$ (without energy and potential). The particle p is represented by a binary string of cell states.

Figure 15.16 shows two typical scenarios where particles p_f and p_s travel in a CA cyclotron. The first scenario (Fig. 15.16a) shows two particles travelling in opposite directions; these particles collide one with another. Their collision site (contact point) is shown by a dark circle in Fig. 15.16a. The second scenario demonstrates a beam routing where a fast particle p_f eventually catches up with a slow particle p_s at a collision site (Fig. 15.16b). If the particles collide like solitons, then the faster particle p_f simply overtakes the slower particle p_s and continues its motion (Fig. 15.16c).

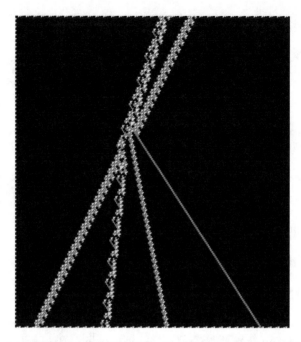

Fig. 15.17 Particle collision in rule 110. Particle $p_{\bar{B}}^-$ collides with particle $p_{\bar{G}}^-$ giving rise to three new particles—p_F^-, $p_{D_2}^+$, and $p_{A^3}^+$, and preserving the $p_{\bar{B}}^-$ particle—that are generated as a result of the collision

Typically, we can find all types of particles in complex CA, including particles with positive p^+, negative p^-, and neutral p^0 displacements, and composite particles assembled from elementary localizations. A sample coding and colliding particles is shown in Fig. 15.17, which displays a typical collision between two particles in rule 110. As a result of the collision one particle is split into three different particles (for full details please see [35]). The previous collision positions of particles determines the outcomes of the collision. Particles are represented now with orientation and name of the particle in rule 110 as follows: $p_G^{+,-,0}$.

To represent particles on a given beam routing scheme (see Fig. 15.16), we do not consider the periodic background configuration in rule 110 because essentially this does not affect on collisions. Figure 15.18 displays a 1D configuration where two particles collide repeatedly and interact as solitons so that the identities of the particles are preserved in the collisions. A negative particle p_F^- collides with and overtakes a neutral particle $p_{C_1}^-$. First cyclotron (Fig. 15.18a) presents a whole set of cells in state 1 (dark points) evolving with the periodic background. By applying a filter we can see better these interactions (Fig. 15.18b).[4] Typical space-time configurations of a CA exhibiting a collision between p_F^- and $p_{C_1}^-$ particles are shown in Fig. 15.18c.

---

[4]Cyclotron evolution was simulated with DDLab software, available at http://www.ddlab.org.

(a)   **(b)**   **(c)**

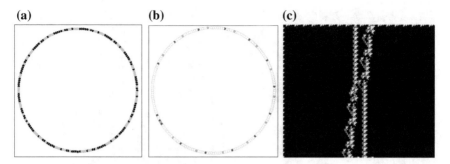

Fig. 15.18   A soliton-type interaction between particles in rule 110: **a**, **b** two steps of beam routing, **c** exact configuration at the time of collision

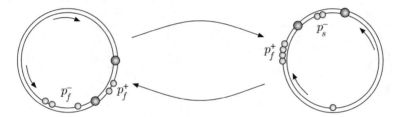

Fig. 15.19   Transition between two beam routing synchronising multiple reactions. When the first set of collisions is done a new beam routing is defined with other set of particles, so that when the second set of collisions is done then first beam returns to its original state

15.6   Beam Routings and Computations

We examine beam routing based on particle-collisions. We will show how the beam routing can be used in designs of computing based-collisions connecting cyclotrons. Figure 15.19 shows a beam routing design, connecting two of beams and then creating a new beam routing diagram where edges represent a change of particles and collisions. In such a transition, new particles emerge and collide to return to the first beam. The particles oscillate between these two beam routing indefinitely.

To understand how dynamics of a double beam differs from a conventional 1D evolution space we provide Fig. 15.20. There we can see multiple collisions between particles from first beam routing and trains particles. Exactly, we have that

$$p_A^+, p_A^+ \leftrightarrow p_{\bar{B}}^-, p_{\bar{B}}^-, p_{\bar{B}}^-$$

changes to the set of particles derived in the second beam routing:

$$p_{A^4}^+ \leftrightarrow p_E^+, p_{\bar{E}}^+.$$

This oscillation determines two beam routing connected by a transition of collisions as:

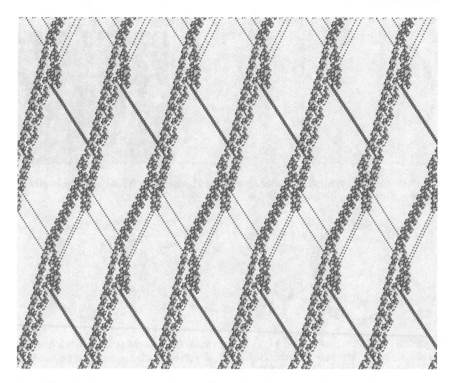

Fig. 15.20 Synchronisation of multiple collisions in rule 110 on a ring of 1,060 cells in 1,027 generations, starting with 50 particles from its initial condition

$$(p_A^+, p_A^+ \leftrightarrow p_{\bar{B}}^-, p_B^-, p_B^-) \to (p_{A^4}^+ \leftrightarrow p_E^+, p_{\bar{E}}^+), \text{ and}$$

$$(p_{A^4}^+ \leftrightarrow p_E^+, p_{\bar{E}}^+) \to (p_A^+, p_A^+ \leftrightarrow p_{\bar{B}}^-, p_B^-, p_B^-).$$

We can see that a beam routing representation allows for a design of collisions in cyclotrons. We employ the beam routing to implement the cyclic tag system in the CA rings. A construction of the cyclic tag system in rule 110 consists of three components (as was discussed in Sect. 15.4.2):

- The *left periodic part*, controlled by trains of 4_A^4 particles. This part is static. It controls the production of 0's and 1's.
- The *centre*, determining the initial value in the tape.
- The *right periodic part*, which has the data to process, adding a leader component which determines if data will be added or erased in the tape.

Left periodic part is defined by four trains of A^4 (Fig. 15.21c), trains of A^4 have three phases. The key point is to implement these components defining both distances and phases, because a distinct phase or a distance induces an undesirable reaction.

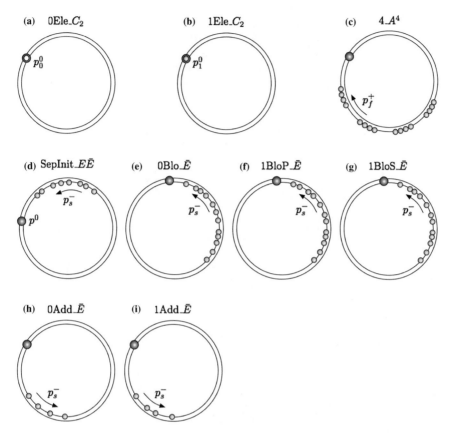

Fig. 15.21 The whole set of beam routing codification representing train of particles, to simulate a cyclic tag system. Each global state represents every component (set of particles) described in Sect. 15.4.1

The *central part* is represented by one value '1' on the tape across a train of four C_2 particles. The component $1Ele\_C_2$ (Fig. 15.21b) represents '1' and the component $0Ele\_C_2$ (Fig. 15.21a) represents '0' on the tape. The component $0Blo\_\bar{E}$ is formed by $12\bar{E}$ particles. The construct includes two components to represent the state '1': $1BloP\_\bar{E}$ (Fig. 15.21f) named *primary* and $1BloS\_\bar{E}$ (Fig. 15.21g) named *standard*. A leader component $SepInit\_E\bar{E}$ (Fig. 15.21d) is used to separate trains of data and to determine their incorporation into of the tape.

The components $1Add\_\bar{E}$ (Fig. 15.21i) and $0Add\_\bar{E}$ (Fig. 15.21h) are produced by two previous different trains of data. The component $1Add\_\bar{E}$ must be generated by a block $1BloP\_\bar{E}$ or by $1BloS\_\bar{E}$. This way, both components can yield the same element. The component $0Add\_\bar{E}$ is generated by a component $0Blo\_\bar{E}$ (Fig. 15.21e). For a complete and full description of such reproduction by phases $f_{i\_}1$, see [38].

To get a cyclic tag system emulation in rule 110 by beam routings, we will use connections between beam routings as a finite state machine represented in Fig. 15.22.

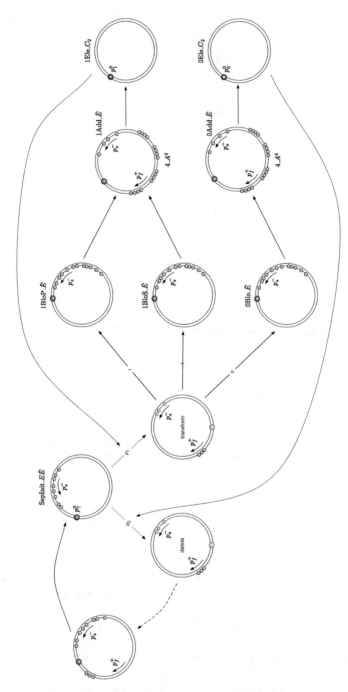

Fig. 15.22 Beam routing finite state machine simulating the cyclic tag system by state of cyclotrons representation

Transitions between beam routings means a change of state (transition function). Initial state is represented by the component 1Ele_C_2. A final state is not specified because it is determined by the state of the computation, i.e., a halt condition. Components 1Ele_C_2 and 0Ele_C_2 are compressed and shown as a dark circle, which represents the point of collision. Both components are made of four C_2 particles being at different distances. When a leader component (SepInit_$E\bar{E}$) is transformed, given previous binary value on the tape, it collides with $p_?^0$ component, i.e., a p_1^0 or p_0^0 element. If $p_?^0$ is '0', then a cascade of collisions starts to *delete* all components that come with three particles successively. If $p_?^0$ is '1' then a cascade of *transformations* dominated by additional particles p^0 is initiated, in order to reach the next leader component. Here, we have more variants because pre-transformed train of particles is encoded into binary values that are then written on the machine tape. If a component of particles is 1BloP_$\bar{E}$ or 1BloS_$\bar{E}$ this means that such a component will be transformed to one 1Add_$\bar{E}$ element. If a component of particles is 0Blo_$\bar{E}$, then such a component will be transformed to 0Add_$\bar{E}$ element. At this stage, when both components are prepared then a binary value is introduced on the tape, a 1Add_$\bar{E}$ element stores a 1 (1Ele_C_2), and a 0Add_$\bar{E}$ element stores a 0 (0Ele_C_2), which eventually will be deleted for the next leader component and starts a new cycle in the cyclic tag system. In bigger spaces these components will be represented just as a point in the evolution space: we describe this representation in the next section.

15.7   Cyclotrons

We use cyclotrons to explore large computational spaces where exact structures of particles are not relevant but only the interactions between the particles. There we can represent the particles as points and trains of particles as sequences of points. A 3D representation is convenient to understand the history of the evolutions, number of particles, positions, and collisions. Figure 15.23 shows a cyclotron evolving from a random initial configuration with 20,000 cells. Three stages are initialised in the evolution and the particles undergo successions of collisions in few first steps of evolution. The evolution is presented in a vertical orientation rotated 90 degrees. The present state shown is a front and its projection in three dimensions unveils the history and the evolution. Following this representation we can design a number of initial conditions to reproduce periodic patterns.[5]

Figure 15.24 shows a basic flip-flop pattern. We synchronise 16 particles $p_F \leftarrow p_B$, the basic collision takes place for two pairs of particles, a p_{D_1} particle and a train of p_{A^2} particles. The distance is determined by a factor of mod 14. A second reaction is synchronised with $p_{D_1} \leftarrow p_{A^2}$ to return back to the initial p_F and p_B particles. All 16 particles are forced in the same phase to guarantee an adequate distance, this distance is fixed in 64 copies of 14 cells (ether). Finally eight collisions are controlled every time simultaneously on an evolution space with 7,464 cells.

---

[5]The simulations are done in *Discrete Dynamics Lab* (DDLab, http://www.ddlab.org/) [59].

Fig. 15.23 ECA rule 110 particles traveling and colliding inside a cyclotron in a evolution space of 20,000 cells. A filter is selected for a better view of particles, each cyclotron initial stage in the history (three dimensional projection) is restarted randomly to illustrate the complex dynamics and variety of particles and collisions

Fig. 15.24 Basic flip-flop oscillator implemented in a cyclotron with 7,464 cells in 25,000 generations. 16 particles $p_F \leftarrow p_B$ were coded

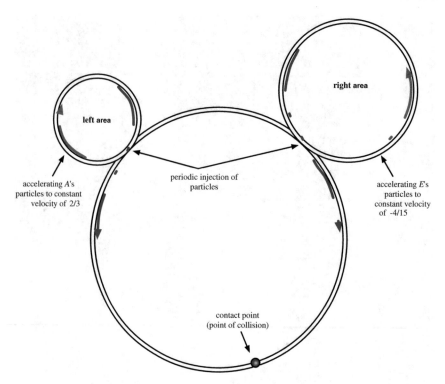

Fig. 15.25 Collider diagram

15.8 Collider Computing

A cyclic tag system consists of three main components. Each stage of computation can be represented with a cyclotron. A synchronisation of these cyclotrons injects beams of particles to a central main collider to obtain the collisions that will simulate a computation. The periodic representations of left and right cyclotrons are fixed. Diagram in Fig. 15.25 shows the dynamics of particles in a collider.

Left part   Periodic area handle beams of three trains of four p_{A^4} particles, travelling from the left side with a constant velocity of 2/3. This ring has 30,640 cells, the minimum interval between trains of particles is 649 copies of ether. Each beam of p_{A^4} can have three possible phases. The sequence of phases is periodic and fixed sequentially: $\{649e\text{-}4A^4(F_i)\}^*$, for $1 \leq i \leq 3$ (Fig. 15.25 left area). Figure 15.26 shows a simulation of these periodic beams of $4p_{A^4(F_i)}$ particles.

Right part   Periodic area handle beams of six trains of 12E's particles ($p_{E^n}, p_{\bar{E}}$), travelling from the right side with a constant velocity of $-4/15$. There are 12 particles related to a perfect square with $13{,}500^2$ possibilities to arrange inputs into the main collider. Interval between 12 particles is mod 14. Figure 15.27

Fig. 15.26   Three beams of $4p_{A^4(F_i)}$ particles. Simulation is displayed in a vertical position to get a better view of particles' trajectories

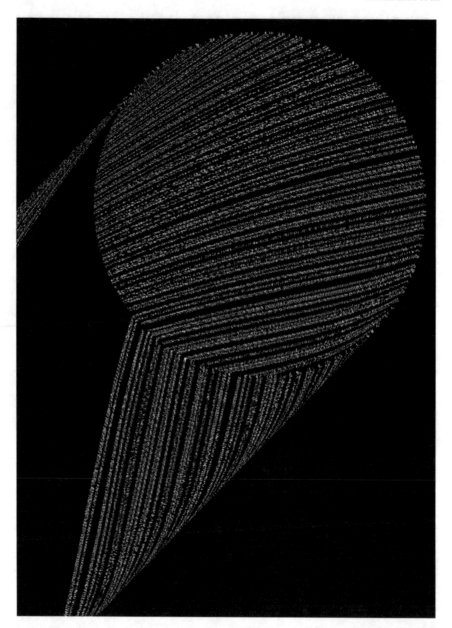

Fig. 15.27 A beam composed of six $12p_{Es}$ particles. Simulation is shown in a vertical position to get a better view of particles' trajectories. Interval between first and last particles can be any number mod 14

Fig. 15.28 First stage of collisions of the cyclic tag system. Solitonic interactions take place $4p_{A^4(F_3)}$ and two $p_{\bar{E}}$ particles. First symbol '1' on the type is deleted (*center*). The first separator is read and deleted

shows the whole set of 72 p_{Es} particles. The set contains leaders and separator components, and beams of particles that introduce '0's and '1's on the tape.

Center   Initial state of particles starts with a '1' on the type of the cyclic tag system. Figure 15.28 shows the first stage of the collider. The system start with one '1' in the type (four vertical p_{Cs} particles), they are static particles that wait for the first beam of p_{Es} particles to arrive at the right side to delete this input and decode the next inputs. In this process two solitons emerge, but they do not affect the system and the first beam of $4p_{A^4(F_3)}$ particles without changing their states.

Figure 15.29 shows how a second symbol '1' is introduced in the collider. A leader component is deleted and the second binary data is prepared to collide later with the first beam of $4p_{A^4(F_3)}$ particles. Finally, the second '1' is represented for the vertical particles, as shown at the bottom of Fig. 15.29.

Fig. 15.29 This snapshot shows when a '1' is introduced in the type. A second beam of $12\,p_{\bar{E}}$ particles is coming to leave just spaced four $p_{\bar{E}}$ particles, these particles collide with one of $4p_{A^4(F_3)}$ particles. The result is four $4p_{C_5}$ particles at the bottom of the simulation that represent one '1' in the cyclic tag system type

Figure 15.30 shows how further symbols '0' and '1' are introduced in the system. They are coded with $p_{\bar{E}_5}$ particles. Before the current '1' is introduced with $4p_{A^4(F_3)}$ particles, the next set of $4p_{A^4(F_3)}$ particles is prepared in advance.

Figure 15.31 shows the largest stage of the collider's working. A second beam of $4p_{A^4(F_1)}$ arrives. More beams of p_{E_5} particles are introduced. Figure 15.32 displays a full cycle of beams of p_A and p_{E_5} particles. All operations are performed at least once. The next set of particles is ready to continue with the next stage of the computation.

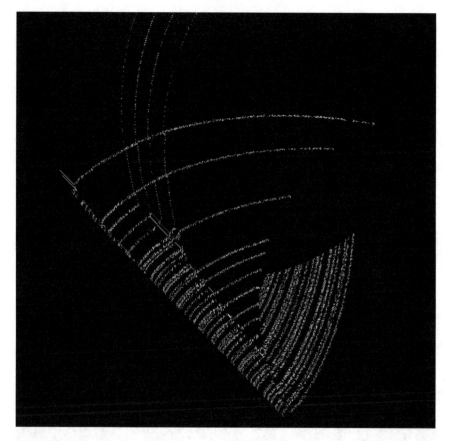

Fig. 15.30   This snapshot shows how a sequence of values '0' and '1' is precoded. You can see sequences of '0's and '1's, and p_{E_s} particles travelling to from the *left* to the *right*

15.9   Discussion

The CA collider is a viable prototype of a collision-based computing device. It well compliments existing models of computing circuits based on particle collisions [15, 18, 23, 42, 45, 56, 60]. How complex is our design? With regarding to time complexity, rule simulates Turing machine a polynomial time and any step of rule 110 can be predicted in a polynomial time [46]. As to space complexity, left cyclotron in the collider is made of 30,640 cells and the right cyclotron of 5,864 cells. The main collider should have 61,280 cells to implement a full set of reactions; however, it is possible to reduce the number of cells in the main collider, because the first train of $4p_{A^4(F_i)}$ particles needs just 10,218 cells; and subsequent trains can be prepared while initial data are processed. Thus, the simulated collider have just thousands of cells not millions. The space complexity of the implemented cyclic tag systems has been reduced substantially [11, 12, 58].

Fig. 15.31 This snapshot shows from another angle how binary values are introduced in the cyclic tag system. We can also see how a number of values are prepared to collide with beams of $4p_{A^4(F_i)}$ particles at the end of simulation

What are chances of implementing the CA collider model in physical substrates? A particle, or gliders, is a key component of the collider. The glider is a finite-state machine implementation of a propagation localisation. A solitary wave, or an impulse, propagating in a polymer chain could be a phenomenologically suitable analog of the glider. A wide range of polymer chains, both inorganic and organic, support solitons [1–3, 6, 9, 13, 14, 17, 19, 50, 52]. We believe actin filaments could make the most suitable substrate for implementation of a cyclic tag system via linked rings of CA colliders.

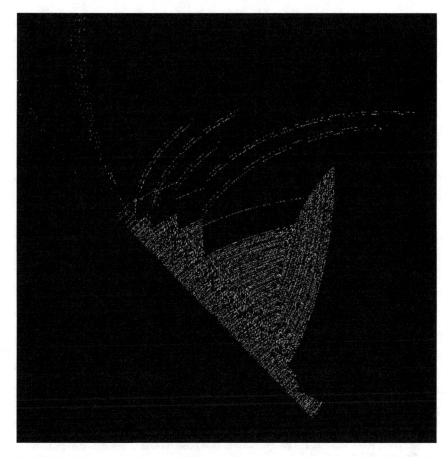

Fig. 15.32 This evolution displays a full cycle of beams of p_A and p_{Es} particles. In this snapshot we can see all necessary operations in the cyclic tag system: input values, deleting block of values, particles like solitons, and the next stage of the collider

An actin filament is a double spiral helix of globular protein units. Not only actin is a key element of a cell skeleton, and is responsible for a cell's motility, but actin networks is a sensorial, information processing and decision making system of cells. In [4] we proposed a model of actin filaments as two chains of one-dimensional binary-state semi-totalistic automaton arrays. We show that a rich family of travelling localisations is observed in automaton model of actin, and many of the localisation observed behave similarly to gliders in CA rule 110. The finite state machine model has been further extended to a quantum cellular automata model in [48]. We have shown that quantum actin automata can perform basic operations of Boolean logic, and implemented a binary adder. To bring more 'physical' meaning in our actin-computing concept we also employed the electrical properties of imitated actin filaments—resistance, capacitance, inductance — and found that it is possible to implement logical gates via interacting voltage impulses [49]; voltage impulses in non-linear transmission wires are analogs of gliders in 1D CA. Clearly, having just actin is not enough: we must couple rings together, arrange physical initiation of solitons and their detection, and solve myriad of other experimental laboratory problems. That will be a scope of further studies.

References

1. Adamatzky, A.: Computing in Nonlinear Media and Automata Collectives. Institute of Physics Publishing, Bristol (2001)
2. Adamatzky, A. (ed.): Collision-Based Computing. Springer, London (2002)
3. Adamatzky, A.: Unconventional Computing. Human Brain Project Magazine (2015)
4. Adamatzky, A., Mayne, R.: Actin automata: phenomenology and localizations. Int. J. Bifurc. Chaos 25(02), 1550030 (2015)
5. Arbib, M.A.: Theories of Abstract Automata. Prentice-Hall Series in Automatic Computation, Michigan (1969)
6. Bandyopadhyay, A., Pati, R., Sahu, S., Peper, F., Fujita, D.: Massively parallel computing on an organic molecular layer. Nat. Phys. 6, 369–375 (2010)
7. Banks, E.R.: Information and transmission in cellular automata. PhD Dissertation. Massachusetts Institute of Technology, Cambridge (1971)
8. Berlekamp, E.R., Conway, J.H., Guy, R.K.: Winning Ways for your Mathematical Plays, vol. 2, Chap. 25, Academic Press, Cambridge (1982)
9. Bredas, J.L., Street, G.B.: Polarons, bipolarons, and solitons in conducting polymers. Acc. Chem. Res. 18(10), 309–315 (1985)
10. Codd, E.F.: Cellular Automata. Academic Press, Inc., New York (1968)
11. Cook, M.: Universality in elementary cellular automata. Complex Syst. 15(1), 1–40 (2004)
12. Cook, M.: A concrete view of Rule 110 computation. In: Neary, T., Woods, D., Seda, A.K., Murphy, N. (eds.) The Complexity of Simple Programs, pp. 31–55 (2008)
13. Davydov, A.S.: Solitons and energy transfer along protein molecules. J. Theor. Biol. 66(2), 379–387 (1977)
14. Davydov, A.S.: Solitons in Molecular Systems. Springer, Heidelberg (1990)
15. Fredkin, E., Toffoli, T.: Design principles for achieving high-performance submicron digital technologies. In: Adamatzky, A. (ed.) Collision-Based Computing, pp. 27–46. Springer, London (2002)
16. Grünbaum, B., Shephard, G.C.: Tilings and Patterns. W. H. Freeman, New York (1986)
17. Heeger, A.J., Kivelson, S., Schrieffer, J.R., Su, W.P.: Solitons in conducting polymers. Rev. Mod. Phys. 60(3), 781 (1988)
18. Hey, A.J.G.: Feynman and computation: exploring the limits of computers. Perseus Books, New York (1998)
19. Jakubowski, M.H., Steiglitz, K., Squier, R.: Computing with solitons: a review and prospectus. Multiple-Valued Logic 6(5–6), 439–462 (2001)
20. Kudlek, M., Rogozhin, Y.: New small universal post machine. Lect. Notes Comput. Sci. 2138, 217–227 (2001)
21. Kudlek, M., Rogozhin, Y.: Small universal circular post machine. Comput. Sci. J. Moldova. 9(25), 34–52 (2001)
22. Lindgren, K., Nordahl, M.G.: Universal computation in simple one-dimensional cellular automata. Complex Syst. 4, 229–318 (1990)
23. Lu, Y., Sato, Y., Amari, S.: Traveling bumps and their collisions in a two-dimensional neural field. Neural Comput. 23(5), 1248–1260 (2011)
24. Margolus, N.H.: Physics-like models of computation. Physica D. 10(1–2), 81–95 (1984)

25. Margolus, N.H.: Crystalline computation, In: Hey, A.J.G. (ed.) Feynman and computation: exploring the limits of computers, pp. 267–305. Perseus Books, New York (1998)
26. Margolus, N.H.: Universal cellular automata based on the collisions of soft spheres. In: Adamatzky, A. (ed.) Collision-Based Computing, pp. 107–134. Springer, London (2002)
27. Martínez, G.J., Adamatzky, A., Chen, F., Chua, L.: On soliton collisions between localizations in complex elementary cellular automata: rules 54 and 110 and beyond. Complex Syst. **21**(2), 117–142 (2012)
28. Martínez, G.J., Adamatzky, A., McIntosh, H.V.: Computing on rings. In: Zenil, H. (ed.) A Computable Universe: Understanding and Exploring Nature as Computation, pp. 283–302. World Scientific, Singapore (2012)
29. Martínez, G.J., Adamatzky, A., McIntosh, H.V.: Computing with virtual cellular automata collider. In: IEEE Proceedings of Science and Information Conference, pp. 62–68. London (2015). doi:10.1109/SAI.2015.7237127
30. Martínez, G.J., Adamatzky, A., Stephens, C.R., Hoeflich, A.: Cellular automaton supercolliders. Int. J. Mod. Phys. C. **22**(4), 419–439 (2011)
31. McIntosh, H.V.: Linear Cellular Automata Via de Bruijn diagrams, http://delta.cs.cinvestav.mx/~mcintosh/cellularautomata/Papers_files/debruijn.pdf. Cited 10 August 1991
32. McIntosh, H.V.: Rule 110 as it Relates to the Presence of Gliders, http://delta.cs.cinvestav.mx/~mcintosh/comun/RULE110W/RULE110.html. Cited 14 May 2001
33. McIntosh, H.V.: A Concordance for Rule 110, http://delta.cs.cinvestav.mx/~mcintosh/cellularautomata/Papers_files/ccord.pdf. Cited 14 May 2002
34. McIntosh, H.V.: One Dimensional Cellular Automata. Luniver Press, Bristol (2009)
35. Martínez, G.J., McIntosh, H.V.: ATLAS: Collisions of Gliders like Phases of Ether in Rule 110, http://uncomp.uwe.ac.uk/genaro/Papers/Papers_on_CA_files/ATLAS/bookcollisions.html. Cited 14 August 2001
36. Martínez, G.J., McIntosh, H.V., Seck, J.C.S.T.: Gliders in Rule 110. Int. J. Unconv. Comput. **2**(1), 1–49 (2006)
37. Martínez, G.J., McIntosh, H.V., Mora, J.C.S.T., Vergara, S.V.C.: Determining a regular language by glider-based structures called phases fi_1 in Rule 110. J. Cell. Automata **3**(3), 231–270 (2008)
38. Martínez, G.J., McIntosh, H.V., Mora, J.C.S.T., Vergara, S.V.C.: Reproducing the cyclic tag system developed by Matthew Cook with Rule 110 using the phases f_1_1. J. Cell. Automata **6**(2–3), 121–161 (2011)
39. Martínez, G.J., McIntosh, H.V., Mora, J.C.S.T., Vergara, S.V.C.: Rule 110 objects and other collision-based constructions. J. Cell. Automata **2**(3), 219–242 (2007)
40. Martínez, G.J., Seck-Tuoh-Mora, J.C., Zenil, H.: Computation and Universality: Class IV versus Class III Cellular Automata. J. Cell. Automata **7**(5–6), 393–430 (2013)
41. Mills, J.W.: The nature of the extended analog computer. Physica D. **237**, 1235–1256 (2008)
42. Minsky, M.: Computation: Finite and Infinite Machines. Prentice Hall, Upper Saddle River (1967)
43. Morita, K.: Simple universal one-dimensional reversible cellular automata. J. Cell. Automata **2**, 159–166 (2007)
44. Morita, K.: Simulating reversible Turing machines and cyclic tag systems by one-dimensional reversible cellular automata. Theor. Comput. Sci. **412**, 3856–3865 (2011)
45. Margolus, N., Toffoli, T., Vichniac, G.: Cellular-automata supercomputers for fluid dynamics modeling. Phys. Rev. Lett. **56**(16), 1694–1696 (1986)
46. Neary, T., Woods, D.: P-completeness of cellular automaton Rule 110. Lect. Notes Comput. Sci. **4051**, 132–143 (2006)
47. Ninagawa, S., Martínez, G.J.: Compression-based analysis of cyclic tag system emulated by Rule 110. J. Cell. Automata **9**(1), 23–35 (2014)
48. Siccardi, S., Adamatzky, A.: Actin quantum automata: communication and computation in molecular networks. Nano Commun. Netw. **6**(1), 15–27 (2015)
49. Siccardi, S., Tuszynski, J. A., Adamatzky, A.: Boolean gates on actin filaments. Phys. Lett. A (2015)

50. Scott, A.C.: Dynamics of Davydov solitons. Phys. Rev. A **26**(1), 578 (1982)
51. Smith III, A.R.: Simple computation-universal cellular spaces. J. Assoc. Comput. Mach. **18**, 339–353 (1971)
52. Toffoli, T.: Non-conventional computers. In: Webster, J. (ed.) Encyclopedia of Electrical and Electronics Engineering, vol. 14, pp. 455–471. Wiley, New York (1998)
53. Toffoli, T.: Symbol super colliders. In: Adamatzky, A. (ed.) Collision-Based Computing, pp. 1–23. Springer, London (2002)
54. von Neumann, J.: Theory of Self-reproducing Automata (edited and completed by A.W. Burks), University of Illinois Press, Urbana and London (1966)
55. Voorhees, B.H.: Computational analysis of one-dimensional cellular automata. In: World Scientific Series on Nonlinear Science, Series A, vol. 15. World Scientific, Singapore (1996)
56. Wolfram, S.: Cellular automata supercomputing. In: Wilhelmson, R.B. (ed.) High Speed Computing: Scientific Applications and Algorithm Design. pp. 40–48. University of Illinois Press, Champaign (1988)
57. Wolfram, S.: Cellular Automata and Complexity. Addison-Wesley Publishing Company, Colorado (1994)
58. Wolfram, S.: A New Kind of Science. Wolfram Media Inc, Champaign (2002)
59. Wuensche, A.: Exploring Discrete Dynamics. Luniver Press, Bristol (2011)
60. Zenil, H. (ed.): A Computable Universe. World Scientific Press, Singapore (2012)

Chapter 16
Quantum Queries Associated with Equi-partitioning of States and Multipartite Relational Encoding Across Space-Time

Karl Svozil

Abstract In the first part of this paper we analyze possible quantum computational capacities due to quantum queries associated with equi-partitions of pure orthogonal states. Special emphasis is given to the parity of product states and to functional parity. The second part is dedicated to a critical review of the relational encoding of multipartite states across (space-like separated) space-time regions; a property often referred to as "quantum nonlocality".

16.1 Unconventional Properties for Unconventional Computing

At the heart of any unconventional form of computation (information processing) appears to be some (subjectively and "means relative" to the current canon of knowledge) strange, mind boggling, unexpected, stunning, surprising, hard to believe, feature or capacity of Nature. That is, in order to search for potentially unconventional information processing, we have to parse for empirical patterns and behaviour as well as for theoretical predictions which, relative to our expectations, go beyond our everyday "classical" experience of the world. "Unconventional" always is "means relative" and has to be seen in a historic context; that is, relative to our present means and capacities which we consider consolidated and conventional.

For the sake of some example, take the transmission of data from one point to another via satellite links or cables; or take (gps) navigation by time synchronization; or take the prediction of all sorts of phenomena, including weather or astronomical events. All these capacities appear conventional today, but would have been unconventional, or even magical, only 200 years ago.

K. Svozil (✉)
Institute for Theoretical Physics, Vienna University of Technology,
Wiedner Hauptstraße 8-10/136, 1040 Vienna, Austria
e-mail: svozil@tuwien.ac.at

© Springer International Publishing Switzerland 2017   429
A. Adamatzky (ed.), *Advances in Unconventional Computing*,
Emergence, Complexity and Computation 22,
DOI 10.1007/978-3-319-33924-5_16

So what are the new frontiers? In what follows I shall mainly concentrate on some quantum physical capacities which are widely considered as potential resources for presently "unconventional" computation.

Before we begin the discussion, a *caveat* is in order. First, we should not get trapped by inappropriate yet convenient formal assumptions which have no operational consequences. For instance, all kinds of "infinity processes" have no direct empirical correspondence. In particular, classical continua abound in physics, but they need to be perceived rather as convenient though metaphysical "completion" of processes and entities which are limited by finite physical means.

We should also not get trapped by what Jaynes called *Mind Projection Fallacy* [1, 2], pointing out that *"we are all under an ego-driven temptation to project our private thoughts out onto the real world, by supposing that the creations of one's own imagination are real properties of Nature, or that one's own ignorance signifies some kind of indecision on the part of Nature."* Instead we should attempt to maintain a curiosity with *evenly-suspended attention* outlined by Freud [3] against *"temptations to project, what [the analyst] in dull self-perception recognizes as the peculiarities of his own personality, as generally valid theory into science."*

The postulate of "true," that is, ontological, randomness in Nature is such a fallacy, in both ways mentioned in the *caveat:* it assumes infinite physical resources (or maybe rather *ex nihilo* creation), as well as our capacity to somehow being able to "prove" this—a route blocked by recursion theory; in particular, by reduction to the Halting problem.

16.2   Quantum Speedups by Equi-Decomposition of Sets of Orthogonal States

One of the mind boggling features of quantum information is that, unlike classical information, it can "reside," or be encoded into, the *relational properties* of multiple quanta [4, 5]. For instance, the singlet Bell state of, say, two electrons is defined by the following property (actually, two orthogonal spatial directions would suffice): if one measures the spin properties of these particles along some arbitrary spatial direction, then the spin value observed on one particle turns out to be always the negative spin value observed on the other particle—their relative spin value is negative—that is, either of "+" sign for the first particle and of "−" sign for the second one; or *vice versa*. The "rub," or rather compensation, for this fascinating encoding of information "across" particles appears to be that none of these individual particles has a definite individual spin value before the (joint) measurement. That is, all information encodable into them is exhausted by these relational specifications. This well recognized capacity of quantum mechanics could be conceived as the "essence of entanglement" [6].

Besides entanglement there is another capacity which is not directly related to relational properties of multipartite systems; yet it shares some similarities with the latter: the possibility to organize elementary, that is, binary (or, in general, d-ary) quantum queries resolving properties which can be encoded into (equi-)partitions of some set of pure states. If such partitions are feasible, then it is possible to obtain one bit (or, in general, dit) of information by staging such a single query *without* knowledge in what particular state the quantized system is.

From a different perspective any such (binary or d-ary) observable is related to a *partial* (i.e. incomplete) state identification [7–9]. Many of the fast quantum algorithms discussed in the literature depend on incomplete state identification.

Note that, in the binary case, any complete state identification—that is, setting up a complete set of quantum observables or queries capable to discriminate between and "locating" all single states—could be seen as the dual (observable) side of what can be considered an arbitrary state preparation for multipartite systems. This latter state preparation also features entanglement by allowing appropriate relational properties among the constituent quanta.

16.2.1 Parity of Two-Partite Binary States

For the sake of a demonstration of the "unconventional" quantum speedup achievable through partial (incomplete) state identification, consider the four two-partite binary basis states $|00\rangle$, $|01\rangle$, $|10\rangle$, and $|11\rangle$. Suppose we are interested in the even parity of these states. Then we could construct a *even parity operator* $\mathbf{P}$ *via* a spectral decomposition; that is,

$$\mathbf{P} = 1 \cdot \mathbf{P}_- + 0 \cdot \mathbf{P}_+, \text{ with}$$
$$\mathbf{P}_- = |01\rangle\langle 01| + |10\rangle\langle 10|, \tag{16.1}$$
$$\mathbf{P}_+ = |00\rangle\langle 00| + |11\rangle\langle 11|,$$

which yields even parity "0" on $|00\rangle$ as well as $|11\rangle$, and even parity "1" on $|01\rangle$ as well as $|10\rangle$, respectively. Note that $\mathbf{P}_-$ as well as $\mathbf{P}_+$ are projection operators, since they are idempotent; that is, $\mathbf{P}_-^2 = \mathbf{P}_-$ and $\mathbf{P}_+^2 = \mathbf{P}_+$.

Thereby, the basis of the two-partite binary states has been effectively equi-partitioned into two groups of even parity "0" and "1;" that is,

$$\{\{|00\rangle, |11\rangle\}, \{|01\rangle, |10\rangle\}\}. \tag{16.2}$$

The states associated with the propositions corresponding to the projection operators $\mathbf{P}_-$ for even parity one and $\mathbf{P}_+$ for even parity zero of the two bits are entangled; that is, this information is only expressed in terms of a *relational property*—in this case parity—of the two quanta between each other [4, 5].

16.2.2 Parity of Multi-partite Binary States

This equi-partitioning strategy [7, 10] to determine parity with a single query can be generalized to determine the parity of multi-partite binary states. Take, for example, the even parity of three-partite binary states definable by

$$\mathbf{P} = 1 \cdot \mathbf{P}_- + 0 \cdot \mathbf{P}_+, \text{ with}$$

$$\mathbf{P}_- = |001\rangle\langle001| + |010\rangle\langle010| + |100\rangle\langle100| + |111\rangle\langle111|, \tag{16.3}$$

$$\mathbf{P}_+ = |000\rangle\langle000| + |011\rangle\langle011| + |101\rangle\langle101| + |110\rangle\langle110|.$$

Again, the states associated with the propositions corresponding to the projection operators $\mathbf{P}_-$ for even parity one and $\mathbf{P}_+$ for even parity zero of the three bits are entangled. The basis of the three-partite binary states has been equi-partitioned into two groups of even parity "0" and "1;" that is,

$$\{\{|000\rangle, |011\rangle, |101\rangle, |110\rangle\},$$
$$\{|001\rangle, |010\rangle, |100\rangle, |111\rangle\}\}. \tag{16.4}$$

16.3 Parity of Boolean Functions

It is well known that Deutsch's problem—to find out whether the output of a binary function of one bit is constant or not; that is, whether the two outputs have even parity zero or one—can be solved with one quantum query [11, 12]. Therefore it might not appear totally unreasonable to speculate that the parity of some Boolean function—a binary function of an arbitrary number of bits—can be determined by a single quantum query. Even though we know that the answer is negative [13] it is interesting to analyze the reason why this parity problem is "difficult" even for quantum resources, in particular, quantum parallelism. Because an answer to this question might provide us with insights about the (in)capacities of quantum computations in general.

Suppose we define the functional parity $P(f_i)$ of an n-ary function $f_i = (g_i + 1)/2$ *via* a function $g_i(x_1, \ldots, x_n) \in \{-1, +1\}$ and

$$P(g_i) = \prod_{x_1, \ldots, x_n \in \{0,1\}} g_i(x_1, \ldots, x_n). \tag{16.5}$$

Let us, for the sake of a direct approach of functional parity, consider all the 2^{2^n} Boolean functions $f_i(x_1, \ldots, x_n)$, $0 \le i \le 2^{2^n} - 1$ of n bits, and suppose that we can represent them by the standard quantum oracle

$$U_i(|x_1, \ldots, x_n\rangle|y\rangle)$$
$$= |x_1, \ldots, x_n\rangle|y \oplus f_i(x_1, \ldots, x_n)\rangle \tag{16.6}$$

as a means to cope with possible irreversibilities of the functions f_i. Because $f_i \oplus f_i = 0$, we obtain $U_i^2 = \mathbf{I}$ and thus reversibility of the quantum oracle. Note that all the resulting $n + 1$-dimensional vectors are not necessarily mutually orthogonal.

For each particular $0 \leq i \leq 2^{2^n} - 1$, we can consider the set

$$F_i = \{f_i(0, \ldots, 0), \ldots, f_i(1, \ldots, 1)\} \tag{16.7}$$

of all the values of f_i as a function of all the 2^n arguments. The set

$$V = \{F_i \mid 0 \leq i \leq 2^{2^n} - 1\}$$
$$= \{\{f_i(0, \ldots, 0), \ldots, f_i(1, \ldots, 1)\} \mid 0 \leq i \leq 2^{2^n} - 1\} \tag{16.8}$$

is formed by all the $2^{2^n + n}$ Boolean functional values $f_i(x_1, \ldots, x_n)$. Moreover, for every one of the 2^{2^n} different Boolean functions of n bits the 2^n functional output values characterize the behavior of this function completely.

In the next step, suppose we equi-partition the set of all these functions into two groups: those with even parity "0" and "1," respectively. The question now is this: can we somehow construct or find two mutually orthogonal subspaces (orthogonal projection operators) such that all the parity "0" functions are represented in one subspace, and all the parity "1" are in the other, orthogonal one? Because if this would be the case, then the corresponding (equi-)partition of basis vectors spanning those two subspaces could be coded into a quantum query [7] yielding the parity of f_i in a single step.

We conjecture that involvement of one or more additional auxiliary bits (e.g., to restore reversibility for nonreversible f_i's) cannot improve the situation, as any uniform (over all the functions f_i) and non-adaptive procedure will not be able to generate proper orthogonality relations.

We know that for $n = 1$ this task is feasible, since (we re-coded the functional value "0" to "-1")

f_i	$P(f_i)$	$f_i(0)$	$f_i(1)$
f_0	0	-1	-1
f_1	0	$+1$	$+1$
f_2	1	-1	$+1$
f_3	1	$+1$	-1

$$(16.9)$$

and the two parity cases "0" and "1," are coded into orthogonal subspaces spanned by $(1, 1)$ and $(-1, 1)$, respectively.

This is no longer true for $n = 2$; due to an overabundance of functions, the vectors corresponding to both parity cases "0" and "1" span the entire Hilbert space:

f_i	$P(f_i)$	$f_i(00)$	$f_i(01)$	$f_i(10)$	$f_i(11)$
f_0	0	-1	-1	-1	-1
f_1	0	-1	-1	$+1$	$+1$
f_2	0	-1	$+1$	-1	$+1$
f_3	0	-1	$+1$	$+1$	-1
f_4	0	$+1$	-1	-1	$+1$
f_5	0	$+1$	-1	$+1$	-1
f_6	0	$+1$	$+1$	-1	-1
f_7	0	$+1$	$+1$	$+1$	$+1$
f_8	1	-1	-1	-1	$+1$
f_9	1	-1	-1	$+1$	-1
f_{10}	1	-1	$+1$	-1	-1
f_{11}	1	-1	$+1$	$+1$	$+1$
f_{12}	1	$+1$	-1	-1	-1
f_{13}	1	$+1$	-1	$+1$	$+1$
f_{14}	1	$+1$	$+1$	-1	$+1$
f_{15}	1	$+1$	$+1$	$+1$	-1

$$(16.10)$$

16.3.1 Proper Specification of State Discrimination

The results of this section are also relevant for making precise Zeilinger's *foundational principle* [4, 5] claiming that an n-partite system can be specified by exactly n bits (dits in general). The issue is what exactly is a "specification?"

We propose to consider a specification appropriate if it can yield to an equi-partitioning of all pure states of the respective quantized system. That is, to give an example, the parity of states could serve as a proper specification, but functional parity in general (for more that two quanta) cannot.

16.4 Relativity Theory *Versus* Quantum Inseparability

Let us turn our attention to another "unconventional" quantum resource, which is mostly encountered at (but not restricted to) spatially separated entangled states: the so-called "quantum nonlocality;" and, in particular, on the paradigm shift of our perception of physical space and time.

First, let us keep in mind that, in the historic perspective it is quite evident why our current theory of space-time, relativity theory [14, 15], does not directly refer to quanta: it was created when quantum mechanics was "unborn," or at least in its early infancy. Indeed, in 1905 it was hardly foreseeable that Planck's self-denominated [16, p. 31] *"Akt der Verzweiflung"* ("act of desperation")—committed five years ago in 1900 for the sake of theoretically accommodating precision measurements of the blackbody radiation—would be extended into one of the most powerful physical theories imagined so far. Therefore it should come as no surprise that all operationalizations and conventions implemented by relativity theory, in particular, simultaneity, refer to classical, pre-quantum, physics.

Besides its applicability and stunning predictions and consequences (such as, for instance "$E = mc^2$," as well as the unification of classical electric and magnetic phenomena) the triumph of special relativity resides in its structural as well as formal clarity: by adopting certain conventions (which were essentially adopted from railroad traffic [17, 18] and are also used by *Cristian's Algorithm* for data network synchronization), and by fixing the speed of electromagnetic radiation for all reference frames (together with the requirement of bijectivity), the Lorentz transformations result from theorems of incidence geometry [19, 20]. Beyond formal conventions, the physical content resides in the form invariance of the equations of motion under such transformations.

In view of these sweeping successes of classical relativity theory it might not be surprising that Einstein, one of the creators of quantum mechanics, never seriously considered the necessity to adapt the concepts of space-time to the new quantum physics. On the contrary—Einstein seemed to have prioritized relativity over quantum theory; the latter one he critically referred to as [21, p. 113] *"noch nicht der wahre Jakob" ("not yet the true [final] answer")*. Time and again Einstein came up with predictions of quantum mechanics which allegedly discredited the (final) validity of quantum theory.

In a letter to Schrödinger dated June 19th, 1935 [22, 23] Einstein concretized and clarified his uneasiness with quantum theory previously published in a paper with Podolsky and Rosen [24] (*"written by Podolsky after many discussions"* [22]). In this communication Einstein insisted that the wave function of a subsystem A of (entangled) particles cannot depend on whatever measurements are performed on its spatially separated (i.e. separated by a space-like interval) "twin" subsystem B: in his own (translated from German[1]) words: *"The true state of B cannot depend on which measurement I perform on A."* Pointedly stated, the *"separability principle"* asserts *that any two spatially separated systems possess their own separate real state* [23].

The separability principle is not satisfied for entangled states [25, 26]; in particular, if general two-partite state

$$|\Psi\rangle = \sum_{i,j\in\{-,+\}} \alpha_{ij}|ij\rangle, \text{ with } \sum_{i,j\in\{-,+\}} |\alpha_{ij}|^2 = 1$$

does not satisfy factorizability [12, p. 18] requiring $\alpha_{--}\alpha_{++} = \alpha_{+-}\alpha_{-+}$. That is, if $\alpha_{--}\alpha_{++} \neq \alpha_{+-}\alpha_{-+}$, then $|\Psi\rangle$ cannot be factored into products of single particle states.

Even in his later years Einstein was inclined to take relativistic space-time as the primary framework; thereby prioritizing it over fundamental quantum mechanical inseparability; in particular, when it comes to multipartite situations [23, 27].

---

[1]Einstein's (underlined) original German text: *"Der wirkliche Zustand von B kann nicht davon abhängen, was für eine Messung ich an A vornehme"*.

16.5 Proximity and Apartness in Quantum Mechanics

In what follows we propose that, when it comes to microphysical situations, in particular, when entanglement is involved, the provenance of classical relativity theory over quantum mechanics has to be turned upside down: while entangled quanta may epistemically (and for many practical purposes [28]) appear "separated," or "apart," or "distinct" to a classical observer ignorant of their relational properties (cf. earlier discussion in Sect. 16.2) encoded "across these quanta," quantum mechanically they are treated holistically "as one."

The pretension of any such observer, or the possibility to actually perceive entangled quanta as being "spatially separated" (by disregarding their correlations) should not be seen as a principal property, but rather as a *"means relative"* one.

For the sake of an example, take the two-particle singlet Bell state $|\Psi^-\rangle = (1/\sqrt{2})(|+-\rangle - |-+\rangle)$, which, by identifying $|-\rangle \equiv (0, 1)$ and $|+\rangle \equiv (1, 0)$, can be identified with the four-dimensional vector whose components in tuple form are $|\Psi^-\rangle \equiv (1/\sqrt{2})[(1, 0) \otimes (0, 1) - (0, 1) \otimes (1, 0)] = (1/\sqrt{2})(0, 1, -1, 0)$. The separability principle is not satisfied, since $0 \cdot 0 \neq 1 \cdot (-1)$. So, from the point of view of those entangled state observables, the quanta appear inseparable.

And yet, the same quanta can be perfectly localized and distinguished by resolving them spatially. This situation—the occurrence of both inseparability and (spatial) distinguishability – has caused a lot of confusion. This is particularly serious if one of these distinct viewpoints on the quantized system, say, spatial separability and locatedness of the particles, is meshed with the inseparability of the spin observables when the latter ones are relationally defined. An yet, we might envision that, with this dual situation we could get a handle on quantum inseparability (*via* encoding of relational information) by spatially separated detectability of the quanta forming this entangled state. Alas this is impossible, because the relational properties do not reveal themselves by individual outcomes—only when all these (relational) outcomes are considered together do the relational properties reveal themselves.

Of course, this would be totally different if it would be possible to wilfully *force* any particular handle or side or component of the entangled state, thereby effectively forcing the respective (relational property on the other handle or side or component. So far, despite speculative attempts to utilize stimulated emission [29], there is no indication that this might be physically feasible.

16.6 Summary

The first part of this chapter has been dedicated to quantum queries relating to properties which can be encoded in terms of (equi-)partitioning of states. We have been particularly interested in the parity of products of binary states, and also in the parity of Boolean functions; that is, dichotomic functions of bits. Thereby we have presented criteria for the (non-) existence of quantum oracles.

In the second part of this chapter we have argued that, instead of perceiving entangled quanta in an a priori "space-time theater," space-time is a secondary, derived concept of our mind which needs to be operationally constructed by conventions and observations. This is particularly true for multipartite entangled states, and their spatio-temporal interconnectedness. Such an approach leaves no room for any hypothetical inconsistency in quantum space-time, and no mind-boggling "peaceful coexistence" with relativity theory.

Acknowledgments This research has been partly supported by FP7-PEOPLE-2010-IRSES-269151-RANPHYS.

References

1. Jaynes, E.T.: Clearing up mysteries—the original goal. In: Skilling, J. (ed.) Maximum-Entropy and Bayesian Methods: Proceedings of the 8th Maximum Entropy Workshop, held on August 1–5, 1988, in St. John's College, Cambridge, England, pp. 1–28. Kluwer, Dordrecht (1989)
2. Jaynes, E.T.: Probability in quantum theory. In: Zurek, W.H. (ed.) Complexity, Entropy, and the Physics of Information: Proceedings of the 1988 Workshop on Complexity, Entropy, and the Physics of Information, held May - June, 1989, in Santa Fe, New Mexico, pp. 381–404. Addison-Wesley, Reading, MA, (1990)
3. Freud, S.: Ratschläge für den Arzt bei der psychoanalytischen Behandlung. In: Anna Freud, E. Bibring, W. Hoffer, E. Kris, and O. Isakower, editors, *Gesammelte Werke. Chronologisch geordnet. Achter Band. Werke aus den Jahren 1909–1913*, pages 376–387, Frankfurt am Main, 1999. Fischer
4. Zeilinger, A.: Quantum teleportation and the non-locality of information. Philos. Trans. R. Soc. Lond. A **355**, 2401–2404 (1997)
5. Zeilinger, A.: A foundational principle for quantum mechanics. Found. Phys. **29**(4), 631–643 (1999)
6. Brukner, Č., Zukowski, M., Zeilinger, A.: The essence of entanglement. Translated to Chinese by Qiang Zhang and Yond-de Zhang, New Advances in Physics (Journal of the Chinese Physical Society) (2002)
7. Niko, D., Svozil, K.: Finding a state among a complete set of orthogonal ones. Phys. Rev. A **65**, 044302 (2002)
8. Donath, N., Svozil, K.: Finding a state among a complete set of orthogonal ones. Virtual J. Quantum Inf. **2** (2002)
9. Svozil, K., Tkadlec, J.: On the solution of trivalent decision problems by quantum state identification. Nat. Comput., in print (2009)
10. Svozil, K.: Quantum information in base n defined by state partitions. Phys. Rev. A **66**, 044306 (2002)
11. Nielsen, M.A., Chuang, I.L.: Quantum Comput. Quantum Inf. Cambridge University Press, Cambridge (2000)
12. David, N.: Mermin Quantum Computer Science. Cambridge University Press, Cambridge (2007)
13. Farhi, E., Goldstone, J., Gutmann, S., Sipser, M.: Limit on the speed of quantum computation in determining parity. Phys. Rev. Lett. **81**, 5442–5444 (1998)
14. Poincaré, Henri: La Science et l'hypothése. Flammarion, Paris (1902)
15. Einstein, Albert: Zur Elektrodynamik bewegter Körper. Annalen der Physik **322**, 891–921 (1905)
16. Hermann, A.: *Frühgeschichte der Quantentheorie (1899-1913)*. Physik Verlag, Mosbach in Baden, 1969. Habilitationsschrift, Naturwissenschaftliche Fakultät der Universität München

17. Galison, P.L.: Einstein's clocks: the place of time. Crit. Inquiry **26**(2), 355–389 (2000)
18. Galison, P.L.: Einstein's clocks, Poincaré's maps: Empires of Time. W.W. Norton & Company, New York (2003)
19. Lester, J.A.: Distance preserving transformations. In: Buekenhout, F. (ed.) Handbook of Incidence Geometry, pp. 921–944. Elsevier, Amsterdam (1995)
20. Naber, G.L.: The Geometry of Minkowski Spacetime. In: Applied Mathematical Sciences. ANU Quantum Optics, 2nd edn, vol. 92. New York (2012)
21. Born, M.: Physics in My Generation, 2nd edn. Springer, New York (1969)
22. Einstein, A.: Letter to Schrödinger. Old Lyme, dated 19.6.35, Einstein Archives 22–47 (searchable by document nr. 22-47) (1935)
23. Howard, D.: Einstein on locality and separability. Stud. Hist. Philos. Sci. Part A **16**(3), 171–201 (1985)
24. Einstein, A., Podolsky, B., Nathan, R.: Can quantum-mechanical description of physical reality be considered complete? Phys. Rev. **47**(10), 777–780 (1935)
25. Schrödinger, E.: Discussion of probability relations between separated systems. Math. Proc. Camb. Philos. Soc. **31**(04), 555–563 (1935)
26. Schrödinger, E.: Die gegenwärtige Situation in der Quantenmechanik. *Naturwissenschaften*, **23**, 807–812, 823–828, 844–849, 1935
27. Einstein, A.: Quanten-Mechanik und Wirklichkeit. Dialectica **2**(3–4), 320324 (1948)
28. Bell, J.S.: Against 'measurement'. Phys. World **3**, 33–41 (1990)
29. Svozil, K.: What is wrong with SLASH? eprint arXiv:quant-ph/0103166 (1989)

Chapter 17
Solving the Broadcast Time Problem Using a D-wave Quantum Computer

Cristian S. Calude and Michael J. Dinneen

Abstract We illustrate how the D-Wave Two quantum computer is programmed and works by solving the Broadcast Time Problem. We start from a concise integer program formulation of the problem and apply some simple transformations to arrive at the QUBO form which can be run on the D-Wave quantum computer. Finally, we explore the feasibility of this method on several well-known graphs.

17.1 Introduction

D-Wave machines are "the world's first commercially available quantum computers." D-Wave Two is an adiabatic machine which operates on 512 qubits. Various teams based in academia and major companies like Lockheed Martin, NASA, Google, are exploring the computational capability of this machine as well as its potential applications.

Programming D-Wave is radically different from programming classical machines, like one's traditional home computer. In this paper we use the Broadcast Time Problem, an NP-complete problem, to illustrate how to develop programs for the D-Wave Two machine and how it operates. We start from a concise integer program formulation of our optimization problem and, through several intermediate phases, we convert it to the QUBO form, the formulation which can be run on the D-Wave computer. Finally we explore the feasibility of this method on several well-known graphs.

C.S. Calude (✉) · M.J. Dinneen
Department of Computer Science, University of Auckland, Auckland, New Zealand
e-mail: cristian@cs.auckland.ac.nz

M.J. Dinneen
e-mail: mjd@cs.auckland.ac.nz

© Springer International Publishing Switzerland 2017
A. Adamatzky (ed.), *Advances in Unconventional Computing*,
Emergence, Complexity and Computation 22,
DOI 10.1007/978-3-319-33924-5_17

17.2 Adiabatic Computing

The adiabatic model of quantum computing uses the propensity of physical systems—classical or quantum—to minimize their free energy. Quantum annealing is free energy minimization in a quantum system.

Its mathematical precursors are the Monte Carlo methods [1]—in which a problem is solved by generating repeated random samplings from a probability distribution, performing simple deterministic computations and aggregating the results—and the Metropolis-Hastings algorithm—a Markov chain Monte Carlo method useful when direct sampling is difficult [2]. In 1983 an analogy between minimizing the cost function of a combinatorial optimization problem—solved efficiently with the Metropolis-Hastings algorithm—and the slow cooling of a solid until it reaches its low energy ground state was discovered in [3]. The proposed method—called *simulated annealing* [3, 4]—is very simple: substitute the cost for energy and run the Metropolis-Hastings algorithm in a sequence of slowly decreasing temperature values, which makes the system progress through various energy states till, hopefully, it finds a global optimal answer.

An *adiabatic quantum computation* (AQC) is an algorithm that computes an exact or approximate solution of an optimization problem encoded in the ground state—its lowest-energy state—of a Hamiltonian (the operator corresponding to the total energy of the system). The algorithm starts at an initial state H_I that is easily obtained, then evolves adiabatically, i.e. by slowly changing to the Hamiltonian H_P. An example of evolution is $H = (1 - t)H_I + t H_P$ as the time t increases monotonically from 0 to 1. During the entire computation, the system must stay in a valid ground state. If the system can reach its ground state we get an exact solution; if it can only reach a local minimum, then we get an approximate solution. The slower the evolution process the better the approximate (possibly exact) solution is obtained.

The adiabatic quantum computing model shares the same paradigm as simulated annealing. The main difference is that simulated annealing is based on "thermodynamic energy" and quantum annealing is based on "quantum fluxuations" during a cooling process. One suggested advantage of quantum annealing is the ability to "quantum tunnel" out of some local optimal states,[1] as illustrated in Fig. 17.1.

AQC is based on the Born–Fock adiabatic theorem [5] which accounts for the adiabatic evolution of quantum states when the change in the time-dependent Hamiltonian is sufficiently slow [6]:

A physical system remains in its instantaneous eigenstate if a given perturbation is acting on it slowly enough and if there is a gap between the eigenvalue and the rest of the Hamiltonian's spectrum.

The quantum adiabatic computation model and the gate quantum computation model—probably the most studied model of quantum computing—are polynomially time equivalent [7].

---

[1] D-Wave Two is capable of using it.

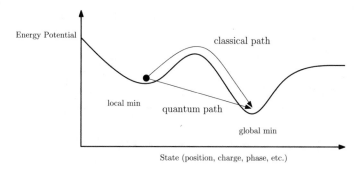

Fig. 17.1 Annealing with quantum tunneling

17.3 D-Wave Computers

The D-Wave computers are produced by the Canadian company D-Wave Systems: D-Wave One (2011) operates on a 128-qubit chipset; D-Wave Two (2013) works with 512 qubits [8]. D-Wave computers use quantum annealing to improve convergence of the system's energy towards the ground state energy of a *Quadratic Unconstrained Binary Optimization* (QUBO) problem. QUBO is an NP-hard mathematical problem consisting in the minimization of a quadratic objective function $f(\mathbf{x}) = \mathbf{x}^T Q \mathbf{x}$, where $\mathbf{x}$ is a n-vector of binary variables and Q is a symmetric $n \times n$ matrix:

$$x^* = \min_{\mathbf{x}} \sum_{i \geq j} x_i Q_{(i,j)} x_j, \text{ where } x_i \in \{0, 1\}.$$

The computer architecture consists of qubits arranged with a host configuration as a subgraph of a *Chimera graph*. A Chimera graph consists of an $M \times N$ two-dimensional lattice of blocks, with each block consisting of $2L$ vertices (a complete bipartite graph $K_{L,L}$), in total $2MNL$ variables. The D-Wave One has $M = N = L = 4$ for a maximum of 128 qubits. D-Wave qubits are loops of superconducting wire, the coupling between qubits is magnetic wiring and the machine itself is supercooled.

To index a qubit we use four numbers (i, j, u, k), where (i, j) indexes the (row, column) of the block, $u \in \{0, 1\}$ is the left/right bipartite half of $K_{L,L}$ and $0 < k < L$ is the offset within the bipartite half. Qubits indexed by (i, j, u, k) and (i', j', u', k') are neighbors if and only if

1. $i = i'$ and $j = j'$ and $[(u, u') = (0, 1)$ or $(u, u') = (1, 0)]$ or
2. $i = i' \pm 1$ and $j = j'$ and $u = u'$ and $u = 0$ and $k = k'$ or
3. $i = i'$ and $j = j' \pm 1$ and $u = u'$ and $u = 1$ and $k = k'$.

Figure 17.2 shows for $L = N = 4$ (and $M > 2$) the structure of an initial part of a Chimera graph where the two half partitions of the bipartite graphs $K_{L,L}$ (blocks) are drawn horizontally and vertically, respectively. The linear index (qubit id of the vertices) from the four tuple (i, j, u, k) is the value $2NLi + 2Lj + Lu + k$.

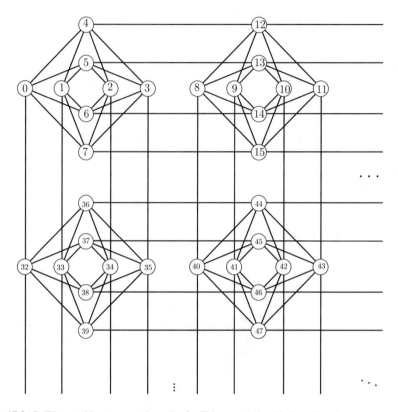

Fig. 17.2 D-Wave architecture: a subgraph of a Chimera graph with $L = N = 4$

17.4 The Broadcast Time Problem

Broadcasting concerns the dissemination of a message originating at one node of a network to all other nodes [9, 10]. This task is accomplished by placing a series of calls over the communication lines of the network between neighboring nodes. Each call requires a unit of time, a call can involve only two nodes and a node can participate in only one call per time step.

A *broadcast tree* for a vertex v (called the originator) of an undirected graph $G = (V, E)$ is an implicit rooted tree based on a sequence $V_0 = \{v\}, E_1, V_1, E_2, \ldots, E_t,$ $V_t = V$ (*of broadcast height t*) such that each $V_i \subseteq V$, each E_i is an oriented subset of E, and for every $1 \leq i \leq t$: (1) each arc (u, w) in E_i has only the source u endpoint in V_{i-1}, (2) no two arcs in E_i share a common endpoint, and (3) $V_i = V_{i-1} \cup \{w \mid (u, w) \in E_i\}$.

The *Broadcast Time Problem* is the following: *Given a connected graph $G = (V, E)$, originator $v \in V$ and integer t, is there a broadcast tree T_v rooted at v with the height of T_v at most t?* This is a well-known NP-complete problem (see [ND49] of [11]), even for graphs of maximum vertex degree 3 (see [12]). The optimization

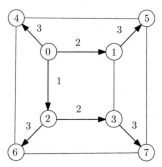

Fig. 17.3 The graph Q_3 with broadcast time 3

version of this problem is approximable within $O(\log^2 |V| / \log\log |V|)$, but is not expected to have a polynomial-time approximation scheme [13].

In the example shown in Fig. 17.3 (hypercube Q_3) an optimal broadcast tree is illustrated for the originator vertex 0.

The development of a quantum solution will be presented in a sequence of four phases, which are described in Sects. 17.5–17.8. We summarize this general solution approach for Q_3 in Sect. 17.9 before giving a detailed implementation of our broadcast time solution for K_2 in Sect. 17.10. Finally, we present some general experimental results for several other small common graphs (Sect. 17.11) and concluding comments (Sect. 17.12).

17.5 Integer Programming Formulation

In the first phase we present a simple formulation (i.e. polynomial-time reduction) of the Broadcast Time Problem with the originator fixed[2] at $v = 0$ as an *Integer Programming (IP) Optimization Problem* (see [14]). The input is a connected graph $G = (V = \{0, 1, \ldots, n-1\}, E)$ representing a network with $n = |V|$ vertices and $m = |E|$ edges. For the graph G, we use the following $n + 2m + 1$ variables:

- t is the required time to complete a broadcast,
- v_i is the time in $\{0, 1, \ldots, t\}$ in which the vertex $i \in V$ receives the message, $0 \le i < n$,
- $b_{i,j}$ is a binary variable which is 1 if and only if the vertex i broadcasts to the vertex j (for each $\{i, j\} \in E$).

The objective function for our optimization problem is $\min(t)$, or equivalently, $\max(n - t)$.

Several families of constraints on the variables are now presented. First, the time t must be at most $n - 1$:

---

[2] Solving the problem for other originators can be easily done by relabelling the vertices of the graph or doing obvious modifications in the formulation below.

$$0 \leq t \leq n - 1. \tag{17.1}$$

Every vertex receives the message at a time step at most t:

$$0 \leq v_i \leq t, \text{ for all } i \in V. \tag{17.2}$$

The originator vertex has no parent and every other vertex must have exactly one parent in the broadcast tree:

$$\sum_{j \neq 0} b_{j,0} = 0, \tag{17.3}$$

$$\sum_{j \neq i} b_{j,i} = 1, \text{ for all } i \in V \setminus \{0\}. \tag{17.4}$$

There are no broadcast cycles, that is for a child vertex, the informed time of the parent must be strictly less than its message received time:

$$b_{i,j}(1 + v_i - v_j) \leq 0, \text{ for all } \{i, j\} \in E. \tag{17.5}$$

Finally, every two child vertices $\{j, k\}$ informed by the same parent i must occur at different times:

$$b_{i,j} + b_{i,k} - (v_j - v_k)^2 \leq 1, \text{ for all } \{i, j\} \in E, \{i, k\} \in E \text{ with } j \neq k. \tag{17.6}$$

17.6 Binary Integer Programming Formulation

Next we convert all non-binary variables (in IP formulation) into binary variables. The following simple procedure converts an integer constrained variable $0 \leq x \leq D$ into a set of $O(\log D)$ binary variables $x_0, x_1, \ldots, x_c$ representing its binary representation:

$$x = x_0 + 2x_1 + 4x_2 + \cdots + 2^c x_c = \sum_{i=0}^{c} 2^i x_i,$$

where $x_i \in \{0, 1\}$ and $2^c \leq D < 2^{c+1}$. Each constraint of the form $x \leq D$ is replaced by the following equivalent constraint:

$$\sum_{i=0}^{c} 2^i x_i \leq D. \tag{17.7}$$

17.7 Linear Binary Integer Programming Formulation

Using standard techniques (e.g., see [15]) we convert the above quadratic binary IP formulation into a linear formulation. Each occurrence of a product of two binary variables xy is replaced by a new variable z_{xy} and the following two linear constraints:

$$0 \le x + y - z_{xy} \le 1, \tag{17.8}$$

$$-1 \le 2z_{xy} - (x + y) \le 0, \tag{17.9}$$

enforce $z_{xy} = xy$.

We can reduce the number of "product" binary variables by observing that for Eq. (17.5) we do not need to consider $j = 0$ and for Eq. (17.6) we can only consider vertices $j > 0$ and $k > 0$ with a common neighbor.

Note that the above reduction was automated and the Sage Mixed Integer Program Solver [16] was used to verify correctness of many small graphs [17].

17.8 QUBO Formulation

The first step to converting the current binary linear IP formulation to QUBO is to use a "standard form," where all inequalities are replaced with equalities by introducing slack variables [14].

The next step is to build an equivalent QUBO of the IP formulation and add rules to force all linear equation constraints to be satisfied when assigning 0/1 to the binary variables. Consider a linear equality constraint C_k of the form $\sum_{i=1}^{n} c_{(k,i)} x_i = d_k$ for $x_i \in \{0, 1\}$ with fixed integer constants $c_{(k,i)}$ and d_k. This equation is satisfied if and only if $\sum_{i=1}^{n} c_{(k,i)} x_i - d_k = 0$, or equivalently, if $\langle \mathbf{c}_k, \mathbf{x} \rangle - d_k = 0$, where $\mathbf{c}_k = (c_{(k,i)}, c_{(k,2)}, \ldots, c_{(k,n)})$ and $\langle \mathbf{c}_k, \mathbf{x} \rangle$ is the product of the vectors $\mathbf{c}_k$ and $\mathbf{x}$. If $\langle \mathbf{c}_k, \mathbf{x} \rangle - d_k$ is not zero we need to have a penalty greater than the maximum feasible value of t, which is n. Thus, we can construct the following QUBO that is equivalent to the IP formulation of the Broadcast Time Problem:

$$x^* = \min_{\mathbf{x}} \left(t + n \cdot \sum_{k} (\langle \mathbf{c}_k, \mathbf{x} \rangle - d_k)^2 \right), \quad \text{where } x_i \in \{0, 1\}.$$

Note first that the variable t is obtained from the variables used in Eq. (17.1) of Sect. 17.5 and is added to the other QUBO entries in Q from the set of linear constraints C_k. The QUBO constants for the binary variables representing t will be powers of 2, as given by Eq. (17.7). Second, any term d_k^2 in the square terms of $(\langle \mathbf{c}_k, \mathbf{x} \rangle - d_k)^2$ which does not involve a variable x_i can be ignored since those additive terms are independent of any assignment of variables (i.e. we have a fixed additive QUBO *offset* to the objective solution). Third, since variables x_i are binary, we have $x_i^2 = x_i$ and the constants for those terms are included in the main diagonal entries of Q. Finally, the conversion from an arbitrary Broadcast Time Problem instance to QUBO was automated [17].

17.9 Q_3 Example

In this section we illustrate the quantum solution phases for Q_3. In the first phase (IP formulation) we get 33 integer variables (the variable t, eight variables of the type v_i, and 24 variables of the type $b_{i,j}$) and 65 quadratic constraints. These constraints are shown in the appendix of this chapter.

The conversion to the binary formulation results in 51 ($= 33 + 2 \cdot 9$) binary variables as we need three binary variables for each of the previous integer variables $t, v_0, \ldots, v_7$. The number of constraints stays the same but each gets expanded with more variables. For example, $x_1 \leq x_0$ becomes $-x_0' + x_3' - 2x_1' + 2x_4' - 4x_2' + 4x_5' \leq 0$.

The next conversion (Sect. 17.7) produces 447 binary variables and 851 linear constraints. Finally, the conversion to QUBO generated 999 slack variables, so in total 1446 binary variables: they represent the number of logical qubits for our QUBO formulation. Full details for our phases 2 through 4 may be found in [17].

To be able to solve this QUBO problem on D-Wave we need one more step to encode the theoretical problem on physical hardware (see [18]) which will be illustrated in the next section with a feasible example for the D-Wave Two.

17.10 K_2 Example

We present both the final IP formulation (see Table 17.1) and QUBO matrix Q (see Table 17.2) for the Broadcast Time Problem for the graph K_2 of one edge. The total number of binary variables is 22 (13 of them are slack variables) and the QUBO offset is 12. When run on the D-Wave simulator [18] (without embedding onto the hardware, which has limited qubit connections) we get this expected result:

```
answer={ 'energies': [-11.0],
         'solutions': [[1,0,0,1,1,0,0,1,0,0,1,1,0,0,0,
                        0,1,0,1,0,0,0]] }
```

When we add the offset 12 to the minimum energy state we get our expected broadcast time of 1. We can also see that $t = x_1 = 1$, $b_{0,1} = x_7 = 1$ and $b_{1,0} = x_8 = 0$, which indicates a valid broadcast tree from the obtained solution x^*.

To actually run this QUBO instance on the D-Wave machine we need to find a minor-containment embedding on the actual physical qubit hardware (the Chimera graph). One valid heuristic is to map each logical qubit to a path of physical qubits.

Table 17.1 Final Binary Integer Program for broadcasting in K_2

Integer program constraints	Comments
$x_0 + x_1 = 1$	x_0 is objective variable t and x_1 is a slack variable
$-x_0 + x_2 + x_3 = 0$	x_2 is vertex variable v_0; Eq. (4)
$-x_0 + x_4 + x_5 = 0$	x_4 is vertex variable v_1; Eq. (4)
$x_6 = 0$	x_6 is broadcast variable $b_{1,0}$; Eq. (5)
$x_7 = 1$	x_7 is broadcast variable $b_{0,1}$; Eq. (6)
$x_2 + x_7 - x_8 + x_9 = 1$	x_8 is for product $b_{0,1}v_0$ with Eq. (10)
$-x_2 - x_7 + 2x_8 + x_{10} = 0$	Equation (7) with (11)
$x_4 + x_7 - x_{11} + x_{12} = 1$	x_{11} is for product $b_{0,1}v_1$
$-x_4 - x_7 + 2x_{11} + x_{13} = 0$	Equation (7)
$x_4 + x_6 - x_{14} + x_{15} = 1$	x_{14} is for product $b_{1,0}v_1$
$-x_4 - x_6 + 2x_{14} + x_{16} = 0$	Equation (7)
$x_2 + x_6 - x_{17} + x_{18} = 1$	x_{17} is for product $b_{1,0}v_0$
$-x_2 - x_6 + 2x_{17} + x_{19} = 0$	Equation (7)
$x_7 + x_8 - x_{11} + x_{20} = 0$	Equation (8)
$x_6 + x_{14} - x_{17} + x_{21} = 0$	Equation (8)

Table 17.2 Final QUBO matrix Q for broadcasting in K_2

3	4	−4	−4	−4	−4	0	0	0	0	0	0	0	0	0	0	0	0	0	0	0	0
4	−2	0	0	0	0	0	0	0	0	0	0	0	0	0	0	0	0	0	0	0	0
−4	0	2	4	0	0	8	8	−12	4	−4	0	0	0	0	0	0	−12	4	−4	0	0
−4	0	4	2	0	0	0	0	0	0	0	0	0	0	0	0	0	0	0	0	0	0
−4	0	0	0	2	4	8	8	0	0	0	−12	4	−4	−12	4	−4	0	0	0	0	0
−4	0	0	0	4	2	0	0	0	0	0	0	0	0	0	0	0	0	0	0	0	0
0	0	8	0	8	0	4	0	0	0	0	0	0	0	−8	4	−4	−16	4	−4	0	4
0	0	8	0	8	0	0	0	−8	4	−4	−16	4	−4	0	0	0	0	0	0	4	0
0	0	−12	0	0	0	0	−8	16	−4	8	−4	0	0	0	0	0	0	0	0	4	0
0	0	4	0	0	0	0	4	−4	−2	0	0	0	0	0	0	0	0	0	0	0	0
0	0	−4	0	0	0	0	−4	8	0	2	0	0	0	0	0	0	0	0	0	0	0
0	0	0	0	−12	0	0	−16	−4	0	0	16	−4	8	0	0	0	0	0	0	−4	0
0	0	0	0	4	0	0	4	0	0	0	−4	−2	0	0	0	0	0	0	0	0	0
0	0	0	0	−4	0	0	−4	0	0	0	8	0	2	0	0	0	0	0	0	0	0
0	0	0	0	−12	0	−8	0	0	0	0	0	0	0	16	−4	8	−4	0	0	0	4
0	0	0	0	4	0	4	0	0	0	0	0	0	0	−4	−2	0	0	0	0	0	0
0	0	0	0	−4	0	−4	0	0	0	0	0	0	0	8	0	2	0	0	0	0	0
0	0	−12	0	0	0	−16	0	0	0	0	0	0	0	−4	0	0	16	−4	8	0	−4
0	0	4	0	0	0	4	0	0	0	0	0	0	0	0	0	0	−4	−2	0	0	0
0	0	−4	0	0	0	−4	0	0	0	0	0	0	0	0	0	0	8	0	2	0	0
0	0	0	0	0	0	0	4	4	0	0	−4	0	0	0	0	0	0	0	0	2	0
0	0	0	0	0	0	4	0	0	0	0	0	0	0	4	0	0	−4	0	0	0	2

One such example is given below where our 22 logical qubits, labeled 0 to 21, become 50 active hardware qubits on D-Wave Two's Chimera graph with $L = 4$, $N = M = 8$.

```
'embedding=': [ 0=[224, 226, 228], 1=[230], 2=[276,
283, 284, 288, 292], 3=[290], 4=[227, 291, 348, 355,
356, 357], 5=[229], 6=[336, 338, 341, 347, 349], 7=
[293, 297, 301, 361], 8=[294, 296, 302], 9=[289], 10=
[300], 11=[298, 362, 365], 12=[359, 367], 13=[364],
14=[345, 351],15=[344], 16=[346], 17=[275, 277, 281,
285, 339], 18=[272], 19=[274], 20=[303], 21=[343] ]
```

This best energy solution of -11 is also obtained when we run it on an actual D-Wave Two machine. This optimal answer occurs about 33 % of the time for our trials of about 1000 runs. In the other cases, the machine did not converge to the optimal ground-state energy.

17.11   Experimental Results

We have produced QUBO representations of the Broadcast Time Problem for several small common graphs using the above IP formulation procedure. Tables 17.3 and 17.4 summarize them for some small common graph families and known special graphs (all graphs can be obtained from Sage [16, 17]). Recall that for non-symmetric graphs we initiate the broadcast at vertex labeled 0, using the vertex labels given by Sage's adjacency lists. In these tables, columns 2 and 3 (Integer Variables and Quadratic Constraints) indicate the size of the IP formulation presented in Sect. 17.5. Next, columns 4 and 5 (Binary Variables and Binary Constraints) indicate the size of the IP formulation given in Sect. 17.7. Finally, columns 6–8 (Slack Variables, Logical Qubits and Chimera/Physical Qubits) indicate the size of the final QUBO representation described in Sect. 17.8. Using this approach, the number of logical qubits equals the number of binary variables plus the number of slack variables.

17.12   Conclusions

In this paper we have shown the process of converting a well-known combinatorial optimization problem, the Broadcast Time Problem, to a QUBO form that can be solved on an adiabatic quantum computer like the D-Wave Two. Our procedure of using an integer programming formulation (e.g., standard polynomial-time reduction) can be easily applied to other hard problems. However, this straightforward approach does require a large number of qubits for relatively small input graph instances. Future work is required to reduce this overhead. One area of study is

Table 17.3 Number of qubits required for some small graphs families

Graph	Order	Size	Integer variables	Quadratic constraints	Binary variables	Binary constraints	Slack variables	Logical qubits	Chimera qubits
C3	3	3	10	16	50	86	96	146	394
C4	4	4	13	21	74	131	146	220	662
C5	5	5	16	26	178	324	366	544	3258
C6	6	6	19	31	240	443	495	735	4164
C7	7	7	22	36	311	580	642	953	
C8	8	8	25	41	391	735	807	1198	
C9	9	9	28	46	778	1484	1608	2386	
C10	10	10	31	51	944	1809	1948	2892	
C11	11	11	34	56	1126	2166	2320	3446	
C12	12	12	37	61	1324	2555	2724	4048	
Grid2 × 3	6	7	21	37	254	472	543	797	4306
Grid3 × 3	9	12	34	65	832	1597	1816	2648	
Grid3 × 4	12	17	47	93	1414	2745	3084	4498	
Grid4 × 4	16	24	65	133	2420	4737	5252	7672	
Grid4 × 5	20	31	83	173	5537	10909	11815	17352	
K2	2	1	5	7	9	15	13	22	47
K3	3	3	10	16	50	86	96	146	394
K4	4	6	17	33	94	171	202	296	1378
K5	5	10	26	61	248	469	606	854	7973
K6	6	15	37	103	366	713	981	1347	
K7	7	21	50	162	507	1014	1482	1989	
K8	8	28	65	241	671	1375	2127	2798	
K9	9	36	82	343	1264	2591	4200	5464	
K10	10	45	101	471	1574	3279	5588	7162	
K2 × 1=P2	3	2	8	12	36	59	64	100	170
K1 × 2=S2	3	2	8	12	40	68	76	116	238
K2 × 2=C4	4	4	13	21	74	131	146	220	662
K2 × 3	5	6	18	32	192	353	414	606	4823
K3 × 3	6	9	25	49	282	529	633	915	
K3 × 4	7	12	32	69	381	727	894	1275	
K4 × 4	8	16	41	97	503	973	1227	1730	
K4 × 5	9	20	50	129	976	1906	2432	3408	
K5 × 5	10	25	61	171	1214	2391	3124	4338	
K5 × 6	11	30	72	118	1468	2914	3896	5364	
K6 × 6	12	36	85	277	1756	3511	4804	6560	
S2=K1 × 2	3	2	8	12	40	68	76	116	238
S3	4	3	11	18	64	114	130	194	505
S4	5	4	14	25	164	301	354	518	3711
S5	6	5	17	33	226	423	501	727	5120
S6	7	6	20	42	297	564	672	969	
S7	8	7	23	52	377	724	867	1244	
S8	9	8	26	63	760	1471	1736	2496	
S9	10	9	29	75	926	1803	2132	3058	
S10	11	10	32	88	1108	2168	2568	3676	

Table 17.4 Number of qubits required for hypercubes and some other small known graphs

Graph	Order	Size	Integer variables	Quadratic constraints	Binary variables	Binary constraints	Slack variables	Logical qubits	Chimera qubits
Q1 = K2	2	1	5	7	9	15	13	22	47
Q2 = C4	4	4	13	21	74	131	146	220	662
Q3	8	12	33	65	447	851	999	1446	
Q4	16	32	81	193	2564	5045	5860	8424	
BidiakisCube	12	18	49	97	1432	2779	3124	4556	
Bull	5	5	16	28	178	324	366	544	3523
Butterfly	5	6	18	33	192	353	414	606	5927
Chvatal	12	24	61	145	1540	3013	3604	5144	
Clebsch	16	40	97	273	2708	5373	6628	9336	
Diamond	4	5	15	27	84	151	174	258	742
Dinneen	9	21	52	142	994	1950	2552	3546	
Dodecahedral	20	30	81	161	5515	10855	11645	17160	
Durer	12	18	49	97	1432	2779	3124	4556	
Errera	17	45	108	320	4480	8900	10890	15370	
Frucht	12	18	49	97	1432	2779	3124	4556	
GoldnerHarary	11	27	66	209	1414	2814	3792	5206	
Grotzsch	11	20	52	118	1288	2508	2968	4256	
Heawood	14	21	57	113	1894	3691	4100	5994	
Herschel	11	18	48	101	1252	2429	2800	4052	
Hexahedral	8	12	33	65	447	851	999	1446	
Hoffman	16	32	81	193	2564	5045	5860	8424	
House	5	6	18	32	192	353	414	606	4176
Icosahedral	12	30	73	205	1648	3257	4164	5812	
Krackhardt	10	18	47	114	1088	2116	2548	3636	
Octahedral	6	12	31	73	324	619	795	1119	
Pappus	18	27	73	145	4514	8869	9575	14089	
Petersen	10	15	41	81	1034	1995	2276	3310	
Poussin	15	39	94	276	2446	4863	6152	8598	
Robertson	19	38	96	229	5211	10287	11570	16781	
Shrikhande	16	48	113	369	2852	5715	7508	10360	
Sousselier	16	27	71	154	2474	4849	5452	7926	
Tietze	12	18	49	97	1432	2779	3124	4556	
Wagner	8	12	33	65	447	851	999	1446	

to exploit the problem's characteristics for a possible direct encoding into QUBO form. Also substantial work is needed to reduce the blowup in embedding the logical qubits to physical qubits, which is required for embedding into the target machine's architecture.

Acknowledgments This work was supported in part by the Quantum Computing Research Initiatives at Lockheed Martin. We thank A. Fowler for comments which improved the presentation.

Appendix: Quadratic IP Formulation for Broadcasting in Q_3

The output of our integer programming formulation from Sect. 17.5 with the hypercube Q_3 as input is given below.

$x_0 \leq 7$ (1) time t

$x_1 \leq x_0$ (2) vertices $v_0 \ldots v_7$ informed times $\leq t$

$x_2 \leq x_0$

$x_3 \leq x_0$

$x_4 \leq x_0$

$x_5 \leq x_0$

$x_6 \leq x_0$

$x_7 \leq x_0$

$x_8 \leq x_0$

$x_9 + x_{10} + x_{11} \leq 0$ (3) originator has no parent

$x_{12} + x_{13} + x_{14} \leq 1$ (4) other vertices have one parent

$x_{15} + x_{16} + x_{17} \leq 1$

$x_{18} + x_{19} + x_{20} \leq 1$

$x_{21} + x_{22} + x_{23} \leq 1$

$x_{24} + x_{25} + x_{26} \leq 1$

$x_{27} + x_{28} + x_{29} \leq 1$

$x_{30} + x_{31} + x_{32} \leq 1$

$x_{12} + x_{12} * x_1 - x_{12} * x_2 \leq 0$ (5) parent time less than child time

$x_{15} + x_{15} * x_1 - x_{15} * x_3 \leq 0$

$x_{21} + x_{21} * x_1 - x_{21} * x_5 \leq 0$

$x_9 + x_9 * x_2 - x_9 * x_1 \leq 0$

$x_{18} + x_{18} * x_2 - x_{18} * x_4 \leq 0$

$x_{24} + x_{24} * x_2 - x_{24} * x_6 \leq 0$

$x_{10} + x_{10} * x_3 - x_{10} * x_1 \leq 0$

$x_{19} + x_{19} * x_3 - x_{19} * x_4 \leq 0$

$x_{27} + x_{27} * x_3 - x_{27} * x_7 \leq 0$

$x_{13} + x_{13} * x_4 - x_{13} * x_2 \leq 0$

$x_{16} + x_{16} * x_4 - x_{16} * x_3 \leq 0$

$x_{30} + x_{30} * x_4 - x_{30} * x_8 \leq 0$

$x_{11} + x_{11} * x_5 - x_{11} * x_1 \leq 0$

$x_{25} + x_{25} * x_5 - x_{25} * x_6 \leq 0$

$x_{28} + x_{28} * x_5 - x_{28} * x_7 \leq 0$

$x_{14} + x_{14} * x_6 - x_{14} * x_2 \leq 0$

$x_{22} + x_{22} * x_6 - x_{22} * x_5 \leq 0$

$x_{31} + x_{31} * x_6 - x_{31} * x_8 \leq 0$

$x_{17} + x_{17} * x_7 - x_{17} * x_3 \leq 0$

$x_{23} + x_{23} * x_7 - x_{23} * x_5 \leq 0$

$x_{32} + x_{32} * x_7 - x_{32} * x_8 \leq 0$

$x_{20} + x_{20} * x_8 - x_{20} * x_4 \leq 0$

$x_{26} + x_{26} * x_8 - x_{26} * x_6 \leq 0$

$$x_{29} + x_{29} * x_8 - x_{29} * x_7 \leq 0$$
$$x_{12} + x_{15} - sqr(x_2 - x_3) \leq 1 \quad (6) \text{ each child with different times}$$
$$x_{12} + x_{21} - sqr(x_2 - x_5) \leq 1$$
$$x_{15} + x_{21} - sqr(x_3 - x_5) \leq 1$$
$$x_9 + x_{18} - sqr(x_1 - x_4) \leq 1$$
$$x_9 + x_{24} - sqr(x_1 - x_6) \leq 1$$
$$x_{18} + x_{24} - sqr(x_4 - x_6) \leq 1$$
$$x_{10} + x_{19} - sqr(x_1 - x_4) \leq 1$$
$$x_{10} + x_{27} - sqr(x_1 - x_7) \leq 1$$
$$x_{19} + x_{27} - sqr(x_4 - x_7) \leq 1$$
$$x_{13} + x_{16} - sqr(x_2 - x_3) \leq 1$$
$$x_{13} + x_{30} - sqr(x_2 - x_8) \leq 1$$
$$x_{16} + x_{30} - sqr(x_3 - x_8) \leq 1$$
$$x_{11} + x_{25} - sqr(x_1 - x_6) \leq 1$$
$$x_{11} + x_{28} - sqr(x_1 - x_7) \leq 1$$
$$x_{25} + x_{28} - sqr(x_6 - x_7) \leq 1$$
$$x_{14} + x_{22} - sqr(x_2 - x_5) \leq 1$$
$$x_{14} + x_{31} - sqr(x_2 - x_8) \leq 1$$
$$x_{22} + x_{31} - sqr(x_5 - x_8) \leq 1$$
$$x_{17} + x_{23} - sqr(x_3 - x_5) \leq 1$$
$$x_{17} + x_{32} - sqr(x_3 - x_8) \leq 1$$
$$x_{23} + x_{32} - sqr(x_5 - x_8) \leq 1$$
$$x_{20} + x_{26} - sqr(x_4 - x_6) \leq 1$$
$$x_{20} + x_{29} - sqr(x_4 - x_7) \leq 1$$
$$x_{26} + x_{29} - sqr(x_6 - x_7) \leq 1$$

References

1. Tee, G.J.: The Monte Carlo Method. Pergamon Press, Oxford and New York (1966)
2. Metropolis, N.C., Rosenbluth, A., Rosenbluth, M., Teller, A., Teller, E.: J. Chem. Phys. **21**(6), 1087 (1953)
3. Kirkpatrick, S., Gelatt, C.D., Vecchi, M.P.: Science **220**(4598), 671 (1983)
4. Wikipedia, Simulated annealing. http://en.wikipedia.org/wiki/Simulated_annealing (2014). Accessed 30 Oct 2014
5. Born, M., Fock, V.: Zeitschrift für Physik **51**(3–4), 165 (1928). doi:10.1007/BF01343193
6. Farhi, E., Goldstone, J., Gutmann, S., Sipser, M.: Quantum computation by adiabatic evolution (2000). arXiv:quant-ph/0001106v1
7. Aharonov, D., Dam, W.v., Kempe, J., Landau, Z., Lloyd, S., Regev, O.: Adiabatic quantum computation is equivalent to standard quantum computation (2005). arXiv:quant-ph/0405098
8. D-Wave. D-Wave overview: a brief introduction to D-Wave and quantum computing (2013). http://www.dwavesys.com/sites/default/files/D-Wave-brochure-102013F-CA.pdf
9. Farley, A., Hedetniemi, S., Mitchell, S., Proskurowski, A.: Discret. Math. **25**, 189 (1979)
10. Hedetniemi, S.M., Hedetniemi, S.T., Liestman, A.L.: Networks **18**, 319 (1998)
11. Garey, M.R., Johnson, D.S.: Computers and Intractability: A Guide to the Theory of NP-Completeness. W.H. Freeman & Co., New York (1979)

12. Dinneen, M.J.: The complexity of broadcasting in bounded-degree networks. Technical Report Combinatorics report LACES-[05C-94-31], Los Alamos National Laboratory (1994). arXiv:math/9411222
13. Ravi, R.: In: Proceedings of the 35th Annual Symposium on Foundations of Computer Science, FOCS'94, pp. 202–213. IEEE Computer Society Press (1994)
14. Wikipedia, Integer programming. http://en.wikipedia.org/wiki/Integer_programming (2015). Accessed 30 Oct 2014
15. Khosravani, M.: Searching for optimal caterpillars in general and bounded treewidth graphs. Ph.D. dissertation, University of Auckland, Auckland, New Zealand (2011)
16. Stein, W., et al.: Sage Mathematics Software (Version 6.3). The Sage Development Team (2014). http://www.sagemath.org
17. Calude, C.S., Dinneen, M.J.: Solving the broadcast time problem using a D-wave quantum computer. Tech. Rep. Report CDMTCS-473, Centre for Discrete Mathematics and Theoretical Computer Science, University of Auckland, Auckland, New Zealand (2014). Paper URL http://www.cs.auckland.ac.nz/research/groups/CDMTCS/researchreports/index.php?download&paper_file=526. Data URL http://www.cs.auckland.ac.nz/research/groups/CDMTCS/researchreports/index.php?download&data_file=7
18. D-Wave, Programming with QUBOs. Technical report, D-Wave Systems, Inc. (2013). Python Release 1.5.1-beta4 (for Mac/Linux), 09-1002A-B

Chapter 18
The Group Zoo of Classical Reversible Computing and Quantum Computing

Alexis De Vos and Stijn De Baerdemacker

Abstract By systematically inflating the group of $n \times n$ permutation matrices to the group of $n \times n$ unitary matrices, we can see how classical computing is embedded in quantum computing. In this process, an important role is played by two subgroups of the unitary group $U(n)$, i.e. $XU(n)$ and $ZU(n)$. Here, $XU(n)$ consists of all $n \times n$ unitary matrices with all line sums (i.e. the n row sums and the n column sums) equal to 1, whereas $ZU(n)$ consists of all $n \times n$ diagonal unitary matrices with upper-left entry equal to 1. As a consequence, quantum computers can be built from NEGATOR gates and PHASOR gates. The NEGATOR is a 1-qubit circuit that is a natural generalization of the 1-bit NOT gate of classical computing. In contrast, the PHASOR is a 1-qubit circuit not related to classical computing.

18.1 Introduction

Often, in the literature, conventional computers and quantum computers are discussed like belonging to two separate worlds, far from each other. Conventional computers act on classical bits, say 'pure zeroes' and 'pure ones', by means of Boolean logic gates, such as AND gates and NOR gates. The operations performed by these gates are described by truth tables. Quantum computers act on qubits, say complex vectors, by means of quantum gates, such as ROTATOR gates and T gates [1]. The operations performed by these gates are described by unitary matrices.

Because the world of classical computation and the world of quantum computation are based on such different science models, it is difficult to see the relationship (be it analogies or differences) between these two computation paradigms. In the present

A. De Vos (✉)
Vakgroep elektronika en informatiesystemen, Universiteit Gent,
Sint Pietersnieuwstraat 41, B-9000 Gent, Belgium
e-mail: alex@elis.UGent.be

S. De Baerdemacker
Vakgroep anorganische en fysische chemie, Universiteit Gent, Krijgslaan 281 - S3,
B-9000 Gent, Belgium
e-mail: Stijn.DeBaerdemacker@UGent.be

© Springer International Publishing Switzerland 2017
A. Adamatzky (ed.), *Advances in Unconventional Computing*,
Emergence, Complexity and Computation 22,
DOI 10.1007/978-3-319-33924-5_18

chapter, we bridge the gap between the two sciences. For this purpose, a common language is necessary. The common tool we have chosen is the representation by square matrices and the construction of matrix groups.

18.2 Reversible Computing

The first step in bridging the gap between classical and quantum computation is replacing (or better: embedding) conventional classical computing in reversible classical computing. Whereas conventional logic gates are represented by truth tables with an arbitrary number w_i of input columns and an arbitrary number w_o of output columns, reversible logic gates are described by truth tables with an equal number w of input and output columns. Moreover, all output rows are different, such that the 2^w output words are merely a permutation of the 2^w input words [2–5]. Table 18.1 gives an example of a conventional gate (i.e. an AND gate, with two input bits A and B and one output bit R), as well as an example of a reversible gate (i.e. a TOFFOLI gate, a.k.a. a controlled controlled NOT gate, with three input bits A, B, and C and three output bits P, Q, and R). The reader may verify that the irreversible AND function is embedded in the reversible TOFFOLI function, as presetting in Table 18.1b the input C to logic 0 leads to the output R being equal to A AND B, as is highlighted by boldface. In the general case, any irreversible truth table can be embedded in a reversible truth table with $w = w_i + w_o$ or less bits [6].

Table 18.1 Truth table of two basic Boolean functions: (a) the AND function, (b) the TOFFOLI function

(a)

$A\,B$	R
0 0	0
0 1	0
1 0	0
1 1	1

(b)

$A\,B\,C$	$P\,Q\,R$
0 0 0	**0 0 0**
0 0 1	0 0 1
0 1 0	**0 1 0**
0 1 1	0 1 1
1 0 0	**1 0 0**
1 0 1	1 0 1
1 1 0	**1 1 1**
1 1 1	1 1 0

The next step in the journey from the conventional to the quantum world, is replacing the reversible truth table by a permutation matrix. As all eight output words $000, 001, \ldots,$ and 110 are merely a permutation of the eight input words $000, 001, \ldots,$ and 111, Table 18.1b can be replaced by an 8×8 permutation matrix, i.e.

$$
\begin{pmatrix}
1 & 0 & 0 & 0 & 0 & 0 & 0 & 0 \\
0 & 1 & 0 & 0 & 0 & 0 & 0 & 0 \\
0 & 0 & 1 & 0 & 0 & 0 & 0 & 0 \\
0 & 0 & 0 & 1 & 0 & 0 & 0 & 0 \\
0 & 0 & 0 & 0 & 1 & 0 & 0 & 0 \\
0 & 0 & 0 & 0 & 0 & 1 & 0 & 0 \\
0 & 0 & 0 & 0 & 0 & 0 & 0 & 1 \\
0 & 0 & 0 & 0 & 0 & 0 & 1 & 0
\end{pmatrix} .
$$

An arbitrary classical reversible circuit, acting on w bits, is represented by a permutation matrix of size $2^w \times 2^w$. In contrast, a quantum circuit, acting on w qubits, is represented by a unitary matrix of size $2^w \times 2^w$. Both kind of matrices are depicted by symbols with w input lines and w output lines:

Invertible square matrices, together with the operation of ordinary matrix multiplication, form a group. The finite matrix group $P(2^w)$ consisting solely of permutation matrices is a subgroup of the continuous group $U(2^w)$ of unitary matrices. In the present chapter, we show a natural means how to enlarge the subgroup to its supergroup, in other words: how to upgrade a classical computer to a quantum computer.

18.3 NEGATORs and PHASORs

For the purpose of upgrading the permutation group $P(n)$ to the unitary group $U(n)$, we introduce two subgroups [7–9] of $U(n)$:

the subgroup XU(n)
consists of all $n \times n$ unitary matrices with all line sums (i.e. the n row sums and the n column sums) equal to 1

Fig. 18.1 Venn diagram of
the Lie groups U(n), XU(n),
and ZU(n) and the finite
groups P(n) and **1**(n). *Note*:
the trivial group **1**(n) is
represented by the bullet.

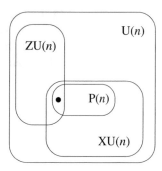

and

> **the subgroup ZU(n)**
>
> consists of all $n \times n$ diagonal unitary matrices with upper-left entry equal to 1.

Whereas U(n) is an n^2-dimensional Lie group, XU(n) is only $(n-1)^2$-dimensional and ZU(n) is only $(n-1)$-dimensional. The two subgroups are quite distinct: their intersection is the trivial group **1**(n), consisting of a single matrix, i.e. the $n \times n$ unit matrix. We note that all P(n) matrices are in XU(n). See Venn diagram in Fig. 18.1.

Why exactly these two groups? The reason becomes clear by looking at the case $n = 2$. There exist only two classical reversible circuits acting on a single bit. They are represented by the two P(2) matrices: $\left(\begin{smallmatrix} 1 & 0 \\ 0 & 1 \end{smallmatrix}\right)$ for the IDENTITY gate and $\left(\begin{smallmatrix} 0 & 1 \\ 1 & 0 \end{smallmatrix}\right)$ for the NOT gate. The latter is also known as the X gate. In contrast, there exists a 4-dimensional infinity of quantum circuits acting on a single qubit. They are represented by the U(2) matrices.

In order to upgrade the group P(2), we construct a unitary interpolation between its two permutation matrices. The interpolation

$$(1-t) \begin{pmatrix} 1 & 0 \\ 0 & 1 \end{pmatrix} + t \begin{pmatrix} 0 & 1 \\ 1 & 0 \end{pmatrix}$$

is unitary if and only if $t = (1 - e^{i\theta})/2$, where θ is an arbitrary angle. We thus obtain a 1-dimensional generalization of the NOT matrix:

> **the NEGATOR gate:**
>
> $$N(\theta) = \frac{1}{2} \begin{pmatrix} 1 + e^{i\theta} & 1 - e^{i\theta} \\ 1 - e^{i\theta} & 1 + e^{i\theta} \end{pmatrix},$$
>
> where θ is an arbitrary angle.

Because U(2) is 4-dimensional, we need some extra building block to generate the full U(2). For this purpose, it suffices to introduce a second 1-dimensional subgroup of U(2):

the PHASOR **gate**:

$$\Phi(\theta) = \begin{pmatrix} 1 & 0 \\ 0 & e^{i\theta} \end{pmatrix},$$

where θ is an arbitrary angle.

Analogously as the NEGATOR is the 1-dimensional generalization of the $\left(\begin{smallmatrix} & 1 \\ 1 & \end{smallmatrix}\right)$ matrix or X gate, the PHASOR can be considered as the 1-dimensional generalization of the $\left(\begin{smallmatrix} 1 & \\ & -1 \end{smallmatrix}\right)$ matrix, a.k.a. the Z gate. The two 1-dimensional subgroups XU(2) and ZU(2) suffice to generate the whole 4-dimensional group U(2). We say: the closure of XU(2) and ZU(2) is U(2). Indeed, an arbitrary matrix U from U(2) can be written as a finite product of matrices from XU(2) and matrices from ZU(2):

$$U(\alpha, \varphi, \psi, \chi) = e^{i\alpha} \begin{pmatrix} \cos(\varphi)e^{i\psi} & \sin(\varphi)e^{i\chi} \\ -\sin(\varphi)e^{-i\chi} & \cos(\varphi)e^{-i\psi} \end{pmatrix}$$
$$= N(\pi)\,\Phi(\alpha + \varphi + \psi)\,N(\pi)\,\Phi(\alpha + \varphi - \chi + \pi/2)\,N(\varphi)\,\Phi(-\psi + \chi - \pi/2).$$

We use the following symbols for the NEGATOR and PHASOR gates:

$$-\boxed{N(\theta)}- \quad \text{and} \quad -\boxed{\Phi(\theta)}- \quad ,$$

respectively. In the literature [5, 10–12], some of these gates have a specific notation:

$$N(0) = \text{I}$$
$$N(\pi/4) = \text{W}$$
$$N(\pi/2) = \text{V}$$
$$N(\pi) = \text{X}$$
$$N(2\pi) = \text{I}$$
$$\Phi(0) = \text{I}$$
$$\Phi(\pi/4) = \text{T}$$
$$\Phi(\pi/2) = \text{S}$$
$$\Phi(\pi) = \text{Z}$$
$$\Phi(2\pi) = \text{I}.$$

In particular, the V gate is known as 'the square root of NOT' [13–16].

18.4 Controlled NEGATORs and Controlled PHASORs

Two-qubit circuits are represented by matrices from U(4). We may apply either the NEGATOR gate or the PHASOR gate from the previous section to either the first qubit or the second qubit. Here are two examples:

i.e. a NEGATOR acting on the second qubit and a PHASOR acting on the first qubit, respectively. These circuits are represented by the 4×4 unitary matrices

$$\frac{1}{2} \begin{pmatrix} 1 + e^{i\theta} & 1 - e^{i\theta} & 0 & 0 \\ 1 - e^{i\theta} & 1 + e^{i\theta} & 0 & 0 \\ 0 & 0 & 1 + e^{i\theta} & 1 - e^{i\theta} \\ 0 & 0 & 1 - e^{i\theta} & 1 + e^{i\theta} \end{pmatrix} \text{ and } \begin{pmatrix} 1 & 0 & 0 & 0 \\ 0 & 1 & 0 & 0 \\ 0 & 0 & e^{i\theta} & 0 \\ 0 & 0 & 0 & e^{i\theta} \end{pmatrix}, \tag{18.1}$$

respectively.

However, we also introduce more sophisticated gates: the so-called 'controlled PHASORs' and 'controlled NEGATORs'. Two examples are

i.e. a positive-polarity controlled NEGATOR acting on the first qubit, controlled by the second qubit, and a negative-polarity controlled PHASOR acting on the second qubit, controlled by the first qubit, respectively. The former symbol is read as follows: 'if the second qubit equals 1, then the NEGATOR acts on the first qubit; if, however, the second qubit equals 0, then the NEGATOR is inactive, i.e. the first qubit undergoes no change'. The latter symbol is read as follows: 'if the first qubit equals 0, then the PHASOR acts on the second qubit; if, however, the first qubit equals 1, then the PHASOR is inactive, i.e. the second qubit undergoes no change'. The 4×4 matrices representing these two circuit examples are:

$$\begin{pmatrix} 1 & 0 & 0 & 0 \\ 0 & \frac{1}{2}(1 + e^{i\theta}) & 0 & \frac{1}{2}(1 - e^{i\theta}) \\ 0 & 0 & 1 & 0 \\ 0 & \frac{1}{2}(1 - e^{i\theta}) & 0 & \frac{1}{2}(1 + e^{i\theta}) \end{pmatrix} \text{ and } \begin{pmatrix} 1 & 0 & 0 & 0 \\ 0 & e^{i\theta} & 0 & 0 \\ 0 & 0 & 1 & 0 \\ 0 & 0 & 0 & 1 \end{pmatrix}, \tag{18.2}$$

respectively.

We now give two examples of a 3-qubit circuit:

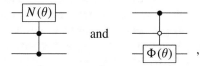

i.e. a positive-polarity controlled NEGATOR acting on the first qubit and a mixed-polarity controlled PHASOR acting on the third qubit. The 8×8 matrices representing these two circuit examples are:

$$
\begin{pmatrix}
1\,0\,0 & 0 & 0\,0\,0 & 0 \\
0\,1\,0 & 0 & 0\,0\,0 & 0 \\
0\,0\,1 & 0 & 0\,0\,0 & 0 \\
0\,0\,0 & \frac{1}{2}(1+e^{i\theta}) & 0\,0\,0 & \frac{1}{2}(1-e^{i\theta}) \\
0\,0\,0 & 0 & 1\,0\,0 & 0 \\
0\,0\,0 & 0 & 0\,1\,0 & 0 \\
0\,0\,0 & 0 & 0\,0\,1 & 0 \\
0\,0\,0 & \frac{1}{2}(1-e^{i\theta}) & 0\,0\,0 & \frac{1}{2}(1+e^{i\theta})
\end{pmatrix}
\quad \text{and} \quad
\begin{pmatrix}
1\,0\,0\,0\,0\,\,0\,\,0\,0 \\
0\,1\,0\,0\,0\,\,0\,\,0\,0 \\
0\,0\,1\,0\,0\,\,0\,\,0\,0 \\
0\,0\,0\,1\,0\,\,0\,\,0\,0 \\
0\,0\,0\,0\,1\,\,0\,\,0\,0 \\
0\,0\,0\,0\,0\,\,e^{i\theta}\,0\,0 \\
0\,0\,0\,0\,0\,\,0\,\,1\,0 \\
0\,0\,0\,0\,0\,\,0\,\,0\,1
\end{pmatrix}, \quad (18.3)
$$

respectively. In each of the expressions (18.1)–(18.3), we note the following properties:

- the former matrix has all row sums and all column sums equal to 1;
- the latter matrix is diagonal and has upper-left entry equal to 1.

Because the multiplication of two square matrices with all line sums equal to 1 automatically yields a third square matrix with all line sums equal to 1, we can easily demonstrate that an arbitrary quantum circuit like

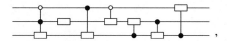

consisting merely of uncontrolled NEGATORs and controlled NEGATORs is represented by a $2^w \times 2^w$ unitary matrix with all line sums equal to 1, i.e. an XU(2^w) matrix. A laborious proof [17] demonstrates that the converse theorem is also valid: any member of XU(2^w) can be synthesized by an appropriate finite string of (un)controlled NEGATORs.

Because the multiplication of two diagonal square matrices yields a third diagonal square matrix and because the multiplication of two unitary matrices with first entry equal to 1 yields a third unitary matrix with first entry equal to 1, we can easily demonstrate that an arbitrary quantum circuit like

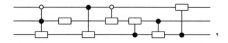

consisting merely of uncontrolled PHASORs and controlled PHASORs is represented by a $2^w \times 2^w$ unitary diagonal matrix with first entry equal to 1, i.e. a $ZU(2^w)$ matrix. It can easily be seen that the converse theorem is also true: any member of $ZU(2^w)$ can be synthesized by an appropriate finite string of (un)controlled PHASORs.

We conclude that the study of NEGATOR and PHASOR circuits automatically leads to the introduction of the two subgroups $XU(2^w)$ and $ZU(2^w)$ of the unitary group $U(2^w)$.

18.5 The FUF Matrix Decomposition

While studying the properties of the XU and ZU groups, a pivotal role is played by the $n \times n$ discrete Fourier transform, i.e. the following unitary matrix:

$$
F = \frac{1}{\sqrt{n}}
\begin{pmatrix}
1 & 1 & 1 & 1 & \ldots & 1 \\
1 & \omega & \omega^2 & \omega^3 & \ldots & \omega^{n-1} \\
1 & \omega^2 & \omega^4 & \omega^6 & \ldots & \omega^{2(n-1)} \\
\vdots & & & & \\
1 & \omega^{n-1} & \omega^{2(n-1)} & \omega^{3(n-1)} & \ldots & \omega^{(n-1)(n-1)}
\end{pmatrix},
$$

where ω is the primitive nth root of unity, i.e. $e^{i\,2\pi/n}$. We have:

the FUF theorem: Any matrix X from $XU(n)$ can be written as the following product [17]:

$$
X = F \begin{pmatrix} 1 & \\ & U \end{pmatrix} F^{-1},
$$

where

- F is the $n \times n$ discrete Fourier transform and
- U is a matrix from $U(n-1)$.

The proof is constructive, i.e. by computation of the matrix product, taking into account the properties of the Fourier matrix. The relationship is a one-to-one mapping. In other words: with one X corresponds one U and with one U corresponds one X. As a result, the group $XU(n)$ is isomorphic to the unitary group $U(n-1)$ and thus has $(n-1)^2$ dimensions. Here is an example from $U(2)$ and $XU(3)$:

$$\frac{1}{\sqrt{3}} \begin{pmatrix} 1 & 1 & 1 \\ 1 & \omega & \omega^2 \\ 1 & \omega^2 & \omega \end{pmatrix} \frac{1}{2} \begin{pmatrix} 2 & 0 & 0 \\ 0 & -1+i & 1+i \\ 0 & 1-i & 1+i \end{pmatrix} \frac{1}{\sqrt{3}} \begin{pmatrix} 1 & 1 & 1 \\ 1 & \omega^2 & \omega \\ 1 & \omega & \omega^2 \end{pmatrix}$$

$$= \frac{1}{6} \begin{pmatrix} 4+2i & -(\sqrt{3}-1)+i\,(\sqrt{3}-1) & \sqrt{3}+1-i\,(\sqrt{3}+1) \\ -(\sqrt{3}-1)-i\,(\sqrt{3}+1) & \sqrt{3}+1+2i & 4+i\,(\sqrt{3}-1) \\ \sqrt{3}+1+i\,(\sqrt{3}-1) & 4-i\,(\sqrt{3}+1) & -(\sqrt{3}-1)+2i \end{pmatrix},$$

where ω is the primitive cubic root of unity, i.e. $\omega = e^{i\,2\pi/3} = -\frac{1}{2} + i\,\frac{\sqrt{3}}{2}$.

We have two special cases of the FUF theorem. The first involves a subgroup of $XU(n)$:

the subgroup $CXU(n)$
consists of all circulant $XU(n)$ matrices.

For such matrices holds

the FZF theorem:
Any matrix C from $CXU(n)$ can be written as follows:

$$C = FZF^{-1},$$

where

- F is the $n \times n$ discrete Fourier transform and
- Z is a matrix from $ZU(n)$.

Similarly, we can write any $ZU(n)$ as an FCF^{-1} product. These two relationships constitute a one-to-one mapping between $ZU(n)$ and $CXU(n)$. The two $(n-1)$-dimensional groups thus are isomorphic. An example for $CXU(4)$ and $ZU(4)$ is

$$\frac{1}{8} \begin{pmatrix} 1+i & 7+i & -1-i & 1-i \\ 1-i & 1+i & 7+i & -1-i \\ -1-i & 1-i & 1+i & 7+i \\ 7+i & -1-i & 1-i & 1+i \end{pmatrix} =$$

$$\frac{1}{2} \begin{pmatrix} 1 & 1 & 1 & 1 \\ 1 & i & -1 & -i \\ 1 & -1 & 1 & -1 \\ 1 & -i & -1 & i \end{pmatrix} \begin{pmatrix} 1 & 0 & 0 & 0 \\ 0 & i & 0 & 0 \\ 0 & 0 & -1 & 0 \\ 0 & 0 & 0 & (1-i)/2 \end{pmatrix} \frac{1}{2} \begin{pmatrix} 1 & 1 & 1 & 1 \\ 1 & -i & -1 & i \\ 1 & -1 & 1 & -1 \\ 1 & i & -1 & -i \end{pmatrix}.$$

The second special case of the FUF theorem is

---

the FXF theorem:
For any matrix X from $\mathrm{XU}(n)$, the following property holds:

$$F \begin{pmatrix} 1 \\ & X \end{pmatrix} F^{-1} = \begin{pmatrix} 1 \\ & X' \end{pmatrix},$$

where

- F is the $(n+1) \times (n+1)$ discrete Fourier transform and
- X' is a matrix from $\mathrm{XU}(n)$.

Proof of this theorem is quite simple. Suffice it to note two facts:

- $F \begin{pmatrix} 1 \\ & x \end{pmatrix}$ is a matrix with an upper row consisting of $n+1$ entries all equal to $1/\sqrt{n+1}$, such that $F \begin{pmatrix} 1 \\ & x \end{pmatrix} F^{-1}$ is a matrix with an upper-left entry equal to 1;
- $F \begin{pmatrix} 1 \\ & x \end{pmatrix} F^{-1}$ is of the form $F \begin{pmatrix} 1 \\ & u \end{pmatrix} F^{-1}$, and therefore an $\mathrm{XU}(n+1)$ matrix, by virtue of the FUF theorem.

A matrix with these two properties is necessarily of the form $\begin{pmatrix} 1 \\ & x' \end{pmatrix}$ with X' a member of $\mathrm{XU}(n)$.

For $n = 2$, the matrix X' is equal to the matrix X. For $n > 2$, usually, the matrix X' is different from X. The relationship thus is a one-to-one mapping from $\mathrm{XU}(n)$ to itself. We give an example from $\mathrm{XU}(3)$:

$$\frac{1}{2}\begin{pmatrix} 1 & 1 & 1 & 1 \\ 1 & i & -1 & -i \\ 1 & -1 & 1 & -1 \\ 1 & -i & -1 & i \end{pmatrix} \frac{1}{2}\begin{pmatrix} 2 & 0 & 0 & 0 \\ 0 & 1 & i & 1-i \\ 0 & -i & 1 & 1+i \\ 0 & 1+i & 1-i & 0 \end{pmatrix} \frac{1}{2}\begin{pmatrix} 1 & 1 & 1 & 1 \\ 1 & -i & -1 & i \\ 1 & -1 & 1 & -1 \\ 1 & i & -1 & -i \end{pmatrix} =$$

$$\frac{1}{4}\begin{pmatrix} 4 & 0 & 0 & 0 \\ 0 & 3 & -1-i & 2+i \\ 0 & -1+i & 2 & 3-i \\ 0 & 2-i & 3+i & -1 \end{pmatrix}.$$

Note that here the Fourier matrices are of larger size than the XU matrices. A relationship like $FXF^{-1} = X'$ with F, X, and X' of the same size does not hold.

18.6   The ZXZ Matrix Decomposition

A quite different theorem is

the ZXZ theorem:
Any matrix U from U(n) can be written as follows [18, 19]:

$$U = aZ_1XZ_2 , \qquad (18.4)$$

where

- a is a member of U(1), i.e. a complex number with unit modulus,
- X is a member of XU(n), and
- both Z_1 and Z_2 are member of ZU(n).

The proof of the theorem is non-constructive and based on symplectic topology [19]. We note that the sum of 1 (number of parameters in a), $n-1$ (parameters in Z_1), $(n-1)^2$ (in X) and $n-1$ (in Z_2) equals the dimensionality n^2 of U(n):

$$1 + (n-1) + (n-1)^2 + (n-1) = n^2 .$$

Thus the number of degrees of freedom in aZ_1XZ_2 exactly matches the number of degrees of freedom in U. This might suggest that the decomposition is unique. However, this is not true: unlike the FUF theorem, the ZXZ theorem is not a one-to-one relationship. As an example, we give here the same matrix from U(2) as in the illustration of the FUF theorem. It has two (and only two) ZXZ decompositions:

$$\frac{1}{2}\begin{pmatrix} -1+i & 1+i \\ 1-i & 1+i \end{pmatrix} = (-1) \begin{pmatrix} 1 & 0 \\ 0 & i \end{pmatrix} \frac{1}{2}\begin{pmatrix} 1-i & 1+i \\ 1+i & 1-i \end{pmatrix}\begin{pmatrix} 1 & 0 \\ 0 & -1 \end{pmatrix}$$

$$= i \begin{pmatrix} 1 & 0 \\ 0 & -i \end{pmatrix} \frac{1}{2}\begin{pmatrix} 1+i & 1-i \\ 1-i & 1+i \end{pmatrix}\begin{pmatrix} 1 & 0 \\ 0 & 1 \end{pmatrix} .$$

Usually a U(n) matrix has a discrete number of decompositions. But sometimes there are as many as a noncountable infinity of decompositions, as is illustrated by another U(2) matrix:

$$\begin{pmatrix} 0 & -i \\ i & 0 \end{pmatrix} = e^{i\beta}\begin{pmatrix} 1 & 0 \\ 0 & ie^{-i\beta} \end{pmatrix}\begin{pmatrix} 0 & 1 \\ 1 & 0 \end{pmatrix}\begin{pmatrix} 1 & 0 \\ 0 & -ie^{-i\beta} \end{pmatrix},$$

where the angle β is allowed to have any value.

Except in some special cases, no analytical method is known to find the ZXZ decompositons of U. Only a numerical procedure is known. It yields one of the solutions with an arbitrarily small error. Which solution is found, depends on the starting conditions of the numerical algorithm [18].

We thus can conclude that any quantum computer looks like

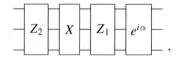

i.e. the cascade of an overall phase factor, an input section consisting merely of (un)controlled PHASORs, a core section consisting merely of (un)controlled NEGATORs, and an output section consisting merely of (un)controlled PHASORs.

By combining the FUF, the ZXZ, the FZF, and the FXF theorems, we can prove the following decomposition of an XU(n) matrix:

the CXC theorem:
For any matrix X from XU(n), the following property holds:

$$X = C' \begin{pmatrix} 1 & \\ & X' \end{pmatrix} C'', \qquad (18.5)$$

where

- X' is a member of XU($n-1$) and
- both C' and C'' are member of CXU(n).

Indeed:

$$X = F \begin{pmatrix} 1 & \\ & U \end{pmatrix} F^{-1} = F \begin{pmatrix} 1 & \\ & aZ'X''Z'' \end{pmatrix} F^{-1} = F \begin{pmatrix} 1 & \\ & aZ' \end{pmatrix} \begin{pmatrix} 1 & \\ & X'' \end{pmatrix} \begin{pmatrix} 1 & \\ & Z'' \end{pmatrix} F^{-1}$$

$$= F \begin{pmatrix} 1 & \\ & aZ' \end{pmatrix} F^{-1} F \begin{pmatrix} 1 & \\ & X'' \end{pmatrix} F^{-1} F \begin{pmatrix} 1 & \\ & Z'' \end{pmatrix} F^{-1} = C' \begin{pmatrix} 1 & \\ & X' \end{pmatrix} C''.$$

Because the ZXZ decomposition is not unique, also the CXC decomposition is not unique.

By applying the CXC theorem again and again, we find the following decomposition of an arbitrary element X of XU(n):

$$X = C'_n \begin{pmatrix} 1 & \\ & C'_{n-1} \end{pmatrix} \cdots \begin{pmatrix} \mathbf{1}_{n-3} & \\ & C'_3 \end{pmatrix} \begin{pmatrix} \mathbf{1}_{n-2} & \\ & C_2 \end{pmatrix} \begin{pmatrix} \mathbf{1}_{n-3} & \\ & C''_3 \end{pmatrix} \cdots \begin{pmatrix} 1 & \\ & C''_{n-1} \end{pmatrix} C''_n,$$

$$(18.6)$$

where $\mathbf{1}_k$ is a short-hand notation for the $k \times k$ unit matrix, and where all C'_k and all C''_k are CXU(k) matrices and C_2 is a CXU(2) matrix. We conclude: any matrix from XU(n) can be decomposed as a product of $2n - 2$ matrices of the form $\begin{pmatrix} \mathbf{1}_{n-k} & \\ & C_k \end{pmatrix}$. We note that a similar reasoning is applicable to permutation matrices, i.e. to classical computation. See Appendix 1.

18.7 The U Circuit Synthesis

The phase factor $a = e^{i\alpha}$ in the ZXZ product may be decomposed into two NEGATOR circuits and two uncontrolled PHASORs. Indeed, if $n = 2^w$, then n is even. If n is even, then we note the following diagonal-matrix property:

$$\text{diag}(a, a, a, a, a, \ldots, a, a)$$
$$= P_0 \, \text{diag}(1, a, 1, a, 1, \ldots, 1, a) \, P_0^{-1} \, \text{diag}(1, a, 1, a, 1, \ldots, 1, a) \, ,$$

where P_0 is the $n \times n$ (circulant) permutation matrix

$$\begin{pmatrix} 0 & 1 & 0 & 0 & \ldots & 0 & 0 \\ 0 & 0 & 1 & 0 & \ldots & 0 & 0 \\ 0 & 0 & 0 & 1 & \ldots & 0 & 0 \\ & \vdots & & & & & \\ 0 & 0 & 0 & 0 & \ldots & 0 & 1 \\ 1 & 0 & 0 & 0 & \ldots & 0 & 0 \end{pmatrix} ,$$

a.k.a. the cyclic-shift matrix, which can be implemented with classical reversible gates (i.e. one NOT and $w - 1$ controlled NOTs [4, 20]). We thus can rewrite (18.4) as a decomposition containing exclusively XU and ZU matrices:

$$U = P_0 Z_0 P_0^{-1} Z_1' X Z_2 \, ,$$

where $Z_0 = \text{diag}(1, a, 1, a, 1, \ldots, 1, a)$ is a ZU matrix which can be implemented by a single (uncontrolled) PHASOR gate and where Z_1' is the product $Z_0 Z_1$:

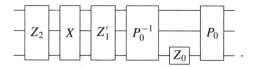

Because both P_0 and P_0^{-1} belong to XU(n), we conclude that any matrix from U(n) can be synthesized by a cascade of XU(n) blocks and ZU(n) blocks. In group-theoretical terms: the closure of XU(n) and ZU(n) is U(n). We note that this circuit decomposition is not unique, because the ZXZ matrix decomposition is not unique.

In the next two sections, we will look in detail to the synthesis of the ZU block and the XU blocks.

18.8 The ZU Circuit Synthesis

The decomposition of an arbitrary member of ZU(n) is straightforward. Indeed, for even n, the matrix can be written as the following product of four matrices:

$$\text{diag}(1, a_2, a_3, a_4, a_5, a_6, \ldots, a_n) =$$

$$\text{diag}(1, a_2, 1, a_4, 1, a_6, \ldots, 1, a_n) \; P_0 \; \text{diag}(1, 1, 1, a_3, 1, a_5, \ldots, 1, a_{n-1}) \; P_0^{-1} \,,$$

where a_j is a short-hand notation for $e^{i\alpha_j}$. If n equals 2^w, then the diagonal matrix $\text{diag}(1, a_2, 1, a_4, 1, a_6, \ldots)$ represents 2^{w-1} PHASORs, each controlled $(w-1)$ times, and the diagonal matrix $\text{diag}(1, 1, 1, a_3, 1, a_5, \ldots)$ represents $2^{w-1} - 1$ PHASORs, each controlled $(w-1)$ times. E.g. for $w = 3$, we obtain

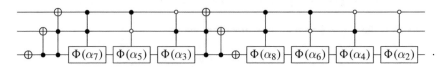

We thus have a total of $2^w - 1$ controlled PHASORs. According to Lemma 7.5 of Barenco et al. [21], each multiply-controlled gate $\Phi(\alpha)$ can be replaced by classical gates and three singly-controlled PHASORs $\Phi(\pm \alpha/2)$. According to De Vos and De Baerdemacker [7], each singly-controlled PHASOR $\Phi(\beta)$ can be decomposed into two controlled NOTs and three uncontrolled PHASORs $\Phi(\pm \beta/2)$. We thus obtain a circuit with a total of $9(2^w - 1)$ uncontrolled PHASORs.

We conclude that any matrix from ZU(n) can be synthesized by a cascade of P(n) blocks and ZZU(n) blocks. Here, ZZU(n) denotes the 1-dimensional subgroup of ZU(n) consisting of all $n \times n$ diagonal unitary matrices with all diagonal elements equal to 1, except the lower-right entry. It is isomorphic to ZU(2) and thus to U(1).

18.9 The *XU* Circuit Synthesis

Because of (18.6), the synthesis of an XU circuit is reduced to the synthesis of matrices consisting of two blocks on the diagonal: a unit submatrix and a CXU submatrix. We will call such matrices block-circulant, as they are composed of two circulant blocks.

18.10 The *CXU* Circuit Synthesis

In spite of the fact that CXU(n) is isomorphic to ZU(n), its synthesis is not as straightforward. The group ZU(n) is isomorphic to U(1)$^{n-1}$. Therefore, the group CXU(n) is equally isomorphic to the direct product U(1)$^{n-1}$. The $n - 1$ generators of ZU(n) are the matrices

$$g_k = \begin{pmatrix} \mathbf{0}_{k-1} & & \\ & 1 & \\ & & \mathbf{0}_{n-k} \end{pmatrix},$$

where $2 \leq k \leq n$ and $\mathbf{0}_j$ is a short-hand notation for the $j \times j$ zero matrix. As an example, we give here the two generators of ZU(3):

$$g_2 = \begin{pmatrix} 0\,0\,0 \\ 0\,1\,0 \\ 0\,0\,0 \end{pmatrix} \text{ and } g_3 = \begin{pmatrix} 0\,0\,0 \\ 0\,0\,0 \\ 0\,0\,1 \end{pmatrix}.$$

For each set $\{j, k\}$, we have that the two generators g_j and g_k commute:

$$[g_j, g_k] = g_j g_k - g_k g_j = 0.$$

That is exactly the reason why ZU(n) is a direct product of its $n - 1$ subgroups isomorphic to ZZU(n) and thus to U(1). Each 1-dimensional subgroup is of the form

$$m_k(\theta) = \begin{pmatrix} \mathbf{0}_{k-1} & & \\ & e^{i\theta} & \\ & & \mathbf{0}_{n-k} \end{pmatrix}.$$

By Fourier conjugating the ZU(n) generators, we find the CXU(n) generators. They look like:

$$g_k = \frac{1}{n} \begin{pmatrix} 1 & \Omega & \Omega^2 & \dots & \Omega^{n-1} \\ \Omega^{-1} & 1 & \Omega & \dots & \Omega^{n-2} \\ \vdots & & & & \\ \Omega^{-n+1} & \Omega^{-n+2} & \Omega^{-n+3} & \dots & 1 \end{pmatrix},$$

where Ω is a short-hand notation for ω^{1-k}. All $n - 1$ generators of CXU(n) equally commute. Whereas in general a single generator g generates a 1-dimensional matrix group given by the matrix exponentiation $m(\theta) = e^{ig\theta}$, in this particular case (because of the property $g_k^2 = g_k$), the generated matrices have the simple expression

$$m_k(\theta) = \mathbf{1}_n + (e^{i\theta} - 1)g_k.$$

As an example, we give here the two generators of CXU(3):

$$g_2 = \frac{1}{3} \begin{pmatrix} 1 & \omega^2 & \omega \\ \omega & 1 & \omega^2 \\ \omega^2 & \omega & 1 \end{pmatrix} \text{ and } g_3 = \frac{1}{3} \begin{pmatrix} 1 & \omega & \omega^2 \\ \omega^2 & 1 & \omega \\ \omega & \omega^2 & 1 \end{pmatrix}.$$

Each generates a 1-dimensional subgroup of CXU(3):

$$m_2(\theta) = \frac{1}{3} \begin{pmatrix} 2+x & \omega^2(x-1) & \omega(x-1) \\ \omega(x-1) & 2+x & \omega^2(x-1) \\ \omega^2(x-1) & \omega(x-1) & 2+x \end{pmatrix} \text{ and}$$

$$m_3(\theta) = \frac{1}{3} \begin{pmatrix} 2+x & \omega(x-1) & \omega^2(x-1) \\ \omega^2(x-1) & 2+x & \omega(x-1) \\ \omega(x-1) & \omega^2(x-1) & 2+x \end{pmatrix},$$

where x is a short-hand notation for $e^{i\theta}$.

The $n \times n$ matrices $m_k(\theta)$ are a generalization of the 2×2 NEGATOR $N(\theta)$: they are a unitary interpolation between the $n \times n$ unit matrix $m_k(0)$ and the $n \times n$ generalized NOT matrix $m_k(\pi) = \mathbf{1}_n - 2g_k$. We have only one 2×2 generalized NOT, i.e. the classical NOT:

$$m_2(\pi) = \begin{pmatrix} 0 & 1 \\ 1 & 0 \end{pmatrix};$$

we have two 3×3 generalized NOTs:

$$m_2(\pi) = \frac{1}{3} \begin{pmatrix} 1 & -2\omega^2 & -2\omega \\ -2\omega & 1 & -2\omega^2 \\ -2\omega^2 & -2\omega & 1 \end{pmatrix} \text{ and } m_3(\pi) = \frac{1}{3} \begin{pmatrix} 1 & -2\omega & -2\omega^2 \\ -2\omega^2 & 1 & -2\omega \\ -2\omega & -2\omega^2 & 1 \end{pmatrix};$$

we have three 4×4 generalized NOTs:

$$m_2(\pi) = \frac{1}{2} \begin{pmatrix} 1 & -i & 1 & i \\ i & 1 & -i & 1 \\ 1 & i & 1 & -i \\ -i & 1 & i & 1 \end{pmatrix}, \ m_3(\pi) = \frac{1}{2} \begin{pmatrix} 1 & 1 & -1 & 1 \\ 1 & 1 & 1 & -1 \\ -1 & 1 & 1 & 1 \\ 1 & -1 & 1 & 1 \end{pmatrix}, \text{ and}$$

$$m_4(\pi) = \frac{1}{2} \begin{pmatrix} 1 & i & 1 & -i \\ -i & 1 & i & 1 \\ 1 & -i & 1 & i \\ i & 1 & -i & 1 \end{pmatrix};$$

etcetera.

By applying the KAK decomposition [22, 23] of U(3), it is proved in [17] that any XU circuit can be decomposed into a cascade of

- uncontrolled NEGATORs,
- singly controlled V gates, and
- doubly controlled NOTs.

E.g. the above CXU(3) matrix $m_3(\theta)$ has the following XU(3) KAK decomposition:

$$V_0(\theta_0) V_3(\theta_1) V_2(\theta_2) V_3(\theta_3) = V_0(\theta/2) V_3(\pi - \theta/2) V_2(\pi) V_3(0)$$

$$= \frac{1}{3} \begin{pmatrix} 1+2y & 1-y & 1-y \\ 1-y & 1+2y & 1-y \\ 1-y & 1-y & 1+2y \end{pmatrix} \frac{1}{3} \begin{pmatrix} 1-2c & 1+c+\sqrt{3}s & 1+c-\sqrt{3}s \\ 1+c-\sqrt{3}s & 1-2c & 1+c+\sqrt{3}s \\ 1+c+\sqrt{3}s & 1+c-\sqrt{3}s & 1-2c \end{pmatrix} \frac{1}{3} \begin{pmatrix} -1 & 2 & 2 \\ 2 & -1 & 2 \\ 2 & 2 & -1 \end{pmatrix}.$$

where we follow the notations V_0, V_1, V_2, and V_3 of Appendix A of Reference [17] and where c and s are short-hand notations for $\cos(\theta/2)$ and $\sin(\theta/2)$, respectively, and y is $c + is = \sqrt{x}$. Subsequently, we can apply to each of these three matrices the XU(4) KAK decomposition. E.g. the rightmost matrix appears in the 2-qubit block-circulant matrix

$$\frac{1}{3}\begin{pmatrix} 3 & 0 & 0 & 0 \\ 0 & -1 & 2 & 2 \\ 0 & 2 & -1 & 2 \\ 0 & 2 & 2 & -1 \end{pmatrix}$$

with decomposition

$$V_0(\theta_0)V_3(\theta_1)V_2(\theta_2)V_3(\theta_3)V_5(\theta_4)V_3(\theta_5)V_2(\theta_6)V_3(\theta_7)V_8(\theta_8)$$
$$= V_0(0)V_3(0)V_2(7\pi/4)V_3(0)V_5(\varphi)V_3(0)V_2(\pi/4)V_3(0)V_8(0)$$

$$= \frac{1}{4}\begin{pmatrix} 2-\sqrt{2} & 2+\sqrt{2} & \sqrt{2} & -\sqrt{2} \\ 2+\sqrt{2} & 2-\sqrt{2} & -\sqrt{2} & \sqrt{2} \\ -\sqrt{2} & \sqrt{2} & 2-\sqrt{2} & 2+\sqrt{2} \\ \sqrt{2} & -\sqrt{2} & 2+\sqrt{2} & 2-\sqrt{2} \end{pmatrix}$$

$$\frac{1}{3}\begin{pmatrix} 1 & -\sqrt{2} & 2 & \sqrt{2} \\ \sqrt{2} & 1 & -\sqrt{2} & 2 \\ 2 & \sqrt{2} & 1 & -\sqrt{2} \\ -\sqrt{2} & 2 & \sqrt{2} & 1 \end{pmatrix} \frac{1}{4}\begin{pmatrix} 2+\sqrt{2} & 2-\sqrt{2} & \sqrt{2} & -\sqrt{2} \\ 2-\sqrt{2} & 2+\sqrt{2} & -\sqrt{2} & \sqrt{2} \\ -\sqrt{2} & \sqrt{2} & 2+\sqrt{2} & 2-\sqrt{2} \\ \sqrt{2} & -\sqrt{2} & 2-\sqrt{2} & 2+\sqrt{2} \end{pmatrix},$$

where, this time, we follow the notations V_0, V_1, ..., and V_8 of Appendix B of Reference [17] and where φ is the angle $\pi + \text{Arccos}(1/3)$. This matrix decomposition leads to the circuit synthesis with

- six uncontrolled NEGATORs,
- six controlled V gates, and
- three controlled NOTs:

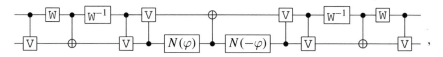

where $\text{W}^{-1} = \text{XVW} = N(7\pi/4)$.

Fig. 18.2 Hierarchy of the
Lie groups U(n), XU(n),
ZU(n), CXU(n), and ZZU(n)
and the finite groups P(n),
CP(n), and **1**(n)

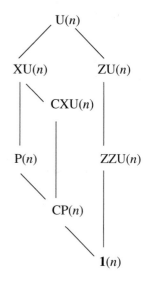

18.11   Conclusion

With the help of truth tables, we have demonstrated that conventional Boolean computation can be embedded in classical reversible computation. With the help of square matrices, we have demonstrated that classical reversible computation is a subspace of quantum computation. Classical reversible computing relates to quantum computing like permutation matrices relate to unitary matrices. The permutation matrix group P(n) forms a subgroup of the unitary matrix group U(n). The main leap from P(n) matrices to U(n) matrices happens by interpolation between two or more permutation matrices, thus enlarging the finite group P(n) to the infinite group XU(n). Figure 18.2 shows in detail the hierarchy of groups and subgroups, revealing the relationship between the finite group P(n) and the infinite group U(n). The decomposition of a given U(2^w) matrix into ZU(2^w) and CXU(2^w) matrices leads to a w-qubit synthesis of the U(2^w) circuit with PHASOR and NEGATOR building blocks.

Appendix

For any matrix P from the group P(n), the following property holds:

$$P = C \begin{pmatrix} 1 & \\ & P' \end{pmatrix},$$

where

- P' is a member of P($n-1$) and

- C is a member of CP(n).

Here, CP(n) denotes the group of $n \times n$ circulant permutation matrices. It is a group isomorphic with the cyclic group $\mathbf{Z}_n$, a finite group of order n. Remarkable is the fact that here $\left(\begin{smallmatrix} 1 & \\ & P' \end{smallmatrix}\right)$ is only multiplied to the left with a circulant matrix, whereas in decomposition (18.5), the matrix $\left(\begin{smallmatrix} 1 & \\ & X' \end{smallmatrix}\right)$ is multiplied both to the left and to the right with a circulant matrix.

The decomposition algorithm is very straightforward: suffice it to choose C such that it has the same leftmost column as the given matrix P. Subsequently, the matrix $\left(\begin{smallmatrix} 1 & \\ & P' \end{smallmatrix}\right)$ follows automatically by computing $C^{-1}P$. Here follows an example from P(4):

$$
\begin{pmatrix} 0\,0\,0\,1 \\ 1\,0\,0\,0 \\ 0\,0\,1\,0 \\ 0\,1\,0\,0 \end{pmatrix} = \begin{pmatrix} 0\,0\,0\,1 \\ 1\,0\,0\,0 \\ 0\,1\,0\,0 \\ 0\,0\,1\,0 \end{pmatrix} \begin{pmatrix} 1\,0\,0\,0 \\ 0\,0\,1\,0 \\ 0\,1\,0\,0 \\ 0\,0\,0\,1 \end{pmatrix}.
$$

By applying the theorem again and again, we find the following decomposition:

$$
P = C_n \begin{pmatrix} 1 & \\ & C_{n-1} \end{pmatrix} \cdots \begin{pmatrix} \mathbf{1}_{n-3} & \\ & C_3 \end{pmatrix} \begin{pmatrix} \mathbf{1}_{n-2} & \\ & C_2 \end{pmatrix}. \tag{18.7}
$$

where all C_k are CP(k) matrices. We conclude: any $n \times n$ permutation matrix can be decomposed as a product of $n - 1$ matrices of the form $\left(\begin{smallmatrix} \mathbf{1}_{n-k} & \\ & C_k \end{smallmatrix}\right)$. We give an example:

$$
\begin{pmatrix} 0\,0\,0\,1 \\ 1\,0\,0\,0 \\ 0\,0\,1\,0 \\ 0\,1\,0\,0 \end{pmatrix} = \begin{pmatrix} 0\,0\,0\,1 \\ 1\,0\,0\,0 \\ 0\,1\,0\,0 \\ 0\,0\,1\,0 \end{pmatrix} \begin{pmatrix} 1\,0\,0\,0 \\ 0\,0\,0\,1 \\ 0\,1\,0\,0 \\ 0\,0\,1\,0 \end{pmatrix} \begin{pmatrix} 1\,0\,0\,0 \\ 0\,1\,0\,0 \\ 0\,0\,0\,1 \\ 0\,0\,1\,0 \end{pmatrix}.
$$

Decomposition (18.6) is not a straightforward generalization of (18.7). This constitutes an illustration of the fact that, in spite of an overall similarity between the group P(n) and the group XU(n), literal translations from P(n) properties to XU(n) properties sometimes fail [24].

References

1. Nielsen, M., Chuang, I.: Quantum Computation and Quantum Information. Cambridge University Press, Cambridge (2000)
2. Fredkin, E., Toffoli, T.: Conservative logic. Int. J. Phys. **21**, 219 (1982)
3. De Vos A., Van Rentergem, Y.: From group theory to reversible computers. In: Adamatzky, A., Teuscher, C. (eds.) From Utopian to Genuine Unconventional Computers, pp 183-208. Luniver Press, Frome (2006)
4. De Vos, A.: Reversible Computing. Wiley-VCH, Weinheim (2010)
5. Wille, R., Drechsler, R.: Towards a Design Flow for Reversible Logic. Springer, Dordrecht (2010)

6. Soeken, M., Wille, R., Keszöcze, O., Miller, M., Drechsler, R.: Embedding of large Boolean functions for reversible logic. *ACM J. Emerg. Technol. Comput. Syst.* **12**, 41 (2015)
7. De Vos, A., De Baerdemacker, S.: The decomposition of U(n) into XU(n) and ZU(n). J. Mult. Val. Log. Soft Comp. **26**, 141 (2016)
8. De Vos, A., De Baerdemacker, S.: Matrix calculus for classical and quantum circuits. ACM J. Emerg. Technol. Comput. Syst. **11**, 9 (2014)
9. De Vos, A., De Baerdemacker, S.: On two subgroups of U(n), useful for quantum computing. In: Proceedings of the 30th International Colloquium on Group-theoretical Methods in Physics, Gent, 14–18 July 2014. J. Phys. Conf. Ser. **597**, 012030 (2015)
10. Sasanian, Z., Miller, D.: Transforming MCT circuits to NCVW circuits. In: Proceedings of the 3 rd International Workshop on Reversible Computation, pp 163–174. Gent (2011)
11. Amy, M., Maslov, D., Mosca, M.: Polynomial-time T-depth optimization of Clifford+T circuits via matroid partitioning. In: IEEE Trans. on Comput-Aided Des. Integr. Circuits Syst. **33**, 1486 (2013)
12. Selinger, P.: Efficient Clifford+T approximations of single-qubit operators. Quantum Inform. Comput. **15**, 159 (2015)
13. Deutsch, D.: Quantum computation. Phys. World **5**, 57 (1992)
14. Galindo, A., Martín-Delgado, M.: Information and computation: classical and quantum aspects. Rev. Moder. Phys. **74**, 347 (2002)
15. De Vos, A., De Beule, J., Storme, L.: Computing with the square root of NOT. Serdica J. Comput. **3**, 359 (2009)
16. Vandenbrande, S., Van Laer, R., De Vos, A.: The computational power of the square root of NOT. In: Proceedings of the 10 th Interntional Workshop on Boolean Problems, pp 257–262. Freiberg, 19-21 September 2012
17. De Vos, A., De Baerdemacker, S.: The NEGATOR as a basic building block for quantum circuits. Open Syst. Inf. Dyn. **20**, 1350004 (2013)
18. De Vos, A., De Baerdemacker, S.: Scaling a unitary matrix. Open Syst. Inf. Dyn. **21**, 1450013 (2014)
19. Idel, M., Wolf, M.: Sinkhorn normal form for unitary matrices. Lin. Algeb. Appl. **471**, 76 (2015)
20. Beth, T., Rötteler, M.: Quantum algorithms: applicable algebra and quantum physics, In: Alber, G., Beth, T., Horodecki, M., Horodecki, P., Horodecki, R., Rötteler, M., Weinfurter, H., Werner, R., Zeilinger, A. (eds.) Quantum Information, pp 96–150. Springer, Berlin (2001)
21. Barenco, A., Bennett, C., Cleve, R., DiVincenzo, D., Margolus, N., Shor, P., Sleator, T., Smolin, J., Weinfurter, H.: Elementary gates for quantum computation. Phys. Rev. A **52**, 3457 (1995)
22. Hermann, R.: Lie Groups for Physicists. Benjamin Inc., New York (1966)
23. Bullock, S., Markov, I.: An arbitrary two-qubit computation in 23 elementary gates. Phys. Rev. A **68**, 012318 (2003)
24. De Vos, A., De Baerdemacker, S.: Symmetry groups for the decomposition of reversible computers, quantum computers, and computers in between. Symmetry **3**, 305 (2011)

Chapter 19
Fault Models in Reversible and Quantum Circuits

Martin Lukac, Michitaka Kameyama, Marek Perkowski,
Pawel Kerntopf and Claudio Moraga

Abstract In this chapter we describe faults that can occur in reversible circuit as compared to faults that can occur in classical irreversible circuits. Because there are many approaches from classical irreversible circuits that are being adapted to reversible circuits, it is necessary to analyze what faults that exists in irreversible circuits can appear in reversible circuit as well. Thus we focus on comparing faults that can appear in classical circuit technology with faults that can appear in reversible and quantum circuit technology. The comparison is done from the point of view of information reversible and information irreversible circuit technologies. We show that the impact of reversible computing and quantum technology strongly modifies the fault types that can appear and thus the fault models that should be considered. Unlike in the classical non-reversible transistor based circuits, in reversible circuits it is necessary to specify what type of implementation technology is used as different technologies can be affected by different faults. Moreover the level of faults and their analysis must be revised to precisely capture the effects and properties of quantum gates and quantum circuits that share several similarities with reversible circuits. By not doing so the available testing approaches adapted from classical circuits would not be able to properly detect relevant faults. In addition, if the classical faults are directly

M. Lukac (✉)
School of Science and Technology, Nazarbayev University, Astana, Kazakhstan
e-mail: martin.lukac@nu.edu.kz

M. Kameyama
Graduate School of Information Sciences, Tohoku University, Sendai, Japan
e-mail: kameyama@ecei.tohoku.ac.jp

M. Perkowski
Department of Computer and Electrical Engineering,
Portland State University, Portland, USA
e-mail: mp8h@pdx.edu

P. Kerntopf
Department of Physics and Applied Informatics, University of Lodz, Lodz, Poland
e-mail: pawel.kerntopf@ii.pw.edu.pl

C. Moraga
TU Dortmund University, Dortmund, Germany
e-mail: claudio.moraga@tu-dortmund.de

© Springer International Publishing Switzerland 2017   475
A. Adamatzky (ed.), *Advances in Unconventional Computing*,
Emergence, Complexity and Computation 22,
DOI 10.1007/978-3-319-33924-5_19

applied without revision and modifications, the presented testing procedure would be testing for such faults that cannot physically occur in the given implementation of reversible circuits. The observation and analysis of these various faults presented in this chapter clearly demonstrates what faults can occur and what faults cannot occur in various reversible technologies. Consequently the results from this chapter can be used to design more precise tests for reversible logic circuits. Moreover the clearly described differences between faults occurring in reversible and irreversible circuits means that new algorithms for fault detection should be implemented specifically for particular reversible technologies.

19.1 Introduction

Reversible circuit representation can be used to debug and test the correctness of the hardware implementation. Unlike the non-reversible representation currently available for the CMOS VLSI, the reversible circuit representation is not commonly associated with one particular technology. Classical irreversible circuits representation is in general associated with a transistor based circuit technology while reversible circuits can represent quantum technology (ion trap, NMR, cavity QED, quantum optics), switch based technology, CMOS adiabatic or simply CMOS implementation of reversible circuits. Moreover, because *quantum circuit model* for true quantum computer has not been constructed (as compared to currently available classical desktops or supercomputers and in contrast to adiabatic quantum computers) multiple technologies might be combined in one reversible computer to achieve the desired functional reversibility and attractive power savings.

Thus when testing a particular reversible circuit one should specify not only what technology is targeted but also what component of a general computer is considered. For instance testing a reversible bus implemented in the switch technology will require different tests than testing the attached reversible circuits that are using a quantum optics implementation.

Currently there have been two distinct approaches to testing of reversible and quantum circuits. On one hand the reversible community has been applying methods used in classical circuits to test and evaluate reversible circuits on the functional level [9, 10, 21, 22, 24, 26, 29–31]. On the other hand, the approach to test quantum circuits with respect to the underlying technology has been exploring the possible physical faults that can really occur in a reversible circuits [6, 13, 23]. Finally, some researchers concluded that testing reversible circuits hardware from the functional point of view is too simplistic [1].

An alternative to testing quantum and reversible circuits for faults, is to build them as fault tolerant. This approach has seen a large momentum in both the theoretical as well as in the engineering communities due to the fact that quantum—as the main implementation technology of reversible circuits—is very noise sensitive and subject to constant perturbations. Consequently, protocols intended to provide the least noise logic operations and computational states have been devised [5, 13, 14,

Fig. 19.1 Information in transistor based circuits is represented by discrete values of voltage. The information is propagated from storage device to storage device by using a clock

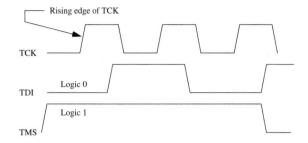

20, 25]. However, not all faults can be captured by the fault tolerant implementation of quantum circuits. In particular, faults that occur in real time and are caused by environmental factors that are beyond the reasonable limit of fault tolerant implementations. Moreover, the fault tolerant implementation addresses the problem of noisy gates and states in general while testing permits to determine the faults more precisely and categorize them by type.

In this chapter we look closer at what types of faults one should be testing when new testing methods for reversible circuits will be developed. We do not analyze faults particular to quantum reversible circuits; starting from faults in information reversible circuits (simulation of reversible circuits) we discuss what faults are worth of testing and what faults are not useful when dealing with energy reversible circuits technology. We show that for energy reversible implementation only certain faults can appear. We also show that other faults exclusively available in information reversible technology must be redefined in order to describe the features of quantum and other energy reversible technologies. Thus, the main contribution of this chapter is a precise characterization of faults that one should consider when designing algorithms for energy reversible circuit testing such as implemented in quantum technologies.

The chapter is organized as follows. Section 19.2 describes the information representation in both the information and energy reversible circuits. Section 19.3 shows faults for information reversible technology while Sect. 19.4 describes the faults of energy conservative circuits. Finally, Sect. 19.5 discusses the results and concludes the chapter.

19.2   Information in Circuits

Classical information in currently available circuit technology is represented by a voltage/current level. In binary circuits the logic value 0 is in general represented by a *low* voltage and logic value 1 is represented by a *high* voltage (see Fig. 19.1 the curves "TDI" and "TMS", respectively). Between two logic circuits, a clock is connected to the registers that during each clock cycle propagates the signal from one register, transfers the signal through a logic circuit and stores it in the next register. Example of a clock signal is shown in Fig. 19.1 by the curve "TCK".

Fig. 19.2 Example of a classical circuit on transistor level implementing the NAND logic gate

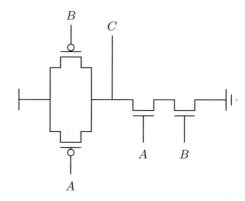

Fig. 19.3 Example of a classical circuit

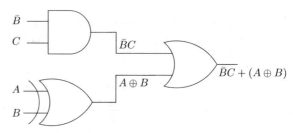

The information is propagated in a switching circuit by a sequential/parallel turning on(off) of transistors in the circuit. For instance, Fig. 19.2 shows the universal logic gate called NAND. It contains two n-MOS and two p-MOS transistors that will produce a *high* output on C when any of the control signals A or B have value "0".

Notice that transistor technology is inherently irreversible without additional overhead because the input signals (A and B) only turn on (off) transistors that either connect the power supply or the ground to the output wire C. Thus the energy forming the output values is not the energy from the inputs. Circuits built from such irreversible gates are usually represented using the gate level schema as shown in Fig. 19.3.

19.2.1 Information Reversible Circuits

Using the transistor technology various reversible circuits implementations have been proposed. One of the oldest is the dual line reversible circuit [8, 27, 28]. An example of the implementation is shown in Fig. 19.4 which presents an information reversible implementation of the CNOT ($A \otimes (A \oplus B)$) logic gate. This approach is using dual lines (the inputs are always provided in double complementary values).

The circuit operates in such a manner that certain variables are used to turn on (off) transistors or transmission gates allowing other signals to pass from input to output.

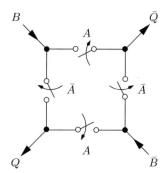

Fig. 19.4 Example of information reversible circuit implementation of the $A \oplus B = Q$ using transistor technology. (The *arrows* across the switches indicate the state when the corresponding variable is in state "1")

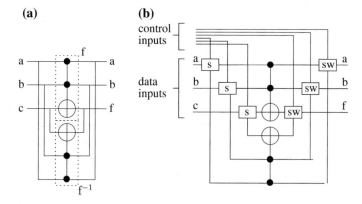

Fig. 19.5 The schematic diagram of **a** a Universal Reversible Cell (URC), **b** with switches required for connections with other URCs

For instance, the gate in Fig. 19.4 will close half of the switches if the variable A takes value "1"and the other half if it takes value "0". The result is then read from the output lines Q and $\bar{Q}$. Naturally, in order to ensure the information reversibility all signals that control the switches must be also propagated to the outputs. Other existing approaches to the implementation of information reversible logic circuits include reversible circuits based on the principles of adiabatic charging and discharging [4], information reversible CMOS Toffoli cell [17] as well as the MEMS implemented reversible circuits [3, 12].

Another type of reversible circuit implementation in classical technology was shown in [15] where a logically reversible circuit (information reversible) was simulated by CMOS cells that could either transmit or store information. In this model the information is only logically retrievable while the energy is lost due to the CMOS transistor implementation. Example of a reversible cell is shown in Fig. 19.5.

Thus by information reversible circuits we mean any circuits that allow for information reversibility without being energy reversible. This includes circuits that use contemporary transistor technology but implement reversible circuits. These approaches are in general intended to simulate and study properties of reversible circuits. However, as the technology used is *classical*, these circuits are not energy reversible. Consequently, we can define the Information Reversible Circuits.

Definition 1 (*Information Reversible Circuits*) These circuits represent a reversible bijection, i.e. such function $f : I \rightarrow O$ for which inverse function $f^{-1} : O \rightarrow I$ exists.

19.2.2 Energy Reversible Circuits

Definition 2 (*Energy reversible circuits*) The circuits in which the final energy state ρ_f permits to recover the initial energy state ρ_i are called energy reversible circuits. An example is the adiabatic charging or quantum computational model.

In information reversible and energy conservative circuits such as implemented in quantum technologies, information is represented using a single elementary particle or a group of elementary particles such as photons, electrons, ions etc. In the information reversibility the logic states are obtained by unitary transformations and the energy states are obtained by discrete energy control signals. Unlike in non-quantum computation, the quantum state is information reversible and energy conservative. The information is represented in three dimensions by a qubit shown in Fig. 19.6. The Bloch sphere shows that the quantum information state is represented as a vector in a

Fig. 19.6 *Bloch sphere*: a symbolic representation of a single qubit. Multiple qubits cannot be represented in this manner due to superposition and entanglement

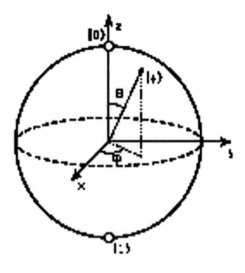

Fig. 19.7 Example of an
implementation of the CNOT
gate using the Hadamard and
CZ quantum gates

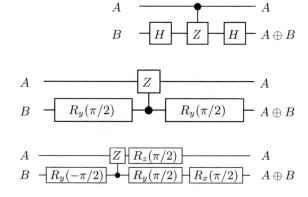

Fig. 19.8 Example of one
possible implementation of
the CNOT gate in the Ion
Trap technology

Fig. 19.9 Example of
CNOT gate built using the
Ising model in NMR
technology

3-dimensional space and specified by an amplitude and a phase. The manipulation of
the quantum state (quantum information) is done through a series of unitary rotations
on any of the three axes R_x, R_y and R_z.

The information reversibility and energy conservation is preserved theoretically
in quantum systems only up to the measurement.

Definition 3 (*Quantum Destructive Measurement*) (QDM) is a non-unitary oper-
ation, that when used on a quantum system described by state $|\tau\rangle$ it projects
$M\tau =_\perp |\rho\rangle$ to one of the orthonormal bases $|\rho\rangle$.

Practically this means that the measurement of a quantum state destroys the quantum
state and only the information captured by the projection remains known. This is
not problematic in most of the cases, but in the case of quantum entanglement the
measured qubits do not allow the recovery of their initial state once they have been
measured.

Quantum circuits are built from quantum non-permutative primitives that are
rarely equivalent to the reversible and permutative circuits building blocks.

For example, Fig. 19.7 shows the quantum realization of the CNOT logic gate
using the CZ (Controlled-Z) and the Hadamard quantum gates. The lines represents
quantum wires, qubits that carry quantum states that represent corresponding logic
states. Gates are in general single qubit rotations or conditional rotations. A single
qubit rotation is such a gate that will indiscriminately rotate a qubit independently
from what state it is in. A conditional rotation is such a logic gate that will rotate a
qubit if and only if the so called control qubit is set to a particular value. More details
on the principles of quantum circuits and quantum computing the interested reader
can for instance obtain from [19]. Another realization of the same gate is shown in
Fig. 19.8 where the Hadamard gates are replaced by the Pauli rotations around the y
of the qubit $|B\rangle$ [19].

Figure 19.9 shows the implementation of the CNOT gate using the Ising model
used in the NMR quantum computing [19].

A final example of implementation of a quantum circuit is shown in Fig. 19.10.
It is the Toffoli gate built using linear quantum optics which has been presented

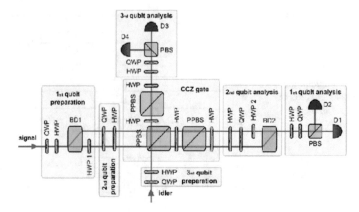

Fig. 19.10 Example of Toffoli gate built using Quantum Linear Optics (Copyright by APL)

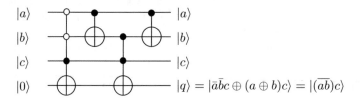

Fig. 19.11 Example of a reversible circuit

recently in [11, 18] and Fig. 19.10 boxes labeled with D represent single photons detectors, BD1 and BD2 are beam displacers, GP glass plates, PBS and PPBS are polarizing beam splitters and partially polarizing beam splitters and HWP and QWP are half and quarter wave plates respectively.

Using reversible primitives such as shown in Figs. 19.7 to 19.10 logically reversible circuits are built. An example of a logical reversible circuit representation is given in Fig. 19.11. Each of the reversible gates is built using a set of quantum realizable gates or a set of information reversible gates. Thus the logical primitives used to design reversible circuits differ substantially from the component reversible gates (Toffoli, CNOT, Fredkin, etc.).

19.2.3   Reversibility and Implementation

In the former section we presented examples of information and energy reversible implementations of reversible logic circuits. We also showed that there are several technologies that guarantee the logic reversibility while the energy reversibility requires a technology that is reversible naturally. However we are looking at the reversible circuits only from the logic side. Moreover the logic reversible side can be

Fig. 19.12 a CNOT circuit, **(a)**   **(b)**
b permutative matrix
representing the circuit

$|a\rangle$ —●— $|a\rangle$

$|b\rangle$ —⊕— $|b\rangle$ if $|a\rangle = |0\rangle$

or $CNOT = \begin{pmatrix} 1 & 0 & 0 & 0 \\ 0 & 1 & 0 & 0 \\ 0 & 0 & 0 & 1 \\ 0 & 0 & 1 & 0 \end{pmatrix}$

$|\bar{b}\rangle$ if $|a\rangle = |1\rangle$

hiding irreversible technology. In this section we have a brief look on the reversible function representation and what it means to observe faults in reversible logic.

For instance, the information reversible technology shown in Fig. 19.4 would be also energy reversible if the components and the power technology to which it was connected were naturally reversible. Because this is not the case the illustrated switch-based technology remains classified only as information reversible.

Consequently when talking about detecting faults in logically reversible circuits it is required to consider what faults can actually appear in a reversible circuit. Depending on the technology some faults might not be physically possible to appear while others can be natural. Thus an analysis of what is possible and what is not can help to design testing algorithms for reversible circuits.

Reversible functions are represented by permutative matrices. For instance, Fig. 19.12a shows the circuit representing the CNOT gate and Fig. 19.12b shows the matrix associated with it. The function implemented on the qubit $|b\rangle$ is Boolean function EXOR: $(a \oplus b)$ if the input states are pure orthonormal states. Thus the logic operation performed on the qubit $|b\rangle$ can also be written as $|b\rangle = |a \oplus b\rangle$ with a and b being Boolean logic values.

The input information considered in this paper is strictly limited to pure states. This means that any quantum or other state specified in various forms must always have a Boolean equivalent. Here the information is represented in a vector form, for instance $|ab\rangle = |11\rangle = \begin{bmatrix} 0 \\ 0 \\ 0 \\ 1 \end{bmatrix}$. To apply the CNOT operation, a multiplication with the permutative matrix representing the gate or the circuit is performed as shown in Eq. 19.1.

$$|a\bar{b}\rangle = \begin{bmatrix} 1 & 0 & 0 & 0 \\ 0 & 1 & 0 & 0 \\ 0 & 0 & 0 & 1 \\ 0 & 0 & 1 & 0 \end{bmatrix} \begin{bmatrix} 0 \\ 0 \\ 0 \\ 1 \end{bmatrix} = \begin{bmatrix} 0 \\ 0 \\ 1 \\ 0 \end{bmatrix} \tag{19.1}$$

A more complex reversible and also universal (logically universal) function—the Toffoli gate—is shown in Fig. 19.13a. Figure 19.13b shows how a reversible function is represented in a truth-tabular form. The inputs to the function are specified by letters a, b, c. The output function, in this case $f = \bar{c}$ iff $a = b = 1$ is shown in the columns under headings a', b' and c' in the table.

Fig. 19.13 **a** The Toffoli
universal gate and **b** Toffoli
gate value table

(a)

$$|a\rangle = a \quad\bullet\quad |a'\rangle = a'$$

$$|b\rangle = b \quad\bullet\quad |b'\rangle = b'$$

$$|c\rangle = c \quad\oplus\quad |c'\rangle = c'$$

(b)

abc	a'b'c'
000	000
001	001
010	010
011	011
100	100
101	101
110	111
111	110

19.3 Information Reversible Circuits Faults

The faults occurring in information reversible circuits are similar to the faults occurring in transistor based circuits.

The permutative nature of the reversible function is only simulated in classical circuit implementation but it is a natural property of truly information and energy reversible circuits: an energy reversible circuit must implement a reversible function. This means that while in the case of the circuits with only information reversibility the faults can affect the unitary and permutative property of reversible function implementation, energy reversible technology always implement a reversible function unless special faults can exist.

19.3.1 Stuck-At Faults at the Output

The well known *stuck-at fault* model means that no matter what input signal is applied to the wire or to the gate the output will be either constant 0 or constant 1. A stuck-at fault is in general called Stuck-at-0 or Stuck-at-1 fault. In transistor based circuit this means that the circuit behaves as if either the drain or the source were connected to the output wire.

If a stuck-at fault occurs in a gate, it means that for any possible input combination the output will be stuck at a value. There are two cases of the *stuck-at fault* that can be observed. In the first case, the qubit can have its state dumped by a non-unitary error resulting in the qubit to be always in a particular state.

$$\text{CNOT with Stuck\_at\_1}(|b\rangle) = \begin{pmatrix} 0 & 0 & 0 & 0 \\ 1 & 1 & 0 & 0 \\ 0 & 0 & 0 & 0 \\ 0 & 0 & 1 & 1 \end{pmatrix} \tag{19.2}$$

Fig. 19.14 Example of a circuit affected by the stuck at fault on a qubit

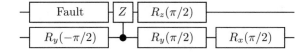

For instance, the stuck-at-1 fault on the input qubit $|a\rangle$ in the CNOT gate from Fig. 19.12b is shown in Eq. 19.2. The circuit for such a fault can be represented as shown in Fig. 19.14 where the fault is simulated by a transformation on the qubit before entering the logic state manipulation. This fault will be called state-fault for the rest of the paper; the faulty qubit is independently changed to a stuck-at state and thus becomes a constant value. Depending on the location of the fault in the quantum gate, the output of the faulty qubit can be probabilistic or constant Fig. 19.2. Notice that the stuck-at fault changed the unitary matrix to a non-unitary matrix; the resulting function is not a reversible function. This is an important fact that limits the existence of this fault to only information reversible circuits.

In particular observe that if the same logic operation as in Eq. 19.1 is applied using the matrix from Eq. 19.2 the result would be as follows (Eq. 19.3).

$$\begin{bmatrix} 0\,0\,0\,0 \\ 1\,1\,0\,0 \\ 0\,0\,0\,0 \\ 0\,0\,1\,1 \end{bmatrix} \begin{bmatrix} 0 \\ 0 \\ 0 \\ 1 \end{bmatrix} = \begin{bmatrix} 0 \\ 0 \\ 0 \\ 1 \end{bmatrix} = |ab\rangle \qquad (19.3)$$

In the second case all qubits have their states changed by sometimes faulty inter-actions in such a manner that the faulty qubit will not be observable stuck-at fault.

For instance, Eq. 19.4 shows a stuck-at-1 fault on the qubit $|B\rangle$.

$$\text{CNOT with Stuck}\_\text{at}\_1(|a\rangle) = \begin{pmatrix} 0\,1\,0\,0 \\ 1\,0\,0\,0 \\ 0\,0\,0\,1 \\ 0\,0\,1\,0 \end{pmatrix} \qquad (19.4)$$

As shown in Eq. 19.4 the fault is now located on such a component of the circuit which does not show the fault directly. Instead the unitary matrix can now be decomposed to $I \otimes X$).

19.3.2 Delay Faults

Delay faults occur when the timing specifications of a gate or a wire differ from the timing exhibited during the usage. Such faults can result in more severe computational faults due to the timing requirements of both classical synchronous and asynchronous circuits. However, delay faults are not discussed here as they are required to be tested using specific model of possible faults not considered here.

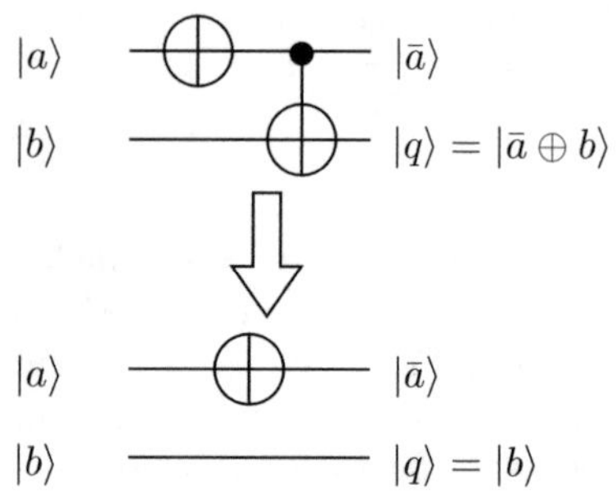

Fig. 19.15 Example of missing gate fault in a reversible circuit

19.3.3 Missing Gate Faults

A missing gate fault is a very general class of faults representing a problem resulting in a circuit that produces an output consistent with its function as if one or more of its component gates were missing. Missing gate faults change the values of the unitary matrix but do not change its unitary property. For instance, consider removing the CNOT gate from the circuit shown in Fig. 19.15.

The permutative unitary matrices representing the function before and after the missing gate fault was inserted are as follows (Eq. 19.5).

$$
f = \begin{bmatrix} 0 & 0 & 1 & 0 \\ 0 & 0 & 0 & 1 \\ 0 & 1 & 0 & 0 \\ 1 & 0 & 0 & 0 \end{bmatrix} \quad f' = \begin{bmatrix} 0 & 0 & 1 & 0 \\ 0 & 0 & 0 & 1 \\ 1 & 0 & 0 & 0 \\ 0 & 1 & 0 & 0 \end{bmatrix} \tag{19.5}
$$

19.3.4 Wrong Gate Faults

Similarly to missing gate fault the wrong gate fault represents a circuit that has the same number of logic gates as the circuit specified but does not generate the correct logic result. The matrices representing the function before and after the wrong gate fault was inserted as follows (Eq. 19.6) (Fig. 19.16).

$$
f = \begin{bmatrix} 0 & 0 & 1 & 0 \\ 0 & 0 & 0 & 1 \\ 0 & 1 & 0 & 0 \\ 1 & 0 & 0 & 0 \end{bmatrix} \quad f' = \begin{bmatrix} 1 & 0 & 0 & 0 \\ 0 & 1 & 0 & 0 \\ 0 & 0 & 0 & 1 \\ 0 & 0 & 1 & 0 \end{bmatrix} \tag{19.6}
$$

Fig. 19.16 Example of
wrong gate fault in a
reversible circuit

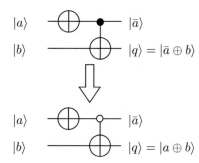

Fig. 19.17 Example of
bridging faults, **a** the original
circuit and **b** the circuit with
bridging fault that does not
alter the unitary property of
the circuit

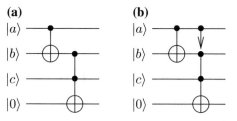

19.3.5   Bridging Faults

The bridging faults are a super category of faults which describe any faults resulting
from a wrong connection between elements that should not be connected. Unlike
stuck-at faults, they can generate fan-outs, fan-ins missing controls in gates or other
faults affecting gate functionality. In some cases the bridging faults will alter the
permutative properties of the unitary matrix of the circuit but in many cases they
only alter the gate functionality.

An example of a bridging fault that can occur as a wrong bridging between two
terminals is shown in Fig. 19.17. Let $|a\rangle$, $|b\rangle$ and $|c\rangle$ be pure states (i.e. a, b and c
represent Boolean values). The original circuit from Fig. 19.17 produces the function
$|q\rangle = |(a \oplus b)c\rangle$. After the bridging fault indicated as a line with arrow in Fig. 19.17
the circuit produces $|q\rangle = |(a \oplus b)ac\rangle = |(1 \oplus b)ac\rangle = |a\bar{b}c\rangle$.

Notice that the bridging fault preserves the unitary property of the circuit. Let F
denote the Feynman gate and T_n denote the Toffoli gate with n controls, then the
original function shown in Fig. 19.17a is given by $(I \otimes T_2) \times (F \otimes I \otimes I)$ while the
function with the bridging fault shown in Fig. 19.17b is given by $T_3 \times (F \otimes I \otimes I)$.

19.4   Energy Reversible Circuits Faults

Energy reversible circuit technology includes but is not limited to quantum circuits such as implemented in Ion Trap, Quantum Optics, NMR, cavity QED or solid state implementations. Another approach to building quantum computers includes the adiabatic computer as implemented by D-Wave [2, 7].

There are only two possible phenomenological faults (expressed by a faulty quantum phenomena) in a truly quantum circuit: value fault and phase fault. Value fault is a phenomenon that will generate a faulty output for a desired input. This is in general a consequence of applying a wrong gate or a gate with some internal fault. The second type is the phase fault and it represents a signal that carries a wrong phase. Depending on the implementation of a quantum circuit, the value fault is as crucial as the phase fault. Moreover, even if the circuit output is value oriented a phase fault can result in a faulty value.

This can be observed in an implementation of CNOT that uses the circuit shown in Fig. 19.18. Notice that if the gate in the middle, called Controlled-Z, changes the output as shown in Eq. 19.7 the output magnitude value is correct but the phase is modified. This has for consequence which the function that the entire circuit from Fig. 19.18 implements is completely wrong since it would produce a two-input identity gate rather than the CNOT gate. (Recall that Hadamard matrix is its own inverse).

$$CZ = \begin{bmatrix} 1 & 0 & 0 & 0 \\ 0 & 1 & 0 & 0 \\ 0 & 0 & 1 & 0 \\ 0 & 0 & 0 & -1 \end{bmatrix} \quad CZ' = \begin{bmatrix} 1 & 0 & 0 & 0 \\ 0 & 1 & 0 & 0 \\ 0 & 0 & 1 & 0 \\ 0 & 0 & 0 & 1 \end{bmatrix} \tag{19.7}$$

19.4.1   Missing Gate Faults

Similarly to the information-only reversible circuits the missing gate fault is such that one full gate is missing and the result is a circuit generating the incorrect result. In quantum circuits, however, the basic gates used to implement reversible functions are rarely directly realizable (Figs. 19.7 to 19.9). This means that while testing a circuit design to output pure state representing logical values of a reversible circuit, the probability of a logical primitive being completely missing is nearly 0. Rather, component gates such as single qubit rotations or interactions could be missing completely.

Fig. 19.18 Realization of CNOT gate using Controlled-Z and two Hadamard gates

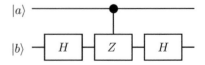

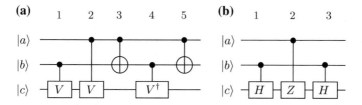

Fig. 19.19 Two realizations of Toffoli gate. **a** Using the $CV/CV^\dagger/$CNOT gates, **b** using the CH/CZ gates [16]

Consequently, the missing gate fault should be considered carefully as depending on the implementation technology the gate itself can have specific meaning. For instance, in quantum optics the missing gate fault means that a particular mirror, splitter or wave plate has been forgotten. A wrong gate fault means that the used components are wrong or the calibration of a non-linear medium has been incorrectly set up. On the other hand, in ion trap technology, the missing gate simply means that one or more laser pulses have not been performed.

Moreover, unlike in classical circuits, a missing gate fault thus simply modifies the logical primitive without really making it disappear. Thus, a missing gate can result in a loss of control or target qubit in a more complex gate. For instance, Fig. 19.19a shows the implementation of a Toffoli gate using the $CV/CV^\dagger/$CNOT gates. This is still not the truly quantum level as each of the gates is in general composed of smaller quantum primitives. However, observe that if the CNOT gate (third from the left) is missing the result is a CV gate controlled by qubit $|a\rangle$. Thus not only a control bit is lost but the output of the obtained gate has become probabilistic. Depending on the realization, the resulting probabilistic behavior will vary as the component quantum primitives will be different.

19.4.2 Wrong Gate Faults

Both the missing gate and the wrong gate fault model should be considered on the truly quantum level which depends on the implementation. As shown in the missing gate faults both of these faults will only very rarely be in the same form as could be found in a classical circuit. In a classical circuit each gate is built from transistors that have fixed properties. In quantum reversible circuits each gate is similarly built from a set of sub-gates that are single qubit rotations or two qubit interactions. The difference between quantum and transistor-like implementation is that the change in one of the components in a classical gate will generate a finite number of possible faults while in quantum there is an infinite amount of possible probabilistic states. A wrong gate means that a pulse of a wrong wavelength and of wrong duration has been applied.

Fig. 19.20 Example of possible faults in the CNOT gate implemented in Ion Trap technology

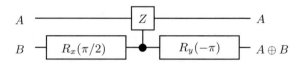

On one hand, the circuit from Fig. 19.8 can have some of its gates removed or simply change the parameters of the angle of rotation. Each such change will result in either a change in the gate output or in the gate simply performing the identity operation (missing gate). This is shown in Fig. 19.20 where the left single qubit rotation was changed from R_y to R_x and the right single qubit rotation had the rotation angle changed.

On the other hand a fault in one of the transistors in a gate such as depicted in Fig. 19.2 will result in either the gate being completely non functioning or producing the wrong logical output.

The wrong gate faults can occur in both reversible or irreversible circuits. Most important observation is that the missing or the wrong gate fault model will never result in such an error that produces functions that cannot be implemented with the available logic gates.

19.4.3 Bridging Faults

As shown in Sect. 19.3 it is not possible to have bridging faults in energy reversible per se because such faults (similarly to stuck-at faults) would violate reversibility. However, there are several faults arising from faults in interaction between qubits. Quantum phenomenological faults such as those described in [6] can be considered due to interaction problems. For instance, the *lost phase fault* is due to a problem of applying the Controlled-Z gate to two qubits. If the gate does not rotate both qubits as desired the result will be equivalent to a faulty gate. The result will be a gate that is missing one or more if its component qubits (control or target).

A fault type that belongs to this category is called missing-control fault (also called *faded control fault*) in which the quantum gate *looses* one or more of its control bits. The missing control fault represents the fact that a logic gate such as Toffoli, Fredkin or Feynman gate loses one or more controls and thus its functionality is changed. While testing for missing control in Toffoli gates is logically sound in quantum realized reversible circuits the result can be often misleading. Consider the two Toffoli gates shown in Fig. 19.19.

Without specifying the realization, exclusively testing for the missing control implies the following:

- if the Toffoli gate is realized as shown in Fig. 19.19a, a missing control of any of the component gates will result in a partially probabilistic gate. This can be seen if any of the two-qubit gates has its control removed. The partially probabilistic gate will result in a probabilistic behavior for some values of the qubits and will work properly for others.

- if the Toffoli gate is realized as shown in Fig. 19.19b, a missing control will either result in partially probabilistic gate or an Identity gate.

19.4.4 Granularity of Quantum Faults

One of the most important differences between testing classical faults and faults in quantum circuits is the granularity. In classical circuits testing for logical faults is adequate because the logical gates are used as components of the circuit. In quantum circuits, however, testing for logical faults does not test directly for the components of the circuit themselves but rather for macros composed of such components. The result is that while testing for a missing control qubit, it is possible to determine in which macro the missing qubit is located but does not allow to determine it precisely enough for diagnosis as there can be multiple problems (multiple fault types or same fault on multiple locations) leading to the same result.

The testing faults in reversible circuits should and must take into account the viability of the faults under test. Reversible circuits can be implemented in various technologies and testing such circuits for similar faults and from a similar point of view as classical circuits might seem a good starting point but can lead to wrong conclusions about the circuit testability. In particular, it should be noted that:

- Bridging faults represent a large category that can be described as interaction faults: such faults are in general the result of problem of transmitting information between two qubits.
- Missing gate and wrong gate faults should be considered on finer grain than directly testing for faulty logical primitives such as Toffoli gates.
- There are no stuck-at faults. Such faults cannot exists because the principle of stuck-at fault would violate the reversibility.

Finally, there are additional faults that exists in quantum circuits and systems: the initialization and the measurement error. However, we did not discuss them on purpose as they have no direct counter part when analyzing faults starting from classical circuits. Moreover, in general initialization and measurement are not part of the logical component of the circuit itself: they represent the ability to initialize the circuit to desired inputs and read the generated outputs. Thus for the purpose of logical analysis of quantum circuits we did not analyze this type of faults.

19.5 Conclusion

In this chapter we presented arguments and examples of faults that should and that should not be tested in quantum realization of reversible circuits. In particular, we showed that quantum circuits are very restricted as to the type of faults that can exist in such circuits. From the high level point of view, testing circuits for a disappearance of a Toffoli gate or for faulty logic gate is not very meaningful as there is a much larger possibility that such a gate will be faulty on more elementary, quantum level. Such faults cannot be directly and exactly diagnosed while testing for logical faults in permutative reversible circuits and require different methods.

References

1. Alves, N.: Detecting errors in reversible circuits with invariant relationships. arXiv:0812.3871v1 (2008)
2. Amin, M.H.S., Omelyanchouk, A.N., Rashkeev, S.N., Coury, M., Zagoskin, A.M.: Quasiclassical theory of spontaneous currents at surfaces and interfaces of d-wave superconductors. Phys. B **318**, 162 (2002)
3. Anantharam, V., He, M., Natarajan, K., Xie, H., Frank, M.P.: Driving fully-adiabatic logic circuits using custom high-q mems resonators. In: Workshop on Methodologies for Low Power Design, part of the ESA 04 (Embedded Systems and Applications) (2004)
4. Athas, W., Jr, Svensson, L., Koller, J.G., Tzartzanis, N., Ying-Chin Chou, E.: Low-power digital systems based on adiabatic-switching principles. IEEE Trans. Very Larg. Scale Integr. (VLSI) Syst. **2**, 398–407 (1994)
5. Bennett, H.C., Brassard, G., Popescu, S., Schumacher, B., Smolin, J.A., Wootters, W.K.: Purification of noisy entanglement and faithful teleportation via noisy channels. Phys. Rev. Let. **76**, 722–725 (1996)
6. Biamonte, J., Allen, J., Perkowski, M.: Fault models for quantum mechanical switching networks. J. Electron. Test. **26**, 499–511 (2010)
7. Blais, A., Zagoskin, A.M.: Operation of universal gates in a solid state quantum computer based on clean Josephson junctions between d-wave superconductors. Phys. Rev. A **61**, 042308 (2000)
8. Desoete, B., De Vos, A.: A reversible carry-look-ahead adder using control gates. Integr. VLSI J. **33**, 89–104 (2002)
9. Farazmand, N., Zamani, M., Tahoori, M.: Online fault testing of reversible logic using dual rail coding. In: Proceedings of the IEEE International On-Line Testing Symposium, pp. 204–205. (2010)
10. Farazmand, N., Zamani, M., Tahoori, M.: Online multiple fault detection in reversible circuits. In: Proceedings of the IEEE International Symposium on Defect and Fault Tolerance in VLSI Systems, pp. 429–437 (2010)
11. Fiurášek, J.: Linear optical Fredkin gate based on partial-SWAP gate. Phys. Rev. A **78**, 032317 (2008)
12. He, M., Frank, M.P., Xie, H.: CMOS-MEMS resonator as a signal generator for fully-adiabatic logic circuits. Devices, and Systems, In: SPIE Proceedings of Smart Structures (2005)
13. Jones, N.C.: Logic Synthesis for Fault-Tolerant Quantum Computers. PhD thesis (2013)
14. Knill, E.: Quantum computing with realistic noisy devices. (2005)
15. Lukac, M., Kameyama, M., Hiura, K.: Natural image understanding using algorithm selection and high level feedback. In: SPIE Intelligent Robots and Computer Vision XXX: Algorithms and Techniques, vol. 8662, pp. 86620D (2013)
16. Lukac, M., Perkowski, M., Kameyama, M.: Evolutionary quantum logic synthesis of boolean reversible logic circuits embedded in ternary quantum space using structural restrictions. In: Proceedings of the IEEE World Congress on Evolutionary Computation (2010)
17. Lukac, M., Shuai, B., Kameyama, M., Miller, M.: Information preserving logic - using logical reversibility to reduce the CPU-memory bottleneck. In: IEEE International Symposium on Multiple-Valued Logic on CD (2011)

18. Micuda, M., Sedlak, M., Straka, I., Mikova, M., Dusek, M., Jezek, M., Fiurasek, J.: Efficient experimental estimation of fidelity of linear optical quantum toffoli gate. Phys. Rev. Lett. **111**, 160407 (2013)
19. Nielsen, M.A., Chuang, I.L.: Quantum Computation and Quantum Information. Cambridge University Press, Cambridge (2000)
20. Paetznick, A., Reichardt, B.W.: Universal fault-tolerant quantum computation with only transversal gates and error correction. Preprint arXiv:1304.3709v1 (2013)
21. Paler, A., Polian, I., Hayes, J.P.: Detection and diagnosis of faulty quantum circuits. In: Proceedings of the Asia and South Pacific Design Automation Conference, pp. 181–186 (2012)
22. Polian, I., Hayes, J.P.: Advanced modeling of faults in reversible circuits. In: Proceedings of the IEEE East-West Design and Test Symposium, pp. 376–381 (2010)
23. Puzzuoli, D., Granade, C., Haas, H., Criger, B., Magesan, E., Cory, D.G.: Tractable simulation of error correction with honest approximations to realistic fault models. arXiv:1309.4717v2 (2013)
24. Rice, J.E.: An overview of fault models and testing approaches for reversible logic. In: Proceedings of the Pacific Rim Conference on Communications, Computers and Signal Processing (PACRIM), p. 6 (2013)
25. Steane, A.M.: Error correcting codes in quantum theory. **77**, 793–797
26. Tague, L., Soeken, M., Minato, S., Drechsler, R.: Debugging of reversible circuits using π DDs. In: IEEE Proceedings of the IEEE International Symposium on Multiple-Valued Logic, pp. 316–321 (2013)
27. Van Rentergem, Y., De Vos, A.: Optimal design of a reversible full adder. Int. J. Unconv. Comput. **1**, 339–355 (2005)
28. Van Rentergem, Y., De Vos, A., Storme, L.: Implementing an arbitrary reversible logic gate. J. Phys. A Math. Gener. Inst. Phys. **38**, 3555–3577 (2005)
29. Wille, R., Zhang, H., Drechsler, R.: Fault ordering for automatic test pattern generation of reversible circuits. In: Proceedings of the IEEE International Symposium on Multiple-Valued Logic, pp. 29–34 (2013)
30. Zamani, M., Tahoori, M.M.: Online missing/repeated gate faults detection in reversible circuits. In: Proceedings of the IEEE International Symposium on Defect and Fault Tolerance in VLSI and Nanotechnology Systems (DFT), pp. 435–442 (2011)
31. Zamani, M., Tahoori, M.B., Chakrabarty, K.: Ping-pong test: Compact test vector generation for reversible circuits. In: Proceedings of the 30th VLSI Test Symposium, pp. 164–169 (2012)

Chapter 20
A Class of Non-optimum-time FSSP Algorithms

Hiroshi Umeo

Abstract Synchronization of large-scale networks is an important and fundamental computing primitive in parallel and distributed systems. The synchronization in cellular automata, known as the firing squad synchronization problem (FSSP), has been studied extensively for more than fifty years, and a rich variety of synchronization algorithms has been proposed. In the present chapter, we give a survey on a class of non-optimum-time $3n$-step FSSP algorithms for synchronizing one-dimensional (1D) cellular automata of length n in $3n \pm O(\log n)$ steps and present a comparative study of a relatively large-number of their implementations. We also propose two smallest-state, known at present, implementations of the $3n$-step algorithm. This chapter gives the first complete transition rule sets implemented on finite state automata for the class of non-optimum-time $3n$-step FSSP algorithms developed so far.

20.1 Introduction

The synchronization in ultra-fine-grained parallel computational model of cellular automata has been known as the firing squad synchronization problem (FSSP) since its development, in which it was originally proposed by J. Myhill in the book edited by Moore [10] to synchronize all/some parts of self-reproducing cellular automata.

In the present chapter, we give a survey on recent developments in a class of non-optimum-time FSSP algorithms for one-dimensional (1D) cellular automata. Here we focus our attention to the 1D FSSP algorithms having $3n \pm O(\log n)$ time complexities and present a comparative study of a relatively large-number of their

A part of this work has been presented at 13th International Conference on Parallel Computing Technologies, PaCT 2015 (chaired by Victor Malyshkin), held on Aug. 31–Sept. 4, 2015, in Petrozavodsk, Russia.

H. Umeo (✉)
University of Osaka Electro-Communication, Neyagawa-shi, Hastu-cho, 18-8,
Osaka 572-8530, Japan
e-mail: umeo@cyt.osakac.ac.jp

© Springer International Publishing Switzerland 2017
A. Adamatzky (ed.), *Advances in Unconventional Computing*,
Emergence, Complexity and Computation 22,
DOI 10.1007/978-3-319-33924-5_20

495

implementations. We also propose two smallest state implementations, known at present, included in the same class of the algorithms. A class of the 3n-step FSSP algorithms is an interesting class of synchronization algorithms among many variants of FSSP algorithms due to its simplicity and straightforwardness and it is important in its own right in the design of cellular algorithms. The first optimum-time FSSP algorithm designed by Goto [4] uses a 3n-step algorithm in its synchronization phase. This chapter gives the first complete transition rule sets implemented on finite state automata for the class of non-optimum-time 3n-step FSSP algorithms developed so far.

Specifically, we attempt to answer the following questions:

- First, what is the local transition rule set for those FSSP algorithms?
- Are all previously presented transition rule sets correct?
- Do these rule sets contain redundant rules? If so, what is the exact rule set?
- How do the algorithms compare with each other?
- Are there still any new implementations of the non-optimum-time FSSP algorithms?
- What is the state-change complexity in those algorithms?

In Sect. 20.2 we give a description of the 1D FSSP and present a small overview of some basic results on 1D FSSP algorithms. See Umeo [14] for details of the FSSP. Section 20.3 gives a survey on non-optimum-time FSSP algorithms. We make implementations of those algorithms on a computer, check and compare their transition rule sets.

20.2   A Class of 3n-Step FSSP Algorithms

20.2.1   Firing Squad Synchronization Problem

Figure 20.1 shows a finite one-dimensional (1D) cellular array consisting of n cells. Each cell is an identical (except the border cells) finite-state automaton. The array operates in lock-step mode in such a way that the next state of each cell (except border cells) is determined by both its own present state and the present states of its left and right neighbors. All cells (*soldiers*), except the left end cell (*general*), are initially in the quiescent state at time $t = 0$ with the property that the next state of a quiescent cell with quiescent neighbors is the quiescent state again. At time $t = 0$, the left end cell C_1 is in the *fire-when-ready* state, which is the initiation signal for the array. The firing squad synchronization problem is to determine a description (state set and next-state function) for cells that ensures all cells enter the *fire* state

Fig. 20.1 A one-dimensional (1D) cellular automaton

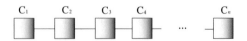

at exactly the same time and for the first time. The set of states and the next-state function must be independent of n.

A formal definition of the FSSP is as follows: A 1D cellular automaton $\mathcal{M}$ is a pair $\mathcal{M} = (Q, \delta)$, where

1. Q is a finite set of states including the following three distinguished states G, Q, and F. G is an initial general state, Q is a quiescent state, and F is a firing state, respectively.
2. δ is a next state function such that $\delta : (Q \cup \{*\}) \times Q \times (Q \cup \{*\}) \to Q$. The state $* \notin Q$ is a pseudo state of the border of the array.
3. The quiescent state Q must satisfy the following conditions:

$$\delta(Q, Q, Q) = \delta(*, Q, Q) = \delta(Q, Q, *) = Q.$$

A 1D cellular automaton of length n, $\mathcal{M}_n$, consisting of n copies of $\mathcal{M}$, is a 1D array of $\mathcal{M}$, numbered from 1 to n. Each $\mathcal{M}$ is referred to as a cell and denoted by C_i, where $1 \le i \le n$. We denote a state of C_i at time (step) t by S_i^t, where $t \ge 0, 1 \le i \le n$. A *configuration* of $\mathcal{M}_n$ at time t is a function $C^t : [1, n] \to Q$ and it is denoted as $S_1^t S_2^t \dots S_n^t$. A *computation* of $\mathcal{M}_n$ is a sequence of configurations of $\mathcal{M}_n$, C^0, C^1, C^2, …, C^t, …, where C^0 is an initial given configuration such that: $C^0 = G \overbrace{Q, \dots, Q}^{n-1}$. The configuration at time $t + 1$, C^{t+1}, is computed by simultaneous applications of the next transition function δ to each cell of $\mathcal{M}_n$ in C^t such that:

$$S_1^{t+1} = \delta(*, S_1^t, S_2^t), \ S_i^{t+1} = \delta(S_{i-1}^t, S_i^t, S_{i+1}^t), 2 \le i \le n - 1, \text{ and } S_n^{t+1} = \delta(S_{n-1}^t, S_n^t, *).$$

A *synchronized configuration* of $\mathcal{M}_n$ at time t is a configuration C^t, $S_i^t = F$, for any $1 \le i \le n$.

The FSSP is to obtain an $\mathcal{M}$ such that for a certain time function $T(n) \ (\ge 2n - 2)$ and any $n \ge 2$,

1. A synchronized configuration at time $t = T(n)$, $C^{T(n)} = \overbrace{F, \dots, F}^{n}$ can be computed from an initial configuration $C^0 = G \overbrace{Q, \dots, Q}^{n-1}$.
2. $S_i^t \ne F$, for any t, i, such that $1 \le t \le T(n) - 1, 1 \le i \le n$.

No cells fire before time $t = T(n)$. We say that $\mathcal{M}_n$ is synchronized at time $t = T(n)$ and the function $T(n)$ is a time complexity for the synchronization.

20.2.2 A Class of 3n-Step FSSP Algorithms

A class of 3n-step FSSP algorithms is an interesting class of synchronization algorithms among many variants of the FSSP algorithms due to its simplicity and straight-

Fig. 20.2 A space-time
diagram for a class of
3n-step FSSP algorithms and
their design parameters:
thread-width and *Zone T* in
the space-time diagram

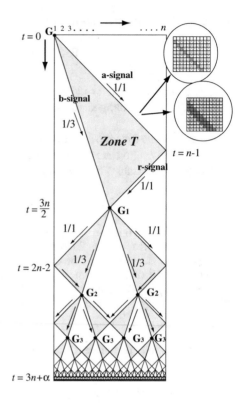

forwardness and it is important in its own right in the design of cellular algorithms.
Figure 20.2 shows a space-time diagram for the well-known 3n-step FSSP algorithm.
The synchronization process can be viewed as a typical *divide-and-conquer strategy*
that operates in parallel in the cellular space. An initial *general* G, located at left end
of the array of size n, generates two special signals, referred to as *a-signal* and *b-
signal*, which propagate in the right direction at a speed of $1/1$ and $1/3$, respectively.
The a-signal arrives at the right end at time $t = n - 1$, reflects there immediately,
and continues to move at the same speed in the left direction. The reflected signal is
referred to as *r-signal*. The b- and the r-signals meet at one or two center cells of the
arry, depending on the parity of n. In the case that n is odd, the cell $C_{\lceil n/2 \rceil}$ becomes a
general at time $t = 3\lceil n/2 \rceil - 2$. The *general* is responsible for synchronizing both
its left and right halves of the cellular space. Note that the *general* is shared by the
two halves. In the case that n is even, two cells $C_{\lceil n/2 \rceil}$ and $C_{\lceil n/2 \rceil + 1}$ become the next
general at time $t = 3\lceil n/2 \rceil$. Each *general* is responsible for synchronizing its left
and right halves of the cellular space, respectively.

Thus, at time

$$t_{\text{center}} = \begin{cases} 3\lceil n/2 \rceil - 2 & n : \text{odd} \\ 3\lceil n/2 \rceil & n : \text{even,} \end{cases} \tag{20.1}$$

the array knows its center point and generates one or two new *general(s)* G_1. The new *general(s)* G_1 generates the same 1/1- and 1/3-speed signals in both left and right directions simultaneously and repeat the same procedures as above. Thus, the original synchronization problem of size n is divided into two sub-problems of size $\lceil n/2 \rceil$. In this way, the original array is split into equal two, four, eight, …, subspaces synchronously. Note that the first general generated at the center G_1 itself is synchronized at time $t = t_{center}$, and the second general G_2 are also synchronized, and the generals generated afterwards are also synchronized at its corresponding step. In the last, the original problem of size n can be split into small sub-problems of size 2. In this way, by increasing the synchronized generals step by step, the initial given array is synchronized. Most of the $3n$-step synchronization algorithms developed so far in Fischer [2], Herman [6], Minsky [9], Umeo, Maeda, and Hongyo [16], and Yunès [22–24] are more or less based on the similar scheme. It can be seen that, by measuring the length of the diagonal path of the b-signal with/without 1 step delay at the center points at each halving iteration in the space-time diagram, the time complexity $T(n)$ for synchronizing n cells is $T(n) = 3n \pm O(\log n)$.

A question is "How can we implement the synchronization diagram above in terms of a small-state finite state automaton?".

The three signals: a-, b-, and r-signals in the space-time diagram in Fig. 20.2 play an important role in finding the center cell(s). A triangle area circled by these three signals is also important in its implementation. We call the area *zone* $\mathcal{T}$.

G State
1: G Y Y → Z
2: G G Q → Y
3: R G Q → G
4: L G R → Y
5: Y G G → Z
6: Q G G → Y
7: Q G L → G
8: Q G Q → G
9: Z G Z → F
10: * G Q → B

A State
11: Y A Q → B
12: Y A Q → B

B State
13: Y B L → X
14: Y B Q → C
15: Q B L → X
16: Q B Q → C
17: * B R → C
18: * B Z → F

C State
19: Y C Q → Q
20: Q C Q → Q
21: * C Q → Q

a State
22: Q a Y → b
23: Q a Q → b

b State
24: R b Y → X
25: R b Q → X
26: Q b Y → c
27: Q b Q → c

c State
28: Q c Y → Q
29: Q c Q → Q

R State
30: G R Y → Q
31: G R Q → Q
32: B R Q → Q
33: R R b → X
34: R R Y → Z
35: R R Q → Q
36: Y R G → L
37: Y R R → Q
38: Y R Y → Q
39: Y R Q → Q
40: Q R b → X
41: Q R Y → Q
42: Q R Q → Q
43: K R b → X
44: K R Q → R
45: * R R → Q
46: * R Q → R

L State
47: G L Y → R
48: G L * → Z
49: B L L → X
50: B L Q → X
51: B L K → X
52: L L Y → Q
53: L L * → Q
54: Y L G → Q
55: Y L L → Z
56: Y L Y → Q
57: Y L Q → Q
58: Q L G → Q
59: Q L L → Q
60: Q L Y → Q
61: Q L Q → Q
62: Q L K → L
63: Q L * → L

X State
64: X X Y → G
65: X X Q → G
66: Y X X → G
67: Q X X → G

Y State
68: G Y G → Y
69: G Y Y → Y
70: a Y A → Y
71: a Y Y → Y
72: b Y B → Y
73: b Y Y → Y
74: c Y C → Y
75: c Y Y → Y
76: R Y L → K
77: R Y Y → K
78: L Y R → Y
79: L Y Y → Y
80: X Y X → Y
81: X Y Y → Y
82: Y Y G → Y
83: Y Y A → Y
84: Y Y B → Y
85: Y Y C → Y
86: Y Y R → Y
87: Y Y L → K
88: Y Y X → Y
89: Y Y Y → Z
90: Y Y Q → Y
91: Y Y Z → F
92: Y Y K → Y
93: Q Y Y → Y
94: Q Y Q → Y
95: Z Y Y → F
96: Z Y Z → F
97: Z Y K → F
98: K Y Y → F
99: K Y Z → F

Q State
100: G Q Y → R
101: G Q Q → R
102: G Q * → Z
103: A Q R → Q
104: A Q L → L
105: A Q Q → Q
106: A Q K → L
107: B Q Q → Q
108: C Q R → A
109: C Q L → G
110: C Q Q → Q
111: a Q Y → Q
112: a Q Q → Q
113: b Q Y → Q
114: b Q Q → Q
115: c Q Y → Q
116: c Q Q → Q
117: R Q a → R
118: R Q c → G
119: R Q Y → R
120: R Q Q → R
121: R Q * → L
122: L Q a → Q
123: L Q Y → a
124: L Q Q → Q
125: X Q Y → Q
126: X Q Q → Q
127: X Q * → Q
128: Y Q G → L
129: Y Q A → Q
130: Y Q B → Q
131: Y Q C → Q
132: Y Q R → A
133: Y Q L → L
134: Y Q X → Q
135: Y Q Q → Q
136: Y Q K → Y
137: Q Q G → L
138: Q Q A → Q
139: Q Q B → Q
140: Q Q C → Q
141: Q Q a → Q
142: Q Q b → Q
143: Q Q c → a
144: Q Q R → Q
145: Q Q L → L
146: Q Q X → Q
147: Q Q Y → Q
148: Q Q Q → Q
149: Q Q K → L
150: Q Q * → Q
151: Z Q * → F
152: K Q a → R
153: K Q Y → Q
154: K Q Q → R
155: * Q G → Z
156: * Q A → Q
157: * Q B → Q
158: * Q C → Q
159: * Q L → R
160: * Q X → Q
161: * Q Q → Q
162: * Q Z → F

Z State
163: G Z * → F
164: B Z * → F
165: Y Z Y → F
166: Z Z Z → F
167: Y Z K → F
168: Y Z * → F
169: Q Z K → F
170: Z Z Y → F
171: K Z Y → F
172: K Z Q → F
173: * Z G → F
174: * Z Y → F

K State
175: L K R → Y
176: L K K → Y
177: Y K Y → Z
178: Y K Z → F
179: Y K K → Z
180: Q K Q → K
181: Q K K → K
182: Z K Y → F
183: Z K Z → F
184: Z K K → F
185: K K R → Y
186: K K Y → Z
187: K K Q → K
188: K K Z → F

Fig. 20.3 A 15-state transition table of the Fischer [2] algorithm

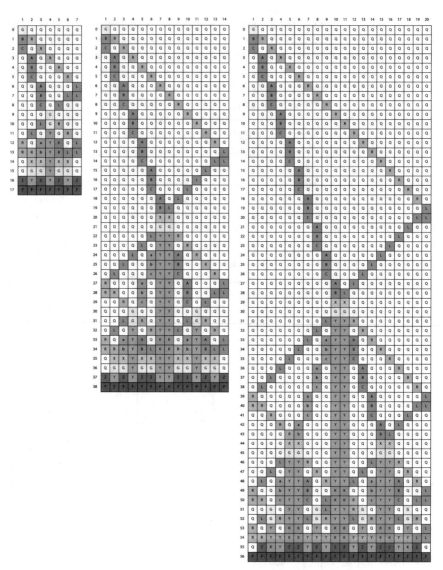

Fig. 20.4 Snapshots of the synchronization process of the Fischer [2] algorithm on 7, 14, and 20 cells, respectively

20.2.3 Complexity Measures and Properties for Synchronization Algorithms

- **Time Complexity**

  Any solution to the original FSSP with a general at one end can be easily shown

Fig. 20.5 A 13-state transition table of the Minsky-McCarthy [9] algorithm

——A State——
1:I A L → I
2:I A Q → B
3:Q A L → I
4:Q A Q → B

——B State——
5:Q B Q → C

——C State——
6:Q C L → I
7:Q C Q → Q

——a State——
8:R a I → I
9:R a Q → I
10:Q a I → b
11:Q a Q → b

——b State——
12:Q b Q → c

——c State——
13:R c Q → I
14:Q c Q → Q

——R State——
15:L R a → Q
16:L R I → F
17:L R Q → Q
18:X R Q → Q
19:X R * → L
20:I R L → L
21:I R Q → Q
22:I R * → L
23:Q R a → Q
24:Q R c → I
25:Q R L → L
26:Q R Q → Q
27:Q R * → L
28:* R a → Q
29:* R I → F
30:* R Q → Q

——L State——
31:A L R → Q
32:A L Q → Q
33:A L * → Q
34:C L Q → I
35:R L I → R
36:R L Q → R
37:X L * → I
38:I L R → F
39:I L * → F
40:Q L R → Q
41:Q L I → Q
42:Q L Q → Q
43:Q L * → Q
44:* L I → R
45:* L Q → R

——X State——
46:X X I → Y
47:I X X → Y
48:I X I → Y
49:Q X Q → Y
50:* X R → X
51:* X L → I
52:* X Q → Q

——Y State——
53:Y Y I → F
54:Y Y * → F
55:I Y I → F
56:Q Y Q → Q

——I State——
57:a I A → Q
58:a I I → Q
59:a I Q → Q
60:R I L → F
61:R I Y → F
62:R I I → F
63:L I R → I
64:L I X → I
65:L I I → I
66:X I R → I
67:X I X → I
68:X I I → I
69:Y I L → F
70:Y I Y → F
71:Y I I → F
72:I I A → Q
73:I I R → I
74:I I L → F
75:I I X → I
76:I I Y → F
77:I I I → X
78:I I Q → I
79:I I * → F
80:Q I A → Q
81:Q I I → I
82:Q I Q → I
83:* I I → F
84:* I Q → X

——Q State——
85:A Q R → Q
86:A Q X → Q
87:A Q Q → Q
88:B Q L → L
89:B Q Y → L
90:B Q Q → Q
91:C Q Q → A
92:a Q Q → Q
93:b Q B → Q
94:b Q Q → Q
95:c Q C → Q
96:c Q Q → Q
97:R Q b → R
98:R Q L → X
99:R Q Q → R
100:R Q * → R
101:L Q a → Q
102:L Q R → Q
103:L Q I → a
104:L Q Q → Q
105:L Q * → Q
106:X Q a → Q
107:X Q R → A
108:X Q I → Q
109:X Q Q → Q
110:Y Q b → R
111:Y Q I → I
112:Y Q Q → R
113:I Q R → A
114:I Q X → Q
115:I Q Y → I
116:I Q I → X
117:I Q Q → R
118:I Q * → R
119:Q Q A → Q
120:Q Q B → Q
121:Q Q C → Q
122:Q Q a → Q
123:Q Q b → Q
124:Q Q c → a
125:Q Q R → Q
126:Q Q L → L
127:Q Q X → Q
128:Q Q Y → L
129:Q Q I → L
130:Q Q Q → Q
131:Q Q * → Q
132:* Q A → Q
133:* Q B → Q
134:* Q C → Q
135:* Q R → Q
136:* Q L → L
137:* Q I → L
138:* Q Q → Q

to require $2n - 2$ steps for synchronizing n cells, since signals on the array can propagate no faster than one cell per one step, and the time from the general's instruction until the final synchronization must be at least $2n - 2$.

Theorem 1 (Goto [4]) The minimum time in which the firing squad synchronization could occur is $2n - 2$ steps, where the general is located on the left end.

Theorem 2 (Goto [4]) There exists a cellular automaton that can synchronize any 1D array of length n in optimum $2n - 2$ steps, where the general is located on the left end.

- **Number of States**

  The following three distinct states: the *quiescent* state, the *general* state, and the *firing* state, are required in order to define any cellular automaton that can solve the FSSP. Note that the boundary state for C_0 and C_{n+1} is not generally counted as an internal state in the study of the FSSP. Balzer [1] and Sanders [12] showed that no four-state optimum-time solution exists. Umeo and Yanagihara [18], Yunès [25], Umeo, Kamikawa, and Yunès [15], and Ng [11] gave several 5- and 4-state *partial* solutions that can solve the synchronization problem for infinitely many sizes n, but not all, respectively. The solution is referred to as *partial* solution, which is compared with usual *full* solution that can solve the problem for all cells.

Theorem 3 (Balzer [1], Sanders [12]) There is no four-state *full* solution that can synchronize n cells.

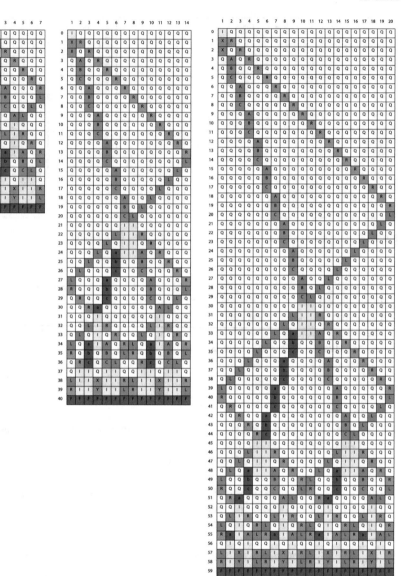

Fig. 20.6 Snapshots of the synchronization process of the Minsky-McCarthy [9] algorithm on 7, 14, and 20 cells, respectively

Yunès [25], Umeo, Kamikawa, and Yunès [15], and Ng [11] developed 4-state partial solutions based on Wolfram's rules 60 and 150. They can synchronize any array/ring of length $n = 2^k$ for any positive integer k. Details can be found in Yunès [25], Umeo, Kamikawa, and Yunès [15], and Ng [11].

——— S State ———				
1: S	S	S	→	S
2: S	S	I	→	I
3: S	S	J	→	S
4: S	S	U	→	S
5: S	S	W	→	U
6: S	S	R	→	I
7: S	S	X	→	S
8: S	S	V	→	S
9: S	S	G	→	S
10: S	S	*	→	S
11: I	S	S	→	I
12: I	S	V	→	I
13: I	S	G	→	I
14: I	S	*	→	I
15: J	S	S	→	S
16: J	S	V	→	S
17: J	S	G	→	S
18: J	S	S	→	S
19: U	S	S	→	S
20: W	S	S	→	U
21: W	S	J	→	I
22: R	S	S	→	I
23: R	S	V	→	I
24: X	S	G	→	S
25: X	S	S	→	S
26: V	S	S	→	S
27: V	S	I	→	S
28: V	S	J	→	S
29: V	S	R	→	I
30: G	S	S	→	S
31: G	S	I	→	I
32: G	S	J	→	S
33: G	S	X	→	S
34: *	S	S	→	S
35: *	S	I	→	I
36: *	S	J	→	S
37: *	S	X	→	S

——— I State ———				
38: S	I	I	→	G
39: S	I	J	→	J
40: S	I	R	→	J
41: S	I	X	→	G
42: S	I	G	→	J
43: I	I	S	→	G
44: I	I	X	→	G
45: J	I	S	→	J
46: J	I	U	→	I
47: J	I	W	→	I
48: J	I	G	→	R
49: I	I	*	→	R
50: U	I	J	→	I
51: U	I	R	→	I
52: W	I	R	→	I
53: W	I	S	→	J
54: R	I	U	→	I
55: R	I	W	→	I
56: X	I	S	→	G
57: X	I	I	→	G
58: G	I	S	→	J
59: G	I	J	→	R
60: G	I	G	→	G
61: G	I	*	→	G
62: *	I	S	→	J
63: *	I	J	→	R
64: *	I	G	→	G

——— J State ———				
65: S	J	S	→	S
66: S	J	I	→	S
67: I	J	S	→	S
68: I	J	U	→	S
69: I	J	G	→	U
70: U	J	I	→	S
71: G	J	I	→	U
72: *	J	I	→	U

——— U State ———				
73: U	U	X	→	V
74: I	U	X	→	I
75: J	U	G	→	V
76: R	U	G	→	G
77: X	U	S	→	V
78: X	U	I	→	I
79: G	U	J	→	V
80: G	U	R	→	G
81: *	U	J	→	V
82: *	U	R	→	G

——— W State ———				
83: S	W	X	→	X
84: S	W	G	→	X
85: I	W	X	→	X
86: I	W	G	→	X
87: X	W	S	→	X
88: G	W	S	→	X
89: G	W	I	→	X
90: *	W	S	→	X
91: *	W	I	→	X

——— R State ———				
92: S	R	G	→	R
93: S	R	*	→	R
94: I	R	G	→	S
95: I	R	*	→	S
96: U	R	G	→	G
97: U	R	*	→	G
98: G	R	S	→	R
99: G	R	I	→	S
100: G	R	U	→	G
101: *	R	S	→	R
102: *	R	I	→	S
103: *	R	U	→	G

——— X State ———				
104: S	X	I	→	I
105: S	X	U	→	X
106: S	X	W	→	S
107: S	X	V	→	X
108: I	X	S	→	I
109: I	X	G	→	I
110: U	X	S	→	X
111: U	X	G	→	X
112: W	X	S	→	X
113: W	X	G	→	S
114: V	X	S	→	X
115: V	X	G	→	X
116: G	X	I	→	I
117: G	X	U	→	X
118: G	X	W	→	S
119: G	X	V	→	X
120: *	X	I	→	I
121: *	X	U	→	X
122: *	X	W	→	S
123: *	X	V	→	X

——— V State ———				
124: S	V	X	→	W
125: S	V	G	→	W
126: X	V	S	→	W
127: G	V	S	→	W
128: *	V	S	→	W

——— G State ———				
129: S	G	S	→	G
130: S	G	G	→	G
131: I	G	I	→	G
132: I	G	G	→	G
133: J	G	J	→	G
134: J	G	G	→	G
135: U	G	U	→	G
136: U	G	G	→	G
137: W	G	W	→	G
138: W	G	G	→	G
139: R	G	R	→	G
140: R	G	G	→	G
141: X	G	X	→	G
142: X	G	G	→	G
143: V	G	V	→	G
144: V	G	G	→	G
145: G	G	S	→	G
146: G	G	I	→	G
147: G	G	J	→	G
148: G	G	U	→	G
149: G	G	W	→	G
150: G	G	R	→	G
151: G	G	X	→	G
152: G	G	V	→	G
153: G	G	G	→	F
154: G	G	*	→	F
155: *	G	G	→	F

Fig. 20.7 A 10-state transition table of the Herman [6] algorithm

Theorem 4 (Ng [11], Umeo, Kamikawa, and Yunès [15], Yunès [25]) There exist 4-state *partial* solutions to the FSSP.

Concerning the optimum-time full solutions, Waksman [21] presented a 16-state optimum-time synchronization algorithm. Afterward, Balzer [1] and Gerken [3] developed an eight-state algorithm and a seven-state synchronization algorithm, respectively, thus decreasing the number of states required for the synchronization. Mazoyer [8] developed a six-state synchronization algorithm which, at present, is the algorithm having the fewest states for 1D arrays.

Theorem 5 (Mazoyer [8]) There exists a 6-state *full* solution to the FSSP.

- **Number of Transition Rules**

  Any k-state (excluding the boundary state) transition table for the synchronization has at most $(k-1)k^2$ entries in $(k-1)$ matrices of size $k \times k$. The number of transition rules reflects a complexity of synchronization algorithms.

- **Filled-In Ratio**

  To measure the density of entries in transition table, we introduce a measure *filled-in ratio* of the state transition table. The filled-in ratio of the state transition table $\mathcal{A}$ is defined as follows: $f_{\mathcal{A}} = e/e_{total}$, where e is the number of exact entries of the next state defined in the table $\mathcal{A}$ and e_{total} is the number of possible entries defined such that $e_{total} = (k-1)k^2$, where k is the number of internal states of the table $\mathcal{A}$.

- **Symmetry versus Asymmetry**

  Herman [5, 6] investigated the computational power of symmetrical cellular automata, motivated by a biological point of view. Szwerinski [13] and Kobuchi [7]

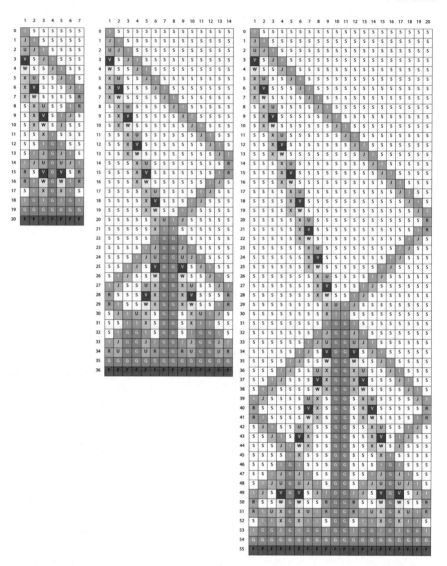

Fig. 20.8 Snapshots of the synchronization process of the Herman [6] algorithm on 7, 14, and 20 cells, respectively

considered a computational relation between symmetrical and asymmetrical CAs with von Neumann neighborhood. A transition table is said to be *symmetric* if and only if the transition table $\delta : Q^3 \rightarrow Q$ such that $\delta(x, y, z) = \delta(z, y, x)$ holds, for any state x, y, z in Q. A symmetrical cellular automaton has a property that the next state of a cell depends on its present state and the states of its two neighbors,

Q State					Z State					A State					d State				

——— Q State ———

1: Q Q Q → Q
2: Q Q G → Z
3: Q Q Z → Z
4: Q Q A → Q
5: Q Q C → A
6: Q Q d → Q
7: Q Q * → Q
8: G Q Q → Z
9: G Q d → Z
10: G Q * → Z
11: Z Q Q → Z
12: Z Q Z → Z
13: Z Q d → Z
14: Z Q * → Z
15: A Q Q → Q
16: A Q Z → Z
17: A Q d → Q
18: C Q Q → Z
19: d Q Q → Q
20: d Q G → Z
21: d Q Z → Z
22: d Q A → Q
23: d Q d → Q

24: d Q * → Q
25: * Q Q → Q
26: * Q Z → Z
27: * Q A → Q
28: * Q d → Q

——— G State ———

29: Q G Q → G
30: Q G G → G
31: G G Q → G
32: G G Z → Q
33: G G A → G
34: G G C → C
35: G G d → d
36: Z G G → Q
37: Z G Z → Q
38: A G G → G
39: A G A → G
40: C G G → C
41: C G C → C
42: d G G → d
43: d G d → d
44: * G Q → G
45: * G Z → C

——— Z State ———

46: Q Z Q → d
47: Q Z G → d
48: Q Z Z → d
49: Q Z d → d
50: Q Z * → d
51: G Z Q → d
52: G Z A → A
53: G Z C → Z
54: G Z * → C
55: Z Z Q → d
56: Z Z d → G
57: A Z G → d
58: A Z d → d
59: A Z * → A
60: C Z G → Z
61: C Z d → Z
62: C Z * → C
63: d Z Q → d
64: d Z Z → G
65: d Z A → A
66: d Z C → Z
67: d Z d → G
68: d Z * → Z

69: * Z Q → d
70: * Z A → A
71: * Z C → C
72: * Z d → Z

——— A State ———

73: Q A A → d
74: Q A C → Q
75: Q A d → d
76: G A d → C
77: Z A d → C
78: A A Q → C
79: C A Z → C
80: C A d → A
81: C A * → C
82: d A Q → A
83: d A G → C
84: d A Z → C
85: d A d → d
86: * A A → d
87: * A C → C

——— C State ———

88: Q C d → d
89: G C G → C
90: G C A → C
91: G C G → C
92: Z C C → C
93: Z C d → Z
94: A C G → C
95: A C A → C
96: A C d → Z
97: C C G → C
98: C C Z → C
99: C C C → F
100: C C d → d
101: C C * → F
102: d C Q → d
103: d C Z → Z
104: d C C → d
105: d C d → d
106: * C A → A
107: * C C → F
108: * C d → C

——— d State ———

109: Q d G → C

110: Q d Z → Q
111: Q d A → A
112: Q d C → Q
113: Q d d → C
114: G d Q → C
115: Z d Q → Q
116: Z d Z → C
117: Z d A → Q
118: Z d C → A
119: Z d d → C
120: Z d * → Q
121: A d Q → d
122: A d Z → Q
123: A d A → G
124: A d d → G
125: C d Q → Q
126: C d Z → A
127: C d C → Q
128: C d d → Q
129: d d Q → d
130: d d Z → C
131: d d A → G
132: d d C → Q
133: * d Z → Q
134: * d C → Q

Fig. 20.9 A 7-state transition table of the Yunès [22] algorithm

but it is same if the states of the left and right neighbors are interchanged. Thus, the symmetrical CA has no ability to distinguish between its left and right neighbors.

- **State-Change Complexity**

  Vollmar [20] introduced a *state-change complexity* in order to measure the efficiency of cellular automata, motivated by energy consumption in certain physical memory systems. The state-change complexity is defined as the sum of *proper* state changes of the cellular space during the computations. A formal definition is as follows: Consider an FSSP algorithm operating on n cells. Let $T(n)$ be synchronization steps of the algorithm. We define a matrix C of size $T(n) \times n$ ($T(n)$ rows, n columns) over $\{0, 1\}$, where each element $c_{i,j}$ on ith row, jth column of the matrix is defined:

$$c_{i,j} = \begin{cases} 1 & S_i^j \neq S_i^{j-1} \\ 0 & \text{otherwise.} \end{cases} \tag{20.2}$$

The state-change complexity $SC(n)$ of the FSSP algorithm is the sum of 1's elements in C defined as:

$$SC(n) = \sum_{j=1}^{T(n)} \sum_{i=1}^{n} c_{i,j}. \tag{20.3}$$

Vollmar [20] showed that $\Omega(n \log n)$ state-changes are required by the cellular space for the synchronization of n cells in $2n - 2$ steps. Gerken [3] presented an optimum-time $\Theta(n \log n)$ minimum-state-change FSSP algorithm.

Theorem 6 (Vollmar [20]) $\Omega(n \log n)$ state-change is necessary for synchronizing n cells in $2n - 2$ steps.

Theorem 7 (Gerken [3]) $\Theta(n \log n)$ state-change is sufficient for synchronizing n cells in $2n - 2$ steps.

Fig. 20.10 Snapshots of the synchronization process of the Yunès [22] algorithm on 7, 14, and 20 cells, respectively

Q State					
1: Q Q Q → Q	25: d Q * → Q	48: * G A → G	71: C Z * → G	C State	114: G d G → Z
2: Q Q G → Q		49: * G C → G	72: d Z Q → d		115: G d Z → Q
3: Q Q Z → Z	G State	50: * G d → G	73: d Z G → Z	93: Q C G → d	116: G d A → A
4: Q Q A → Q	26: Q G G → Q		74: d Z A → Z	94: Q C d → d	117: G d C → Q
5: Q Q C → A	27: Q G G → G	Z State	75: d Z C → Z	95: G C Q → d	118: Z d Q → Q
6: Q Q d → Q	28: G G Q → G		76: d Z * → Z	96: G C Z → Q	119: Z d G → Q
7: Q Q * → Q	29: G G G → F	51: Q Z Q → d	77: * Z Q → d	97: Z C G → Q	120: Z d Z → C
8: G Q Q → Q	30: G G Z → G	52: Q Z G → d		98: Z C Z → G	121: Z d A → Q
9: G Q Z → Z	31: G G A → G	53: Q Z A → d	A State	99: Z C C → G	122: Z d C → A
10: G Q A → Q	32: G G C → G	54: Q Z d → d		100: Z C d → A	123: Z d d → C
11: G Q d → Q	33: G G d → G	55: Q Z * → d	78: Q A G → C	101: C C Z → G	124: Z d * → Q
12: Z Q Q → Z	34: G G * → F	56: G Z Q → d	79: Q A Z → Z	102: C C d → G	125: A d Q → d
13: Z Q Q → Z	35: Z G G → G	57: G Z Z → G	80: Q A A → d	103: d C Q → d	126: A d G → d
14: Z Q A → Z	36: Z G Z → G	58: G Z A → Z	81: Q A d → d	104: d C Z → A	127: A d Z → Q
15: Z Q d → Z	37: Z G d → d	59: G Z C → G	82: G A Q → C	105: d C C → G	128: C d Q → Q
16: Z Q * → Z	38: A G G → G	60: G Z d → Z	83: G A Z → Z	106: d C d → G	129: C d G → Q
17: A Q Q → Q	39: A G A → G	61: Z Z Q → d	84: G A A → d	107: * C Z → G	130: C d Z → A
18: A Q G → Q	40: C G G → G	62: Z Z G → d	85: G A d → A	108: * C d → G	131: d d Q → d
19: A Q Z → Z	41: C G C → G	63: Z Z * → G	86: Z A Q → Z		132: d d G → d
20: A Q d → Q	42: d G G → G	64: A Z Q → d	87: Z A A → Z	d State	133: d d Z → C
21: C Q Q → A	43: d G A → G	65: A Z G → Z	88: Z A d → G	109: Q d Z → Q	134: * d Z → C
22: d Q Q → Q	44: d G d → G	66: A Z Z → d	89: A A Q → C	110: Q d A → A	
23: d Q G → Q	45: * G Q → G	67: A Z d → Z	90: d A Q → A	111: Q d C → Q	
24: d Q A → Q	46: * G G → F	68: A Z * → Z	91: d A A → A	112: Q d d → C	
	47: * G Z → G	69: C Z G → G	92: d A Z → Z	113: G d Q → Z	
		70: C Z d → Z			

Fig. 20.11 A 7-state transition table of the Yunès [22] algorithm

20.2.4   A Brief History of the Developments of the 3n-Step FSSP Algorithms and Their Implementations

The 3n-step algorithm is a simple and straightforward one that exploits a parallel divide-and-conquer strategy based on an efficient use of 1/1- and 1/3-speed of signals. After Minsky and McCarthy (See Minsky [9]) gave an idea for designing the 3n-step synchronization algorithm, Fischer [2] also presented a 3n-step synchronization algorithm, yielding a 15-state implementation, respectively. Herman [6] implemented the 3n-step algorithm in terms of 10-state finite state automaton. Yunès [22] developed two seven-state synchronization algorithms. His algorithms were interesting in that he decreased the number of internal states of each cellular automaton by extending the width of signal threads in the space-time diagram. Umeo, Maeda, and Hongyo [16] presented a 6-state 3n-step algorithm. Afterward, Yunès [25] also presented a 6-state 3n-step algorithm.

20.3   Implementations of the 3n-Step FSSP Algorithms

The non-optimum-time 3n-step FSSP algorithms that we discuss in this chapter are as follows:

- **Fischer [2] algorithm,**
- **Minsky-McCarthy [9] algorithm,**
- **Herman [6] 10-state algorithm,**
- **Yunès [22] 7-state algorithm,**
- **Umeo-Maeda-Hongyo [16] 6-state algorithm,**
- **Yunès [25] 6-state algorithm,**
- **Two 6-state algorithms proposed in this chapter,** and
- **Umeo-Yanagihara [18] 5-state algorithm.**

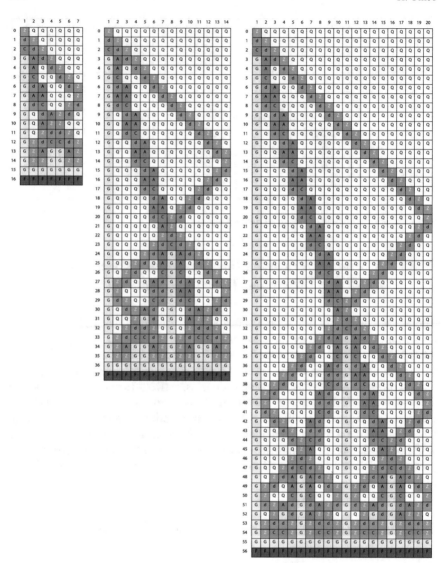

Fig. 20.12 Snapshots of the synchronization process of the Yunès [22] algorithm on 7, 14, and 20 cells, respectively

In this section, we examine the state transition rule sets for these FSSP protocols developed so far above. A transition rule is expressed in a usual 4-tuple style: W X Y → Z which represents the state transition rule that an automaton in currently in state X, with its left neighbor in state W and the right neighbor in state Y will enter state Z at the next step. The state "*" that appears in the state transition table is a border state for

```
──────Q State──────  │ 15:* Q P → P │ 28:R P * → Z │ 41:P R M → M │ 54:Q Z Z → P │ 70:M Z M → Q
                     │ 16:* Q R → Q │ 29:Z P Q → R │ 42:R R Q → R │ 55:Q Z M → Q │ 71:M Z * → Q
 1:Q Q Q → Q         │ 17:* Q Z → Q │ 30:Z P P → Z │ 43:R R P → M │ 56:P Z P → Z │ 72:* Z P → Z
 2:Q Q P → P         │              │ 31:Z P Z → Z │ 44:R R M → M │ 57:P Z Z → Z │ 73:* Z R → Q
 3:Q Q R → Q         │ ───P State── │ 32:Z P * → Z │ 45:R R Q → P │ 58:R Z Q → Q │ 74:* Z Z → F
 4:Q Q Z → Q         │              │ 33:* P Q → Z │ 46:Z R P → R │ 59:R Z R → Q │ 75:* Z M → Q
 5:Q Q * → Q         │ 18:Q P Q → Z │ 34:* P R → Z │ 47:Z R M → R │ 60:R Z Z → Q │
 6:P Q Q → P         │ 19:Q P P → Z │ 35:* P Z → Z │ 48:M R Q → Z │ 61:R Z * → Q │ ───M State───
 7:P Q P → P         │ 20:Q P R → R │              │ 49:M R P → M │ 62:Z Z Q → P │
 8:P Q * → P         │ 21:Q P Z → R │ ───R State── │ 50:M R R → M │ 63:Z Z P → Z │ 76:M M R → R
 9:R Q Q → Q         │ 22:P P Q → Z │              │ 51:M R Z → R │ 64:Z Z R → Q │ 77:R M Z → Z
10:R Q R → Q         │ 23:P P R → Z │ 36:Q R R → R │ 52:M R M → M │ 65:Z Z Z → F │ 78:Z M R → Z
11:Z Q Q → Q         │ 24:P P Z → Z │ 37:Q R Z → P │              │ 66:Z Z M → Q │
12:Z Q Z → Q         │ 25:R P Q → R │ 38:Q R M → Z │ ───Z State── │ 67:Z Z * → F │
13:Z Q * → Q         │ 26:R P P → Z │ 39:P R R → M │              │ 68:M Z Q → Q │
14:* Q Q → Q         │ 27:R P R → Z │ 40:P R Z → R │ 53:Q Z R → Q │ 69:M Z Z → Q │
```

Fig. 20.13 A 6-state transition table of the Umeo, Maeda, and Hongyo [16] algorithm

the left and right end cells. It is noted that, according to the conventions in FSSP, the border state "*" is not counted as the number of states for the synchronizer. We have tested the validity of those tables for any array of length n such that $2 \leq n \leq 500$. It reveals that all of the rule tables tested in this chapter include no redundant rules.

20.3.1   Fischer Algorithm: $\mathcal{A}_1$

Fischer [2] firstly presented an idea for synchronizing any 1D array in non-optimum-time. We implemented a space-time diagram (Fig. 1 in Fischer [2]) for the synchronization in terms of a finite state automaton with 15 states. The set Q of the internal states for the Fischer's algorithm is $Q = \{$G, Q, A, B, C, a, b, c, R, L, X, Y, Z, K, F$\}$, where the state G is the initial *general* state, Q is the *quiescent* state, and F is the *firing* state, respectively. The table itself, consisting of 188 4-tuple rules, is constructed newly in this chapter. See Fig. 20.3. The readers can find that the table is very sparse in a sense that each table has many empty entries. The filled-in ratio of the implementation is $f_{\text{Fischer}[2]} = 188/14 \times 15 \times 15 = 5.9$ (%). The time complexity for synchronizing any array of length n is $3n - 4$. Figure 20.4 shows some snapshots of the synchronization process of the Fischer's algorithm on 7, 14, and 20 cells. The thread width of the a-, b-, and r-signals implemented is one, respectively.

20.3.2   Minsky–McCarthy Algorithm: $\mathcal{A}_2$

Minsky and McCarthy (See Minsky [9]) also presented an idea for designing the $3n$-step synchronization algorithm. Yunès [22] gave an implementation of the algorithm for 14 cells in terms of a 13-state finite state automaton. Figure 20.5, consisting of 138 rules, is the transition table constructed in this chapter based on Fig. 2 in Yunès [22]. The set Q of internal states for the Minsky-McCarthy algorithm is $Q=\{$I, Q, A, B, C, a, b, c, R, L, X, Y, F$\}$, where the state I is the initial *general* state, Q is the *quiescent* state, and F is the itfiring state, respectively. The filled-in ratio of

Fig. 20.14 Snapshots of the synchronization process of the Umeo, Maeda, and Hongyo [16] algorithm on 7, 14, and 20 cells, respectively

Q State

#				
1:	Q Q Q	→	Q
2:	Q Q A	→	C
3:	Q Q B	→	B
4:	Q Q C	→	Q
5:	Q Q D	→	Q
6:	Q Q *	→	Q
7:	A Q Q	→	C
8:	A Q A	→	Q
9:	A Q B	→	B
10:	A Q D	→	C
11:	A Q *	→	C
12:	B Q Q	→	B
13:	B Q A	→	B
14:	B Q B	→	B
15:	B Q C	→	A
16:	B Q *	→	B
17:	C Q Q	→	Q
18:	C Q A	→	C
19:	C Q C	→	Q
20:	C Q D	→	Q
21:	C Q *	→	Q
22:	D Q Q	→	Q
23:	D Q A	→	C
24:	D Q C	→	Q
25:	D Q D	→	Q
26:	D Q *	→	Q
27:	* Q Q	→	Q
28:	* Q A	→	C
29:	* Q B	→	B
30:	* Q C	→	Q
31:	* Q D	→	Q

A State

#				
32:	Q A A	→	B
33:	Q A C	→	A
34:	A A Q	→	B
35:	A A A	→	E
36:	A A B	→	A
37:	A A C	→	C
38:	A A D	→	C
39:	A A *	→	E
40:	B A A	→	A
41:	B A B	→	A
42:	B A D	→	A
43:	C A Q	→	A
44:	C A A	→	C
45:	C A C	→	C
46:	C A D	→	C
47:	D A A	→	C
48:	D A B	→	A
49:	D A C	→	C
50:	D A D	→	C
51:	D A *	→	C
52:	* A Q	→	B
53:	* A A	→	E
54:	* A B	→	A
55:	* A C	→	C
56:	* A D	→	C

B State

#				
57:	Q B Q	→	D
58:	Q B A	→	C
59:	Q B B	→	D
60:	Q B C	→	D
61:	Q B D	→	D
62:	Q B *	→	D
63:	A B Q	→	C
64:	A B A	→	A
65:	A B B	→	A
66:	A B D	→	A
67:	A B *	→	A
68:	B B Q	→	D
69:	B B A	→	A
70:	B B C	→	A
71:	B B D	→	B
72:	C B Q	→	D
73:	C B B	→	A
74:	C B C	→	A
75:	C B *	→	A
76:	D B Q	→	D
77:	D B A	→	A
78:	D B B	→	B
79:	D B B	→	B
80:	D B *	→	B
81:	* B Q	→	D
82:	* B A	→	A
83:	* B C	→	A
84:	* B D	→	B

C State

#				
85:	Q C A	→	D
86:	Q C B	→	B
87:	Q C D	→	C
88:	A C Q	→	D
89:	A C A	→	D
90:	A C B	→	D
91:	A C C	→	D
92:	B C Q	→	B
93:	B C A	→	D
94:	B C B	→	B
95:	B C C	→	B
96:	B C *	→	B
97:	C C A	→	D
98:	C C B	→	B
99:	C C D	→	C
100:	D C Q	→	C
101:	D C C	→	C
102:	D C D	→	C
103:	* C A	→	D
104:	* C B	→	B
105:	* C D	→	C

D State

#				
106:	Q D A	→	Q
107:	Q D B	→	Q
108:	Q D C	→	A
109:	A D Q	→	Q
110:	A D A	→	Q
111:	A D C	→	B
112:	A D D	→	Q
113:	A D *	→	Q
114:	B D Q	→	Q
115:	B D B	→	Q
116:	B D D	→	Q
117:	B D *	→	Q
118:	C D Q	→	A
119:	C D A	→	B
120:	C D D	→	A
121:	D D A	→	Q
122:	D D B	→	Q
123:	D D C	→	Q
124:	* D A	→	Q
125:	* D B	→	Q

Fig. 20.15 A 6-state transition table of the Yunès [25] algorithm

the implementation is $f_{\text{Minsky-McCarthy}[9]} = 138/12 \times 13 \times 13 = 6.8$ (%). The time complexity for synchronizing any array of length n is $3n + O(\log n)$. Figure 20.6 shows some snapshots of the synchronization process of the Minsky–McCarthy [9] algorithm on 7, 14, and 20 cells, respectively. Note that the thread width of the three signals implemented is one.

20.3.3   Herman's 10-State Algorithm: $\mathcal{A}_3$

Herman [6] also gave a 10-state implementation for the $3n$-step synchronization algorithm. Figure 20.7, consisting of 155 rules, is the transition table constructed in this chapter based on Herman [6] (Figs. 3, 4, and 5 in Herman [6]). The set Q of the internal states for the Herman [6] algorithm is $Q=\{$I, S, J, U, W, R, X, V, G, F$\}$, where the state I is the initial *general* state, S is the *quiescent* state, and F is the *firing* state, respectively. The filled-in ratio of the implementation is $f_{\text{Herman}[6]} = 155/9 \times 10 \times 10 = 17.2$ (%). The time complexity for synchronizing any array of length n is $3n + O(\log n)$. Figure 20.8 shows some snapshots of the synchronization process of the Herman [6] algorithm on 7, 14, and 20 cells, respectively. Note that the thread width of the a-, b-, and r-signals implemented is two, respectively. Herman [6] is the first paper which introduced wider threads to describe those signals, yielding the decrease of the number of states required.

20.3.4   Yunès Seven-State Algorithm: $\mathcal{A}_4$

Yunès [22] presented two 7-state implementations for the $3n$-step FSSP algorithms and decreased the number of states required. The set Q of internal states for the first Yunès algorithm is $Q=\{$G, Q, A, C, d, Z, F$\}$, where the state G is the initial *general* state, Q is the *quiescent* state, and F is the *firing* state, respectively. The

Fig. 20.16 Snapshots of the synchronization process of the Yunès [25] algorithm on 7, 14, and 20 cells, respectively

——— Q State ———		——— Z State ———	——— A State ———	——— C State ———	——— d State ———
1: Q Q Q → Q	24: * Q d → Q	46: * Z C → Z	68: * A Z → d	90: * C C → F	112: * d A → A
2: Q Q Z → Z		47: Q A Z → Z	69: * A C → Q	91: Q d Z → Q	113: * d C → Z
3: Q Q A → Q	25: Q Z Q → A	48: Q A A → d	70: Q C A → d	92: Q d A → A	114: * d d → C
4: Q Q C → A	26: Q Z Z → A	49: Q A C → Q	71: Q C d → d	93: Q d C → Q	
5: Q Q d → Q	27: Q Z A → C	50: Q A d → d	72: Z C Z → A	94: Q d d → C	
6: Q Q * → Q	28: Q Z C → A	51: Z A Q → Z	73: Z C A → A	95: Z d Q → Q	
7: Z Q Q → Z	29: Q Z d → d	52: Z A Z → d	74: Z C C → A	96: Z d Z → Q	
8: Z Q Z → Z	30: Z Z Q → A	53: Z A A → d	75: Z C d → A	97: Z d d → Q	
9: Z Q d → C	31: Z Z A → d	54: Z A C → A	76: Z C * → A	98: Z d * → Q	
10: Z Q * → Z	32: Z Z C → Q	55: Z A d → C	77: A C Q → d	99: A d Q → d	
11: A Q Q → Q	33: A Z Q → C	56: Z A * → d	78: A C Z → A	100: A d A → A	
12: A Q Z → C	34: A Z Z → d	57: A A Q → C	79: A C d → Z	101: A d d → A	
13: A Q A → Q	35: A Z A → d	58: A A Z → d	80: C C Z → A	102: C d Q → Q	
14: A Q C → A	36: A Z d → d	59: A A C → Q	81: C C C → F	103: C d C → Z	
15: C Q A → A	37: A Z * → d	60: A A d → C	82: C C d → Z	104: C d d → Z	
16: C Q A → A	38: C Z Q → C	61: C A Q → Q	83: C C * → F	105: d d Q → d	
17: d Q Q → Q	39: C Z Z → Z	62: C A Z → A	84: d C Q → A	106: d d Z → Z	
18: d Q A → Q	40: C Z C → Z	63: C A A → Q	85: d C Z → A	107: d d A → A	
19: d Q d → Q	41: C Z * → Z	64: C A C → Q	86: d C A → Z	108: d d C → Z	
20: d Q * → Q	42: d Z Q → d	65: d A Q → A	87: d C C → Z	109: d d d → C	
21: * Q Q → Q	43: d Z A → d	66: d A Z → C	88: d C d → Z	110: d d C → C	
22: * Q A → Q	44: * Z Q → Z	67: d A A → C	89: * C Z → A	111: * d Z → Q	
23: * Q A → Q	45: * Z A → d				

Fig. 20.17 A 6-state transition table of the new algorithm

following Fig. 20.9, consisting of 134 rules, is the transition table. The filled-in ratio of the implementation is $f_{Yunès[22]} = 134/6 \times 7 \times 7 = 45.6\,(\%)$. The time complexity for synchronizing any array of length n is $3n + O(\log n)$. Figure 20.10 shows some snapshots of the synchronization process of the Yunès [22] algorithm on 7, 14, and 20 cells, respectively. Note that the width of the signals implemented is two. Note that Yunès [22] also represented a-, b-, and r-signals as threads of width 2, thus decreasing the number of states to seven.

Yunès [22] also gave a different 7-state implementation for the $3n$-step FSSP algorithm. The set Q of internal states for the Yunès algorithm is $Q=\{G, Q, A, C, d, Z, F\}$, where the state Z is the initial *general* state, Q is the *quiescent* state, and F is the *firing* state, respectively. The following Fig. 20.11, consisting of 134 rules, is the transition table. The filled-in ratio of the implementation is $f_{Yunès[22]} = 134/6 \times 7 \times 7 = 45.6\,(\%)$. The time complexity for synchronizing any array of length n is $3n + O(\log n)$. Figure 20.12 shows some snapshots of the synchronization process of the Yunès [22] algorithm on 7, 14, and 20 cells, respectively. Note that the width of the signals implemented is also two. A major difference between these two implementations is a center marking for each splitting.

20.3.5   Umeo, Maeda, and Hongyo's 6-State Algorithm: $\mathcal{A}_5$

Umeo, Maeda, and Hongyo [16] presented a 6-state $3n$-step FSSP algorithm. The implementation was quite different from previous designs. The set Q of internal states for the algorithm is $Q=\{P, Q, R, Z, M, F\}$, where the state P is the initial *general* state, Q is the *quiescent* state, and F is the *firing* state, respectively. The following Fig. 20.13, consisting of 78 rules, is the transition table. The filled-in ratio of the implementation is $f_{Umeo, Maeda, and Hongyo[16]} = 78/5 \times 6 \times 6 = 52.0\,(\%)$. The time complexity for synchronizing any array of length n is $3n + O(\log n)$. Figure 20.14 shows some snapshots of the synchronization process of the algorithm on 7, 14

Fig. 20.18 Snapshots of the synchronization process of the new 6-state algorithm on 7, 14, and 20 cells, respectively

and 20 cells, respectively. Umeo, Maeda, and Hongyo [16] improved the seven-state implementation developed by Yunès [22]. The number six is the smallest one known in the class of $3n$-step synchronization algorithms. An important key idea is to increase the number of cells being active during their computation in the *zone* $\mathcal{T}$, shown in Fig. 20.2. The width of the thread is unbounded. It is seen that the algorithm

G State		L State		X State	Q State
	19: Q G X → L	36: X M M → M	53: M L * → M	70: G X X → Q	87: M Q * → Q
	20: * G L → M	37: X M L → M	54: L L G → X	71: M X M → M	88: L Q L → X
1: G G L → L		38: Q M G → Q	55: L L M → M	72: M X X → M	89: L Q Q → L
2: G G X → L	M State	39: Q M M → L	56: L L X → M	73: L X L → Q	90: L Q * → L
3: G G Q → G		40: Q M L → Q	57: L L Q → M	74: L X X → Q	91: X Q X → M
4: M G L → G	21: G M G → Q	41: * M G → Q	58: X L G → M	75: L X Q → M	92: Q Q G → Q
5: M G X → G	22: G M M → Q	42: * M M → F	59: X L L → M	76: X X G → Q	93: Q Q M → Q
6: M G Q → L	23: G M Q → Q	43: * M L → Q	60: X L Q → X	77: X X M → M	94: Q Q L → L
7: L G G → L	24: G M * → Q		61: Q L G → G	78: X X L → Q	95: Q Q Q → Q
8: L G M → G	25: M M G → Q	L State	62: Q L M → G	79: X X X → M	96: Q Q * → Q
9: L G L → L	26: M M M → F		63: Q L L → M	80: X X Q → M	97: * Q G → Q
10: L G X → L	27: M M L → M	44: G L G → G	64: Q L X → X	81: Q X L → M	98: * Q M → Q
11: L G Q → M	28: M M X → M	45: G L M → M	65: Q L Q → M	82: Q X X → M	99: * Q L → L
12: X G G → L	29: M M Q → L	46: G L L → X	66: * L G → M		100: * Q Q → Q
13: X G M → G	30: M M * → F	47: G L X → M	67: * L M → M	Q State	
14: X G L → L	31: L M M → M	48: G L Q → G	68: * L Q → G		
15: X G Q → L	32: L M L → M	49: G L * → M		83: G Q G → Q	
16: Q G X → G	33: L M X → M	50: M L G → M	X State	84: G Q Q → Q	
17: Q G M → L	34: L M Q → Q	51: M L L → M		85: M Q M → Q	
18: Q G L → M	35: L M * → Q	52: M L Q → G	69: G X G → X	86: M Q Q → Q	

Fig. 20.19 A transition table of a new 6-state $O(n^2)$-state-change implementation

has $O(n^2)$ state-change complexity. The algorithm can be extended to a new non-trivial symmetrical six-state $3n$-step generalized FSSP algorithm. See Umeo, Maeda, and Hongyo [16] for details.

20.3.6   Yunès 6-State Algorithm: $\mathcal{A}_6$

Yunès [25] presented a 6-state implementation for the $3n$-step FSSP algorithm. His implementation is based on wider threads. The set Q of the internal states for the Yunès [25] algorithm is $Q = \{A, Q, B, C, D, E\}$, where the state A is the initial *general* state, Q is the *quiescent* state, and E is the *firing* state, respectively. The following Fig. 20.15, consisting of 125 rules, is the transition table. The filled-in ratio of the implementation is $f_{\text{Yunès}[25]} = 125/5 \times 6 \times 6 = 69.4\,(\%)$. The time complexity for synchronizing any array of length n is $3n + O(\log n)$. Figure 20.16 shows some snapshots of the synchronization process of the Yunès's algorithm on 7, 14, and 20 cells, respectively. Note that the width of the a- and r-signals implemented is two, but for the b-signal is a combination of two and three.

20.3.7   A New 6-State Algorithm: $\mathcal{A}_7$

Here we present a new 6-state implementation for the $3n$-step algorithm. The set Q of internal states for the algorithm is $Q = \{Z, Q, A, C, d, F\}$, where the state Z is the initial *general* state, Q is the *quiescent* state, and F is the *firing* state, respectively. The following Fig. 20.17, consisting of 114 rules, is the 6-state transition table. The filled-in ratio of the implementation is $f_{\text{This chapter}} = 114/5 \times 6 \times 6 = 63.3\,(\%)$. The time complexity for synchronizing any array of length n is $3n + O(\log n)$. Figure 20.18 shows some snapshots of the synchronization process of the algorithm on 7, 14, and 20 cells, respectively. Note that the width of the implemented a-signal is three and those for the b- and r-signals are two.

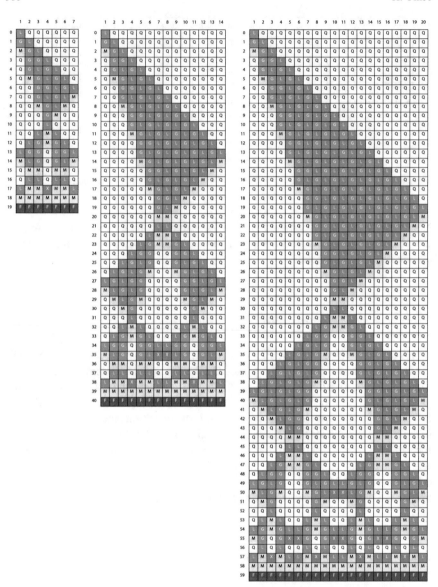

Fig. 20.20 Snapshots of the synchronization process of a new 6-state $O(n^2)$-state-change implementation

Q State								
	13: S Q Q → Q	24: R R R → R	L State	49: S L Q → R	60: L S Q → S
1: Q Q Q → Q	14: S Q R → S	25: R R L → L		50: S L R → F	61: L S S → S
2: Q Q R → Q	15: * Q Q → Q	26: R R * → L	37: Q L Q → L	51: S L * → F	62: S S Q → Q
3: Q Q L → L	16: * Q R → Q	27: L R Q → S	38: Q L R → S	52: * L L → R	63: S S R → S
4: Q Q S → Q	17: * Q L → L	28: L R L → Q	39: Q L L → Q	53: * L S → R	64: S S L → F
5: Q Q * → Q	18: * Q S → Q	29: L R S → F	40: Q L S → Q		65: * S Q → Q
6: R Q Q → R		30: S R Q → Q	41: R L Q → Q	S State	66: * S R → S
7: R Q R → R	R State	31: S R R → R	42: R L R → Q		67: * S L → F
8: R Q * → R		32: S R L → L	43: R L L → R	54: Q S Q → Q	
9: L Q Q → Q	19: Q R Q → R	33: S R * → L	44: R L S → R	55: Q S R → S	
10: L Q L → L	20: Q R R → R	34: * R Q → S	45: R L * → Q	56: Q S L → L	
11: L Q S → S	21: Q R L → Q	35: * R L → Q	46: L L Q → L	57: Q S S → Q	
12: L Q * → Q	22: Q R S → L	36: R S * → F	47: L L L → L	58: R S Q → R	
	23: R R Q → Q		48: L L S → L	59: R S S → F	

Fig. 20.21 A transition table of the 5-state Umeo and Yanagihara [18] algorithm

20.3.8   A New 6-State Algorithm: $\mathcal{A}_8$

We also present a new 6-state $O(n^2)$-state-change implementation for the $3n$-step FSSP algorithm. The implementation is quite similar to the algorithm $\mathcal{A}_5$. The set Q of internal states for the algorithm is $Q=\{L, Q, G, M, X, F\}$, where the state L is the initial *general* state, Q is the *quiescent* state, and F is the *firing* state, respectively. The following Fig. 20.19, consisting of 100 rules, is the transition table. The table is nearly symmetric. The filled-in ratio of the implementation is $f_{\text{This Chapter}} = 100/5 \times 6 \times 6 = 55.6$ (%). The time complexity for synchronizing any array of length n is $3n + O(\log n)$. Figure 20.20 shows some snapshots of the synchronization process of the algorithm on 7, 14, and 20 cells, respectively. The state-change complexity is $O(n^2)$.

20.3.9   Umeo-Yanagihara 5-State Algorithm: $\mathcal{A}_9$

Umeo and Yanagihara [18] presented a 5-state implementation for the $3n$-step FSSP algorithm. The solution is a *partial* one that can synchronize any array of length n such that $n = 2^k, k = 1, 2, 3, ...,$. The set Q of internal states for the implementation is $Q=\{R, Q, S, L, F\}$, where the state R is the initial *general* state, Q is the *quiescent* state, and F is the *firing* state, respectively. The following Fig. 20.21, consisting of 125 rules, is the transition table. The filled-in ratio of the implementation is $f_{\text{Umeo and Yanagihara[18]}} = 67/4 \times 5 \times 5 = 67.0$ (%). The time complexity for synchronizing any array of length n is $3n - 3$. Figure 20.22 shows some snapshots of the synchronization process of the algorithm on 8, 16, and 32 cells, respectively. Note that the state change complexity is $O(n^2)$.

20.3.10   State-Change Complexity

Concerning the state-change complexity, the following theorems are established.

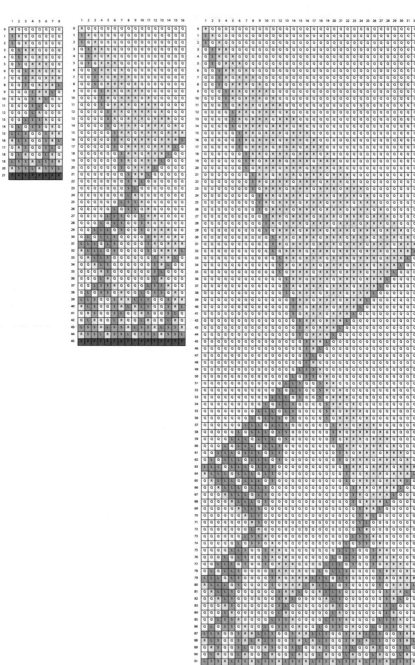

Fig. 20.22 Snapshots of the synchronization process of the 5-state Umeo and Yanagihara [18] algorithm on 8, 16, and 32 cells, respectively

Table 20.1 Quantitative comparison of transition rule sets for non-optimum-time FSSP algorithms

Algorithm	# States	# Rules	Time complexity	State-change complexity	Generals's position	Type	Thread width	Filled-in ratio (%)
Fischer [2]	15	188*	$3n - 4$	$O(n \log n)$	left	thread	1	5.9
Minsky–McCarthy [9]	13	138*	$3n + O(\log n)$	$O(n \log n)$	left	thread	1	6.8
Herman [6]	10	155*	$3n + O(\log n)$	$O(n \log n)$	left	thread	2	17.2
Yunès [22]	7	134	$3n \pm O(\log n)$	$O(n \log n)$	left	thread	2	45.6
Yunès [22]	7	134	$3n \pm O(\log n)$	$O(n \log n)$	left	thread	2	45.6
Umeo et al. [16]	6	78	$3n + O(\log n)$	$O(n^2)$	left	plane	–	43.3
Umeo et al. [16]	6	115	$\max(k, n - k + 1) + 2n + O(\log n)$	$O(n^2)$	arbitrary	plane	–	63.9
Umeo and Yanagihara [18]	5	67	$3n - 3$ $n = 2^k, k = 1, 2, \ldots$	$O(n^2)$	left	plane	short	67.0
Yunès [25]	6	125	$3n + \lceil \log n \rceil - 3$	$O(n \log n)$	left	thread	2, 3	69.4
This chapter	6	114	$3n + O(\log n)$	$O(n \log n)$	left	thread	2, 3	63.3
This chapter	6	100	$3n + O(\log n)$	$O(n^2)$	left	plane	–	55.6

The "*" symbol in the table shows the correction and reduction of transition rules implemented in this chapter

Theorem 8 *The non-optimum-time algorithms developed by Fischer [2], Minsky–McCarthy [9], Herman [6], Yunès [22], Yunès [25], and a new 6-state algorithm in this chapter have $O(n \log n)$ optimum state-change complexity for synchronizing n cells in $3n \pm O(\log n)$ steps.*

Theorem 9 *The non-optimum-time algorithms developed by Umeo–Maeda–Hongyo [16], a new one in this chapter, and Umeo–Yanagihara [18] have $O(n^2)$ state-change complexity for synchronizing n cells in $3n \pm O(\log n)$ steps.*

20.4 Discussions

We have given a survey on a class of non-optimum-time FSSP algorithms for one-dimensional (1D) cellular automata, focusing our attention to the 1D FSSP algorithms having $3n \pm O(\log n)$ time complexities. Here, we present a Table 20.1 based on a quantitative comparison of non-optimum-time synchronization algorithms and their transition tables discussed above.

References

1. Balzer, R.: An 8-state minimal time solution to the firing squad synchronization problem. Inf. Control **10**, 22–42 (1967)
2. Fischer, P.C.: Generation of primes by a one-dimensional real-time iterative array. J. ACM **12**(3), 388–394 (1965)
3. H.-D. Gerken: Uber Synchronisations-Probleme bei Zellularautomaten. pp. 50. Diplomarbeit Institut fur Theoretische Informatik, Technische Universitat Braunschweig, (1987)
4. E. Goto: A minimal time solution of the firing squad problem. Dittoed course notes for Applied Mathematics vol. 298, pp. 52-59. Harvard University, (1962)
5. Herman, G.T.: Models for cellular interactions in development without polarity of individual cells. I. General description and the problem of universal computing ability. Int. J. Syst. Sci. **2**(3), 271–289 (1971)
6. Herman, G.T.: Models for cellular interactions in development without polarity of individual cells. II. Problems of synchronization and regulation. Int. J Syst. Sci. **3**(2), 149–175 (1972)
7. Kobuchi, Y.: A note on symmetrical cellular spaces. Inform. Process. Lett. **25**, 413–415 (1987)
8. Mazoyer, J.: A six-state minimal time solution to the firing squad synchronization problem. Theor. Comput. Sci. **50**, 183–238 (1987)
9. Minsky, M.: Computation: Finite and infinite machines. pp. 28–29. Prentice Hall, Upper Saddle River (1967)
10. Moore, E.F.: The firing squad synchronization problem. In: Moore, F. (ed.) Sequential Machines, Selected Papers (E), pp. 213–214. Addison-Wesley, Reading MA (1964)
11. W. L. Ng: Partial solutions for the firing squad synchronization problem on rings. pp.363. UMI Dissertation Publishing, Ann Arbor, MI (2011)
12. Sanders, P.: Massively parallel search for transition-tables of polyautomata. In: Jesshope, C., Jossifov, V., Wilhelmi, W. (eds.) Proceedings of the VI International Workshop on Parallel Processing by Cellular Automata and Arrays, pp. 99–108. Akademie (1994)
13. Szwerinski, H.: Symmetrical one-dimensional cellular spaces. Inform. Control **67**, 163–172 (1982)

14. H. Umeo: Firing squad synchronization problem in cellular automata. In: R. A. Meyers (Ed.) Encyclopedia of Complexity and System Science, vol.4, pp. 3537–3574. Springer, Heidelberg (2009)
15. Umeo, H., Kamikawa, N., Yunès, J.B.: A family of smallest symmetrical four-state firing squad synchronization protocols for ring arrays. Parallel Process. Lett. **19**(2), 299–313 (2009)
16. Umeo, H., Maeda, M., Hongyo, K.: A design of symmetrical six-state $3n$-step firing squad synchronization algorithms and their implementations. Proceedings of 7th International Conference on Cellular Automata for Research and Industry, ACRI, LNCS vol. 4173, pp. 157-168.(2006)
17. Umeo, H., Maeda, M., Sousa, A., Taguchi, T.: A class of non-optimum-time $3n$-step FSSP algorithms-a survey. In: Proceedings of the 13th International Conference on Parallel Computing Technologies, PaCT: Petrozavodsk, LNCS, vol. 9251. pp. 231–245. Springer, Heidelberg (2015)
18. Umeo, H., Yanagihara, T.: A smallest five-state solution to the firing squad synchronization problem. In: Proceedings of 5th International Conference on Machines, Computations, and Universality, MCU, LNCS, vol. 4664 (2997), pp.292-302. (2007)
19. Umeo, H., Yunès, J.-B., Kamikawa, N., Kurashiki, J.: Small non-optimum-time firing squad synchronization protocols for one-dimensional rings. In: Proceedings of the 2009 International Symposium on Nonlinear Theory and its Applications, NOLTA'09, pp. 479–482. (2009)
20. Vollmar, R.: Some remarks about the "Efficiency" of polyautomata. Int. J. Theor. Phys. **21**(12), 1007–1015 (1982)
21. Waksman, A.: An optimum solution to the firing squad synchronization problem. Inf. Control **9**, 66–78 (1966)
22. Yunès, J.-B.: Seven states solutions to the firing squad synchronization problem. Theor. Comput. Sci. **127**(2), 313–332 (1994)
23. Yunès, J.-B.: An intrinsically non minimal-time Minsky-like 6-states solution to the firing squad synchronization problem. Theor. Inf. Appl. **42**(1), 55–68 (2008)
24. Yunès, J.-B.: Simple new algorithms which solve the firing squad synchronization problem: a 7-state $4n$-step solution. In: Durand, J., Margenstern, M. (eds.) Proceedings of the International Conference on Machines, Computations, and Universality, MCU, LNCS, vol. 4664, pp.316–324. Springer, Berlin (2007)
25. J. B. Yunès: A 4-states algebraic solution to linear cellular automata synchronization. Inf. Process. Lett. **19**(2), 71–75 (2008)

Chapter 21
Universality of Asynchronous Circuits Composed of Locally Reversible Elements

Jia Lee

Abstract Reversible computing reflects a fundamental law of microscopic physics, which usually attempts to develop an equivalence between the global reversibility and local reversibility in computational systems. Hitherto the equivalence assumes implicitly that the underlying systems are synchronously timed. Alternative systems include the delay-insensitive circuits, a special type of asynchronous circuits, of which the operations are robust to arbitrary delays involved in the transmissions of signals. This chapter aims at exploring the universal input and output behavior of delay-insensitive circuits composed by reversible elements, with the globally non-reversible behavior arising from the locally invertible operations.

21.1 Introduction

Local reversibility reflects a fundamental law of physics at microscopic scale, which in general is likely insufficient to ensure the macroscopic reversibility in physical systems [8]. Nevertheless, logical reversibility potentially promises to develop computers with zero power consumption [1], and is also a precondition for quantum computation [4]. For this reason, extensive efforts have been made to facilitate the constructions of reversible computing models, such as reversible cellular automata [9, 11], reversible logical circuits [5, 6], billiard-ball models [3, 17] and quantum computers [4, 10].

Since reversibility, in general, is connected with determinism, the dynamical behavior of a reversible model can be drawn as a linear graph without branching, where each vertex represents a global state of the model, and each arrow pointing from a vertex to another expresses a global transition via which the former state can be transformed to the latter. In this case, a valid equivalence between the global reversibility and local reversibility not only provides an efficient way to test for the reversibility of a computational model, but also can substantially reduce the

J. Lee (✉)
College of Computer Science, Chongqing University, Chongqing 400044, China
e-mail: lijia@cqu.edu.cn

© Springer International Publishing Switzerland 2017
A. Adamatzky (ed.), *Advances in Unconventional Computing*,
Emergence, Complexity and Computation 22,
DOI 10.1007/978-3-319-33924-5_21

construction of the inverse of a reversible model to direct reversing the model's local function that is invertible [18]. The inverse of a reversible model is also reversible which can exactly exhibit the time-reversed dynamics of the original model.

However, the equivalence between global and local reversibility implicitly assumes that the underlying reversible model is synchronously timed, i.e., each global transition comprises the local transitions of all elements that are performed in lock step in accordance with a central clock signal. Removal of central clocks may lead to models that can work in asynchronous timings. Delay-insensitive (DI) circuits are one of the most typical examples of asynchronous models, in which signals may be subject to arbitrary delays but without violating the external input and output behavior of the circuits, thereby no need a central clock to synchronize the local operations of circuit elements or transmission of signals.

As microscopic physical processes are asynchronous in nature, it makes sense to investigate the reversibility at local and global levels in asynchronous models [2, 15, 20]. Local reversibility can be defined on primitive elements in DI-circuits [13, 18], each of which takes the same number of input and output lines. In spite of the randomness associated with asynchronous operations, local reversibility in logical elements can directly lead to the global reversibility in a DI-circuit composed of such elements, provided that both the circuit as well as all elements process at most one signal at any time [13, 14, 16], i.e., they are serial. Accordingly, every serial DI-circuit composed of serial reversible elements exhibits linearity in its graph of all configurations, though each transition between a pair of configurations may take even unpredictable time due to the lack of central clocks.

A more general and practical case allows multiple signals to be processed simultaneously by reversible elements as well as the circuits composed of them. In such a case, the concurrency among signal transmissions tends to cause forking and merging of paths in the graph of all possible states of a DI-circuit, whereby the global behavior of the circuit might be regarded as incompatible with the reversibility of conventional logical circuits. This chapter demonstrates that locally reversible elements can be used to construct the whole class of DI-circuits, with the universality emerging from the concurrency among reversible operations at local level. As microscopic physical interactions are inherently asynchronous, the emergence of non-reversible global behavior from reversible local operations in DI-circuits might be consistent with the macroscopic consequence of microscopic reversible dynamics in physics.

Section 21.2 gives definitions on local reversibility in delay-insensitive circuits. Section 21.3 describes the constructions of DI-circuits using locally reversible elements. This chapter finishes with the conclusion given in Sect. 21.4.

21.2   Delay-Insensitive Circuits and Reversible Elements

A *delay-insensitive* (DI) circuit is a modular system with a finite number of input and output lines that interface with the outside world. Communications between the circuit and the outside world are done via exchanging signals through the input and output lines. Signals are one-valued and are denoted by a token on a line.

A DI module can be formalized as $(I, O, \Lambda, \lambda, R)$ where I and O are finite sets of input and output lines ($I \cap O = \emptyset$), respectively. Λ is a finite set of the states of the module, as well as λ being the initial state ($\lambda \in \Lambda$). Also, $R \subseteq (2^I \setminus \{\emptyset\}) \times \Lambda \times (2^O \setminus \{\emptyset\}) \times \Lambda$ is a finite set of transitions. For any $\gamma \subseteq I, \delta \subseteq O$, and $s, s' \in \Lambda$, a transition $\gamma; s \rightarrow \delta; s' \in R$ corresponds to a situation in which the module, when in state s with one signal on each input line $I_i \in \gamma$, will be activated to assimilate a signal from I_i, produce a signal on each output line $O_j \in \delta$, and finally switch its state to s'. A module's operation, therefore, depends merely on the combination of input signals, rather than on the arrival order of the signals or on their timings of arrival.

Keller [12] formulated several operating conditions for DI-circuits, under which the external input and output behavior of any circuit composed of DI modules can be characterized itself as a DI module, i.e., the circuit is delay-insensitive. In particular, one of Keller's conditions prohibits more than one signal to appear on an interconnection line simultaneously. Thus, when two successive signals appear on an input line of a DI module, there must be at least one signal on an output line of the module in response to the first signal, before the next signal is put on the same line.

Assume a DI module $(I, O, \Lambda, \lambda_0, R)$. The module is called *conservative* if each transition $\gamma; s \rightarrow \delta; s'$ in R satisfies $|\gamma| = |\delta|$, i.e., no transition results in the increment or decrement of the number of input signals. Every conservative DI module can be realized using a DI-circuit composed by a fixed set of primitive DI modules (elements), as given in Fig. 21.1, i.e., these elements are universal [15, 19].

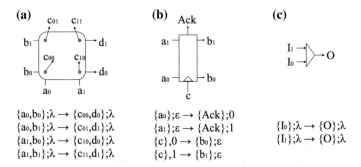

Fig. 21.1 **a** 2×2-**SJoin** $= (\{a_0, a_1, b_0, b_1\}, \{c_{ij} \mid 0 \leq i, j \leq 1\} \cup \{d_0, d_1\}, \{\lambda\}, \lambda, R_{s22})$: Assume $i, j \in \{0, 1\}$. A signal arriving on input line a_i, together with a signal arriving on line b_j, are assimilated and result in one signal on each of the output lines c_{ij} and d_j. **b** **Arbiter** $= (\{c, a_0, a_1\}, \{b_0, b_1, Ack\}, \{\epsilon, 0, 1\}, \epsilon, R_A)$: Assume $i \in \{0, 1\}$. When the element is in state ϵ, a token arriving on input line a_i changes the state to i, and results in an output token on line Ack. When the element is in state i, an input token on line c changes the state to ϵ, and gives rise to a token on output line b_i. Moreover, when the state of the element is ϵ, simultaneous arrivals of signals on a_0 and a_1 are allowed, and in that case only one of them (may be chosen arbitrarily) is processed. The signal on a_0 or a_1 that was not assimilated keeps pending on the same input line until the element's state reverts to ϵ. **c** **Merge** $= (\{I_0, I_1\}, \{O\}, \{\lambda\}, \lambda, R_M)$: A signal arriving on input line I_0 or I_1 is assimilated and gives rise to an output signal to line O

A *serial* DI module is a special type of conservative modules in which each transition $\gamma; s \rightarrow \delta; s' \in R$ satisfies $|\gamma| = |\delta| = 1$. Moreover, let

$$\theta_1 = \gamma_1; s_1 \rightarrow \delta_1; s_1'$$
$$\theta_2 = \gamma_2; s_2 \rightarrow \delta_2; s_2'$$

be any two transitions in R (θ_1 and θ_2 may be the same). The module is called *deterministic* if

$$s_1 = s_2 \wedge \gamma_1 \subseteq \gamma_2 \implies \theta_1 = \theta_2.$$

Moreover, the module is *reversible* if it is deterministic and conservative, and

$$s_1' = s_2' \wedge \delta_1 \subseteq \delta_2 \implies \theta_1 = \theta_2.$$

Thus, each transition of a reversible DI module does not overlap on the left-hand side or on the right-hand side with any other transitions. It can be verified that the 2×2-SJoin and Arbiter in Fig. 21.1 are reversible elements, whereas the Merge is serial but not reversible.

Figure 21.2 gives some typical reversible elements [13, 15, 18] that can be used to construct the whole class of conservative DI-circuits, as we will show in the next section, including the non-reversible Merge in Fig. 21.1c.

21.3   Constructing Conservative Delay-Insensitive Circuits by Reversible Elements

Because the DI-modules given in Fig. 21.1 are capable of constructing all conservative DI-circuits, universality of the reversible elements in Fig. 21.2 can be shown by decomposing each module into the elements. For this purpose, Fig. 21.3 illustrates the construction of an 2×2-SJoin in Fig. 21.1a using the RE, CDE and Join elements.

To verify how the construction in Fig. 21.3 works, suppose that the 2×2-SJoin receives two signals with one on input line a_i and another one on input line b_j ($i, j \in \{0, 1\}$). In this case, the signal from input line a_i changes the state of the CDE A to i and results in a signal to an input line of the Join. Likewise, the signal from input line b_j changes in turn the states of the RE R_{0j} and R_{1j} from V to H, after which it will change the CDE B's state to j and be transferred to the Join. The Join, therefore, outputs a signal back to each of the CDEs A and B. As a result, the CDE B reverts its state from j to 0 and generates a signal on the output line b_j of the construction. In addition, the CDE A reverts its state from i to 0 and transfers a signal to the CDE R_{i0}, which will eventually change the states of the R_{0j} and R_{1j} to V and give rise to a signal on output line c_{ij} of the 2×2-SJoin.

The inverse of an reversible DI module can be obtained by simply exchanging the input and output lines of the module, as well as the left-hand and right-hand sides

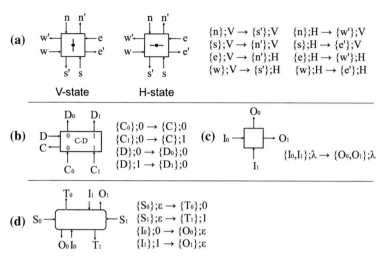

$$\{n\};V \rightarrow \{s'\};V \quad \{n\};H \rightarrow \{w'\};V$$
$$\{s\};V \rightarrow \{n'\};V \quad \{s\};H \rightarrow \{e'\};V$$
$$\{e\};V \rightarrow \{n'\};H \quad \{e\};H \rightarrow \{w'\};H$$
$$\{w\};V \rightarrow \{s'\};H \quad \{w\};H \rightarrow \{e'\};H$$

(a) V-state   H-state

(b)
$$\{C_0\};0 \rightarrow \{C\};0$$
$$\{C_1\};0 \rightarrow \{C\};1$$
$$\{D\};0 \rightarrow \{D_0\};0$$
$$\{D\};1 \rightarrow \{D_1\};0$$

(c)
$$\{I_0,I_1\};\lambda \rightarrow \{O_0,O_1\};\lambda$$

(d)
$$\{S_0\};\varepsilon \rightarrow \{T_0\};0$$
$$\{S_1\};\varepsilon \rightarrow \{T_1\};1$$
$$\{I_0\};0 \rightarrow \{O_0\};\varepsilon$$
$$\{I_1\};1 \rightarrow \{O_1\};\varepsilon$$

Fig. 21.2 Reversible DI modules and their respective sets of transitions. **a RE** (Rotary Element) = $(\{n, s, e, w\}, \{n', s', e', w'\}, \{V, H\}, V, R_{RE})$: The local states H and V are displayed by horizontal and vertical bars, respectively. Roughly speaking, when a token comes from a direction parallel to the bar, then the token will pass straight through to the opposite output line without changing the direction of the bar. When a token comes from a direction orthogonal to the bar, the token will be deflected to the right and rotate the bar 90°. **b CDE** (Coding-Decoding Element) = $(\{C_0, C_1, D\}, \{D_0, D_1, C\}, \{0, 1\}, 0, R_{CD})$: Assume $i \in \{0, 1\}$. When the element is in state 0, a token arriving on input line C_i updates the state to i, and gives rise to a token on output line C. A token arriving on input line D reverts the state to 0, and results in a token on output line D_i if the element is in state i. **c Join** = $(\{I_0, I_1\}, \{O_0, O_1\}, \{\lambda\}, \lambda, R_{CJ})$: Two signals, with one arriving on input lines I_0 and another one arriving on line I_1, are assimilated and produce one signal on each of the output lines O_0 and O_1. **d Multiplexer** = $(\{S_0, S_1, I_0, I_1\}, \{T_0, T_1, O_0, O_1\}, \{\varepsilon, 0, 1\}, \varepsilon, R_{ML})$: Assume $i \in \{0, 1\}$. When the module is in state ε, a signal arriving on input line S_i changes the state to i and gives rise to an output line on line T_i. When the module is in state i, a signal arriving on input line I_i is processed and gives rise to an output signal on line O_i; in this case, the state of the module is changed to ε. When the state is ε, simultaneous arrivals of a signal on each of the input lines S_0 and S_1 are allowed. In this case, only one of the two signals (may be arbitrarily chosen) is processed, whereas the other signal that was not assimilated keeps pending on the line until the module reverts to state ε

of each transition. Obviously, the inverse of an RE in Fig. 21.2a is still an RE. The cases for other reversible elements in Fig. 21.2 are the same. A remarkable feature of the DI-circuit in Fig. 21.3 is that reversing the directions of each interconnection line in the construction, and replacing each element with its inverse, directly give rise to the inverse of an 2 × 2-SJoin. This feature also carried over to the construction of an Arbiter (Fig. 21.1b) via the reversible elements: CDE and Multiplexer, as shown in Fig. 21.4. In particular, when a signal arrives on each of the input lines a_0 and a_1 at the same time, the Multiplexer in the construction will arbitrate between them so as to allow only one signal to be processed at a time, whereas the other signal is left waiting on the corresponding input line of the Arbiter. This behavior, therefore, reveals the arbitrating ability of the Arbiter's construction in Fig. 21.4.

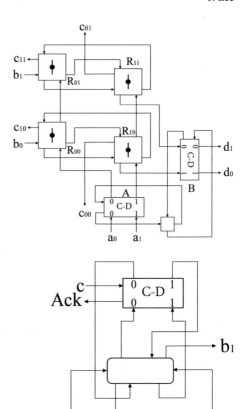

Fig. 21.3 Construction of 2×2-SJoin using RE, CDE and Join. Initially, all CDEs are in state V

Fig. 21.4 Construction of Arbiter using CDE and Multiplexer

The Merge in Fig. 21.1c is the simplest and most typical DI module that is not reversible. This module is serial and works as a fan-in element, such that a signal arriving on either input line I_0 or I_1 is transferred to the output line O. In particular, a Merge can not be realized by the use of only serial reversible elements, such as the RE in Fig. 21.2a [15, 18]. Hence, reversible elements that are able to process multiple input signals must be employed. Figure 21.5 shows a circuit scheme for constructing a Merge out of the CDE, TJoin and Multiplexer, in which two signals that are assigned initially play a key role in accomplishing the non-reversible functionality of a Merge [15]. For simplicity, assume a signal arrives on an input line I_i of the Merge in Fig. 21.5 with $i \in \{0, 1\}$. The input signal, together with a pending signal to the Join μ_i, is processed by μ_i and results in two signals running around in the construction. The resulting signals change the states of the Multiplexer and both of the CDEs to i, and eventually give rise to a signal appearing on the output line O of the Merge, and another signal being left in the circuit. After that, the latter signal will revert both CDEs as well as the Multiplexer to their initial states respectively, and finally return to an input line of the Join μ_i and wait there until a new signal is received from the input line I_i of the construction.

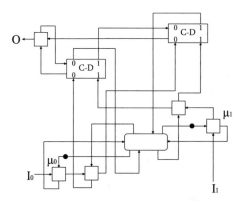

Fig. 21.5 Construction of Merge using CDE, Join and Multiplexer. A signal, denoted by a black blob, is assigned in advance on an input line of each of the Joins μ_0 and μ_1

As soon as a DI-circuit outputs all signals in response to a valid set of inputs, the circuit must get ready to receive new input signals [12]. The DI-circuit, therefore, must be stabilized before actually processing the new inputs. A straightforward way to accomplish this is to ensure that the circuit contains no signal on any interconnection line in the absence of any input signal, for example, as the constructions of an 2×2-SJoin and Arbiter given in Figs. 21.3 and 21.4, respectively. The lack of additional signals to be assigned initially on some interconnection lines, however, may probably bring the difficulty in designing a conservative DI-circuit to conduct some intended tasks [15]. In the case of using additional signals, there is always an identical number of signals to be left in the circuit, after every valid set of input signals is processed and gives rise to corresponding output signals.

As mentioned above, after a signal appears on the output line O, a signal may run around within the DI-circuit in Fig. 21.5, i.e., the construction is not stabilized. To cope with such a situation, a Multiplexer as well as its connected four Joins are employed in the construction of a Merge [15]. As a result, as soon as an input signal results in a signal on output line O, the Multiplexer and the Joins enable the Merge's construction in Fig. 21.5 to receive a new signal from either of its input lines, and prevent the new input from being actually processed before the construction is stabilized.

Because the 2×2-SJoin, Arbiter and Merge modules in Fig. 21.1 are universal for the whole class of conservative DI-circuits [15, 19], all constructions in Figs. 21.3, 21.4 and 21.5 demonstrate that any arbitrary conservative DI module can be decomposed into a DI-circuit composed merely of the reversible elements: RE, CDE, Join and Multiplexer. Asynchronous operations of the reversible elements, therefore, tend to actually boost their constructability and computing power [15].

The RE and CDE can be substituted for each other, because either of them can be constructed from the other [13, 14]. In addition, both of them can be decomposed into a pair of mutually reversed elements, as given in Fig. 21.6. Called RT and IRT, these elements have simpler functionality than CDE and RE, and take a minimal

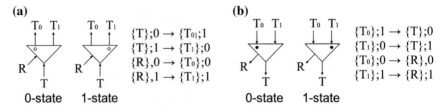

Fig. 21.6 a RT (Reading Toggle) = ({T, R}, {T_0, T_1}, {0, 1}, 0, Ψ_{RT}): When the element is in state 0 (or 1), a token arriving on input line T changes the state to 1 (resp. 0), and results in a token on output line T_0 (resp. T_1); a token arriving on input line R does not change the state and gives rise to a token on output line T_0 (resp. T_1). **b IRT** (Inverse Reading Toggle) = ({T_0, T_1}, {T, R}, {0, 1}, 0, Ψ_{IRT}): When the element is in state 0 (or 1), a token arriving on input line T_1 (resp. T_0) changes the state to 1 (resp. 0), and results in a token on output line T; a token arriving on input line T_0 (resp. T_1) does not change the state, and gives rise to a token on output line R

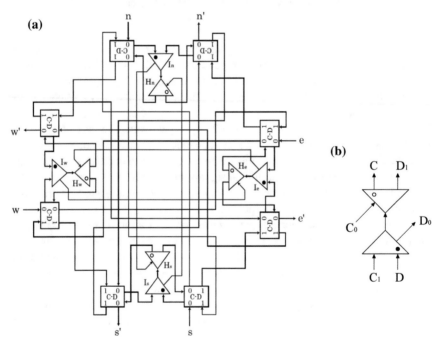

Fig. 21.7 a Construction of RE using RT, IRT and CDE (reprinted from [13], © Springer). **b** Construction of CDE using RT and IRT

number of input and output lines. Figure 21.7 illustrates the constructions of an RE and CDE using RT and IRT [13], in accordance with which the RT, IRT, Join and Multiplexer are capable of realizing any arbitrary conservative DI-circuit, thereby forming a universal set of reversible elements.

21.4 Conclusion

Logical reversibility promises in principle the possibility of computers with zero dissipation, by preventing entropy loss during computation [1]. Likewise, asynchronous operations tend to reduce power consumption and energy dissipation, by keeping all idle elements in a sleeping state [7]. Thus, it makes sense to include the reversibility into asynchronous systems. This chapter provided definitions on local reversibility in delay-insensitive circuits, and showed that local reversibility, in general, is unable to ensure the global reversibility in DI-circuits. On the contrary, reversible elements are capable of constructing any arbitrary DI-circuit, i.e., they are universal [15]. It is well known that local reversibility reflects the time-reversible nature of microscopic dynamics of particles and fields. Macroscopic physics, however, are likely to exhibit non-reversibility as described by the Second Law of Thermodynamics [8]. Hence, the emergence of universal global behavior from reversible local operations in DI-circuits might be consistent with the macroscopic consequence of microscopic reversible dynamics in physics.

Acknowledgments The author is grateful to Professor Kenichi Morita, Dr. Katsunobu Imai at Hiroshima University, Japan, and Dr. Ferdinand Peper at National Institute of Information and Communications Technology, Japan, for the fruitful discussions. This research work was supported by NSFC (No. 61170036).

References

1. Bennett, C.H.: The thermodynamics of computation–a review. Int. J. Theor. Phys. **21**, 905–940 (1982)
2. Danos, V., Krivine, J.: Reversible communicating systems. In: Proceedings of 15th International Conference on Concurrency Theory, CONCUR 2004, LNCS, vol. 3170, pp. 292–307. (2004)
3. Durand-Lose, J.: Computing Inside the Billiard Ball Model, Collision-Based Computing, (Ed. A. Adamatzky), pp. 135–160, Springer, Berlin (2002)
4. Feynman, F.: Quantum mechanical computers. Found. Phys. **16**, 507–531 (1986)
5. Fredkin, E., Toffoli, T.: Conservative logic. Int. J. Theor. Phys. **21**, 219–253 (1982)
6. Gupta, P., Agrawal, A., Jha, N.K.: An algorithm for synthesis of reversible logic circuits. IEEE Trans. CAD **25**, 2317–2330 (2006)
7. Hauck, S.: Asynchronous design methodologies: an overview. Proc. IEEE **83**, 69–93 (1995)
8. Holian, B.L., Hoover, Wm.G., Posch, H.A.: Resolution of Loschmidt's Paradox: the origin of irreversible behavior in reversible atomistic dynamics. Phys. Rev. Lett. **59**, 10–13 (1987)
9. Imai, K., Morita, K.: A computation-universal two-dimensional 8-state triangular reversible cellular automaton. Theor. Comput. Sci. **231**, 181–191 (2000)
10. Kane, B.E.: A silicon-based nuclear spin quantum computer. Nature **393**, 133–137 (1998)
11. Kari, J.: Reversible cellular automata. Dev. Lang. Theory, LNCS **3572**, 2–23 (2005)
12. Keller, R.M.: Towards a theory of universal speed-independent modules, IEEE Trans. Comput. **C-23**, 21–33 (1974)
13. Lee, J., Peper, F., Adachi, S., Morita, K.: An asynchronous cellular automaton implementing 2-state 2-input 2-output reversed-twin reversible elements, ACRI 2008. LNCS **5191**, 67–76 (2008)
14. Lee, J., Yang, R.L., Morita, K.: Design of 1-tape 2-symbol reversible Turing machines based on reversible logic elements. Theor. Comput. Sci. **460**, 78–88 (2012)

15. Lee, J., Adachi, S., Xia, Y.N., Zhu, Q.S.: Emergence of universal global behavior from reversible local transitions in asynchronous systems. Inf. Sci. **282**, 38–56 (2014)
16. Tang, M.X., Lee, J., Morita, K.: General design of reversible sequential machines based on reversible logic elements. Theor. Comput. Sci. **568**, 19–27 (2015)
17. Margolus, N.: Physics-like models of computation. Phys. D **10**, 81–95 (1984)
18. Morita, K.: Reversible computing and cellular automata–a survey. Theor. Comput. Sci. **395**, 101–131 (2008)
19. Patra, P., Fussell, D.S.: Conservative delay-insensitive circuits, Workshop on Physics and Computation, pp. 248–259 (1996)
20. Phillips, I., Ulidowski, I.: Reversibility and models for concurrency. Electron. Notes Theor. Comput. Sci. **192**, 93–108 (2007)

Chapter 22
Reservoir Computing as a Model for *In-Materio* Computing

Matthew Dale, Julian F. Miller and Susan Stepney

Abstract Research in substrate-based computing has shown that materials contain rich properties that can be exploited to solve computational problems. One such technique known as *Evolution-in-materio* uses evolutionary algorithms to manipulate material substrates for computation. However, in general, modelling the computational processes occurring in such systems is a difficult task and understanding what part of the embodied system is doing the computation is still fairly ill-defined. This chapter discusses the prospects of using *Reservoir Computing* as a model for *in-materio* computing, introducing new training techniques (taken from Reservoir Computing) that could overcome training difficulties found in the current *Evolution-in-Materio* technique.

22.1 Introduction

Biological organisms vastly outperform classical/conventional computing paradigms in many respects, from possessing inherent fault-tolerance to constructing highly parallel machines. Much of this is achieved by exploiting physicality and by sharing and distributing computational effort throughout the spatial system. As such, they exploit physical interactions through feedback with the real-world, utilising features such as their own morphology.

Many of these systems comprise relatively simple elements that emerge and coalesce into more complex, but robust, structural layers across different scales. Such

M. Dale (✉) · S. Stepney
Department of Computer Science, University of York, York, UK
e-mail: md596@york.ac.uk

J.F. Miller
Department of Electronics, University of York, York, UK
e-mail: julian.miller@york.ac.uk

S. Stepney
York Centre for Complex Systems Analysis, University of York, York, UK
e-mail: susan.stepney@york.ac.uk

© Springer International Publishing Switzerland 2017   533
A. Adamatzky (ed.), *Advances in Unconventional Computing*,
Emergence, Complexity and Computation 22,
DOI 10.1007/978-3-319-33924-5_22

grounding properties (and many more) have enabled these complex systems to thrive and evolve, adapting and co-evolving in real-time with their local ecosystem.

The conventional von Neumann computing architecture, although expertly refined over time, poses some fundamental inefficiencies. For example, classical computers require the transformation between high-level languages to low-level machine code, a process that requires layers of conversion through a compiler stack, making it computationally costly, slow, and highly susceptible to faults and errors. These systems typically succumb to many issues in speed, from both an inability to deal with concurrent computations and the bottleneck created by the transfer of data between separate memory and processing entities. Because of these architectural weaknesses other intertwined discrepancies arise, such as an increase in power consumption, system size, and top-down design complexity.

Unconventional computing tries to address some of these limitations by attempting to provide alternative architectures and systems that typically exploit the underlying physics and many-scale interactions of the real-world. Many forms of unconventional systems have been explored in recent years. For example, quantum computing is one such system, where two-state quantum bits, typically described by electron or photon spin/polarisation, can be exploited to perform large numbers of parallel computations (i.e. $2^{no. of qubits}$). This is achieved through the principles of superposition, enabling a qubit to be in both states simultaneously (see [104] for a review). Another example is reaction-diffusion computing [2], which performs computation through local chemical reactions and diffusion. By using chemical processes, the system can execute highly parallel computations, performed by the complex interactions of propagated waves of information caused by local disturbances. Physarum polycephalum (slime mould) is currently under investigation as an excitable, reaction-diffusion medium that could form the basis of a programmable amorphous biological computer [3, 123].

For the purpose of this chapter we discuss research into configurable *in-materio* systems (typically on the nanoscale), with the intention of identifying a computational model that could suit and conform to many *in-materio* systems.

This chapter begins with an introduction to the concepts of material computation with a brief mention of the criticality hypothesis (Sect. 22.2). Next, the field of Evolution-*in-materio* as a current investigation into the theory of material computation is reviewed, discussing its methodology and current research (Sect. 22.3). Reservoir Computing and its potential as an *in-materio* model is examined in Sects. 22.4 and 22.5. To further aid in identifying "good" reservoirs some evaluation techniques are highlighted in Sect. 22.6. Lastly, an argument is made in Sect. 22.7 to the possible benefits reservoir computing could have on the design of computational substrates.

22.2 Material Computation

This section discusses some of the principal concepts of material computation, such as some informal definitions, the role critical dynamics has in complexity and where it may fit within other computing paradigms.

Material computation is defined to be a computational process that occurs in a sufficiently complex dynamical system realised in the form of a physical material substrate.

This definition relies on the idea that matter contains physical information and that systems of collective matter can dynamically modulate and redistribute information to change state; therefore, implied here, to perform some form of computation. A simple example of this would be the physical process in which the state or phase of matter changes in relation to energy, which could (loosely) in some sense be viewed as a computational mechanism.

Stepney et al. [111] provide a more contextual definition of material computation as: "computation directly by physical and chemical processes of a complex substrate, with little or no abstraction to a virtual machine". From this definition we can categorise material computation as an analogue process that utilises physical constructs, the tangible medium itself, and meta-processing to do computation. This "physicality" may be defined and observed as the structural topology, characteristic behaviours and information processing associated with the many-scale interactions occurring in that system.

As a concept, material computation is still in its developmental or conceptual stage, with early experiments supporting, or working in tandem with, current digital technology to form hybrid, and potentially very powerful, machines. A configurable substrate can be used to transfer some of the computational burden from the digital system to the material [111]. As such, the material can endow the system with many of the properties and advantages of analogue systems, such as speed and concurrency, in a device where memory and processing are not separable. To achieve this, engineers attempt to exploit processes and behavioural phenomena that naturally occur, properties that are governed directly by the underlying physics and chemistries of the substrate.

Exploiting computation directly from materials offers many potential advantages over classical systems where the computation performed does not depend so much on the details of the materials used. As a paradigm, substrate computing potentially offers vast amounts of computational power by capitalising on the massive parallelism and natural constraints of these rich systems. Such properties are suggested to have the potential to provide solutions "for free", or at least computationally cheaper, and provide a rich explorable state space, aligning computation to particular trajectories [108].

Much of the current interest in material computation is to abstract a model (or models) of computation from what the substrate does naturally. It has been proposed that the first step to producing a potential unified theory of material computation, i.e. a theory that better understands what computation is and how it occurs in materials, should take place in "primitive" (un-evolved) substrates, where the general principles are in plain sight [108]. After this, material computation could emerge with some supportive reasoning to a better understanding of computation in biological substrates.

22.2.1 The Effects of Criticality on Complexity

It is hypothesised that there is a critical state in which a system can exhibit maximal computational power, and therefore where maximal complexity can be acquired, labelled a region near (or at) the *edge of chaos* [60]. This concept may have a direct relationship to material computation, whereby a material can exhibit "richness" [75], and therefore be exploitable, by operating close to or within this region.

This *edge of chaos* represents the transitional border between ordered dynamics, where perturbations to the system quickly fade and the system settles, and chaotic behaviour, where perturbations significantly affect long-term stability and the system becomes unpredictable. This critical landscape can be observed by looking at the systems trajectory in the phase space by monitoring the convergence towards or away from a steady state, and thus highlighting a system's sensitivity to initial conditions. Both behaviours are thought to be necessary to gain maximal complexity, using some ordered behaviour to maintain memory, and some chaotic behaviour to enable processing.

Langton [60] observed the effects and advantages of systems working in this transitional region, using cellular automata. At a critical point, Langton observed that a cellular automaton could optimally perform computations, imitating complex life-like behaviour. Earlier, Packard [86] observed another unique property: that genetic algorithms tend to evolve populations in these critical regions, suggesting that adaptability was therefore optimised close to the edge of chaos. Similar conclusions are proposed and demonstrated in neural networks, where vast computational power and capability in this region is described through network connectivity. Bertschinger and Natschläger [11] demonstrate these relationships in input-driven networks, by accurately determining the position of the critical line with respect to structural parameters.

It has also been suggested that *living* neural networks support the "criticality hypothesis". Beggs [9] discusses this notion by looking at how the power-law distribution of neuronal avalanche sizes (a cascade of bursts of activity) suggests operation near a critical point. Beggs further explains, that the implications of these avalanche size distributions implies that information transmission, information storage, computational power and stability could all simultaneously optimise at the edge of chaos.

22.2.2 Configuration and Structure

Conventional programs and algorithms represent idealised mathematical objects, irrespective of their underlying hardware. In a physical system (say a biological system) computation is embodied, behaviour may not be completely captured by a closed mathematical model. As such, trying to program these embodied systems requires different techniques. The programming and manipulation of materials requires (to some extent) a complex understanding of the properties and interactions within that system. We therefore require either some convoluted top-down "programming" approach (in the traditional sense) or an alternative mechanism (e.g. through training, learning or evolution). Whichever method is applied, this "program" would alter the details of system in a controlled manner, for example, using controlling fields that affect structure and dynamics.

22.3 Evolution-In-Materio

Evolution-*in-materio* (EIM) is a term coined by Miller and Downing [75] to refer to the means by which a physical system, a complex material, could be manipulated by computer controlled evolution (CCE) to perform useful computation.

The idea of using unconstrained evolution as a search method in physical media is deep rooted in the field of Evolvable Hardware (EH) [34, 36]. Most evolved configurations in EH lead to digital products or components later embedded into physical artefacts. For example, evolving simulated models and optimisation programs, or, designing a physical system that can be manufactured after evolution; like antennas [65], robots [64, 90] or chemical systems made of oil droplets [35].

Miller argues that Evolution-*in-materio* sits between full embodiment and the realised evolved EH solutions described above [76]. In this form, physical artefacts are configured (or conceptually created) during the process and assessed/controlled by simulated Darwinian evolution. This can also be seen in some EH systems, but, typically such systems are limited to constrained silicon hardware, e.g. electronic circuits evolved on Field Programmable Gate Arrays (FPGA) [43]. EIM relies on a hybrid analogue/digital architecture where the evolutionary process (encapsulated on a digital computer) controls the writing/reading of physical signals to/from an analogue material. As such, the directed search tries to exploit the dynamics of the material by evaluating the performance of individual test configurations. Physical realisations are therefore embodied in the search process but evaluated externally. A system that operates in this manner is theoretically very powerful, allowing the manipulation of physical properties which are hitherto unknown.

An early example of EIM can be found in Thompson's work with FPGAs [114]. Thompson attempted to evolve a frequency discriminator by allowing evolution to reconfigure circuit elements on the FPGA. In the process, he discovered that evolution had in fact used subtle electrical variations in the underlying material to form a solution. It was only made evident when evolved configurations no longer solved the problem when moved around the FPGA and when areas not directly involved somehow contributed to the overall operation. Thompson's work led to an explosion of interest [36, 76]. Miller and Downing [75] label this work as a "starting point" for the exploration of intrinsic evolution as a means by which to exploit the natural properties of materials for computation.

22.3.1 Computation in Liquid Crystal

Miller and Downing [75] discuss several interesting materials that could possess the desired characteristics needed for both computation and evolution. These include, liquid crystal, conducting and electro-active polymers, voltage controlled colloids, nanoparticle suspensions and irradiated or damaged semiconductors [75]. These materials are exploited using a device that can alter the material's function through configuration parameters (discrete signals), using a configurable analogue processor (CAP) (see Fig. 22.1).

Liquid Crystal (LC) has a number of advantages for readily applying the theory. LC contains several key features including, wide availability, addressable using digital voltages, exhibits emergent behaviour, has a unique mesomorphic structure between ordered and disordered, and can relax to an initial base state. Harding and Miller [39] adopt liquid crystal as a basis material and construct a bespoke platform to solve multiple computational problems. The hardware houses a liquid crystal display (LCD) and an array of dynamically selectable input/output connections to both the LCD and external measurement devices. They demonstrate liquid crystal as an efficient evolvable material where relatively small numbers of generations can produce effective solutions. Over the course of their investigation the LC system has been applied to three separate tasks; tone discrimination [39], creating logic gates [41] and a real-time robot controller [40].

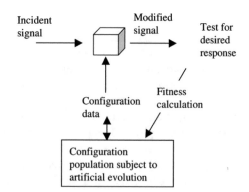

Fig. 22.1 Configurable Analogue Processor (CAP): using an evolutionary guided search the material is reconfigured to solve some computational task through applied input signals [75]

Harding and Miller found that a *rich* substrate of liquid crystal supplied many more exploitable properties compared to conventional silicon hardware (i.e. Thompson's FPGA). This in turn increased the diversity of solutions, thus increasing its evolvability. Harding and Miller's work demonstrated the advantages of emergent design by configuring intractable characteristic properties with no knowledge of their existence. But, the experiments also raised other fundamental questions on applying techniques from the Evolution-*in-materio* paradigm. For example, the length of time needed to "program" materials, i.e. how long does the search need to be and what constraints are there when transitioning into the physical world? what are the difficulties in defining the boundaries of the system under evolution, i.e. what is actually doing the computation? and what effects come with non-isolated systems embedded in a physical environment? is the system/evolution utilising environmental conditions and sources of noise? and what are the affects on system/solution reproducibility? what are the consequences of varying conditions on replicating the solution? Many of these questions and more are discussed in [76].

22.3.2   NASCENCE Project: Carbon Nanotube Substrates

As part of the European FP7-ICT research project NASCENCE[1] (NAnoSCale Engineering for Novel Computation using Evolution) further materials were considered, along with a new bespoke hardware platform [15]. The present hardware iteration, known as the Mecobo board [70], forms another hybrid hardware architecture to integrate digital computers with experimental materials. The system interfaces with materials placed on micro-electrode arrays (MEA) (Fig. 22.2) using a similar premise to Harding and Miller's liquid crystal system, the CAP.

The substrates currently under analysis consist of Single-Walled Carbon Nanotubes (SWCNT) mixed with either a polymer or liquid crystal. Another substrate consisting of randomly-dispersed gold nanoparticles is also under investigation, but requires temperatures of less than a few Kelvin to function. The polymer/LC mixtures disperse the nanotubes into random static networks, forming varying connection topologies and conductive pathways, possibly forming something akin to a random electrical circuit. Carbon nanotubes are used as they can exhibit either metallic or semi-conducting behaviour and contain other unique properties (e.g. ballistic conduction, thermal conductivity, self-assembly via van der Waals forces), whilst the mixing material is believed to create isolating regions, forming an insulator between elements and creating network structure.

A number of recent investigations have demonstrated the capabilities of SWCNT/ polymer mixtures, in particular Poly-methyl-methacrylate (PMMA) and Poly-butyl-methacrylate (PBMA), as a potentially rich, evolvable and ubiquitous substrate. These investigations include: solving classification and optimisation problems such as frequency classification [78]; classifying various data instances [23, 80]; solving

---

[1]NASCENCE homepage: nascence.no.

Fig. 22.2 Micro-electrode array used in the NASCENCE project to contain, stimulate and record activity in a carbon nanotube-based substrate. Computer controlled evolution is used to select active electrodes and mode (i.e. record or stimulate)

small numbers of cities instances of the travelling salesman problem (TSP) [22]; and applied to the (NP-hard) bin packing problem [79]. Early evidence highlights and supports the plausibility and potential of the methodology, but, in some respects it still lacks competitive results and still exhibits some of the issues raised at the end of Sect. 22.3.1.

PBMA appears to show greater stability than PMMA. The electrical percolation threshold of PBMA is said to occur around a concentration of 1 % (w.r.t. polymer weight), forming a useful mixture of short and long-conductive pathways. After this, adding more nanotubes to the mixture is said to provide a negligible computational advantage as a suitable network appears to already exist. Although, interestingly, higher concentrations do demonstrate more non-linear current-voltage (I-V) behaviour in comparison. At less than 1 %, the nanotube networks become fairly sparse and are argued to have reduced computational performance (and potentially more linear properties) [74]. The PMMA polymer was investigated towards the beginning of the NASCENCE project (see [56]) and is reported to have a percolation threshold between 0.17 and 0.70 % in the literature, as explained in [74], but more importantly, a direct comparison is difficult to make between the two polymers as they come under different polymer groups, i.e. contain different chain lengths.

Investigations into SWCNT/liquid crystal mixtures has shown some promise, for example, non-linear I-V behaviour appears to be more prominent. It has also been demonstrated that conductivity and orientation can be changed by an in-plane electric field. But, LC has been shown to experience a longer configuration time, in-terms of LC molecule and SWCNT alignment, due to LC molecules being smaller than SWCNT ribbons, and associated relaxation times. Other issues include; Long-term stability and reconfigurability, and the exact role of the liquid crystal in nanotube alignment [121].

New engineering technologies, adaptations in the search method (or fitness criteria), changes in hardware (number of electrodes, different pitch sizes, etc.) and new materials could all have a significant effect on the field. The key components to the success of EIM lay in improvements to the fabrication of materials and the interfacing system used for stimulation and observation. This early work has also only highlighted one of many means of "programming" a material via evolution (i.e. through discrete voltage inputs). Other controlling fields, or even a combination of fields, could be utilised to manipulate/configure different materials—hopefully further separating the distinction between configuration and input signals. The ideal scenario for this field would be to pave the way for cheap, small, easily reconfigured and manufactured, multi- or single-purpose standalone computational devices.

22.4 Reservoir Computing

22.4.1 What Is Reservoir Computing?

Reservoir Computing is the unification of three individually conceived methods for creating and training artificial recurrent neural networks (RNN): Echo State Networks (ESN) [44], Liquid State Machines (LSM) [72] and the Backpropagation-Decorrelation (BPDC) on-line learning rule [105].

A typical RNN model consists of a system of three layers: an input layer, a hidden layer (the core network), and an output layer (see Fig. 22.3). The hidden layer contains processing elements (neurons) that are interconnected through weighted synapses (connection weights). The input and output layers are connected to the hidden layer, again through weighted synapses. Variations on the types of connectivity, e.g. feedback from the output to hidden layer or input layer to output layer, depends on the task and method. For simplicity, we examine a simple input-to-hidden-to-output system encompassing a recurrent network in the hidden layer.

Fig. 22.3 A typical three-layer recurrent neural network. The input and output layer are connected to the hidden layer via weighted connections. The connections between neurons in the hidden layer are also weighted

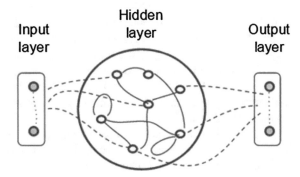

When driven by an input, neurons activate, propagating information through the network to other neurons through varying connection strengths. The presence of recurrent connections can produce self-sustained activations, preserving a dynamic memory in the network's internal state. Networks such as this have been shown to be theoretically very powerful and can be both Turing equivalent [53] and good universal approximators of dynamical systems [30]. However, making the most of RNNs comes at a price, as they suffer from many training difficulties, such as the computational expense of updating large networks, bifurcation points, and sometimes falling into inescapable local optima when using gradient-descent.

Reservoir Computing offers an alternative training technique. It reduces the computational cost and removes the problem of degenerative gradient information that leads to poor convergence. Also, the concepts of reservoirs go beyond traditional neural networks and encompass (to some extent) more general dynamical systems. Evidence of this can be seen in the following implementations: in electronic circuits [95, 98], a bucket of water [28], Gene Regulation Networks (GRN) of *E. coli* bacteria [24, 52], deoxyribozyme oscillators (referred to as "DNA reservoir computing") [33] and a cat's primary visual cortex [81].

22.4.2 Reservoir Types

There are many "flavours" of reservoir, originating from two separate research fields of machine learning and computational neuroscience. The first focuses on training dynamical systems for temporal learning tasks using artificial recurrent neural networks. The second aims at realistically modelling the computational properties of neural microcircuits. We give a summary of the two main branches of RC, and reservoirs in unconventional hardware. For more types and variations see [68].

22.4.2.1 Echo State Network

The *Echo State Network* (ESN) is a discrete-time neural network constructed from a sparse, random collection of analogue neurons (Fig. 22.4). The typical neuron model employed uses the sum of its weighted inputs, applied to a sigmoid function (generally a hyperbolic-tangent), to give the neuron state $x(n)$ at time n. The state activations $x(n)$ of these neurons are termed as echo states [44], i.e. echoes of the input history. To propagate and hold this history the network requires the *echo state property*, or more generally speaking, a fading memory. The property itself is provided by the characteristic dynamics of the system.

In ESNs, different scaling parameters, and in particular the *spectral radius* $\rho(.)$, influence these dynamics. These parameters fundamentally alter and control the amount of memory and non-linearity present in the system. The $\rho(.)$ parameter is used to scale the weight matrix W so that the largest absolute eigenvalue satisfies

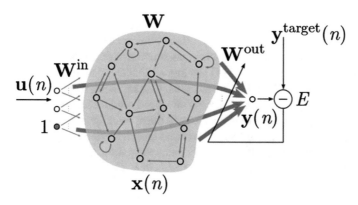

Fig. 22.4 Echo State Network: A random, static, recurrent network of sigmoidal neurons. The input driven system projects $u(n)$ into the reservoir network. Each node possesses a one-to-one weighted connection to all inputs (via W^{in}), outputs (via W^{out}) and other nodes (via W). The extra input (labelled "1" in the figure) is provided as a bias W_{bias}. Training occurs on the output weights W^{out} (in relation to reservoir states $x(n)$) by reducing the error between $y(n)$ and $y^{target}(n)$ [66]

$\rho(W) < 1$ (typically, but not always; see [68]). Within this region the echo state property is said to be assured.

Another variant of the ESN model is the leaky-integrator neuron model, using a neuron that possesses some form of memory of previous activations. These neurons contain a leaking rate, or decay parameter α, which can control the speed of the reservoirs update dynamics (Eq. 22.1).

As Jaeger [44] describes, each neuron acts like a digital low-pass filter enabling a discrete network to approximate the dynamics of a continuous network (variations and uses can be seen in [44, 66, 68]). Dynamical systems have a natural time-scale; understanding the time-scale on which the input is changing compared to the time-scale of the system dynamics can be difficult. The leaky parameter α helps control and mediate any differences in input time-scales.

Putting together these components gives the neuron state update equation:

$$x(n) = (1 - \alpha)x(n-1) + \alpha f(W_{in}u(n) + Wx(n-1)) \qquad (22.1)$$

22.4.2.2 Liquid State Machine

The *Liquid State Machine* (LSM) model arose as a method for defining the computational properties and power of neural microcircuits, "an alternative to Turing and attractor-based models in dynamical systems" [71]. The LSM model represents a competitive model for describing computations in biological networks of neurons. The LSM attempts to model cortical micro-columns in the neocortex, structured in cortical layers of randomly created spiking neurons based on a spatial embedding. Among other things, it has been described as a possible process used by mammalian brains in speech recognition [27] and has been verified for cortical microcircuits in the primary visual cortex [81] and the primary auditory cortex [54].

Networks based on LSM use continuous streams of data (spike trains) to achieve real-time computations. Maass [71] has argued that classical models cannot handle real-time data streams based on spike trains. Unlike ESNs, they are generally more adaptive systems, supporting additional advanced readout features such as parallel perceptrons [72], although, in many cases, a linear readout is preferred.

Investigations using Liquid State Machines has highlighted the potential for more abstract unconventional applications, for example, pattern recognition using a physical medium (water) [28], and imitating LSMs in *E. Coli* [52].

22.4.2.3 Unconventional Hardware: Single Non-linear Dynamical Node with Delayed Feedback

Recent experimental applications of reservoir computing in optoelectronics and photonics [6, 61, 87] have demonstrated a new way of constructing a pseudo-reservoir system. Using *delay systems* theory, a system can imitate the characteristics of a recurrent network without being one. Delay systems represent a class of dynamical systems which incorporate non-linear systems with some form of delayed feedback and/or delayed coupling.

The key feature of this new RC flavour is to replace a physical network of nodes (often large in size) with a single non-linear node and a delay line. The delay system mimics a large interconnected network by creating a topology of *virtual* nodes in the delay line. This is achieved by applying time-multiplexing techniques on the input, i.e. through a combination of sample-and-hold operations mixed with an additional input mask. The sample-and-hold operation creates a stream $I(t)$ which defines the state update determined by the delay τ in the feedback loop. An additional mask M is created to represent coupled weights between stream $I(t)$ and the virtual nodes. The matrix M was initially randomly created at first, motivated by the random connectivity in reservoirs; but later optimised masks were proposed [7].

The number of virtual nodes in the system is defined by N equidistant points separated in time ($\theta = \tau/N$) along the delay interval τ. The resulting time-multiplexed input sequence becomes $J(t) = M \times I(t)$ which is then fed into the non-linear node (Fig. 22.5). Once the system has updated after time τ, the output nodes access the states in the delay line using $\sum_{i=1}^{N} w_i x(t - \frac{N}{\tau}(N - i))$. For more detail on the masking process see the supplementary information provided for [6, 7].

The single non-linear node scheme has many interesting implications for designing hardware reservoir computers. A clear advantage is that the overall architecture needed is very simple. But using a delay loop implies a serial process (in contrast to the parallel feeding of nodes in RC). Therefore, the speed of information processing depends on the state update given by time τ. A few suggestions have been given to compensate for such discrepancies and to increase computational capability, e.g. adding additional delay lines (increasing memory capacity) and finding an optimal number of nodes that can reasonably be implemented physically: adding more nodes

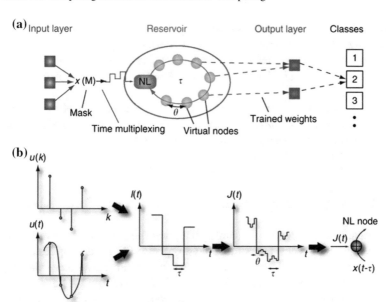

Fig. 22.5 a Single non-linear node with delayed feedback. A new system derived from delay systems theory where virtual nodes are created in the delay loop τ to represent the hidden layer. The number of virtual nodes N is separated by time θ given by $\theta = \tau/N$, **b** time-multiplexing of the input stream $I(t)$ (sampled from discrete $u(k)$ or continuous time $u(t)$) and a randomly created mask M that creates the input sequence $J(t)$. $J(t)$ is then added to the delayed state of the reservoir $x(t - \tau)$ before being fed back into the non-linear node [6]

will decrease τ, but having too many nodes is physically impractical given the hardware involved. Future work using this technique looks promising and could produce some ultra-high-speed computing solutions for specific tasks.

22.4.3 *Reservoirs and Kernels*

A reservoir can be interpreted as possessing kernel-like properties. A *kernel* acts as a pre-processor, embedding input data into a vector space known as a *feature* space. It is understood in many statistical machine learning methods that combining this feature space with a simple linear discriminant algorithm can enable the learning of complex non-linear functions.

In kernel methods this is achieved by projecting the input space $u(n)$ into a high-dimensional (possibly infinite-dimensional) feature space $x(n)$ *without* paying the price of its explicit computation, referred to as the *kernel trick* [92]. A kernel can therefore be expressed as the expansion function $x(u(n))$. However, there are two significant differences between reservoirs and kernels: reservoirs ***do*** explicitly compute the input transformation, i.e. do not possess the kernel trick; and kernels are typically ill-equipped to handle temporal signals. To tackle temporal tasks the learned function $x(.)$ requires some form of memory of previous inputs. Reservoirs solve this by utilising the network's recurrent topology, which creates memory by retaining previous state activations. The final expansion function of the reservoir can therefore be represented as $x(n) = x(x(n-1), u(n))$.

Reservoirs use this pre-processing technique to map temporal features of the input into a spatially defined feature map (the network). The desired features can then be extracted, or combined, in a linear fashion to create the output $y(n)$:

$$y(n) = W^{out} x[u(n)] \qquad (22.2)$$

where $W^{out} \in \mathbb{R}^{N_y \times N_x}$, and N_y, N_x are the number of output nodes and internal nodes respectively.

This enables reservoirs to tackle many temporal and dynamic real-world tasks not possible using simple non-temporal kernels. Equation (22.2) also implies the system contains a clear separation between the reservoir and the linear readout, separating the training procedure from the hidden layer, i.e. only W^{out} is trained. As such, the kernel representation offers a much faster and better converging mechanism compared to other RNN models, as it does not suffer from vanishing gradients. This representation also classes reservoirs as powerful adaptive filters. For more information on kernels see [100].

22.4.4   Reservoirs and Criticality

To design an optimal reservoir one should find a good trade-off between: (i) the transformation of the input that optimally boosts the linear classifiers capability, later referred to as the "*quality*" of the kernel; and (ii) a sufficiently-long (fading) memory based on the input history. These two properties often conflict; to obtain a useful memory requires ordered behaviour and a rich transformation requires dynamic behaviour. Legenstein and Maass [62] have shown that optimal reservoirs tend to experience the best trade-offs at a critical point near the "edge of chaos".

Dynamic networks are said to exhibit emergent criticality and self-organising properties [110]. A novel example of this can be seen [10], where a self-organising structure of carbon nanotubes evolves to produce maximum entropy given a strong applied electric field. Moreover, it has been observed that dynamic networks can self-organise into critical regions where they can perform interesting computations.

These systems are characterised by motion in the phase space described as *trajectories* or state transitions. A trajectory may converge towards (i.e. be attracted to) a stable or unstable steady state; an attractor. Attractors vary from *point*—a point in the state space that attracts trajectories into its basin—to *strange* and chaotic—attracting trajectories, but inside diverge exponentially. These systems are very robust, small perturbations in the trajectory will tend to converge towards the same attractor. But, both external inputs and parameter changes in the system can drastically alter the shape of the phase space. Such changes may perturb a trajectory, moving or "clamping" the system between different attractor basins. As a result, clamping may even create, remove or change existing attractors and thus alter the initial phase space created by the natural dynamics of the system [109]. The overall dynamics of the system can therefore, to some extent, be controlled.

As discussed in Sect. 22.2, certain critical regions have been identified to display interesting properties for computation (e.g. maximum complexity and performance). Although the reasons for maximised performance is not entirely understood, a series of quantitative measures have been used to observe the effects of phase transition on computational capability.

Early measurements of dynamics in reservoir-like systems can be found in [11] where, using a similar technique to Derrida and Pomeau [26], dynamic behaviour is measured using the Hamming distances between two output states. By observing a growth in the distance between states it can be determined that chaotic behaviour is present; a decrease in distance would indicate more ordered behaviour. This concept is similar to computing the characteristic Lyapunov Exponent for a dynamical system (see Sect. 22.6.2.2), with both measures analysing the sensitivity to differences in initial conditions.

Bertschinger and Natschläger [11] highlight two fundamental properties required for these networks: a *fading memory* and a "network mediated" separation. A fading memory is indicative of an ordered phase (memory) with some dynamics (fading). The same property is found in both Liquid State Machines and Echo State Networks, referred to as the "echo state property" in the latter. This allows the readout function to use information from recent inputs and derive functions of those inputs from the network state. Network mediated separation is deemed fundamentally important for input time-series networks, with similarities to the separation property in LSMs (see [72]). The property requires (ideally a large) diversity in network states that is the result of differences in inputs alone, allowing characteristic features to be identified in the input, and that any changes in state should not directly be a result of chaotic dynamics which could produce the same phenomena. As such, this property enables a readout function to respond effectively to any variation in the inputs.

Legenstein and Maass [63] propose two critical elements that characterise the computational capabilities of a complex dynamical system (cortical neural microcircuits in this case). These new measures, or properties, are proposed because Lyapunov exponents are only useful for analysing a systems dynamics, and are not necessarily helpful in predicting good parameter regions that create high computational performance. These measures are the *kernel-quality* and the *generalisation-capability*. The kernel-quality refers to the linear separation property found in kernels (Sect. 22.4.3). An empirical measure of this property is achieved by examining the complexity of functions that can be carried out on the inputs that boost the classification power of a subsequent linear layer. The generalisation-capability quantifies a system's capability to generalise any learned behaviour to a new input.

Boedecker et al. [13] extend these ideas to ESNs, and create a general framework for direct and localised measurements for each neuron. Boedecker et al. give measurements indicating the memory of each neuron and the transfer of information between each neuron. This work highlights some interesting and relevant points for all systems: a network does not necessarily need to be at the edge of chaos to do computation, the measured region where a system is at the edge of chaos is not universal for all tasks, and a critical state may maximise computational capabilities, but, such criticality may also be unnecessary or detrimental to certain tasks.

22.4.5 Training Reservoirs

This section describes the training procedure and some of the available techniques one can use to train both the linear outputs and the reservoir itself. The two methods for training linear readouts described here are: "off-line" *batch-mode* training, using simple linear or ridge regression techniques (done once all reservoir states are collected into matrix X for training length T), and "on-line" training, often gradient descent-based, Recursive Least Squares algorithm (a useful extensive investigation of RLS-type training is shown in [57]). We then describe some *pre*-processing training techniques that can be used on the reservoir itself, methods that have been identified as useful in creating tailored/optimal reservoirs.

There are many training techniques available in this diverse field. Here we just discuss methods perceived as "classical" training methods. For more examples of training techniques see [47, 66, 68].

22.4.5.1 Off-Line Training

Reservoirs are traditionally trained in a supervised manner where the temporal input $u(n)$ and coupled target output $y^{target}(n)$ are provided. Given a desired output the system can learn input-output behaviour by minimising the error (Eq. 22.3) between system output and desired output.

$$E(y, y^{target}) = \sqrt{\frac{1}{T} \frac{\sum_{t=1}^{T} (y(n) - y^{target}(n))^2}{\sigma^2(y^{target}(n))}} \tag{22.3}$$

To evaluate if the learned behaviour generalises accordingly, new input data is tested and the error between the two are again compared.

Reservoir computers are conceptually viewed as recurrent neural networks incorporating the three-layered topology of N_x hidden nodes (neurons), N_u inputs and N_y outputs. As discussed in Sect. 22.4.2, variations on how these are implemented are also possible, but, essentially the system still adheres to the same structural layers.

The general update state equations for most systems are defined in Eqs. (22.4) and (22.5) for, discrete time $n = 1, \ldots, T$, internal state $x(n)$ and output $y(n)$:

$$x(n) = f(W^{in}u(n) + Wx(n-1) + W^{fb}y(n-1) + W^{bias}) \tag{22.4}$$

$$y(n) = f^{out}(W^{out}[x(n); u(n)]) \tag{22.5}$$

The function f is commonly represented by a sigmoid, typically a hyperbolic tangent. In echo state networks this represents a basic *tanh* neuron, but varies depending on the application. In other networks, f can be designed to form linear nodes, threshold logic gates or spiking neurons. In regards to the output $y(n)$, f^{out} may also be a non-linear function (sigmoid) but tends to be identity in most cases.

The weight matrices; input weights $W^{in} \in \mathbb{R}^{N_u \times N_x}$, reservoir/hidden layer weights $W \in \mathbb{R}^{N_x \times N_x}$ and feedback weights (from the output to the reservoir, if needed) $W^{fb} \in \mathbb{R}^{N_x \times N_y}$ are all drawn randomly from a uniform distribution at creation and remain static. The output weight matrix $W^{out} \in \mathbb{R}^{N_y \times (N_x + N_u)}$ includes weights for the inputs as they act as additional states (hence the concatenation of $[x(n); u(n)]$). Typically, the W matrix forms a sparse network with many of the weights set to zero, the other W^{in} and W^{fb} matrices can either be dense or sparse. Additional scaling parameters might also be applied to the matrices to govern properties such as non-linearity, stability and global dynamics. Techniques that can optimise/adapt each matrix (on-line or pre-processing) will be discussed later in the section.

The bias W^{bias} can be used to counteract training issues and large weights by adding noise, acting as a *regularisation* parameter, or to push the *tanh* neuron to a particular state, creating a smoothing effect.

Applying feedback W^{fb} can be useful, or detrimental, to certain tasks. Some tasks might not be learnt to a reasonable degree without feedback, or, certain systems may require dynamics beyond what is supplied by the driven input to construct a suitable output. Adding feedback comes with its own risks, feedback will ultimately change the stability of the system and requires adaptations in the training procedure. It is often advised only to use feedback when necessary. For more information see [66].

The *off-line* technique is completed in one training cycle T after the system has computed all states for the given inputs. It provides a very fast training mechanism as it essentially computes a linear model given by the known output Y, collected reservoir states X and desired output Y^{target}:

$$Y = W^{out} X \tag{22.6}$$

The collected state matrix $X \in \mathbb{R}^{N_x \times T}$ is created when the input $u(n)$ is run through the reservoir states $x(n)$. To avoid initial transients created by an initial zero state $x(0)$, a section at the beginning of the training sequence is discarded in the state matrix X. Essentially, the system goes through a "warming-up" process where states bounce around rather than returning to the equilibrium output, i.e. the system is too chaotic to retrieve any useful information about the input.

Given Eq. (22.6), we can find the optimal weights that minimise the error between $y(n)$ and $y^{target}(n)$ by solving the overdetermined system:

$$Y^{target} = W^{out} X \tag{22.7}$$

Equation (22.7) can be solved for W^{out} using linear regression. The simplest method is to use Ordinary Least-Squares (Eq. 22.8), but typically this method succumbs to stability issues when inverting (XX^T).

$$W^{out} = Y^{target} X^T (XX^T)^{-1} \tag{22.8}$$

Lukoševičius [66] recommends using either ridge regression (regression with Tikhonov regularisation β) (Eq. 22.9) or the Moore–Penrose pseudo-inverse (Eq. 22.10). Ridge regression is a stable and effective solution and is generally advised. Adding a regularisation parameter counteracts the problems of producing very large output weights, which often indicates very sensitive and unstable solutions.

$$W^{out} = Y^{target} X^T (XX^T + \beta I)^{-1} \tag{22.9}$$

where I is the identity matrix and β the regularisation parameter.

Setting $\beta = 0$ gives the same method for solving linear regression in Eq. (22.8). It is therefore recommended to use a logarithmic scale for selecting β where it never reaches zero [68].

The pseudo-inverse is applied in some cases typically because it is straightforward to implement in certain programming environments (e.g. MATLAB). However this comes at a price. The pseudo-inverse method is computationally expensive for large matrices of X and typically overdetermined. However, in most cases the network is made up of relatively small matrices and over-fitting depends on the difficulty of the task.

$$W^{out} = Y^{target} X^+ \tag{22.10}$$

22.4.5.2 On-Line Training

Some tasks require an on-line training method that adapts with time, minimising the error at each time step. This implicitly turns W^{out} into an adaptive linear combiner. The Recursive Least Squares (RLS) algorithm (Eq. 22.11) is more commonly applied as it overcomes the severely impaired convergence performance of the Least Means Square (LMS) algorithm [46].

$$E(y, y^{target}, n) = \frac{1}{N_y} \sum_{i=1}^{N_y} \sum_{j=1}^{n} \lambda^{n-j} (y_i(j) - y_i^{target}(j))^2 \tag{22.11}$$

Using RLS comes at a cost: the number of weights is quadratic rather than linear, and it can still be numerically unstable in some cases. Other powerful on-line methods may be useful to a practitioner, particularly in the presence of feedback connections, such as Backpropagation-Decorrelation (BPDC) [105].

The RLS training procedure is described here, derived from [25]. First, set the error forgetting parameter λ close to but less than one; the forgetting parameter controls the contribution of previous samples. Next, initialise the autocorrelation matrix $\rho(0) = I/\delta$, with δ being a small constant and I the identity matrix. At each time step compute the output weights using the following steps:

Step 1: Calculate the reservoir state $x(n)$ and output signal $y(n)$ for input $u(n)$.

Step 2: Calculate the error between target output $y^{target}(n)$ and system output given previous output weights:

$$e(n) = y^{target}(n) - W^{out}(n-1)x(n) \tag{22.12}$$

Step 3: Update the gain vector $K(n)$:

$$K(n) = \frac{\rho(n-1)x(n)}{\lambda + x^T(n)\,\rho(n-1)x(n)} \tag{22.13}$$

Step 4: Update the autocorrelation matrix $\rho(n)$:

$$\rho(n) = \frac{1}{\lambda}[\rho(n-1) - K(n)x^T(n)\,\rho(n-1)] \tag{22.14}$$

Step 5: Assign new output weights $W^{out}(n)$ using (22.12) and (22.13):

$$W^{out}(n) = W^{out}(n-1) + K(n)e(n) \tag{22.15}$$

For more readout training methods including feedback training (such as FORCE training), supervised, unsupervised, reinforcement learning, etc. consult the excellent review [68] and practical aid [66].

22.4.5.3 Adaptation and Pre-training

Adaptive reservoirs, ones that change weights or configuration, are inspired by natural adaptation in biological neurons. These adaptive processes are the result of persistent changes in a neuron's electrical properties, governed by unsupervised *local* adaptation rules often referred to as *Intrinsic Plasticity* (IP). These rules represent a homeostatic mechanism in which neurons self-modify their intrinsic activity (i.e. excitability). Using such learning rules has shown to increase robustness and performance when pre-training reservoirs [96, 106]. For an overview of recent investigations including both local and global adaptation schemes, see [67, 68].

In classical RC, reservoirs are generated randomly, hence the performance of each reservoir varies on creation. Reservoir computing boasts its training performance on the separation between the reservoir and readout. The readout training, at its core, is quite inexpensive, allowing the possibility of other forms of reservoir *pre-training*, i.e.

generating reservoirs deterministically for each task. Even a crude experiment such as selecting a reservoir which produces the smallest error from a pool of randomly-created reservoirs highlights the advantages of pre-training. Evolutionary algorithms are a potential search strategy for pre-training.

Investigations using evolutionary optimisation for pre-training are well documented. Many strategies have been attempted, including evolving topologies (i.e. network size), weight matrices (such as W^{in}, W, W^{fb}), global parameters (e.g. spectral radius), connection density, adapting slopes of the activation function $f(.)$ and even training w^{out} when no target signal is available. Other interesting methods include *EvoLino*—evolving hidden connections to gradient-based long-short term memory (LSTM) RNNs [91] and using *Neuro-Evolution of Augmented Topologies* (NEAT) as a meta-search algorithm for evolving ESNs [20] (the "related work" section also discusses other neuro-evolution methods for constructing ESNs). All of these methods have shown great potential, highlighting the performance increases and optimisations reservoir pre-training can create for specific tasks. Pre-training and adaptation appears to be one of many key branches under investigation in the field of reservoir computing.

22.5 Modelling Materials with Reservoir Computing

Any high-dimensional dynamical system with an observable state $x(n)$ that is a result of input $u(n)$ could form the basis of a reservoir according to [68]. This implies that any material that can exhibit sufficient dynamics and a fading memory could therefore, theoretically, be adopted as a reservoir.

Given that reservoir computing is based on artificial recurrent neural networks, one implementation route would be to design hardware substrates modelled on simplified neural network-like structures, e.g. large coupled networks of non-linear elements with varying densities of connectivity. Using this structural model, semi-isolated regions of activity may form exploitable meta-states for the trainable readout. Various implementations of hardware-based artificial neural networks have existed for some time; see [77] for a review of HNNs. More recently, there has been increased popularity towards applying memristive components to neuromorphic circuits, see [58] (a review CMOS/memristor hybrids) and [101].

Kulkarni and Teuscher [59], among others, have examined and demonstrated reservoir networks built from memristor devices. Memristors appear to be ideal components for building reservoirs: they exhibit non-linear properties and maintain a memory of previous inputs. In the Kulkarni and Teuscher experiment, simulated circuits are randomly-created from networks of memristors, then a subsequent readout layer is trained using a genetic algorithm to solve some computational task. Other simulated memristor reservoirs include: simple-cycle reservoirs [16] i.e. memristive networks that form nodes instead of analogue neurons (see [88]), training more realistic volatile memristor models [19], and variation-tolerant reservoirs [17].

Fabricating random, highly-interconnected networks and devices from nanoscale switching elements is a challenging task. At these scales, features of self-assembly and self-organisation are essential: characteristics that might only be achievable through unconventional methods and materials. Konkoli and Wendin [55] provide a brief review of some non-CMOS devices and discuss the suitability of the RC model on such unconventional devices. These devices include, molecular electronic networks, memristor networks, and other substrates outlined as part of the SYMONE project [122]. At present, the SYMONE project is investigating networks of organic transistors (NOMFETs) and self-assembled networks of gold nanoparticles that could feature functionalised memristive junctions [55].

Two examples that demonstrate physical, *in-materio* reservoir computing systems are silicon-based photonic chips (based on nanophotonic crystal cavities) [29, 117], and Atomic Switch Networks (ASNs) [103, 112, 113].

The *photonics chip* primarily exploits the propagation of light through silicon. Inside these chips are photonic crystals that remove the propagation of certain frequencies of light, known as the band gap. Adding a line defect to a crystal produces a photonic crystal waveguide, effectively a process by which light is forced between the defect. Cavities are then created along the line defect to create an optical "resonator" which traps light, increasing the power inside the resonator. These resonators then form a optical reservoir which can be trained and manipulated using different types of readouts, e.g. [117] creates a linear system, and the non-linearity is added at the readout through the inherent non-linearity of the measuring equipment. Methods such as this propose an interesting optical alternative to hardware-based reservoir computing.

The *atomic switch network* approach focuses on the electrical and chemical properties of a material. These networks attempt to mimic the vast complexity, emergent dynamics, and connectivity of the brain. Highly-interconnected networks are constructed by bottom-up self-assembly of silver nanowires. Through a triggered electro-chemical reaction, coated copper seed nucleation sites spawn large quantities of silver nanowires of various lengths, from nano- to millimetre scale. Large random networks are formed, creating crossbar-like junctions between nanowires, and when exposed to gaseous sulphur create $Ag|Ag_2S|Ag$ nanoscale metal-insulator-metal (MIM) junctions. Applying external activation (a bias voltage) to these junctions creates a temporary resistance drop, leading to functional memristive junctions called Atomic Switches. Applying this construction and functionalisation process the ASN method offers some unique properties, such as scalability and practicality in creating highly-complex nanoscale substrates.

The emergent behaviour and dynamics of ASNs can be observed through fluctuations in network conductivity, a result of spontaneous switching between discrete metastable resistance states, where locally excited regions cause cascading changes in resistance throughout the system. As such, the non-linear responses to resistive switching are reported to result in higher harmonic generation (HHG), also suggested as a useful measure for quantitatively evaluating reservoir dynamics [103].

A clear advantage of the ASN technique is that it allows some degree of regulation in fabrication and further control through "resistance control" training [113].

For example, varying the parameters of the nucleation sites (copper seeds) can alter the network structure and therefore the substrates dynamics. The relative size, numbers and pitch of copper seeds can determine the length of wires, forming distant connections or confining spatial regions to dendrite-like structures, thus ultimately defining the density of connections [103].

ASNs appear to be natural candidates for reservoir computing, producing a high-dimensional recurrent network that does not require the manipulation of individual elements. ASNs have been applied to one derived RC task; the waveform generation task [103, 113]. This is a simple analogue task which measures a reservoir's ability to construct a desired output waveform from an independent input waveform using network generated harmonics. For example, given an input sine-wave, can a trained reservoir construct a sawtooth, square-wave, or any other periodic function at the output (essentially a Fourier series task using harmonic analysis). These initial experiments have proven ASNs to be capable reservoirs and has also highlighted HHG as a potential metric for evaluating reservoir properties.

22.6   Evaluating the Characteristics of Reservoirs

Creating a random reservoir (in simulation at least) is fairly straightforward, but designing one with the right properties, using the large parameter space available, is a challenging task. In many cases parameter selection is done by hand, through trial and error, and with expert knowledge of the desired characteristics. So how can we better understand and evaluate reservoirs?

One approach is to simplify its construction. In doing so, one could provide a more theoretical understanding of what makes reservoirs useful/successful. Reference [88] explores this idea by addressing three issues: (i) what is the minimal complexity of topology and parameters that produce comparable performance to standard models? (ii) what degree of randomness is needed to construct competitive reservoirs? and (iii) how do completely deterministic reservoirs compare? These are good questions for understanding underlying RC principles, but may be impractical to investigate given an already created (maybe static) physical substrate. Instead, we desire more experimental quantitative measures that individually describe the reservoir and its qualities as an efficient kernel.

Determining, or evaluating, reservoir quality and performance can be achieved in two ways, either through direct measurement of performance on a given task, or by cumulatively assessing individual properties of the reservoir. Using the latter method provides a mechanism in which performance could be partly predicted for any task. As [82] explains, a good reservoir that scores well in all properties may be able to facilitate the process of "learning transfer", where the reservoir can be trained to some objective function that will increase its capability without seeing any output task. The objective function, in the mentioned case, measures how well a reservoir (an LSM) separates different classes of inputs into different reservoir states. As such, it was shown that improving separability and instilling an adaptive balance between chaotic and ordered behaviour (through changes in structure) a reservoir can increase its performance across different artificial problems.

In this section we discuss some of the measures proposed in the reservoir computing literature, accompanied by a variety of different benchmark tasks used to assess performance. Such measures and benchmarks may be an effective method for evaluating the potential of substrate-based reservoirs.

22.6.1 Kernel Quality and Separation

22.6.1.1 Kernel Quality

Kernel quality evaluates a reservoir's ability to produce diverse and complex mappings (functions) of the input stream u that can consequently increase the classification performance of a linear decision-hyperplane [62].

Kernel quality, also known as the *linear separation property*, was first introduced by Legenstein and Maass [62], along with an accompanying metric referred to as the *generalisation capability* of a reservoir. The two properties are closely coupled and can be measured using a similar ranking mechanism. The first, kernel quality, measures a reservoir's ability to produce diverse reservoir states given significantly different input streams. The second measures the reservoir's capability to generalise given similar input streams with respect to a target function. Both measurements can be carried out using the method in [18], by computing the rank r of an $n \times m$ matrix M, with the two methods differing only in the selection of m input streams $u_i, \ldots, u_m$, i.e. input streams being largely different or of similar type/class.

The rank is assessed as follows; Given the input stream u_i the reservoir state vector x_{u_i} of length n is collected to form each column in the matrix M. The rank r of each matrix is then estimated by Singular Value Decomposition (SVD). Büsing et al. [18] explains that a good reservoir should possess a high kernel quality and a low generalisation rank, and also identify a correlation in the measurement to the reservoir's dynamics. For example, reservoirs in ordered regimes produce low ranking values in both measures, and in chaotic regimes produces high values in both measures. A similar connection is also observed in [21], where the Lyapunov exponent and the kernel quality strongly correlate.

22.6.1.2 Class Separation

Class separation is a metric that corresponds directly to different classes of input stimuli. Demonstrations of class separation can be found in [21, 31, 82]. Separation is measured as the average distance between resulting states, once again, given the assumption that significantly different inputs should generate significantly different reservoir states. To calculate separation requires the division of the input and state vectors into discrete classes; [31] provides an alternative measure characterised on the original assumption. For example, given two different input vectors $u_j(n)$ and $u_k(n)$ the euclidean distance between inputs should be large and positive, as described by D:

$$D := \|u_j(n) - u_k(n)\| \tag{22.16}$$

If the reservoir exhibits a good separation property the reservoir states ($x_j(n)$ and $x_k(n)$) should increase in distance, or be equal to the original distance:

$$D \leq \|x_j(n) - x_k(n)\| \tag{22.17}$$

which can be represented as the ratio:

$$\frac{\|x_j(n) - x_k(n)\|}{\|u_j(n) - u_k(n)\|} \geq 1 \tag{22.18}$$

This simplified measure has been extended [31] into *Separation Ratio Graphs* to produce a visual representation of separation and the phase transition of correlated dynamic behaviour (see Fig. 22.6).

Konkoli and Wendin [55] offer another comparable method for identifying reservoir quality in memristor networks. This metric is again based on the assumption that quality can be measured by observing the reservoir's ability to generate different dynamic states at the output. In this case, it is observed by measuring the dissimilarity between output states and a linear combination of the inputs, i.e. determining if the non-linear frequency response of a network cannot be approximated by a linear mixture of delayed inputs. Dissimilarity is measured in the Fourier space (ω) between outputs $o(n)$ and a linear combination of the time shifted inputs $z(n)$, given by:

Fig. 22.6 Separation Ratio Graph [31]. Graphical representation of the phase transition between chaos and order. Systems in the target zone are said to possess both a good separation property and ideal dynamic behaviour to produce optimal reservoirs

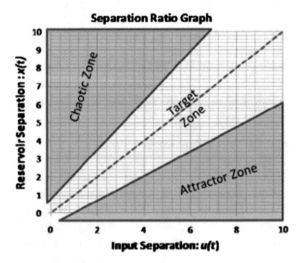

$$\delta = \frac{\| o(\omega) - z(\omega) \|}{\| o(\omega) \|} \tag{22.19}$$

A large dissimilarity (large δ) is ideal in a reservoir as it describes a complex projection of the input. A small δ on the other hand may simply describe a linear propagation of the input, highlighting the absence of richness in the reservoir.

22.6.2 Quantifying Dynamics

Producing a reservoir with rich dynamics is evidently desirable, but how can we quantify dynamical behaviour? how can we pin-point the critical region at the edge of chaos? In the literature, there are several prominent themes in classifying and measuring dynamics, including observation of trajectories and attractor behaviour, internal scaling of input-driven activity, memory and retention, and higher-harmonic generation, particularly in reservoirs comprising resistive switches.

22.6.2.1 Spectral Radius

In Echo State Networks, the spectral radius $\rho(.)$ (the largest absolute eigenvalue) of the internal weight matrix W is used to determine and control reservoir dynamics [44]. The parameter ρ globally scales the internal weights, moving the system between different regimes, altering the non-linearity and impulse response of the reservoir. Effectively, the scaling parameter alters the internal time-scales of the system, providing the *echo state property*.

Parameter selection of the spectral radius is typically centred around a value of one; smaller than one is attributed to a stable regime (a fading response to input stimuli), if larger than one, a system will typically be unstable, causing undesirable interference to new inputs.

22.6.2.2 Lyapunov Exponents

A popular measurement for criticality, or chaotic behaviour, in a reservoir is the empirical estimation of the Lyapunov exponents (LE) for a dynamical system [11, 21, 31, 63, 118]. To calculate the Lyapunov Exponents, and quantify a system's criticality, we measure the divergence between two close trajectories due to some small perturbation. For example, if an orbit is close to an attractor, in an ordered phase, small changes should dissipate over time. In a chaotic phase, an applied change will diverge exponentially from an orbit, persisting or increasing over time.

Gibbons [31] provides a simple approximation of the largest (maximal) Lyapunov exponent (derived from [89]), as the largest tends to dominate. Various formats and interpretations exist in calculating different Lyapunov Exponents due to different approximation methods (see [51, 63, 118]). For example, Verstraeten et al. [118] examines the local Lyapunov spectrum and the Jacobian of the reservoir, suggesting performance can be better predicted by the maximum of the minimal Lyapunov Exponent. Boedecker et al. [13] takes a similar approach to Gibbons, estimating the mean exponential rate of divergence of the trajectories for an n-dimensional phase space.

A system with a maximal Lyapunov exponent $\lambda \approx 0$ is somewhere near the *edge of chaos*. A chaotic system is present at $\lambda > 0$ and an ordered, or sub-critical, system at $\lambda < 0$.

Another suggested measure of criticality is the measure of instantaneous entropy of reservoir states, defined as the *average state entropy* (ASE) in [85]. Applying Rényi quadratic entropy, one can measure the distribution of instantaneous amplitudes in reservoir states, providing some measure of the "richness" of dynamics. The entropy measurement is associated with an expectation that increased diversity of amplitudes, i.e. increased spread of amplitudes away from some concentrated point, will increase the readout's ability construct the desired response.

22.6.2.3 Memory Capacity

Measuring the short-term memory capacity of a reservoir was first outlined by Jaeger [45] as a quantitative measurement to observe the echo state property (fading memory). To determine the memory capacity of a reservoir we simply measure how many delayed versions of the input $u(n - k)$ the outputs can recall or recover with precision. As Jaeger describes, using the equation in (22.20), we can measure memory capacity by how much variance of the delayed input can be recovered, summed over all delays [45]. This is carried out by training individual output units to recall the input at time k with the maximum capacity MC of an N node reservoir typically bounded by its size, i.e. $MC \leq N$.

$$MC = \sum_{k=1}^{\infty} MC_k = \sum_{k=1}^{\infty} \frac{cov^2(u(n - k), y(n))}{\sigma^2(u(n))\sigma^2(y(n))} \qquad (22.20)$$

This measurement has direct connection to the dynamic behaviour of a system. It can be helpful in identifying the boundaries between static structure that provides memory, and the point of complex dynamics that gives us processing. As such, one might expect a chaotic system to lose information regarding previous inputs at a faster rate and a more ordered regime to increase (to some extent) input retention.

22.6.2.4 Harmonic Generation

The generation of higher harmonics in Atomic Switch Networks has been identified as a technique for examining emergent behaviour and network connectivity

[103, 113]. In ASNs, higher harmonic generation (HHG) is attributed to the non-linear frequency response of the system due to both input amplitude and memristive "hard" switching behaviour. To examine HHG, one can plot the frequency response of the system, which can be used to identify connectivity and system dynamics, i.e. changes in network response (an on-set of HHG) is typically due to an increase in "hard" switching memristive connections past a percolation threshold [103]).

22.6.3 Evaluation Through Benchmark Tasks

As a means of direct assessment, we can measure performance of reservoirs and their subsequent readouts by applying them to specific tasks. Reservoir computing (and neural networks as a whole) possesses an abundance of benchmark tasks, from simple classification and time-series prediction to robot navigation [5, 25] and non-linear channel equalisation [87, 88].

In this section we discuss benchmark tasks that are the most prevalent in the reservoir computing literature.

22.6.3.1 Waveform Generation

This task requires a rich transformation of a temporal input waveform (a periodic signal) to create an entirely new waveform. It is based on Sillin et al. [103]'s physical adaptation of Jaeger's [47] sine-wave generator task and is linked directly to Fourier series/analysis. The task is to train the system to produce three different waveforms, given an input sine wave. In [103, 113], this is achieved by applying a 10 Hz input sine-wave (at one electrode in the ASN) to produce a 10 Hz square-wave, and sawtooth, and a 20 Hz sine-wave at the output $(y(n))$, via the combination of other weighted electrode readings (e.g. recorded states $x(n)$). The task is said to be an excellent precursor to testing potential reservoir substrates on more difficult temporal problems [103], as it highlights an abundance of temporal features (phase shifts, delays, harmonic generation, recurrence etc.).

A similar task is the continuous-time multiple superimposed oscillator (MSO) task. In this benchmark, the reservoir's role is to predict the evolution of, and generate a superposition of, two or more sinusoidal waves with different frequencies.

$$s(t) = \sin(\omega_1 t) + \sin(\omega_2 t) \qquad (22.21)$$

where $\omega_1 = 0.2$ and $\omega_2 = 0.311$. The task has been demonstrated in photonics experiments [29] and other non-traditional reservoirs [107].

22.6.3.2 Time-Series Prediction and Generation

Two prominent benchmark tasks in reservoir computing are Non-linear Auto-regressive Moving Average (NARMA) dynamical system modelling and the Mackey–Glass chaotic time-series prediction task.

The NARMA task originates from Atiya and Parlos [8]'s work on training recurrent network; its goal is to evaluate a reservoir's ability to model an nth order, highly non-linear dynamical system where the system state depends on the incoming input as well as its own history. The challenging aspect of the NARMA task is that it contains long-term dependencies created by the nth order time-lag. Typically, the benchmark is carried out on 10th and 30th ordered systems [6, 46, 47, 106].

A description of the 10th ordered task is as follows; Given white noise $u(n)$ from a uniform distribution of interval $[0, 0.5]$, the reservoir should predict an output $y(n)$ close to the target $y(n + 1)$, calculated by:

$$y(n + 1) = 0.3y(n) + 0.05y(n)\left(\sum_{i=0}^{9} y(n - i) \right) + 1.5u(n - 9)u(n) + 0.1$$

$$(22.22)$$

Mackey–Glass chaotic time-series prediction is another common benchmark, where the system is trained to predict one time-step into the future (for examples see [44, 49, 118]).

The system is described by the Mackey–Glass delay differential equation:

$$\dot{y}(n) = \alpha y(n - \tau)/(1 + y(n - \tau)^{\beta}) - \gamma y(n) \qquad (22.23)$$

As [44] explains, parameters for the MG task are typically set to $\alpha = 0.2$, $\beta = 10$ and $\gamma = 0.1$ with the parameter τ set to 17 to produce a mildly chaotic attractor: the system has a chaotic attractor for $\tau > 16.8$.

Other time-series prediction benchmarks are summarised in [88], including predicting laser activations in the Santa Fe Laser dataset (originally used in [50]), predicted next output in the Hénon Map dataset, and IPIX Radar and Sunspot series datasets.

22.6.3.3 Classification Tasks

Simple classification problems are wide and varied in the field of machine learning, some of which can be seen in both the RC and EIM literature. For example, typical tasks for EIM are tone and frequency discriminators, and Iris and Lenses dataset classification [38, 78, 80]. Examples in reservoir computing include, signal classification (discriminating between two waveforms) [87], various n-bit parity problems [11, 25, 94, 97] and other time-independent classification tasks [4].

Possibly the most adopted classification task in RC involves the recognition of isolated digits from multiple speakers [61, 87, 88, 93, 96, 119, 120]. Taken from a subset of the T146 speech corpus dataset, the task uses a total of 500 speech fragments collected from five different participants listing the digits zero to nine (repeated ten times). The reservoir interprets these speech fragments through a preprocessing filter in the form of a digital model of the human cochlea. Linear classifiers are then trained to be sensitive to individual digits, with a final classification being made on the temporal mean of the output. Performance on the task is measured using cross-validation by calculating the number of misclassified digits using the Word Error Rate (WER).

Reservoir computing is said to be very competitive to, or outperform, many of the state-of-the-art approaches on these tasks.

22.6.4 Material Properties and Considerations

In order to perform material computation we require some ability to manipulate and control certain aspects of physical structure and behaviour. To observe such effects requires a means of observation and measurement. In the two examples presented (ASNs and EIM), we have discussed one method by which this can be achieved, through the application and recording of electrical voltages to a micro-electrode array. There are other possible methods, for example: optical stimulus/measurement and other stimuli across the electromagnetic spectrum (e.g. optoelectronic and photonic reservoirs); image recognition for observation (e.g. a method also used in Fernando and Sojakka [28]'s bucket of water); control and observation through magnetic fields (e.g. manipulating ionised gases or observing nuclear magnetic resonance (NMR) [108]); chemical excitation and reaction (e.g. reaction-diffusion computers and slime mould [1, 2]).

There are many physical properties and considerations that require discussion when talking about using any novel material for computation. Here we focus on four key factors that possess some relevance to substrate-based reservoirs: (i) a means by which to observe network connectivity and activity, (ii) assuring non-linearity is present in the system, (iii) methods for modelling activity and behaviour, and (iv) the impact of environmental factors.

22.6.4.1 Network Connectivity and Activity

The computational capability of a material is often directly related to conductivity and the density of connections inside it. Variations in these concentrations can have an adverse or favourable effect on conductivity and task performance. For example, to optimise an ASN one can control the densities of silver nanowires [103]. Massey et al. [74] identify a similar relationship, where nanotube concentration directly alters the conductivity and computational performance of SWCNT/PBMA composites.

How can we measure connectivity in materials and analyse distributed activity? One possible method is demonstrated in [112], where ultra-sensitive infra-red (IR)

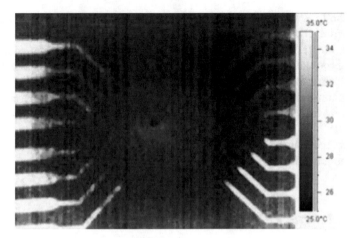

Fig. 22.7 Infrared image of carbon nanotube/polymer mixture. Infrared has been identified as a useful method for observing current flow and local regions of activity in both EIM and ASNs [73]

imaging is used to help identify network conductivity and behaviour in ASNs. IR imaging was used to observe and measure possible dominant conductive pathways and identify the activity of local regions to stimulus by thermal emission. Similar experiments have also been carried out on carbon nanotube/polymer substrates within the NASCENCE project (see Fig. 22.7).

Other forms of network observation include: observation of power-law scaling in the power spectral density [103], optical microscopy such as examining structure through an optical microscope (an example can be seen in Fig. 22.8 with a SWCNT/LC substrate), electron microscopy and scanning-probe microscopy such as scanning electron microscopy (SEM) [32] and scanning-force microscopy (SFM) [14], observing current flow in polymer substrates by electron-beam-induced-current (EBIC) [37], investigating structural properties using Fourier transform infrared and Raman spectroscopy (FTIR) (a method used to observe the interaction of molecules in liquid crystal/SWCNT composites in [121]), and other absorption and emission spectroscopy techniques [99].

22.6.4.2 Non-linearity

Non-linearity within a material can be measured through current-voltage (I-V) characteristics. In ASNs, non-linearity is observed by the presence of pinched hysteresis curves as a function of input amplitude (produced by applying slow voltage sweeps). This non-linear I-V characteristic is said to be the result of changing switch behaviour (in this case towards a "hard-switching" regime) and increased harmonic generation [103].

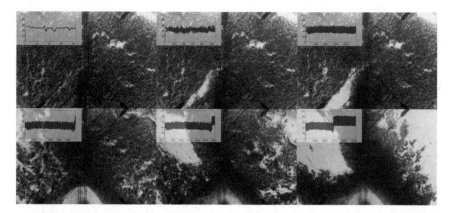

Fig. 22.8 Image of SWCNT/LC behaviour to voltage stimulus, observed through a microscope. Observation at this scale shows detail in the complexity of local interactions and carbon nanotube alignment to electric fields [73]

Similar I-V measurements have also been used in both SWCNT/PBMA and SWCNT/LC composites proposed in the NASCENCE project [74, 121].

22.6.4.3 Modelling

Simulating the internal properties of specific materials would be desirable, but it is typically impractical to create an exact representation of one individual material. Instead, we could use an abstraction, a model, of what the material is doing. Some methods that model specific features of the system, and others that are potential accompaniments to the reservoir model, include: the modelling of electron transport using a Monte Carlo approach, percolation visualisation for electrode pathways, Voronoi diagrams for visualising electrode activity [102], Volterra series and Wiener series models, using Random Boolean Networks as a model, abstract neural networks for modelling global activity, NARMAX (non-linear autoregressive moving average model with exogenous inputs) modelling, and using Cellular Automata as a reservoir/substrate model [83, 84].

22.6.4.4 Environment and Noise

The local environment, thermal noise, other noise related fluctuations and quantum effects are tangible concerns in the physical domain. Sensitive systems require good isolation and compensation techniques to reduce the effects of both internal and external noise. Effectively, the systems susceptibility to noise will determine its robustness and adaptability. In many cases, noise and unwanted variability could be filtered using conventional techniques, but requires some care in implementation (i.e.

does not impede on the boundaries of the system or unduly affect what is doing the computation). Evolution-*in-materio* on the other hand has been found to exploit such external influences. Using an evolutionary search in physical systems has shown an uncanny ability to utilise strange characteristics that are clearly attributed to external influences. For example, Bird and Layzell [12]'s evolvable motherboard was directly influenced by local laboratory equipment in producing evolved oscillators.

22.7   Feedback Design: Designing Substrates to Be Good Reservoirs

Beyond simply applying the reservoir model, could we utilise some of its highlighted features to aid in the design of *in materio* systems?

Often the difficulty in designing and engineering computational substrates is knowing what range of features are necessary, or even exploitable. Some aspects of the physical system are unknown. There are clearly some attributes that are desirable (for evolution at least): reconfigurable, input-driven, reproducible, some stability, etc. Can we use known computational phenomena from reservoir computing to help reduce the search space of suitable materials? Could the abstract model guide material design, and in return, serve to create a more realistic and efficient model? This could create a virtuous feedback-loop in design, where new novel materials can be evaluated for computational capability and reservoir quality.

Going further, could multiple smaller reservoirs reproduce the overall dynamics of a larger network? As stated in [66], a powerful extension of ESN consists of many small ESNs in parallel, where an averaged output has drastically improved performance, but, at a price in terms of memory capacity, i.e. fewer nodes typically equals less memory. Another approach would be to use multiple reservoirs to overcome hardware limitations, such as the number of electrode contacts available. These networked reservoirs could, as a by-product, provide increased robustness (further distributing computation) and added variability in states, i.e. promote a good overall separation property. Additionally, such systems could be constructed from various materials, each with different kernel properties and time-scales, allowing a global weighting system or training method (possibly backpropagation) to optimally choose which material to use for certain tasks. Moreover, this could lead to a more generic reservoir network suitable to multiple tasks.

Hierarchical reservoirs represent an up-and-coming, possibly highly advantageous, avenue worth pursuing [68]. A hierarchical system attempts to overcome some of difficulties found in classical reservoirs such as scalability, learning complex intelligent tasks and working with multiple time-scales simultaneously. Some early examples of these architectures can be found in [48] where high-level reservoirs extract features from low-level ones using a "feature-voting" system. Another can be seen in a system comprised of decoupled sub-reservoirs with inhibitory connections [124] where the inhibitory connections predict the activation of sub-reservoirs. Other examples include, hierarchically clustered ESNs and the impact substructures have on stability [51], increased performance on speech recognition tasks [115] and acoustic modelling [116], and a hierarchical architecture used for hand-writing recognition to sort delivery parcels [69].

22.8 Conclusion

We believe that reservoir computing can be used as a stepping stone to modelling and training the currently somewhat static in structure *in-materio* systems. Reservoir computing is considered mostly on the basis that the systems themselves may not be able to undergo "on-line" training/adaptation (manually or autonomously). Thus, the material is perceived as an initially-configurable black-box, or a suitably rich kernel, removed from the final training process.

The reservoir method provides many advantages when dealing with unknown or intractable properties, but, it is also limited by the complexity of the task it is required to learn. For example, a good reservoir requires a high-dimensional expansion of the input that can be exploited by a subsequent readout layer. This implies that for more complex tasks the reservoir may require an exponential growth in exploitable features from an exceptionally large feature space, which itself is potentially impractical, or at least very difficult to extract useful features from. Harnessing these features then becomes a significant challenge when implemented in a physical device. Standard artificial neural networks try to overcome this problem by building the required non-linear features internally. This is achieved through internal training, which is traditionally achieved through backpropagation.

To apply backpropagation in a physical system requires a material to contain the right attributes for internal training and manipulation, however. A recent example of a physical system that does use error-backpropagation as a training method is demonstrated in [42]. The experiment and implementation is carried out on an acoustic system, using the propagation of sound waves between a speaker and microphone, and on an electro-optical system, harnessing the reciprocal transmission of light through an optical circuit. This experiment provides a unique insight into backpropagation applied to physical systems and presents a potentially competitive alternative to digital neural networks.

The research areas discussed in this chapter, if combined, provide a rich avenue to explore. Reservoir computing presents us with a convenient theoretical model. It also provides some indication of what properties are required to increase the performance: the right dynamic criticality, good kernel projection, etc.

Optimisation and pre-training can improve performance in reservoirs, by allowing a reservoir to be *pre-set* with some properties that effectively increase the performance of the main training process. Evolution as a form of pre-training in the physical domain may provide a crucial and efficient technique for manipulating/configuring matter into suitable reservoirs, e.g. through structural alignment/deformation, or creating rich local regions of varying activity. As is described in the Evolution-*in-materio*

M. Dale et al.

doctrine, evolution may be the best practical technique to exploring and exploiting properties that are currently intractable or hitherto unknown, properties that could produce the most interesting and competent physical reservoirs.

Acknowledgments Matthew Dale is funded by a Defence Science and Technology Laboratory (DSTL) Ph.D. studentship.

References

1. Adamatzky, A.: Physarum Machines: Computers from Slime Mould, vol. 74. World Scientific, Singapore (2010)
2. Adamatzky, A., Costello, B., Asai, T.: Reaction-Diffusion Computers. Elsevier, Amsterdam (2005)
3. Adamatzky, A., Erokhin, V., Grube, M., Schubert, T., Schumann, A.: Physarum chip project: growing computers from slime mould. IJUC **8**(4), 319–323 (2012)
4. Alexandre, L.A., Embrechts, M.J., Linton, J.: Benchmarking reservoir computing on time-independent classification tasks. In: International Joint Conference on Neural Networks IJCNN 2009, pp. 89–93. IEEE (2009)
5. Antonelo, E.A., Schrauwen, B., Van Campenhout, J.: Generative modeling of autonomous robots and their environments using reservoir computing. Neural Process. Lett. **26**(3), 233–249 (2007)
6. Appeltant, L., Soriano, M.C., Van der Sande, G., Danckaert, J., Massar, S., Dambre, J., Schrauwen, B., Mirasso, C.R., Fischer, I.: Information processing using a single dynamical node as complex system. Nature Commun. **2**, 468 (2011)
7. Appeltant, L., Van der Sande, G., Danckaert, J., Fischer, I.: Constructing optimized binary masks for reservoir computing with delay systems. Sci. Rep. **4**(3629) (2014)
8. Atiya, A.F., Parlos, A.G.: New results on recurrent network training: unifying the algorithms and accelerating convergence. IEEE Trans. Neural Netw. **11**(3), 697–709 (2000)
9. Beggs, J.M.: The criticality hypothesis: how local cortical networks might optimize information processing. Philos. Trans. R. Soc. Lond. A: Math., Phys. Eng. Sci. **366**(1864), 329–343 (2008)
10. Belkin, A., Hubler, A., Bezryadin, A.: Self-assembled wiggling nano-structures and the principle of maximum entropy production. Sci. Rep. **5**(8323) (2015)
11. Bertschinger, N., Natschläger, T.: Real-time computation at the edge of chaos in recurrent neural networks. Neural Comput. **16**(7), 1413–1436 (2004)
12. Bird, J., Layzell, P.: The evolved radio and its implications for modelling the evolution of novel sensors. In: Proceedings of the Congress on Evolutionary Computation CEC'02, vol. 2, pp. 1836–1841. IEEE (2002)
13. Boedecker, J., Obst, O., Lizier, J.T., Mayer, N.M., Asada, M.: Information processing in echo state networks at the edge of chaos. Theory Biosci. **131**(3), 205–213 (2012)
14. Bose, S.K., Lawrence, C.P., Liu, Z., Makarenko, K.S., van Damme, R.M.J., Broersma, H.J., van der Wiel, W.G.: Evolution of a designless nanoparticle network into reconfigurable boolean logic. Nature Nanotechnol. (2015). doi:10.1038/nnano.2015.207
15. Broersma, H., Gomez, F., Miller, J., Petty, M., Tufte, G.: Nascence project: nanoscale engineering for novel computation using evolution. Int. J. Unconv. Comput. **8**(4), 313–317 (2012)
16. Bürger, J., Goudarzi, A., Stefanovic, D., Teuscher, C.: Composing a reservoir of memristive networks for real-time computing. arXiv:1504.02833 (2015)
17. Burger, J., Teuscher, C.: Variation-tolerant computing with memristive reservoirs. In: 2013 IEEE/ACM International Symposium on Nanoscale Architectures (NANOARCH), pp. 1–6. IEEE (2013)

18. Büsing, L., Schrauwen, B., Legenstein, R.: Connectivity, dynamics, and memory in reservoir computing with binary and analog neurons. Neural Comput. **22**(5), 1272–1311 (2010)
19. Carbajal, J.P., Dambre, J., Hermans, M., Schrauwen, B.: Memristor models for machine learning. Neural Comput. **27**(3), 725–747 (2015)
20. Chatzidimitriou, K.C., Mitkas, P.A.: A NEAT way for evolving echo state networks. In: ECAI 2010, pp. 909–914. IOS Press (2010)
21. Chrol-Cannon, J., Jin, Y.: On the correlation between reservoir metrics and performance for time series classification under the influence of synaptic plasticity. PloS One **9**(7), e101792 (2014)
22. Clegg, K.D., Miller, J.F., Massey, M.K., Petty, M.: Travelling salesman problem solved 'in materio' by evolved carbon nanotube device. In: Parallel Problem Solving from Nature–PPSN XIII, pp. 692–701. Springer, Heidelberg (2014)
23. Clegg, K.D., Miller, J.F., Massey, M.K., Petty, M.C.: Practical issues for configuring carbon nanotube composite materials for computation. In: IEEE International Conference on Evolvable Systems, ICES 2014, pp. 61–68. IEEE (2014)
24. Dai, X.: Genetic regulatory systems modeled by recurrent neural network. In: Advances in Neural Networks-ISNN 2004, pp. 519–524. Springer, Heidelberg (2004)
25. Dasgupta, S., Wörgötter, F., Manoonpong, P.: Information theoretic self-organised adaptation in reservoirs for temporal memory tasks. In: Engineering Applications of Neural Networks, pp. 31–40. Springer, Heidelberg (2012)
26. Derrida, B., Pomeau, Y.: Random networks of automata: a simple annealed approximation. EPL (Europhys. Lett.) **1**(2), 45 (1986)
27. Dominey, P.F.: Complex sensory-motor sequence learning based on recurrent state representation and reinforcement learning. Biol. Cybern. **73**(3), 265–274 (1995)
28. Fernando, C., Sojakka, S.: Pattern recognition in a bucket. In: Advances in Artificial Life, pp. 588–597. Springer, Heidelberg (2003)
29. Fiers, M., Van Vaerenbergh, T., Wyffels, F., Verstraeten, D., Dambre, J., Schrauwen, B., Bienstman, P.: Nanophotonic reservoir computing with photonic crystal cavities to generate periodic patterns. IEEE Trans. Neural Netw. Learn. Syst. **25**(2), 344–355 (2014)
30. Funahashi, K., Nakamura, Y.: Approximation of dynamical systems by continuous time recurrent neural networks. Neural Netw. **6**(6), 801–806 (1993)
31. Gibbons, T.E.: Unifying quality metrics for reservoir networks. In: IJCNN 2010, The International Joint Conference on Neural Networks, pp. 1–7. IEEE (2010)
32. Goldstein, J., Newbury, D.E., Echlin, P., Joy, D.C., Romig, A.D Jr., Lyman, C.E., Fiori, C., Lifshin, E.: Scanning Electron Microscopy and X-ray Microanalysis: A Text for Biologists, Materials Scientists, and Geologists. Springer Science & Business Media, Heidelberg (2012)
33. Goudarzi, A., Lakin, M.R., Stefanovic, D.: DNA reservoir computing: a novel molecular computing approach. In: DNA Computing and Molecular Programming, pp. 76–89. Springer, Heidelberg (2013)
34. Greenwood, G.W., Tyrrell, A.M.: Introduction to Evolvable Hardware: A Practical Guide for Designing Self-Adaptive Systems, vol. 5. Wiley, New York (2006)
35. Gutierrez, J.M., Hinkley, T., Ward Taylor, J., Yanev, K., Cronin, L.: Evolution of oil droplets in a chemorobotic platform. Nature Commun. **5** (2014)
36. Haddow, P.C., Tyrrell, A.M.: Challenges of evolvable hardware: past, present and the path to a promising future. Genet. Program. Evolvable Mach. **12**(3), 183–215 (2011)
37. Hanoka, J.I., Bell, R.O.: Electron-beam-induced currents in semiconductors. Ann. Rev. Mater. Sci. **11**(1), 353–380 (1981)
38. Harding, S., Miller J.F.: Evolution in materio: a tone discriminator in liquid crystal. In: CEC 2004, Congress on Evolutionary Computation, vol. 2, pp. 1800–1807. IEEE (2004)
39. Harding, S., Miller J.F.: Evolution in materio: initial experiments with liquid crystal. In: 2004 NASA/DoD Conference on Evolvable Hardware, pp. 298–305. IEEE (2004)
40. Harding, S., Miller J.F.: Evolution in materio: a real-time robot controller in liquid crystal. In: 2005 NASA/DoD Conference on Evolvable Hardware, pp. 229–238. IEEE (2005)

41. Harding, S., Miller J.F.: Evolution in materio: evolving logic gates in liquid crystal. In: ECAL 2005 Workshop on Unconventional Computing: From cellular automata to wetware, pp. 133–149. Beckington, UK (2005)
42. Hermans, M., Burm, M., Van Vaerenbergh, T., Dambre, J., Bienstman, P.: Trainable hardware for dynamical computing using error backpropagation through physical media. Nature Commun. **6**, (2015)
43. Higuchi, T., Iwata, M., Kajitani, I., Yamada, H., Manderick, B., Hirao, Y., Murakawa, M., Yoshizawa, S., Furuya, T.: Evolvable hardware with genetic learning. In: IEEE International Symposium on Circuits and Systems, ISCAS'96, vol. 4, pp. 29–32. IEEE (1996)
44. Jaeger, H.: The "echo state" approach to analysing and training recurrent neural networks-with an erratum note. Bonn, Germany: German National Research Center for Information Technology GMD Technical Report **148**, 34 (2001)
45. Jaeger, H.: Short term memory in echo state networks. GMD-Forschungszentrum Informationstechnik (2001)
46. Jaeger, H.: Adaptive nonlinear system identification with echo state networks. In: Advances in Neural Information Processing Systems, pp. 593–600 (2002)
47. Jaeger, H.: Tutorial on training recurrent neural networks, covering BPPT, RTRL, EKF and the "echo state network" approach. GMD-Forschungszentrum Informationstechnik (2002)
48. Jaeger, H.: Discovering multiscale dynamical features with hierarchical echo state networks. Technical report No. 9 (2007)
49. Jaeger, H., Haas, H.: Harnessing nonlinearity: predicting chaotic systems and saving energy in wireless communication. Science **304**(5667), 78–80 (2004)
50. Jaeger, H., Lukoševičius, M., Popovici, D., Siewert, U.: Optimization and applications of echo state networks with leaky-integrator neurons. Neural Netw. **20**(3), 335–352 (2007)
51. Jarvis, S., Rotter, S., Egert, U.: Extending stability through hierarchical clusters in echo state networks. Front. Neuroinformatics **4**, (2010)
52. Jones, B., Stekel, D., Rowe, J., Fernando, C.: Is there a liquid state machine in the bacterium escherichia coli? In: IEEE Symposium on Artificial Life, 2007. ALIFE'07, pp. 187–191. IEEE (2007)
53. Kilian, J., Siegelmann, H.T.: The dynamic universality of sigmoidal neural networks. Inf. Comput. **128**(1), 48–56 (1996)
54. Klampfl, S., David, S.V., Yin, P., Shamma, S.A., Maass, W.: Integration of stimulus history in information conveyed by neurons in primary auditory cortex in response to tone sequences. In: 39th Annual Conference of the Society for Neuroscience, Program, vol. 163 (2009)
55. Konkoli, Z., Wendin, G.: On information processing with networks of nano-scale switching elements. Int. J. Unconv. Comput. **10**(5–6), 405–428 (2014)
56. Kotsialos, A., Massey, M.K., Qaiser, F., Zeze, D.A., Pearson, C., Petty, M.C.: Logic gate and circuit training on randomly dispersed carbon nanotubes. Int. J. Unconv. Comput. **10**(5–6), 473–497 (2014)
57. Küçükemre, A.U.: Echo state networks for adaptive filtering. Ph.D. thesis, University of Applied Sciences (2006)
58. Kudithipudi, D., Merkel, C., Soltiz, M., Garrett, S.R., Robinson, E.P.: Design of neuromorphic architectures with memristors. In: Network Science and Cybersecurity, pp. 93–103. Springer, Heidelberg (2014)
59. Kulkarni, M.S., Teuscher, C.: Memristor-based reservoir computing. In: IEEE/ACM International Symposium on Nanoscale Architectures, NANOARCH, 2012, pp. 226–232. IEEE (2012)
60. Langton, C.G.: Computation at the edge of chaos: phase transitions and emergent computation. Phys. D: Nonlinear Phenom. **42**(1), 12–37 (1990)
61. Larger, L., Soriano, M.C., Brunner, D., Appeltant, L., Gutiérrez, J.M., Pesquera, L., Mirasso, C.R., Fischer, I.: Photonic information processing beyond turing: an optoelectronic implementation of reservoir computing. Opt. Express **20**(3), 3241–3249 (2012)
62. Legenstein, R., Maass, W.: Edge of chaos and prediction of computational performance for neural circuit models. Neural Netw. **20**(3), 323–334 (2007)

63. Legenstein, R., Maass, W.: What makes a dynamical system computationally powerful. In: New Directions in Statistical Signal Processing: From Systems to Brain, pp. 127–154 (2007)
64. Lipson, H., Pollack, J.B.: Automatic design and manufacture of robotic lifeforms. Nature **406**(6799), 974–978 (2000)
65. Lohn, J.D., Linden, D.S., Hornby, G.S., Kraus, W.F., Rodriguez-Arroyo, A.: Evolutionary design of an X-band antenna for NASA's space technology 5 mission. In: NASA/DoD Conference on Evolvable Hardware, pp. 155–155. IEEE (2003)
66. Lukoševičius, M.: A practical guide to applying echo state networks. In: Neural Networks: Tricks of the Trade, pp. 659–686. Springer, Heidelberg (2012)
67. Lukoševicius, M., Jaeger, H.: Overview of reservoir recipes. Technical report 11, Jacobs University Bremen (2007)
68. Lukoševičius, M., Jaeger, H.: Reservoir computing approaches to recurrent neural network training. Comput. Sci. Rev. **3**(3), 127–149 (2009)
69. Lukoševičius, M., Jaeger, H., Schrauwen, B.: Reservoir computing trends. KI-Künstliche Intelligenz **26**(4), 365–371 (2012)
70. Lykkebø, O.R., Harding, S., Tufte, G., Miller, J.F.: Mecobo: A hardware and software platform for in materio evolution. In: Unconventional Computation and Natural Computation, pp. 267–279. Springer, Heidelberg (2014)
71. Maass, W.: Liquid state machines: motivation, theory, and applications. In: Computability in Context: Computation and Logic in the Real World, pp. 275–296 (2010)
72. Maass, W., Natschläger, T., Markram, H.: Real-time computing without stable states: a new framework for neural computation based on perturbations. Neural Comput. **14**(11), 2531–2560 (2002)
73. Massey, M.K.: Presentation at NASCENCE Consortium Progress Meeting. Totnes, UK (2015)
74. Massey, M.K., Kotsialos, A., Qaiser, F., Zeze, D.A., Pearson, C., Volpati, D., Bowen, L., Petty, M.C.: Computing with carbon nanotubes: optimization of threshold logic gates using disordered nanotube/polymer composites. J. Appl. Phys. **117**(13), 134903 (2015)
75. Miller, J.F., Downing, K.: Evolution in materio: looking beyond the silicon box. In: NASA/DoD Conference on Evolvable Hardware 2002, pp. 167–176. IEEE (2002)
76. Miller, J.F., Harding, S., Tufte, G.: Evolution-in-materio: evolving computation in materials. Evol. Intell. **7**(1), 49–67 (2014)
77. Misra, J., Saha, I.: Artificial neural networks in hardware: a survey of two decades of progress. Neurocomputing **74**(1), 239–255 (2010)
78. Mohid, M., Miller, J.F., Harding, S., Tufte, G., Lykkebo, O.R., Massey, M.K., Petty, M.C.: Evolution-in-materio: a frequency classifier using materials. In: International Conference on Evolvable Systems, ICES 2014, pp. 46–53. IEEE (2014)
79. Mohid, M., Miller, J.F., Harding, S., Tufte, G., Lykkebo, O.R., Massey, M.K., Petty, M.C.: Evolution-in-materio: solving bin packing problems using materials. In: International Conference on Evolvable Systems, ICES 2014, pp. 38–45. IEEE (2014)
80. Mohid, M., Miller, J.F., Harding, S., Tufte, G., Lykkebø, O.R., Massey, M.K., Petty, M.C.: Evolution-in-materio: solving machine learning classification problems using materials. In: PPSN XIII, Parallel Problem Solving from Nature, pp. 721–730. Springer, Heidelberg (2014)
81. Nikolić, D., Haeusler, S., Singer, W., Maass, W.: Temporal dynamics of information content carried by neurons in the primary visual cortex. In: Advances in Neural Information Processing Systems, pp. 1041–1048 (2006)
82. Norton, D., Ventura, D.: Improving liquid state machines through iterative refinement of the reservoir. Neurocomputing **73**(16), 2893–2904 (2010)
83. Ozgur, Y.: Reservoir computing using cellular automata. arXiv:1410.0162 [cs.NE] (2014)
84. Ozgur, Y.: Connectionist-symbolic machine intelligence using cellular automata based reservoir-hyperdimensional computing. arXiv:1503.00851 [cs.ET] (2015)
85. Ozturk, M.C., Xu, D., Príncipe, J.C.: Analysis and design of echo state networks. Neural Comput. **19**(1), 111–138 (2007)
86. Packard, N.H.: Adaptation toward the edge of chaos. In: Kelso, J.A.S., Mandell, A.J., Shlesinger, M.F. (eds.) Dynamic Patterns in Complex Systems, pp. 293–301. World Scientific, Singapore (1988)

87. Paquot, Y., Duport, F., Smerieri, A., Dambre, J., Schrauwen, B., Haelterman, M., Massar, S.: Optoelectronic reservoir computing. Sci. Rep. **2**, (2012)
88. Rodan, A., Tino, P.: Minimum complexity echo state network. IEEE Trans. Neural Netw. **22**(1), 131–144 (2011)
89. Rosenstein, M.T., Collins, J.J., De Luca, C.J.: A practical method for calculating largest Lyapunov exponents from small data sets. Phys. D: Nonlinear Phenom. **65**(1), 117–134 (1993)
90. Samuelsen, E., Glette, K.: Real-world reproduction of evolved robot morphologies: automated categorization and evaluation. In: Applications of Evolutionary Computation, vol. 9028. LNCS, pp. 771–782. Springer, Heidelberg (2015)
91. Schmidhuber, J., Wierstra, D., Gagliolo, M., Gomez, F.: Training recurrent networks by evolino. Neural Comput. **19**(3), 757–779 (2007)
92. Scholkopf, B., Smola, A.J.: Learning with Kernels: Support Vector Machines, Regularization, Optimization, and Beyond. MIT press (2001)
93. Schrauwen, B., Defour, J., Verstraeten, D., Van Campenhout, J.: The introduction of timescales in reservoir computing, applied to isolated digits recognition. In: Artificial Neural Networks–ICANN 2007, pp. 471–479. Springer, Heidelberg (2007)
94. Schrauwen, B., Büsing, L., Legenstein, R.A.: On computational power and the order-chaos phase transition in reservoir computing. In: Advances in Neural Information Processing Systems, pp. 1425–1432 (2008)
95. Schrauwen, B., D'Haene, M., Verstraeten, D., Van Campenhout, J.: Compact hardware liquid state machines on fpga for real-time speech recognition. Neural Netw. **21**(2), 511–523 (2008)
96. Schrauwen, B., Wardermann, M., Verstraeten, D., Steil, J.J., Stroobandt, D.: Improving reservoirs using intrinsic plasticity. Neurocomputing **71**(7), 1159–1171 (2008)
97. Schumacher, J., Toutounji, H., Pipa, G.: An analytical approach to single node delay-coupled reservoir computing. In: Artificial Neural Networks and Machine Learning–ICANN 2013, pp. 26–33. Springer, Heidelberg (2013)
98. Schürmann, F., Meier, K., Schemmel, J.: Edge of chaos computation in mixed-mode vlsi-a hard liquid. In: Advances in Neural Information Processing Systems, pp. 1201–1208 (2004)
99. Shah, J.: Ultrafast Spectroscopy of Semiconductors and Semiconductor Nanostructures, vol. 115. Springer Science & Business Media, Heidelberg (1999)
100. Shawe-Taylor, J., Cristianini, N.: Kernel Methods for Pattern Analysis. Cambridge University Press, Cambridge (2004)
101. Sheridan, P., Ma, W., Lu, W.: Pattern recognition with memristor networks. In: IEEE International Symposium on Circuits and Systems, ISCAS 2014, pp. 1078–1081. IEEE (2014)
102. Sillin, H.O.: Neuromorphic hardware: the investigation of atomic switch networks as complex physical systems. Ph.D. thesis, University of California, Los Angeles (2015)
103. Sillin, H.O., Aguilera, R., Shieh, H., Avizienis, A.V., Aono, M., Stieg, A.Z., Gimzewski, J.K.: A theoretical and experimental study of neuromorphic atomic switch networks for reservoir computing. Nanotechnology **24**(38), 384004 (2013)
104. Steane, A.: Quantum computing. Rep. Prog. Phys. **61**(2), 117 (1998)
105. Steil, J.J.: Backpropagation-decorrelation: online recurrent learning with o (n) complexity. In: 2004 IEEE International Joint Conference on Neural Networks, vol. 2, pp. 843–848. IEEE (2004)
106. Steil, J.J.: Online reservoir adaptation by intrinsic plasticity for backpropagation-decorrelation and echo state learning. Neural Netw. **20**(3), 353–364 (2007)
107. Steil, J.J.: Several ways to solve the MSO problem. In: ESANN, pp. 489–494 (2007)
108. Stepney, S.: The neglected pillar of material computation. Phys. D: Nonlinear Phenom. **237**(9), 1157–1164 (2008)
109. Stepney, S.: Nonclassical computation: a dynamical systems perspective. In: Rozenberg, G., Bäck, T., Kok, J.N. (eds) Handbook of Natural Computing, vol. 4, pp. 1979–2025. Springer, Heidelberg (2012)
110. Stepney, S., Braunstein, S.L., Clark, J.A., Tyrrell, A., Adamatzky, A., Smith, R.E., Addis, T., Johnson, C., Timmis, J., Welch, P.: Journeys in non-classical computation I: a grand challenge for computing research. Int. J. Parallel, Emergent Distrib. Syst. **20**(1), 5–19 (2005)

111. Stepney, S., Abramsky, S., Adamatzky, A., Johnson, C., Timmis, J.: Grand challenge 7: Journeys in non-classical computation. In: Visions of Computer Science, London, UK, September 2008, pp. 407–421. BCS (2008)
112. Stieg, A.Z., Avizienis, A.V., Sillin, H.O., Martin-Olmos, C., Aono, M., Gimzewski, J.K.: Emergent criticality in complex Turing B-type atomic switch networks. Adv. Mater. **24**(2), 286–293 (2012)
113. Stieg, A.Z., Avizienis, A.V., Sillin, H.O., Aguilera, R., Shieh, H., Martin-Olmos, C., Sandouk, E.J., Aono, M., Gimzewski, J.K.: Self-organization and emergence of dynamical structures in neuromorphic atomic switch networks. In: Memristor Networks, pp. 173–209. Springer, Heidelberg (2014)
114. Thompson, A.: An evolved circuit, intrinsic in silicon, entwined with physics. In: Evolvable Systems: From Biology to Hardware, pp. 390–405. Springer, Heidelberg (1997)
115. Triefenbach, F., Jalalvand, A., Schrauwen, B., Martens, J.: Phoneme recognition with large hierarchical reservoirs. In: Advances in Neural Information Processing Systems, pp. 2307–2315 (2010)
116. Triefenbach, F., Jalalvand, A., Demuynck, K., Martens, J.: Acoustic modeling with hierarchical reservoirs. IEEE Trans. Audio, Speech, Lang. Process. **21**(11), 2439–2450 (2013)
117. Vandoorne, K., Mechet, P., Van Vaerenbergh, T., Fiers, M., Morthier, G., Verstraeten, D., Schrauwen, B., Dambre, J., Bienstman, P.: Experimental demonstration of reservoir computing on a silicon photonics chip. Nature Commun. **5**, (2014)
118. Verstraeten, D., Schrauwen, B.: On the quantification of dynamics in reservoir computing. In: Artificial Neural Networks–ICANN 2009, pp. 985–994. Springer, Heidelberg (2009)
119. Verstraeten, D., Schrauwen, B., Stroobandt, D., Van Campenhout, J.: Isolated word recognition with the liquid state machine: a case study. Inf. Process. Lett. **95**(6), 521–528 (2005)
120. Verstraeten, D., Schrauwen, B., d'Haene, M., Stroobandt, D.: An experimental unification of reservoir computing methods. Neural Netw. **20**(3), 391–403 (2007)
121. Volpati, D., Massey, M.K., Johnson, D.W., Kotsialos, A., Qaiser, F., Pearson, C., Coleman, K.S., Tiburzi, G., Zeze, D.A., Petty, M.C.: Exploring the alignment of carbon nanotubes dispersed in a liquid crystal matrix using coplanar electrodes. J. Appl. Phys. **117**(12), 125303 (2015)
122. Wendin, G., Vuillaume, D., Calame, M., Yitzchaik, S., Gamrat, C., Cuniberti, G., Beiu, V.: Symone project: synaptic molecular networks for bio-inspired information processing. Int. J. Unconv. Comput. **8**(4), 325–332 (2012)
123. Whiting, J., de Lacy Costello, B., Adamatzky, A.: Slime mould logic gates based on frequency changes of electrical potential oscillation. Biosystems **124**, 21–25 (2014)
124. Xue, Y., Yang, L., Haykin, S.: Decoupled echo state networks with lateral inhibition. Neural Netw. **20**(3), 365–376 (2007)

Chapter 23
On Reservoir Computing: From Mathematical Foundations to Unconventional Applications

Zoran Konkoli

Abstract In a typical unconventional computation setup the goal is to exploit a given dynamical system, which cannot be easily adjusted or programmed, for information processing applications. While one has some intuition of how to use the system, it is often the case that it is not entirely clear how to achieve this in practice. Reservoir computing represents a set of approaches that could be useful in such situations. As a paradigm, reservoir computing harbours enormous technological potential which can be naturally released in the context of unconventional computation. In this chapter several key concepts of reservoir computing are reviewed, re-interpreted, and synthesized to aid in realizing the unconventional computation agenda, and to illustrate what unconventional computation might be. Some philosophical approaches are discussed too, e.g. the strongly related implementation problem. The focus is on understanding reservoir computing in the classical setup, where a single fixed dynamical system is used: To this end, mathematical foundations of reservoir computing are presented, in particular the Stone-Weierstrass approximation theorem, with a mixture of rigor, and intuitive explanations. To make the synthesis it was crucial to thoroughly analyze both the differences and similarities between Liquid State Machines and Echo State Networks, and find a common context insensitive base. The result of the synthesis is the suggested Reservoir Machine model. The model could be used to analyze how to build unconventional information processing devices and to understand their computing capacity.

23.1 Introduction

In a typical information processing setup a *reservoir computing device* implements a mapping $\mathcal{F}$ from the space of time series input data Ω to the space of time series output data Ω',

Z. Konkoli (✉)
Department of Microtechnology and Nanoscience - MC2,
Chalmers University of Technology, SE-412 Gothenburg, Sweden
e-mail: zorank@chalmers.se

© Springer International Publishing Switzerland 2017
A. Adamatzky (ed.), *Advances in Unconventional Computing*,
Emergence, Complexity and Computation 22,
DOI 10.1007/978-3-319-33924-5_23

$$\mathcal{F} : \Omega \to \Omega' \tag{23.1}$$

The computation is performed in the on-line mode: At each time instance the previously "seen" time series is "inspected" by the device and an output is assigned. In signal engineering such an object is referred to as a *filter*. As the time instance moves, an input data series is converted into the output data series.

The term "reservoir computing" describes a related set of ideas for performing computation with non-linear dynamical systems as filters. Surprisingly, despite the fact that the idea originates from a solid mathematical background, the theory of reservoir computing is still phenomenological in many ways. There are essentially two perspectives that emphasize the use of either a class of systems or a single system. These perspectives are strongly related, and in the literature are often treated as one concept. This can cause a great deal of confusion since implicit context dependent assumptions are frequently made. For example, when envisioning applications, one often assumes the one-system perspective and, yet, when discussing the expressive power of the same system the other perspective is assumed (without rigorous justification). This is one of the reasons why the theory of reservoir computing is still phenomenological.

In the following, the distinction between the two perspectives will be kept explicit. The one-system setup will be referred as the classical setup of reservoir computing, since this perspective is often (implicitly) assumed in the literature when applications are discussed:

> A reservoir computer consists of a dynamical system, a reservoir, and an interface that extracts the information stored in the internal states of the system, a readout layer. If the configuration space of the system is complex enough, various inputs to the system can drive the system to different regions of the configuration space, which represents computation. The result of the computation is extracted by the readout layer.

As a paradigm, reservoir computing addresses the natural question whether and under which conditions an arbitrary system can be used for advanced information processing applications. The outcome depends on which choices are available, e.g. whether it is possible to choose among many systems or, if only one system is available and how adjustable it actually is. The key foundation of the reservoir computing field is an insight that this indeed can be done, regardless of the perspective assumed. If the following conditions are met the reservoir computer could be used to compute in principle anything[1]:

---

[1] While such machines are expected to be able to implement a broad class of information processing tasks, they are not Turing universal. There is no obvious way of establishing that such machines have infinite expressive power, at least not in the strict sense of the word. For example, one would have to provide a construct to realize a universal Turing machine using reservoir computing. The abstract concepts such as "the tape" or "the reading head" are not easily realized in this context.

> The properties that guarantee that the system can be used for reservoir computing are usually referred to as (i) *the separability*, (ii) *the echo state*, and (iii) *the fading memory* property. Such system are referred to as *good reservoirs*. Further, the readout layer should possess the (iv) *the approximation* property.

The goal will be to re-interpret the above requirements by adopting a practical point of view assuming that the aim is to actually construct a reservoir device from a given dynamical system. *This classical scenario is ubiquitous in unconventional computation.*

The Promise of Reservoir Computing

As a paradigm, reservoir computing might have an enormous technological potential. Property (iv), the approximation property, might be hard to realize for readout layer implementations that are not done in-silico. However, properties (i-iii) appear generic. Many systems might exhibit such behaviors.[2] Should this be really true, the idea of reservoir computing could have enormous practical implications for a range of information processing technologies.

> It is possible that there are many systems that can be used for advanced information processing in the reservoir computing context, but are simply overlooked since they have never been studied that way.

However, it is not yet clear whether the field of reservoir computing can live up to its technological promise. There are two pertaining issues. First, to verify rigorously whether the system of interest (a reservoir) has properties (i–iii) is a highly non-trivial task. This is a serious obstacle towards understanding which systems might be used for reservoir computing. Second, the theory of reservoir computing has not been developed enough.

> The requirements (i–iii) cannot be expressed in clear engineering terms since they are a direct translation of the related mathematical formulations which are simply impractical for engineering applications. This is another reason why the theory of reservoir computing appears phenomenological. It is necessary to understand and re-interpret the existing reservoir computing mathematical background to bridge towards the engineering side.

A brute-force strategy for finding suitable reservoir systems is to simply start checking for every conceivable system whether the properties exist. Eventually, by

---

[2]For example, systems with the dynamics that is (a) chaotic-like or input sensitive (to ensure the separation property), or (b) with some sort of friction or energy dissipation (to ensure the echo state or equivalently the fading memory property).

studying many such systems one should be able to generalize and gain an understanding of which systems might have such properties. The main problem with the strategy is that it is in general very hard to judge whether an arbitrary system has indeed the required properties. There are several reasons for that. From the strict mathematical point it is possible to rigorously define when the system possess the separation property. However, despite the presence of the mathematical rigor, the validity check is not straight forward since there is simply no universal procedure for performing the check in practice. Further, when engineering reservoir computing systems, one can ask: how much should the system separate inputs? Is there a tolerance range (a resolution) which is acceptable? In general, the issues related to accuracy, tolerance to damage or noise, have not been extensively addressed in the reservoir computing literature.

The goal of this chapter is to provide the necessary overview of key mathematical concepts, and take them as a starting point to develop a suitable theory and the related strategies for building reservoir computing systems, which could provide some generic guidance for the related engineering efforts. Several practical principles will be discussed throughout the text. The text is organized as follows.

- A brief history of reservoir computing is presented in Sect. 23.2, together with some key mathematical concepts. The section contains a discussion about the two most common up to date implementations of the idea: Liquid State Machines and Echo State Networks.
- Section 23.3 contains the definition of the Reservoir Machine concept. The definition is a mathematical formalisation of what reservoir computing in the classical setup is. The related mathematical concepts serve as the foundation for the analysis in the sections that follow.
- The technological promise of reservoir computing is discussed in two subsequent Sects. 23.4 and 23.5, each inspired by two different perspectives:

  - Section 23.4 contains a discussion on the relationship between reservoir computing and philosophy of computation, and in particular the implementation problem.
  - Section 23.5 analyzes the mathematical foundation of reservoir computing, the Stone-Weierstrass approximation theorem. This section contains a formal mathematical background that is necessary for understanding reservoir computing on one hand, and understanding how to build reservoir computers on the other. The section also contains some examples of how the theorem can be used.

- Section 23.6 contains a discussion of how the technological potential of reservoir computing could be realized in practice. It contains a set of practical guidelines that, if adhered to, could enable us to build powerful reservoir computers.
- The concluding Sect. 23.7 contains a brief discussion of several key theoretical concepts that one should learn to command (understanding, implementation, and exploitation). If we are able to do that, the "prophecy" of reservoir computing might be fulfilled.

23.2   A Brief History of Reservoir Computing

Reservoir computing was independently suggested in [1, 2] and [3–5]: These studies introduced the concepts of Liquid State Machines (LSMs) and Echo State Networks (ESNs) respectively. Both focused on explaining some specific features of neural network dynamics. The LSM and ESN concepts were further developed and exploited thereafter, e.g. as discussed in the reviews [6, 7].

After being initially suggested, the concept has been exploited in a range of applications. The web-site [8] dedicated to reservoir computing contains an excessive list of references. Since this book chapter is not a review paper the reader is directed to these sources for additional information. To illustrate the flexibility of the concept, a few more recent examples that exploit different reservoirs can be mentioned: reservoirs made of memristor networks have been considered in for pattern recognition [9, 10] or harmonics generation [11], photonic systems as reservoirs have been discussed in [12–14], etc. For a more detailed discussion on various reservoirs that have been considered please see [15] and references therein. The field seems to be exploding and reviewing all possibilities that have been investigated in the literature is out of the scope of this book chapter.

Surprisingly, since the original publications, three are very few studies that address the fundamental (conceptual) side of the problem. For example, one might wonder, what are the limits of reservoir computing? or what is the computing capacity of such devices? and ultimately, how to increase it? A few examples of such studies can be found in [12, 16–20] but there seems to be a gap in the literature, as this fundamental side of the problem is not attracting extensive attention. In particular, there seem to be a lack of interest in the mathematical foundations of reservoir computing, which is strange given that these play a prominent role in formulating the concept. One of the goals of this book chapter is to remedy this situation by re-visiting the mathematical foundations of reservoir computing.

For historical reasons, in the literature, the reservoir computing concept is assumed to be synonymous with Liquid State Machines and Echo State Networks. Despite being related, these two models emphasize slightly different perspectives. The LSM model is formulated for a class of systems, while the ESN model emphasizes the classical (one system) perspective. Both approaches assume a strict separation between the dynamical system (the reservoir) and the readout layer (the interface), where the reservoir carries the full burden of computation.

Liquid State Machine(s)

Liquid State Machine is a model of computation with, in principle, the expressive power of the Turing machine. The expressive power follows directly from the use of the Stone-Weierstrass approximation theorem and the assumption that there is whole class of devices (machines) to choose from that implement the assumptions of the theorem. The Stone-Weierstrass approximation theorem strongly emphasizes the concept of the algebra of functions. In some sense, the LSM model is the most direct implementation of the Stone-Weierstrass approximation theorem in the context of time series information processing. This is exactly the reason why one assumes the existence of a class of machines (the base filters) that realize an algebra of continuous filters.

Echo State Network(s)

The echo state network approach strongly adopts the classical perspective of reservoir computing with a focus on a chosen dynamical system. The concept was suggested to be able to explain the observation that random networks can be made to perform computation by using a somewhat "lighter" training procedure. This insight is a result of numerous numerical investigations of recurrent neural networks. The background to the echo state network idea is the realization that it is not necessary to train the full network, but only a smaller part consisting of the outer (interface) layer of neurons. To train the interface part it is sufficient to use very simple methods, e.g. the linear regression or similar.

Are These Two Approaches Similar?

In the literature no formal distinction is made between the two approaches, which might seem rather confusing: Strictly speaking the two perspectives are not identical. However, there is a deep connection between the two perspectives due to the fact that a single complex dynamical system can harbor many smaller subsystems. If these small subsystems are coupled, their dynamics will be also complex. In this way a single complex system can implicitly represent a class of smaller "mini/micro-reservoirs". A typical example is an artificial neural network with many neurons or groups of neurons.

23.3   One-System Reservoir Computing: Reservoir Machine

The classical reservoir computing setup, where the goal is exploit a *single* dynamical system for computation, can be mathematically formalized as follows. The dynamical system is normally referred to as a reservoir $\mathcal{R}$. The reservoir needs to be equipped with a readout layer ψ, which is used to probe the states of the reservoir. The reservoir and the readout layer define a reservoir device.

To be able to describe information processing features in exact mathematical terms the phrase *reservoir (computing) machine* will be used to indicate that a reservoir has been assigned a readout layer (possibly optimized for a specific information processing task). Mathematically, a reservoir machine $\mathcal{M}$ is an ordered pair that consists of the reservoir and a readout layer

$$\mathcal{M} = (\mathcal{R}, \psi) \tag{23.2}$$

In this construction the only variable (adjustable) part of the machine is the readout layer ψ. From the engineering perspective a key challenge is to actually construct a suitable readout layer ψ and train (adjust) it. If the readout layer is simple in some sense, it should be easily trainable.[3]

23.3.1 Reservoir $\mathcal{R}$

A reservoir $\mathcal{R}$ is a dynamical system that evolves in time and can respond to external inputs. The input is assumed to exert a direct influence on the internal (microscopic, mesoscopic, or macroscopic) degrees of freedom of the reservoir. For theoretical modeling, the time variable t can be both discrete, $t \in \mathbb{Z}$, or continuous, $t \in \mathbb{R}$, depending on which type of dynamics seems more appropriate.

The input consists of time data series. Formally, a data series u is a mapping

$$u : \mathbb{R} \to \mathbb{U} \text{ or } u : \mathbb{Z} \to \mathbb{U}$$

that assigns to each time instance t a value taken from a set $\mathbb{U}$. Typically, the set $\mathbb{U}$ is taken to be a bounded subset of $\mathbb{R}^n$, e.g. a point in the set is given by $u \equiv (u_1, u_2, \ldots, u_n)$ where each $u_i \in [u_{\min}, u_{\max}]$ is taken from an interval on the real line; $u_{\min}, u_{\max} \in \mathbb{R}$. The time variable can be used to "index" all values in the series by evaluating $u(t)$. Further, let

$$x \in \mathcal{S}$$

denote an internal degree of freedom (a state) of the reservoir, where $\mathcal{S}$ denotes the space of all possible states of the system. The variable x could be treated as an observable: It does not necessarily have to represent a microscopic degree of freedom, like a position of a molecule. It could be also a variable like the temperature, or the concentration profile.

The dynamics of the system is governed by an evolution operator that describes how the internal state of the system $x(t)$ changes in time under the influence of the external input $u(t)$. A typical continuous model is stated as

$$\frac{dx(t)}{dt} = H(x(t), u(t)) ; \quad t \in \mathbb{R} \tag{23.3}$$

and a typical discrete model is given by

$$x_t = H(x_{t-1}, u_t) ; \quad t \in \mathbb{Z} \tag{23.4}$$

---

[3]Naturally, this cannot be taken for granted as the computational simplicity does not necessarily imply that the system is easily adjusted in engineering terms.

where $H(x, u)$ denotes a multi-variable function that governs the dynamics of the system. To make the notation simpler the same symbol H has been used in both equations.

Note the difference in notation for the continuous and the discrete time dynamics, e.g. $u(t)$ versus u_t, and $x(t)$ versus x_t. In the following the symbols $q(t)$ with $q \in \{x, u\}$ will be used when discussing both continuous and discrete systems. The symbols q_t with $q \in \{x, u\}$ will be used exclusively for discrete systems.

The machine performs computation by changing its internal state under the influence of the input. The internal state of the system at each time instance depends on the whole history of previous inputs. In signal engineering, such a system is referred to as a *filter*. It maps input time series $u(t)$ into another times series $x(t)$. Thus the reservoir can be seen as an implementation of the filter

$$u \rightarrow x = \mathcal{R}(u) \tag{23.5}$$

where the individual sequence values can be inspected as $x(t) = \mathcal{R}(u)(t)$.[4] The same symbol is used to denote the physical reservoir and the mapping (the filter) it realizes.

23.3.2 Readout-Layer ψ

Every reservoir is expected to be equipped with a readout layer ψ. The readout layer should not process information in any substantial way. It should be only used to assess the information stored in the dynamical system.

What is the complexity of ψ that we can allow and still be able to claim that ψ is not doing any substantial computation? This is of course a complicated mathematical and philosophical problem. Please see [21] for a possible answer to this question. In the reservoir computing context it is naturally resolved by claiming that the computation should be done instantaneously. The readout function should not be a filter in the strict engineering sense. The function per se should not accumulate any memory. It should only "see" history through x.[5] Mathematically, these principles are expressed as follows. The readout layer is used to extract the output of computation $o(t)$, i.e. by "inspecting" the value of $x(t)$ at each time instance t:

$$o(t) = \psi(x(t)) \tag{23.6}$$

The symbol $\psi(x)$ denotes a multi-variable function that provides a mathematical description of the readout layer.

---

[4]Note that it would be wrong to use $\mathcal{R}(u(t))$.

[5]For complex systems, $x(t)$ is expected to contain a very long list of values and, in principle, a lot of history could be stored in x.

23.3.3 Reservoir Machine as a Model of Computation

The Filter Implemented by the Machine $\mathcal{M}$

Taken together, each machine $\mathcal{M}$ realizes the related function (filter) $\mathcal{F}$ that maps an input sequence u into the output sequence o:

$$u \rightarrow o = \mathcal{F}(u) \tag{23.7}$$

The values of these sequences can be inspected at each time instance t as $o(t) = \mathcal{F}(u)(t) \equiv \psi(\mathcal{R}(u)(t))$.[6] In what follows no formal distinction will be made between the machine and the filter it implements, and $\mathcal{M}$ will be used in place of $\mathcal{F}$ except in situations when a confusion might arise.

Configurability of the Readout Layer Generates a Class of Machines

It will be useful to considers a collection of all machines that are obtained from a fixed but otherwise arbitrary reservoir $\mathcal{R}$ and all possible readout layers:

$$\mathfrak{M} = \{\mathcal{M} \equiv (\mathcal{R}, \psi) : \psi \in \Psi\} \tag{23.8}$$

The collection of machines $\mathfrak{M}$ is generated by the reservoir and will be referred to as a *programmable reservoir machine*. Once a particular readout layer has been chosen (engineered) it will be referred to as a *programmed reservoir machine*, or just a *reservoir machine*. Thus Reservoir Machine can be viewed as a model of computation. It is clearly not a universal model of computation. Its computing power comes mostly from the reservoir, which is "frozen": In the LSM construct it is always allowed to choose among different reservoirs. In the ESN model, this is sometimes allowed, and sometimes forbidden. In the Reservoir Machine model it is *always* forbidden.

23.4 The Technological Potential of Reservoir Computing: Lessons from Philosophy of Computation

Hilary Putnam suggested a construction of how to use *any* object to implement *any* finite state automaton [22]. The statement is a paradox: even a rock should be able to compute anything. The recent advances in understanding the relatively novel approach to information processing, the reservoir computing paradigm, seem to indicate that this seemingly absurd idea is not without merits.

---

[6]Note that here we do not use $\psi(\mathcal{R}(u))(t)$. That would be incorrect, given the assumption that the readout layer ψ should not accumulate any information from the past. The notation $\Psi(x)$ implies that Ψ acts as a filter, which is clearly not the case according to the definition.

23.4.1 Putnam's Construction

Understanding which dynamic properties guarantee information processing ability is a highly complex problem. As an example, to appreciate the difficulties associate with the problem, consider the following example. Can a rock compute? An engineer's answer would very likely be "no", which is also the common intuitive expectation. Interestingly, a philosopher's answer is "yes". This begs a question: Which one is correct? Probably both.

One of the key issues that the philosophy of mind tries to address is whether the human brain implements an automaton. If yes, then different states of mind are just different states of the automaton, e.g. as summarized in [23]. As a response to this thesis Hilary Putnam [22] provided a construction through which it can be shown that even a simple object as a rock can be made to implement any finite state automaton. This rather absurd conclusion was the very reason why Putnam considered the construction. He used it to show that the notion of computation needs to be restrained when discussing philosophy of mind. Clearly, just the ability to compute cannot be used to define what a mind is. In this context, the notion of computation is simply too broad, as any object seems to be able of performing it. Nevertheless, the statement is a paradox that illustrates an important principle, and a way of thinking about computation, and especially unconventional computation.

23.4.2 A Thought Experiment

Putnam's construction and the reservoir computing idea are strongly related. What would happen if Putnam's idea were taken out of context and applied in the context of reservoir computing? Assume that the goal is to use the rock as a reservoir. Note that Putnam's rock is taken grossly out of context, but such a possibility can be envisioned, and servers the purpose of illustrating an important point.

The Rock can Compute but an Auxiliary Interface must be Used

Putnam's construction has been criticized from many angles (cf. [24–26] and references therein). The most common argument against the construction states that to implement a complex automaton on a rock, large auxiliary equipment would be needed. At the end, the actual computing would be done by the auxiliary equipment and not by the rock. This is certainly a valid argument and extremely useful for understanding the reservoir computing idea.

Rock can be used for computation in principle, but this is hard to achieve in practice. This particular line of criticism against Putnam's construction indicates that when discussing computing ability of a dynamical system, it is very important to distinguish between the system per se and the interface that the system might be equipped with to achieve information processing.

The Interface is Equally Important as the Reservoir

This further implies that it is very important to understand how to interface the system properly, and how to mathematically describe the interface and gauge its computing power versus the computing power of the system. This line of reasoning has been explored in [21] which contains an example of how these ideas can be formalized. Interpreted in this vein, the thesis of reservoir computing states implicitly that the interface cannot be too complex. This is exactly the reason why either a simpler linear readout algorithm or an easily trained perceptron networks is used as the interface in the context of reservoir computing. Clearly, by using an auxiliary equipment the rock could be turned into the system that has the desired properties, but this equipment, very likely, will never be classified as a "simple readout layer".

How to Judge Whether a Given System has the Key Properties?

The properties seem so generic that it seems reasonable to assume that many physical systems can perform computation, even the systems which were not specifically designed by humans for that purpose. In that sense the original construction by Putnam has some practical relevance. This connection between Putnam's paradox and the idea of reservoir computing has been already pointed out in [27]. The rock is certainly not one of these systems. From the reservoir computing perspective, a rock cannot compute, as it lacks these key properties.[7]

23.5 The Technological Potential of Reservoir Computing: Lessons from Mathematics

The problem of a suitable approximation is probably as old as mathematics itself. In the particular case that deals with the problem of approximating an arbitrary function with a fixed class of functions, the approximation problem has a solution with a surprisingly elegant formulation. The theorem is one of the foundations of reservoir computing, and the structure of the theorem will be discussed in the following. An attempt will be made to present a deeply mathematical subject by emphasizing intuitive reasoning.

The theorem is normally referred to as the Stone-Weierstrass approximation (SWA) theorem (SWAT). Behind the simple formulation of the theorem hides an enormous application potential. The theorem has been used frequently in both mathematics and engineering to analyse approximation properties (the computing capacity) for plethora of systems. It is somewhat surprising that is has not been used excessively in unconventional computation too.

---

[7]Note that this statement might be an oversimplification. For example, imagine that there is simple readout layer that can access the internal states of the rock. Then the rock might still have the desired properties.

First, the theorem will be stated in its most general abstract form. The initial abstract formulation is useful when proving the theorem and, most importantly, it is useful for analysing unconventional computation applications, as will be illustrated in the following sections. Perhaps one of the reasons why the theorem has not found its way into unconventional computation is that it is very likely incomprehensible to somebody not versed in advanced analysis. Accordingly, the abstract formulation will be augmented by discussing the key components of the theorem to provide more intuitive formulations and facilitate its use in the unconventional computation context. Second, a version of the theorem will be provided as used by engineers. Third, while being an extremely useful tool, the theorem has its limitations, which will be discussed too.

A note on notation: In this section various metric spaces are discussed. Naturally, the most important "metric" space of interest is the space of input sequences Ω.[8] The elements of this space should be labeled by the symbol u. However, in this section a temporary change of notation is made, to make the connection with the existing mathematical literature more explicit: It is a custom to label elements of a metric space by using the symbol x.[9]

23.5.1 A Rigorous Mathematical Formulation of the Theorem

Theorem 1 (Stone-Weierstrass—the first version) *Let A be an algebra of continuous functions that map from a compact metric space Ω to the set of real numbers $\mathbb{R}$. Let elements of A separate points, and let there be a constant function in the algebra. Let $C(\Omega, \mathbb{R})$ denote the set of all continuous functions that map from Ω to $\mathbb{R}$. Then, A is dense in $C(\Omega, \mathbb{R})$.*

Some of the concepts used in the above formulation of the theorem are explained below. The explanations that follow are not provided solely to educate the reader regarding the theorem per se, but they are essential for understanding the reservoir computing idea, and for practical applications of the theorem. In particular, from the discussions that follow it should be clear which mathematical properties a physical system must possess that would render it useful for computation. Later on, these abstract mathematical properties will be discussed from a practical point of view (e.g. when building such devices). Several examples will be provided that illustrate how these mathematical concepts might manifest themselves in applications.

Definition 1 When does the set of functions form an algebra? Formally, this is stated as follows. Let A and Ω be as in the theorem above. The set of functions

---

[8]Here the quotation marks are used since Ω is not a metric space for any norm. To turn it into a metric space a special norm has to be used, as discussed later.

[9]The reader can substitute u in place of x whenever in doubt regarding how mathematical statements discussed in this section relate to the discussion on reservoir computing contained in other sections.

$\mathcal{A} = \{a | a : \Omega \to \mathbb{R}\}$ forms an algebra when for any pair of elements $a, b \in \mathcal{A}$, the combinations $a + b$, ab, and wa with $w \in \mathbb{R}$ are also in $\mathcal{A}$, where $(a + b)(x) \equiv a(x) + b(x)$, $(wa)(x) \equiv wa(x)$, and $(ab)(x) \equiv a(x)b(x)$ for every $x \in \Omega$.

An algebra of functions is a slight extension of the concept of the vector space of functions. In addition to the usual vector operations (the addition and multiplication by a constant), to form an algebra, the set of functions needs to be closed with regard to the pairwise multiplication operation as well.[10] Further, it is somewhat surprising that the presence of a constant function is explicitly required. An example will be provided later on to illustrate why this property is important.

> This closure property is an important property of an algebra that makes it useful for analysing expressive power of the algebra (e.g. it approximation power, in the mathematical sense of the word), and for analysing unconventional computing devices (e.g. their computing capacity).

Definition 2 What does it mean for an algebra to separate points? Let $\mathcal{A}$ and Ω be as in the theorem above. Algebra of functions $\mathcal{A}$ separates points if for every two elements $x_1, x_2 \in \Omega$ that are different, $x_1 \neq x_2$, an element in the algebra g can be found such that $g(x_1) \neq g(x_2)$.

This is another important property of the algebra that is necessary for approximation purposes. Note the particular form of the requirement. For example, it is not required that two distinct elements from the algebra are found such that $g_1(x_1) \neq g_2(x_2)$, or such that $g_1(x_1) \neq g_2(x_1)$ and $g_1(x_2) \neq g_2(x_2)$. The condition used in the definition is a rather mild in the sense that it is not too restrictive on the algebra.

> The separation property should be checked for a given pair of points, one at a time. Once the pair has been fixed, the user of the theorem needs only to find an element in the algebra that will satisfy the condition. In principle, if different pairs of points require different elements is of no concern. However, there must be enough elements in the algebra: The real difficulty is in checking that this can be done for every pair of points. Later on, it will be shown that this transforms into an important engineering requirement.

Definition 3 What does it mean that $\mathcal{A}$ is dense in $C(\Omega, \mathbb{R})$? By definition, it means that $\bar{\mathcal{A}} = C(\Omega, \mathbb{R})$, where the symbol $\bar{\mathcal{A}}$ denotes the closure of the algebra $\mathcal{A}$. The closure of $\mathcal{A}$ is defined as the smallest closed set that contains $\mathcal{A}$.

For someone without a background in point-set topology this statement might not make much sense. Thus alternative more intuitive definition will be provided.

---

[10]Note that we do not require that the set is closed with respect to the composition operation $a \circ b$ with $(a \circ b)(x) = a(b(x))$. In here, ab does not refer to $a \circ b$.

Definition 4 Intuitively, one should think of $\bar{A}$ as the set of all functions that can be approximated by elements in A. Now it becomes clear that the statement $\bar{A} = C(\Omega, \mathbb{R})$ implies that under the assumed conditions the algebra can approximate any function.

Thus there are two definitions of the closure, as the smallest closed set containing A and as a set of functions that can be approximated by elements in A. To see that they are equivalent requires a bit of work.

First, it is necessary to assume that A forms a metric space. For any two elements a and b in A one must be able to compute the distance between these two elements $\rho(a, b)$. Any algebra of *continuous* functions supports the natural sup norm metric,

$$\rho(a, b) \equiv \sup_{x \in \Omega} |a(x) - b(x)| \qquad (23.9)$$

This is a very sensitive measure of similarity, as the difference is checked at every point. With the ability to measure the distance between any two elements one can define what a convergence is by using the standard arguments: A sequence $a_1, a_2, a_3, \ldots, a_n, \cdots$ of elements in A converges to a_* if for every $\varepsilon > 0$, an index $n(\varepsilon)$ exists such that $n > n(\varepsilon)$ guarantees that $\rho(a_*, a_n) < \varepsilon$.

Second, if A is a metric space, its closure $\bar{A}$ can be defined as the set of all possible limits that can be obtained by considering all converging sequences made by using elements from A,

$$\bar{A} \equiv \{a_* | \lim_{n \to \infty} a_n = a_*, a_n \in A\} \qquad (23.10)$$

This is a rigorous theorem in mathematics. It is not obvious, and it requires some thinking to show that this is true (regardless of the fact that the proof consists of few lines [28]). The statement above implies that if a function f is in $\bar{A}$, then for every $\varepsilon > 0$ there is an element a in A such that

$$|a(x) - f(x)| < \varepsilon \qquad (23.11)$$

The approximation is uniform, it holds for every $x \in \Omega$. Equation (23.11) follows from the definition of convergence and (23.10). For any $f \in \bar{A}$ a convergent sequence $\{a_n\}_{n=0,\infty}$ can be constructed. Take the first element in the sequence for which $\rho(a_n, f) < \varepsilon$. This element defines the a in the equation above.

The above formulation of the SWA theorem does not feature the interface (the readout layer). However, the interface that is used to read internal states is an important part of the device. Alternatively, to emphasize the presence of the interface the following formulation of the theorem might be more useful.

Theorem 2 (Stone-Weierstrass—the second version) *Let B be an algebra of continuous functions that map from a compact metric space Ω to the set of real numbers $\mathbb{R}$. Let elements of B separate points. Let P denote the algebra of multivariate polynomials with d variables. Let $C(\Omega, \mathbb{R})$ denote the set of all continuous functions that map from Ω to $\mathbb{R}$. Then, for every accuracy requirement $\varepsilon > 0$, and every continuous* *function* $f \in C(\Omega, \mathbb{R})$ *elements* $b_1, b_2, \ldots, b_d \in \mathcal{B}$ *and the polynomial* $p \in \mathcal{P}$ *can be found such that* $|f(x) - p(b_1, b_2, \ldots, b_d)(x)| < \varepsilon$ *for every* $x \in \Omega$.[11]

This formulation is the one that is often used in the context of reservoir computing, since in some situations this formulation is more practical. Note that in the above statement of the theorem there are less restrictions on the algebra since only separation property is required, no constants are needed. The constants are, of course, included in the algebra of polynomial functions.

Further, the two versions of the theorem point to the issue that was discussed previously. How to balance the computing power of the reservoir, versus the computing power of the interface? For example, one might focus on linear interfaces instead, then the question is whether a linear combination of d elements in $\mathcal{A}$ can be found in place of a in Eq. (23.11), e.g. $a(x) = \sum_{i=1}^{d} w_i a_i(x)$. As ε is made smaller the number of elements d that have to be taken in the weighted sum are expected to increase.

> In brief, the Stone-Weierstrass approximation theorem states that if an algebra of continuous functions that maps from a compact metric space Ω to real numbers separates points and contains constant element $\mathbf{1}$, then it can approximate any continuous function on Ω uniformly.

Proof The proof of the SWA theorem is a constructive proof, and can be found in many textbooks, e.g. cf. [29]. It usually extends over several pages of text (if all concepts are defined from scratch). The proof will not be presented in here. However, a few interesting steps, from the reservoir computing perspective, will be commented upon. These will be returned to later in the text, but from a more engineering-like perspective.

The proof of the theorem outlines a procedure (an algorithm essentially) of how to approximate an arbitrary continuous function using the elements in the algebra. As one goes through the procedure it is important to check that every step in the algorithm is valid, which is ensured by the assumed conditions in the theorem. More specifically, the recipe works by showing that any $f \in C(\Omega, \mathbb{R})$ can be approximated at two arbitrary points, and exploiting this fact to show than a uniform approximation can be found everywhere. For this procedure to work, the following requirements must be met:

- The two point approximation procedure discussed above: This is the reason why the presence of the constant function in the algebra is important. It is needed to show that any continuous function can be approximated at any two points by the elements from the algebra by exploiting a linear-like interpolation and requiring that $w_1 \mathbf{1}(x) + w_2 g(x) = f(x)$ at $x = x_1, x_2$ where $\mathbf{1}$ denotes the constant function.
- All functions of the algebra must map from a *compact domain*. This is used to claim that various parts of the domain can be covered by a finite number of open

---

[11] Here naturally, the expression $p(b_1, b_2, \ldots, b_d)(x)$ is interpreted as $p(b_1(x), b_2(x), \ldots, b_d(x))$.

sets. This step is necessary to be able to cover all Ω space by patches where the pairwise approximations work well.

- The algebra property, in particular the closure with regard to multiplications, is used to show that $\bar{A}$ forms a lattice (in the mathematical sense of the word), i.e. that min and max operations are possible on functions. These are used to switch to different functions when the patches are changed to stay ε-close to the target function.   □

23.5.2 A Few Application Examples of the Theorem

To illustrate the power of the theorem several applications of the first version of the theorem will be illustrated. Each example was chosen to illustrate a particular aspect of the theorem.

The use of the theorem can appear at odds with its formulation, in the sense that it is hard to build an intuition of what is going on behind the scenes when the conditions of the theorem are being checked. The examples below illustrate the fact that the procedure of checking for the separation property lies completely outside of the theorem, and is a challenging problem on its own.

Example: All Polynomials on a Finite Interval $\Omega = [0, 1]$

Consider all continuous functions on a finite interval $\Omega = [0, 1]$, and an algebra of polynomials on that interval. The goal is to see whether it is possible to approximate any continuous function on the interval by using polynomials. It is a well-known result that this is possible (the Weierstrass approximation theorem). It is straight forward to show this using the SWA theorem. The set of all polynomials clearly forms an algebra. The algebra contains a constant element (in fact infinitely many such elements). The algebra also separates points, since any linear polynomial, e.g. $p(x) = x$, works for every pair of points.

Example: The First Ten Legendre Polynomials (Defined on $\Omega = [-1, 1]$

This set does not form an algebra. For example, the product of the tenth polynomial with itself is not in the algebra. This algebra cannot approximate all functions since it is not closed with respect to the multiplication operation.

Example: Polynomials without Constant Term on a Finite Interval $\Omega = [0, 1]$

Note that every such polynomial vanishes at $x = 0$. It can be shown that such polynomials can approximate all continuous functions on Ω that vanish at the origin.

However, such polynomials cannot approximate all continuous functions on Ω. This example illustrates why the presence of the constant function 1 in the algebra is important.

Example: An Algebra of Predicate Based Functions on the Interval $\Omega = [0, 1]$

Assume that all functions in the algebra are defined by using logical predicates. Every predicate is a Boolean formula that features x and some constants. For example, a predicate function could be the Boolean expressions $\pi_1(x) = x > 1$, $\pi_2(x) = \sin^2(x) + \cos(x) == 0$, or $\pi_3 = \pi_1(x) \wedge \pi_2(x)$. Let the elements in the algebra a be defined by considering all possible predicate functions. Each predicate function defines the algebra element as $a(x) = 0$ (1) when $\pi(x) = F$ (T) where F and T stand for false and true logical values. Can this algebra be used to approximate all continuous functions on the interval? Unfortunately, the theorem cannot be used since this algebra of functions is not continuous. Clearly, in the present form, the theorem has its limitations.

Example: Polynomials on $\Omega = (-\infty, t]$ with $t \in \mathbb{R}$

In this case the domain is not compact. The theorem cannot be used. It does not matter which algebra of functions is considered. We shall return later on to this example when the fading memory property will be discussed.

23.6   Realizing the Technological Potential: From Mathematical Concepts to Practical (Engineering) Guidelines

How to build reservoir computers with powerful information processing abilities? The elegance of the LSM formalism is an illustration of how the SWA theorem can be used to understand the expressive power of a class of machines. How can one do the same in the classical reservoir computing setup? when only one system can be used to build a reservoir computer.

Motivated by the LSM and ESN setups it is tempting to consider two options: (i) build a class of dynamical systems that can be combined, or (ii) use a sufficiently complex single dynamical system.[12] Both approaches have its merits. The success of the CMOS technology is strong evidence in favor of the first approach. The key technological feature of the CMOS paradigm is that one can exercise full control of the construction process, down to the tiniest component that contributes to the overall information processing ability of the system. However, this control cannot be often exercised in the context of unconventional (natural) computation, where one can only control some selected parts of the system but not all its components.

---

[12]Used in this way, the LSM and ESN concepts are taken slightly out of context. They were introduced as models of computation, a tools to study specific features of neural network dynamics.

For example, in molecular computing it is very challenging to control individual molecules. For unconventional computing applications the second option seems more relevant. Accordingly, the following text emphasizes the classical (one-) reservoir computing paradigm.

This section discusses how to bridge from the abstract mathematical context of the SWA theorem towards a more engineering like setup when the goal is to build an actual device, a reservoir machine. This is done by carefully analyzing how the conditions of the SWA theorem can be met for an arbitrary reservoir machine, and how these conditions can be engineered in practice. This section is not meant to be a historical overview of the reservoir computing method (though a part of the text follows the historical development of the field), but rather aims to provide a synthesis of it from a practical point of view.

> If one were to interpret the Stone-Weierstrass theorem in a broader, more engineering like context, it would appear that the following reasoning and the resulting hypothesis seem feasible: If a physical system can realize an algebra of functions that separate points, then it should be possible to use the system to compute in principle anything. It is possible that both requirements are naturally realized by physical systems at microscopic level, and for complex systems at even higher levels (meso-, macro-scales). The hypothesis is that the technological potential of reservoir computing can be indeed "released": Provided there are readout layers that can resolve such microstates, there are no a priori reasons why powerful reservoir computing devices could not be realized.

The possibility that the hypothesis is actually true is too important to be ignored. It might change the way we think about information processing and have profound impact on information processing engineering. It is important to understand whether this agenda can be realized in practice, and if not, where the limitations are. This section is an interpretation of the SWA theorem in the technological context of reservoir computing in the classical setup. The goal is to provide a set of broad guidelines of how reservoir computers could be engineered and which requirements should be met in order to turn them into powerful information processing devices. Some open problems are pointed out too.

23.6.1 Existence of the Filter

The first and the most important question is whether any dynamical system realizes a filter. This cannot be taken for granted since the system has to be started from an initial state, and deciding whether the initial state matters or not is a highly non-trivial issue. The echo state property and the fading memory property have been suggested specifically to address this issue.

23.6.1.1 The Fading Memory Requirement

To be able to use a system as a filter, in the on-line computation context, the present state of the system should be weakly dependent on distant inputs. Many dynamical systems found in nature often equilibrate, and have the potential to act as filters. But there are also systems that do not equilibrate easily, e.g. chaotic systems. It is important to be able to distinguish these two classes of systems. There are several ways of formalizing mathematically the condition that the dynamics is insensitive to the initial condition. The most common definition is as follows.

Definition 5 (*Fading memory*) For reservoirs with the fading memory property, the dynamics of the reservoir should not be influenced by a too distant past. Two input time series that differ in the distant past should lead to roughly the same output: For every u and $\varepsilon > 0$, there exist a $\delta(u, \varepsilon) > 0$ and an interval $[t_0 - T, t_0]$ such that $|(\mathcal{F}u)(t_0) - (\mathcal{F}v)(t_0)| < \varepsilon$ for every input v that is δ-close to u on that interval; $|u(t) - v(t)| < \delta$ for $t \in [t_0 - T, t_0]$.

Note that this resembles the definition of continuity at a point (not uniform continuity). It was shown that the systems with fading memory have unique steady states (that lock-onto the input, for a proof see section XIII, Theorem 6, in [30]).

Fading Memory Leads to a Special form of Continuity

Interestingly, the fading memory property ensures some useful mathematical properties of the filter realized by the system. Since Ω is the space of infinite time series it is not automatically compact. The Arzelà-Ascoli theorem states that to make the space of functions compact, one would have to, at least, limit the time frame by considering only a finite time window $t \in [t_0 - \tau, t_0]$. However, there are two immediate problems, what should one choose for the reference (computation) time t_0 (the reference issue) and the interval length τ (the length issue)?

The compactness problem can be solved as follows. If the distance between two time series is defined by using the weighting functions construct, which tend to favor more recent values in time,

$$\rho^{(t)}(u, v) \equiv \sup_{k \leq t} w(t - k)|u(k) - v(k)| \tag{23.12}$$

where $w(k) \to 0$ with $k \to \infty$, then the space Ω with this metric is compact.

Interestingly, once the compactness is in place, the fading memory ensures that the mapping realized by the filter is continuous. Any filter with fading memory is continuous in the metric $\rho^{(t)}$:

$$\rho^{(t)}(u, v) < \delta \Rightarrow |(\mathcal{F}u)(t) - (\mathcal{F}v)(t)| < \varepsilon \tag{23.13}$$

This result was stated as a theorem in [31].[13] This has been also used as an alternative definition of fading memory (e.g., see section III, the definition 3.1 in [30]).

23.6.1.2 The Echo State Property

The echo state property is a result of a direct attempt to deal with the filter existence issue. In echo state networks the present state of the network is an "echo" of the input history. The initial state of the system can be forgotten if the system has been exposed to the input for a sufficiently long time. The rest is a mathematical formulation of the idea.

Historically, it has been realized that the echo state property is crucial if the network can be trained by adjusting its output weights only. Quoting from [32]:

> For the supervised learning algorithms which are used with Echo State Networks (Edit: citing the original technical report [4], and a later review [7]) it is crucial that the current network state x_k is uniquely determined by any left-infinite input sequence $u_{-\infty}, \ldots, u_{k-1}, u_k$.

Such behavior guarantees that the on-line computation is possible, i.e. that the dynamics of the system does not depend on the initial state of the device. The original definition of the echo state property is as follows [5].

Definition 6 (*Echo state, discrete dynamics*) Assume a fixed time instance t. Let $q[-\infty : t] \equiv (\ldots, q_{t-2}, q_{t-1}, q_t)$ denote a left infinite sequence obtained by truncating $(\ldots, q_{t-1}, q_t, q_{t+1}, \cdots)$ at t. For any input $u[-\infty : t]$ that has been used to drive the system (for an infinitely distant past until the time t), and for any two trajectories $x[-\infty : t]$ and $x'[-\infty : t]$ that are consistent with the dynamic mapping (23.4),[14] it must be that $x_t = x'_t$.

Figure 23.1 is a graphical illustration of this property. The echo state property is equivalent to stating that there *exists* an input echo function $\mathcal{E}$ such that if the system has been exposed to the infinite input sequence, its current state is given by

$$x_t = \mathcal{E}(\ldots, u_{t-2}, u_{t-1}, u_t) \qquad (23.14)$$

This notation is somewhat uncomfortable since it involves a function with an infinite list of arguments. However, the statement implied by Eq. (23.14) is useful from an engineering perspective, this particular definition emphasizes the existence of a filter. The original Definition 6 is more suitable for mathematical analysis, e.g. for

---

[13]Note that the key property is the fading memory. The particular definition of the metric is "for free" (it can be always made). There is a nice alignment with the assumptions of the SWA theorem: the domain of the mapping implemented by the filter Ω should be compact, and the mapping should be continuous. Then the filter realized by the reservoir maps from a compact metric space and is continuous. These properties are useful for establishing the expressive power of the filter.

[14]The consistency is expressed as requiring that for a given sequence x it is true that $x_k = H(x_{k-1}, u_k)$ for every k. The same must hold for the other sequence, i.e. $x'_k = H(x'_{k-1}, u_k)$ for every k.

(a)

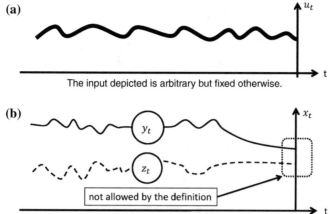

The input depicted is arbitrary but fixed otherwise.

(b)

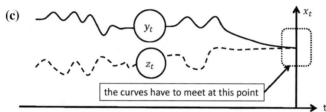

This is situation is not allowed to occur for systems with echo state property. If an input is found, and this situation occurs, the system does not have the echo state property.

(c)

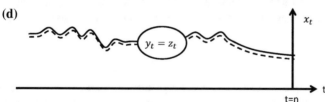

This situation must hold if the system has the echo state property. The depicted situation is the one stated (required) by the definition.

(d)

An example of the situation not required by the definition, though this can happen accidentally. It is not required that the two trajectories that are compatible with the input are identical. This can happen, but this is not required by the definition.

Fig. 23.1 An illustration of the echo state property (backward-oriented perspective). *Panel* **a** Depicts a fixed but otherwise arbitrary input. For a given input, *panels* **b–d** depict various situations. The definition of the echo state specifies exactly which situation is allowed or forbidden. A much more intuitive, the forward-oriented definition, is illustrated in Fig. 23.2

identifying whether the system has such a property. An alternative (and equivalent) definition is listed in the appendix that might be even better suited to that end.

There are systems for which the echo state function $\mathcal{E}$ does not exist, and a few examples will be provided (assuming the discrete dynamics). The examples are chosen to illustrate that instead of (23.14) the following expression should be used

$$x_t = \mathcal{E}(\dots, u_{t-2}, u_{t-1}, u_t | \mathfrak{x}) \qquad (23.15)$$

where $\mathfrak{x}$ denotes the initial state of the system in the infinite past. The initial state can exert an influence on the dynamics for an infinitely long time. Several systems that behave like that are listed below:

Example 1

As the first negative example, consider the system with the discrete dynamics that strongly depends on the initial condition:

$$H_1(x, u) = x \qquad (23.16)$$

The system "remembers" the initial condition forever. In fact it is insensitive to the input which is a rather special case. This system does not exhibit the echo state property since it is possible that two separate trajectories exist for a given input u.[15]

Example 2

The following example exhibits a less trivial dynamics, where the system can be influenced by the input:

$$H_2(x, u) = x + \lambda \tanh u \qquad (23.17)$$

This mapping resembles a discrete version of the random walk where the spatial increment at each time step depends on the input received by the walker. The input is wrapped by the tanh function just to limit the size of individual steps, which is controlled by λ. Note that for any choice of λ the initial state in the infinite past influences the dynamics.

Example 3

As the last example, consider any system that has at least two quasi[16] stable states with two (or more) basins of attraction. The key feature of the dynamics is that for "weak" inputs the system never crosses the basins of attraction. Transitions from one basin of attraction to the other can only happen under the influence of "strong" inputs. For strong inputs it is not true that (A) for two sequences of states x and x' that are consistent with the dynamics it follows that (B) $x_t = x'_t$. Note that the echo state Definition 6 requires $A \Rightarrow B$. It is possible to find trajectories where A is true but B is false leading to $A \nRightarrow B$.[17] Thus the system under consideration does not exhibit the echo state property.

---

[15]Note that such a pair of trajectories can be found for any input. For example, consider two trajectories that started from x_0 and x'_0 with $x_0 \neq x'_0$ in the infinite past.

[16]Here the term "quasi" indicates that the states are easily "disturbed" by the external input.

[17]It is sufficient to consider two initial conditions that start from different basins of attraction.

23.6.1.3   Summary of the Existence Issue

It is clear that the fading memory and echo state properties are strongly related. It has been proved that if a system has the echo state property then it also has the fading memory property [5]. A version of the (likely[18]) proof that the fading memory implies echo state can be found in [30] (Sect. 8.2, Theorem 6). However, despite being strongly related, both concepts do emphasize slightly different aspects of the problem. For example, the fading memory definition emphasizes the possibility to perform on-line computation, while the echo state property emphasizes the existence of the time invariant filter.

The main problem with both definitions is that it is hard to check whether a given system has either of these. There are some results regarding the existence of the echo state property for a particular class of dynamic functions $H(x, u)$ [3–5]. Clearly, from the above discussion one can see that it is very hard to make generic statements regarding the systems which exhibit these properties. Such analysis has to be done for every system of interest.

23.6.2   The Expressive Power of Reservoir Machine

The notion of the expressive power of the reservoir is important for realizing the advocated technological goals. While it is clear that the echo state mapping exists, provided the specific requirements are met, it is less clear what one can actually do with such devices. If we knew which features of the device influence it expressive power, we would have a theory for building powerful reservoir computers, and we would also understand the limits of the approach.

For discrete reservoirs with the echo state property the input echo function involves an infinite list of arguments. If the system has natural relaxation time (or past forgetting time) τ_* then this limits the number of arguments: Instead of (23.14) the following description of the filter might be closer to the truth,

$$x(t) \approx \mathcal{E}_*(u_{t-\tau_*}, \ldots, u_{t-2}, u_{t-1}, u_t) \tag{23.18}$$

For such systems, how do we realize filters with very short or very long list of arguments? Both situations are problematic. For example, assume that the goal is to realize a filter that takes only the two immediate inputs, $o(t) = (\mathcal{F}u)(t)$, that should be trained to return the sum $(\mathcal{F}u)(t) = u(t-1) + u(t)$. It is not at all clear how to get rid of the dependence on the remaining inputs further away in the past. Filtering longer signals is equally problematic since it is not clear where the distant

---

[18] "Likely" is emphasized due to the following. First, the proof assumes continuous dynamics. Can it be generalized to discrete dynamics? Second, the proof does not lead straightly to Definition 5, but to a continuous formulation of the echo state that is possibly equivalent to the original echo state definition.

information should be stored. To realize both scenarios would likely require more powerful reservoirs, but there is only one reservoir to choose from.

In the following a brief summary of what is known about the expressive power of liquid state machines and echo state networks is discussed. The expressive power of Reservoir Machine is discussed after that.

23.6.2.1   Lessons from the Past: The Expressive Power of LSM and ESN

A lot of research effort has been spent to understand which dynamic properties of the system can guarantee the echo state property. Much less effort has been spend on understanding how to exploit such a property if it exists. This is still an open problem.

LSM is indeed Turing universal (in the fading memory sense), but only provided that a whole class of machines is considered (the base filters). Not surprisingly, this class should have exactly the same properties as the ones stated in the second version of the SWA theorem. In fact, the LSM model originated from a direct application of the second version of the SWA theorem on filters.

In contrast to LSM, in the literature there seems to be a lack of precision in stating the expressive power of ESN, and possibly some misconceptions. This very likely results from using the work by Maass et al. on LSM in imprecise way without specifying the context properly (e.g. one system, or a class of systems). For example, consider the following quote from [33]:

> Universal computation and approximation properties. ESNs can realize every nonlinear filter with bounded memory arbitrary well. This line of theoretical research has been started and advanced in the file of LSM (Edit: quoting the LSM work by Maass et al. [1] and [34])

In the quote, there are several important assumptions that are implicit. Consider a few examples of what might go wrong if such a claim were made without specifying a proper context.

For example, the first implicit assumption is that one should consider a class of networks, and not a single system (network). The second implicit assumption is that this class should have the properties required by the SWA theorem, be an algebra and separate points (which sould be proven, however). These implicit assumptions are very fragile since their validity is context sensitive. Let us discuss each in turn.

The ESN idea, by construction, emphasizes the use of a fixed network, and in this classical setup the quote is most certainly incorrect (as will be explained later on). But, if the network is complex, then it might implicitly harbor a class of smaller subsystems (see Sect. 3.1 in [5] for an illustrative neural network example). In that sense, the assumption that possibly a large class of echo state mappings is available for a single system is justified. However, while these subsystems might exhibit the echo state property (i.e. realize a set of filters), do these subsystems form an algebra? In fact, they very likely do not form an algebra since the number of elements is finite. But should one worry about it, i.e. is it necessary to require that they do? e.g. as in the

LSM model. Subsystems might exhibit some separation features, but most certainly not the exact separation property.

Another strongly related misconception regards the issue of the expressive power of the classical reservoir computer (since it is often associated with the ESN concept). It is implicitly assumed that RC has an infinite computing power, e.g. as in the following quote from [7]:

> Modeling capacity. RC is computationally universal for continuous-time, continuous-value real-time systems modeled with bounded resources (including time and value resolution (Edit: quoting the work by Maass et al. [35] and [36])

This is simply not true a priori for the same reasons as discussed above. If one-system is used as a reservoir, a great caution should be exercised in making such a claim. The main motivation behind suggesting the Reservoir Machine model was to make these issues explicit so that they can be rigorously addressed. Understanding the expressive power of the model is a highly non-trivial and still an open problem, as discussed below.

The SWA theorem states clearly which properties a collection of filters must possess if used in the information processing context: realize an algebra with constant element that separate points, the algebra must realize mappings from a compact space. In the following, each of the requirements of this generic theorem will be revisited an interpreted in practical (engineering) terms.

23.6.2.2 The Burden of Realizing an Algebra Can Be Taken by the Readout Layer

The key lesson from the SWAT is that machines $\mathcal{M}$ should form an algebra in some sense. Do they, and can we engineer them in such a way in the context of unconventional computation?

At this stage the mathematical concept of an algebra needs to be converted into an engineering one. The requirement that the collection of machines described by Reservoir Machine forms an algebra implies that they can be combined in some way, or adjusted, to obtain new, hopefully more powerful, machines. Any two machines $\mathcal{M}$ and $\mathcal{M}'$ might be combined into a larger machine $\mathcal{M} \odot \mathcal{M}'$ to compute another function (filter). Here the symbol $\odot$ denotes an engineering operation on the machines. The filter being realized this way should be exactly the one that is obtained by applying the algebraic operations on the respective output values $o(t)$ and $o'(t)$, as $o(t)o'(t)$.

The above considerations put some constraints on how the engineering operation should be implemented

$$u(t) \to (\mathcal{M} \odot \mathcal{M}')(u)(t) = \psi(\mathcal{R}(u)(t))\psi'(\mathcal{R}(u)(t)) \qquad (23.19)$$

Note that the reservoir is not changing. Only the readout function is allowed to change. *Conceptually, the only freedom we have when designing new machines is to combine their readout layers.* Mathematically, the engineering challenge can be represented as

$$\mathcal{M}\odot\mathcal{M}' \equiv (\mathcal{R}, \psi)\odot(\mathcal{R}, \psi') = (\mathcal{R}, \psi \odot \psi') \tag{23.20}$$

A direct comparison between (23.19) and (23.20) shows that if we can design a new readout layer ψ'' such that $\psi''(x) = \psi(x)\psi'(x)$ for every $x \in \mathcal{S}$ then it is possible to engineer an algebra of machines. In the similar way it can be shown that engineering the vector space operations, multiplication by scalar $w\odot\mathcal{M}$ or addition $\mathcal{M} + \mathcal{M}'$, can be transferred to the readout layer.

23.6.2.3   Engineering the Readout Layer ψ

The preceding discussion shows that the first technological requirement towards building a reservoir machine is to ensure that the set of readout layers Ψ forms an algebra that can be realized. Of course, for in-silico implementations that is not an issue, but in any other setup it has to be addressed explicitly.

The readout layer, or the interface, is one of the key concepts of the reservoir computing. Intuitively, since the goal is to use the system as is, any auxiliary equipment that has to be added onto the system to turn it into an information processing unit should be as simple as possible. Typically, when performing mathematical analysis the readout function ψ is assumed to be a multivariate polynomial or even a simple linear polynomial. How complex the readout layer should be?

Again, this is an instance where the SWA theorem needs to be reinterpreted in an engineering context. The second version of the SWA theorem indicates that any class of functions which have the universal approximation property will do.[19] However, this might be an unnecessarily strong requirement. In the literature, several implementations of the readout layer have been suggested ranging from a simple linear readout towards a more complicated simple perceptron network. An extensive list of various readout layers can be found in [7] (in Sect. 8).

The question is why should one engineer more complex readout layers if simpler ones will do? and what is the simplest readout layer that can be used? To answer these questions is an extremely challenging problem that has been grossly overlooked in the literature.

Since the reservoir is frozen, and all the burden of the algebraic properties rests on the readout layer, it is clear that the required complexity of the readout layer is conditioned on the complexity of the dynamics of the reservoir. It is hard to make generic statements without knowing what the reservoir looks like in a bit more detail. Understanding this interplay is still an open problem. There is simply no available theory that could be used to address the issue.

---

[19]Multivariate polynomials have this universal approximation property. Any function that behaves in the same way can be used instead.

From the practical point of view, it is implicit in the construction that such readout needs to be constructed, and it should be kept be kept as simple as possible. The computing that can be done by the reservoir should not be done by the readout layer. To understand how much of the burden needs to be put on the readout layer one should really start from basics, i.e. the first (abstract) version of the SWA theorem, as discussed below.

23.6.2.4   Separating Points with One Reservoir Is Problematic but Still Possible

The algebra of machines does not automatically satisfy the requirements of SWAT for one obvious reason. There is only one reservoir to choose from. However, it is possible that complex reservoirs can be used to implement such an algebra.

A Problem

Let us see whether the algebra of reservoir machines separates points. For every pair of time series u and v we must find a reservoir machine $\mathcal{M}$ such that $\mathcal{M}(u)(t) \neq \mathcal{M}(v)(t)$ where t stands for the observation (reference) time. This is the place where the construct with a single reservoir breaks: in contrast to the LSM model, there are simply no reservoirs to choose from, only one is available. What choices can one actually make? In principle, there are two. First, while a single system cannot exhibit infinite "separation" power, it might exhibit some if it is complex enough, i.e. if it contains many components that are "addressable". Second, by assumption, there is a class of readout layers to choose from. However, by construction, these cannot contribute to realizing the separation property. Thus, strictly speaking, there is only one choice to be made, one has to "dig" into the system.

A Possibility

If complex enough, the reservoir could be divided into smaller parts: A truly complex reservoir with many components realizes many filters, where the number of filters is equal to the number of microscopic components N:

$$x \equiv (x_1, x_2, \ldots, x_i, \ldots, x_N) \tag{23.21}$$

where each component realizes one filter

$$u \to x_i(u) \tag{23.22}$$

This possibility has been already pointed out in [5] (Sect. 3.1). For example, for a discrete system it implies that input echo functions (filters) exist, such that

$$x_t^{(1)} = \mathcal{E}^{(1)}(\ldots, u_{t-2}, u_{t-1}, u_t)$$
$$x_t^{(2)} = \mathcal{E}^{(2)}(\ldots, u_{t-2}, u_{t-1}, u_t)$$
$$\ldots$$
$$x_t^{(i)} = \mathcal{E}^{(i)}(\ldots, u_{t-2}, u_{t-1}, u_t) \tag{23.23}$$
$$\ldots$$
$$x_t^{(N)} = \mathcal{E}^{(N)}(\ldots, u_{t-2}, u_{t-1}, u_t)$$

Thus, possibly, there is a very large class of mappings to choose from.

The filters realized by such sub-systems might or might not form an algebra. However, this feature is *not* an engineering requirement: The readout layer carries the burden of realizing an algebra! What is important is that all these filters separate points: In some sense all these filters should be "different".

The output of the computation is given by

$$o(t) = \psi(x_1(u)(t), x_2(u)(t), \ldots, x_i(u)(t), \ldots, x_N(u)(t)) \tag{23.24}$$

At this stage it is important to formalize what a sub-system might be, i.e. *what the variables x_i actually represent*. In this context, the concept of observable in statistical physics is extremely useful, and in particular the observables that are relevant for information processing as discussed in [21]. Such observables could be referred to as "information processing observables". While formalizing (23.24) in terms of information processing observables would make the discussion more complete it would also make it a bit more technical. We just leave it at claiming, as in the reservoir dynamics Sect. 23.3.1, that x_i does not necessarily describe microscopic degrees of freedom but can also refer to features on larger scales.

In practice, it might be hard to access all subsystems (even if they are not microscopic). From the engineering point of view, a more reasonable assumption is that the readout layer will have only access to a limited number of components. Thus the equation for the output should read instead

$$o(t) = \psi(x_{i_1}(u)(t), x_{i_2}(u)(t), \ldots, x_{i_d}(u)(t)) \tag{23.25}$$

where it has been assumed that the readout layer can only access d components. The multiple $(i_1, i_2, \ldots, i_d)$ denotes a particular choice of component filters being indexed. In practice the number of accessible filters will be such that $d \ll N$.

One Should Resist the Temptation to Divide into too Many Parts

It is tempting to increase N by considering smaller and smaller systems. This would make the collection of the filters generated by the subsystems very large and increase the resolution of the algebra. Note that this is somewhat equivalent to the assumption that x describes microscopic states. It also strongly parallels Putnam's construction (the assumption that every micro-state is accessible). However, there is a fundamental problem with realizing such an agenda. The dynamics becomes noisy.

For observables that refer to very small sub-systems the dynamic laws expressed in (23.3) and (23.4) do not longer apply. For such systems a noise term should be added in the equations of motion. For example, it is a well-known fact that the dynamics of a single molecule in the sea of solvent molecules kept at a finite temperature should be modelled by using a stochastic differential equation. For observables that are microscopic in nature one would have to assume the following dynamical law

$$\frac{dx(t)}{dt} = \tilde{H}(x(t), u(t)) + \eta(t) \tag{23.26}$$

where η is a stochastic variable with a given mean and a finite variance and $\tilde{H}$ indicates that H needs to be modified for the effects of friction. For example, the form of the noise that describes experiments that involve diffusion is $\langle \eta(t) \rangle = 0$ and $\langle \eta(t)\eta(t') \rangle = \gamma \delta(t - t')$ where $\delta(t)$ denotes the Dirac delta function and γ is a temperature dependent parameter.

The Expressive Power of a Reservoir Machine is Limited

Without knowing more details about the component filters $x_i(t); i = 1, 2, \ldots, N$, it is impossible to make further progress. This is exactly the reason why one-reservoir construct is hard to analyze. However, it is clear that if the component filters separate inputs, and if the readout functions can approximate sufficiently well, then it should be possible to tune the reservoir machine towards arbitrary information processing task, but there are clearly limits to what can be computed. The expressive power of a programmable reservoir machine is not infinite.

23.7 Conclusions

Perhaps it is fair to say that reservoir computing is more an insight about computing than an approach to computing. The insight is about the possibility to use an arbitrary dynamical system for computation without elaborate re-configuring (training) procedures. In this chapter the existing problems with realizing this classical setup were discussed, together with what could be done to address these problems, and what could be gained by doing so. The three big questions that were addressed were:

Which features of the construct are hard to engineer? How to engineer such features in principle? What is the expected technological impact of such devices?

> In this classical setup, reservoir computing is also a "call" to study dynamical systems, and in particular unconventional computation, in a new way. The Reservoir Machine concept was suggested to make this point of view explicit.

In the literature Echo State Networks and Liquid State Machines are often treated as one concept. Admittedly, they are strongly related but they are not identical. Any interpretation of these concepts is strongly context dependent, as illustrated in the text. This lack of clarity might obscure further progress, and the goal was to, first, point out the key differences between the concepts and then, second, provide a synthesis of the ideas they represent. This was done in the context of the classical setup of reservoir computing. The *Reservoir Machine* concept is the result of the attempted synthesis.

> Reservoir Machine has been introduced to emphasize the fact that, ulti- mately, in the worst case scenario, one is facing the problem of using a fixed dynamical system for computation with a very little freedom of tuning the system. This worst case scenario is ubiquitous in situations when in-silico solutions are not possible, and in particular in the context of unconventional computation.

The explicit formulation of Reservoir Machine established a clear starting point for the analysis of several aspects of the problem: The mathematical foundation of reservoir computing, the SWA theorem, has been re-interpreted to provide strate- gic guidelines for building powerful unconventional reservoir machines and to aid in understanding their computing power. The most important observations are as follows.

1. There is a strong connection between Putnam's construction and the reservoir computing idea in the classical setup. Putnam's construction addresses nearly the same problem.[20] The notion of the microscopic degree of freedom and the ability to access such states is crucial to establish the connection. This line of thinking puts a clear emphasis on the need to understand how to build good readout layers. This aspect of the problem should be given a serious consideration. While being extremely "passive" in mathematical terms, in the engineering sense the readout layer is as important as the reservoir.
2. It has been shown that, in the Reservoir Machine setup, the readout layer takes the burden of realizing the algebra, while the reservoir takes the burden of separating

---

[20]It is tempting to argue that the separation property being emphasized in reservoir computing is, in some sense, related to Putnam's requirement of non-periodic behavior and the idea of the internal dial [24].

points. This is in contrast with Liquid State Machines where base filters realize an algebra and the readout layer plays a somewhat passive role. The interplay between the formal requirement to realize an algebra and the requirement that the algebra separates points should be understood better.[21]

3. Given that one can build powerful readout layers, it is important to be aware of the fact that making the sub-systems too small results in noisy dynamics. It is possible that for some applications the presence of noise is not an issue, and might even be desirable, but in majority of cases noise is probably a nuisance. In general, the effects of noise in the reservoir computing setup still need to be understood.

4. There seems to be no theory of reservoir computing that might be directly relevant for engineering applications. The Reservoir Machine construct provides an illustration of how to approach the problem of constructing such a theory.[22]

Perhaps the suggested reservoir machine model could be taken as a starting point for constructing a theory that is relevant from a practical point of view, as it clearly points out the relevant set of issues. There are several options for constructing a better theory of reservoir computing.

1. The SWA theorem might not be particularly useful in the engineering context, after all. The theorem is simply an existence statement. It states that an approximation can be found under given conditions. It does not address the accuracy or tolerance issues, or gives any bounds. One might try to constructs and prove a completely different version of the SWA theorem, in a form that is more useful for engineers.

2. The second option is to re-work some ingredients of the existing theorem towards a more engineering like setup. For example, the separation property seems to be crucial. The separation property is a mathematical formulation of an intuitive understanding that the system must have some "resolution power" (of inputs). It could be useful to re-phrase the separation criterion in less absolute terms by, e.g., requiring not strict separation of inputs, but separation up to a certain accuracy (resolution).[23]

---

[21]Intuitively, one expects that the expressive power of the algebra is strongly related to the "richness" of the algebra, provided such a concept can be defined, e.g. by using the number of sub-filters as a measure. The readout layer is not the source of that richness and, yet, takes the burden of realizing the algebra. This is a result of the systematic analysis that has been undertaken, and should be taken as such.

[22]A lot has been achieved since the reservoir computing concept has been originally suggested. However, there are many issues that are still open, as pointed out throughout the text. For example, in the classical setup, reservoir computing does not have the universal computing power, not even in the fading memory (on-line computation) sense. There is still no sufficient understanding of the expressive power of reservoir computing in the classical (reservoir machine) setup. The understanding provided by the use of SWA theorem is an important first step (e.g. LSM and ESN studies) but to understand the computing capacity of a reservoir machine we clearly need a much better theory of reservoir computing.

[23]To a mathematician, an interesting question, perhaps, might be: given that a class of functions forms an algebra and separates inputs up to a certain accuracy, is there any way to characterize a class of functions that can be represented that way? Which types of theorems could be proven?

3. The third option is to further explore and refine the fading memory and the echo state concepts, despite the fact that, perhaps, they are the part of the reservoir computing theory that has been mostly developed and investigated. For one thing, the equivalence between the two concepts has been proven, but under very specific conditions. It should be clear from the presented discussion that these concepts are strongly related. Accordingly, it is surprising that the equivalence has not been proven under more generic conditions.

To conclude, as a technological platform, the single reservoir perspective featured in Reservoir Machine should be less restrictive when it comes to practical implementations. Admittedly, it is also less expressive, but might have a larger technological impact. It seems that the envisioned reservoir machine technology might solve complicated unconventional information processing problems. As an illustration of how the Reservoir Machine concept can be applied in the context of material computation (e.g. for building material machines) see [15]. Reservoir Machine might be suitable as a platform for improving the existing, and realizing new unconventional computing scenarios. It is perfectly possible that important information processing applications can be realized by using relatively simple reservoirs, since not all relevant information processing tasks are complicated.[24] For example, one can envision plethora of applications of reservoir computing in situations where *in-silico* realization is not feasible, e.g. in medical sensing applications when bio-compatibility is an issue rather than the computing capacity. This suggests that while reservoir computing machines might be used for high-performance computing, their natural zone of application is very likely elsewhere. Such machines could be used for the information processing tasks that require deep integration of the information processing equipment and biological systems.

Acknowledgments This work was supported by Chalmers University of Technology and by the European Commission under the contracts FP7-FET-318597 SYMONE and HORIZON-2020-FET-664786 RECORD-IT.

Appendix

An Alternative Echo State Property Definition

Intuitively, the following definition of the echo state property might be easier to understand. [5] The following notation is useful to rephrase the original definition in the ambience of this chapter. Let $x_t = \mathcal{R}(u|x_n)(t)$ denote the configuration of

---

[24]For example, consider the problem of on-line time series data analysis and pattern recognition. This is typical example where the separation property is not a strong requirement. What is needed is that only particular input patterns drive the system to a different regions of the configuration space when compared to the background input. It is possible that relatively simple reservoir machines could handle such tasks.

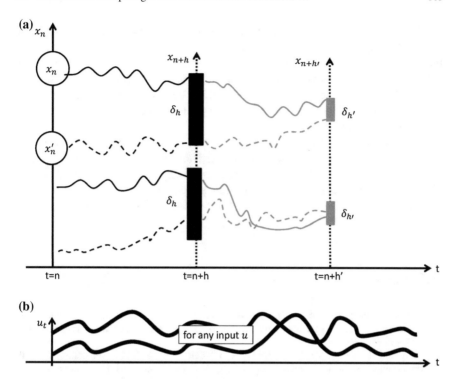

Fig. 23.2 An illustration of the state contracting property. Any pair of trajectories start aligning if one waits long enough. For example, for a fixed pair of initial conditions the trajectories after time h can be both "covered" with the *black box* by moving it vertically. Note that the height of the box δ_h is fixed beforehand and depends only on h. Likewise, if one waits a bit longer, until $t = n + h'$ with $h' > h$ then the size of the box becomes smaller. Note that the *gray box* is smaller than the *black box*. There is always a tendency for each pair of curves to start aligning. The more one waits the more aligned they become. **a** The desired behaviour for any pair of trajectories. **b** The behavior must hold for any input

the system that was at time n in state x_n and that has been exposed to the input u after that, during the time interval $[n : t] \equiv (n, n + 1, n + 2, \ldots, t - 2, t - 1, t)$. To emphasize that the system has been exposed to the input during h time steps we write $\mathcal{R}(u|x_n)(n + h)$. Further, let $\|x_t - x'_t\|_S$ denote the distance between any two elements x_t and x'_t in S for arbitrary t.[25]

Definition 7 (*state contracting*) The network is uniformly state contracting iff there exists a null sequence[26] of bounds δ_h with $h = 0, 1, 2, \ldots, \infty$ such that for every input sequence u and every pair of initial conditions $x_n, x'_n \in S$ the chosen pair of trajectories approach each other, i.e. $\|\mathcal{R}(u|x_n)(n + h) - \mathcal{R}(u|x'_n)(n + h)\|_S < \delta_h$.

---

[25] The notation is implicitly suggesting that S is a vector space but this need not be the case. This form is used to make such expressions more readable.

[26] A null sequence is a sequence of positive numbers that converges to zero.

Note that, in contrast to Definition 6, this definition does not feature any knowledge of an infinite past. For example, there is no need to know what the input looked like in the interval $(-\infty : n) \equiv (\ldots, n-3, n-2, n-1)$. Accordingly, the above definition is easier to understand since it is "forward oriented" and more aligned with the human intuition of how dynamic systems behave. Figure 23.2 is a graphical representation of this property. It describes a system that equilibrates in some sense, i.e. by "locking" onto the input.

The fact that the definition above is equivalent to Definition 6 was proven rigorously in the original publication by Jaeger from 2001 [4]. Note that Figs. 23.1 and 23.2 are different. They depict a priori genuinely different behaviors. Thus a mathematical proof that these two behaviors are equivalent was indeed necessary. The above definition might be more useful if one wants to check whether a given physical system has the echo state property.

References

1. Maass, Wolfgang, Natschläger, Thomas, Markram, Henry: Real-time computing without stable states: a new framework for neural computation based on perturbations. Neural Comput. **14**(11), 2531–2560 (2002)
2. Markram, H., Natschlger, T., Maass, W.: The "liquid computer": A novel strategy for real-time computing on time series (special issue on foundations of information processing). TELEMATIK, **8**, 39–43 (2002)
3. Jaeger, Herbert, Haas, Harald: Harnessing nonlinearity: predicting chaotic systems and saving energy in wireless communication. Science **304**(5667), 78–80 (2004)
4. Jaeger., H.: The "echo state" approach to analysing and training recurrent neural networks. Technical Report GDM Report 148 (contains errors), German national research center for information technology (2001)
5. Jaeger, H.: The "echo state" approach to analysing and training recurrent neural networks - with an erratum note. Technical Report erratum to GDM Report 148, German national research center for information technology (2010)
6. Jaeger, H., Lukoöevicius, M., Schrauwen, B.: Reservoir computing trends. KI - Konstliche Intelligenz, **26**, 365–371 (2012)
7. Lukoöevicius, M., Jaeger, H.: Reservoir computing approaches to recurrent neural network training. Comput. Sci. Rev. **3**, 127–149 (2009)
8. ORGANIC-EU-FP7. Reservoir Computing: Shaping Dynamics into Information (2009)
9. Kulkarni, M.S., Teuscher, C.: Memristor-based reservoir computing. In: 2012 IEEE/ACM International Symposium on Nanoscale Architectures (NANOARCH), pp. 226–232 (2012)
10. Carbajal, J.P., Dambre, J., Hermans, M., Schrauwen, B.: Memristor models for machine learning. Neural Comput. **27**, 725–747 (2015)
11. Zoran, Konkoli, Goran, Wendin: On information processing with networks of nano-scale switching elements. Int. J. Unconv. Comput. **10**(5–6), 405–428 (2014)
12. Appeltant, L., Soriano, M.C., Van der Sande, G., Danckaert, J., Massar, S., Dambre, J., Schrauwen, B., Mirasso, C.R., Fischer, I.: Information processing using a single dynamical node as complex system. Nat. Commun. **2**, 468 (2011)
13. Larger, L., Soriano, M.C., Brunner, D., Appeltant, L., Gutierrez, J.M., Pesquera, L., Mirasso, C.R., Fischer, I.: Photonic information processing beyond turing: an optoelectronic implementation of reservoir computing. Opt. Express **20**(3), 3241–3249 (2012)

14. Mesaritakis, C., Bogris, A., Kapsalis, A., Syvridis, D.: High-speed all-optical pattern recognition of dispersive fourier images through a photonic reservoir computing subsystem. Opt. Lett. **40**, 3416–3419 (2015)
15. Konkoli, Z., Stepney, S., Dale, M., Nichele, S.: Reservoir computing with computational matter. In: Amos, M., Rasmussen, S., Stepney, S. (eds.) Computational Matter. Springer, Heidelberg (2016)
16. Dambre, J., Verstraeten, D., Schrauwen, B., Massar, S.: Information processing capacity of dynamical systems. Sci. Rep. **2**, 514 (2012)
17. Massar, M., Massar, S.: Mean-field theory of echo state networks. Phys. Rev. E **87** (2013)
18. Goudarzi, A., Stefanovic, D.: Towards a calculus of echo state networks. Procedia Comput. Sci. **41**, 176–181 (2014)
19. Soriano, M.C., Brunner, D., Escalona-Moran, M., Mirasso, C.R., Fischer, I.: Minimal approach to neuro-inspired information processing. Front. Comput. Neurosci. **9**, 68 (2015)
20. Bennett, C., Jesorka, A., Wendin, G., Konkoli, Z.: On the inverse pattern recognition problem in the context of the time-series data processing with memristor networks. In: Adamatzky, A. (ed.) Advances in Unconventional Computation. Springer, Heidelberg (2016)
21. Zoran, K.: A perspective on Putnam's realizability theorem in the context of unconventional computation. Int. J. Unconv. Comput. **11**, 83–102 (2015)
22. Putnam, H.: Representation and Reality. MIT Press, Cambridge (1988)
23. Chalmers, D.J.: A computational foundation for the study of cognition. J. Cogn. Sci. **12**, 325–359 (2011)
24. Chalmers, D.J.: Does a rock implement every finite-state automaton? Synthese **108**, 309–333 (1996)
25. Scheutz, M.: When physical systems realize functions. Minds Mach. **9**, 161–196 (1999)
26. Joslin, D.: Real realization: Dennett's real patterns versus Putnam's ubiquitous automata. Minds Mach. **16**, 29–41 (2006)
27. Kirby, K.: Nacap 2009 Extended Abstract: Putnamizing the Liquid State (2009)
28. Rudin, W.: Principles of Mathematical Analysis. McGraw-Hill (1976)
29. Dieudonne, J.: Foundations of Modern Analysis. Read Books (2008)
30. Boyd, S., Chua, L.O.: Fading memory and the problem of approximating nonlinear operators with Volterra series. IEEE Trans. Circuits Syst. **32**, 1150–1161 (1985)
31. Maass, W., Markram, H.: On the computational power of circuits of spiking neurons. J. Comput. Syst. Sci. **69**, 593–616 (2004)
32. Yildiz, I.B., Jaeger, H., Kiebel, S.J.: Re-visiting the echo state property. Neural Netw. **35**, 1–9 (2012)
33. Jaeger, H.: Echo state network. Scholarpedia **2**, 2330 (2007)
34. Maass, W., Joshi, P., Sontag, E.D.: Computational aspects of feedback in neural circuits. Plos Comput. Biol. **3**, 15–34 (2007)
35. Maass, W., Natschlger, T., Markram, H.: A model for real-time computation in generic neural microcircuits. In: Becker, S., Thrun, S.., Obermayer, K. (eds.) NIPS (Advances in Neural Information Processing Systems 15), pp. 229–236. MIT Press, Cambridge (2003)
36. Maass, W., Joshi, P., Sontag, E.D.: Principles of real-time computing with feedback applied to cortical microcircuit models. In: Weiss, Y., Schölkopf, B., Platt, J.C. (eds.) NIPS (Advances in Neural Information Processing Systems 18), pp. 835–842. MIT Press, Cambridge (2006)

Chapter 24
Computational Properties of Cell Regulatory Pathways Through Petri Nets

Paolo Dini

Abstract The paper develops a Petri net model of a negative feedback oscillator, Case 2a from Tyson et al. (Curr. Opin. Cell Biol. 15, 221–231, (2003), [48]), in order to be able to perform the holonomy decomposition of the automaton derived from its token markings and allowed transitions for a given initial state. The objective is to investigate the algebraic structure of the cascade product obtained from its holonomy components and to relate it to the behaviour of the physical system, in particular to the oscillations. The analysis is performed in two steps, first focusing on one of its component systems, the Goldbeter–Koshland ultrasensitive switch (Case 1c from Tyson et al. Curr. Opin. Cell Biol. 15, 221–231, (2003), [48]), in order to verify the validity of its differential model and, from this, to validate the corresponding Petri net through a stochastic simulation. The paper does not present new original results but, rather, discusses and critiques existing results from the different points of view of continuous and discrete mathematics and stochasticity. The style is one of a review paper or tutorial, specifically to make the material and the concepts accessible to a wide interdisciplinary audience. We find that the Case 2a model widely reported in the literature violates the assumptions of the Michaelis–Menten quasi-steady-state approximation. However, we are still able to show oscillations of the full rate equations and of the corresponding Petri net for a different set of parameters and initial conditions. We find that even the automata derived from very coarse Petri nets of Case 1c and Case 2a, with places of capacity 1, are able to capture meaningful biochemical information in the form of algebraic groups, in particular the reversibility of the phosphorylation reactions. Significantly, it appears that the algebraic structures uncovered by holonomy decomposition are a larger set than what may be relevant to a specific physical problem with specific initial conditions, although they are all physically possible. This highlights the role of physical context in helping select which algebraic structures to focus on when analysing particular problems. Finally, the interpretation of Petri nets as positional number systems provides an additional perspective on the computational properties of biological systems.

P. Dini (✉)
Royal Society Wolfson Biocomputation Research Lab, School of Computer
Science, University of Hertfordshire, Hatfield, UK
e-mail: p.dini@herts.ac.uk

© Springer International Publishing Switzerland 2017
A. Adamatzky (ed.), *Advances in Unconventional Computing*,
Emergence, Complexity and Computation 22,
DOI 10.1007/978-3-319-33924-5_24

24.1 Introduction

This paper builds on the research of the BIOMICS project,[1] whose main objective
is to map the spontaneous order-construction ability of biology to computer science
in the form of a new model of computation that we call Interaction Computing
(IC). The paper provides a brief summary of the conceptual framework for IC as
motivation for what has turned out to be a very challenging effort in interdisciplinary
theory construction involving biology, physics, mathematics and computer science
that, at this point, is only partly completed. The paper does not contain new or
original research results in any of the disciplines it touches, but develops an in-
depth critique and comparison of some of the models and methods used in these
different disciplinary domains for the analysis of two cellular pathways. Thus, the
paper is in part a review of some systems biology work and in part a contribution
to interdisciplinary dialogue that we hope will motivate further research towards the
realization of IC.

By spontaneous order construction we refer to ontogenetic processes rather than to
phylogenetic, evolutionary processes, even though the former clearly evolved from
the latter. The motivation comes from Stuart Kauffman's observation that natural
selection by itself does not seem powerful enough to explain the bewildering variety
of complex life forms we observe, and that some other source of order must be at
play ([25]: Preface). The order construction ability of biological organisms appears
to depend on their ability to maintain "dynamically stable" behaviour in the presence
of external inputs and internal propagation of signals and materials at multiple scales.
The ability of biological organisms to react to external signals in a meaningful and
useful way is a consequence of coevolution with the environment itself, implying that
not *any* environment will do. Adami and co-workers expressed this in an alternative
way through the concept of physical complexity, i.e. the "coding" of the environment
into at least one part of the DNA [1, 2]. From the viewpoint of statistical physics self-
organization can be understood as the minimization of free energy, which is consistent
with the open-system architecture the above description implies. The 'self' in self-
organization is merely the fall towards equilibrium of a system that is continually kept
away from equilibrium by a flow of high-free energy nutrients. That is, in biology
equilibrium is death.

In our view, therefore, self-organization depends on open systems interacting with
a compatible environment in such a way that the inputs keep them away from thermodynamic equilibrium, but also that they remain within a dynamically stable range of
possible behaviours. Furthermore, the interactions relate also to the components internal to the system, whose number can change at run-time depending on the needs of
the moment [38]. Although the concept of 'dynamical stability' has not been defined
mathematically and formally, yet, we assume it is related in some way to invariant or
conserved properties of the non-linear system comprising the organism and, in some
cases, also its environment. This explains the reliance on *algebraic* methods and

---

[1]http://www.biomicsproject.eu.

theories in BIOMICS research. The point is that, since algebraic structure underpins order in physics and since its tell-tale symptom—symmetry—is routinely used as a "compass" in the development of mathematical models of physical phenomena, we wonder what computational properties such symmetry may map to. The challenge so far has been, however, that whereas uncovering conserved quantities in *physical* systems is often relatively straightforward, doing so in *biological* systems is rather more difficult.

The focus in the BIOMICS project has been on cellular pathways as a useful starting point given the wealth of pathways available for study at various levels of detail and accuracy, and of analytical, numerical, and stochastic methods to study their mathematical models. This paper focuses on the systems biology perspective as advanced by Uri Alon [3] and John Tyson,[2] who advocate a modular approach whereby cell metabolic and regulatory pathways are understood to be built up by simpler building blocks or modules, justifying an emphasis on the latter before attempting to understand the global behaviour of the former.

24.1.1   Biological Background

The paper builds on BIOMICS report D1.1.2 [11] and focuses on the analysis of Case 2a from [48], a system of three ordinary differential equations (ODEs) that model the Cyclin mitotic oscillator as a negative feedback oscillator with limit cycle behaviour. Cyclins are proteins whose concentration regulates various phases of the cell cycle. The type at the root of the Case 2a model is Cyclin B and it is associated with mitosis, the splitting of the nucleus and of the chromosomes it contains into two identical copies at the end of the cell division process. Strictly speaking, for mitosis to take place only *one* oscillation in Cyclin B concentration is required. The mechanism by which Cyclin B is created and depleted, however, has been modelled by a system of equations that can as well start over at the end of the first oscillation, even though in nature this does not happen (at least, not until the same point in the next cell cycle is reached, minutes or hours later). As a result, periodic oscillations can be observed in the model.

The model provided in [48] was originally proposed by Goldbeter in 1991 [21], based on experimental work by Félix et al. [16] where the presence of a negative feedback loop had been postulated. Such a loop is set up through the interaction of three reactions:

- Cyclin B is produced in response to a signal connected to the overall cell cycle.
- The production of Cyclin triggers, after some concentration threshold is reached, the phosphorylation of a cyclin-dependent kinase, named 'cdc2' for 'cell division cycle protein 2'. This intermediate process is also modelled by Tyson et al. as Case 1c, and was already analysed in BIOMICS report D1.1.1 with the method of Lie groups [9].

---

[2]http://mpf.biol.vt.edu/lab_website/.

- When the phosphorylated cdc2 kinase reaches another threshold it, in turn, activates the phosphorylation of a Cyclin protease that accelerates the degradation of Cyclin B, thereby ending the first oscillation. This third sub-system is another copy of the Case 1c system.

Negative feedback is one of the main examples of stability in a dynamical system that is achieved through the interaction between different sub-systems rather than through the structure of a single component (such as the return force of a spring in a spring-mass system). What makes this system and other similar regulatory cycles of the cell interesting is their autonomous character, which appears to originate from their being embedded in a hierarchy of other regulatory dynamical processes connected to one another by various feedbacks at the same scale and between different scales. For example, this Cyclin oscillator is itself embedded in, and triggered by, the cell cycle. Thus, systems such as Case 2a, which is the result of combining 3 sub-systems, can be seen as a recursive step in the construction of a model of dynamic biological structure (i.e. whose size as measured by the number of states or number of subsystems is not constant) as a nested hierarchy of regulatory and metabolic systems. The foundational work towards the formalization of such a hierarchy is presented in [38].

We draw a distinction between 'dynamical', a term used in dynamical systems theory, a sub-discipline of physics and mathematics, and 'dynamic', which is associated with computer science. Whereas the former indicates the study of the relationship between forces (causes) and motion (effect)—as opposed to, for example, just the formal description of motion as in kinematics—the latter indicates more generally any system or effect that depends on time in some way. Although 'dynamical stability' has not been defined mathematically, yet, we believe it to be intimately associated with non-linear behaviour. As a consequence, both the 'dynamical' and the 'dynamic' concepts are important to BIOMICS and are discussed in this paper.

24.1.2 Petri Nets

Although it is possible to derive a finite-state automaton directly from a given metabolic or regulatory pathway (see, for example, [14]), in this paper we rely exclusively on Petri nets to effect this mapping. Petri nets have been a reliable and widely used analytical and modelling tool in systems biology for many years [45, 50]. We emphasize that Petri nets are not the ideal modelling tool for arriving at IC, since they tend to model systems in a monolithic way. As will be discussed in Sect. 24.5, Abstract State Machines (ASMs) [4] are much better-suited for this purpose. Since ASMs can be seen as generalizations of Petri nets and since Petri nets have been used very successfully to model (bio)chemical reactions, however, they are an ideal stepping stone in the development of a bridging theory between biology and computer science.

Petri nets are bipartite graphs of 'places' and 'transitions'. Each place corresponds to a chemical species and can hold an integer number of 'tokens' which represent discrete quanta of the species in question. In some modelling contexts one token can represent a single molecule of that species, and in this paper we will rely on both interpretations. A distribution of tokens over the (ordered) places of a given Petri net is called a 'marking'. Each transition corresponds to a chemical reaction in a particular direction (i.e. a single arrow in a chemical reaction equation). Places and transitions are connected by directed arcs whose weights correspond to the stoichiometry of the reaction being modelled. If the weight of each of the arcs feeding a transition is less than or equal to the number of tokens in the place from which that arc departs, the transition is 'enabled'. When an enabled transition 'fires' tokens disappear from the places from which the arcs depart, according to the weights of the arcs in question; and tokens appear in the places on the other side of the transition that fired, again according to the weights of the outgoing arcs. As long as the weights respect the stoichiometry, conservation of mass in the different atomic elements will be satisfied, but Petri nets are more abstract and their definition does not itself require conservation of mass—so some care must be exercised when modelling physical systems.

As a given transition fires, the distribution of tokens over the places of the Petri net changes. With these definitions, the construction of the corresponding finite-state automaton is straightforward [13], i.e. each marking of a Petri net is a state of its corresponding automaton. We use the GAP [47] package pn2a [12] to generate the automaton composed of all the states reachable from a given initial state or, in some cases, all the possible states of a given Petri net.

Using mass-action kinetics, it is straightforward to draw the Petri net corresponding to a given system of chemical reactions in a way that respects the different rate constants as well as the stoichiometry [45]. At the same time, the same chemical reactions can be used to derive the ordinary differential equations (ODEs) for the time rates of change of the various chemical species, also using mass-action kinetics. Based on Gillespie's work [19, 20], we know that the token distributions in the places of a Petri net over a fixed time window, calculated with a stochastic simulation method and averaged over r runs, converge to the numerical solution of the corresponding system of ODEs as $r \to \infty$. However, it is a known fact that ODE models derived from mass-action kinetics cannot describe systems with low copy numbers or when particles mix poorly. While the Petri net approach advocated here does not suffer from the poor mixing (by construction), the low copy number problem might be still an issue. In this context, it might be possible to use ODE models, but these have to be extended in a special way to account for the presence of noise and low copy numbers. For an example of how this can be done see [26–30] and references therein. Although in this paper we do not address this point in detail, we do point out where a generalization based on stochastic methods appears to be necessary to obtain more accurate ODE-based systems biology models. We use our own stochastic simulation method coded in *Mathematica*, based on [44], which has been verified with the program Netbuilder [6] and was previously utilized for a similar problem [8].

24.1.3 Computational Motivation

The motivation for studying this problem is to see whether the dynamical properties of the biological system chosen can be related to the computational properties of its discrete, finite-state automaton model. This can be understood in two conceptual steps of increasing abstraction. First, we construct a Petri net of the system and regard the places of the Petri net as the digits of a positional number system whose base is the capacity of the Petri net (assumed constant for all places).[3] Each number in such a system is given by a different marking of the Petri net, so that transitions between markings can be seen as elementary addition and subtraction operations.

Second, this then prepares the ground for a much more abstract conceptual and formal step, i.e. the study of the algebraic properties of the same finite-state automaton model through its holonomy decomposition [51], which was inspired by the more general Krohn–Rhodes theorem [31]. To calculate the holonomy decomposition we use the GAP-based [47] software package SgpDec [15]. Although the proof of the holonomy (or of the Krohn–Rhodes) theorem provides one form of "understanding", and arguably the deepest possible, from an applied perspective it is useful to focus on any *computational* properties that the algebraic structure might uncover. Here we encounter a significant challenge, since either theorem provides complete structural information about a cascade of machines[4] that emulates the original automaton, but does not provide any dynamic information about possible algorithms such a cascade might compute. This is not surprising, given the generality of the result: if the cascade emulates the original automaton, clearly it can emulate *any* of its algorithms. However, it leaves unanswered the biologically, physically, and computationally interesting question whether or not the observed behaviour of the system the automaton models is related in any way to the algebraic structure of that automaton.

Further motivation for this question is provided when we dig deeper into what kind of structure the holonomy theorem uncovers. As discussed more fully in [42], each component of the decomposition is either a set of irreversible resets[5] or a group of (by definition invertible) permutations, both acting on subsets of states. The fact that in the decomposition of any automaton the number of groups is usually much smaller than the number of (sets of) irreversible transformations matches our experience of biological systems, with reversible processes and symmetries providing a "backbone" of ordered structure around which open-system dissipative and entropic processes are organized. In computing, too, we find mostly irreversible processes such as memory

---

[3]More precisely, if the capacity of the Petri net is 3, for example, meaning that each place can hold up to 3 tokens, then the base is $3 + 1 = 4$ since the absence of tokens in a given place corresponds to a 0 digit for that place. So for a constant-capacity Petri net the base for this scheme is always given by the capacity $+1$.

[4]Either theorem shows that more than one cascade of machines can emulate the same automaton, i.e. the decomposition is not unique. Uniqueness up to isomorphism is only guaranteed in the analogous and immensely simpler case of the decomposition of finite groups, known as the Jordan–Hölder theorem of elementary finite group theory.

[5]A reset is a constant map from two or more states to a single state.

overwrites interspersed with some symmetry, but we are still far short of being able to emulate the spontaneous order-construction ability of biology.

The irreversible resets and reversible groups acting on subsets of states as established by the holonomy theorem can be further decomposed into two-states resets (flip-flops or 1-bit memory resets) and simple non-abelian groups (SNAGs), thus recovering the more fundamental Krohn–Rhodes theorem [31]. Since flip-flops and SNAGs cannot be decomposed further, they play an analogous role in the decomposition of automata to the role that prime numbers play in the factorization of the integers—hence the 'prime decomposition' name of the theorem.

Focusing for convenience on holonomy decomposition, the question of its computational significance for a given automaton has been discussed many times previously (see for instance [8, 34–36, 42]) but only some partial conclusions have been reached so far. For example, in the 3-Queens puzzle presented and analysed in detail in [42], the higher levels of the decomposition embody the history of previous moves, whereas the lowest level encodes the number of queens on the board for the current state. This paper aims to add to that discussion by invoking an intriguing perspective that was proposed by Rhodes in the 1960s [40] and according to which either kind of decomposition is referred to as a 'coordinatization' of the given problem. This term makes sense when viewed from the physics or engineering perspectives, where it is well-known that the choice of variables can make a huge difference in the complexity of its mathematical model, and therefore also in whether a solution can be found or not.[6] The fact that automata decompositions by Krohn–Rhodes theory are not unique does not give much reassurance in how easy or difficult the best coordinatization of a problem might be. However, it still makes it possible to relate this rather abstract theory to the first conceptual step in our discussion, above. Namely, the holonomy decomposition of an automaton can be interpreted as an "expansion" of its states into an abstract generalization of a positional number system of variable base, again implying that a change of state of a physical system can be viewed as an elementary "addition" in this abstract positional "number" system.

The computational motivation, therefore, is to see whether algebraic properties can provide a bridge between dynamical and algorithmic behaviour of the same system as modelled through the different viewpoints of continuous and discrete mathematics, and to use the coordinatization or number system perspective as a guide and formal context for this exploration. The paper reports on the partial progress achieved so far towards the analysis of Case 2a. Namely, whereas Case 2a is too large for the discrete algebraic analysis, this could be performed for Case 1c. Since no conclusive general results have been found so far by relating these different views of the system, they should be seen mainly as insights of pedagogical relevance for further study and that may help students and interdisciplinary researchers relate physical behaviour to computational behaviour at different levels of abstraction.

---

[6]The search for the right variables of a problem was vigorously pursued in the empirical discipline of hydraulics and hydrodynamics in the 19th Century, which led to method of similarity analysis which was then later shown to be a branch of group theory. In other words, looking for the right variables of a problem and looking for its symmetries are often one and the same problem.

24.1.4   Overview

The paper summarizes the analysis of Case 1c and Case 2a, which D1.1.1 [9] and D1.1.2 [11] provide in full detail. For Case 2a we encountered a problem in that the ODEs used in [48] (originally derived in [21]) are not an accurate model of the biochemistry for the physical concentrations reported in [48]. These equations were derived using the Michaelis–Menten (MM) approximation, but the numerical values of the variables reported for this system in [48] violate the conditions of validity of this approximation. Since the MM approximation is very commonly used in systems biology, a significant amount of effort was therefore spent investigating its limits of validity, in an attempt to gauge the quality of the approximation afforded by the model used in [48] and, secondly, to understand its relationship to the Petri net model of the same system.

The limits of the validity of the MM approximation were investigated for the simpler Case 1c system. We found that the quasi-steady-state (QSS) approximation used in [48] to derive the three Case 2a ODEs can be significantly improved by employing the 'total QSS' approximation, or tQSS [5]. The tQSS ODEs are more complex algebraically than the QSS version, but still simpler as a dynamical system than the full rate equations for this problem. Since the Petri net models can be related directly to the full rate equations, if follows that for Case 2a the discrete behaviour can only be compared to the solution of the full rate equations or, at most, of the tQSS model, but certainly not to the model provided by [48].

The paper then goes on to analyse the full Case 2a problem with a Petri net that builds on the insights gained in the Case 1c analysis. We do not replicate the linear stability analysis of Case 2a, which can be found in [11] and which does not offer any surprises: the Lyapunov exponent is negative, compatibly with stable limit-cycle behaviour. Finally, the paper uses the Petri nets of the two systems to obtain the corresponding automata. We find that even a very coarse Petri net with places of capacity 1 (whose markings correspond to binary numbers whose number of digits equals the number of places of such a net) is able to capture meaningful biochemical information. Significantly, it appears that only a very small subset of the algebraic structures found through holonomy decomposition corresponds to the behaviour normally observed for this reaction.

24.2   Case 1c

In preparation of the analysis of Case 2a from Tyson et al. [48], this section analyses a system that is formally identical to Case 1c, which was already analysed numerically and analytically via Lie groups in BIOMICS report D1.1.1 [9]. This is the Goldbeter and Koshland's (G–K) ultrasensitive switch [23], which has been analysed extensively since it was first published in 1981 [7, 22, 24, 48].

Although a Lie symmetry was found that integrates the differential equation for this system, there are two reasons for not reporting that result in this paper. First, the equation we integrated is the form given by Tyson et al. and implied by Goldbeter and Koshland, which relies on the MM approximation. Since Case 2a violates the conditions required by this approximation, the analysis of Case 1c subject to this approximation is not relevant to Case 2a. Second, in the application of the Lie group method the last step involves the inversion of a function $t(y)$ to obtain the desired $y(t)$, where t is time and y is the dependent variable of the ODE in question. The function $t(y)$ given by the symmetry we found for this problem is not invertible analytically, meaning that we could not find a full analytical solution in closed form even for Case 1c under the MM approximation.

Therefore, in the first part of the paper we examine the simplifying assumptions commonly made for MM systems as they apply to Case 1c. The careful retracing, discussion, and double-checking of the derivations that is presented in detail in [9, 11] and summarized here is a consequence of some minor inconsistencies that we found in some of these publications [7, 48].[7]

After a recap of Petri nets, Sect. 24.2.1 provides some context for the problem being studied in the form of background on the analysis of the two coupled enzymatic reactions from the literature and within BIOMICS. Section 24.2.2 explains the two main approximations for this system, the Michaelis-Menten quasi-steady-state approximation and the 'total quasi-steady-state' approximation. In order to show their effects in detail, Sect. 24.2.2.1 derives both approximations for the single-enzyme system. The next two Sects. 24.2.2.2 and 24.2.2.3, do the same for the 2-enzyme system, and Sect. 24.2.3 presents a comparison between the various models and the stochastic Petri net simulations for Case 1c.

24.2.1 Problem Context

The G–K switch [23] is based on the following two coupled enzymatic reactions for the phosphorylation and dephosphorylation of a protein Y:

$$
\begin{aligned}
[Y] + [X] &\underset{d_1}{\overset{a_1}{\rightleftharpoons}} [YX] \xrightarrow{k_1} [Y_p] + [X] \\
[Y_p] + [Z] &\underset{d_2}{\overset{a_2}{\rightleftharpoons}} [ZY_p] \xrightarrow{k_2} [Y] + [Z],
\end{aligned}
\tag{24.1}
$$

where the subscript 'p' indicates the phosphorylated state, X and Z denote the enzymes for the forward and backward reactions, respectively, a_i is the 'association' rate constant, d_i is the 'dissociation' rate constant, and k_i is the 'catalysis' rate constant. Other symbols often used for these constants are shown in Table 24.1. In

---

[7]One of these inaccuracies is present in our own work: it appears in [9] and was explained and rectified in [11].

Table 24.1 Some of the common symbols for the Michaelis-Menten chemical reaction rate constants

	Notation 1	Notation 2	Notation 3	Notation 4	Etc.
Association rate	a	k_+	k_1	k_1	$\ldots$
Dissociation rate	d	k_-	k_{-1}	k_2	$\ldots$
Catalytic rate	k	k_{cat}	k_2	k_3	$\ldots$

this paper we use consecutive number subscripts such as Notation 4 (but not necessarily starting at 1) because it makes it easier to relate the chemical reaction equations to the Petri net diagrams.

As discussed in D1.1.1 [9], the switch behaviour refers to the rapid rise in concentration exhibited by the phosphorylated variable Y_p when the ratio of the total concentration of the forward enzyme X_T to the reverse reaction enzyme Z_T increases beyond a certain threshold. The so-called signal-response curve is a steep sigmoid when the enzymes are saturated, i.e. when their concentration is much smaller than the substrate's. The signal-response curve corresponds to a *family* of steady-state responses for this system, since the total amounts of enzyme are constant for any one analysis, but it is useful to think of this system as a module in a larger system in which one or both of the enzymes are synthesized by some other module. This is precisely what Case 2a does, where the switch-like behaviour serves to amplify the growth of the compounds in the downstream modules, which are arranged in a closed loop. The result is a sustained oscillation or limit cycle—at least in the case of the ODE model used in [48] for Case 2a.

However, in the analysis of Case 2a, we found that the Petri net does not oscillate for the same initial conditions and parameter values that cause the ODE system to oscillate, and exhibits a damped rather than sustained oscillation for a different set of values (shown and discussed below). This is worrying, because it implies that the oscillation may be an artifact that partly depends on the approximations made to arrive at the Case 2a ODEs. This is what motivated a careful analysis of the approximations that have been developed for these kinds of systems, especially in light of the fact that they are widely used in systems biology [17, 39].

Figure 24.1 shows the Petri net corresponding to this example. We constructed this Petri net starting from the diagram presented by Nabli et al. [33] to model the Michaelis-Menten reaction. Starting from the rate equations obtained from Eq. (24.1) and retracing the derivation outlined by Goldbeter and Koshland [23], in D1.1.1 [9] we showed how to derive the single ODE which in [48] appears as Case 1c (although with different variable names). The full rate equations derivable from Eq. (24.1) using mass-action kinetics and corresponding to Fig. 24.1 are:

$$\dot{X} = -k_1 XY + (k_2 + k_3)[YX] \qquad\qquad X(0) = X_T \qquad (24.2)$$

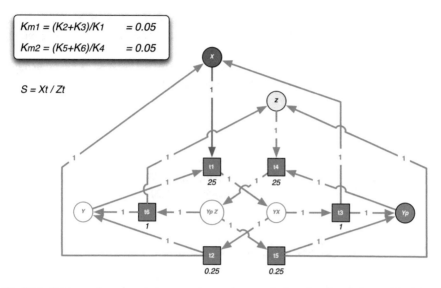

Fig. 24.1 Petri net of two coupled enzymatic reactions catalyzing phosphorylation and dephosphorylation of a protein Y. This system is one component of the more complex Case 2a system

$$\dot{Y} = -k_1 XY + k_2 [YX] + k_6 [Y_P Z] \qquad Y(0) = 0 \qquad (24.3)$$

$$[\dot{YX}] = k_1 XY - (k_2 + k_3)[YX] \qquad [YX](0) = 0 \qquad (24.4)$$

$$\dot{Z} = -k_4 ZY_p + (k_5 + k_6)[Y_p Z] \qquad Z(0) = Z_T \qquad (24.5)$$

$$\dot{Y}_p = k_3 [YX] - k_4 ZY_p + k_5 [Y_p Z] \qquad Y_p(0) = Y_T \qquad (24.6)$$

$$[\dot{Y_p Z}] = k_4 ZY_p - (k_5 + k_6)[Y_p Z], \qquad [Y_p Z](0) = 0 \qquad (24.7)$$

where we have dropped most of the square brackets for notational expediency since all variables are assumed to be in units of molar concentration (moles/unit volume). For the enzyme-substrate complexes we do retain the square brackets, to distinguish them from the product of the two species they are composed of.

Using the notation used here, Eq. (3.128) in [9], which we derived from the above rate equations, takes the following form:

$$\dot{Y}_p = k_1 X_T \frac{Y}{k_{m_1} + Y} - k_2 Z_T \frac{Y_p}{k_{m_2} + Y_p}, \qquad (24.8)$$

where the subscript 'T' (or 't' in Fig. 24.1) indicates total amount (in this case of each enzyme), and the MM constants are given by

$$k_{m_i} = \frac{d_i + k_i}{a_i}, \qquad i = 1, 2. \qquad (24.9)$$

The same constants are shown in Fig. 24.1 in terms of the notation of that figure, where the subscript of each rate constant equals the number of the corresponding transition.

Correcting a minor oversight in Tyson et al.'s paper [48] that is discussed in depth in [11], the ODE for Y_p that should have been used for Case 1c in [48] is

$$\dot{Y}_p = k_1 X_T \frac{Y_T - Y_p}{k_{m_1} + Y_T - Y_p} - k_2 Z_T \frac{Y_p}{k_{m_2} + Y_p}, \qquad (24.10)$$

which we use in the rest of this report. We now discuss briefly the two approximations of the full rate equations (24.2)–(24.7) found in the literature and used to arrive at Eq. (24.10) (QSS) and to an intermediate system (tQSS).

24.2.2 Assumptions and Approximations

In the derivation leading up to Eq. (24.10) the same two assumptions usually used to derive the MM equation are made. The first is the 'standard' *quasi-steady state* (QSS$_1$) assumption, namely that the rate of change of the enzyme-substrate complex is zero. Referring to Fig. 24.1, this means that the "outflow", $(k_2 + k_3)[YX]$, equals the "inflow", $k_1 XY$; and likewise for the reverse (dephosphorylation) reaction. The second assumption is not generally considered part of QSS but we introduce it explicitly here because the form of the ODEs used for Case 1c and Case 2a in [48] requires it. The assumption is that the amount of substrate-enzyme complex is so small relative to the amount of substrate to be negligible (QSS$_2$).

The justification for QSS$_1$ depends on the requirement that the enzyme is saturated, i.e. that the concentration of substrate is much higher than the enzyme concentration. Clearly the same condition also renders QSS$_2$ valid. In fact, under this condition the available enzymes will be quickly sequestered by the substrate to form the complex, after which it (the complex) will not be able to vary much since, as soon as it splits into a product plus enzyme (catalytic reaction) or back into a substrate plus enzyme (dissociation reaction), the enzyme will quickly be sequestered again by some other substrate molecule.

Segel [46] rationalized QSS$_1$ in terms of the different time-scales at which the different reactions of the MM system progress. That work was further developed a few years later by Borghans et al. [5] who introduced a better approximation, called the *total quasi-steady state* (tQSS) approximation. The QSS$_1$ and tQSS approximations were then examined in detail again by Tzafriri [49], and a few years later by Ciliberto et al. [7], who discuss specifically the G–K [23] ultrasensitive switch.

Interestingly, in their formulation of the model corresponding to the QSS approximation Ciliberto et al. [7] do not make use of the second assumption. As a consequence, their QSS model is different from the form normally found in the literature. Therefore, we now analyse in detail the QSS and tQSS approximations as they apply

to the *single-enzyme* MM system before discussing these approximations for the full
rate equations (24.2)–(24.7) of the 2-enzyme G–K ultrasensitive switch.

24.2.2.1   QSS and tQSS Approximations for the Single-Enzyme Reaction

QSS Derivation. In the standard MM derivation, the rate equations for the substrate
and the enzyme-substrate complex are:

$$\dot{X} = -k_1 XY + (k_2 + k_3)[YX] \qquad\qquad X(0) = X_T \qquad (24.11)$$

$$\dot{Y} = -k_1 XY + k_2[YX] \qquad\qquad Y(0) = Y_T \qquad (24.12)$$

$$[\dot{YX}] = k_1 XY - (k_2 + k_3)[YX] \qquad\qquad [YX](0) = 0 \qquad (24.13)$$

$$\dot{Y_p} = k_3[YX] \qquad\qquad Y_p(0) = 0 \qquad (24.14)$$

Adding (24.11) and (24.13) leads to the conservation law for the enzyme:

$$X_T = X + [YX]. \qquad (24.15)$$

Adding (24.12)–(24.14) leads to the conservation law for the substrate:

$$Y_T = Y + [YX] + Y_p. \qquad (24.16)$$

Assuming that (QSS$_1$)

$$[\dot{YX}] = 0 \qquad (24.17)$$

in (24.13), leads to the familiar MM expression for the complex:

$$[YX] = X_T \frac{Y}{k_m + Y}, \qquad (24.18)$$

such that (24.14) becomes

$$\dot{Y_p} = k_3 X_T \frac{Y}{k_m + Y}. \qquad (24.19)$$

Differentiating (24.16) and making use of the first assumption (24.17) again leads to

$$\dot{Y} = -k_3 X_T \frac{Y}{k_m + Y}, \qquad Y(0) = Y_T. \qquad (24.20)$$

This is all that's needed for the model of the MM system with the standard QSS$_1$
approximation. However, if we want to calculate the time-evolution of the phospho-

rylated product, then we need to express (24.19) in terms of Y_p only. Solving for Y from (24.16) does not work since $[YX]$ appears on the right-hand side (see below), so we introduce the second assumption, QSS$_2$. If $[YX]$ is negligible compared to Y_T, then from (24.16) we have

$$Y = Y_T - Y_p \qquad (24.21)$$

such that the MM time-evolution of the product under QSS$_1$ + QSS$_2$ can be found by integrating:

$$\dot{Y}_p = k_3 X_T \frac{Y_T - Y_p}{k_m + Y_T - Y_p}, \qquad Y_p(0) = 0. \qquad (24.22)$$

Following [46] or [5, 11] shows that the most stringent condition to guarantee the validity of the QSS approximation is

$$\frac{\Delta Y}{Y_T} = \varepsilon = \frac{X_T}{Y_T + k_m} \ll 1. \qquad (24.23)$$

tQSS Derivation. An intuitively accessible derivation of the tQSS approximation can be obtained by noting that when $[YX]$ is not negligible we can use (24.16) to eliminate Y from (24.18) in favour of Y_p and $[YX]$:

$$[YX] = X_T \frac{Y_T - Y_p - [YX]}{k_m + Y_T - Y_p - [YX]}, \qquad (24.24)$$

which leads to a quadratic for $[YX]$:

$$[YX]^2 - (k_m + X_T + Y_T - Y_p)[YX] + X_T(Y_T - Y_p) = 0. \qquad (24.25)$$

Its solution is:

$$[YX] = \frac{1}{2}(k_m + X_T + Y_T - Y_p)$$
$$- \frac{1}{2}\sqrt{(k_m + X_T + Y_T - Y_p)^2 - 4X_T(Y_T - Y_p)}, \qquad (24.26)$$

where we have used the negative branch of the square-root [5]. This expression for $[YX]$ is the analogue for the tQSS approximation of Eq. (24.18) for the QSS approximation.[8] In fact, we can now use (24.26) in (24.14) to obtain the ODE that corresponds to the tQSS approximation:

_____

[8] Strictly speaking, for the QSS$_2$ assumption, since both QSS and tQSS rely on the QSS$_1$ assumption.

$$\dot{Y}_p = k_3 \frac{1}{2}(k_m + X_T + Y_T - Y_p)$$
$$- k_3 \frac{1}{2}\sqrt{(k_m + X_T + Y_T - Y_p)^2 - 4X_T(Y_T - Y_p)}. \qquad (24.27)$$

We can see where the name 'tQSS' comes from by following Borghans et al. [5], who derive (24.27) by introducing a new variable:

$$\hat{Y} = Y + [YX], \qquad (24.28)$$

which lumps together the variable that changes on a fast time-scale ($[YX]$) and one of those that changes on a slow time-scale (Y) in such a way that the total substrate $\hat{Y}$ cannot be depleted by the formation of the complex. As explained by Borghans et al. 'Because the validity of the classical QSSA depends strongly on negligible initial depletion of substrate [...], this simple variable change is expected to have an important effect' [5].

Because the new approximation that ensues involves the total amount of substrate, it was dubbed 'total QSS' or tQSS.[9] Using (24.15) and (24.28), Eqs. (24.12) and (24.13) become

$$\dot{\hat{Y}} - [\dot{YX}] = -k_1(X_T - [YX])(\hat{Y} - [YX]) + k_2[YX] \qquad \hat{Y}(0) = Y_T \qquad (24.29)$$
$$[\dot{YX}] = k_1 \left[(X_T - [YX])(\hat{Y} - [YX]) - k_m[YX] \right] \qquad [YX](0) = 0. \qquad (24.30)$$

Adding,

$$\dot{\hat{Y}} = -k_3[YX] \qquad \hat{Y}(0) = Y_T. \qquad (24.31)$$

It is worth noting that so far we have only introduced a new variable but have not changed the character of the equations. The functions Y and $[YX]$ obtainable from the solution of (24.31) and (24.30) are no different from those obtained with the full rate equations (24.11)–(24.14).

Setting (24.30) equal to zero (as for QSS$_1$) we obtain a quadratic in $[YX]$:

$$[YX]^2 - (k_m + X_T + \hat{Y})[YX] + X_T\hat{Y} = 0, \qquad (24.32)$$

whose solution is

$$[YX] = \frac{1}{2}(k_m + X_T + \hat{Y}) - \frac{1}{2}\sqrt{(k_m + X_T + \hat{Y})^2 - 4X_T\hat{Y}}, \qquad (24.33)$$

---

[9]Confusion may be caused by the use of 'total' in the name of this approximation to indicate only $Y + [YX]$, whereas in (24.16) we used Y_T for all the species that involve Y in some way. The former is a subjective choice for the name of a variable that makes sense in reactions where the product is actually a different molecule rather than recognizably the same molecule in a phosphorylated state, as here.

Note that, using (24.16),

$$Y_T - Y_p = Y + [YX] = \hat{Y}, \tag{24.34}$$

such that (24.33) is the same expression as (24.26). Substituting (24.33) into (24.31) we obtain:

$$\dot{\hat{Y}} = -k_3 \frac{1}{2}(k_m + X_T + \hat{Y})$$
$$+ k_3 \frac{1}{2}\sqrt{(k_m + X_T + \hat{Y})^2 - 4X_T\hat{Y}}, \quad \hat{Y}(0) = Y_T, \tag{24.35}$$

from whose solution $[YX](t)$, $Y(t)$, and $Y_p(t)$ can be recovered using (24.33), (24.28), and (24.34), respectively. Differentiating (24.34) makes it clear that (24.35) is the same ODE as (24.27).

There is one remaining point to be clarified, in that the initial condition for Eq. (24.35) is inconsistent with the value of $[YX]$ given by (24.33) at $t = 0$, which is not zero when $\hat{Y} = Y_T$. This can be explained by noting that the initial condition for the ODE (24.35) is derivable from the full rate equations under the change of variable (24.28) and is therefore correct. It is (24.33) that is an approximation which, although better than the QSS equivalent (24.18), gets worse the closer one gets to $t = 0$.

Figure 24.2 shows a comparison of the solution found with the full rate equations with these two approximations. As the conditions approach the requirement (24.23) the QSS$_1$ and tQSS approximations converge to the exact solution. When the enzyme concentration is of the same magnitude as the substrate's, on the other hand, the tQSS approximation does a little better. Figure 24.3 shows the same trend for the Y_p product, for which the QSS$_2$ assumption was also used in order to arrive at Eq. (24.22).

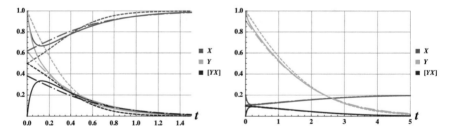

Fig. 24.2 Numerical solution of the time evolution of the substrate, enzyme, and substrate-enzyme complex for the single-enzyme Michaelis-Menten system for two different initial enzyme/substrate ratios. *Left* $\varepsilon = X_T/(Y_T + k_m) = 0.5$. *Right* $\varepsilon = X_T/(Y_T + k_m) = 0.1667$. *Solid* full rate equations (Eqs. (24.12)–(24.14)); *dashed* QSS$_1$ approximation (Eq. (24.20)); *dot-dashed* tQSS approximation (Eq. (24.35)). $k_1 = 10$, $k_2 = 5$, $k_3 = 5$, $k_m = 1$, $Y_T = 1$

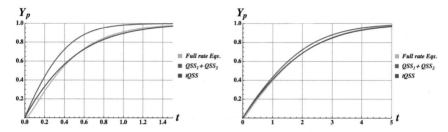

Fig. 24.3 Numerical solution of the time evolution of the phosphorylated product for the single-enzyme Michaelis-Menten system for two different initial enzyme/substrate ratios. *Left* $\varepsilon = X_T/(Y_T + k_m) = 0.5$. *Right* $\varepsilon = X_T/(Y_T + k_m) = 0.1667$. Full rate equations: Eqs. (24.12)–(24.14); QSS$_1$ + QSS$_2$: Eq. (24.22); tQSS: Eq. (24.27). $k_1 = 10$, $k_2 = 5$, $k_3 = 5$, $k_m = 1$, $Y_T = 1$

24.2.2.2 QSS Approximation for Coupled 2-Enzyme Reactions

For the 2-enzyme ultrasensitive switch, the ODE for Y_p that corresponds to the QSS$_1$ + QSS$_2$ approximation, shown above as Eq. (24.10), is obtained by using (24.21) in (24.8). This ODE corresponds to Fig. 24.1 and Eq. (24.1), but only as long as (24.23) is true. Since in Case 2a the concentrations of the enzymes are *not* much smaller than the concentrations of the substrates, this approximation is clearly not justified. It was because of the fact that the Case 2a system of ODEs relies on both the QSS assumptions when clearly neither is justified that the problem of verifying the Petri net of Fig. 24.1, which corresponds to the full rate equations (24.2)–(24.7), became more challenging than it had seemed at first.

Ciliberto, Capuani and Tyson [7] cite the above form of the Goldbeter and Koshland ODE for Y_p (Eq. (24.10)) but then, oddly, use a *different* form for what they call the QSS approximation of the full rate equations (24.2)–(24.7). Their derivation can be reconstructed as follows. Starting with the following equation for the conservation law of the substrate,

$$Y_T = Y + [YX] + Y_p + [Y_p Z], \tag{24.36}$$

which is obtained by adding (24.3), (24.4), (24.6), and (24.7), it can be used to eliminate Y from the equation for the $[YX]$ complex,

$$[YX] = X_T \frac{Y}{k_{m_1} + Y} = X_T \frac{Y_T - [YX] - Y_p - [Y_p Z]}{k_{m_1} + Y_T - [YX] - Y_p - [Y_p Z]}, \tag{24.37}$$

as we did above for (24.18) to obtain (24.24). However, if Y_p is eliminated from the corresponding equation for $[Y_p Z]$,

$$[Y_p Z] = Z_T \frac{Y_p}{k_{m_2} + Y_p}, \tag{24.38}$$

also using (24.36), the result will be a system of two simultaneous *quadratic* equations each involving both complexes $[YX]$ and $[Y_pZ]$, whose explicit solution does not appear possible. Therefore, Ciliberto et al. eliminated just Y but kept (24.38) as it is. In fact, their QSS model appears to model tQSS in the forward direction (for Y) and QSS in the backward direction (for Y_p). The result is more accurate than QSS as generally defined (i.e. Eq. (24.10)) but less accurate than tQSS. The governing equations are algebraic except for the ODE for Y_p:

$$X = X_T - [YX] \tag{24.39}$$

$$Y = Y_T - [YX] - Y_p - [Y_pZ] \tag{24.40}$$

$$[YX] = \frac{1}{2}\left\{(k_{m_1} + X_T + Y_T - Y_p - [Y_pZ])\right.$$
$$\left. - \sqrt{(k_{m_1} + X_T + Y_T - Y_p - [Y_pZ])^2 - 4X_T(Y_T - Y_p - [Y_pZ])}\right\} \tag{24.41}$$

$$Z = Z_T - [Y_pZ] \tag{24.42}$$

$$\dot{Y}_p = k_3[YX] - k_6[Y_pZ] \tag{24.43}$$

$$[Y_pZ] = Z_T \frac{Y_p}{k_{m_2} + Y_p}, \tag{24.44}$$

where the initial condition for Y_p is the same as for Eq. (24.6). The initial values for the other variables are given by substituting $Y_p(0) = Y_T$ in the algebraic equations shown.

To make sure we understand the approximations discussed so far, Fig. 24.4 shows a comparison between the solution to the full rate equations and the system above. In reference to the switching behaviour, this example corresponds to the very middle of the switch, i.e. to the case $S = X_T/Z_T = 1$. Thus, at equilibrium the product Y_p is neither high nor low. Most of the solution curves calculated with Ciliberto et al.'s

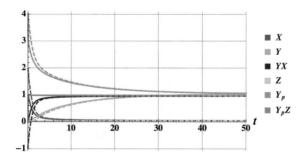

Fig. 24.4 Comparison of the numerical solutions of the time evolution of all the variables of the G–K switch (Case 1c) as calculated with the full rate equations (24.2)–(24.7) (*solid*) and the Ciliberto et al.'s [7] system (24.39)–(24.44) (*dashed*). $k_1 = 25$, $k_2 = 0.25$, $k_3 = 1$, $k_4 = 25$, $k_5 = 0.25$, $k_6 = 1, k_{m_1} = 0.05, k_{m_2} = 0.05, Y(0) = 0, X(0) = X_T = 1, [YX](0) = 0, Y_p(0) = Y_T = 4, Z(0) = Z_T = 1, [Y_pZ](0) = 0$

system are close to the exact solution from the full rate equations, as we might expect given that the ratio $\varepsilon_1 = X_T/(Y_T + k_{m_1}) = \varepsilon_2 = Z_T/(Y_T + k_{m_2}) = 1/4.05 = 0.247$. The curves for $[YX]$ and, to a much smaller extent, Y start negative, which is a bit worrying. However, any approximation that relies on the QSS_1 assumption in some way is not expected to be accurate in the initial transient region of the system's response. Finally, the fact that Eq. (24.36) is satisfied by both solutions is easy to see as $t \to \infty$ $(1 + 1 + 1 + 1 = 4)$.

Figure 24.5 shows a comparison of three sets of solution curves, for a completely different set of parameters: the full rate equations (24.2)–(24.7), Ciliberto et al.'s [7] system (24.39)–(24.44), and the standard $QSS_1 + QSS_2$ ODE (24.10).

This example is interesting because the values of the k_m constants in the two directions are different. In the forward direction, $\varepsilon_1 = X_T/(Y_T + k_{m_1}) = 20/(50 + 1) = 0.392$; in the backward direction, $\varepsilon_2 = Z_T/(Y_T + k_{m_2}) = 200/(50 + 100) = 1.333$, so we would expect the QSS approximations not to be very accurate, especially in the backward direction. This example corresponds to the OFF position of the switch, with $X_T/Z_T = 1/10$, as can be seen by the low equilibrium value of Y_p. The curves corresponding to the QSS approximation are significantly different from the exact (full rate equations) solution in the approach to equilibrium, although with the exception of the curve for Y they converge to the correct values. Ciliberto's et al.'s solutions similarly show significant deviations in the initial transient region, especially $[YX]$'s initial negative value, but they all converge to the correct equilibrium values. The gross inaccuracy of the QSS Y curve is simply a consequence of the QSS_2 assumption, embodied in Eq. (24.21), and was probably the motivation for using the tQSS approximation in the forward direction.

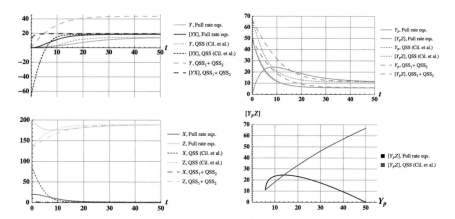

Fig. 24.5 Comparison of the numerical solutions of the time evolution of all the variables of the G–K switch (Case 1c). The full rate equations (24.2)–(24.7) are compared to the standard $QSS_1 + QSS_2$ approximation from (24.10) and to the QSS approximation (24.39)–(24.44) from [7], which amounts to QSS_1 only in our notation and from which we have taken the initial conditions and parameter values: $k_1 = 0.1, k_2 = 0.05, k_3 = 0.05, k_4 = 0.0009, k_5 = 0.005, k_6 = 0.085, k_{m_1} = 1, k_{m_2} = 100$, $Y(0) = 0, X(0) = X_T = 20, [YX](0) = 0, Y_p(0) = Y_T = 50, Z(0) = Z_T = 200, [Y_pZ](0) = 0$

The last plot, on the bottom right, is a 'phase diagram' because it involves only the dependent variables. The curves are parametrized by time, with $t = 0$ on the right end of both curves shown and the common equilibrium point on the far left. The black curve is the exact solution, whereas the blue curve is the solution obtained with Eqs. (24.39)–(24.44). The latter shows how the QSS approximation models the fast initial transient as a discontinuous vertical jump, followed by a gradual descent towards equilibrium, increasingly approaching the exact solution. The same diagram in Ciliberto et al.'s article [7] shows that the time-steps of these two curves are not uniform, with the solution spending most of the time on the far left, which is consistent with the other diagrams in this figure. Interestingly, the curve for Ciliberto et al.'s solution is indistinguishable from the one for the QSS solution, which is why only one of them is drawn here. The reason can be surmised by looking at the plot immediately above the phase diagram: here we can see that even though the two solutions (dashed and long-dashed) for $[Y_pZ]$ and Y_p are different, their *ratios* appear also to the naked eye to be very similar.

We now derive the ODEs for the tQSS approximation of the rate equations (24.2)–(24.7) for the chemical reaction equations (24.1).

24.2.2.3 tQSS Approximation for Coupled 2-Enzyme Reactions

The power of the tQSS approximation developed by Borghans et al. [5] can now be appreciated better. Define

$$\hat{Y} = Y + [YX] = Y_T - \hat{Y}_p \qquad (24.45)$$

$$\hat{Y}_p = Y_p + [Y_pZ] = Y_T - \hat{Y}, \qquad (24.46)$$

where Y_T is given by (24.36). As in the derivation that led to ODE (24.35) for the single-enzyme reaction, if we set $[\dot{Y}X] = 0$ and $[\dot{Y}_pZ] = 0$, we get the quadratics

$$[YX]^2 - (k_{m_1} + X_T + \hat{Y})[YX] + X_T\hat{Y} = 0 \qquad (24.47)$$

$$[Y_pZ]^2 - (k_{m_2} + Z_T + \hat{Y}_p)[Y_pZ] + Z_T\hat{Y}_p = 0, \qquad (24.48)$$

which are now decoupled and from which we can therefore obtain the functions for the complexes:

$$[YX] = \frac{1}{2}(k_{m_1} + X_T + \hat{Y}) - \frac{1}{2}\sqrt{(k_{m_1} + X_T + \hat{Y})^2 - 4X_T\hat{Y}}, \qquad (24.49)$$

$$[Y_pZ] = \frac{1}{2}(k_{m_2} + Z_T + \hat{Y}_p) - \frac{1}{2}\sqrt{(k_{m_2} + Z_T + \hat{Y}_p)^2 - 4Z_T\hat{Y}_p}. \qquad (24.50)$$

Replicating the derivation of Eq. (24.35), the two ODEs for the tQSS approximation are obtained as

$$\dot{\hat{Y}} = -k_3[YX] + k_6[Y_pZ] \qquad\qquad \hat{Y}(0) = 0 \qquad\qquad (24.51)$$

$$\dot{\hat{Y}}_p = -k_6[Y_pZ] + k_3[YX] \qquad\qquad \hat{Y}_p(0) = Y_T. \qquad\qquad (24.52)$$

Substituting (24.49) and (24.50) into Eqs. (24.51) and (24.52) gives two coupled ODEs for $\hat{Y}$ and $\hat{Y}_p$. The remaining two equations for the tQSS approximation are just the conservation laws for the two enzymes:

$$X_T = X + [YX] \qquad\qquad (24.53)$$

$$Z_T = Z + [Y_pZ]. \qquad\qquad (24.54)$$

The tQSS model used by Ciliberto et al. [7] looks different but in fact is the same as the above. The difference is that instead of using the ODE for $\hat{Y}$, Eq. (24.51), they use the conservation law, Eq. (24.36). This is equivalent since adding (24.51) and (24.52) gives (24.36) after an integration step. The rest of Ciliberto et al.'s equations for the tQSS model are the same as the above.

Finally, (24.44) is indeed a QSS and not a tQSS approximation, which can be seen by substituting $Y_p + [Y_pZ]$ for $\hat{Y}_p$ in (24.48):

$$[Y_pZ]^2 - (k_{m_2} + Z_T + \hat{Y}_p)[Y_pZ] + Z_T\hat{Y}_p = 0$$

$$[Y_pZ]^2 - (k_{m_2} + Z_T + Y_p + [Y_pZ])[Y_pZ] + Z_T(Y_p + [Y_pZ]) = 0$$

$$-k_{m_2}[Y_pZ] - Y_p[Y_pZ] + Z_TY_p = 0$$

$$[Y_pZ] = Z_T\frac{Y_p}{k_{m_2} + Y_p},$$

which is the same as (24.44).[10] In other words, $\hat{Y}_p$ hides a "piece" of $[Y_pZ]$, and the "remainder" of this variable happens to satisfy a quadratic. When this piece is released, the function for $[Y_pZ]$ reverts to (24.44). In fact, (24.44) is the origin of the negative values for $[YX]$ and Y in the transient region, under the model (24.39)–(24.44), if we trace its effect through the other equations of that model.

Figures 24.6 and 24.7 show that the tQSS approximation is practically indistinguishable from the exact solution. The plots are shown to different magnifications to make it easier to see the very slight difference between the curves. The details of the $[Y_pZ]$ are not visible in Fig. 24.7b, but its behaviour is practically a mirror image of Z about 0.5.

Figure 24.8 shows a comparison between the solutions to the full rate equations and the QSS equations, as in Fig. 24.5, and the tQSS equations (24.45)–(24.50). Consistently with the different values of the QSS test (24.23) in the two directions, the agreement in the forward direction is significantly better than in the backward direction.

---

[10]As expected, the same result is obtained by substituting $Y_p + [Y_pZ]$ for $\hat{Y}_p$ in (24.50).

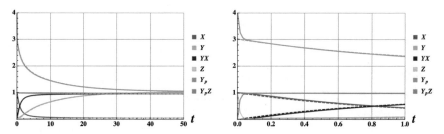

Fig. 24.6 Comparison of the numerical solutions of the time evolution of all the variables of the G–K switch (Case 1c) as calculated with the full rate equations (24.2)–(24.7) (*solid*) and the tQSS approximation (24.49)–(24.54) (*dashed*). The plot on the *right* shows the same data but only in the first second, to show how tQSS approximates the rapid decrease in Y_p by a discontinuous jump at $t = 0$. $k_1 = 25, k_2 = 0.25, k_3 = 1, k_4 = 25, k_5 = 0.25, k_6 = 1, k_{m_1} = 0.05, k_{m_2} = 0.05, Y(0) = 0$, $X(0) = X_T = 1, [YX](0) = 0, Y_p(0) = Y_T = 4, Z(0) = Z_T = 1, [Y_pZ](0) = 0$

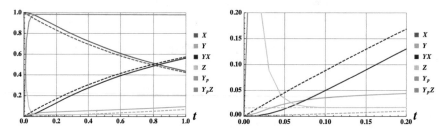

Fig. 24.7 Same conditions as previous figure but at greater vertical and horizontal magnification to show the level of accuracy of the tQSS approximation in the transient region

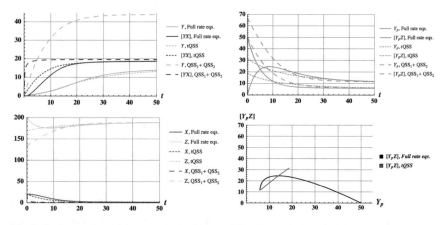

Fig. 24.8 Comparison of the numerical solutions of the time evolution of all the variables of the G–K switch (Case 1c). The full rate equations (24.2)–(24.7) are compared to the standard QSS$_1$ + QSS$_2$ approximation from (24.10) and to the tQSS approximation (24.49)–(24.54). $k_1 = 0.1, k_2 = 0.05$, $k_3 = 0.05, k_4 = 0.0009, k_5 = 0.005, k_6 = 0.085, k_{m_1} = 1, k_{m_2} = 100, Y(0) = 0, X(0) = X_T = 20$, $[YX](0) = 0, Y_p(0) = Y_T = 50, Z(0) = Z_T = 200, [Y_pZ](0) = 0$

24.2.3   Petri Net and Numerical Analysis Results

As a first step in verifying the stochastic simulation of the Petri net of Fig. 24.1, Fig. 24.9 shows that the conservation laws are satisfied.

Figure 24.10 shows a comparison of the stochastic simulation of the Petri net of Fig. 24.1 with the numerical solution of the full rate equations (24.2)–(24.7), for $S = 0.5$. The curves shown in the left diagram are the distributions of tokens for the 6 places of this Petri net, averaged over 1000 runs over the same time window and starting from the same initial conditions. They all show a rapid transient and they all reach equilibrium. The agreement with the numerical solution of the full rate equations is quite good, especially in the equilibrium region. The only variable that deviates a little from the numerical solution, which we take as the correct or 'exact' solution here, is the equilibrium value of Y_p, which is a bit higher than 0 in the Petri net simulation. The reason is that since we are using discrete tokens to model concentration occasional transitions of tokens into the Y_p place will cause its average value over many runs to be greater than zero. The substrate concentration is 4, whereas the X and Z enzymes have initial values of 1 and 2, respectively, yielding

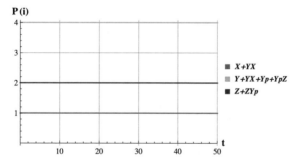

Fig. 24.9 Verification of the conservation laws in the stochastic simulation of the Petri net of Fig. 24.1, averaged over 1000 runs and for $S = 0.5$. Initial conditions: $(X, Y, [YX], Y_p, Z, [Y_pZ]) = (1, 0, 0, 4, 2, 0)$

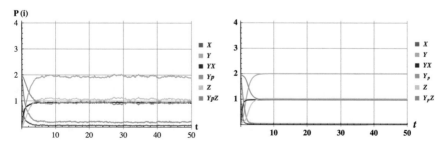

Fig. 24.10 Comparison of the Stochastic simulation averaged over 1000 runs of the Petri net of Fig. 24.1 (*left*) and the numerical solution of the full rate equations (24.2)–(24.7) (*right*). $S = 0.5$; initial conditions: $(X, Y, [YX], Y_p, Z, [Y_pZ]) = (1, 0, 0, 4, 2, 0)$

a value for the signal S of 0.5. The value of $\varepsilon = 1/4.05 = 0.247$ from Eq. (24.23) is not so small, and yet the agreement of the Y_p curve (dark green) with the lowest, red curve in Fig. 3.15a of D1.1.1 [9] is pretty good.

Figures 24.11 and 24.12 show two more cases for the same system, for $S = 1.0$ and $S = 2.0$. Also here the behaviour is as expected and qualitatively consistent with the numerical solution of D1.1.1. There is a noticeable mismatch in the rate at which Y and Y_p reach equilibrium. It is not clear how this can be accounted for. One possibility is that the observed discrepancy is caused by the usual mismatch between the simulation and the ODE approach when particle copy numbers are low. See [26–30] for further explanation of the effect.

The next set of Figs. 24.13, 24.14 and 24.15 compares the tQSS approximation to the QSS approximation. Whereas the tQSS approximation is almost identical to the full system of rate equations (24.2)–(24.7), the QSS does quite poorly, especially for Y, as we already saw above. It seems, therefore, that to validate the Petri net for a set of chemical reactions one should use either the original rate equations or the tQSS approximation, but not the QSS. This would seem to be particularly important when the amount of enzyme is of the same order of magnitude as or larger than the substrate.

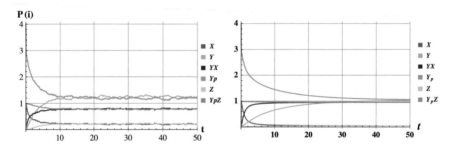

Fig. 24.11 Comparison of the Stochastic simulation averaged over 1000 runs of the Petri net of Fig. 24.1 (*left*) and the numerical solution of the full rate equations (24.2)–(24.7) (*right*). $S = 1.0$; initial conditions: $(X, Y, [YX], Y_p, Z, [Y_pZ]) = (1, 0, 0, 4, 1, 0)$

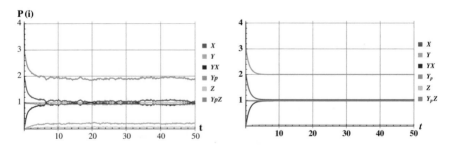

Fig. 24.12 Comparison of the Stochastic simulation averaged over 1000 runs of the Petri net of Fig. 24.1 (*left*) and the numerical solution of the full rate equations (24.2)–(24.7) (*right*). $S = 2.0$; initial conditions: $(X, Y, [YX], Y_p, Z, [Y_pZ]) = (2, 0, 0, 4, 1, 0)$

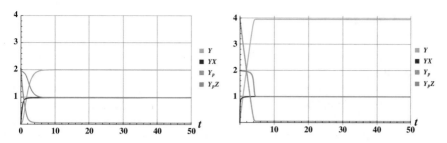

Fig. 24.13 Comparison of the tQSS (*left*) and the QSS (*right*) approximations of Eqs. (24.2)–(24.7). $S = 0.5$; initial conditions: $(X, Y, [YX], Y_p, Z, [Y_pZ]) = (1, 0, 0, 4, 2, 0)$

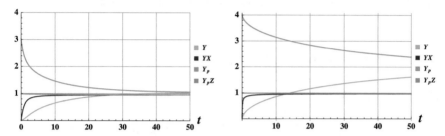

Fig. 24.14 Comparison of the tQSS (*left*) and the QSS (*right*) approximations of Eqs. (24.2)–(24.7). $S = 1.0$; initial conditions: $(X, Y, [YX], Y_p, Z, [Y_pZ]) = (1, 0, 0, 4, 1, 0)$

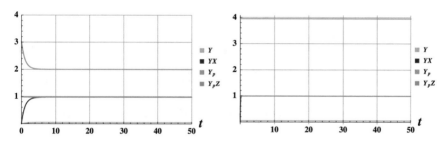

Fig. 24.15 Comparison of the tQSS (*left*) and the QSS (*right*) approximations of Eqs. (24.2)–(24.7). $S = 2.0$; initial conditions: $(X, Y, [YX], Y_p, Z, [Y_pZ]) = (2, 0, 0, 4, 1, 0)$

These results show that the best approximation is tQSS, and that the Petri net matches the full rate equations. Thus, in the remainder of the paper we will rely on the full rate equations, and we will use the Petri net with some confidence that it does model the physics of the problem. A somewhat surprising finding, however, is that our faith in the full rate equations is only justified if the number of molecules is large. For small numbers, the Petri net simulations turn out to be the more "physical".

24.3 Case 2a

Tyson et al. took this system from [21]. Both models have the same problem, in that both use the QSS approximation without satisfying the assumptions upon which it is based. As a consequence, the full rate equations for the same parameters and a compatible set of initial conditions do not show oscillations, and the same applies to the Petri net model. Using a different set of parameters, the numerical solution of the full rate equations shows weakly damped oscillations. The verification of the Petri net is not as convincing as for Case 1c since it can't reproduce the same amplitude of the oscillations as the numerical solution's. It appears that this may be caused, again, by the mass-action assumption and will require further analysis in future work.

24.3.1 Full Rate Equations and Petri Net

Figure 24.16 shows the Case 2a pathway, which was first modelled by Goldbeter [21] based on the empirical work of Félix et al. [16].

The analysis begins with the chemical reaction equations. These are not given in [48] but are easily inferrable from the pathway and a knowledge of the Case 1c 2-enzyme reaction:

$$S \xrightarrow{k_1} [X] \tag{24.55}$$

$$[R_p] + [X] \xrightarrow{k'_2} [R_p] + (X \text{ depleted}) \tag{24.56}$$

$$[Y] + [X] \underset{k_4}{\overset{k_3}{\rightleftharpoons}} [YX] \xrightarrow{k_5} [Y_p] + [X] \tag{24.57}$$

$$[Y_p] + [Z] \underset{k_7}{\overset{k_6}{\rightleftharpoons}} [ZY_p] \xrightarrow{k_8} [Y] + [Z] \tag{24.58}$$

$$[R] + [Y_p] \underset{k_{10}}{\overset{k_9}{\rightleftharpoons}} [RY_p] \xrightarrow{k_{11}} [R_p] + [Y_p] \tag{24.59}$$

$$[R_p] + [W] \underset{k_{13}}{\overset{k_{12}}{\rightleftharpoons}} [WR_p] \xrightarrow{k_{14}} [R] + [W], \tag{24.60}$$

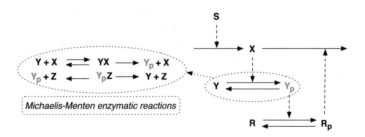

Fig. 24.16 Pathway of the Cyclin (X) oscillator [21] or Case 2a [48]

where S is the 'signal' which controls the rate of production of X, and the presence of R_p accelerates the depletion of X, Eqs. (24.55) and (24.56). Y and R are proteins being phosphorylated to produce Y_p and R_p. X and Z act as enzymes for the forward and backward reactions for Y_p, (24.57) and (24.58), and Y_p and W serve the same purpose for the reactions for R_p, (24.59) and (24.60).

The corresponding rate equations are:

$$\dot{X} = k_1 S - k_2' X R_p - k_3 XY + (k_4 + k_5)[YX] \tag{24.61}$$

$$\dot{Y} = -k_3 XY + k_4[YX] + k_8[Y_p Z] \tag{24.62}$$

$$[\dot{YX}] = k_3 XY - (k_4 + k_5)[YX] \tag{24.63}$$

$$\dot{Y}_p = k_5[YX] - k_6 ZY_p + k_7[Y_p Z] - k_9 Y_p R + (k_{10} + k_{11})[RY_p] \tag{24.64}$$

$$\dot{Z} = -k_6 ZY_p + (k_7 + k_8)[Y_p Z] \tag{24.65}$$

$$[\dot{Y_p Z}] = k_6 ZY_p - (k_7 + k_8)[Y_p Z] \tag{24.66}$$

$$\dot{R} = -k_9 Y_p R + k_{10}[RY_p] + k_{14}[R_p W] \tag{24.67}$$

$$[\dot{RY}_p] = k_9 Y_p R - (k_{10} + k_{11})[RY_p] \tag{24.68}$$

$$\dot{R}_p = k_{11}[RY_p] - k_{12} W R_p + k_{13}[R_p W] \tag{24.69}$$

$$\dot{W} = -k_{12} W R_p + (k_{13} + k_{14})[R_p W] \tag{24.70}$$

$$[\dot{R_p W}] = k_{12} W R_p - (k_{13} + k_{14})[R_p W]. \tag{24.71}$$

The Petri net that can be derived from either set of equations is shown in Fig. 24.17.

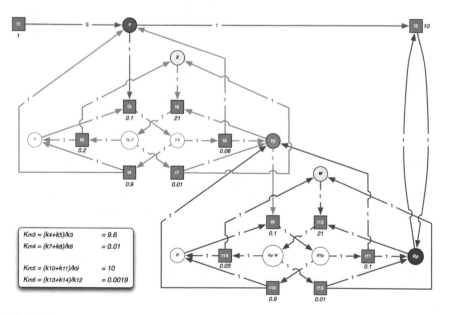

Fig. 24.17 Petri net corresponding to the full rate equations of the negative feedback oscillator

24.3.2   Comparison with Tyson et al.'s Case 2a Model

It took a long time and many attempts to arrive at this Petri net. As discussed in the previous section, much effort went into establishing and validating the Petri net for one of the components of the Case 2a system, the ultrasensitive Goldbeter–Koshland (G–K) switch [23]. The main problem was that the Petri net shown in Fig. 24.17 just wouldn't oscillate using the parameters and initial values provided by Tyson et al. [48]. We now briefly retrace these steps.

Tyson et al.'s Case 2a is given as the following set of equations

$$\dot{X} = k_0 + k_1 S - k_2 X - k'_2 R_p X \tag{24.72}$$

$$\dot{Y}_p = \frac{k_5 X (Y_T - Y_p)}{K_{m_3} + Y_T - Y_p} - \frac{k_8 Y_p}{K_{m_4} + Y_p} \tag{24.73}$$

$$\dot{R}_p = \frac{k_{11} Y_p (R_T - R_p)}{K_{m_5} + R_T - R_p} - \frac{k_{14} R_p}{K_{m_6} + R_p}. \tag{24.74}$$

As we showed in the previous section, the ODEs for Y_p and R_p in this set can be derived from the full rate equations (24.61)–(24.71) using the QSS (QSS$_1$ + QSS$_2$) approximation in both directions for each module. We can see that X plays the same role in the ODE for Y_p as S did for Case 1c. Similarly, Y_p, in turn, plays the role of the enzyme in the ODE for R_p. However, in order to make these equations dimensionally consistent, the total amount of reverse enzyme in the second and third equations (Z and W in the full rate equations) must be assumed to have concentration equal to 1. The numbering of the rate constants has been changed to match the more general full rate equations model. On the other hand, the numbering of the Michaelis-Menten constants K_{m_i} has been left unchanged since they do not appear explicitly in the full rate equations model.

Table 24.2 lists all the parameters relevant to the two models. The constants of the full rate equations model were chosen to match the corresponding constants in the Tyson et al. model and so that the Michaelis-Menten constants K_{m_i}, also shown in the table, would come out with the same values (see Fig. 24.17 for the expressions used to calculate them). We should point out that the degradation constant k_2, which equals 0.01 in Tyson et al.'s model, was set equal to 0 in the full rate equation model. This is because numerical analysis showed that its presence or absence did not make any difference to the solutions, but adding another transition to the Petri net would have made the automaton, to be analysed in a later section, even more complex.

Some additional difficult modelling choices had to be made. Referring to Fig. 24.16, we can see that the action of the species X and Y_p, which serve as enzymes in the forward reactions for Y_p and R_p, respectively, are drawn as dotted arrows rather than continuous arrows. This means 'promotion' of the reaction the arrow points to by the presence of the compound the arrow starts from,[11] and does not necessarily

---

[11] Or could mean inhibition if the dotted line ends with a flat segment instead of an arrow.

Table 24.2 Parameter values and initial conditions for Tyson et al.'s Case 2a model [48] and for the full rate equations (24.61)–(24.71), for a few different cases.

Parameter values	S	k_0	k_1	k_2	k_2'	k_3	k_4	k_5	k_6	k_7	k_8
From [48] Fig. 24.18	2	0	1	0.01	10	–	–	0.1	–	–	0.2
Equations (24.61)–(24.71), Fig. 24.19	2	0	1	0	10	11	0.01	0.1	21	0.01	0.2
Figures 24.17, 24.21, 24.22 and 24.24	1	0	1	0	10	0.1	0.9	0.06	21	0.01	0.2
Parameter values	k_9	k_{10}	k_{11}	k_{12}	k_{13}	k_{14}		K_{m_3}	K_{m_4}	K_{m_5}	K_{m_6}
From [48] Fig. 24.18	–	–	0.1	–	–	0.05		0.01	0.01	0.01	0.01
Equations (24.61)–(24.71), Fig. 24.19	11	0.01	0.1	6	0.01	0.05		0.01	0.01	0.01	0.01
Figures 24.17, 24.21, 24.22 and 24.24	0.1	0.9	0.1	21	0.01	0.03		9.60	0.01	10.0	0.0019
Initial conditions	X	Y	[YX]	Y_p	Z	$[Y_pZ]$	R	$[RY_p]$	R_p	W	R_pW
From [48] Fig. 24.18	5	–	–	1	–	–	–	–	0.5	–	–
Equations (24.61)–(24.71), Fig. 24.19	5	0	0	1	1	0	0	3	1	1	0
Figures 24.17 and 24.21	5	0	0	10	1	0	0	3	5	3	0
Figures 24.17 and 24.22	50	0	0	100	20	0	0	30	50	20	0
Figures 24.17 and 24.24	500	0	0	1000	200	0	0	300	500	200	0

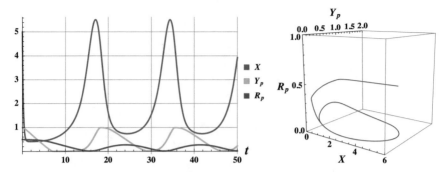

Fig. 24.18 Numerical solution of the negative feedback oscillator according to the Tyson et al. Case 2a ODE model [48], Eqs. (24.72)–(24.74)

imply mass flow. However, since we had already verified and validated the Petri net for Case 1c in the previous section, using the same "wiring" seemed justified. Indeed, the arrows issuing, for example, from t_4 and t_5 replenish X, consistently with enzymatic reactions.

The action of R_p on t_2, on the other hand, was not as straightforward. The best behaviour was obtained by introducing two arcs between them, in opposite directions. In this manner two desired effects are obtained:

- An increase in the concentration of R_p increases the probability of t_2 firing, consistently with the effect of a promotor.
- While the firing of t_2 has the desired effect of depleting X, the arrow pointing back towards R_p replenishes it. Therefore R_p is not depleted when this transition fires, consistently with the semantics of the dotted promotion arrow and with the absence of the term $-k_2' X R_p$ in Eq. (24.69).

Figure 24.18 shows the numerical solution of the Tyson et al. Case 2a model, according to Eqs. (24.72)–(24.74). There is a slight inconsistency in [48]: the initial condition for R_p is stated as $R_p(0) = 1$. However, the plot of the numerical solution provided in this article shows an initial value of $R_p(0) = 0.1$, approximately. We were not able to obtain exactly the same plot for the three solution curves as shown in this reference; however, with $R_p(0) = 0.5$ Fig. 24.18 is very close. The figure also shows the phase-space trajectory, with the expected limit cycle.

Figure 24.19 shows the numerical solution of the full rate equations *for the same parameter values and initial conditions*, given in Table 24.2. Unfortunately, the solution curves are totally different and do not show any oscillations at all. Or, rather, the response seems heavily overdamped.

Such a mismatch between these models is not surprising given that the parameters and initial conditions used by Tyson et al.'s model violate its assumptions. The implication is that the oscillations that this model generates for these parameter values and initial conditions are artifacts of the approximation and are not physical.

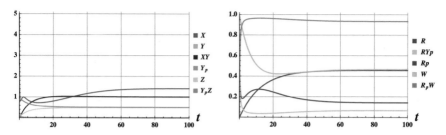

Fig. 24.19 Numerical solution of the negative feedback oscillator according to the full rate equations, Eqs. (24.61)–(24.71)

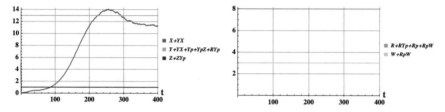

Fig. 24.20 Verification of the conservation laws of the negative feedback oscillator as calculated with the stochastic Petri net simulator. As expected X is not conserved

However, it still makes physical sense that the stacked components of Case 2a should oscillate, as long as there is some time delay between them. Therefore, it seemed that *some* combination of parameters and initial conditions should work. This is in fact what we found, although not in as clear-cut a way as we would have wished.

Figure 24.20 shows that the conservation laws are respected, so the implementation appears to be correct. Unlike Case 1c, the "enzyme" X in this case is not conserved, but neither do we expect it to be since it is being created and destroyed at different rates by t_1 and t_2.

After many attempts working with the numerical solution of the full rate equations we finally found a combination of parameters and initial conditions that gave convincing oscillations, as shown in Fig. 24.21, although they are weakly damped. The turning point that allowed finding this response was the realization that biochemical oscillations, being based on 1st-order and not 2nd-order systems, are not characterized by an equilibrium position in the centre of the oscillation, an inertia that overshoots equilibrium, and a return force that brings the mass back towards equilibrium. There is no inertia, and it is not accurate to talk about a "return force". Rather, we posit that the behaviour of either of the 2-enzyme G–K components can be characterized as follows:

- Something akin to "equilibrium" appears to exist (or in any case can be arbitrarily defined) in the dephosphorylated state of the protein (either Y_p or R_p, the argument applies to

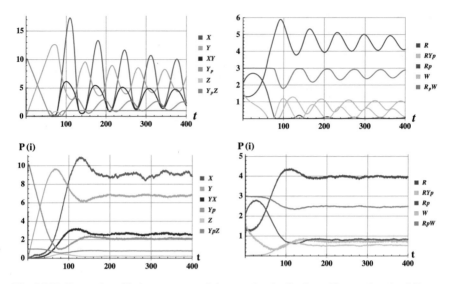

Fig. 24.21 Damped oscillation response of the negative feedback oscillator using the full rate equations (*top*) and a stochastic simulation of the Petri net shown in Fig. 24.17, averaged over 3000 runs (*bottom*). The parameters and initial conditions used are listed in Table 24.2

both).[12] Therefore, referring to Fig. 24.17, this state can be visualized as being "on the left" rather than "in the centre" of the oscillation, i.e. with Y "full" and Y_p "empty".

- The system (meaning both G–K components) should have a very fast return action towards the "equilibrium" dephosphorylated state. As discussed by Goldbeter and Koshland [23], this steep 'ultrasensitive' behaviour is achieved when the enzyme is saturated with substrate (QSS conditions) and the Michaelis-Menten constant K_m is very small, e.g. 0.01.
- A property of such a system that introduces something analogous to a delay is to make the forward (phosphorylation) action much slower, i.e. *not* ultrasensitive. This is achieved by making K_m fairly large (3 orders of magnitude larger than the reverse K_m in our example).

Figure 24.21 also shows the result of the stochastic simulation of the Petri net, averaged over 3000 runs. The oscillations are visible, but much more damped than the full rate equations solution. At first we thought that a possible cause for this 'washed out' behaviour of the oscillations in the stochastic simulation was that the onset of oscillations takes place at some point after the system is started. If this occurrence signals the loss of stability of the system, it is possible that the stochastic simulation is not as sensitive at picking up loss of stability as other features of the dynamics. However, the behaviour of the other variables in the same initial region is captured very well compared to the oscillations that occur later. Therefore, the fault lies elsewhere.

---

[12]Rather than a statement of biochemistry, this should be seen as a dynamical systems interpretation of the observed behaviour.

According to Konkoli's work [26–30], the reason is that the mass-action assumption, upon which the full rate equations are based, breaks down for small numbers of molecules. Therefore, the stochastic simulation shows the correct behaviour and the fault lies with the ODE solution, whose marked oscillations are not "physical".

A simple way to test this claim is to compare the full rate equations to the stochastic simulation for a larger system, i.e. a system with more molecules. This is shown in Figs. 24.22 and 24.23, where the initial conditions are changed slightly to obtain more marked oscillations in the ODE solution and then multiplied by 10 without changing the rate constants (see Table 24.2). In Figs. 24.24 and 24.25 the initial conditions are then multiplied by 10 again. In this case, although the two sets of solution curves share the same geometric features, the frequency of the oscillations as calculated by the ODE model is markedly higher. For all these cases the 'correct' solution is the stochastic, but the ODE solution improves the larger the system becomes, supporting the claim.

For the purposes of this paper it is not necessary to derive the tQSS approximation for this system since we already know that it is at best as accurate as the full rate equations. Rather, we now switch to the discrete analysis of these systems.

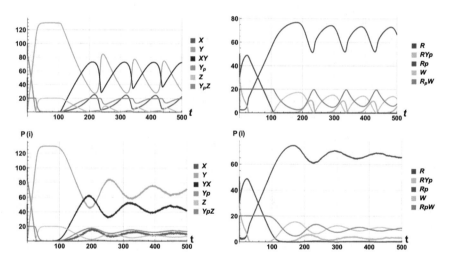

Fig. 24.22 Response of the negative feedback oscillator using the full rate equations (*top*) and a stochastic simulation of the Petri net shown in Fig. 24.17, averaged over 1000 runs (*bottom*). The parameters and initial conditions used are listed in Table 24.2

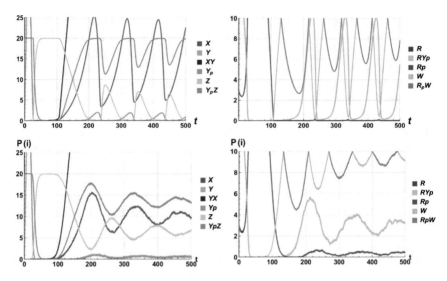

Fig. 24.23 Expanded vertical scale for Fig. 24.22, to resolve the variables with smaller values

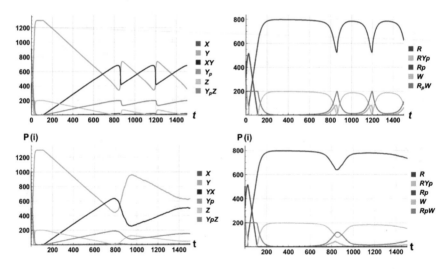

Fig. 24.24 Response of the negative feedback oscillator using the full rate equations (*top*) and a stochastic simulation of the Petri net shown in Fig. 24.17, averaged over 500 runs (*bottom*). The parameters and initial conditions used are listed in Table 24.2. As shown in [44], a higher number of tokens leads to shorter stochastic time-steps; thus, for this simulation there were approximately 1,000,000 time-steps over the time-window shown and for each of the 500 runs, requiring approximately 116 h of CPU time on a 2.7 GHz MacBook Pro Intel Core i7 in total to run the simulation (our *Mathematica* code only uses one core)

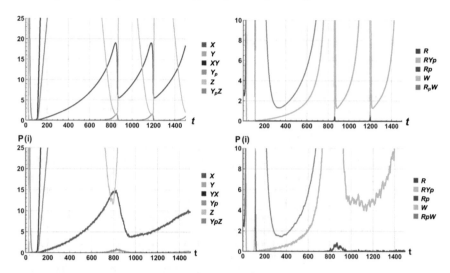

Fig. 24.25 Expanded vertical scale for Fig. 24.24, to resolve the variables with smaller values

24.4 Discrete Analysis of Case 1c and Case 2a

In this section we start analysing discrete models of the systems analysed in the previous sections. The emphasis is on Petri nets as a bridge between the continuous and the discrete views of these systems. We should clarify that we are just beginning to figure out how to extract structural and dynamical information from the discretized finite-state automaton corresponding to a biochemical system, based on its hierarchical (Krohn-Rhodes or holonomy) decomposition. Whether or not this information can be obtained any other way is not yet clear, nor how useful it is from a computational point of view. The reasons for this—and what could be done to address the challenge—are slowly becoming apparent the more we work with ASMs [4], and are discussed at the end of this section. Before we can compare different views on the modelling of continuous and discrete dynamical systems, however, we need to develop some understanding of the kind of information that each view provides. This section aims to give a sense of the kind of information that algebraic automata theory provides, and how it might be related to some of the physical properties of the systems discussed in the previous sections. We first provide a brief conceptual summary of holonomy decomposition.

24.4.1 Holonomy Decomposition

As discussed in detail and through several examples in [10, 42], the holonomy decomposition (HD) of an automaton is visualized by means of the 'skeleton diagram'.

However, the skeleton diagram is only a shorthand of the algebraic structure of an automaton understood as its HD. Here we can only provide a telegraphic summary of its main properties.

- The skeleton, obtained with the GAP [47] program SgpDec [15], shows the holonomy [51] rather than the Krohn-Rhodes [31] decomposition of an automaton.
- The HD involves a loop-free cascade of permutation-reset automata rather than of simple non-abelian groups (SNAGs) and (banks of) 2-state resets (flip-flops).
- Permutation groups can be decomposed using the Jordan-Hölder theorem into a composition series involving SNAGs, and constant maps (resets) can similarly be decomposed into 2-state resets. Since SNAGs and 2-state resets cannot be further decomposed, the HD can be regarded as an intermediate step to the full decomposition of an automaton into its 'prime' components. However, it is conceptually easier to think about the HD, and the only available implementation [15] was developed for the HD.
- More precisely, the HD of an automaton is a multi-stage wreath product between two "orthogonal" classes of transformation semigroups, i.e., permutation groups and constant maps or resets. As such, it is a loop-free hierarchy of so-called permutation-reset automata[13] acting on image sets of the state set of the original automaton in the form of macro-states of decreasing size (increasing resolution) as one progresses down the hierarchy to the singletons at the bottom.
- These image sets are called 'tiles', but in the sense of roof tiles rather than bathroom tiles since they can overlap on one or more states.[14]
- At any one level of the decomposition there are usually many isomorphic copies of a given group or set of constant maps, so they are arranged in equivalence classes.
- In the skeleton diagram, the presence of a group at a given level and acting on tiles at one or more levels below is indicated by drawing the equivalence class as a long rectangle with dark background, whereas the equivalence classes associated with the constant maps (and trivial groups, i.e. identities) have a white background.
- The skeleton diagram shows only one of the isomorphic copies of a permutation group or constant map, which is called the 'representative' (in analogy with coset representatives from group theory). The tiles that are the images of a given representative and upon which the group or constant map acts are called 'tiles of the representative' or 'rep-tiles'.
- In the skeleton diagram, the rep-tiles can be recognized because they are image sets of the representative they come from and they are joined to it by a solid (or sometimes dotted) line.
- A dotted line from a representative to one of its rep-tiles indicates a 'Garden of Eden' state, meaning a state which once left can never be recovered (revisited) by the system.
- Although the HD is a much larger computational object than the original automaton, one of its sub-structures, the 'cascade product', emulates the behaviour of the original automaton.
- A cascade of permutation-reset automata is composed by one representative from each equivalence class and the (macro)states upon which it acts (rep-tiles). Together, the representatives and their rep-tiles form the 'holonomy components'.

The appeal of the HD analysis is that it identifies reversible groups among (usually many more) sets of irreversible constant maps—both kinds of objects as subsemi-

---

[13] A permutation-reset automaton is an automaton whose action is either a permutation of the state set or a constant map or reset (i.e. a map from the state set to a single state).

[14] This metaphor is due to Attila Egri-Nagy.

groups of the semigroup of an automaton. A drawback is that the HD is not unique, a point to which we return at the end of this section. Be that as it may, the objective here is to achieve some insight into the physical—and computational—interpretation of the groups, when they occur. This is because the presence of groups suggests that some quantities are being conserved. BIOMICS started from the assumption that when such structures occur in automata derived from biological systems they (the groups) may have something to do with the biological systems' self-organizing properties. As a final 'health warning', it helps to distinguish in our mind the algebraic view, for which the action is on the whole state set, from the physical or computational view, for which the action is sequential from one state to the next.

24.4.2   Single Component: The Coupled 2-Enzyme System

We start with one of the three components of the Case 2a oscillator, i.e. Case 1c from [48], shown in Fig. 24.26 along with the automaton generated from its markings. The skeleton diagram shows the presence of two groups, C_3 and C_2.

This system is interesting because it allows us to compare easily physical behaviour to group action. However, this is not as straightforward as we might expect. For example, the skeleton diagram in Fig. 24.26 tells us that one of the components of the holonomy cascade is a permutation group that consists of a group C_2 acting on states

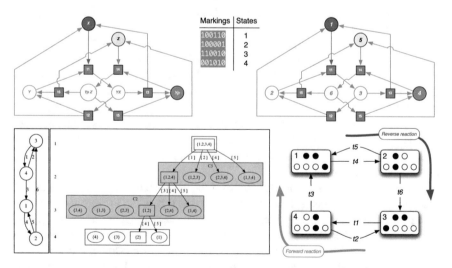

Fig. 24.26 Analysis of a Capacity-1 discrete model of the 2-enzyme Goldbeter–Koshland ultrasensitive switch [23] (Case 1c in [48]). *Top row* Petri net showing biochemical species as place names; table showing how Petri net markings map to the states of the automaton; Petri net with explicit place numbering to facilitate the interpretation of the markings. *Bottom row* automaton and skeleton diagrams generated by SgpDec [15]; automaton using a visualization of the markings and indicating the two opposite directions of the enzymatic reactions

Table 24.3 Action of individual transitions making up C_2 element to permute states 1 and 2

	Starting state	t_6	t_1	t_4	t_3	Ending state
Action on state 1	1	1	1	2	2	2
Action on state 2	2	3	4	4	1	1

1 and 2. From a physical point of view, the natural inference is that this action is related to the formation and the dissociation of the enzyme-substrate complex [YX]. However, the transitions between these two states are effected by *different* arrows and, therefore, cannot result from the action of C_2. In fact, in the permutation group C_2 acting on a set of two states, the non-identity element permutes the states, but here t_4 only moves 1–2, it does not affect 2; similarly, t_5 moves 2–1 but does not affect 1.

Rather, the algebraic 'pool of reversibility' [40] that the presence of this group implies is related to the physical reversibility of the enzymatic reactions. In other words, the permutation action is effected through a roundabout sequence of transitions: the word provided by SgpDec for permuting states 1 and 2 involves t_4 but not t_5, and is actually $t_6 t_1 t_4 t_3$.[15] As discussed in [10] and proved in [13], the only way to obtain permutation of states in an automaton generated from a Petri net without inhibition arcs is to allow 'mute' transitions—in other words, to treat input symbols to the automaton that are not relevant to the current state as equivalent to an Assembly language NOP or to the identity mapping. Table 24.3 shows the action of this group element on states 1 and 2.

We should note that there are actually 6 isomorphic copies of C_2 acting on 6 different pairs of singleton states (4 choose 2 = 6). Of these, {1, 2} and {3, 4} have a similar physical interpretation. Another pair of states that is permuted is 1 and 3, by the word $t_1 t_4 t_6 t_3$. The physical significance in this case is that 1 and 3 are the opposite ends of the oscillation of the G–K switch when it is embedded in the Case 2a system. In the present case, i.e. when the system is on its own, it does not oscillate, but each state is reachable from the other. The action of C_3 is rather more complicated given that now what is permuted is 3 objects, each of which is a set of 2 states, as shown by the skeleton diagram.

Going on to the whole automaton shown in the lower-right corner of Fig. 24.26, the visualization of the states as place markings makes it possible to "see" the chemical reactions as the system moves from one state to the next along the sequence $t_1 t_3 t_4 t_6$. From a dynamical systems point of view, this sequence of reactions corresponds to the system starting with a maximum concentration of substrate, moving to the opposite extreme of zero substrate and maximum phosphorylated protein, and then coming back to the starting point. Of course, since there is no analogue of inertia or a "spring" return force in this 1st-order system the oscillation is actually not possible, but the reversibility appears to be captured by the algebraic analysis.

---

[15]Transformation semigroup elements composed of Petri net transitions are normally assumed to act on the left. This is done so that a string of transitions can be written and read from left to right.

A more fruitful point of view is to focus on the fundamental properties of groups, i.e. their association with symmetries. Paraphrasing Ian Stewart, (a) a symmetry is an invertible transformation that preserves some aspect of the structure of a mathematical object; and (b) it is easily proven that the set of all such symmetries always forms a group. The converse is what we need here: the presence of a group always implies that something is being conserved. Referring again to the lowest level of the skeleton shown in Fig. 24.26, in transitioning from State 1 to State 2, and vice versa, it appears that Places 1, 2, and 3 are not affected by these transformations. So the "state" of these three places is the invariant. Between States 1 and 3, on the other hand, the two enzymes and the two complexes (Places 1, 3, 5, and 6) remain constant. At the level above, we find that the rep-tiles shown preserve only the substrate, i.e. Place 2. The isomorphic copy of this action that corresponds to States $\{2, 3, 4\}$ preserves the phosphorylated protein Y_p. The remaining two tiles $\{1, 2, 3\}$ and $\{1, 3, 4\}$ are similar in preserving Places 1 & 3 and 5 & 6, respectively. The expectation is that for more complex systems, and in particular for integrable systems of ODEs, the quantities preserved may provide more insight into their dynamics. The task, then, will be to see whether the invariants of the Lie symmetries may be physically related to the invariants of the discrete automata models of the same systems.

Figure 24.27 shows the automaton that is generated when allowing for all possible initial conditions. For this system, the result is 11 different and disjoint automata (rather than a single large automaton), only one of which corresponds to the case

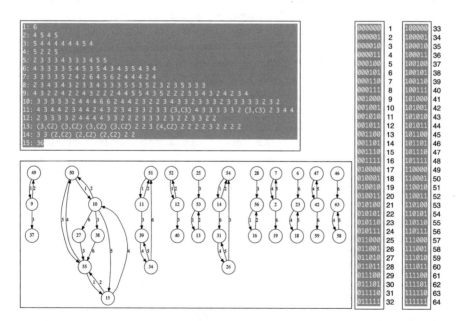

Fig. 24.27 *Bottom* disconnected automata generated using all possible initial conditions Capacity-1 discrete model of the 2-enzyme Goldbeter–Koshland ultrasensitive switch [23] (Case 1c in [48]); *top* holonomy components; *right* state numbering

just analysed in Fig. 24.26. The figure also shows the map of the markings to the state numbers. Since in this case we are analysing the global automaton and since all the places in this Petri net of 6 places had capacity 1, each place can only have zero or 1 token, so the total number of states is simply 2^6, and the mapping is just the binary representation of the state numbers. The state numbering changes with each analysis, so state 1 in the previous example does not correspond to state 1 in this example, and similarly for the others. Finally, the figure also shows the holonomy components of the global automaton.

In this case the skeleton diagram is very large indeed and cannot fit in a single A4 page, so the holonomy components are a useful shorthand. Each component shown corresponds to an equivalence class whose size is not shown, i.e. each component corresponds to a class representative (picked arbitrarily by SgpDec). The plain number 6 at the top level, for example, means (1) that the top component is a set of resets and identity maps (because the component is shown as a simple integer), and (2) that the set of resets acts on a set of 6 'points'. In other words, the permutation-reset automaton at the top level of the cascade consists in a set of 6 elements of the semigroup (and trivial groups or identities) that act as resets, i.e. that map a set of 6 points to itself by constant maps. The 6 'points' are actually subsets of the state set, also known as 'macrostates' in computer science. Although not relevant for the emulation of the original automaton, these 6 rep-tiles are obtained as image-sets of the set of all 64 states of this automaton under the action of 6 different elements of the semigroup of the automaton on the whole state set (these maps are *not* the same as the resets, which *are* relevant to the emulation function of the cascade).

In the lower levels we can see many more integers next to each other at the same level. Each such integer indicates a different equivalence class, and other sets of as many rep-tiles that also map to themselves through resets and identities. In a holonomy decomposition the only other possibility for the components of the original automaton is the presence of permutation groups. The first such group is seen at Level 11, and it is denoted as $(3, C_3)$, indicating the C_3 cyclic group acting on 3 rep-tiles.

Figure 24.28 shows a visualization of one of each of the different types of automaton shown in the previous figure. This helps us see that of all these types only the one we have just seen in Fig. 24.26 is clearly related to this physical problem, although it is not the only one that is physically feasible. For example, the automaton at the top-right of Fig. 24.28 is not likely to be observed for this problem because two of the states show that Y and Y_p are full at the same time. However, with such a coarse resolution it is not possible to distinguish between the case where both species have maximum concentration and the case where they are both at 0.5 concentration. This latter case corresponds to $S = 1$ in Case 1c, i.e. in the equilibrium state of the switch in the middle of the sigmoid curve, which is certainly feasible. The same system with greater capacity would need to be analysed to see what concentration levels these places actually have. The largest automaton has the same problem for one of its states. The two small automata are both physically plausible, but their dynamics are not very interesting.

Figure 24.29 shows the analysis of the Capacity-2 version of the Petri net for this system, i.e. where each place can hold zero, 1 or 2 tokens. For the initial condition

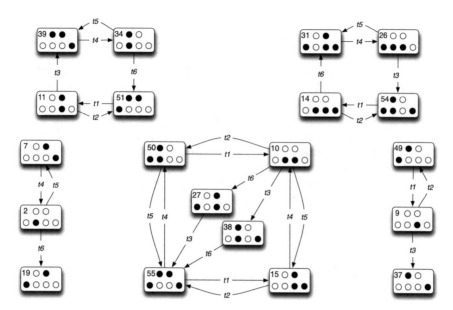

Fig. 24.28 Visualization of some of the automata shown in Fig. 24.27. The automaton at the *top-left* is the same as that shown in Fig. 24.26

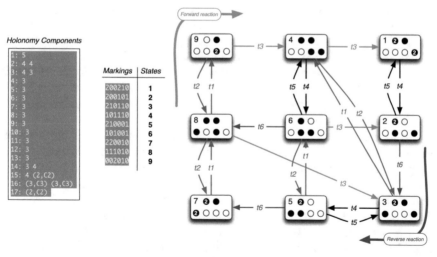

Fig. 24.29 Visualization of the automaton corresponding to the Petri net shown in Fig. 24.26 for Capacity-2 places. The automaton is comprised of the 9 states that are reachable from the initial state, shown as State 1. Places with two tokens are shown explicitly, places with 1 token are *black*, empty places are *white*. The "oscillation" between Y_p and Y that is shown in Fig. 24.26 applies also in this higher-resolution automaton, following the outer states in the clock-wise direction from State 1

$((p_1, p_2, p_3, p_4, p_5, p_6) = (2, 0, 0, 2, 1, 0))$, shown in the figure as State 1, only 8 states are reachable of a possible $3^6 = 729$. These are all physically plausible states, as can be easily verified by following the visualization in the figure. The states are arranged in a way to highlight that the "oscillation" between the two extremes $(Y_p = 2, Y = 0)$ (State 1) and $(Y_p = 0, Y = 2)$ (State 7) corresponds to traversing the automaton around its outer states in the clockwise direction, just as in Fig. 24.26. Other paths (state traces) are also admissible. Finally, the holonomy components show a similar structure to the lower-resolution, Capacity-1 case just analysed.

The algebraic regularities of this discrete structure apply to a much larger set of states. However, some of these are not necessarily helpful or meaningful in a specific applied context. For example, it is certainly possible physically for both Y and Y_p to be at maximum concentration simultaneously, but it is not likely to be observed in situations where this pathway acts like a switch or is part of an oscillator. This highlights the importance of physical context in "making sense" of the algebra.

24.4.3 Negative Feedback Oscillator (Case 2a)

In this section we analyse the automaton derived from the simplest possible Petri net model of the full Case 2a system, applying the insights gained in the previous section to find a possible interpretation for at least some aspects of its algebraic structure. Figure 24.30 shows the automaton generated by SgpDec [15] along with a possible 3-D visualization. This is done to highlight how the state transitions defined as transformations between Petri net markings correspond to movements along the independent dimensions of $\mathbb{R}^n$, where n is the number of transitions. So what is shown here is a 3-D projection of an object in $\mathbb{R}^{14}$ (actually $\mathbb{R}^{13}$ since t_2 is not firing in this specific case), which however is already much more expressive than the normal 2-D automaton representation.

Figure 24.31 shows the same automaton but with the additional visualization of the token markings. The number key for the Petri net places at the bottom of this figure refers to the columns of the states-markings table on the left and to the geometry of the Petri net for this system as shown in Fig. 24.17. Figure 24.32 shows the holonomy components for this system. The cascade product has 76 levels. We notice the presence of some large groups in the higher levels: D_{12}, S_7 and S_6.

If we allow t_2 to fire, the automaton acquires 10 additional states, as shown in Fig. 24.33. It is worth noting that 16 of the new set of 26 states are identical to the previous case, as are the transitions between them. This demonstrates how adding a transition adds a dimension to the space of the automaton and creates a larger automaton that is a superset of the original. On the other hand, adding capacity to a place changes the number of states *along a particular dimension*, without changing the dimension of the space in which the automaton is embedded. Thus, the resolution of the discretization increases.

Figure 24.34 shows the same automaton but with the visualization of the Petri net markings corresponding to each state. This helps us see the chemical processes

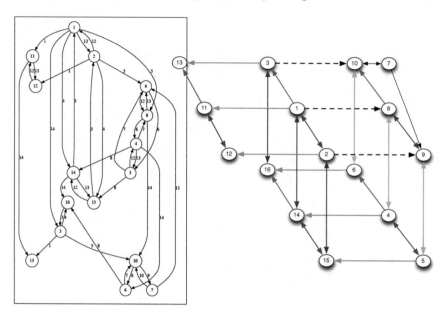

Fig. 24.30 3-D visualization of the automaton generated by `SgpDec` [15] for the Case 2a Petri net with Capacity 1 and t_2 disabled. State 9 is the starting state: $\{X, Y, [YX], Y_p, Z, [Y_pZ], R, [Y_pR], R_p, W, [WR_p]\} = \{1, 0, 0, 1, 1, 0, 0, 0, 1, 1, 0\}$

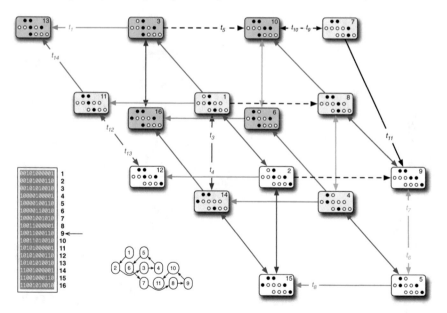

Fig. 24.31 Markings-to-states mapping and visualization of the same automaton as in the previous figure, this time using the token visualization of each state. The number key at the *bottom* refers to the places shown in Fig. 24.17. Transition numbers are drawn near the head of the respective *arrows*

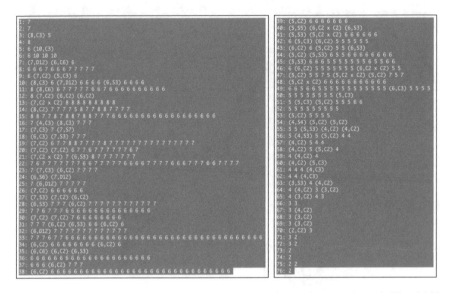

Fig. 24.32 Holonomy components of the decomposition of the automaton shown in Figs. 24.30 and 24.31

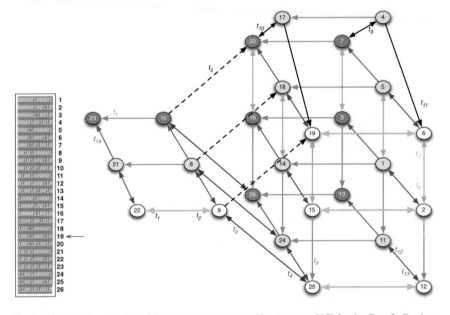

Fig. 24.33 3-D visualization of the automaton generated by SgpDec [15] for the Case 2a Petri net with Capacity 1. State 19 is the starting state: $\{X, Y, [YX], Y_p, Z, [Y_pZ], R, [Y_pR], R_p, W, [WR_p]\} = \{1, 0, 0, 1, 1, 0, 0, 0, 1, 1, 0\}$

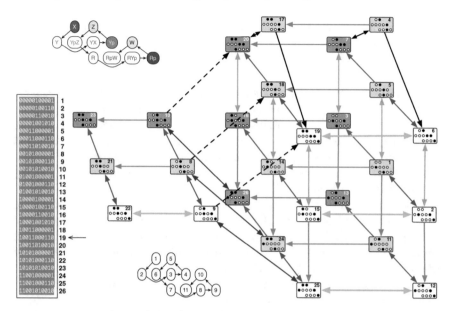

Fig. 24.34 3-D visualization of the automaton generated by `SgpDec` [15] for the Case 2a Petri net with Capacity 1, with Petri net visualization. The key is provided to help relate the column numbers of the place markings with the relative locations of the places in the Petri net. State 19 is the starting state: $\{X, Y, [YX], Y_p, Z, [Y_pZ], R, [Y_pR], R_p, W, [WR_p]\} = \{1, 0, 0, 1, 1, 0, 0, 0, 1, 1, 0\}$

corresponding to different paths followed along the arrows. As the arrows define transitions, these are essentially different "algorithms" (see D3.3.2 [10]). As in the previous figure, different shading is used for the states to highlight the depth of the figure. Thus, darker states are in the back vertical plane, light gray in the middle plane, and white states in the front. This representation of the automaton allows us to see the oscillations of the individual components. In particular, movement along each dimension is associated with a different chemical process. These state traces are not all linear because we are dealing with a fairly drastic projection onto $\mathbb{R}^3$:

- Starting at State 19, if we move into the page horizontally towards the darker states, i.e. towards States 18 and 20, then we are following the dephosphorylation reaction that starts with R_p and ends with R. This is Eq. (24.60). Note that this same reaction is enacted by *all* triplets of states that are parallel to these three.
- Starting with the same State 19, if we move vertically down from the "roof" to the "basement" of the house, then we are enacting Eq. (24.58), the dephosphorylation of Y_p. Again, this same process applies to *all* vertical triplets of states. It is important to note that both this and the previous bullet involve a reversible step and an irreversible step, just like the original chemical equations.
- If we start at State 25, move diagonally up to State 9, and then up again to State 19, we are following the phosphorylation of Y, Eq. (24.57).
- Equation (24.55) is modelled by moving towards the left between the "front" of the house on the far right and the plane just to the left of it, or from States 8, 9, or 10 towards 21, 22, or 23.

- The bright green arrows in the front plane, on the other hand, model both Eqs. (24.55) and (24.56).
- Finally, Eq. (24.59) is "on the roof": States 20, 17, 19 (or 7, 4, 6), replicating the triangular pattern of Eq. (24.57).

This modelling effort has been successful, so far, in that it has yielded an automaton that is easily interpretable in terms of the Petri net and of the original chemical reactions. The individual oscillations of the second and third components are modelled by going around the two kinds of triangles: (19, 15, 25, 9, 19) for R and (19, 18, 20, 17, 19) for Y. The effectiveness of these diagrams is apparently a consequence of the fact that three dimensions are sufficient for drawing any graph of any topology without crossing arcs.

Unfortunately, it was not possible to generate the HD of the larger automaton, as the computer ran out of its 64GB of RAM after reaching the 99th level of the decomposition. So although this case could be run on a supercomputer with much larger RAM, it is better to focus first on simpler problems the analytical solution of whose ODE systems can be found explicitly (with e.g. Lie groups), so as to try to relate the groups observed in the HD to any invariants in the solution of the corresponding ODE systems. A potentially fruitful direction of analysis, for example, will be to examine one or more of the first integrals of the fully integrated differential systems, since the solution curves, by definition, are level sets of such quantities, which are functions of the dependent variables and time. Knowing which functions of the dependent variables are conserved along solutions may give us valuable explanatory insights in the dynamics of these systems.

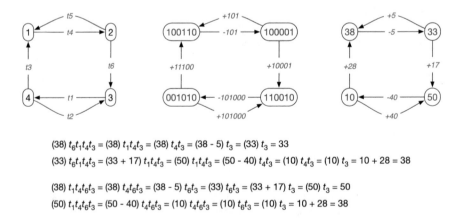

$(38) \, t_6 t_1 t_4 t_3 = (38) \, t_1 t_4 t_3 = (38) \, t_4 t_3 = (38 - 5) \, t_3 = (33) \, t_3 = 33$

$(33) \, t_6 t_1 t_4 t_3 = (33 + 17) \, t_1 t_4 t_3 = (50) \, t_1 t_4 t_3 = (50 - 40) \, t_4 t_3 = (10) \, t_4 t_3 = (10) \, t_3 = 10 + 28 = 38$

$(38) \, t_1 t_4 t_6 t_3 = (38) \, t_4 t_6 t_3 = (38 - 5) \, t_6 t_3 = (33) \, t_6 t_3 = (33 + 17) \, t_3 = (50) \, t_3 = 50$

$(50) \, t_1 t_4 t_6 t_3 = (50 - 40) \, t_4 t_6 t_3 = (10) \, t_4 t_6 t_3 = (10) \, t_6 t_3 = (10) \, t_3 = 10 + 28 = 38$

Fig. 24.35 *Top row, left* automaton from Fig. 24.26. *Middle* markings of the Petri net shown in Fig. 24.26 interpreted as expansions in a number system given by its places and with base equal to its capacity $+1$. *Right* same numbers expressed in base 10. *Bottom row* permutations (1 2) and (1 3) are shown as the action of words composed of transitions on these two subsets of states

24.4.4   Petri Nets as Number Systems

We now look again at the Petri net for the Case 1c system, this time interpreting the states as numbers in a positional number system defined by the places of the Petri net itself, and transitions as elementary addition and subtraction operations between these numbers. Figure 24.35 shows that, if we treat each place as a bit in a base-2 number, with position of each bit given by the place number given in Fig. 24.26, each state can be interpreted as a number whose decimal expansion is also given in the figure. Transitions then become additions and subtractions as shown. Figure 24.35 also shows the two permutations discussed earlier in this section, of states 1 & 2 and 1 & 3.

24.5   Discussion and Conclusions

The starting assumption of the BIOMICS project was that the order-constructing behaviour of biological systems could be captured mathematically through appropriate formalization of the relevant physical laws. Although the analysis we could perform of Case 2a, the negative feedback oscillator, was limited, we were able to establish the accuracy of the continuous and discrete models of Case 1c, the Goldbeter-Koshland ultrasensitive switch, and uncovered some algebraic structure in the latter.

The situation at this point appears to be that there is structure, but no preferred dynamics: i.e., no preferred "direction" of time-evolution from a state of disorder to a state of order. To be fair, the systems we have been able to analyse so far have perhaps been too simple to be able to exhibit such sophistication in their behaviour.

At an epistemological level, it seems worthwhile to explore the deeply interesting idea of Rhodes's that the Krohn-Rhodes (or the holonomy) decomposition of an automaton as a loop-free cascade of simpler machines can be thought of as an "expansion", for each automaton state, into an abstract generalization of the concept of positional number system, where each level in the cascade corresponds to a position in such system. This raises the question of the ontological primacy of the concept of 'state' vs. that of 'number'. An initial discussion is provided in [8], where it is shown how, for example, each digit in a binary number is simply a binary counter that acts as a submachine at that position or level in the binary expansion of a given number. In this paper we have shown how also the global marking of a Petri net can be seen as a p-digit number in base $c + 1$, where p is the number of places and c is the capacity of the Petri net, assumed equal for all places. Although this representation of the system does not explain anything we did not already know, hopefully it makes it easier to relate to Rhodes's much more general abstract number system idea.

Put in these terms, the problem of modelling biological behaviour becomes "merely" one of encoding it into such an abstract number system. But computer science is familiar with this process, and it is called *programming*, with the support of a *compiler*. We are back at the starting point, therefore, of whether mathematical structure derived from physical law can in some way and to some extent replace top-down engineering thinking in the creation of computational systems that are able to perform useful tasks compatibly with an "order construction design pattern" based on formalized structural and dynamical biological properties.

Although no conclusive way to encode ordered biological behaviour into computational systems has so far been found, in the BIOMICS project we have begun to work with the Abstract State Machines (ASM) software engineering framework [4]. It is helpful to think of ASMs as abstract and much more powerful generalizations of Petri nets. For example, in place of the functions that determine the numbers of tokens for each place, any (dynamic) function at all can be defined; more importantly, the output of one ASM can be used as the input of another, enabling interaction; the 'abstract state' of an ASM is somewhat analogous to the global marking of a Petri net, and a state trace is a sequence of such abstract states; an algorithm or program for an ASM is analogous to the rules according to which a given Petri net transition is enabled to fire or not; and so forth. Most importantly, ASMs are based on a rigorous mathematical and logic foundation, a fact that suggests that it will be relatively straightforward to map the formalization of conserved quantities in biological systems to suitable function definitions for a set of interacting ASMs, via category theory and coalgebraic specification [37, 38, 41, 43].

In principle, what ASMs enable is the definition of data structures that "fit" the problem at hand 'like a glove'. The result of using ASMs for requirements specification and iterative refinement, therefore, is not only a reliable software application but also a more natural and easier programming task. But this property of modelling frameworks is very well-known in physics, and is related to the search for the most appropriate variables for a given problem. For example, the governing equations for the flow in a circular pipe are much easier to work with when expressed in cylindrical than in rectangular coordinates. Carrying this idea further leads to the well-developed field of dimensional and similarity analysis, which is a sub-field of (Lie) group theory [32]. Dimensional and similarity analysis systematize the derivation of the dimensionless form of a given problem by expressing it in terms of dimensionless groupings of variables. Such dimensionless groupings are necessarily invariant with respect to changes of units, which is in fact the condition used to derive them. Crucially, these groupings (obtainable through the Buckingham PI Theorem, for example [18]) are not unique, and may have different impacts on the ease with which the solution to a given problem is obtained. Since there is no general procedure to find the best configuration, some trial and error, intuition, and experience are required.

The result of applying this kind of analysis to empirical data is often to collapse multiple curves onto a single relationship. This useful property, discovered by empirical scientists and engineers in the 19th Century, was later shown to be part of Lie group theory, for example in the reduction of a partial differential equation (PDE) to an ODE. The point is that the search for algebraic invariants usually simplifies the mathematical models of physical problems and makes their solution possible. This is a very broad statement that applies to most of theoretical physics, to the point that it has become part of the language of physics itself and is no longer a very remarkable thing to say.

However, the field of computer science is still relatively young and could in principle benefit from such an algebraic perspective. The search for a formal connection between the symmetry structure of non-linear dynamical systems and the algebraic structure of the HD of the finite-state automata derived from the *same* biochemical systems was in fact was the main motivation behind the development of the BIO-MICS project. Interestingly, although the Jordan-Hölder decomposition of a group is unique up to isomorphism, holonomy (or Krohn-Rhodes) decomposition is not, meaning that some 'encodings' are likely to be more expressive/useful than others.

With the above considerations in mind, therefore, it appears that the best HD for a given problem is not discernible from the algebraic structure alone, since many possible such decompositions are possible for a given automaton. If we can find a way to relate the continuous (Lie) and discrete (HD) algebraic structures of a given problem we may be able to use the former to guide the selection of the latter. Although, as discussed in D1.3.1 and D1.3.2, this is looking harder than we had originally envisaged, some input from physics still seems plausible in some way, as discussed at the beginning of this section. In the meantime, the research is developing the theoretical (coalgebraic and category theory) perspective further, to link ASMs, automata, and biological systems together functorially.

In conclusion, although no new and original results have been presented in this paper, we hope that the comparison of formalisms and methodologies discussed will make it easier for systems biologists, mathematicians, physicists, and computer scientists to continue working together on the very challenging problem of Interaction Computing.

Acknowledgments The author is grateful to Chrystopher L. Nehaniv and Attila Egri-Nagy for helpful discussions and for help with running SgpDec. The author is also grateful to Zoran Konkoli for discussions about the shortcomings of ODE models in the presence of low particle numbers and their generalizations based on stochastic methods, and to the TRUCE EU project that provided the context that enabled such discussions to take place. Finally, the author is grateful to Egon Börger for the many stimulating discussions about the similarities and differences between Petri nets and Abstract State Machines. The work reported in this article was funded by the BIOMICS EU project, contract number CNECT-ICT-318202. Its support is gratefully acknowledged.

References

1. Adami, C., Cerf, N.: Physical complexity of symbolic sequences. Phys. D **69**, 137–162 (2000)
2. Adami, C., Ofria, C., Collier, T.: Evolution of biological complexity. Proc. Natl. Acad. Sci. **97**(9), 4463–4468 (2000)
3. Alon, U.: An Introduction to Systems Biology: Design Principles of Biological Circuits. Chapman and Hall, Boca Raton (2007)
4. Börger, E., Stärk, R.: Abstract State Machines: A Method for High-Level System Design and Analysis. Springer, New York (2003)
5. Borghans, J.A.M., de Boer, R.J., Segel, L.A.: Extending the quasi-steady state approximation by changing variables. Bull. Math. Biol. **58**(1), 43–63 (1996)

6. Brown, C.T., Rust, A.G., Clarke, P.J.C., Pan, Z., Schilstra, M.J., De Buysscher, T., Griffin, G., Wold, B.J., Cameron, R.A., Davidson, E.H., Bolouri, H.: New computational approaches for analysis of cis-regulatory networks. Dev. Biol. **246**(1), 86–102 (2002)

7. Ciliberto, A., Capuani, F., Tyson, J.J.: Modeling networks of coupled enzymatic reactions using the total quasi-steady state approximation. PLoS Comput. Biol. **3**(3), e45 (2007). http://www.ploscompbiol.org/article/info%3Adoi%2F10.1371%2Fjournal.pcbi.0030045

8. Dini, P., Nehaniv, C.L., Egri-Nagy, A., Schilstra, M.J.: Exploring the concept of interaction computing through the discrete algebraic analysis of the Belousov-Zhabotinsky reaction. BioSystems **112**(2), 145–162 (2013)

9. Dini, P., Nehaniv, C.L., Schilstra, M.J., Karimi, F., Horváth, G., Muzsnay, Z., Christodoulides, K., Bonivárt, Á., den Breems, N.Y., Munro, A.J., Egri-Nagy, A.: D1.1.1: tractable dynamical and biological systems for numerical discrete, and lie group analysis. BIOMICS Deliverable, European Commission (2013). http://biomicsproject.eu/file-repository/category/11-public-files-deliverables

10. Dini, P., Rothstein, E.M., Schreckling, D., Nehaniv, C.L., Egri-Nagy, A.: D3.3.2: algebraic analysis of more complex computer science systems with specific security properties. BIOMICS Deliverable, European Commission (2014). http://biomicsproject.eu/file-repository/category/11-public-files-deliverables

11. Dini, P., den Breems, N.Y., Munro, A.J.: D1.1.2: ODE and automata analysis of cell metabolic and regulatory pathways. BIOMICS Deliverable, European Commission (2014). http://biomicsproject.eu/file-repository/category/11-public-files-deliverables

12. Egri-Nagy, A., Nehaniv, C.L.: PN2A: Petri Net Analysis GAP Package. http://sourceforge.net/projects/pn2a/

13. Egri-Nagy, A., Nehaniv, C.L.: Algebraic properties of automata associated to Petri nets and applications to computation in biological systems. BioSystems **94**(1–2), 135–144 (2008)

14. Egri-Nagy, A., Nehaniv, C.L., Schilstra, M.J.: Symmetry groups in biological networks. In: Information Processing in Cells and Tissues, IPCAT09 Conference. Journal preprint, 5–9 April 2009

15. Egri-Nagy, A., Nehaniv, C.L., Mitchell, J.D.: SgpDec – Hierarchical Decompositions and Coordinate Systems, Version 0.7.29 (2014). http://sgpdec.sf.net

16. Félix, M.-A., Labbé, J.-C., Dorée, M., Hunt, T., Karsenti, E.: Triggering of cyclin degradation in interphase extracts of amphibian eggs by cdc2 kinase. Nature **346**, 379–382 (1990)

17. Ferrel, J.E., Tsai, T.Y.-C., Yang, Q.: Modeling the cell cycle: why do certain circuits oscillate? Cell **144**(3), 874–885 (2011)

18. Gibbings, J.C.: Dimensional Analysis. Springer, London (2011)

19. Gillespie, D.T.: A general method for numerically simulating the stochastic time evolution of coupled chemical reactions. J. Comput. Phys. **22**, 403–434 (1976)

20. Gillespie, D.T.: Exact stochastic simulation of coupled chemical reactions. J. Phys. Chem. **81**(25), 2340–2361 (1977)

21. Goldbeter, A.: A minimal cascade model for the mitotic oscillator involving cyclin and cdc2 kinase. Proc. Natl. Acad. Sci. **88**, 9107–9111 (1991)

22. Goldbeter, A.: Biochemical Oscillations and Cellular Rhythms. Cambridge University Press, Cambridge (1996)

23. Goldbeter, A., Koshland, D.E.: An amplified sensitivity arising from covalent modification in biological systems. Proc. Natl. Acad. Sci. **78**(11), 6840–6844 (1981)

24. Goldbeter, A., Koshland, D.E.: Sensitivity amplification in biochemical systems arising from covalent modification in biological systems. Q. Rev. Biophys. **15**(3), 555–591 (1982)

25. Kauffman, S.: The Origins of Order: Self-Organisation and Selection in Evolution. Oxford University Press, Oxford (1993)

26. Konkoli, Z.: Application of Bogolyubov's theory of weakly nonideal Bose gases to the AA, AB, BB reaction-diffusion system. Phys. Rev. E **69**, 011106 (2004)

27. Konkoli, Z.: Diffusion-controlled reactions in small and structured spaces as a tool for describing living cell biochemistry. J. Phys.: Condens. Matter **19**, 065149 (2007)

28. Konkoli, Z.: Modeling reaction noise with a desired accuracy by using the X-level approach reaction noise estimator(XARNES) method. J. Theor. Biol. **305**, 1–14 (2012)
29. Konkoli, Z.: On the relevance of diffusion-controlled reactions for understanding living cell biochemistry. Int. J. Softw. Inf. **7**(4), 675–694 (2013)
30. Konkoli, Z., Johannesson, H.: Two-species reaction-diffusion system with equal diffusion constants: anomalous density decay at large times. Phys. Rev. E **62**(3), 3276–3280 (2000)
31. Krohn, K., Rhodes, J.: Algebraic theory of machines. I. Prime decomposition theorem for finite semigroups and machines. Trans. Am. Math. Soc. **116**, 450–464 (1965)
32. Moran, J.: A unification of dimensional and similarity analysis via group theory. Ph.D. thesis, University of Wisconsin, Madison, WI (1967)
33. Nabli, F., Fages, F., Martinez, T., Soliman, S.: A boolean model for enumerating minimal siphons and traps in Petri nets. In: Proceedings of CP2012, 18th International Conference on Principles and Practice of Constraint Programming. LNCS, vol. 7514, pp. 798–814. Springer (2012)
34. Nehaniv, C.L.: Algebraic engineering of understanding: global hierarchical coordinates on computation for the manipulation of data, knowledge, and process. In: Proceedings of the 18th Annual International Computer Software and Applications Conference (COMPSAC94), pp. 418–425. IEEE Computer Society Press, Taipei, Taiwan (1994)
35. Nehaniv, C.L.: Algebraic models for understanding: coordinate systems and cognitive empowerment. In: Proceedings of the Second International Conference on Cognitive Technology. 1997: Humanizing the Information Age, pp. 147–162, Aizu-Wakamatsu City, Japan (1997)
36. Nehaniv, C.L., Rhodes, J.L.: The evolution and understanding of hierarchical complexity in biology from an algebraic perspective. Artif. Life **6**, 45–67 (2000)
37. Nehaniv, C.L., Karimi, F., Rothstein, E., Dini, P.: D2.2: constrained realization of stable dynamical organizations, logic, and interaction machines. BIOMICS Deliverable, European Commission (2015). http://biomicsproject.eu/file-repository/category/11-public-files-deliverables
38. Nehaniv, C.L., Rhodes, J., Egri-Nagy, A., Dini, P., Rothstein Morris, E., Horváth, G., Karimi, F., Schreckling, D., Schilstra, M.J.: Symmetry structure in discrete models of biochemical systems: natural subsystems and the weak control hierarchy in a new model of computation driven by interactions. Philos. Trans. R. Soc. A **373**, 20140223 (2015). http://dx.doi.org/10.1098/rsta.2014.0223
39. Novak, B., Tyson, J.J.: Design principles of biochemical oscillators. Nat. Rev.: Mol. Cell Biol. **9**(12), 981–991 (2008)
40. Rhodes, J.: Applications of Automata Theory and Algebra via the Mathematical Theory of Complexity to Biology, Physics, Psychology, Philosophy, and Games. World Scientific Press (2010). Foreword by M.W. Hirsch, edited by C.L. Nehaniv (Original version: University of California at Berkeley, Mathematics Library, 1969)
41. Rothstein, E., Schreckling, D.: D4.1: candidate for a (co)algebraic interaction computing specification language. BIOMICS Deliverable, European Commission (2015). http://biomicsproject. eu/file-repository/category/11-public-files-deliverables
42. Rothstein, E.M., Dini, P., Nehaniv, C.L., Schreckling, D., Egri-Nagy, A.: D3.3.1: algebraic analysis of simple computer science systems. BIOMICS Deliverable, European Commission (2013). http://biomicsproject.eu/file-repository/category/11-public-files-deliverables
43. Rothstein Morris, E., Dini, P., Ruzsnavszky, F., Schreckling, D., Li, L., Munro, A.J., Börger, E.: D5.1: requirements collection for an interaction computing execution environment. BIOMICS deliverable, European Commission (2015). http://www.biomicsproject.eu
44. Schilstra, M.J., Martin, S.R.: Simple stochastic simulation. In: Michael, L., Ludwig, B. (eds.) Methods in Enzymology, pp. 381–409. Academic Press (Elsevier), New York (2009)
45. Schilstra, M.J., Martin, S.R., Keating, S.M.: Methods for simulating the dynamics of complex biological processes. In: Correia, J.J., Detrich, H.W. (eds.) Methods in Cell Biology, pp. 807–842. Elsevier, New York (2008)
46. Segel, L.A.: On the validity of the steady state assumption of enzyme kinetics. Bull. Math. Biol. **6**, 579–593 (1988)

47. The GAP Group. Gap–Groups, Algorithms, and Programming, V 4.7.5 (2014). http://www. gap-system.org
48. Tyson, J.J., Chen, K.C., Novak, B.: Sniffers, buzzers, toggles and blinkers: dynamics of regulatory and signaling pathways in the cell. Curr. Opin. Cell Biol. **15**, 221–231 (2003)
49. Tzafriri, A.R.: Michaelis-Menten kinetics at high enzyme concentrations. Bull. Math. Biol. **65**, 1111–1129 (2003)
50. Wingender, E. (ed.): Biological Petri Nets. IOS Press, Amsterdam (2011) (Silver Anniversary Edition)
51. Zeiger, H.P.: Cascade synthesis of finite-state machines. Inf. Control **10**(4), 419–433 (1967). Plus erratum

Chapter 25
Kernel P Systems and Stochastic P Systems for Modelling and Formal Verification of Genetic Logic Gates

Marian Gheorghe, Savas Konur and Florentin Ipate

Abstract *P* systems are the computational models of *membrane computing*, a computing paradigm within natural computing area inspired by the structure and behaviour of the living cell. In this chapter, we discuss two variants of this model, a non-deterministic case, called *kernel P (kP)* systems, and a stochastic one, called *stochastic P (sP)* systems. For both we present specification languages and associated tools, including simulation and verification components. The expressivity and analysis power of these natural computing models will be used to illustrate the behaviour of two genetic logic gates.

25.1 Introduction

Membrane computing is a computational paradigm, within the more general area of natural computing [32], inspired by the structure and behaviour of eukaryotic cells. The formal models introduced in this context are called *membrane systems* or *P systems*. After their introduction [27], membrane systems have been widely investigated for computational properties and complexity aspects, but also as a model for various applications [28]. Many different variants of P systems have been introduced and studied, mainly due to the many theoretical challenges induced by them, but also motivated by the need to model different problems. Most of these variants of P systems consider key features of the biological cell as part of the computational models introduced. In this respect, they deal with either simple bio-chemical elements

M. Gheorghe (✉) · S. Konur
School of Electrical Engineering and Computer Science, University of Bradford,
Bradford BD7 1DP, UK
e-mail: m.gheorghe@bradford.ac.uk

S. Konur
e-mail: s.konur@bradford.ac.uk

F. Ipate
Department of Computer Science, University of Bucharest, Str. Academiei nr. 14,
010014 Bucharest, Romania
e-mail: florentin.ipate@ifsoft.ro

© Springer International Publishing Switzerland 2017
A. Adamatzky (ed.), *Advances in Unconventional Computing*,
Emergence, Complexity and Computation 22,
DOI 10.1007/978-3-319-33924-5_25

(called *objects*) or more complex molecules like DNA strands which are codified as *strings*. Specific molecules like catalysts, activators and inhibitors are also utilised by the models. Chemical interactions within compartments and transmembrane regulations are represented by *rewriting rules* and *communication rules*, respectively. Various biological entities, like, cells, tissues, as well as specialised cells, such as neurons, are described in the membrane systems framework by *cell P systems*, *tissue P systems*, and *neural P systems*, respectively. The combination of these features leads to a rich set of variants of P systems. A thorough presentation of the theoretical developments is provided in [28], whereas various applications of this computing paradigm in modelling problems from various areas, including computer science, graphics and linguistics, can be found in [8]. More recently, it has been applied to systems and synthetic biology [12], optimisations and graphics [18] and synchronisation of distributed systems [10]. Some of the future challenges of the field are presented in [16]. The most up-to-date information on P systems can be found on its website [26].

Every membrane system consists of a set of *compartments* linked together in accordance with certain well-defined system structures, e.g., tree and graph in the case of cell P systems and tissue (or neural) P systems, respectively. Some systems have a static structure, others have a dynamic one. Each compartment contains a multiset of elements, either simple objects or more complex data, *strings*. These are either transformed or transferred between neighbour compartments, due to some *rules* which are specific to each compartment. A membrane system appears to be a computational model of a distributed system, where the structure of the system, the types of objects and transformations matter and collaborate in order to express a certain computation.

Membrane computing has been an umbrella for the proliferation of different variants of membrane systems. Not only studies investigating relationships between different classes of P systems, but also software tools supporting them have been considered. The best known tool that covers the most used P system models has a specification language, known generically as *P–Lingua* [25], which provides adequate syntax for each of the variants of P systems supported. P–Lingua aims to keep the syntax as close as possible to the original models and provides a simulation platform for all these models and a consistent user interface environment, called MeCoSim [24].

An alternative approach has been considered, by defining a more general membrane system model, allowing to relatively easily specify the most utilised P system models. This model is called *kernel P systems* (*kP systems*). A revised version of the model and the specification language can be found in [14] and its usage to specify the 3-colouring problem and a comparison to another solution provided in a similar context [9], is described in [15]. The kP systems have also been used to specify and analyse, through formal verification, synthetic biology systems [21, 22].

Kernel P systems are supported by a software framework, kPWORKBENCH [1, 2], which integrates a set of related simulation and verification methodologies and tools.

All these classes of P systems deal with non-deterministic behaviour, but in various circumstances, especially when biological systems are considered, stochastic systems are more appropriate. In this respect several variants of stochastic P systems have been introduced and utilised to model various problems in systems and synthetic biology [12]. A variant which is also supported by a simulation and verification environment is based on Gillespie approach for executing the system [17]. The tool and some applications are presented in [6].

In this work, we utilise kP systems and stochastic P systems (based on Gillespie approach), together with the corresponding software platforms developed, in order to model and verify certain properties of biological systems. The novelty of the approach is given by (a) the methodology that combines quantitative and qualitative analysis; (b) the modular way of specifying systems very close to the their informal descriptions; and (c) by the power of the verification method, relying on *model checking* techniques, which combines various approaches in order to adequately check the desired properties.

The chapter consists of five sections. Section 25.2 introduces the basic concepts related to kernel P systems and stochastic P systems. Section 25.3 introduces the AND and OR logic gates. Section 25.4 discusses the model described by using stochastic P systems and the modules associated with them. In Sect. 25.5 it is presented the verification methodology combining quantitative and qualitative analysis. Finally, Sect. 25.6 draws conclusions.

25.2 P Systems—Basic Definitions

The reader is assumed to be familiar with basic elements of membrane computing, e.g., from [28]. Some basic concepts utilised in the sequel will be introduced. Let A be an alphabet. An word with elements from A is a sequence containing these elements. The set of all words over A is denoted by A^*; λ denotes the empty word and $A^+ = A \setminus \{\lambda\}$. A multiset w over A is a mapping, $w : A \longrightarrow \mathbb{N}$ and $w(a), a \in A$, defines the number of occurrences of a in the multiset. In the sequel a multiset will be defined by a word where the order of the elements is not considered.

In this section we will introduce two P system models, a non-deterministic version, called kernel P systems, and a stochastic one, called stochastic P systems.

25.2.1 Kernel P Systems

A kP system is made of compartments placed in a graph-like structure. Each compartment C_i, $1 \leq i \leq n$, has a type $t_i = (R_i, \sigma_i)$, $t_i \in T$, where T represents the set of all types, describing the associated set of rules R_i and the execution strategy, σ_i, of that compartment. In the sequel, we will present a simplified version of kP systems—for the full definition we refer to [13].

Definition 1 A *kernel P (kP) system* of degree n is a tuple

$$k\Pi = (A, \mu, C_1, \ldots, C_n, i_0),$$

where A is a finite set of elements called *objects*; μ defines the *membrane structure*, a graph, (V, E), where V are vertices indicating compartments, and E edges; $C_i = (t_i, w_i)$, $1 \le i \le n$, is a *compartment* of the system consisting of a *compartment type*, $t_i \in T$, and an *initial multiset*, w_i over A; i_0 is the *output compartment* where the result is obtained.

Each rule r may have a *guard* g denoted as r $\{g\}$. The rule r is applicable to a multiset w when its left hand side is contained into w and g is true for w. The guards are constructed using multisets over A and relational and Boolean operators. For example, rule $r : ac \to c$ $\{\ge a^3 \wedge < b^5\}$ can be applied to the current multiset, $w = a^5 b^4 c$, as it includes the left hand side of r, i.e., ac and the guard condition is satisfied by w—there are at least 3 a's and no more than 5 b's.

In the sequel, we will present the types of rules utilised by kP systems. In the more general definition of such systems, see [13], there are two main types of rules, *rewriting and communication rules* and *structure changing rules*. The later set of rules is meant to be used when the structure of system, involving both compartments and links, is changing. In this work, we will be using only rewriting and communication rules and the definition below will deal with these types of rules.

Definition 2 A *rewriting and communication* rule has the form: $x \to y$ $\{g\}$, in compartment l_i, where $x \in A^+$ and y has the form $y = (a_1, t_1) \ldots (a_h, t_h), h \ge 0, a_j \in A$ and t_j indicates a compartment type from T—see Definition 1—with instance compartments linked to the current compartment; if a link does not exist (the two compartments are not in E) then the rule is not applied; if a target, t_j, refers to a compartment type that has more than one instance connected to l_i, then one of them will be non-deterministically chosen;

Each compartment has an execution strategy for its set of rules that can be defined as a sequence $\sigma = \sigma_1 \& \sigma_2 \& \ldots \& \sigma_n$, where σ_i denotes an atomic component of the form:

- ε, means *empty* execution strategy—an analogue of a *skip* instruction;
- r, a rule from the set R_t (the set of rules associated with type t), describes the fact that *if r is applicable, then it is executed*; otherwise, the compartment terminates the execution thread for this particular computational step and thus, no further rule will be applied;
- $(r_1, \ldots, r_n)$, with $r_i \in R_t, 1 \le i \le n$, describes a *non-deterministic choice within a set of rules*; one and only one applicable rule will be executed, if such a rule exists, otherwise this is simply skipped;
- $(r_1, \ldots, r_n)^*$, with $r_i \in R_t, 1 \le i \le n$, indicates that the rules $\{r_1, \ldots, r_n\}$ are executed *iteratively an arbitrary number of steps*;

- $(r_1, \ldots, r_n)^\top, r_i \in R_t, 1 \le i \le n$, represents the maximally parallel execution of a set of rules. If no rules are applicable, then the execution proceeds to the subsequent atom in the chain.

We introduce now the concept of a *configuration* of a kP system of degree n as being an n-tuple $\mathcal{K} = (u_1, \ldots, u_n)$, where u_i is a multiset in compartment C_i, $1 \le i \le n$. The initial configuration is $\mathcal{K}_0 = (w_1, \ldots, w_n)$. Starting from $\mathcal{K}_0$ and using rules from $R_1, \ldots, R_n$ in accordance with the execution strategies $\sigma_1, \ldots, \sigma_n$, one gets a sequence of configurations. The process of getting a configuration from another one is called *transition*. A *computation* of Π is a sequence of transitions starting from $\mathcal{K}_0$. If the sequence is finite then this leads to a *halting computation* and the result is read out from i_0. In applications one can consider partial computations and the result is not always restricted to only one single compartment.

We present now an example of a kP system with two compartments by using first the notations introduced so far and then its transcription in a machine readable language, called *kP–Lingua* [11].

Example 1 There are two types $t_1 = (R_1, \sigma_1)$, and $t_2 = (R_2, \sigma_2)$, where $R_1 = \{r_{1,1} : bb \to (a, t_2) \{\ge b^2\}; r_{1,2} : b \to (b, t_1)(b, t_1)\}$, $R_2 = \{r_{2,1} : a \to (c, t_1)(c, t_1)\}$ and $\sigma_1 = (r_{1,1}, r_{1,2}), \sigma_2 = (r_{2,1})$. One can notice that rule $r_{1,1}$ has a guard that requires at least $2b$'s to be present in the current multiset. The execution strategies, σ_1 and σ_2, are non-deterministic choices. The kP system of degree 2 is given by

$$k\Pi_1 = (\{a, b, c, d\}, \mu, C_1, C_2, 1),$$

where μ is a graph with two vertices, C_1, C_2, and an edge between them. The two components are given by $C_1 = (t_1, w_1)$, $C_2 = (t_2, w_2)$, where $w_1 = d^2 b$, $w_2 = d$. The initial configuration of the system is $\mathcal{K}_0 = (d^2 b, d)$. The only possible transition allows only the rule $r_{1,2}$ to be applied in C_1 and consequently the next configuration is $\mathcal{K}_1 = (d^2 b^2, d)$. In this configuration, one can use in C_1 either $r_{1,1}$ or $r_{1,2}$ (both are applicable) and nothing in C_2; hence, one gets either $\mathcal{K}'_2 = (d^2 b^3, d)$ or $\mathcal{K}''_2 = (d^2, da)$, respectively. From $\mathcal{K}'_2$ one can continue with either $r_{1,1}$ or $r_{1,2}$ in C_1 and nothing in C_2. In $\mathcal{K}''_2$ one can only use $r_{2,1}$ in C_2 leading to $\mathcal{K}''_3 = (d^2 c^2, d)$, which is a final configuration and we obtain a halting computation with the result in $C_1, d^2 c^2$.

This example written in kP-Lingua is:

```
type t1 {
    choice {
        >= 2b : 2b ->  a(C2) .
        b ->  2b .
    }
}
type t2 {
    choice {
        a -> {2c}(C1) .
```

```
    }
  }
  C1 {2d, b} (t1) - C2 {d} (t2) .
```

Above, t1, t2 denote two compartment types, which are instantiated as C1, C2, respectively. C1 starts with the initial multiset 2d, b and C2 starts with d. The rules of C1 are chosen non-deterministically, only one at a time—this is achieved by the use of the key word choice. The first rule is fired only when its guard becomes true. This rule also sends an a to the instance of t2 that is linked. In the type t2, there is only one rule to be fired, which happens only when there is an a in the compartment C2.

25.2.2 Stochastic P Systems

In the case of stochastic P systems, constants are associated with rules in order to compute their probabilities. The precise definition is given below. It refers to a class of P systems, called *tissue P systems*, where the system structure is defined as a graph of components—a precise formal definition can be found in [28].

Definition 3 A *stochastic P (sP) system* is a model consisting of a tissue P system

$$sP = (O, L, \mu, M_1, \dots, M_n, R_1, \dots, R_n)$$

where O is a finite set of objects, called *alphabet*, denoting the entities involved in the system; L is a finite set of labels naming compartments; μ is a membrane structure composed of $n \geq 1$ membranes defining the regions or compartments of the system and their connections, forming an arbitrary graph; $M_i = (l_i, w_i)$, $1 \leq i \leq n$, is the initial configuration of the compartment or region defined by the membrane i, where $l_i \in L$ is the label of the compartment and $w_i \in O^*$ is a finite initial multiset of objects; $R_i = \{r_1^i, \dots, r_{m_i}^i\}$, $1 \leq i \leq n$, is a set of multiset rewriting rules, of the form: $r_k^i : [x \rightarrow^{c_k} y]_{l_i}$, where x and y are multisets of objects (y might be empty) over O, representing the molecular species consumed and produced in the corresponding molecular interaction occurring in the compartment labelled l_i. An application of a rule of this form changes the content of the membrane with label l_i by replacing the multiset x with y. The stochastic constant c_k is used by the Gillespie algorithm [17] in order to compute the probabilities associated with the rules.

In each compartment of the sP system the execution strategy is based on Gillespie algorithm. Similarly to kP systems, one can define configurations, transitions and computations. Partial computations are also widely used in this context.

The model of the sP systems has been considered as the basis of the Infobiotics Workbench [6] where this is extended with some modularity features allowing a more flexible specification of a system. Each module has a name and some attributes associated with it. We do not provide the full formal definition of these modules (for a formal approach see [6, 31]), but we prefer to introduce them through some examples utilised later in this work.

The unregulated gene expression module, whereby some genes are expressed constitutively and independently of transcription factors is defined by the module:

$UnReg(\{G, R, P\}, \{c_1, c_2, c_3, c_4\})\{$
$G \rightarrow^{c_1} G + R;$
$R \rightarrow^{c_2} R + P;$
$R \rightarrow^{c_3};$
$P \rightarrow^{c_4}\}$

This module describes the process of transcribing the gene G into its corresponding mRNA, R, which in turn is translated into a protein P. The mRNA and the protein can be degraded. The propensities of these processes are determined by the stochastic coefficients c_i, $1 \leq i \leq 4$. Some variants of this might appear when more than a protein is involved. The module has the form:

$UnRegM(\{G, R, P_1, P_2\}, \{c_1, c_2, c_3, c_4, c_5, c_6\})\{$
$G \rightarrow^{c_1} G + R;$
$R \rightarrow^{c_2} R + P_1;$
$R \rightarrow^{c_3} R + P_2;$
$R \rightarrow^{c_4};$
$P_1 \rightarrow^{c_5};$
$P_2 \rightarrow^{c_6}\}$

When it is assumed that the protein is obtained in one step from the gene, the module is:

$UnRegS(\{G, P\}, \{c_1, c_2\})\{$
$G \rightarrow^{c_1} G + P;$
$P \rightarrow^{c_2}\}$

One can also describe the process of complex formation, when two molecules M_1 and M_2 form a more complex molecule $M_1 \cdot M_2$ and this might be reversible. The module is:

$Comp(\{M_1, M_2\}, \{c_1\})\{$
$M_1 + M_2 \rightarrow^{c_1} M_1 \cdot M_2\}$

A negative regulation of a gene, when a repressor protein R binds reversibly to the gene G preventing it to produce any protein, is given by:

$Neg(\{R, G\}, \{c_1, c_2\})\{$
$R + G \rightarrow^{c_1} R \cdot G;$
$R.G \rightarrow^{c_2} R + G\}$

25.2.3   Tools

The specifications written in kP-Lingua are supported by a software platform, kPWORKBENCH, which integrates a set of tools and translators that bridge several target specifications that we employ for kP system models, written in kP-Lingua. kPWORKBENCH permits *simulation* and *formal verification* of kP system models using several simulation and verification methodologies and tools.

The Infobiotics Workbench (IBW) is an integrated software suite of tools to perform in silico experiments for sP models in systems and synthetic biology [6]. This software platform includes *simulators* and tools for *formal verification* of stochastic models.

The IBW tool is aimed at providing support for quantitative (stochastic) analysis of systems, whereas kPWORKBENCH is meant to help with qualitative analysis. In this respect, systems are specified within IBW, by using modularity features and then analysed with the existing tools of this environment. For qualitative analysis these are then automatically translated into a non-deterministic version of the system which is then analysed within the kPWORKBENCH environment.

25.3   Genetic Logic Gates

Genetic logic gates have been considered in various papers, including [4, 30, 34], using various synthetic biology tools, amongst them GEC [29], Eugene [5] and Proto [3]. In [23, 33], we have studied two basic logic gates, AND and OR, using the IBW tool for quantitative analysis and kPWORKBENCH for qualitative one. Here, we provide a summary of our results.

The genetic parts and designs of these gates are proposed by Beal et al. [4]. Both gates use two inducers, aTc and IPTG, as input and use GFP as output. aTc and IPTG disable the activities of TetR and LacI proteins, respectively.

Figure 25.1a illustrates the genetic design of an AND gate, which receives two input signals: aTc and IPTG. In this system, the transcription factors LacI and TetR are expressed by a gene controlled by the same promoter. The aTc molecules repress TetR, and IPTG molecules repress LacI, to prevent them from inhibiting the production of GFP by binding to the corresponding promoter which up-regulates the expression of GFP. If both IPTG and aTc are set to high, then neither LacI nor TetR can inhibit the GFP production.

Figure 25.1b illustrates a genetic OR gate, comprising two mechanisms. Each mechanism leads to the production of GFP, when it is activated. The first mechanism is repressed by LacI while the second is repressed by TetR. Therefore, GFP can be produced from the former when IPTG is set to high and from the latter when aTc is set to high.

The stochastic model, in each of the two cases, consists of an sP system with one compartment and a set of stochastic rules, governing the kinetic and stochastic behaviour of the system. The rewriting rules and the kinetic constants (taken

(a)

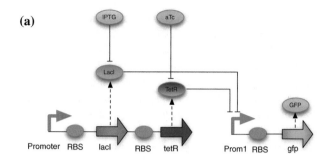

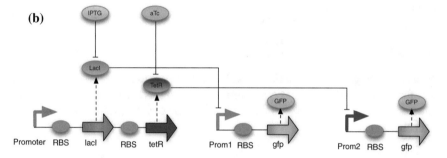

(b)

Fig. 25.1 Beal et. al.'s genetic devices functioning as an AND and OR gate (taken from [33]). **a** AND gate. **b** OR gate

from [4]) of the devices are described in the Table 25.1 (a) AND and (b) OR of the Appendix.

In the next section we will show how the two gates are modelled by using the modular approach provided by the sP systems and available as part of the IBW specification component.

25.4 Modelling with sP Systems

The sP systems model utilised for each of the two logic gates consists of a one compartment system and an alphabet of objects including all the species and molecules that appear in the two sets of reactions listed in Table 25.1 (a) AND and (b) OR of the Appendix. For expressing the behaviour of each of these systems we will be using modules as they are supported by the IBW environment. These specifications are also automatically translated into non-deterministic kP systems and made available to kPWORKBENCH.

The AND logic gate fully described by the reactions listed in Table 25.1 (a) AND can be specified using modules in a manner that maps better the informal specification above. The rules r_1, r_2, r_3 and r_{10}, r_{11}, r_{12} can be embedded into a module $UnRegM$ which expresses the fact that the transcription factors LacI and TetR are expressed by a gene, but also that the mRNA and the transcription factors degrade. Each of the reactions r_4 and r_5 expresses a complex formation, whereby the aTc molecules repress TetR, and IPTG molecules repress LacI, respectively. These are captured by *Comp* modules. The fact that LacI and TetR inhibit the GFP production (rules r_{6a}, r_{6b} and r_{7a}, r_{7b}) is captured by two modules describing this negative regulation, *Neg*. Finally, the GFP production (rules r_8, r_9) is defined by a module *UnRegS*. The complete specification of the AND logic gate using modules is:

$UnRegM$({gene_LacI_TetR, mLacI_TetR, LacI, TetR}, {k_1, k_2, k_3, k_{12}, k_{10}, k_{11}})
$Comp$({LacI, IPTG}, {k_4})
$Comp$({TetR, aTc}, {k_5})
Neg({LacI, gene_GFP}, {k_{6a}, k_{6b}})
Neg({TetR, gene_GFP}, {k_{7a}, k_{7b}})
$UnRegS$({gene_GFP, GFP}, {k_8, k_9})

The OR logic gate is very similar to AND with respect to modelling modules. The first three lines, using modules *UnRegM* and *Comp* twice are the same. As the OR gate uses two mechanisms to produce GFP then the two *Neg* modules of the AND gate are replaced by

Neg({LacI, gene_GFP1}, {k_{6a}, k_{6b}})
Neg({TetR, gene_GFP2}, {k_{7a}, k_{7b}})

and finally, the *UnRegS* of the AND gate is replaced by two *UnRegS* modules responsible for producing GFP.

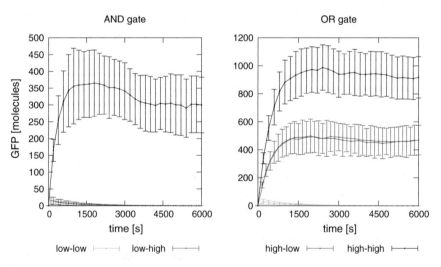

Fig. 25.2 GFP expression in the AND (*left*) and (OR) gate over time for the aTc/IPTG input combinations *low-low*, *low-high*, *high-low*, and *high-high* (taken from [33]). *Error bars* denote the standard deviations of 100 statistically independent samples

Having the systems implemented by using modules one can run various simulations with IBW environment and check their behaviours. The trajectories of both gate dynamics for the four different input combinations of low and high aTc and IPTG concentrations are shown in Fig. 25.2. The graphs presented show that the gates quickly approach a steady state with output concentrations. These show that the models implement the desired Boolean logics. Apart from simulations revealing various aspects of the systems' behaviour the tools discussed earlier provide more insights into the system by revealing certain properties or relationships between different components. All these will be investigated in the next section.

The translation of the sP system into a kP system is obtained automatically by removing the kinetic coefficients of the former. The kP system obtained can be executed in KPWORKBENCH by using the execution strategy corresponding to non-deterministic choice. The simulation of such a non-deterministic system does not bring any new information about the model, but this specification is useful for the formal verification performed in the next section.

25.5 P Systems Verification

In this section we briefly present a methodology for verifying P system models using model checking approaches. This has been developed by looking at quantitative and qualitative results whereby various model checking tools have been used to investigate properties of different types [20, 22]. In this work we will be illustrating the use of two model checkers, PRISM [19] for quantitative analysis and NUSMV [7] for qualitative aspects; they are part of the IBW and kPWORKBENCH platforms, respectively. They can be directly called from these platforms and the queries can be formulated in a natural language format [6].

In standard logic gates, any voltage value above a threshold value, such as 3V, is considered as 1. Since there is no such a standard value for genetic Boolean gates, we propose a threshold for this particular design. To analyse the behaviour of the genetic devices formally, we verify the following property using PRISM:

"What is the likelihood that GFP eventually exceeds the threshold Thr at time t?"

which is expressed in a probabilistic logic (CSL) as

$$P_{=?}\left[F^t \text{ GFP} \geq Thr\right].$$

This query returns different results for different parameter values. For example, if we consider the OR gate, it returns 1 for $Thr = 100, t = 1000$, and aTc = IPTG = 1000.

When models are built, especially when these are complex, with many species and interactions, it is essential that one can verify their correctness. In many cases there are chains of reactions leading to certain results and it is important to check dependabilities and the way they influence certain results. In our case, both logic gates show certain dependabilities. For instance, one can check that GFP is not present initially in the system, but it will eventually appear. One can verify whether GFP will finally appear in the system, by using the following statement in NuSMV:

"There are pathways in the system that eventually lead to the production of GFP"

$$EF (GFP > 0).$$

This property, expressed in CTL, is true, as expected.

We now show how one can check that a certain sequence of events must or might appear in a chain of reactions. This is illustrated by the relationship between the production of LacI and TetR and the complex formation of LacI.IPTG and TetR.aTc, respectively. We illustrate now the chain of events triggered by LacI by using a CTL formula in NuSMV:

"Always the LacI *production might eventually lead to the complex* LacI.IPTG*."*

$$AG (LacI > 0 \Rightarrow EF LacI.IPTG > 0)$$

This property is true, as expected.

25.6 Conclusions

In this paper, we have shown how an unconventional computing paradigm, membrane systems, is utilised to model and analyse various systems, especially biological systems. In particular, we have considered kP systems and stochastic P systems, together with the corresponding software platforms developed, in order to model and verify certain properties of biological systems.

Our approach is novel in the sense that our methodology (i) combines quantitative and qualitative analysis; (ii) is a modular way of specifying systems, (iii) employs simulation methods to analyse system dynamics and (iv) integrates various verification methods to adequately check the desired properties.

We are currently working on the next versions of IBW and kPWORKBENCH tools by incorporating more methods for specifying, modelling, simulating and verifying biological systems.

Acknowledgments MG and SK acknowledge the support provided for synthetic biology research by EPSRC ROADBLOCK (project number: EP/I031812/1). The work of FI was supported by a grant of the Romanian National Authority for Scientific Research, CNCS-UEFISCDI (project number: PN-II-ID-PCE-2011-3-0688).

Appendix: Table

Table 25.1 (a) AND gate by Beal et al. (taken from [33]) and, (b) OR gate by Beal et al. (taken from [33])

(a)

r_{0a}	$\overset{k_{0a}}{\to}$ IPTG	$k_{0a} \in \{0, 1000\}$
r_{0b}	$\overset{k_{0b}}{\to}$ aTc	$k_{0b} \in \{0, 1000\}$
r_1	gene_LacI_TetR $\overset{k_1}{\to}$ gene_LacI_TetR + mLacI_TetR	$k_1 = 0.12$
r_2	mLacI_TetR $\overset{k_2}{\to}$ mLacI_TetR + LacI	$k_2 = 0.1$
r_3	mLacI_TetR $\overset{k_3}{\to}$ mLacI_TetR + TetR	$k_3 = 0.1$
r_4	LacI + IPTG $\overset{k_4}{\to}$ LacI-IPTG	$k_4 = 1.0$
r_5	TetR + aTc $\overset{k_5}{\to}$ TetR-aTc	$k_5 = 1.0$
r_{6a}	gene_GFP + LacI $\overset{k_{6a}}{\to}$ gene_GFP-LacI	$k_{6a} = 1.0$
r_{6b}	gene_GFP-LacI $\overset{k_{6b}}{\to}$ gene_GFP + LacI	$k_{6b} = 0.01$
r_{7a}	gene_GFP + TetR $\overset{k_{7a}}{\to}$ gene_GFP-TetR	$k_{7a} = 1.0$
r_{7b}	gene_GFP-TetR $\overset{k_{7b}}{\to}$ gene_GFP + TetR	$k_{7b} = 0.01$
r_8	gene_GFP $\overset{k_8}{\to}$ gene_GFP + GFP	$k_8 = 1.0$
r_9	GFP $\overset{k_9}{\to}$	$k_9 = 0.001$
r_{10}	LacI $\overset{k_{10}}{\to}$	$k_{10} = 0.01$
r_{11}	TetR $\overset{k_{11}}{\to}$	$k_{11} = 0.01$
r_{12}	mLacI_TetR $\overset{k_{12}}{\to}$	$k_{12} = 0.001$

(b)

$r_0 - r_5$	Same as the rules $r_0 - r_5$ of the AND gate above	
r_{6a}	gene_GFP1 + LacI $\overset{k_{6a}}{\to}$ gene_GFP1-LacI	$k_{6a} = 1.0$
r_{6b}	gene_GFP1-LacI $\overset{k_{6b}}{\to}$ gene_GFP1 + LacI	$k_{6b} = 0.01$
r_{7a}	gene_GFP2 + TetR $\overset{k_{7a}}{\to}$ gene_GFP2-TetR	$k_{7a} = 1.0$
r_{7b}	gene_GFP2-TetR $\overset{k_{7b}}{\to}$ gene_GFP2 + TetR	$k_{7b} = 0.01$
r_8	gene_GFP1 $\overset{k_8}{\to}$ gene_GFP1 + GFP	$k_8 = 1.0$
r_9	gene_GFP2 $\overset{k_9}{\to}$ gene_GFP2 + GFP	$k_9 = 1.0$
$r_{10} - r_{13}$	Same as the rules $r_9 - r_{12}$ of the AND gate	

References

1. Bakir, M.E., Ipate, F., Konur, S., Mierlă, L., Niculescu, I.: Extended simulation and verification platform for kernel P systems. In: 15th International Conference on Membrane Computing. LNCS, vol. 8961, pp. 158–168. Springer, Heidelberg (2014)

2. Bakir, M.E., Konur, S., Gheorghe, M., Niculescu, I., Ipate, F.: High performance simulations of kernel P systems. In: Proceedings of the 2014 IEEE 16th International Conference on High Performance Computing and Communication. HPCC '14, pp. 409–412. France, Paris (2014)
3. Beal, J., Lu, T., Weiss, R.: Automatic compilation from high-level biologically-oriented programming language to genetic regulatory networks. PLoS One **6**(8), e22490 (2011)
4. Beal, J., Phillips, A., Densmore, D., Cai, Y.: High-level programming languages for biomolecular systems. Design and Analysis of Biomolecular Circuits, pp. 225–252. Springer, New York (2011)
5. Bilitchenko, L., Liu, A., Cheung, S., Weeding, E., Xia, B., Leguia, M., Anderson, J.C., Densmore, D.: Eugene - a domain specific language for specifying and constraining synthetic biological parts, devices, and systems. PLoS One **6**(4), e18882 (2011)
6. Blakes, J., Twycross, J., Konur, S., Romero-Campero, F., Krasnogor, N., Gheorghe, M.: Infobiotics workbench: a P systems based tool for systems and synthetic biology. In: Applications of Membrane Computing in Systems and Synthetic Biology. Emergence, Complexity and Computation, vol. 7, pp. 1–41. Springer, Heidelberg (2014)
7. Cimatti, A., Clarke, E., Giunchiglia, E., Giunchiglia, F., Pistore, M., Roveri, M., Sebastiani, R., Tacchella, A.: NuSMV version 2: an open source tool for symbolic model checking. In: Proceedings of the International Conference on Computer-Aided Verification (CAV 2002). LNCS, vol. 2404, pp. 359–364. Springer, Denmark (2002)
8. Ciobanu, G., Păun, Gh, Pérez-Jiménez, M.J. (eds.): Applications of Membrane Computing. Springer, Heidelberg (2006)
9. Díaz-Pernil, D., Gutiérrez-Naranjo, M.A., Pérez-Jiménez, M.J.: A uniform family of tissue P systems with cell division solving 3-COL in a linear time. Theor. Comput. Sci. **404**, 76–87 (2008)
10. Dinneen, M.J., Yun-Bum, K., Nicolescu, R.: Faster synchronization in P systems. Nat. Comput. **11**(4), 637–651 (2012)
11. Dragomir, C., Ipate, F., Konur, S., Lefticaru, R., Mierlă, L.: Model checking kernel P systems. In: 14th International Conference on Membrane Computing. LNCS, vol. 8340, pp. 151–172. Springer, Heidelberg (2013)
12. Frisco, P., Gheorghe, M., Pérez-Jiménez, M.J. (eds.): Applications of Membrane Computing in Systems and Synthetic Biology. Springer, Heidelberg (2014)
13. Gheorghe, M., Ipate, F., Dragomir, C.: Kernel P systems. In: 10th Brainstorming Week on Membrane Computing, pp. 153–170. Fénix Editora, Brazil (2013)
14. Gheorghe, M., Ipate, F., Dragomir, C., Mierlă, L., Valencia-Cabrera, L., García-Quismondo, M., Pérez-Jiménez, M.J.: Kernel P systems - version 1. In: 11th Brainstorming Week on Membrane Computing, pp. 97–124. Fénix Editora, Brazil (2013)
15. Gheorghe, M., Ipate, F., Lefticaru, R., Pérez-Jiménez, M.J., Ţurcanu, A., Valencia-Cabrera, L., García-Quismondo, M., Mierlă, L.: 3-Col problem modelling using simple kernel P systems. Int. J. Comput. Math. **90**(4), 816–830 (2012)
16. Gheorghe, M., Păun, G., Pérez-Jiménez, M.J., Rozenberg, G.: Research frontiers of membrane computing: open problems and research topics. Int. J. Found. Comput. Sci. **24**, 547–624 (2013)
17. Gillespie, D.: A general method for numerically simulating the stochastic time evolution of coupled chemical reactions. J. Comput. Phys. **22**(4), 403–434 (1976)
18. Gimel'farb, G.L., Nicolescu, R., Ragavan, S.: P system implementation of dynamic programming stereo. J. Math. Imaging Vis. **47**(1–2), 13–26 (2013)
19. Hinton, A., Kwiatkowska, M., Norman, G., Parker, D.: PRISM: a tool for automatic verification of probabilistic systems. Proceedings of the TACAS. LNCS, vol. 3920, pp. 441–444. Springer, Heidelberg (2006)
20. Konur, S., Gheorghe, M.: Property-driven methodology for formal analysis of synthetic biology systems. IEEE/ACM Trans. Comput. Biol. Bioinform. **12**(2), 360–371 (2015)
21. Konur, S., Gheorghe, M., Dragomir, C., Ipate, F., Krasnogor, N.: Conventional verification for unconventional computing: a genetic XOR gate example. Fundamenta Informaticae **134**(1–2), 97–110 (2014)

22. Konur, S., Gheorghe, M., Dragomir, C., Mierlă, L., Ipate, F., Krasnogor, N.: Qualitative and quantitative analysis of systems and synthetic biology constructs using P systems. ACS Synth. Biol. **4**(1), 83–92 (2015)
23. Konur, S., Ladroue, C., Fellermann, H., Sanassy, D., Mierla, L., Ipate, F., Kalvala, S., Gheorghe, M., Krasnogor, N.: Modeling and analysis of genetic boolean gates using Infobiotics Workbench. In: Verification of Engineered Molecular Devices and Programs, pp. 26–37. Vienna, Austria (2014)
24. MeCoSim website. http://www.p-lingua.org/mecosim/
25. P-Lingua website. http://www.p-lingua.org
26. P systems website. http://ppage.psystems.eu
27. Păun, G.: Computing with membranes. J. Comput. Syst. Sci. **61**(1), 108–143 (2000)
28. Păun, G., Rozenberg, G., Salomaa, A. (eds.): The Oxford Handbook of Membrane Computing. Oxford University Press, Oxford (2010)
29. Pedersen, M., Phillips, A.: Towards programming languages for genetic engineering of living cells. J. R. Soc. Interface **6**(Suppl 4), S437–S450 (2009)
30. Regot, S., Macia, J., Conde, N., Furukawa, K., Kjellen, J., Peeters, T., Hohmann, S., de Nadal, E., Posas, F., Sole, R.: Distributed biological computation with multicellular engineered networks. Nature **469**(7329), 207–211 (2011)
31. Romero-Campero, F., Twycross, J., Càmara, M., Bennett, M., Gheorghe, M., Krasnogor, N.: Modular assembly of cell systems biology models using P systems. Int. J. Found. Comput. Sci. **20**(3), 427–442 (2009)
32. Rozenberg, G., Bäck, T., Kok, J.N. (eds.): Handbook of Natural Computing. Springer, Heidelberg (2012)
33. Sanassy, D., Fellermann, H., Krasnogor, N., Konur, S., Mierlă, L., Gheorghe, M., Ladroue, C., Kalvala, S.: Modelling and stochastic simulation of synthetic biological boolean gates. In: Proceedings of the 2014 IEEE 16th International Conference on High Performance Computing and Communication. HPCC '14, pp. 404–408. France, Paris (2014)
34. Tamsir, A., Tabor, J.J., Voigt, C.A.: Robust multicellular computing using genetically encoded NOR gates and chemical 'wires'. Nature **469**(7329), 212–215 (2011)

Chapter 26
On Improving the Expressive Power of Chemical Computation

Erik Bergh and Zoran Konkoli

Abstract The term chemical computation describes information processing setups where an arbitrary reaction system is used to perform information processing. The reaction system consists of a set of reactants and a reaction volume that harbours all chemicals. It has been argued that this type of computation is in principle Turing complete: for any computable function a suitable chemical system can be constructed that implements it. Turing completeness cannot be strictly guaranteed due to the inherent stochasticity of chemical reaction dynamics. The computation process can end prematurely or branch off in the wrong direction. The frequency of such errors defines the so-called fail rate of chemical computation. In this chapter we review recent advances in the field, and also suggest a few novel generic design principles which, when adhered to, should enable engineers to build accurate chemical computers.

26.1 Introduction

Chemical computers could be used to achieve energy efficient wireless information processing on small scales. This type of computation can be potentially useful for many technological applications. The idea to use chemical reactions for information processing has been around for quite some time. The concept has been explored from many different angles [1–24].

In a chemical computer, the information about the state of the computation is represented as a collective state of a collection of molecules. The most important property of a chemical computer is that the motion of molecules (transport) occurs spontaneously, even when the reactor is not stirred. In principle, once a system has been prepared, the chemical computer needs no power to operate, apart from the obvious thermodynamic requirement that it has to operate in a finite (room) temperature. This implies that the state of the computer changes without any external influence, and no explicit wiring is needed.

E. Bergh · Z. Konkoli (✉)
Department of Microtechnology and Nanoscience - MC2, Chalmers
University of Technology, SE-412 Göteborg, Sweden
e-mail: zorank@chalmers.se

© Springer International Publishing Switzerland 2017
A. Adamatzky (ed.), *Advances in Unconventional Computing*,
Emergence, Complexity and Computation 22,
DOI 10.1007/978-3-319-33924-5_26

The transport of reactants is powered by thermal fluctuations of the surrounding solvent. Reacting molecules move by diffusion with the diffusion constant given by $D \sim k_B T / \eta$ where k_B is the Boltzmann constant, while T and η denote the temperature and the viscosity of the solvent respectively. Reactions happen spontaneously when reacting molecules are close enough. The energy for a reaction comes either from the thermal motion of the surrounding solvent molecules, or is already stored in the electronic states of the reacting molecules.

Since particles can move by diffusion another key advantage of a chemical computer is that there is no need to implement wires to carry information across the device. This is in a sharp contrast with the traditional silicon-based solutions where correct wiring at small scales is a serious issue. In the standard setup the information carrier is the electrical charge and its motion needs to be directed by using conducting materials.

Several specific applications of a chemical computer have been demonstrated in the context of unconventional computation. [25] The main drive behind these developments was not to build a copy of, or to mimic, the standard silicon solutions (which heavily rely on the use of the von Neumann architecture). Instead, the goal was to find ways of using the chemistry in its natural setup, to perform the so-called natural computation, without elaborate auxiliary engineering. A few cases demonstrated in the laboratory deal with a chemical computer implementation of the Vornoi diagram problem, or the skeletonisation problem. There are numerous examples in nature how chemical computation is used to realize information processing. A typical example is bacterial chemotaxis where the intracellular chemistry is used to modulate the motion of the bacteria towards the food or away from the poison, or various gene regulation mechanisms.

Several attempts have been made to achieve a rigorous mathematical formalisation of the concept of a computing chemical reaction system. The goal is to ignore experimental details and emphasize instead the relevant information processing features of a chemical reaction system. A range of various calculi have been developed to that end. [26–28] Such generic (and often very abstract) formulations are useful when the reaction system of interest is very large, i.e. when it involves many reactions and reactant types, and when there is a need for automation and re-use of models. These formulations are often meant to be used in two ways, to analyze chemical computers in a generic but rigorous mathematical way, or to exploit the formality of the description to build involved and highly optimised simulation tools. It is fair to say that, so far, majority of the activity in the field has dealt with the later. These approaches will not be discussed. Instead, the focus will be on discussing the approaches that specifically focus on understanding chemical computation per se. In particular, one very important issue that can be analyzed using a rigorous mathematical formulation is the property of the Turing universality.

What can one compute using the chemical computer? Several recent studies have addressed the question [21, 24, 29–31]. The main conclusion from these studies is that the chemical computation is Turing complete in principle. However, unpredictable errors in the computation can occur. The origin of such errors is the inherent stochasticity of chemical reaction dynamics, which originates from thermal motion of the solvent.

While the motion of the solvent molecules servers the purpose of powering the computer it also leads to occurrence of errors in chemical computation. It is impossible to separate the two. The frequency of such errors defines the so-called fail rate of computation (or just the fail rate). If we can build a theoretical model to explain what controls the fail rate, we will be able to build chemical computers that would make very few errors.

Formally, the fail rate is defined as the fraction of runs during which the error occurs (assuming the the computation can be repeated infinitely many times). In this context a run should be thought of as a dynamical process where the chemical computer changes its state from the time it has been prepared, to the time when its stops, i.e. when the outputs can be inspected.

In this chapter several features of chemical reaction dynamics will be discussed with the goal of understanding how they can influence the intrinsic stochastic behavior of the chemical computer, and ultimately the fail rate. Based on the analysis of several examples, a few design principles will be suggested that could be adhered to when constructing chemical computers.

The fail rate has been investigated in the well-mixed chemical reactor setup. The main conclusion from these studies is that the fail rate increases with the number of particles in the system. This is a serious issue. For example, imagine a computer that makes much more mistakes when programs with long execution time are being run on it.

The present study aims to further explore the fail rate in much more detail in order to understand how to remedy the above problem and understand how it depends on some generic engineering aspects of building a chemical computer. Specifically, the focus is on understanding how the spatial, or temporal noise, affects the fail rate. Any attempt to miniaturize chemical computers for advanced bio-compatible applications will necessarily result in the following situations: (a) there might be many reactant types but very few copies of each, and (b) external stirring will not be possible. Thus it is highly prudent to understand how these features make an influence on the fail rate. Note that from the technical point of view, these regimes (low particle numbers and weak stirring) are the ones that are hardest to analyze. Several pedagogical examples are discussed in [32] (a review article where spatially extended kinetics is discussed) or [33] (study of the effects related to low particle numbers). An attempt has been made to present an intrinsically technical subject using intuitive yet mathematically rigorous constructs.

The text is organized as follows. First some important properties of the diffusion controlled chemical reactions will be reviewed. The goal will be to point out the most important features that are relevant for information processing. Second, the Bare Bones programming language will be discussed. The language is Turing universal with an advantage that any Bare Bones program can be implemented as an (abstract) chemical computer. The key element of the language is the while loop. It will be explained how the loop can be implemented using chemical reactions. Third, the fail rate of the several loop implementations will be investigated in detail, e.g., by analysing its dependence on the topology of the chemical reaction network, the speed of the reactant transport versus the speed of reactions, the number of reactants in the device, etc.

26.2 Fluctuation-Dominated Kinetics Essentials

There are two major types of noise, spatial noise, and temporal noise. This distinction is not always made explicitly in the literature, but it will be used throughout this chapter. Kinetics is said to be fluctuation-dominated when the presence of noise is overwhelming so that the classical chemical kinetics cannot be used to explain the observed experimental behavior.

The dynamics of chemical reaction systems is intrinsically noisy. This noise cannot be controlled since it is an integral part of the dynamics. The classical (mass action) kinetics formulation is extremely practical and straightforward to use, and it appears as an integral part of many software modeling applications.

The equations of classical chemical kinetics are never exact. These equations do not contain information about the intrinsic chemical noise that is always present in the system. The mass action law formulation can be only used to describe (i) the so called well-stirred chemical reactor and (ii) systems with many particles in the reaction volume.

In practice, this means that the theoretical predictions obtained by using the mass action law agree with experimental results only if the special conditions (i) and (ii) are met. In such situations, noise is present but it never influences the dynamics. However, there are situations when this is not true. The influence of noise can be so strong, so that the theoretical predictions obtained by using the mass action law are qualitatively different from the ones observed in experiments.

Temporal noise is also being referred to as intrinsic noise. Even if the chemical reactor is well-stirred, the chemical reaction dynamics can be stochastic if there are very few particles in the system. A well-stirred chemical reactor is often described as having a dimension zero ($d = 0$). In electronics, this type of noise is referred as shot-noise, but this term will not be used in here. It can be shown that when the number of particles is large then such noise disappears. However, for some systems with few particles inside the reactor, the mass action law predictions are qualitatively wrong when compared to experiments (or simulations). Typical examples are bistable systems. For a review of these topics cf. [33, 34] and references therein.

By spatial noise we mean inhomogeneities in the way particles are distributed in space. Such inhomogeneities always occur spontaneously but are smeared out by either external mixing or the internal motion of the fluid around the reacting particles. However, when there is no external mixing, and when the intrinsic transport of reactants is relatively slow when compared to the speed of the reactions in the system, the inhomogeneities can persist for a very long time. Spatial noise can have an enormous influence on the dynamics of the system. It can render any form of mass action law kinetics qualitatively incorrect, even if there are many particles in the system. Systems where this happens are said to exhibit fluctuation dominated kinetics. Conversely, if the transport of reactants is fast the effects of noise are negligible. Such reaction systems are often described as well-mixed.

Traditionally, in the special case when spatial noise plays a dominant role, and the system cannot be described by using the mass action law, one talks about fluctuation dominated kinetics. The typical trademark of such behavior is that particle concentration is not uniform.

As an illustration of a system that exhibits fluctuation dominated kinetics, consider a reaction system where particles A diffuse in space with the diffusion constant D, and react on contact with a reaction rate λ as $A + A \to P$. Assume that the product molecule P does not have any influence on the dynamics of the system (there is no back-reaction). Further, assume that initially the particle concentration $c(\mathbf{r}, t)$ is spatially homogeneous (no spatial dependence of the concentration): $c(\mathbf{r}, t = 0) = c_0$, where $\mathbf{r}$ denotes a position in space, and t denotes time.

In this situation a typical use of the mass action law would be to assume (correctly) that the particle concentration stays homogeneous is space *in average*, $c(\mathbf{r}, t) = c(t)$, and then use the mass law equation for the $A + A$ reaction

$$\frac{dc(t)}{dt} = -\lambda c(t)^2 \tag{26.1}$$

to show that

$$c(t) = \frac{c_0}{1 + \lambda c_0 t} \tag{26.2}$$

This very simple expression conveys several important messages, but not all of them will be discussed in here. The only reason why the equation above is shown is to illustrate how the mass action law kinetics, or classical chemical kinetics, can fail to describe reality. This is best seen by inspecting the $t \to \infty$ limit of the expression, the so-called asymptotic solution, which behaves as $c(t) \propto 1/t$. Experiments show that indeed in this time regime the concentration stays homogeneous in average, but the mass action prediction is qualitatively wrong in low spatial dimensions [35].

For example, in one dimension (e.g. the particles are forced to move in a narrow cylinder) the correct result is $c(t) \propto 1/\sqrt{t}$. Note that qualitative difference for very large times. Thus the reaction $A + A$ exhibits fluctuation dominated kinetics in one dimension. The same occurs for two dimensions (e.g. all reactants move on the surface) where the correct asymptotic behavior is given by $c(t) \propto t/\ln t$. In three dimensions (reactants move in a finite volume) the mass action law provides qualitatively correct description. In three dimensions the dynamics of $A + A$ reaction is not fluctuation dominated. For a review of these topics cf. [32, 35] and references therein.

In the following the two types of noise discussed above will be jointly referred to as spatio-temporal noise, and the term fluctuation dominated kinetics will be used to describe the systems for which both types of noise are important.

The goal will be to investigate fail rate for the systems with both low copy numbers and inefficient particle mixing. Further, an attempt will be made to systematically exploit such features to gain an understanding of how to build the most accurate chemical computers.

26.3 Catalytic Particle Computer as a Bare Bones Programing Language Implementation

The Bare Bones programming language is Turing complete: it can theoretically describe any computable mathematical function as shown by Brookshear and Brylow [36]. The language is suitable for chemical implementation in the sense that any Bare Bones program can be implemented using a suitable chemical reaction system.

A program in the Bare Bones programming language consists of the description of variables v_1, v_2, etc., and a list of statements being either DECREASE v_i, INCREASE v_j, or a while do loop construct, $while(v \neq 0)...endwhile$. These statements can be arranged in a list, even nested, and are expected to be executed in a well-defined order. Below is a Bare Bones program performing $x = u + v - w$.

> **while**$(u \neq 0)$
>
>   *INCREASEx*, *DECREASEu*
>
> **endwhile**
>
> **while**$(v \neq 0)$
>
>   *INCREASEx*, *DECREASEv*
>
> **endwhile**
>
> **while**$(w \neq 0)$
>
>   *DECREASEx*, *DECREASEw*
>
> **endwhile**

To implement such program using a chemical computer it is necessary to transform it into a form that is more suitable for chemical implementation. Liekens and Fernando [31] suggested the so called catalytic particle computer as a way to implement Bare Bones programs. The catalytic particle computer is a particular implementation, a chemical model, of a Bare Bones program. The model consists of an abstract description of how a chemical machine can be implemented in principle. The specification contains the list of reactant types, and the chemical reactions that simulate the execution of the Bare Bones program.

As the Bare Bones program implementation executes it passes through different states. The state of the execution is encoded using chemicals, e.g. $S_1, S_2, S_3, \ldots, S_n$, and observing their quantities (copy numbers)

$$c^{(1)} \equiv (\#S_1, \#S_2, \#S_3, \ldots, \#S_n) \tag{26.3}$$

For example, in the initial state of the program $c^{(1)} = (1_1, 0_2, 0_3, \ldots, 0_n)$, in the next state of the execution e.g. $c^{(1)} = (0_1, 1_2, 0_3, \ldots, 0_n)$, etc. where 1_i (0_i) indicates that 1 (0) occurs at a position i in the list.

Likewise, variable values $v_1, v_2, \ldots, v_m$ are encoded as copy numbers of the respective chemicals $V_1, V_2, \ldots, V_m$,

$$c^{(2)} \equiv (\#V_1, \#V_2, \#V_3, \ldots, \#V_m) \tag{26.4}$$

Taken together, the configuration state of the system

$$c = c^{(1)} \cup c^{(2)} \tag{26.5}$$

encodes the state of the computation, where the symbol $\cup$ denotes the list concatenation.

As an illustration, the Bare Bones program indicated above can be represented as follows:

S_1 : **while**$(u \neq 0)$
    INCREASEx, DECREASEu
    endwhile
    GOTO S_2
S_2 : **while**$(v \neq 0)$
    INCREASEx, DECREASEv
    endwhile
    GOTO S_3
S_3 : **while**$(w \neq 0)$
    DECREASEx, DECREASEw
    endwhile

The above program code is implemented utilizing catalytic particles as in Fig. 26.1.

In the depicted chemical implementation S_1, S_2 and S_3 represent the particles making up the states $\mathbf{S_1}, \mathbf{S_2}$, and $\mathbf{S_3}$ while catalyzing the creation or conversion of the different particles: U, V, W and X. Here and in the following, the particles are named using capital letters, and their respective copy numbers are denoted by using lower-case letters, e.g. u, v, w and x.

26.4   Mathematical Formalization of the While Loop

The while loop construct is the most important element of the language. The catalytic particle computer implementation described previously is one of many possibilities. In the following other implementations will be discussed.

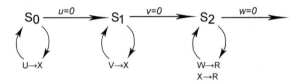

Fig. 26.1 The catalytic particle computer implementation of the Bare Bones program calculating $u + v - w$. The result of the computation is encoded in the final number of X molecules. Note the presence of the state particles which catalyze the chemical reactions. When the condition stated above an *arrow* is fulfilled, the state particle is allowed to undergo the indicated unimolecular reaction (to the next state)

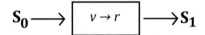

Fig. 26.2 An implementation of the 'CLEAR v' command. A process in the *box* transforms state S_1 to state S_1, and converts every particle V to a particle R

To make the forthcoming discussion easier it is necessary to formalize the while loop concept mathematically. This is done in Eq. (26.6) where the symbol F denotes the mathematical operator that acts on the configuration space of state vectors c and converts a predefined set of copy numbers as shown. This is illustrated in Fig. 26.2.

$$\begin{pmatrix} \#V[begin] \\ \#R[begin] \\ S_0 \end{pmatrix} \equiv \begin{pmatrix} v \\ 0 \\ \#S_1[begin] \\ \#S_2[begin] \\ \vdots \\ \#S_n[begin] \end{pmatrix} \xrightarrow{F} \begin{pmatrix} 0 \\ v \\ \#S_1[end] \\ \#S_2[end] \\ \vdots \\ \#S_n[end] \end{pmatrix} \equiv \begin{pmatrix} \#V[end] \\ \#R[end] \\ S_1 \end{pmatrix} \qquad (26.6)$$

As an example, consider the while loop implementation suggested by Liekens and Fernando. The list of chemical reactions is shown in Eq. (26.7):

$$S_0 + V \xrightarrow{k_1} S_0 + R \qquad (26.7)$$
$$S_0 \xrightarrow{k_2} S_1$$

The state particle S_0 is catalyzing the reaction of particle V to R until there are no V particles left and S_0 changes to S_1 representing another state. This represents a while-loop with the condition that v is non-zero. In the following, this implementation will be referred to as the minimal while loop implementation. The respective F for this case is given by

$$\begin{pmatrix} v \\ 0 \\ s_0[begin] = 1 \\ s_1[begin] = 0 \end{pmatrix} \xrightarrow{F} \begin{pmatrix} 0 \\ v \\ s_0[end] = 0 \\ s_1[end] = 1 \end{pmatrix} \tag{26.8}$$

Note that in (26.8) states S_0 and S_1 are encoded with two copy numbers ($\#S_0$, $\#S_1$) as $S_0 \equiv (1, 0)$ and $S_1 \equiv (0, 1)$.

26.5 The Fail Rate Concept

Since the while loop construct is the key element of the Bare Bones programming language the chance that the error occurs during its execution is of particular importance. This probability will be heavily analyzed in the following.

The only way the while loop execution can fail is that the reaction that leads to the next step of the execution (the one just after the while loop) happens too soon. When this happens molecules V will still be in the system. This would classify as a 'run-time' error, that could propagate further and cause all sorts of inconsistencies in the program execution.

For example, due to the stochastic nature of chemical reactions it is not possible to prevent the second reaction in (26.7), though the first reaction might be much faster. This is illustrated in Fig. 26.3 where arrows represent different reaction pathways. Thus due to the presence of stochastic effects there is a chance that the loop ends prematurely. To analyse the probability that this happens, it is necessary to investigate the probabilities that each reaction will occur, given that the configuration of the system is known.

These probabilities can be computed by using the standard techniques from the theory of Poisson processes [37]:

$$P(S_0 + V \xrightarrow{k_1} S_0 + R) = \frac{k_1 v}{k_1 v + k_2} \tag{26.9}$$

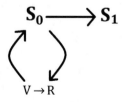

Fig. 26.3 The pathway topology of the minimal while loop implementation in Eq. (26.7) is shown. From state S_0 the system either converts one V particle to R, or changes directly to state S_1 (which would be an error provided this happened when $v \neq 0$)

$$P(S_0 \overset{k_2}{\rightarrow} S_1) = \frac{k_2}{k_1 v + k_2} \qquad (26.10)$$

From Eq. (26.9) it is straightforward to calculate the probability P of not finishing the loop by using the probability of the opposite event, i.e. the probability $\bar{P}$ that there is no error during the loop execution. Due to the independence of the reaction events $\bar{P}$ is given as the product of probabilities that loop performs without error during each iteration (given by Eq. 26.9 for each $v = V_0, V_0 - 1, \ldots, 2, 1$).

$$P = 1 - \prod_{v=1}^{v_0} \frac{k_1 v}{k_1 v + k_2} = 1 - \frac{v_0! \Gamma(\frac{k_2}{k_1} + 1)}{\Gamma(\frac{k_2}{k_1} + v_0 + 1)} \qquad (26.11)$$

26.6 Strategies for Identifying Good Designs

Once the fail rate formula is known it can be used to theoretically analyze fail rate and suggest feasible design on pure theoretical basis. To be able to identify good designs it is essential to have an accurate fail rate formula.

There are many effects that can influence fail rate and to account for them all is rather challenging. It is equally important to be critical of the results. For a given formula it is important to understand under which assumptions it has been derived. For example, Eq. (26.11) has been derived under the assumption of the so called well-mixed regime where the mixing time is considerably faster than the typical reaction time. The analysis of this formula has been performed by Liekens and Fernando [31]. It can be seen immediately that $P \to 0$ for $k_1/k_2 \to \infty$. This shows that to reduce the chance that the chemical computer makes an error the first design rule is that chemical reactions should be used where $k_1 \gg k_2$. This can be achieved in two ways by making k_1 as large as possible, or by making k_2 as small as possible. The second alternative results in a problem, as discussed by Liekens and Fernando [31]. The duration of the loop execution (the run-time of the loop) becomes infinitely long when $k_2 \to 0$. This suggests that designs with $k_1 \gg k_2$ and k_2 sufficiently large are more practical. The question is whether this conclusion holds when other effects are considered.

There are many features that are known to exert influence on kinetics, e.g. reactor-size, reactor-shape, diffusivity and the initial distribution of particles. All these features can be used to achieve better designs (fast and accurate computation) and it is therefore of great interest to understand how they affect the fail rate. Note that even if not exploited explicitly, ignoring these parameters can cause enormous errors when theoretically analyzing the performance of a chemical computer designs (for eventual experimental implementations).

In the following two fail rate shaping features of the system dynamics will be investigated (1) the influence of the speed of mixing is analysed through by modification of the diffusion constants (large diffusion constants corresponding to fast mixing) and (2) the effects of the while-loop reaction network topology (analyzed by investigating the fail rate for several explicit reaction schemes). Before embarking on analysing these effects, the next section summarizes the reaction schemes considered in this work.

26.7 Reaction Network Topologies

In Sect. 26.6 it was shown that the choice of the kinetic (reaction rate) parameters can be exploited to achieve a good performance (with low fail rates). This is by no means the only aspects that can be changed to optimize the performance.

In this section various while loop realization have been suggested, each detailed in a separate subsection. All reaction schemes were chosen based on the expectation that they should improve the accuracy of the related chemical computer implementation, i.e. result in lower fail rates.

In the following the terms 'reaction network topology' or 'reaction system topology' will be used when the goal is to emphasize the structure of the reaction network, i.e. how various reactions couple together. All reaction system topologies comply with the template specified in Eq. (26.6) and can be seen as an implementation of the while-loop operator F. For example, the minimal while loop implementation in Eq. (26.7), shown also in Fig. 26.3, is the simplest possible implementation of the while loop that complies with the template.

26.7.1 Sequential System

To achieve less random and more controlled state transition, the reaction topology is changed so that the change of state is governed by a chain of reactions.

It is assumed that a cluster of S_0 needs to be formed before the next step is allowed to happen. The critical number of S_0 particles that needs to be accumulated will be denoted as ν. By assumption, particles S_0 are collected (adsorbed) by a particle P sequentially. This is illustrated in Fig. 26.4, and shown explicitly in Eq. (26.12).

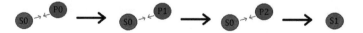

Fig. 26.4 An illustration of the sequential topology dynamics. An example with $\nu = 3$ is depicted. The process involves three S_0 particles: First the P_0 particle reacts with a S_0 particle creating a P_1 complex. This P_1 complex then adsorbs another S_0 particle resulting in a new complex P_2. When this P_2 complex adsorbs yet another S_0 particle, the complex with ν S_0 particles is finally converted into the S_1 particle. This indicates that the state of the computation has changed

$$S_0 + V \xrightarrow{k_1} S_0 + R$$

$$P_n + S_0 \xrightarrow{k_2} P_{n+1} \qquad (26.12)$$

$$P_{\nu-1} + S_0 \xrightarrow{k_3} S_1$$

The F mapping for this case is given by

$$\begin{pmatrix} \nu \\ 0 \\ s_0[begin] = \nu \\ p_0[begin] = 1 \\ p_1[begin] = 0 \\ p_2[begin] = 0 \\ \cdots \\ p_{\nu-1}[begin] = 0 \\ s_1[begin] = 0 \end{pmatrix} \xrightarrow{F} \begin{pmatrix} 0 \\ \nu \\ s_0[end] = 0 \\ p_0[end] = 0 \\ p_1[end] = 0 \\ p_2[end] = 0 \\ \cdots \\ p_{\nu-1}[end] = 0 \\ s_1[end] = 1 \end{pmatrix} \qquad (26.13)$$

A similar mechanism is used in various gene regulation mechanisms in the living cell. Normally, a relatively large cluster of molecules needs to aggregate at specific parts of DNA before genes can be expressed. The required pre-clustering effectively lowers the reaction rate, which indirectly favours larger k_1/k_2 ratios. However, it is not at all clear whether such assumption holds when spatial aspects of the dynamics are taken into account.

26.7.2 Feedback System

A further improved control of the transition to the next state can be achieved if the exit-reaction (to the next step) is inhibited by any existing V particle.

An implementation of this idea is shown in Eq. (26.14). The system in Eq. (26.12) has been extended with a reaction where the P_n complex decays into its components every time the complex encounters a V particle. Then, it decays into one P and n S_0 particles, and hence the chain of the reactions that leads to the necessary accumulation of S_0 particles must be redone. The reactions leading to a state-change are illustrated in Fig. 26.5 where also a branch has been incorporated where the particles collapses due to an encounter with a V particle. The F mapping for this case is given by Eq. (26.13).

$$S_0 + V \xrightarrow{k_1} S_0 + R$$

$$P_n + V \xrightarrow{k_2} P + nS_0 + V \qquad (26.14)$$

$$P_n + S_0 \xrightarrow{k_3} P_{n+1}$$

$$P_{\nu-1} + S_0 \xrightarrow{k_4} S_1$$

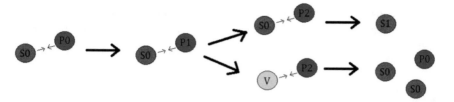

Fig. 26.5 An illustration of the feedback system dynamics with $v = 3$. The process is identical to that in Fig. 26.4 with the only difference being that every P complex can decay under the presence of a V molecule. For example, the P_2 particle instead of reacting with a S_0 particle encounters a V particle causing it to decay into a P_0 particle and two S_0 particles. This forces the system to redo the chain of reactions hence delaying the state change which is of interest when there are remaining V particles

26.7.3   Delayed Feedback System

A problem with the feedback system in Eq. (26.14) is that when P_n collapses the resulting particles are in the same position making it likely for the particles to quickly react back together. To avoid this, the system is extended further, as shown in Eq. (26.15). The reaction systems features a new particle, S_r. When P_n disintegrates, instead of n S_0 particles, n S_r particles are created. The S_r particles must then decay into S_0 before they can react with P. While this is happens, the particles will also diffuse away from each other. Our expectation is that this inhibits the particles to instantly react back together.

$$S_0 + V \xrightarrow{k_1} S_0 + R$$
$$P_n + V \xrightarrow{k_2} P + nS_r + V$$
$$P_n + S_0 \xrightarrow{k_3} P_{n+1} \tag{26.15}$$
$$P_{v-1} + S_0 \xrightarrow{k_4} S_1$$
$$S_r \xrightarrow{k_5} S_0$$

The F mapping for this case is given by

$$
\begin{pmatrix}
v \\
0 \\
s_0[begin] = v \\
p_0[begin] = 1 \\
p_1[begin] = 0 \\
p_2[begin] = 0 \\
s_r[begin] = 0 \\
\cdots \\
p_{v-1}[begin] = 0 \\
s_1[begin] = 0
\end{pmatrix}
\xrightarrow{F}
\begin{pmatrix}
0 \\
v \\
s_0[end] = 0 \\
p_0[end] = 0 \\
p_1[end] = 0 \\
p_2[end] = 0 \\
s_r[end] = 0 \\
\cdots \\
p_{v-1}[end] = 0 \\
s_1[end] = 1
\end{pmatrix}
\tag{26.16}
$$

Note that this particular topology exploits explicitly the spatial aspects of the dynamics. Here, the diffusion of S_r particles is used as a delay mechanism to ensure more accurate transition.

26.8   Methods

The fail rate for the implementations discussed in the previous section has been analyzed numerically using simulation techniques. An analytic, or semi-analytic, investigation might be possible but this would likely complicate the discussions that follow by making it much more technical. This section summarizes how the fail rate was computed, which simulation technique was used, and how the simulation parameters were prepared.

26.8.1   The Fail Rate Computation

To model the dynamics of the system it is possible to use off-lattice or on-lattice simulations techniques. Both have their respective merits and problems. In principle there is no difference between off-lattice and on-lattice simulations provided they are matched appropriately. In the following, for the reasons of the simplicity of implementation, and numerical efficiency, the on-lattice paradigm was used.

The dynamics of each chemical system is modelled by assuming that particles move on a lattice with spacing between lattice sites given by h. Each site is assumed to be the centre of a small well-mixed chemical reactor with volume h^d, where d is the dimension of the system. Particles move among these volumes, and react within each volume. By assumption, each particle can jump and when two or more particles meet on the same lattice site they can react. The probabilities of these events are governed by their respective rates. The content of each lattice site and the reaction rate values determine which reactions are possible. Each rate describes a Poisson process [37]. The initial condition is prepared by placing each particle on the lattice at random.

The simulations were performed using the version of the Gillespie's algorithm suggested in [38]. The algorithm works as follows. Each lattice point has a vector of copy numbers assigned to it representing the number of particles of each species at that point. Rates for all events were calculated for each lattice point based on the copy numbers, with events referring to the diffusion and reaction processes. The process to occur is chosen as follows. Based on the rates a table with probabilities for the occurrence of each process is constructed. Computationally, this is the most expensive step during the algorithm execution. A process is chosen at random according to the weights specified by the table. When the copy numbers of a lattice point is changed, the rates is updated based on the new copy numbers, and the time is updated. In such a way it is possible to obtain a trajectory through the configuration space of the system.

Every trajectory requires a separate simulation. All simulations are performed on a PC with a 64-bit 3.30 GHz 8 core AMD with 16 GB RAM.

Numerically, the fail rate is computed by examining the fraction of runs that resulted in an error:

$$f = \frac{N_{error}}{N_{tot}} \tag{26.17}$$

where N_{tot} denotes the total number of runs and N_{error} the number of runs that resulted in an error. To obtain good statistics many such runs are repeated. The simulation is stopped when the state is changed: Then there can either be no V particles left ($v = 0$), or there can still be remaining V particles (an error has occurred).

26.8.2 The Units Used in Simulations

To obtain agreement with off-lattice dynamics, it is necessary to choose the simulation parameters in a very special way. In what follows all parameters will be specified using an arbitrary set of units. The physical (experimental) time t_{exp} will be expressed in the units of arbitrary time interval τ, $t_{exp} = t\tau$ and the physical length will expressed in the units of lattice spacing h, $l_{exp} = lh$. The bulk diffusion constant (experimentally measurable parameter) can be expressed as

$$D_{exp} = D_0 \frac{h^2}{\tau} \tag{26.18}$$

D_0/τ denotes the rate which with a particle jumps from the current node to a specific neighbouring node. The experimental reaction rates are defined as

$$k_{exp} = k \frac{h^{d(\omega-1)}}{\tau} \tag{26.19}$$

where ω is the overall stoichiometric coefficient of the reaction (e.g. the sum of the exponents in the mass action law). For example, for unary reactions $\omega = 1$ and for binary reactions $\omega = 2$. Note that t, l, D_0, and k are dimensionless. From here on, to simplify the notation, we express all quantities in units of τ and h and use t instead of t_{exp}, k instead of k_{exp}, etc.

By construction, a lattice simulation is less accurate than the off-lattice simulation. They agree when the number of lattice sites becomes infinite. Thus when studying this type of convergence for lattice simulations while keeping the measurable quantities constant, i.e. D_{exp} and k_{exp}, it is necessary to carefully adjust the parameters used in simulations.

For example, assume that the number of lattice sites is increased in the simulation by a factor α (while keeping the experimental parameters constant). This amounts to scaling the lattice spacing h by the factor α^{-1} (follows from $h = l_{exp}/l \sim 1/l$

and $l \sim \alpha$). Further, from Eq. (26.18) one sees that D_0 scales with α^2 (follows from $D_0 = \tau D_{exp}/h^2 \sim h^{-2}$), and from (26.19), in a similar way, one concludes that $k \sim \alpha^{d(\omega-1)}$. Note that the reaction rate for a unary reaction does not scale, while the rate for a binary reaction scales with α^d. For example, doubling of the number of lattice sites would correspond to choosing $\alpha = 2$: D_0 would have to be increased four times, $D_0 \rightarrow 4D_0$, unary reaction rates would be left unchanged, and binary reaction rates would have to be increased by the factor 2^d. The intuitive explanation is that as the size of the lattice site is decreased, the particles must jump more often to maintain the same overall displacement, and since particles do not meet that often (they can be placed on more sites), once they do meet they should react with a larger reaction rate to maintain the initial speed of reaction.

26.9 Results

While reading this section, it is very important to follow the chronological order of the presentation. The presentation is organised around the key hypotheses, or simply results, which were formulated after careful analysis of the simulation data and further generalizations.

R1: Any form of poor-mixing makes the fail rate orders of magnitude higher than what would be predicted by using the assumption of a well-mixed reaction volume.

This can be seen by comparing the fail rates for the situations with good and bad mixing (Sect. 26.9.1). As a reference, a k_1/k_2 ratio was chosen that results in very few errors for well-mixed system, ca. 1ppm. The effects of poor mixing increase the fail rate by orders of magnitude, from 1ppm to the values in the range 40–100 %, depending on the dimensionality of the system and how many particles need to be consumed.

R2: Even if the mixing is poor, the minimal while loop implementation can still be significantly improved by using the topology of the sequential reaction network. In this way, the fail rate can be roughly halved (Sect. 26.9.2, paragraph 'The sequential system').

As an example, the fail rate can be reduced from 100 % to 60 % in one dimension, which is the most pathological case where the worst mixing is expected.

R3: The fail rate can be even reduced further using the feedback system topology, and the rate can be lowered roughly by half (Sect. 26.9.2, paragraph 'The feedback system').

For example, in the most pathological dimension ($d = 1$) the fail rate can be further reduced from 60 % to roughly 20 %, which is a remarkable achievement. By using the delayed feedback system a further decrease is possible, but the reduction is not that dramatic, though it can be significant in some situations. For example, in one dimension, when many loop cycles need to be performed, the reduction is from the fail rate value of 20 % to barely 10 %.

R4: The time it takes the while loop to complete its cycles is changing when different
     topologies of the implementation are considered (Sect. 26.9.3). In general, more
     complex topologies lead to longer execution times. However, a typical increase
     in the program execution time is not dramatic (e.g. no orders of magnitude
     difference). This suggests that the gain in accuracy comes at a balanced increase
     in the program execution time.

26.9.1 The Effects of the Mixing Speed

In order to investigate the effects of space, the fail rate of the minimal while loop
implementation described in Sect. 26.5 was studied for two cases, one with $k_1/k_2 =$
10 and one with $k_1/k_2 = 10^6$. This was done in one ($d = 1$), two ($d = 2$) and three
($d = 3$) dimensions.

In the simulations, the total number of lattice sites was kept constant, 100 cells,
for all dimensions. Thus the lattice sizes were as follows: 100 cells for $d = 1$, 10×10
cells for $d = 2$, and $5 \times 5 \times 4$ cells in $d = 3$. Periodic boundary conditions were
assumed. The following values for the diffusion constants were used: $D_0 = 100$,
1000 and 10000 for $d = 1, 2, 3$ respectively. For every simulation set ten thousand
simulations were performed ($N_{tot} = 10000$) and the error-rate was calculated as
described earlier, i.e. by monitoring the number of runs N_{error} that resulted in an
error (the runs that did not successfully transform all V particles to R before exiting
the while-loop).

Figure 26.9 depicts how the fail-rate of the minimal while-loop implementation
in Eq. (26.7) depends on the spatial aspects of the dynamics. Here, the simulations
were performed with $k_1/k_2 = 10$ to make the overall error large. The analysis of
Fig. 26.9 shows that the speed of mixing can have a strong influence on the fail rate.
For example, in panel (a) one sees that the fail-rate curve strongly differs from the
curve that describes the well-mixed result, even for very large diffusion constant
values with $D_0 = 10000$ for $d = 1$. Panels (b) and (c) suggest that the spatial effects
are weaker for dimensions $d = 2, 3$ since all curves are closer.[1]

The reason behind the higher fail-rate for lower D_0 or lower dimensions can be
explained as follows. Figure 26.6 is an illustration of the most important principles. It
shows a snapshot of the dynamics where the occupancy of each lattice site is shown.
The typical features that signal the fluctuation dominated kinetics are explicit.

Once a particle V has been converted it leaves a void that needs to be filled by
the diffusion. There is always a depletion zone around the molecule S_0, the so called
cavity. If the diffusion is slow, this cavity is long-lived. Further, it is a well-known
result that in higher dimensions these cavities do not have such a strong influence on
the kinetics [32]. Thus the effect should be less pronounced in higher dimensions or
for larger diffusion constants (e.g. see Fig. 26.7).

---

[1]The results for d = 2,3 only qualitative since the results are not fully converged with respect to the
size of the lattice. This analysis is illustrated briefly in Fig. 26.18.

(a)

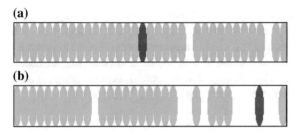

(b)

Fig. 26.6 An illustration of the (spatial) fluctuation dominated kinetics. Snapshots from one dimensional simulations of the minimal while loop-system implementation in Eq. (26.7). A *green oval* symbolizes that the position is occupied by one or several V particles, the *red oval* denotes the S_0 particle, and *white* colour stands for empty sites. Panel **a** depicts the so-called well-mixed system. The rate of diffusion is significantly greater than the reaction rate. This causes an equal distribution of V particles in space. When the S_0 particle consumes an V particle, the other neighbouring V particles quickly fill the void. In panel **b** the reaction rate is significantly higher than the rate of diffusion. Note a void of V particles around the S_0 particle. If a V particle comes in contact with the S_0 particle it instantaneously reacts and because of the low rate it takes longer time for the neighbouring V particles to fill the void

In panel (a) is there a high degree of mixing giving a homogeneous distribution of V particles. In panel (b) is there a low degree of mixing causing a lowering of the concentration of V particles around the S_0 particles. The same is visualized in Fig. 26.7 for two dimensions.

For comparison, Fig. 26.8 shows the fail-rate of an identical systems to that described above but with with $k_1/k_2 = 10^6$. This is a more natural choice of the parameters since it should lead to lower fail rates. Indeed, when compared to the Fig. 26.9 all curves are shifted downwards. The shift of the reference curve is the largest (orders of magnitude larger). The curve for the well-mixed system almost coincides with the abscissa in the graph, while other curves move only slightly. The reference curve has not been plotted since it is so small that it cannot be seen.

This analysis suggests that the natural choice for reaction rate parameters has a drawback. While the natural adjustment of the reaction rates leads to a low reference fail rate, the effect from weak mixing that has the opposite effect (it tends to increase the fail rate) is much more pronounced. There is a competition between these two effects, and the effect of weak-mixing dominates. This suggests that the fail rate of naturally tuned system is much more sensitive to the speed of mixing. The relative increase in the fail rate (when diffusion constant is lower) is much bigger for this system.

The results for $k_1/k_2 = 10^6$ and $k_1/k_2 = 10$ are similar in the sense that the chance of failure increases for weak mixing, e.g. for low values of the diffusion constant the fail rates tend to be generally larger. This can be seen from Table 26.1. For example, in one dimension with $D_0 = 10000$, the fail-rate is 0.21 for $k_1/k_2 = 10^6$. The corresponding value for $k_1/k_2 = 10$ is 0.51. As expected, the fail rate is lower for the $k_1/k_2 = 10^6$ case, assuming everything else the same. However, in relative terms, the effect of weak-mixing is much more pronounced for the natural choice: e.g. the

Fig. 26.7 Same as Fig. 26.6 but for two dimensions. *Green circles* represent a position occupied by one or more V particles with a darker tone indicating a higher number of V particles. Panel **a** shows well-mixed conditions with a homogeneous concentration over the domain while panel **b** shows conditions with low rate of diffusion with a distinct drop in concentration of V particles around the S_0 particle

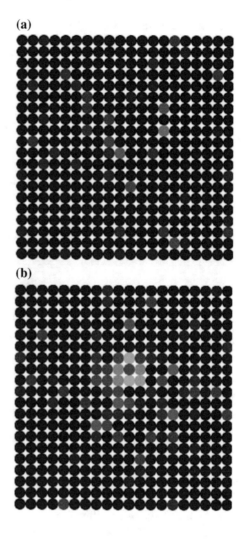

(a)

(b)

fail rates for $k_1/k_2 = 10$ and $k_1/k_2 = 10^6$ are 0.40 and 5.2×10^{-6} respectively for the well-mixed systems, but the respective differences in the fail rates from the weak-mixed systems are much larger: note a jump from the fail rates in $d = 0$ from 0.4 to the $d = 1$ fail rates of 1, 0.93 etc. on one hand, and the respective jumps from 5.2×10^{-6} to 1, 0.88 on the other. This holds uniformly across all considered (physical) dimensions.[2]

In brief, the above analysis suggests that the effects of space play an important role when the reaction rate parameters are tuned as suggested in [31], i.e. when k_1/k_2

---

[2]We believe that the f_2 values for $d = 2$ and $d = 3$ should be actually a bit higher due to lack of convergence for two ($d = 2$) and three ($d = 3$) dimensions with respect to the number of sites (cells) in the lattice.

Fig. 26.8 The fail rate of the minimal while-loop implementation in Eq. (26.7) with $k_1/k_2 = 10^6$ plotted against the initial number of V particles v_0. *Blue* is simulations for $D_0 = 100$, *red* is $D_0 = 1000$ and *green* is $D_0 = 10000$. Panel **a** shows results for simulations in one dimension, panel **b** for two dimensions and panel **c** for three dimensions

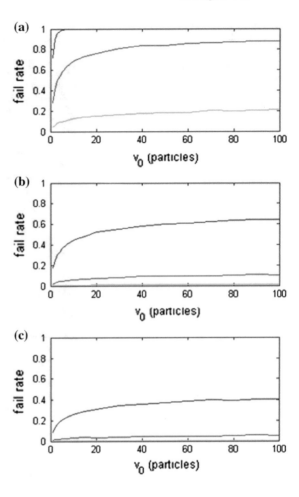

is chosen as large as possible. Further, even for very large diffusion constants the effects of finite mixing speed seem to decide on the value of the fail rate. The spatial features of the dynamics need to be taken into account when analysing the behavior of the fail rate.

26.9.2 The Effects of the Reaction Topology

The fail rates of the several extensions of the minimal while loop implementation have been investigated numerically. In order to reduce the number of parameters, most of the reaction rate parameters were set to very high values, around 10^6, to achieve the diffusion-controlled limit. This is a standard procedure used in these types of simulations (see e.g. [32] and references therein). The remaining parameters were

Fig. 26.9 The fail rate of the minimal while-loop implementation in Eq. (26.7) with $k_1/k_2 = 10$ plotted against the initial number of V particles v_0 for various diffusion constants (each data point represents an average over 10000 runs) and dimensions: panel **a** for $d = 1$, **b** for $d = 2$, and **c** for $d = 3$. The *blue curves* are for $D_0 = 100$, *red* for $D_0 = 1000$, and *green* for $D_0 = 10000$. The *black (dashed) curves* depict the analytic fail rate from Eq. (26.11) representing the $D_0 \rightarrow \infty$ limit

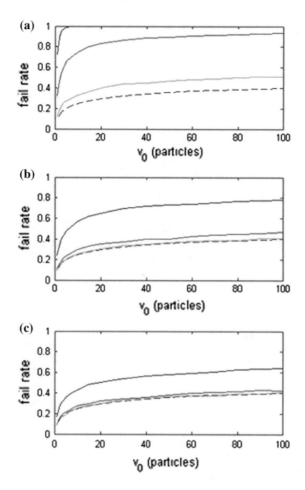

chosen to ensure that the average duration of the loop execution (to be also referred as the mean loop-execution time) is 1 (in units of τ) when $v_0 = 1$, and in a well-mixed regime (zero-dimensional case). This was done to get a fair comparison between the topologies. The kinetic constants used for the simulations are all stated in Table 26.2.

Besides the fail-rate, we also recorded the mean loop execution time of the while-loops that did not result in an error. The diffusion rate was set to $D_0 = 1$ for all particles. The simulations were performed on 16 cells in one and two dimensions, on a 1×16 and a 4×4 domain. For every system simulations were performed for $v = 1, 2, 3, 4$. Note however that the three topologies: sequential, feedback and delayed are identical for $v = 1$. The fail-rate was calculated from an ensemble of 10000 simulations.

Table 26.1 Selected fail rate values with $v_0 = 100$

d	D	f_1 ($k_1/k_2 \sim 1$)	f_2 ($k_1/k_2 \sim \infty$)
0	∞	0.40	5.2×10^{-6}
1	100	1	1
1	1000	0.93	0.88
1	10000	0.51	0.21
2	100	0.78	0.64
2	1000	0.47	0.10
2	10000	0.41	0.01
3	100	0.64	0.40
3	1000	0.42	0.05
3	10000	0.41	0.006

A side by side comparison of fail rates is shown for two systems with $k_1/k_2 = 10$ (the f_1 column) and $k_1/k_2 = 10^6$ (the f_2 column). The second choice (to the right) is referred to as a natural choice: it has been suggested that this choice leads to lower fail rates [31]

Table 26.2 The values of the kinetic parameters used in the simulations of the different reaction topologies

Topology	k_1	k_2	k_3	k_4	k_5
Minimal implementation	10^6	1			
Sequential $v = 1$	10^6	1	1		
Sequential $v = 2$	10^6	1.5	1.5		
Sequential $v = 3$	10^6	1.83	1.83		
Sequential $v = 4$	10^6	2.08	2.08		
Feedback $v = 2$	10^6	10^6	1.5	1.5	
Feedback $v = 3$	10^6	10^6	1.83	1.83	
Feedback $v = 4$	10^6	10^6	2.08	2.08	
Delayed $v = 2$	10^6	10^6	10^6	10^6	0.29
Delayed $v = 3$	10^6	10^6	10^6	10^6	0.265
Delayed $v = 4$	10^6	10^6	10^6	10^6	0.225

All parameters are expressed in arbitrary units, as discussed in Sect. 26.8.2

26.9.2.1 The Sequential System

Figure 26.10 depicts how the fail rate depends on the initial number of V particles, v_0, for the sequential system. When compared to the minimal while loop implementation, the sequential system achieves a lower fail rate. Note that the effect is genuine. For example, it is likely possible that the overall effect of the auxiliary reactions is only to re-normalize the effective rate of the reaction $S_0 \rightarrow S_1$ which could make the ratio k_1/k_2 larger. However, this is not the case since, as discussed above, a care has been taken to tune the reaction parameters to ensure that the duration of the while loop

Fig. 26.10 The fail rate of the sequential system plotted against initial number of V particles v_0 in the regime of poor mixing. The different colored *solid curves* represent different values v and the *dashed black curve* represent the system in Eq. (26.7). Panel **a** shows the results for simulations in one dimension and panel **b** for two dimensions

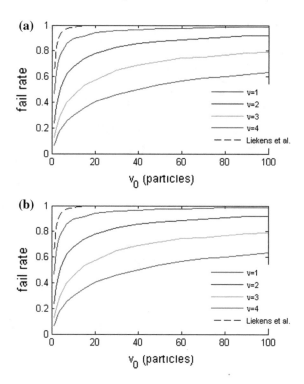

execution is always equal to one. This implies that the effective reaction rate for the $S_0 \rightarrow S_1$ reaction is not changed. An increase in v leads to lower fail rate for all v_0.

26.9.2.2   The Feedback System

Figure 26.11 shows similar results for the feedback-system. The fail rate also decreases with an increasing number v as for the sequential system. It should be noted that the fail rate for the feedback system is lower from the one for the sequential system.

26.9.2.3   The Delayed System

The results for the delayed system are plotted in Fig. 26.12. As for the previous two topologies, the delayed system shows a lower fail-rate with increasing v. However, there are some subtle differences. The delayed system is better for all v_0 except for very small v_0 values. The differences in the fail rate are not that large but do exist, at least for $d = 2$. This behavior could be caused by the lack of convergence with respect to the number of lattice size, but it could be also a genuine effect. This

Fig. 26.11 Solid curves depict the fail rate of the feedback system plotted against initial number of V particles v_0. The *dashed curves* are the fail rate for the sequential system included for comparison. The different colors represent different v values. Panel **a** shows the results for simulations in one dimension and panel **b** for two dimensions

Fig. 26.12 Solid curves are the fail rate of the delayed system plotted against initial number of V particles v_0. The *dashed curves* are the fail rate for the feedback system included for comparison. The different colors represent different values v. Panel **a** shows the results for simulations in one dimension and panel **b** for two dimensions

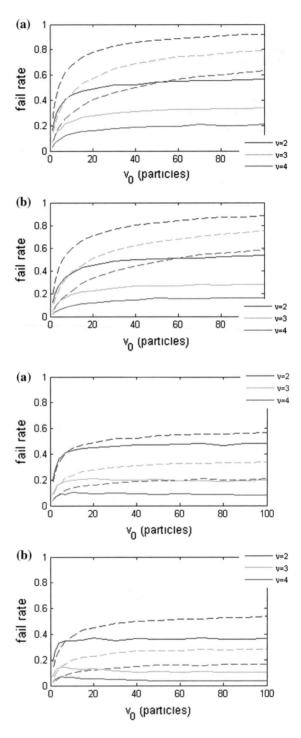

example suggests that, in general, the performance for two different topologies might differ depending on the values of v_0. This in turn implies that it could be possible to optimize the loop topology for various regions.

26.9.2.4 A Summary of Key Findings

A maximum fail-rate with respect to v_0 can be observed for the delayed system seemingly increasing with v. This maximum is more flagrantly seen in two dimensions. Again, this effect as well can be caused by the lack of convergence. However, it is likely genuine since only relative comparisons are made across different values of v_0. This suggests a possibility to fine tune while loop implementations to a particular range of v_0.

The feedback and the delayed system can exit the while-loop prematurely despite the the feedback from V particles. What causes these systems to exit prematurely is visualized with snapshots from simulations of the delayed system. In Fig. 26.13 are snapshots from simulations in one dimension and in Fig. 26.14 from simulations in two dimensions.

26.9.3 Running Time

Figure 26.15 shows the mean of the while-loop duration time, in units of τ, of a successful while loop for $v_0 = 1$ for the sequential system. The reason for focusing

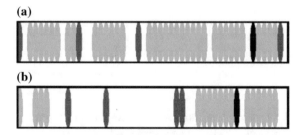

Fig. 26.13 Snapshots from simulations of the feedback system in one dimension. *Green* represent a position occupied by one or several V particles and *red* represent a position occupied by a S_0 particle and *black* represent the position of the P particle. Panel **a** depicts a situation when the system cannot exit the while-loop easily since there are V particle surrounding all S_0 particles. The P particle cannot collect each S_0 without encountering which would force it to release its load of S_0 particles. Panel **b** shows a dangerous situation which can arise when all S_0 particles are in the same cavity (of V particles). Then it is possible for the P particle by entering the cavity to collect all S_0 particles without encountering a V particle. This would cause the system to exit the while-loop even though there are remaining V particles. This is an example of how the effects of fluctuation dominated kinetics can affect the fail rate at low dimensions

Fig. 26.14 Snapshots from simulation as in Fig. 26.13 but in two dimensions with identical color coding. In panel **a** it is impossible for the particle P to collect all S_0 without encounter V particles and decay. In panel **b**, however, there are few V particles left making it possible for the P particle to collect all S_0 by going around the remaining V particles. This of course would make the system prematurely exit the while-loop. Again, this is an additional illustration of the fact that the effects of fluctuation dominated kinetics can exert a strong influence on the computation

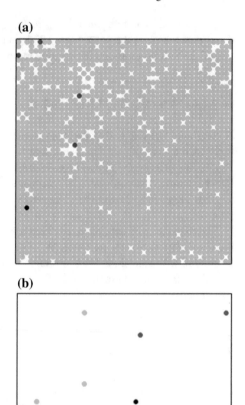

(a)

(b)

on $v_0 = 1$ is that very few while-loops were successful for greater v_0 (which requires an enormous amount of runs to obtain a reasonable statistics).

The mean time grows moderately with v and is larger in the lower dimension. This is expected, since particles do not meet as often due to the recurrence of random walks in low dimensions. As a comparison, the related mean duration time for the minimal while-loop implementation was about 1.43 for $d = 1$ and 1.56 for $d = 2$.[3] (No graphs shown). It is important to point out again that the reaction rate parameters were chosen to ensure that the mean time is 1 for $d = 0$ for every implementation

---

[3]The same problem with few successful while-loops arises. The specified numbers are for $v_0 = 1$.

Fig. 26.15 The mean time
for successful while-loop
runs for the sequential
system with initial number of
V particles $v_0 = 1$ plotted
against v. Panel **a** shows the
results for simulations in one
dimension and panel **b** for
two dimensions

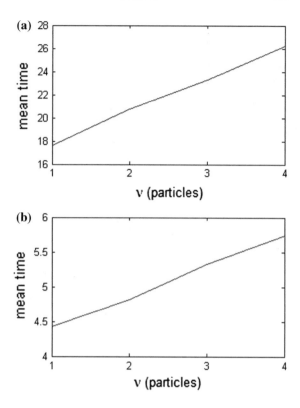

topology. Thus all observed changes are caused by spatial dynamic effects. In general,
we noticed that the inclusion of spatial effects affects the mean execution times of
considered implementation topologies very differently (e.g. compare the previously
mentioned mean time values 1.43 and 1.56 with the ones observed in Figs. 26.15,
26.16, and 26.17).

The dependence of the mean time on v_0 is illustrated in Fig. 26.16 for the feedback
system. By inspecting the mean time data values in Figs. 26.15 and 26.16 (for $v_0 = 1$)
one sees that the mean times are slightly longer for the feedback system. The reason
is that in the feedback system every P_n complex can decay into n S_0 particles and the
aggregation process of assembling P_n again needs to be restarted, which consumes
overall time. The results for the delayed system are similar to the ones for the feedback
system, as shown in Fig. 26.17.

In general, the mean execution time for all three topologies grows with v_0. This is
expected since conversion of each V particles adds a small contribution to the total
duration time. There are differences that are worth discussing. The mean execution
time increases with increasing v for small v_0. Very likely, for higher v there are more

Fig. 26.16 The mean time
for successful while-loop
runs for the feedback system
plotted against initial number
of V particles v_0. The
different *curves* represent
different values of v. Panel **a**
depicts results for $d = 1$ and
panel **b** depicts results for
$d = 2$

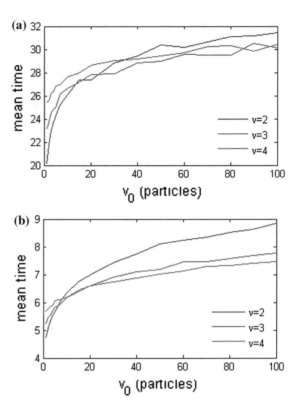

intermediate steps that take more time. Interestingly, for high v_0 the pattern seems to be reversed. There, larger v produce lower mean times. We believe that this somewhat counterintuitive anomaly has to do with the effects of fluctuation dominated kinetics.

Our hypothesis is that for low v_0 the the process that takes the longest time (the rate limiting step) is the formation of the final P_v complex. In the opposite situation when v_0 is high, the rate limiting step is conversion of all V particles. For larger v values there are more catalytic sites performing the conversion, which shortens the execution time.

In brief, the above analysis suggests that the mean duration time of a while loop implementation does depend on the details of the topology. The dimensionality of the system clearly also plays a role. In some cases are there much more differences between the data for different dimensions than for different topologies.

Fig. 26.17 The mean time for successful while-loop runs for the delayed system plotted against initial number of V particles v_0. The different *curves* represent different values of v as indicated in the legend. Panel **a** $d = 1$, panel **b** $d = 2$

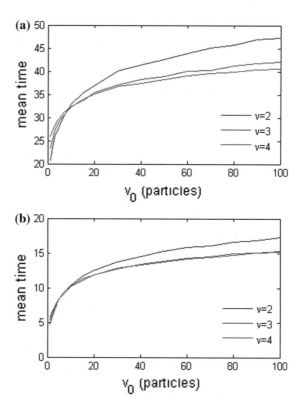

26.10 Conclusions

This chapter reviewed the Bare Bones programming language and a generic chemical implementation by Liekens et al. [31]. The Bare Bones programming language is Turing complete in principle: Any algorithm that can be represented as a Turing machine can be transformed into the corresponding Bare Bones program, which in turn can be implemented using a suitable chemical system by following the recipe suggested in [31].

The while loop construct is the most important part of the language. Any chemical implementation of the while loop is bound to be imperfect. Due to the inherent randomness of chemical reactions, there is always a chance that an error in the computation occurs.

The unavoidable occurrence of such errors is the reason that chemical computers are not Turing universal in the strict sense of the word. The Turing completeness of any chemical computer hinges on our practical ability to reduce the chance that such errors occur. For practical reasons it is extremely important to understand this error. If we could make a theoretical model that could help us understand how to control (minimize) the frequency of such errors it would be possible to build accurate chemical computers. This would enable a range of important technological applications targeting various bio-compatible information processing scenarios, e.g. in medicine or biotechnology.

As an illustration of which types of investigations are necessary, a few possible improvements of the minimal while loop implementation have been investigated in detail. The dynamics of several chemical implementations of the while loop has been simulated. The simulation data were used to analyze the fail rate of the loop and its typical duration time.

The goal of this investigation was to suggest a way of identifying suitable design strategies for building accurate chemical information processing devices. This was done in two stages. First, a thorough investigation has been performed regarding the influence of the spatial aspects of the dynamics on the fail rate. This was followed by an investigation of how the reaction topology influences the fail rate. The focus was not on understanding or optimizing the reaction rate parameters (which would have been a much easier task). Instead, a much more complicated issue has been addressed, regarding how the structure of the chemical reaction system topology affects the fail rate. In doing so, an attempt was made to ensure that a fair comparison was made between different while loop realizations. Thus simulation parameters were carefully chosen to ensure that all while-loop realizations would have a mean time equal to τ in zero dimensions.

The suggested idea to explicitly minimize the fail rate by varying specific design features of a chemical computer seems feasible in the sense that it is possible to draw generic conclusions regarding the optimal designs by studying a limited number of cases. As a proof of the concept, several topological features of the chemical reaction network were varied. It is possible to exploit various topologies to improve the performance in terms of lowering the fail rate. This makes the program run longer, possibly in a balanced way.

The list below contains the final suggestions for efficient design principles for while-loop implementations:

- An implementation that uses the delayed feedback topology seems to exhibit the lowest fail rate. The feedback topology is the second best. The sequential topology comes after that, followed by the minimal while loop implementation.
- The fail rate depends also on the range of v_0 considered. The dependence differs for different implementation topologies. This in turn implies that it could be possible to optimize the loop topology for various regions of v_0.
- Increase in the degree of cooperative binding, v, leads to lower fail rates. Thus implementation with higher degrees of cooperativity should in general have lower fail rates.
- The mean duration time of a while loop implementation does depend on the details of the topology. The observed increase in the program execution time is less than tenfold (no order of magnitude increase) for considered topologies. However, there are examples (not shown) where the increase can be much more dramatic (several orders of magnitude).

- The dimensionality of the system clearly plays a role. There are much more differences between the fail rate data for different dimensions than for different topologies in some cases. It seems that the fail rate and the execution time increase with lowering the dimensionality, but due to the convergence problems this is only stated as a hypothesis. Given the hypothesis holds, extremely low dimensions should be avoided.
- There are typical features that strongly relate to the emergence of fluctuation dominated kinetics which can be exploited to improve the execution runtime.

On the method side, the focus was on using simulation techniques. However, it is possible to perform similar investigation using analytic or semi-analytic techniques. These would very likely provide additional insights to the already established bulk of knowledge.

The reaction systems that were analysed were chosen based on the intuition. As it comes to exploring various possibilities to optimize the experimental design we barely scratched the surface: in principle, the data shows that everything matters (the order of the cooperativity, the effects of fluctuation dominated kinetics, the distinct copy number dependencies). From the practical point of view, it would be useful to make the suggested procedure more automatic. This could be done by implementing search algorithms in the space of reaction topologies, or other features such as the form of the reaction volume, that comply with the structure of the while loop semantics. While this could be hard to do for various reasons, e.g. there is a large number of auxiliary parameters that can be adjusted (e.g. reaction rates or diffusion constants), the potential benefits could be enormous.

Acknowledgments This work has been supported by Chalmers University of Technology. A part of the work has been done for meeting the master's degree requirements at Chalmers University of Technology.

Appendix: Convergence Tests

We studied how the number of lattice sites affects the simulation data. Extensive convergence tests were performed which consisted of analyzing how the graphs presented in this section change when the number of lattice sites is increased. A few examples are shown in Fig. 26.18.

The conclusions are as follows. For one ($d = 1$) the convergence has been achieved and these results are quantitative. The results for two ($d = 2$) dimensions are only qualitative. To obtain better results, the lattice size should be increased. This could not be done due to the usual hardware limitations. In these types of simulations the CPU speed is the most limiting factor.

Fig. 26.18 Convergence
tests of the fail rate versus
number of nodes. The
system is a minimal while
loop implementation with
$k_1/k_2 = 10^6$ and $v_0 = 1$. On
the x-axis is the length in
nodes of the domain. In
panel **a** is the convergence
test for one dimension, this
yields that the x-axis also
correspond to the total
number of nodes. Panel **b**
corresponds to two
dimensions with the total
number of nodes then being
the x-axis value squared.
Panel **c** consequently
corresponds to three
dimensions. Since the
domain used has been
$5 \times 5 \times 4$ for three
dimensions, the total number
of nodes is the x-axis value
cubed times a factor 0.8 due
to the non-symmetry

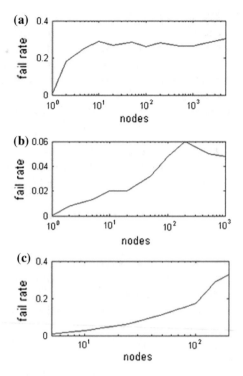

References

1. Conrad, M.: Evolutionary learning circuits. J. Theor. Biol. **46**, 167–188 (1974)
2. Rosen, R.: Pattern generation in networks. Prog. Theor. Biol. **6**, 161–209 (1981)
3. Kampfner, R.R., Conrad, M.: Sequential behavior and stability properties of enzymatic neuron networks. Bull. Math. Biol. **45**, 969–980 (1983)
4. Kirby, K.G., Conrad, M.: Intraneuronal dynamics as a substrate for evolutionary learning. Phys. D **22**, 205–215 (1986)
5. Conrad, M.: Rapprochement of artificial-intelligence and dynamics. Eur. J. Op. Res. **30**, 280–290 (1987)
6. Hjelmfelt, A., Weinberger, E.D., Ross, J.: Chemical implementation of neural networks and turing-machines. Proc. Natl. Acad. Sci. U. S. A. **88**, 10983–10987 (1991)
7. Hjelmfelt, A., Weinberger, E.D., Ross, J.: Chemical implementation of finite-state machines. Proc. Natl. Acad. Sci. U. S. A. **89**, 383–387 (1992)
8. Aoki, T., Kameyama, M., Higuchi, T.: Interconnection-free biomolecular computing. Computer **25**, 41–50 (1992)
9. Conrad, M.: Molecular computing - the lock-key paradigm. Computer **25**, 11–20 (1992)
10. Hjelmfelt, A., Ross, J.: Chemical implementation and thermodynamics of collective neural networks. Proc. Natl. Acad. Sci. U. S. A. **89**, 388–391 (1992)
11. Hjelmfelt, A., Schneider, F.W., Ross, J.: Pattern-recognition in coupled chemical kinetic systems. Science **260**, 335–337 (1993)
12. Hjelmfelt, A., Ross, J.: Mass-coupled chemical-systems with computational properties. J. Phys. Chem. **97**, 7988–7992 (1993)
13. Rambidi, N.G.: Biomolecular computer: roots and promises. Biosystems **44**, 1–15 (1997)

14. Rambidi, N.G., Maximychev, A.V.: Towards a biomolecular computer. information processing capabilities of biomolecular nonlinear dynamic media. Biosystems **41**, 195–211 (1997)
15. Conrad, M., Zauner, K.P.: Dna as a vehicle for the self-assembly model of computing. Biosystems **45**, 59–66 (1998)
16. Hiratsuka, M., Aoki, T., Higuchi, T.: Enzyme transistor circuits for reaction-diffusion computing. IEEE Trans. Circuits Syst. I-Fundam. Theory Appl. **46**, 294–303 (1999)
17. Stange, P., Zanette, D., Mikhailov, A., Hess, B.: Self-organizing molecular networks. Biophys. Chem. **79**, 233–247 (1999)
18. Simpson, M.L., Sayler, G.S., Fleming, J.T., Applegate, B.: Whole-cell biocomputing. Trends Biotechnol. **19**, 317–323 (2001)
19. Winfree, E.: Dna computing by self-assembly. Natl. Acad. Eng.: The Bridge **33**, 31–38 (2003)
20. Lizana, L., Konkoli, Z., Orwar, O.: Tunable filtering of chemical signals in a simple nanoscale reaction-diffusion network. J. Phys. Chem. B **111**, 6214–6219 (2007)
21. Soloveichik, D., Cook, M., Winfree, E., Bruck, J.: Computation with finite stochastic chemical reaction networks. Nat. Comput. **7**, 615–633 (2008)
22. Soloveichik, D., Seelig, G., Winfree, E.: Dna as a universal substrate for chemical kinetics. Proc. Natl. Acad. Sci. **107**, 5393–5398 (2010)
23. Jiang, H., Riedel, M.D., Parhi, K.K.: Digital logic with molecular reactions. 2013 IEEE/Acm International Conference on Computer-Aided Design. ICCAD-IEEE ACM International Conference on Computer-Aided Design, pp. 721–727. IEEE, New York (2013)
24. Cummings, R., Doty, D., Soloveichik D.: Probability 1 computation with chemical reaction networks. In: Murata, S., Kobayashi, S., (eds.) DNA Computing and Molecular Programming, Lecture Notes in Computer Science, vol. 8727, pp. 37–52. Springer International Publishing, Berlin (2014)
25. Sienko, T., Adamatzky, A., Rambidi, N.G., Conrad, M.: Molecular Computing. MIT Press, Cambridge (2005)
26. Cardelli, L.: On process rate semantics. Theor. Comput. Sci. **391**, 190–215 (2008)
27. Krivine, J., Danos, V., Benecke, A.: Modelling epigenetic information maintenance: A kappa tutorial. In: Bouajjani, A., Maler, O. (eds.) Computer Aided Verification. Proceedings, volume 5643 of Lecture Notes in Computer Science, pp. 17–32. Springer, Berlin (2009)
28. Degano, P., Bracciali, A.: Process calculi, systems biology and artificial chemistry. Handb. Nat. Comput. **3**, 1863–1896 (2012)
29. Liekens, A.M.L., Fernando, C.T.: Turing complete catalytic particle computers. In: Costa, F.A.E., Rocha, L.M., Costa, E., Harvey, I., Coutinho, A. (eds.) Advances in Artificial Life, Proceedings, volume 4648 of Lecture Notes in Artificial Intelligence, pp. 1202–1211 (2007)
30. Cardelli, L., Zavattaro, G.: On the computational power of biochemistry. In: Horimoto, K., Regensburger, G., Rosenkranz, M., Yoshida, H. (eds.) Algebraic Biology. Proceedings, volume 5147 of Lecture Notes in Computer Science, pp. 65–80. Springer, Berlin (2008)
31. Liekens, A., Fernando, T: Turing Complete Catalytic Computers. In: Advances in Artificial life, vol. 4648, pp. 1202–1211. Springer, Berlin (2007)
32. Konkoli, Z.: Diffusion controlled reactions, fluctuation dominated kinetics, and living cell biochemistry. Int. J. Softw. Inform. **7**, 675 (2013)
33. Konkoli, Z.: Modeling reaction noise with a desired accuracy by using the x level approach reaction noise estimator (xarnes) method. J. Theor. Biol. **305**, 1–14 (2012)
34. Singh, A., Hespanha, J.P.: A derivative matching approach to moment closure for the stochastic logistic model. Bull. Math. Biol. **69**, 1909–1925 (2007)
35. Privman, V.: Nonequilibrium Statistical Mechanics in One Dimension. Cambridge Univ. Press, Cambridge (1997)
36. Brookshear, G., Brylow, D.: Computer Science: An Overwiew, 11th edn. Addison-Wesley, Boston (2003)
37. Van kampen, N.G.: Stochastic Processes in Physics and Chemistry, 1st edn. Elsevier Science, Amsterdam (2003)
38. Elf, J., Donĉić, A., Ehrenberg, M.: Mesoscopic reaction-diffusion in intracellular signaling. Proc. SPIE **5110**, 114–124 (2003)

Chapter 27
Conventional and Unconventional Approaches to Swarm Logic

Andrew Schumann

Abstract We consider two possible ways of logical formalization of swarm behaviour: the conventional way by classical automata and the unconventional one by labelled transition systems coded by p-adic integers. Swarm intelligence is one of the directions in emergent computing. We show that the computational complexity of conventional way connected to implementations of Kolmogorov–Uspensky machines, Schönhage's storage modification machines, and random-access machines on swarms is very high. The point is that computable functions can be simulated by swarm behaviors with a low accuracy, because of the following two main features: (i) in swarms we observe the propagation in all possible directions; (ii) there are some emergent patterns. These features cannot be defined conventionally by inductive sets. However, we can consider swarms in the universe of streams which is permanently being expended and can be coded by p-adic numbers. In this universe we can define functions and relations for the algorithmization of swarm intelligence in an unconventional way.

27.1 Introduction

In bio-inspired computations there are many researches focused on swarm intelligence and different formalizations of swarm behaviours [6, 16, 38]. An intelligent swarm behaviour is demonstrated not only by animal groups, but also by groups of insects or even bacteria. For example, it was discovered experimentally that swarms of social insects [10] can solve complex computational problems in searching for food and in transporting sources and information due to massive-parallel behaviour with labour divisions.

The majority of researches in swarm intelligence are concentrated on algorithms in simulating and modelling swarms: the Particle Swarm Optimization (PSO) [18], the Bacterial Foraging Optimization Algorithm (BFOA) [21], the Artificial Bee Colony

A. Schumann (✉)
University of Information Technology and Management in Rzeszow, Rzeszow, Poland
e-mail: andrew.schumann@gmail.com

© Springer International Publishing Switzerland 2017
A. Adamatzky (ed.), *Advances in Unconventional Computing*,
Emergence, Complexity and Computation 22,
DOI 10.1007/978-3-319-33924-5_27

711

(ABC) [14], the Cuckoo Optimization Algorithm (COA) [22], the Social Spider Optimization (SSO) [7], the Ant Colony Optimization (ACO) [9], etc. These models combine different mathematical tools including probability theory to propose artificial swarms.

In this paper we try to formulate swarms as labelled transition systems with the same set of labels (events, or actions): direction, splitting, fusion, repelling (Sect. 27.2). Then we show that these labelled transition systems can implement Kolmogorov–Uspensky machines, Schönhage's storage modification machines, and random-access machines, but with a low accuracy because of emergent patterns which occur if we have many states of appropriate transition system (Sect. 27.3). Then we propose p-adic arithmetic and logic to formalize emergent patterns of swarms (Sect. 27.4).

To sum up, we offer a general logical approach to swarm intelligence.

27.2   From Emergent Computing to Swarm Computing

In conventional logic circuits some electrical properties of transistors are used. In particular, the voltage is managed to be in only one of two states: high (if the voltage runs the range from 2.8 to 5.0V) or low (if the voltage is in the range from 0 to 0.8V). The high state of voltage means '1' or logical true. The low state means '0' or logical false. Then the power of the circuit, P, is expressed as follows:

$$P = P_{static} \cdot P_{dynamic},$$

where $P_{static} = V_{CC} \cdot I_{CC}$ and $P_{dynamic} = [(C_{pd} + C_L) \cdot V_{CC}^2 \cdot f] \cdot N_{SW}$, V_{CC} is a supply voltage (V), I_{CC} is a power supply current (A), C_{pd} is a power dissipation capacitance (F), C_L is an external load capacitance (F), f is an operating frequency (Hz), N_{SW} is a total number of outputs switching.

The main idea of electric devices based on electrical properties of transistors is that the Boolean logic can be implemented with a very high accuracy. In this logic any complex logic expression is considered a composition of logical atoms (i.e., of simple logical propositions). In other words, there are no emergent phenomena. Let us recall that the emergence is detected, when new patterns of a highly structured collective behavior appear and these patterns cannot be reduced to a linear composition of simple subsystems.

Notably, emergent phenomena are key phenomena in all self-organizing systems such as collective intelligent behaviors of animal groups: flocks of birds, colonies of ants, schools of fish, swarms of bees, etc. Emergence is observed in the economy as well: macroeconomic fluctuations, traffic jams, hierarchy of cities, motion picture industry and mass protest behavior [20]. There are attempts to formalize the notion of emergence by algorithmic complexity theory. However, the Kolmogorov complexity function is not computable. There is no way to define the emergence by minimum linear compositions. Wolfram proposed a more useful approach in a mathematical definition of emergency [39]. He showed that the behavior of one-dimensional cellular automata is divided into the following four cases:

- there are limit points of the system, i.e. we obtain a homogeneous state of the system;
- there are limit cycles of the system, i.e. we obtain separated periodic structures;
- there are chaotic attractors, i.e. we face chaotic patterns;
- there are complex localized structures.

In the last case we deal with an emergent phenomenon in the true sense. On the basis of the Wolfram's approach the so-called *emergent computing* has developed. In this kind of computing (i) the computation process is distributed over a set of parallel and autonomous processing units; (ii) each unit computes locally and can interact directly only with a small number of other units; (iii) there is a consistency among the processing units; (iv) there are more outputs, than inputs.

One of the most studied instance of emergent computing is represented by *swarm computing* [6, 18]. This computing is based on labor division in animal groups. Any swarm as a result of collective behavior, such as birds flocking or fish schooling, is a self-organizing system, where, on the one hand, each unit responds to local stimuli individually and, on the other hand, all together they accomplish a global task (transporting, eating, self-protecting, etc.).

There were proposed many swarm algorithms to simulate the behavior of insect or animal groups [7, 9, 14, 18, 21, 22]. In the PSO [17, 18] it is assumed that the particles (agents) know (i) their best position 'local best' (*lb*) and (ii) their neighborhood's best position 'global best' (*gb*). The next position is determined by velocity. Let $x_i(t)$ denote the position of particle i in the search space at time step t, where t is discrete. Then the position x_i is changed by adding a velocity to the current position:

$$x_i(t+1) = x_i(t) + v_i(t+1),$$

where $v_i(t+1) = v_i(t) + c_1 r_1 (lb(t) - x_i(t)) + c_2 r_2 (gb(t) - x_i(t))$ and i is the particle index, c_1, c_2 are acceleration coefficients, such that $0 \leq c_1, c_2 \leq 2$, r_1, r_2 are random values (such that $0 \leq r_1, r_2 \leq 1$) regenerated every velocity update.

One of the possible PSO algorithms can be exemplified by the bird flocking [23, 24]. In flocks 'local best' and 'global best' of birds are defined by the following three rules: (i) collision avoidance (birds fly away before they crash into one another); (ii) velocity matching (birds fly about the same speed as their neighbors in the flock); and (iii) flock centering (birds fly toward the center of the flock as they perceive it). So, the position of a bird i at time t is given by its placement x_i at time $t-1$ shifted by its current velocity v_i. This v_i is determined by the rules (i)–(iii).

All the algorithms PSO, BFOA, ABC, COA, SSO, ACO are used to simulate swarms of different insects or animals. Let us try to answer the question, whether swarms can be considered an unconventional computer, i.e. whether swarms can calculate.

Each swarm is a natural transition system

$$TS = (S, E, T, I),$$

where:

- S is the non-empty set of states;
- E is the set of events;
- $T \subseteq S \times E \times S$ is the transition relation;
- $I \subseteq S$ is the set of initial states.

Any transition system is a labeled graph with nodes corresponding to states from S, edges representing the transition relation T, and labels of edges corresponding to events from E.

Let us consider some examples.

27.2.1 Ant Colony Transitions

Any ant colony is a swarm of ants localized first at the nest. This nest can be regarded as an initial state of ant colony transitions. Then ants use a special mechanism called *stigmergy* to build up all transitions. Stigmergy (stigma + ergon) means 'stimulation by work'. This mechanism has the following steps [9]:

- At first ants are looking for food randomly, laying down pheromone trails.
- If ants find food, they return to the nest, leaving behind pheromone trails. So there is more pheromone on the shorter path than on the longer one.
- Ants prefer to go in the direction of the strongest pheromone smell. As a consequence, the concentration of pheromone is so strong on the shorter path, that all the ants prefer this path (it is experimentally proven in [9]).

Thus, stigmergy allows ants to transport food to their nest in a remarkably effective way. Food localizations are considered new states and ant roads to food places are regarded as transitions. Let $P_{ant} = \{n\}$ be an initial state of ant transitions (i.e. the nest), $A_{ant} = \{a_1, a_2, \ldots, a_j\}$ be a set of food pieces (attractants) localized at different places, $V_{ant} = \{r_1, r_2, \ldots, r_i\}$ be a set of ant roads. So the ant colony transition system, $TS_{ant} = (S_{ant}, E_{ant}, T_{ant}, I_{ant})$, can be defined as follows:

- $\sigma : P_{ant} \cup A_{ant} \rightarrow S_{ant}$ assigning a state to each original point of the ant colony as well as to each attractant;
- $\tau : V_{ant} \rightarrow T_{ant}$ assigning a transition to each ant road;
- $\iota : P_{ant} \rightarrow I_{ant}$ assigning an initial state to the nest.

Each event of the set of events E_{ant} is assigned to ant transitions in accordance with the following types of ant expansion:

- *direction* (the ants move from one state/attractant/initial point to another state/ attractant),

- *fusion* (the ants move from different states/attractants to the same one state/attractant),
- *splitting* (the ants move from one state/attractant/initial point to different states/attractants),
- *repelling* (the ants stop to move in one direction).

The system TS_{ant} can be used to solve the Travelling Salesman Problem [9] formulated as follows: given a list of cities and the distances between each pair of cities, we must to define the shortest possible route that visits each city exactly once and returns to the origin city. This problem is NP-hard. Let P_{ant} be considered the origin city, A_{ant} be a set of all other cities, and V_{ant} be a set of all connections between cities. The Travelling Salesman Problem can be solved, because the ants lay down pheromone trails faster on the shortest path so that the shortest path gets reinforced with more pheromone to attract more future ants. As a result, pheromone trails on the edges between cities depend on the distance: more shorter, more attracting. This allows ants to find shorter tours of cities.

27.2.2 Bee Colony Transitions

A bee colony is another example of swarm intelligence [15]. The bee nest is an initial state of bee colony transitions. Any bee colony exploits a mechanism called *waggle dance* to optimize the food transporting to the nest. This mechanism is as follows. In the nest there is an area for communication among bees. At this area the bees knowing, where the food source is precisely, exchange the information about the direction, distance, and amount of nectar on the related food source by a waggle dance. The direction of waggle dancing bees shows the direction of the food source in relation to the Sun, the intensity of the waggles is associated to the distance, and the duration of the dance shows the amount of nectar. Due to this form of communication the bee colony transitions are built up by the following steps [14]:

- There are two kinds of bees: employed and unemployed. Employed bees know exactly, where a particular food source (nectar) is, and visit just this source. Unemployed bees do not know and seek a food source. The unemployed bees are divided into the following two groups: scouts and onlookers. A scout bee carries out search for new food sources without any guidance. An onlooker bee follows the instruction of a waggle dancing bee and visits the food source for the first time. An employed bee visits this source many times. So the first step in constructing the bee transporting system is in sending scout bees.
- Then onlookers are sent.
- At the next step the food source is exploited by employed bees.
- An employed bee tests if the nectar amount of the new food source is higher than that of the previous one. If it is so, the bee memorizes the new place and forgets the old one. If the nectar amount decreased or exhausted and the employed bee dos not know a new place, this bee become an unemployed bee. So at this step

the employed bees exchange the nectar information of the food sources to change their decision.

Assume that $P_{bee} = \{n\}$ is an initial state of bee transitions (i.e. the bee nest), $A_{bee} = \{a_1, a_2, \ldots, a_j\}$ is a set of food sources (attractants) localized at different places, $V_{onlooker} = \{r_1, r_2, \ldots, r_k\}$ is a set of onlooker bee roads, and $V_{employed} = \{r_1, r_2, \ldots, r_l\}$ is a set of employed bee roads. Then the bee colony transition system, $TS_{bee} = (S_{bee}, E_{bee}, T_{bee}, I_{bee})$, can be defined thus:

- $\sigma : P_{bee} \cup A_{bee} \rightarrow S_{bee}$ assigning a state to each original point of the bee colony as well as to each attractant;
- $\tau : V_{onlooker} \cup V_{employed} \rightarrow T_{bee}$ assigning a transition to each bee road;
- $\iota : P_{bee} \rightarrow I_{bee}$ assigning an initial state to the bee nest.

In the set of events E_{bee} there are the following types of labels for transitions:

- *direction* (the onlooker or employed bees move from one state/attractant/initial point to another state/attractant),
- *fusion* (the onlooker or employed bees move from different states/attractants to the same one state/attractant),
- *splitting* (the onlooker or employed bees move from one state/attractant/initial point to different states/attractants),
- *repelling* (the onlooker or employed bees stop to move in one direction).

The system TS_{bee}, if we use only $V_{employed}$ as a set of roads, can solve the Travelling Salesman Problem, also. Thereby, P_{bee} is examined as the origin city, A_{bee} is a set of all other cities, and $V_{employed}$ is a set of all connections between cities. The point is that the greater the number of iterations in sending onlooker or employed bees, the higher the influence of the distance in attracting the bees to appropriate food sources. As a result, the shorter distance seems to be more attracting for employed bees.

There is another NP-hard problem that can be solved by TS_{bee}, the so-called Generalized Assignment Problem formulated as follows: there are a number of agents and a number of tasks, each agent has a budget and each task assumes some cost and profit; we must find an assignment in which all agents do not exceed their budget and total profit of the assignment is maximized. In the case of the bee colony, the bees are regarded as agents, the nectar sources as tasks, the amount of nectar as profit, and the distance as cost. Hence, in this interpretation the bee colony can solve the Generalized Assignment Problem.

27.2.3 Escherichia Coli *Transitions*

The complex group behavior can be observed in the bacterium life also. For instance, *Escherichia coli* bacteria form swarms in semisolid nutrient medium [8, 36]. In these swarms with high bacterial density, large-scale swirling and streaming motions of thousands or millions of cells are observed.

Let us suppose that $P_{one\ E.coli} = \{n_1, n_2, \ldots, n_m\}$ is a set of initial states of *Escherichia coli* transitions, $A_{one\ E.coli} = \{a_1, a_2, \ldots, a_j\}$ is a set of attractants (nutrient gradients), $V_{one\ E.coli} = \{r_1, r_2, \ldots, r_k\}$ is a set of paths for each *Escherichia coli* bacterium. Then the *Escherichia coli* transition system,

$$TS_{one\ E.coli} = (S_{one\ E.coli}, E_{one\ E.coli}, T_{one\ E.coli}, I_{one\ E.coli}),$$

can be defined in the following manner:

- $\sigma : P_{one\ E.coli} \cup A_{one\ E.coli} \to S_{one\ E.coli}$ assigning a state to each original point of the *Escherichia coli* population as well as to each attractant;
- $\tau : V_{one\ E.coli} \to T_{one\ E.coli}$ assigning a transition to each path of each *Escherichia coli* bacterium;
- $\iota : P_{one\ E.coli} \to I_{one\ E.coli}$ assigning initial states to initial positions of *Escherichia coli* bacteria.

There are the four types of labels for transitions in the set of events $E_{one\ E.coli}$:

- *Direction*: the *Escherichia coli* bacterium moves from one state/attractant/initial point to another state/attractant presented by a nutrient gradient. This locomotion is carried out by a set of tensile flagella which move in the counterclockwise direction helping the bacterium to swim very fast.
- *Tumbling*: the *Escherichia coli* bacterium can tumble to change its swim direction in the future. It is a Brownian motion of the bacterium. The tumbling is achieved by the flagella moving in the clockwise direction. In this direction each flagellum operates relatively independently of the others.
- *Repelling*: the *Escherichia coli* bacterium stops to move in one direction, because it avoids noxious environment.
- *Repropuction*: the health bacterium splits into two bacteria while the least healthy bacteria die.

In semisolid nutrient medium *Escherichia coli* bacteria make swarms: they become multinucleate and move across the surface in coordinated packs. However, swarms of *Escherichia coli* are not attracted by chemotaxis [8]. Their dynamics can be explained only mechanistically by collisions with neighbors. This means that *Escherichia coli* swarms can be simulated mathematically (by hydrodynamic interactions) on the basis of $TS_{one\ E.coli}$, but these swarms are not intelligent.

So we can add a set $E_{E.coli\ swarm}$ of swarming transitions defined mathematically to the transition system $TS_{one\ E.coli}$. There are the following four kinds of maneuvers during swarming [8, 36]:

- *Forward motion*: When the majority of *Escherichia coli* bacteria swim in the same way, their swarm moves forward.
- *Lateral motion*: Because of collisions with neighboring cells or motor reversals there is a lateral motion in a swarm with propulsion angles of $>35°$.
- *Reversals*: Reversals occur every 1.5 s and require about 0.1 s for completion, but they do not have a large impact on the average cell behavior.

- *Stalls*: When the bacteria pause, their flagella continue to spin and pump fluid over the agar in front of the swarm. As a result, stalls occur at the swarm edge.

Hence, *Escherichia coli* swarms do not calculate.

27.2.4 Paenibacillus Vortex *Transitions*

Notice that there are bacteria with intelligent and successful swarming strategies such as *Paenibacillus vortex* [5]. These strategies help *Paenibacillus vortex* bacteria to carry out a cooperative colonization of new territories. When *Paenibacillus vortex* is inoculated on hard agar surfaces with peptone, it develops complex colonies of vortices. When it is inoculated on soft agar surfaces, it organizes a special network of swarms with intricate internal traffic. In contrast to *Escherichia coli*, the *Paenibacillus vortex* swarms are sensitive to chemotaxis (attractants). As a consequence, the *Paenibacillus vortex* network is built up on the basis of interactions between swarms allowing them to transport nutrients, spores and other organisms [12, 34].

Let $P_{one\ P.vortex} = \{n_1, n_2, \ldots, n_m\}$ be a set of initial states of *Paenibacillus vortex* transitions, $A_{one\ P.vortex} = \{a_1, a_2, \ldots, a_j\}$ be a set of attractants (nutrient gradients), $V_{one\ P.vortex} = \{r_1, r_2, \ldots, r_k\}$ be a set of paths for each *Paenibacillus vortex* bacterium. Then the *Paenibacillus vortex* transition system, $TS_{one\ P.vortex} = (S_{one\ P.vortex}, E_{one\ P.vortex}, T_{one\ P.vortex}, I_{one\ P.vortex})$, is as follows:

- $\sigma : P_{one\ P.vortex} \cup A_{one\ P.vortex} \rightarrow S_{one\ P.vortex}$ assigning a state to each original point of the *Paenibacillus vortex* population as well as to each attractant;
- $\tau : V_{one\ P.vortex} \rightarrow T_{one\ P.vortex}$ assigning a transition to each path of each *Paenibacillus vortex* bacterium;
- $\iota : P_{one\ P.vortex} \rightarrow I_{one\ P.vortex}$ assigning initial states to initial positions of *Paenibacillus vortex* bacteria.

The four types of labels for transitions in the set of events $E_{one\ P.vortex}$:

- *direction*: the *Paenibacillus vortex* bacterium moves from one state/attractant/ initial point to another state/attractant presented by a nutrient gradient;
- *tumbling*: the *Paenibacillus vortex* bacterium can tumble for a while;
- *repelling*: the *Paenibacillus vortex* bacterium stops to move in one direction, as it avoids some chemical concentrations;
- *repropuction*: the health bacterium splits into two bacteria.

Paenibacillus vortex bacteria can be organized in swarms. The locomotion in a swarm can be explained hydrodynamically by collisions among the bacteria and by the boundary of the layer of lubricant collectively generated by them. However, interactions among swarms can be considered intelligent. Each swarm has a snake-like formation. It looks for food and can cross each other's trail. When food is detected, swarms change their direction. The *Paenibacillus vortex* swarms can split and fuse in accordance with topology of nutrients.

So, for these swarms we can propose another transition system

$$TS_{P.vortex\ swarm} = (S_{P.vortex\ swarm}, E_{P.vortex\ swarm}, T_{P.vortex\ swarm}, I_{P.vortex\ swarm}),$$

where

- $\sigma : P_{P.vortex\ swarm} \cup A_{P.vortex\ swarm} \rightarrow S_{P.vortex\ swarm}$, where $P_{P.vortex\ swarm} = \{n_1, n_2, ..., n_l\}$ is a set of initial states of *Paenibacillus vortex* swarm transitions and $A_{P.vortex\ swarm} = A_{one\ P.vortex}$ is a set of attractants (nutrient gradients). The function σ assigns a state to each original point of the *Paenibacillus vortex* swarms as well as to each attractant;
- $\tau : V_{P.vortex\ swarm} \rightarrow T_{P.vortex\ swarm}$, where $V_{P.vortex\ swarm} = \{r_1, r_2, ..., r_k\}$ is a set of paths for each *Paenibacillus vortex* swarm. The function τ is to assign a transition to each path of each *Paenibacillus vortex* swarm;
- $\iota : P_{P.vortex\ swarm} \rightarrow I_{P.vortex\ swarm}$ is to assign initial states to initial positions of *Paenibacillus vortex* swarms.

Each event of the set of events $E_{P.vortex\ swarm}$ is assigned to swarm motions according to the following types of maneuvers:

- *direction*: the *Paenibacillus vortex* swarm moves from one state/attractant/initial point to another state/attractant,
- *fusion*: the *Paenibacillus vortex* swarms move from different states/attractants/ initial points to the same one state/attractant,
- *splitting*: the *Paenibacillus vortex* swarm moves from one state/attractant/initial point to different states/attractants (for the experimental details see [11]),
- *repelling*: the *Paenibacillus vortex* swarm stops to move in one direction if it faces a repellent.

As we see, the *Paenibacillus vortex* transition system $TS_{P.vortex\ swarm}$ can solve the Travelling Salesman Problem, too.

27.2.5 Conclusion: Towards Physarum machines

In the project *Physarum Chip Project: Growing Computers From Slime Mould* [3, 4, 31] supported by FP7 we are going to design an unconventional computer on plasmodia of *Physarum polycephalum*. Notice that *Physarum polycephalum* is a one-cell organism whose plasmodia behave as a swarm organizing a network for transporting sources and information. In order to simulate *Physarum polycephalum* networks we have proposed *Physarumsoft* [33], a software tool for programming *Physarum* computing and simulating *Physarum* expansions.

Taking into account the fact that any plasmodium can be considered a typical intelligent swarm, we can use *Physarumsoft* for demonstrating computational powers of different swarms: ant colonies, bee colonies, and *Paenibacillus vortex* swarms. Indeed, let

$$T S_{P.polycephalum} = (S_{P.polycephalum}, E_{P.polycephalum}, T_{P.polycephalum}, I_{P.polycephalum})$$

be a transition system for plasmodia and this system is defined standardly. Let f be a mapping from $S_{P.polycephalum}$ to $S_\star$ and from $I_{P.polycephalum}$ to $I_\star$, where $\star \in \{ant, bee, P.vortex\ swarm\}$. Assume that all transitions denoted by $\longrightarrow$ are the same for all $T S_{P.polycephalum}$, $T S_{ant}$, $T S_{bee}$, $T S_{P.vortex\ swarm}$. For example, we have the same direction, fusion, splitting, and repelling. The function f is a homomorphism if and only if

- for all $s \in (S_{P.polycephalum} \cup I_{P.polycephalum})$, if $s \longrightarrow s'$, for some $s' \in S_{P.polycephalum}$, then $f(s) \longrightarrow f(s')$;
- for all $s \in S_{P.polycephalum}$, if $f(s) \longrightarrow t$, for some $t \in S_\star$, then there exists $s' \in S_{P.polycephalum}$ with $s \longrightarrow s'$ and $f(s') = t$.

If $f : S_{P.polycephalum} \cup I_{P.polycephalum} \to S_\star \cup I_\star$ is a homomorphism as well as $f^{-1} : S_\star \cup I_\star \to S_{P.polycephalum} \cup I_{P.polycephalum}$ is a homomorphism, then f is an isomorphism. So, for any $T S_\star$, where $\star \in \{ant, bee, P.vortex\ swarm\}$, there is an isomorphism from $S_{P.polycephalum}$ to $S_\star$ and from $I_{P.polycephalum}$ to $I_\star$. This means that it is enough to study the *Physarum* machine $T S_{P.polycephalum}$ to know computational properties of different swarms: ant colonies, bee colonies, *Paenibacillus vortex* swarms, etc.

According to our previous study of $T S_{P.polycephalum}$ we know that it is impossible to define *Physarum* transitions as atomic acts [30, 32]. For instance, under the same conditions, the plasmodium can follow splitting or direction, fusion or direction, etc. Nevertheless, with a low accuracy we can implement some conventional algorithms in $T S_{P.polycephalum}$.

Notice that $T S_{P.polycephalum}$ cannot be defined as an inductive set because of the absence of atomic acts. Let us define (i) the universe U of all transitions defined in $T S_{P.polycephalum}$ and (ii) functions, F, which take one or more elements from U as arguments and return an element of U. The universe is defined inductively (hence, it is an inductive set) if there is a base set $B \subseteq U$ such that unboundedly repeated compositions of F on B cover the whole U. So, we assume that B contains atomic acts (transitions). All functions $f \in F$ are defined *inductively* if (i) each f is defined on B; (ii) for each f we have $f(a_1, a_2, \ldots, a_n) = f(b_1, b_2, \ldots, b_n)$ for $a_1, a_2, \ldots, a_n \in U$ and $b_1, b_2, \ldots, b_n \in U$ iff $a_i = b_i$ for $i = 1, \ldots, n$; (iii) each f has a range which is disjoint from the ranges of all other functions in F and from B. A function h is called *recursive* if (i) $h(a)$ is defined for all $a \in B$; (ii) for each $f \in F$ the value of $h(f(a_1, a_2, \ldots, a_n))$ is defined in terms of $h(a_1), \ldots, h(a_n)$.

In the very next section we will show how we can consider fragments of U such that there are some recursive functions in them.

27.3 Conventional Logical Approach to *Physarum* Machines

It is known that, theoretically, Turing machines, Kolmogorov–Uspensky machines [19, 37], Schönhage's storage modification machines [26, 27], and random-access machines [35] have the same expressive power. In other words, the class of functions computable by these machines is the same. For the first time A. Adamatzky [2] experimentally showed that the *Physarum* machine $TS_{P.polycephalum}$ can be represented as a kind of Kolmogorov–Uspensky machines. Hence, we can implement conventional algorithms in $TS_{P.polycephalum}$.

27.3.1 Physarum *Kolmogorov–Uspensky Machines*

Let us show that $TS_{P.polycephalum}$ can be considered a Kolmogorov–Uspensky machine. Let $\Gamma = S_{P.polycephalum} \cup I_{P.polycephalum}$ be an alphabet, $k = |E_{P.polycephalum}|$ a natural number. We say that a tree is (Γ, k)-tree, if one of nodes is designated and it is called *root* and all edges are directed. Each node is labelled by one of signs of Γ and each edge from the same node is labelled by different numbers $\{1, \ldots, k\}$ (so, each node has not more than k edges). We see that by this definition of (Γ, k)-tree, the plasmodium grows from the one active zone (so, we simulate the expansion from the one nest of ants or bees, or from the one inoculation of *Paenibacillus vortex* swarms), where all attractants are labelled by signs of Γ and protoplasmic tubes (roads of ants or roads of bees) are labelled by numbers of $\{1, \ldots, k\}$. Thus, $TS_{P.polycephalum}$ (as well as TS_{ant}, TS_{bee}, or $TS_{P.vortex\ swarm}$) can be represented as a (Γ, k)-tree.

(Γ, k)-*Physarum complex* is any initial finite digraph which is connected (i.e. each vertex is accessible from the initial one by a directed path), each node is labelled by one of signs of Γ, and each edge from the same node is labelled by different numbers $\{1, \ldots, k\}$. The set of all vertices of (Γ, k)-*Physarum* complex U is denoted by $v(U)$.

The *r-neighborhood* of (Γ, k)-complex is represented by a (Γ, k)-complex which consists of edges and vertices of initial complex that are accessible from initial vertex by a directed path that is not longer than r. Notice that r can be arbitrary. Any property of (Γ, k)-complex which is dependent just of r-neighborhood is called *r-local property of (Γ, k)-complex*. Hence, we can ever project *Physarum* transitions (using attractants and repellents) for inducing different numbers r and appropriate local properties.

Definition 1 A program of *Physarum* Kolmogorov–Uspensky machine is any r-local action transforming some (Γ, k)-complexes of growing plasmodia into other (Γ, k)-complexes of growing plasmodia:

$$U \to \langle W, \gamma, \delta \rangle,$$

where U, W are (Γ, k)-*Physarum* complexes, γ is a mapping from $v(U)$ to $v(W)$, δ is an injection from $v(U)$ into $v(W)$. The algorithm of transformation complexes $S \to S^*$ is as follows [19, 37]:

- r-Neighborhood of complex S is the same as of U.
- $v(S') = v(S \backslash U) \cup v(W)$.
- If $b \in U$, $a \in S \backslash U$, there is $\langle a, b \rangle$ in S and $\gamma(b)$ is defined, then $\langle a, \gamma(b) \rangle$ is an edge in S' with the same number as $\langle a, b \rangle$.
- If $a \in U$, $b \in S \backslash U$, there is $\langle a, b \rangle$ in S and $\delta(a)$ is defined, then $\langle \delta(a), b \rangle$ is an edge in S' with the same number as $\langle a, b \rangle$ (due to injectivity of δ we have different numbers for different edges from the same vertex).
- The initial vertex of W is an initial vertex of S' and we delete in S' all vertices (with appropriate edges) which are not accessible from the initial one. In this way we obtain S^*.

The simpler version of Kolmogorov–Uspensky machines is represented by Schönhage's storage modification machines [26, 27].

27.3.2 Physarum *Schönhage's Storage Modification Machines*

These machines consist of a fixed alphabet of input symbols, $\Gamma = S_{P.polycephalum} \cup I_{P.polycephalum}$, and a mutable directed graph with its arrows labelled by $E_{P.polycephalum}$ denoted by X for short. The set of nodes Γ, identifying with attractants, is finite. One fixed node $a \in I_{P.polycephalum}$ is identified as a distinguished center node of the graph. It is the first active zone of growing plasmodium (also associated with the one nest of ants or bees, just one inoculum of *Paenibacillus vortex* swarms). The distinguished node a has an edge x such that $x_\gamma(a) = a$ for all $\gamma \in X$. That is, all pointers from the distinguished center node point back to the center node. Each $\gamma \in X$ defines a mapping x_γ from Γ to Γ in accordance with directions of growing plasmodium; $x_\gamma(b)$ is the node found at the end of the edge starting at b labelled by γ. Each word of symbols in the alphabet Γ is a pathway through the machine from the distinguished center node. For example $ABBC$ would translate to taking path A from the start node, then path B from the resulting node, then path B, then path C. With respect to the word $ABBC$, the plasmodium moves.

Schönhage's machine modifies storage by adding new elements and redirecting edges. Its basic instructions are as follows:

- Creating a new node: **new** W. The machine reads the word W, following the path represented by the symbols of W until the machine comes to the last symbol in the word. It causes a new node y associated with the last symbol of W to be created and added to X; its location in relation to the other nodes and pointers is determined by W. If W is the empty string, this has the effect of creating a new center node a, linked to the old a. For example, **new** AB creates a new node that is reached by following the B pointer from the node designated by A. The growing plasmodium from active zone A to active zone B corresponds to this word AB. Adding a new node B means adding a new attractant denoted by B.

- A pointer redirection: **set** W **to** V. This instruction redirects an edge from the path represented by word W to a former node that represents word V. If W is the empty string, then this has the effect of renaming the center node a to be the node indicated by V. Notice that **set** W **to** V means removing nodes and the edges incident to $W \backslash V$. So, we can remove some attractants denoted by $W \backslash V$.
- A conditional instruction: **if** $V = W$ **then instruction** Z. It compares two paths represented by words W and V and if they end at the same node, then we jump to instruction Z else continue. This instruction serves to add edges between existing nodes. It corresponds to the splitting or fusion of *Physarum polycephalum*.

Definition 2 A program of *Physarum* Schönhage's storage modification machine is any action transforming sets X of nodes for growing plasmodia with the alphabet Γ into other sets X' of nodes for growing plasmodia with the same alphabet Γ which carries out by instructions **new** W; **set** W **to** V; **if** $V = W$ **then instruction** Z.

27.3.3 Physarum *Random-Access Machines*

In random-access machines there are registers (defined as simple locations with contents (single natural numbers) labelled by signs of $X = \{1, 2, \ldots, k\}$, where $k = |E_{P.polycephalum}|$. In case of *Physarum polycephalum*, the alphabet

$$\Gamma = S_{P.polycephalum} \cup I_{P.polycephalum}$$

consisting of nodes (attractants) for growing plasmodia can be represented as set of registers. Their contents is defined by the number of protoplasmic tubes located at the place of appropriate register (i.e. by the number of protoplasmic tubes linked to this node). For example, $[\gamma]$ means 'the contents of register with location $\gamma \in \Gamma$'. So, $[\gamma]$ may be equal to 3. Then the node γ has three edges.

Instructions of *Physarum* random-access machines copy the contents of one registers and deposit them into other without destructing or changing registers. To do so we need repellents which stimulate plasmodium active zones to leave attractants.

Definition 3 A program of *Physarum* random-access machine is any action transforming contents of registers from X for the growing plasmodium with the alphabet Γ into other contents of the same registers in Γ for the growing plasmodium which carries out by the following instructions:

- Clear the content of register γ (set it to zero): $CLR([\gamma])$. All active zones are leaving γ due to repellents.
- Increment the contents of register γ: $INC([\gamma])$. The intensity of repellents at γ decreases.

- Decrement the contents of register γ: $DEC([\gamma])$. The intensity of repellents at γ increases.
- Copy the contents of register γ_j to register γ_k leaving the contents of γ_j intact: $CPY([\gamma_j], [\gamma_k])$. Using the *Physarum* transition, called direction, we can transmit active zones located at γ_j to add them to active zones located at γ_k.
- If register γ contains zero then jump to instruction Z else continue in sequence: $JZ([\gamma], Z)$.
- If the contents of register γ_j is equal to the contents of register γ_k then jump to instruction Z else continue in sequence: $JE([\gamma_j], [\gamma_k], Z)$.

Notice that in *Physarum* Kolmogorov–Uspensky machines and Schönhage's storage modification machines, the key role in computation models belongs to attractants, but in *Physarum* random-access machines, this role belongs to repellents.

27.3.4 Conclusion: Problems of Conventional Approach

Unfortunately, the computational complexity of implementations Kolmogorov–Uspensky machines, Schönhage's storage modification machines, and random-access machines on the *Physarum polycephalum* medium is so high. The point is that not every computable functions can be simulated by plasmodium behaviors properly (the more bit function the higher complexity in its computation):

- first, the motion of plasmodia is too slow (several days are needed to compute simple functions such as 5-bit conjunction, 3-bit adder, etc., but the plasmodium stage of *Physarum polycephalum* is time-limited, therefore there is not enough time for realizations, e.g., of thousands-bit functions);
- second, the more attractants or repellents are involved in designing computable functions, the less accuracy of their implementation is, because the plasmodium tries to be propagated in all possible directions and we will deal with indirected graphs, cycles, and other problems;
- third, the plasmodium is an adaptive organism that is very sensitive to environments, therefore it is very difficult to organize the same laboratory conditions for calculating the same k-bit functions, where k is large;
- fourth, the plasmodium has a free will and can make different decisions under the same conditions;
- fifth, the plasmodium follows emergent patterns which are fully eliminated in conventional automata such as Kolmogorov–Uspensky machines, although these patterns are natural for occupying many attractants.

Thus, swarm intelligence can be reduced to conventional automata, but with very low accuracy.

27.4 Unconventional Logical Approach to *Physarum* Machines

In swarm behaviours we can observe the two main features: (i) the propagation in all possible directions; (ii) some emergent patterns. These features show that the swarm dynamics cannot be represented as inductive sets. Hence, the dynamics of swarms is out of conventional tools of computing if we take into account the features mentioned above. We face there streams which refer to mathematical objects which can be generated just as non-well-founded sets [1]. There are some theories of these objects such as coalgebra [13] and coinductive calculus of streams [25]. The universe of these objects is permanently being expended and can be coded by p-adic numbers [28].

27.4.1 p-Adic Universe for **Physarum** *Machines*

Let us take a set of edges, $X = \{0, 1, \ldots, p - 1\}$, where $p - 1 = |E_{P.polycephalum}|$, from each node. The set of alphabet $\Gamma = S_{P.polycephalum} \cup I_{P.polycephalum}$ is identified with attractants. Hence, we have assumed that at each step of plasmodium propagation there are not more than $p - 1$ neighboring attractants which can be directly occupied. This means that our universe is p-adic and the plasmodium transition system can be coded by p-adic integer. Let us remember that the set of p-adic integers is denoted by $\mathbf{Z}_p$ and each p-adic integer $n \in \mathbf{Z}_p$ has the following meaning:

$$n = \sum_{i=0}^{\infty} a_i \cdot p^i,$$

where $a_i \in \{0, 1, \ldots, p - 1\}$, and the following notation:

$$n = \ldots a_i a_{i-1} \ldots a_1 a_0.$$

For each transition $s \longrightarrow s'$, the state $s' \in \Gamma$ is called the child of $s \in \Gamma$. For each two transitions $s \longrightarrow s'$ and $s' \longrightarrow s''$, the state $s'' \in \Gamma$ is called the grandchild of $s \in \Gamma$. Let us consider just strings $\gamma_0 \gamma_1 \ldots \gamma_k$, where γ_1 is a grandchild for γ_0, γ_2 is a grandchild for γ_1, ..., γ_k is a grandchild for γ_{k-1}. Let each string $\gamma_0 \gamma_1 \ldots \gamma_k$ have a numeric value $[\gamma_0 \gamma_1 \ldots \gamma_k] \in \mathbf{Z}_p$ which is defined as follows:

- Let $[\gamma_0]$ denote an integer $\leq p - 1$ for the node $\gamma_0 \in X$. This integer is equal to the number of children for γ_0. If γ_0 has no grandchildren, then its value is coded by a p-adic integer $\ldots 0000[\gamma_0]$, see Figs. 27.1, 27.2 and 27.3.
- Let $[\gamma_0]$ and $[\gamma_1]$ denote some integers $\leq p - 1$ for the nodes $\gamma_0, \gamma_1 \in X$, where γ_1 is a grandchild of γ_0. The integer $[\gamma_0]$ is equal to the number of children for γ_0 and the integer $[\gamma_1]$ is equal to the number of neighbours for γ_1 occupied by

Fig. 27.1 At each step of plasmodium propagation, there are only two attractants which can be occupied. At the first time step the plasmodium expansion is coded by the 3-adic integer …000001

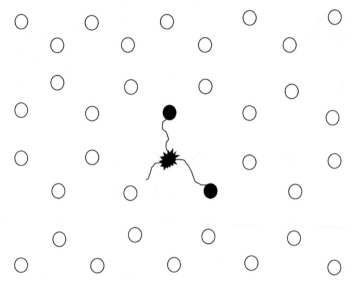

Fig. 27.2 At each step of plasmodium propagation, there are only three attractants which can be occupied. At the first time step the plasmodium expansion is coded by the 4-adic integer …000002

the plasmodium. If γ_1 has no grandchildren, then the value of $\gamma_0 \gamma_1$ is coded by a p-adic integer $\dots 0000[\gamma_1][\gamma_0]$.

- …

- Let $[\gamma_0], [\gamma_1], \dots, [\gamma_k]$ denote some integers $\leq p-1$ for the nodes $\gamma_0, \gamma_1, \dots, \gamma_k \in \Gamma$, respectively, where γ_1 is a grandchild of γ_0, γ_2 is a grandchild of γ_1, …, γ_k is a grandchild of γ_{k-1}. The integer $[\gamma_0]$ is equal to the number of children for γ_0, the integer $[\gamma_1]$ is equal to the number of neighbours for γ_1 occupied by the plasmodium, …, the integer $[\gamma_k]$ is equal to the number of neighbours for γ_k occupied by the plasmodium. If γ_k has no grandchildren, then the value of $\gamma_0 \gamma_1 \dots \gamma_k$ is coded by a p-adic integer $\dots 0000[\gamma_k] \dots [\gamma_1][\gamma_0]$. See Fig. 27.4.

Evidently that according to this definition if in the string $\gamma_0 \gamma_1 \dots \gamma_k$ each γ_i $(0 \leq i \leq k)$ has no neighboring attractants occupied by the plasmodium, then $[\gamma_0 \gamma_1 \dots \gamma_k] = 0 \in \mathbf{Z}_p$ and if in the string $\gamma_0 \gamma_1 \dots \gamma_k$ each γ_i $(0 \leq i \leq k)$ has all neighboring attractants occupied by the plasmodium, then $[\gamma_0 \gamma_1 \dots \gamma_k] = \sum_{i=0}^{k} (p-1) \cdot p^i \in \mathbf{Z}_p$. The strings $[\gamma_0 \gamma_1 \dots \gamma_k] \neq 0$ are called non-empty.

Let us analyze the case when we have two strings $\gamma_0 \gamma_1 \dots \gamma_k$ and $\gamma_0 \gamma_1' \dots \gamma_m'$ started from the same state γ_0. Suppose that γ_{i_k} $(0 \leq i_k \leq k)$ and γ_{i_m}' $(0 \leq i_m \leq m)$

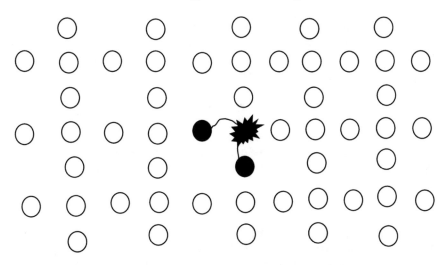

Fig. 27.3 At each step of plasmodium propagation, there are not more than four attractants which can be occupied. At the first time step the plasmodium expansion is coded by the 5-adic integer ...000002

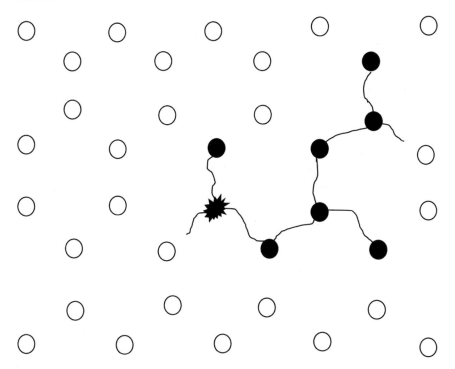

Fig. 27.4 At each step of plasmodium propagation, there are not more than four attractants which can be occupied. At the time step $t > 0$ the plasmodium expansion is coded by the 4-adic integer ...00000232

have some neighboring attractants occupied by the plasmodium. This means that we face a splitting of the plasmodium at the node γ_0. Assume that there is not more splitting for nodes from $\gamma_0 \gamma_1 \ldots \gamma_k$ and $\gamma_0 \gamma'_1 \ldots \gamma'_m$ and only $\gamma_0 \gamma_1 \ldots \gamma_k$ and $\gamma_0 \gamma'_1 \ldots \gamma'_m$ are non-empty. Then the transition system is coded by the set

$$\{[\gamma_0\gamma_1 \ldots \gamma_k], [\gamma_0\gamma'_1 \ldots \gamma'_m]\}.$$

Another similar situation is observed when we have two strings $\gamma_0 \gamma_1 \gamma_2 \ldots \gamma_k$ and $\gamma_0 \gamma_1 \gamma'_2 \ldots \gamma'_m$ started from the same state γ_0 and with the splitting at the node γ_1. If only $\gamma_0 \gamma_1 \gamma_2 \ldots \gamma_k$ and $\gamma_0 \gamma_1 \gamma'_2 \ldots \gamma'_m$ are non-empty, then the transition system is coded by the set

$$[\gamma_0\gamma_1\gamma_2 \ldots \gamma_k], [\gamma_0\gamma_1\gamma'_2 \ldots \gamma'_m].$$

Let A_{γ_0} be a set of all strings started from γ_0 and this set be coded by $[A_{\gamma_0}]$, a set of p-adic integers obtained for each string from A_{γ_0}. Now we can define compositions of two sets A_{γ_0} and $A_{\gamma'_0}$, where $\gamma_0 \neq \gamma'_0$:

- If no strings from A_{γ_0} have joint nodes with some strings from $A_{\gamma'_0}$, then we have $A_{\gamma_0,\gamma'_0} = A_{\gamma_0} \cup A_{\gamma'_0}$ and this system is coded by $[A_{\gamma_0}] \cup [A_{\gamma'_0}]$.
- If some strings from A_{γ_0} have joint nodes with some strings from $A_{\gamma'_0}$, then we have $A_{\gamma_0,\gamma'_0} = A_{\gamma_0} + A_{\gamma'_0}$ and this set contains all strings started from γ_0 and started from γ'_0. The system A_{γ_0,γ'_0} is coded by $[A_{\gamma_0,\gamma'_0}]$.

By induction, we can define sets $A_{\gamma_0,\gamma'_0,\ldots,\gamma''_0}$.

Notably, the plasmodium expansion is time dependent. So we can consider sets $A^t_{\gamma_0,\gamma'_0,\ldots,\gamma''_0}$ at $t = 0, 1, \ldots$ Let us define logical operations over the same sets $A^t_{\gamma_0,\gamma'_0,\ldots,\gamma''_0}$ with different t:

conjunction $A^{t=k}_{\gamma_0,\gamma'_0,\ldots,\gamma''_0} \wedge A^{t=l}_{\gamma_0,\gamma'_0,\ldots,\gamma''_0}$: Notice that strings from $A^{t=k}_{\gamma_0,\gamma'_0,\ldots,\gamma''_0}$ and $A^{t=l}_{\gamma_0,\gamma'_0,\ldots,\gamma''_0}$, where $k \neq l$, are the same, but they can be coded by different p-adic integers at $t = k$ and $t = l$. Let us consider each string $\gamma_0 \gamma_1 \gamma_2 \ldots \gamma_m$. Let $[\gamma_0 \gamma_1 \gamma_2 \ldots \gamma_m]_k$ be a p-adic numerical value of $\gamma_0 \gamma_1 \gamma_2 \ldots \gamma_m$ at $t = k$ and $[\gamma_0 \gamma_1 \gamma_2 \ldots \gamma_m]_l$ be a p-adic numerical value of $\gamma_0 \gamma_1 \gamma_2 \ldots \gamma_m$ at $t = l$. Then we define $\min([\gamma_0 \gamma_1 \gamma_2 \ldots \gamma_m]_k, [\gamma_0 \gamma_1 \gamma_2 \ldots \gamma_m]_l)$ digit by digit:

$$\min(\ldots 000\gamma_{m,k} \ldots \gamma_{2,k}\gamma_{1,k}\gamma_{0,k}; \ldots 000\gamma_{m,l} \ldots \gamma_{2,l}\gamma_{1,l}\gamma_{0,l}) =$$

$$\ldots 000 \min(\gamma_{m,k}, \gamma_{m,l}) \ldots \min(\gamma_{2,k}, \gamma_{2,l}) \min(\gamma_{1,k}, \gamma_{1,l}) \min(\gamma_{0,k}, \gamma_{0,l}).$$

The set $A^{t=k}_{\gamma_0,\gamma'_0,\ldots,\gamma''_0} \wedge A^{t=l}_{\gamma_0,\gamma'_0,\ldots,\gamma''_0}$ contains such minimum for each string.

disjunction $A^{t=k}_{\gamma_0,\gamma'_0,\ldots,\gamma''_0} \vee A^{t=l}_{\gamma_0,\gamma'_0,\ldots,\gamma''_0}$: Let us consider each string $\gamma_0 \gamma_1 \gamma_2 \ldots \gamma_m$. Let $[\gamma_0 \gamma_1 \gamma_2 \ldots \gamma_m]_k$ be a p-adic numerical value of $\gamma_0 \gamma_1 \gamma_2 \ldots \gamma_m$ at $t = k$ and $[\gamma_0 \gamma_1 \gamma_2 \ldots \gamma_m]_l$ be a p-adic numerical value of $\gamma_0 \gamma_1 \gamma_2 \ldots \gamma_m$ at $t = l$. Then we define $\max([\gamma_0 \gamma_1 \gamma_2 \ldots \gamma_m]_k, [\gamma_0 \gamma_1 \gamma_2 \ldots \gamma_m]_l)$ digit by digit:

$$\max(\ldots 000\gamma_{m,k} \ldots \gamma_{2,k}\gamma_{1,k}\gamma_{0,k}; \ldots 000\gamma_{m,l} \ldots \gamma_{2,l}\gamma_{1,l}\gamma_{0,l}) =$$

$$\ldots 000 \max(\gamma_{m,k}, \gamma_{m,l}) \ldots \max(\gamma_{2,k}, \gamma_{2,l}) \max(\gamma_{1,k}, \gamma_{1,l}) \max(\gamma_{0,k}, \gamma_{0,l}).$$

The set $A^{t=k}_{\gamma_0,\gamma_0',\ldots,\gamma_0''} \vee A^{t=l}_{\gamma_0,\gamma_0',\ldots,\gamma_0''}$ contains such maximum for each string.

negation $\neg A^{t=k}_{\gamma_0,\gamma_0',\ldots,\gamma_0''}$: Let us define the universe $\Omega_{\gamma_0,\gamma_0',\ldots,\gamma_0''}$ as a set of all possible strings started from the nodes $\gamma_0, \gamma_0', \ldots, \gamma_0''$. A numerical value of each string $\gamma_0 \gamma_1 \ldots \gamma_m$ from $\Omega_{\gamma_0,\gamma_0',\ldots,\gamma_0''}$ is maximal: $[\gamma_0 \gamma_1 \ldots \gamma_m] = \sum_{i=0}^{m}(p-1)\cdot p^i$. Then $\neg A^{t=k}_{\gamma_0,\gamma_0',\ldots,\gamma_0''} = \Omega_{\gamma_0,\gamma_0',\ldots,\gamma_0''} \cap A^{t=k}_{\gamma_0,\gamma_0',\ldots,\gamma_0''}$ and it is coded by a set of p-adic integers $\sum_{i=0}^{m}(p-1)\cdot p^i - [\gamma_0 \gamma_1 \gamma_2 \ldots \gamma_m]_k$ for each string $\gamma_0 \gamma_1 \gamma_2 \ldots \gamma_m \in \neg A^{t=k}_{\gamma_0,\gamma_0',\ldots,\gamma_0''}$.

By using this logic we can define an expansion strategy of the plasmodium on $\gamma_0 \gamma_1 \gamma_2 \ldots \gamma_m$ in the universe $A_{\gamma_0,\gamma_0',\ldots,\gamma_0''}$:

$$\mathcal{S}(\gamma_0\gamma_1\gamma_2 \ldots \gamma_m \in A_{\gamma_0,\gamma_0',\ldots,\gamma_0''}) = \bigwedge_{k=0}^{k=m} \sum_{t=0}^{\infty}(max([\gamma_k]_t, [\gamma_k]_{t+1})) \cdot p^i.$$

$$\mathcal{S}(A_{\gamma_0,\gamma_0',\ldots,\gamma_0''}) = \{\mathcal{S}(\gamma_0\gamma_1\gamma_2 \ldots \gamma_m): \gamma_0\gamma_1\gamma_2 \ldots \gamma_m \in A_{\gamma_0,\gamma_0',\ldots,\gamma_0''}\}.$$

27.4.2 p-Adic Valued Probability Measures

We know that $A_{\gamma_0,\gamma_0',\ldots,\gamma_0''} \subseteq \Omega_{\gamma_0,\gamma_0',\ldots,\gamma_0''}$. Assume, $A_{\gamma_0,\gamma_0',\ldots,\gamma_0''}$ is finite and the cardinality $|A_{\gamma_0,\gamma_0',\ldots,\gamma_0''}|$ is equal n. Let n be written in a p-adic form. Now, let us define the so-called p-adic cardinality, $\lceil A_{\gamma_0,\gamma_0',\ldots,\gamma_0''}\rceil$, as follows:

$$\lceil A_{\gamma_0,\gamma_0',\ldots,\gamma_0''}\rceil = \sum_{\gamma_0 \gamma_1 \gamma_2 \ldots \gamma_m \in A_{\gamma_0,\gamma_0',\ldots,\gamma_0''}} \frac{\mathcal{S}(\gamma_0 \gamma_1 \gamma_2 \ldots \gamma_m)}{n}$$

Evidently, $\lceil A_{\gamma_0,\gamma_0',\ldots,\gamma_0''}\rceil \leq \lceil \Omega_{\gamma_0,\gamma_0',\ldots,\gamma_0''}\rceil$ even if $A_{\gamma_0,\gamma_0',\ldots,\gamma_0''} = \Omega_{\gamma_0,\gamma_0',\ldots,\gamma_0''}$, as for each $\gamma_0 \gamma_1 \gamma_2 \ldots \gamma_m \in \Omega_{\gamma_0,\gamma_0',\ldots,\gamma_0''}$, $\mathcal{S}(\gamma_0 \gamma_1 \gamma_2 \ldots \gamma_m) = \bigwedge_{k=0}^{k=m} \sum_{t=0}^{\infty}(max([\gamma_k]_t, [\gamma_k]_{t+1})) \cdot p^i = \sum_{t=0}^{\infty}(p-1) \cdot p^i = -1$. Thus,

$$\lceil \Omega_{\gamma_0,\gamma_0',\ldots,\gamma_0''}\rceil = -1$$

Suppose that $A_{\gamma_0,\gamma_0',\ldots,\gamma_0'',\ldots}$ is infinite. We can define the p-adic cardinality, $\lceil A_{\gamma_0,\gamma_0',\ldots,\gamma_0'',\ldots}\rceil$, in the following manner:

$$\lceil A_{\gamma_0,\gamma_0',\ldots,\gamma_0'',\ldots}\rceil = \max_{\gamma_0 \gamma_1 \gamma_2 \ldots \gamma_m \in A_{\gamma_0,\gamma_0',\ldots,\gamma_0'',\ldots}} \mathcal{S}(\gamma_0\gamma_1\gamma_2 \ldots \gamma_m).$$

Let Ω be a set of all strings including infinite ones. Then

$$\mathcal{S}(\Omega) = \{\mathcal{S}(\gamma_0\gamma_1\gamma_2\ldots\gamma_m\ldots) \in \mathbf{Z}_p : \gamma_0\gamma_1\gamma_2\ldots\gamma_m\cdots \in \Omega\}$$

and

$$\lceil\Omega\rceil = \max(\mathcal{S}(\Omega)) = -1.$$

Definition 4 For any $A_{\gamma_0,\gamma_0',\ldots,\gamma_0''} \subseteq \Omega_{\gamma_0,\gamma_0',\ldots,\gamma_0''}$ and $A_{\gamma_0,\gamma_0',\ldots,\gamma_0''\ldots} \subseteq \Omega$, a p-adic valued probability measure, $P(A_{\gamma_0,\gamma_0',\ldots,\gamma_0''})$ and $P(A_{\gamma_0,\gamma_0',\ldots,\gamma_0''\ldots})$, is defined as follows:

- $P(A_{\gamma_0,\gamma_0',\ldots,\gamma_0''}) = \frac{\lceil A_{\gamma_0,\gamma_0',\ldots,\gamma_0''}\cap\Omega_{\gamma_0,\gamma_0',\ldots,\gamma_0''}\rceil}{\lceil\Omega_{\gamma_0,\gamma_0',\ldots,\gamma_0''}\rceil}$ and $P(A_{\gamma_0,\gamma_0',\ldots,\gamma_0''\ldots}) = \frac{\lceil A_{\gamma_0,\gamma_0',\ldots,\gamma_0''\ldots}\cap\Omega\rceil}{\lceil\Omega\rceil}$.
- $P(\Omega_{\gamma_0,\gamma_0',\ldots,\gamma_0''}) = 1$, $P(\Omega) = 1$, and $P(\emptyset) = 0$.
- if $A_{\gamma_0,\gamma_0',\ldots,\gamma_0''} \subseteq \Omega_{\gamma_0,\gamma_0',\ldots,\gamma_0''}$ and $B_{\gamma_0,\gamma_0',\ldots,\gamma_0''} \subseteq \Omega_{\gamma_0,\gamma_0',\ldots,\gamma_0''}$ are disjoint, i.e.

$$\inf(P(A_{\gamma_0,\gamma_0',\ldots,\gamma_0''}), P(B_{\gamma_0,\gamma_0',\ldots,\gamma_0''})) = 0,$$

then $P(A_{\gamma_0,\gamma_0',\ldots,\gamma_0''} \cup B_{\gamma_0,\gamma_0',\ldots,\gamma_0''}) = P(A_{\gamma_0,\gamma_0',\ldots,\gamma_0''}) + P(B_{\gamma_0,\gamma_0',\ldots,\gamma_0''})$. Otherwise,

$$P(A_{\gamma_0,\gamma_0',\ldots,\gamma_0''} \cup B_{\gamma_0,\gamma_0',\ldots,\gamma_0''}) = P(A_{\gamma_0,\gamma_0',\ldots,\gamma_0''}) + P(B_{\gamma_0,\gamma_0',\ldots,\gamma_0''}) -$$

$$\inf(P(A_{\gamma_0,\gamma_0',\ldots,\gamma_0''}), P(B_{\gamma_0,\gamma_0',\ldots,\gamma_0''})) = \sup(P(A_{\gamma_0,\gamma_0',\ldots,\gamma_0''}), P(B_{\gamma_0,\gamma_0',\ldots,\gamma_0''})).$$

All these equalities hold for infinite sets $A_{\gamma_0,\gamma_0',\ldots,\gamma_0''\ldots}$ and $B_{\gamma_0,\gamma_0',\ldots,\gamma_0''\ldots}$, also.
- $P(\neg A_{\gamma_0,\gamma_0',\ldots,\gamma_0''}) = 1 - P(A_{\gamma_0,\gamma_0',\ldots,\gamma_0''})$ for all finite $A_{\gamma_0,\gamma_0',\ldots,\gamma_0''} \subseteq \Omega_{\gamma_0,\gamma_0',\ldots,\gamma_0''}$, where $\neg A_{\gamma_0,\gamma_0',\ldots,\gamma_0''} = \Omega_{\gamma_0,\gamma_0',\ldots,\gamma_0''}\backslash A_{\gamma_0,\gamma_0',\ldots,\gamma_0''}$. $P(\neg A_{\gamma_0,\gamma_0',\ldots,\gamma_0''\ldots}) = 1 - P(A_{\gamma_0,\gamma_0',\ldots,\gamma_0''\ldots})$ for all infinite $A_{\gamma_0,\gamma_0',\ldots,\gamma_0''\ldots} \subseteq \Omega$, where $\neg A_{\gamma_0,\gamma_0',\ldots,\gamma_0''\ldots} = \Omega\backslash A_{\gamma_0,\gamma_0',\ldots,\gamma_0\ldots}''$.
- Relative probability functions $P(A_{\gamma_0,\gamma_0',\ldots,\gamma_0''}|B_{\gamma_0,\gamma_0',\ldots,\gamma_0''}) \in \mathbf{Q}_p$ and $P(A_{\gamma_0,\gamma_0',\ldots,\gamma_0''\ldots}| B_{\gamma_0,\gamma_0',\ldots,\gamma_0''\ldots}) \in \mathbf{Q}_p$ are defined as follows:

$$P(A_{\gamma_0,\gamma_0',\ldots,\gamma_0''}|B_{\gamma_0,\gamma_0',\ldots,\gamma_0''}) = \frac{P(A_{\gamma_0,\gamma_0',\ldots,\gamma_0''} \cap B_{\gamma_0,\gamma_0',\ldots,\gamma_0''})}{P(B_{\gamma_0,\gamma_0',\ldots,\gamma_0''})},$$

where $P(B_{\gamma_0,\gamma_0',\ldots,\gamma_0''}) \neq 0$ and

$$P(A_{\gamma_0,\gamma_0',\ldots,\gamma_0''} \cap B_{\gamma_0,\gamma_0',\ldots,\gamma_0''}) = \inf(P(A_{\gamma_0,\gamma_0',\ldots,\gamma_0''}), P(B_{\gamma_0,\gamma_0',\ldots,\gamma_0''})).$$

All these equalities hold for infinite sets $A_{\gamma_0,\gamma_0',\ldots,\gamma_0''\ldots}$ and $B_{\gamma_0,\gamma_0',\ldots,\gamma_0''\ldots}$, too.

27.4.3 Relations and Functions on p-Adic Valued Strings

Let us consider only sets with non-empty strings:

$$A_{\gamma_0,\ldots,\gamma_0'} = \{\gamma_0\gamma_1\gamma_2\ldots\gamma_m : [\gamma_0\gamma_1\gamma_2\ldots\gamma_m] = \sum_{i=0}^{m}[\gamma]\cdot p^i \wedge [\gamma]_m \neq 0\}.$$

Then the set $R_n(\underbrace{A_{\gamma_0,\ldots,\gamma_0'}, \ldots, B_{\gamma_0'',\ldots,\gamma_0'''}}_{n}) \subseteq \mathcal{P}(\Omega)^n$ is called n-ary relation on p-adic valued strings.

We can define some unusual relations such as vertical equivalence $R_{\parallel}$:

$$R_{\parallel}(A_{\gamma_0,\ldots,\gamma_0'}, B_{\gamma_0'',\ldots,\gamma_0'''}) \quad if \ and \ only \ if \quad P(A_{\gamma_0,\ldots,\gamma_0'}|B_{\gamma_0'',\ldots,\gamma_0'''}) > 0.$$

The vertical equivalence has the following properties:

- *identity*: $R_{\parallel}(A_{\gamma_0,\ldots,\gamma_0'}, A_{\gamma_0'',\ldots,\gamma_0'''})$ for any $A_{\gamma_0,\ldots,\gamma_0'} \subseteq \Omega$, as

$$P(A_{\gamma_0,\ldots,\gamma_0'}|A_{\gamma_0,\ldots,\gamma_0'}) > 0;$$

- *symmetry*: if $R_{\parallel}(A_{\gamma_0,\ldots,\gamma_0'}, B_{\gamma_0'',\ldots,\gamma_0'''})$, then $R_{\parallel}(B_{\gamma_0'',\ldots,\gamma_0'''}, A_{\gamma_0,\ldots,\gamma_0'})$ for any $A_{\gamma_0,\ldots,\gamma_0'}, B_{\gamma_0'',\ldots,\gamma_0'''} \subseteq \Omega$, since from $P(A_{\gamma_0,\ldots,\gamma_0'}|B_{\gamma_0'',\ldots,\gamma_0'''}) > 0$ it follows that

$$P(B_{\gamma_0'',\ldots,\gamma_0'''}|A_{\gamma_0,\ldots,\gamma_0'}) > 0.$$

- *supertransitivity*: if $R_{\parallel}(A_{\gamma_0,\ldots,\gamma_0'}, C_{\gamma_0'''',\ldots,\gamma_0'''''})$, then there exists $B_{\gamma_0'',\ldots,\gamma_0'''}$ for any $A_{\gamma_0,\ldots,\gamma_0'}, C_{\gamma_0'''',\ldots,\gamma_0'''''} \subseteq \Omega$ such that $R_{\parallel}(B_{\gamma_0'',\ldots,\gamma_0'''}, C_{\gamma_0'''',\ldots,\gamma_0'''''})$ and $R_{\parallel}(A_{\gamma_0,\ldots,\gamma_0'}, B_{\gamma_0'',\ldots,\gamma_0'''})$, because from $P(A_{\gamma_0,\ldots,\gamma_0'}|C_{\gamma_0'''',\ldots,\gamma_0'''''}) > 0$ it follows that there exists $B_{\gamma_0'',\ldots,\gamma_0'''}$ such that $P(B_{\gamma_0'',\ldots,\gamma_0'''}|C_{\gamma_0'''',\ldots,\gamma_0'''''}) > 0$ and $P(A_{\gamma_0,\ldots,\gamma_0'}|B_{\gamma_0'',\ldots,\gamma_0'''}) > 0$.

Notably that the supertransitivity is a feature of p-adic valued probabilities. In real probabilities with values in the interval $[0, 1]$ of real numbers the conditional probability $P(A|B) > 0$ defines a standard equivalence relation (called by us horizontal equivalence) that gives a partition of Ω into disjoint subsets. In particular, $P(A|B) > 0$ satisfies the transitivity: if $P(B|C) > 0$ and $P(A|B) > 0$, then $P(A|C) > 0$. In case of p-adic valued probabilities we have no partition in accordance with the horizontal equivalence relation. Nevertheless, we can build a partition of $\mathcal{P}(\Omega)$ in accordance with the vertical equivalence relation:

$$\mathcal{P}(\Omega) = \bigcup_{A_{\gamma_0,\ldots,\gamma_0'} \subseteq \Omega} \{A_{\gamma_0,\ldots,\gamma_0'} : R_{\parallel}(A_{\gamma_0,\ldots,\gamma_0'}, B_{\gamma_0'',\ldots,\gamma_0'''})\},$$

where $\{A_{\gamma_0,\ldots,\gamma_0'} : R_{\parallel}(A_{\gamma_0,\ldots,\gamma_0'}, B_{\gamma_0'',\ldots,\gamma_0'''})\}$ is a class of vertical equivalence. Notice that this partition is not exclusive: two classes of vertical equivalences can be intersected.

We can prove the proposition that a union of two vertical equivalences is a vertical equivalence.

Let Φ and Ψ be two binary relations on p-adic valued strings. Then their composition $\Phi \circ \Psi$ is defined as follows:

$\Phi \circ \Psi = \{\langle A_{\gamma_0,\dots,\gamma_0'}, B_{\gamma_0'',\dots,\gamma_0'''}\rangle: \langle A_{\gamma_0,\dots,\gamma_0'}, C_{\gamma_0'''',\dots,\gamma_0'''''}\rangle \in \Phi$ and $\langle C_{\gamma_0'''',\dots,\gamma_0'''''}, B_{\gamma_0'',\dots,\gamma_0'''}\rangle \in \Psi$ for some $C_{\gamma_0'''',\dots,\gamma_0'''''} \in \Omega\}$.

Also, we can define a diagonal relation in $\mathcal{A}$:

$$1_{\mathcal{A}} = \{\langle A_{\gamma_0,\dots,\gamma_0'}, A_{\gamma_0,\dots,\gamma_0'}\rangle: A_{\gamma_0,\dots,\gamma_0'} \in \mathcal{A}\},$$

and an inverse relation:

$$\Phi^{-1} = \{\langle A_{\gamma_0,\dots,\gamma_0'}, B_{\gamma_0'',\dots,\gamma_0'''}\rangle: \langle B_{\gamma_0'',\dots,\gamma_0'''}, A_{\gamma_0,\dots,\gamma_0'}\rangle \in \Phi\}.$$

A relation Φ is a function from $\mathcal{A}$ to $\mathcal{B}$ if and only if $\Phi \circ \Phi^{-1} \supseteq 1_{\mathcal{A}}$ and $\Phi^{-1} \circ \Phi \subseteq 1_{\mathcal{B}}$.

Formal theories of these functions are proposed in [30]. Reversible logic gates on the basis of non-linear group theory are proposed in [29].

27.5 Conclusions

In any intelligent swarm behaviour there are emergent patterns that cannot be reduced to linear combinations of subsystems. Therefore conventional algorithms such as Kolmogorov-Uspensky machines have very low accuracy of their implementations on swarm systems (Sect. 27.3). Nevertheless, we can define a p-adic valued logic that can describe a massive-parallel behavior of swarms (Sect. 27.4).

Acknowledgments This research is supported by FP7-ICT-2011-8. The ideas of this paper were presented in BICT 2015.

References

1. Aczel, A.: Non-Well-Founded Sets. Stanford University Press, Redwood city (1988)
2. Adamatzky, A.: Physarum machine: implementation of a Kolmogorov–Uspensky machine on a biological substrate. Parallel Process. Lett. **17**, 455–467 (2007)
3. Adamatzky, A.: Physarum Machines: Computers from Slime Mould. World Scientific, Singapore (2010)
4. Adamatzky, A., Erokhin, V., Grube, M., Schubert, T., Schumann, A.: Physarum chip project: Growing computers from slime mould. Int. J. Unconv. Comput. **8**(4), 319–323 (2012)
5. Ariel, G., Shklarsh, A., Kalisman, O., Ingham, C., Ben-Jacob, E.: From organized internal traffic to collective navigation of bacterial swarms. New J. Phys. **15**, 125019 (2013)
6. Bonabeau, E., Dorigo, M., Theraulaz, G.: Swarm Intelligence: From Natural to Artificial Systems. Oxford University Press, Oxford (1999)
7. Cuevas, E., Cienfuegos, M., Zaldivar, D., Perez-Cisneros, M.: A swarm optimization algorithm inspired in the behavior of the social-spider. Expert Syst. Appl. **40**(16), 6374–6384 (2013)

8. Darnton, N.C., Turner, L., Rojevsky, S., Berg, H.C.: Dynamics of bacterial swarming. Biophys. J. **98**, 2082–2090 (2010)
9. Dorigo, M., Stutzle, T.: Ant Colony Optimization. MIT Press, Cambridge (2004)
10. Gordon, D.: The organization of work in social insect colonies. Complexity **8**(1), 43–46 (2003)
11. Ingham, C.J., Ben-Jacob, E.: Swarming and complex pattern formation in paenibacillus vortex studied by imaging and tracking cells. BMC Microbiol. **8**(36), 1–16 (2008)
12. Ingham, C.J., Kalisman, O., Finkelshtein, A., Ben-Jacob, E.: Mutually facilitated dispersal between the nonmotile fungus aspergillus fumigatus and the swarming bacterium paeni bacillus vortex. Proc. Natl. Acad. Sci. U.S.A. **108**(49), 19731–19736 (2011)
13. Jacobs, B., Rutten, J.: A tutorial on (co)algebras and (co)induction. EATCS Bull. **62**, 222–259 (1997)
14. Karaboga, D.: An Idea Based on Honey Bee Swarm for Numerical Optimization. Technical Report-TR06. Engineering Faculty, Computer Engineering Department. Erciyes University (2005)
15. Karaboga, D., Akay, B.: A comparative study of artificial bee colony algorithm. Appl. Math. Comput. **214**(1), 108–132 (2009)
16. Kassabalidis, I., El-Sharkawi, M., II Marks R.J., Arabshahi, P., Gray, A.: Swarm intelligence for routing in communication networks. In: Proceedings of the Global Telecommunications Conference, GLOBECOM'01, vol. 6, pp. 3613–3617 (2001)
17. Kennedy, J., , Eberhart, R.: Particle swarm optimization. In: Proceedings of the 1995 IEEE International Conference on Neural Networks, vol. 4 (1995)
18. Kennedy, J., Eberhart, R.: Swarm Intelligence. Morgan Kaufmann Publishers Inc, Amsterdam (2001)
19. Kolmogorov, A.: On the concept of algorithm. Uspekhi Matematicheskich Nauk **8**(4), 175–176 (1953)
20. Lee, C.: Emergence and universal computation. Metroeconomica **55**(2–3), 219–238 (2004)
21. Passino, K.: Biomimicry of bacterial foraging for distributed optimization and control. Control Syst. **22**(3), 52–67 (2002)
22. Rajabioun, R.: Cuckoo optimization algorithm. Appl. Soft Comput. **11**, 5508–5518 (1987)
23. Reynolds, C.: Flocks, herds, and schools: a distributed behavioral model. Comput. Graph. **21**, 25–34 (1987)
24. Reynolds, R.G.: An introduction to cultural algorithms. In: Proceedings of the Third Annual Conference on Evolutionary Programming, pp. 131–139 (1994)
25. Rutten, J.: A coinductive calculus of streams. Math. Struct. Comput. Sci. **15**(1), 93–147 (2005)
26. Schoenhage, A.: Real-time simulation of multi-dimensional Turing machines by storage modification machines. Project MAC Technical Memorandum 37. MIT Press, Cambridge (1973)
27. Schoenhage, A.: Storage modification machines. SIAM J. Comput. **9**, 490–508 (1980)
28. Schumann, A.: Non-archimedean fuzzy and probability logic. J. Appl. Non-Class. Log. **18**(1), 29–48 (2008)
29. Schumann, A.: Conventional and unconventional reversible logic gates on physarum polycephalum. Int. J. Parallel, Emerg. Distrib. Syst. (2015). doi:10.1080/17445760.2015.1068775
30. Schumann, A.: Towards context-based concurrent formal theories. Parallel Process. Lett. **25**, 1540008 (2015)
31. Schumann, A., Adamatzky, A.: Physarum spatial logic. New Math. Nat. Comput. **7**(3), 483–498 (2011)
32. Schumann, A., Adamatzky, A.: The double-slit experiment with physarum polycephalum and p-adic valued probabilities and fuzziness. Int. J. Gen. Syst. **44**(3), 392–408 (2015)
33. Schumann, A., Pancerz, K.: Towards an object-oriented programming language for physarum polycephalum computing: A petri net model approach. Fundamenta Informaticae **133**(2–3), 271–285 (2014)
34. Shklarsh, A., Finkelshtein, A., Ariel, G., Kalisman, O., Ingham, C., Ben-Jacob, E.: Collective navigation of cargo-carrying swarms. Interface Focus **2**, 689–692 (2012)
35. Tarjan, R.E.: Reference machines require non-linear time to maintain disjoint sets. STAN-CS-77-603 (March 1977)

36. Turner, L., Zhang, R., Darnton, N.C., Berg, H.C.: Visualization of flagella during bacterial swarming. J. Bacteriol. **192**, 3259–3267 (2010)
37. Uspensky, V.U.: Kolmogorov and mathematical logic. J. Symb. Log. **57**, 385–412 (1992)
38. Wang, Y., Li, B., Weise, T., Wang, J., Yuan, B., Tian, Q.: Self-adaptive learning based particle swarm optimization. Inf. Sci. **181**(20), 4515–4538 (2011)
39. Wolfram, S.: Universality and complexity in cellular automata. Physica D **10**, 1–35 (1984)

Chapter 28
On the Inverse Pattern Recognition Problem in the Context of the Time-Series Data Processing with Memristor Networks

Christopher Bennett, Aldo Jesorka, Göran Wendin and Zoran Konkoli

Abstract The implementation problem deals with identifying computations that can be performed by a given physical system. This issue is strongly related to the problem of describing a computing capacity of the system. In this chapter, these issues have been addressed in the context of on-line (real-time) pattern recognition of time series data, where memristor networks are used in the reservoir computing setup to perform information processing. Instead of designing a network that can solve a particular task, an inverse question has been addressed: Given a network of a certain design, which signals might it be particularly adept at recognizing? Several key theoretical concepts have been identified and formalised. This enabled us to approach the problem in a rigorous mathematical way: The problem has been formulated as an optimization problem, and a suitable algorithm for solving it has been suggested. The algorithm has been implemented as computer software: For a given network description the software produces the time-dependent voltage patterns (signals) that can be best recognized by the network. These patterns are found by performing a directed random search in the space of input signals. As an illustration of how to use the algorithm, we systematically investigated all networks containing up to four memristors.

C. Bennett · G. Wendin · Z. Konkoli (✉)
Department of Microtechnology and Nanoscience - MC2, Chalmers University
of Technology, 412 Gothenburg, Sweden
email: zorank@chalmers.se

Present Address:
C. Bennett
Institut d' Electronique Fondamentale, Univ. Paris-Sud, CNRS, 91405 Orsay, France

A. Jesorka
Department of Chemical and Biological Engineering (CBE), Chalmers
University of Technology, 412 Gothenburg, Sweden

© Springer International Publishing Switzerland 2017   735
A. Adamatzky (ed.), *Advances in Unconventional Computing*,
Emergence, Complexity and Computation 22,
DOI 10.1007/978-3-319-33924-5_28

28.1   Introduction

Unconventional or natural computation refers to computing with structures that do not naturally implement Boolean logic. Instead, their information processing abilities are based on particular aspects of their dynamics [1]. The act of computation is often based on rather complex and non-linear dynamics. While some a priori guiding principles are known about how such devices can be built in a few selected areas of application, it is far from clear how to achieve a specific functionality in general. It is even not clear what the computing capacity of such a device would be. This study takes a pragmatic view on these matters. There are technological limitations to what can be built with given hardware and to what can be achieved with these devices.

The conceptual problem of optimizing a device design towards a particular functionality is challenging, but often doable, and has been frequently addressed in the literature. However, we wish to address an inverse problem: given a device, what can it compute?

There are numerous philosophical discussions about this so-called implementation problem (cf. [2] and references therein). For example, it has been argued that even a simple object as a rock can implement any finite state automaton. However, the suggested implementation can be (and has been) criticised on the grounds that the auxiliary hardware necessary to implement an automaton is doing all the computation, and not the rock. To resolve this issue a more precise version of the problem has been formulated, the natural computability problem and a way of solving it, as presented in [3]. The formulation emphasizes practical aspects of implementing a computation: The natural computation can be identified by balancing the costs of implementing the computation with the complexity of the function being computed. The present study adopts a similar approach. The emphasis on practical aspects is expected to help in identifying key conceptual issues that need to be defined in order to solve the generic questions.

In the following, a rather complicated philosophical issue is addressed by studying a concrete example. We specifically focused on the problem of pattern recognition. This is a well-defined computing task with many applications. Likewise, a well-defined but potentially computationally powerful class of devices has been chosen as information processing substrate. The goal is to identify suitable pattern recognition tasks for memristor networks in the context of on-line (real-time) time series pattern recognition. Thus the key question is: Which patterns are naturally recognized by memristor networks without any additional auxiliary equipment? This question is naturally addressed in the context of reservoir computing paradigm which emphasises computing capabilities of the device per se. In the following the problem will be referred to as the inverse pattern recognition problem.

28.2 Reservoir Computing

The key idea behind the reservoir computing (RC) paradigm is that the system performs computation naturally without any specific training. The computation is performed automatically as the system changes states. The only component that is trained is an easily adjustable read-out layer. This is the only auxiliary equipment that is allowed.

It is clear that not every system can perform information processing in such a way. A set of criteria has been formulated that a good reservoir computer must satisfy. In 2002, Maass et al., put forth a computational model called the Liquid State Model (LSM) to describe the information processing capability of neural circuits in *real time*, in contradistinction to traditional, feed-forward artificial neural networks (ANNs) that must be trained over time [4]. They demonstrated that despite its inability to train or alter weights, a simple alterable readout layer can nonetheless parse an extremely non-linear signal input by using the already high dimensional transient set of states (so-called Liquid State) of the reservoir. Independently, Jaeger et al. derived a similar formulation (Echo State Network or ESN), which originally concerned harnessing non-linearity within ANNs [5]. Hereafter, these computing models are referred to as the Reservoir Computing (RC) paradigm.

The idea of exploiting memristor networks in the context of reservoir computing is not new. The work by Kulkarni and Teuscher [6] has demonstrated that pattern recognition is possible using memristor networks. Note that in [6] the network structure has been optimized for a fixed task, whereas we aim to find answers to precisely the opposite question by exploiting the paradigm of reservoir computing, where the only permitted auxiliary equipment is the easily trained linear readout layer. This layer is rather passive and is not doing any substantial computation. It simply extracts the information processed by the system. An attempt to address a similar set of issues has been made in [7].

28.3 Methods: Formalising the Recognition Problem

28.3.1 Choosing an Initial State x_0

Formally, one should distinguish between the off-line and on-line (real-time) computation. For example, assume that a signal that needs to be recognized starts at t_0, has a duration τ, and ends at $t_f = t_0 + \tau$. Without any loss of generality it will be assumed that $t_0 = 0$. It is clear that prior to the occurrence of the signal the system could be in any state x_0. In the off-line computation the initial state of the system could be used as a training parameter which, interestingly, has not been discussed much in the literature, despite being suggested [8]. In the context of on-line (real-time) pattern recognition any initial state is possible in principle, and the initial state depends on the history of the system for $t < t_0$. In the context of reservoir computing this issue is solved by requiring the "fading memory" property (the distant past is successively "forgotten" by the system). Instead of investigating how all possible histories influence the initial state, it will be simply required that the system performs pattern recognition in the interval $[t_0, t_0 + \tau]$ equally well for all x_0. This is a rather strong requirement since not all initial states x_0 might be actually relevant.

28.3.2 The Role of the Background

The history of the system can be defined as a set of patterns that the reservoir has been continuously exposed to until time t_0. These will be referred to as background patterns. A signal that can be recognized by the device is termed a task (signal). Since the history decides the x_0, it is clear that the task signal and the background signals are closely related and, in some sense, inseparable. In the following, this task-background duality will be accounted for explicitly by keeping a separate lists of task and background signals. The set of background signals will be clearly defined and made explicit.

The full set of signals is not countable, but if a sampling strategy is used, this is not a major problem. We always work with finite subsets of signals, each having some pre-determined cardinality which is always maintained. Such sets are constructed iteratively by replacing the task signals with better candidates, until some convergence criteria are met.

28.3.3 Measuring the Total Quality of Recognition: δ

The quality of the pattern recognition is quantified as follows. Assume the set of signals contains a set of tasks, the rest being background

$$\mathfrak{B} \equiv \{u_\alpha; \alpha = 1, 2, \dots, B\} \tag{28.1}$$

where B (base functions) denotes the number of signals considered. What should one choose for base functions? In principle, this set should contain all possible signals. The number of base functions should be as large as possible. However, for practical purposes to demonstrate the approach we focus on several pre-defined classes of signals, e.g. the Legendre polynomials, or square waves. A typical base function Assume that there is a finite set of initial states of the system and these are ordered and numbered as

$$\mathfrak{S} = \{x_0^\mu; \mu = 1, 2, \dots, S\} \tag{28.2}$$

where S (states) counts the total number of initial states considered. Each signal drives the system (memristor values) from its initial state into the final state

$$x_i^{\alpha,\mu} = T_i(\tau|x_0^\mu, u_\alpha) \tag{28.3}$$

where the function T describes the dynamics of the system, and specifies how each memristor value (the reservoir configuration space)

$$\mathfrak{C}(t) = \{x_i(t); i = 1, 2, \ldots, D\} \tag{28.4}$$

changes in time under the influence of an external signal u_α. Note that

$$x_0^\mu \equiv \mathfrak{C}(t = 0) \tag{28.5}$$

D (dimensionality) denotes the number of memristors in the network and corresponds to the dimensionality of the configuration space of the reservoir.

From the information processing point of view such a system is a device that realizes an abstract mapping F from the space of input signals to real numbers: Depending on the initial condition of the device μ, each input signal u_α is converted into a real number $F_\mu(u_\alpha)$ as

$$F_\mu(u_\alpha) = \sum_{i=1}^{D} w_i x_i^{\alpha,\mu} - K; \quad \forall\mu, \forall\alpha \tag{28.6}$$

Note the presence of the initial condition index. If the initial condition of the device cannot be controlled then one must allow for the possibility that these values differ even if the device is always exposed to the same input u_α. Ideally, for on-line computation, the outcome of the computation should depend weakly on the initial condition. However, this cannot be taken for granted.

The pattern recognition is implemented in the standard way where the goal is optimize the device to compute a predefined classifier function f. This function is defined by listing explicitly how each u_α maps to a predefined value $f_\alpha \equiv f(u_\alpha)$ for all α. Values for f_α are user defined, e.g. in our software implementation if a pattern u_α is to be recognized $f_\alpha = 1$, otherwise $f_\alpha = 0$. The goal is to minimize the differences

$$\Delta_{\alpha,\mu}[f] \equiv F_\mu(u_\alpha) - f_\alpha \tag{28.7}$$

The only quantities that can be adjusted in the weighted sum (28.6) are weights w_i and the offset K. These parameters can be exploited by a device design engineer to improve its performance. What works against such efforts is the fact that many initial conditions are possible. Likewise, the device might respond to various patterns u_α in a strongly unbalanced way. For example, while the error can be made small for some patterns it might be intrinsically large for others. Thus, in strict mathematical terms the differences $\Delta_{\alpha,\mu}$ should be made as small as possible for every initial condition x_0^μ and input pattern u_α. The optimization of the weights and the shift factor should be performed in such a way that the recognition works uniformly well across all initial conditions and input signals.

Let δ_f denote the smallest typical error of realising a mapping f across all input signals and initial conditions:

$$\delta_f^2 = \frac{1}{S} \min_{w,K} \sum_{\alpha,\mu} \left(\Delta_{\alpha,\mu}[f] \right)^2 \tag{28.8}$$

Equation (28.8) is an instance of the standard least squares optimization problem; a consequence of not using the standard non-linear wrapper function in (28.6). Since the solution to the least squares problem is widely known, formulas for w and K are not explicitly stated. Both can be readily computed form the values for $x_i^{\alpha,\mu}$ and f_α. The optimized weights and the shift factor will be denoted by w_* and K_*

28.3.4 Measuring the Quality of Recognition for Individual Signals and Signal Sets: δ_α and $\delta(\mathfrak{B})$

It is practical to separate different contributions to the typical error as

$$\delta_f^2 = \frac{1}{S} \sum_{\alpha,\mu} \left(\delta_{\alpha,\mu}[f] \right)^2 \tag{28.9}$$

which defines $\delta_{\alpha,\mu}[f]$ as $\Delta_{\alpha,\mu}[f]$ computed with optimized weights.

$$\delta_{\alpha,\mu} \equiv \left(\Delta_{\alpha,\mu} \right)_{w=w_*, K=K_*} \tag{28.10}$$

The separation of contributions is useful since the signal specific error $\delta_\alpha(\mathfrak{B})$ can be defined as a typical recognition error across all initial conditions (e.g. here defined as a generalized average)

$$\delta_\alpha(\mathfrak{B}) \equiv \left(\langle \delta_{\alpha,\mu}^2 \rangle_\mu \right)^{1/2} = \left(\frac{1}{S} \sum_\mu \delta_{\alpha,\mu}^2 \right)^{1/2} \tag{28.11}$$

In the discussions that follow, this ability to compare accuracy of various signals in a natural way will be extremely important. Note that $\delta_\alpha(\mathfrak{B})$ depends on the whole base set. For example, if u_β with $\beta \neq \alpha$ were replaced by a new signal, the value for $\delta_\alpha(\mathfrak{B})$ might change. The total recognition error for a given base set $\mathfrak{B}$ is naturally expressed as

$$\delta(\mathfrak{B}) = \left(\sum_\alpha \delta_\alpha(\mathfrak{B})^2 \right)^{1/2} \tag{28.12}$$

The parameters B and S should be taken as large as possible since this improves the accuracy of numerical results. In principle, these parameters should be taken infinitely large. However, it is clear that due to the practical limitations of the computer hardware, the parameters B and S cannot be arbitrarily large. The questions are which values are reasonable, and most importantly, how does the error behave as these parameters increase? The error has been defined per initial state, since during the device operation, the device will always start from one state. Furthermore, when the $S \to \infty$ limit is taken, the space of initial conditions is sampled more densely, since all memristance values are bound: $x \in [x_{min}, x_{max}]$. As B is increased it is much more challenging to understand the trend of the total error, as the input signals are not bound by default. An advantage of the present formulation is that the computing capacity of different types of networks can be compared. It is sufficient to keep the values for B and S constant when the comparison is done.

28.4 The Sampling Algorithm

The algorithmic ideas presented in the previous section can be used to address the implementation problem or the natural computability problem rigorously. If the algorithm is implemented, it can be run on a computer to suggest a natural input signal. Note that it would be extremely hard to use analytic approaches to implement the algorithm.

The algorithm has been implemented as follows. The input to the algorithm must consist of the description of the memristor device, and the output consists of the list of patterns which can be recognized together with the associated background. The device description is used to define the mapping function T which defines the dynamics of the network.

At each iteration step of the algorithm, a list of the base input patterns is stored. This list is divided in two parts, a list of signals that are recognized, and the list of signals that are not. The user must specify how many slots should be allotted for each. In our numerical examples the number of background slots was chosen larger than the number of task slots. For a given list the recognition error is computed for each base function. This information is used to decide which signals to keep and which to regenerate. It is expected that after a sufficient number of iterations the signals that are recognized should differ clearly from the background signals. Some of the key elements of the sampling algorithm are detailed in the following subsections.

28.4.1 Memristor Networks

To construct a memristor network a set of nodes is given and connected by memristive links. Each link is modelled by assuming that the memristance x changes according to

Fig. 28.1 Simple
symmetrical memristor
threshold model. Function
$f(V)$ describes how
memristance changes with
applied voltage

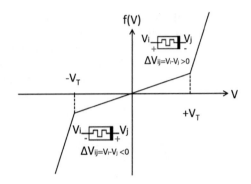

$$\dot{x}(t) = \begin{cases} f(V(t)) & x_{min} \le x(t) \le x_{max} \\ 0 & x(t) < x_{min} \text{ or } x(t) > x_{min} \end{cases} \qquad (28.13)$$

where V denotes the voltage difference on the respective nodes that the link is bridging
and

$$f(V) = b * V + \frac{(a-b)}{2}(|V + V_{th}| - |V - V_{th}|) \qquad (28.14)$$

which is the standard form used in the literature. In the following x is expressed in
ohm, and f in (Fig. 28.1) in ohm/sec. The $f(V)$ in the equation above should not be
confused with $f(u)$ (the classifier function introduced earlier).

This model is a good first-order approximation for a variety of simple, bipolar
memristor media [9] and has already been used successfully to study computational
performance and complexity of memristor networks [10]. For each memristor link the
user defines the following set of parameters: V_{th}, a, and b. To compute the mapping
function for such a model the MENES software has been used [11].

28.4.2 The Turning Power (TP) Concept

By far the biggest challenge was how to choose signals when sampling. Given a
signal, it was not entirely clear which part is definitive for the evolution of the
system. Consider an extreme example of a very long, flat signal, with a large voltage
spike at its end. In this case, only the last part of the signal will be definitive upon
the final state of memristive values, and the signal could be substantially compressed
to the length of just this final pulse. In this way, length τ alone does not capture
the strength of the signal. Then, the question becomes what is a correct length and
strength of a signal that can "move" the system in the most efficient way. To answer
this question rigorously, we have introduced the concept of "turning power".

Turning power corresponds to the largest change of a memristance possible under
a pulse duration. In studying this change, it is important to consider all possible

sub-intervals of the pulse $[t_1, t_2] \subset [0, \tau]$. Over an interval from $[t_1, t_2]$, one could calculate the turning power as the total voltage-controlled effect on memristance. Thus for a given signal

$$\text{turning power} \sim \max_{0 \leq t_1 < t_2 \leq \tau} |x(t_2) - x(t_1)| \qquad (28.15)$$

where here and in the following the symbol $\sim$ indicates that the right hand side of the equation is an estimate of the turning power. Namely, this quantity is only valid for a single memristor (e.g. for a particular initial state), and challenging to define properly on the network level: the voltage difference experienced by every individual memristor is a complicated function of applied voltages. Instead, we suggest a related definition, that is less precise but signal specific:

$$\text{TP}[V] \equiv \max_{0 \leq t_1 < t_2 \leq \tau} \left| \int_{t_1}^{t_2} f(V(t)) dt \right| \sim \max_{0 \leq t_1 < t_2 \leq \tau} \left| \int_{t_1}^{t_2} V(t) dt \right| \qquad (28.16)$$

In strict mathematical terms, the TP represents a functional on the space of input (voltage) signals. An intuitive understanding of this equation is as follows.

The TP value distinguishes between the signals that are strongly oscillating and the signals that do not change sign. For example, imagine the following two input types. First, imagine a series of many alternating spikes each with roughly the same surface under the curve. Alternatively, assume that the spike series is re-ordered, by collecting the spikes with strictly positive and negative voltages in two groups where, e.g. the group with positive voltages arrives first. Neither signal type will change the memristance value much. However, the first signal type will not change the memristance value for intermediate times, while the second signal type will cause substantial changes at the time instances where the series of positive and negative spikes end. The second signal type will have larger TP. Note that the TP value cannot distinguish between a strictly positive signal that is very broad and the related signal that is very narrow but with the same surface under the curve. For information processing purposes, these signals are virtually identical (provided their respective timings are not playing a role).

28.4.3   Signal Construction Algorithm

As an input to our signal searching engine, the user must supply the following parameters: (i) Signal time length τ. (ii) the number of slices n_{slices} in the $V(t)$ graph. This measures the complexity of a signal. (iii) The number of steps n_{steps} that the MENES simulator uses for integration.[1] (iv) A global voltage scale V_{tr}, necessarily $V_{tr} > V_{th}$, that specifies the voltage range of each signal: $|V(t)| \leq V_{tr}$. (v) The num-

---

[1] As a general rule of thumb, we used n_{steps} that were 2–3 times larger than n_{slices} so as to allow the integrator to capture all features in the signal.

ber of possible voltage values in the range V_{values}, and maximum turning power TP. Prior to every run, a certain target vector f is given, and a total simulation time is specified.[2] Once the pool of base signals is defined, individual candidate signals are spawned as follows. At each slice, one value of possible discrete voltages is chosen at random. Then the whole list is interpolated into spike pulses using the lowest order interpolation, producing step-like functions.

The random search implementation: For a given set of base functions $\mathfrak{B}$ the individual recognition errors δ_α with $\alpha = 1, 2, \ldots, B$ are computed. The worst among the recognized signals is replaced, i.e. the one with the largest δ_α. This defines a new set of base functions $\mathfrak{B}'$. If $\delta(\mathfrak{B}) > \delta(\mathfrak{B}')$ then the new set replaces the old, $\mathfrak{B}' \rightarrow \mathfrak{B}$, and the process is repeated.

28.5 Results

28.5.1 Confirming Configuration Space Separation: One-Memristor Case

We preliminarily tested whether input signals $u_\alpha(t)$ can drive the memristance $x(t)$ into separate regions of configuration space $\mathfrak{S}$ by using the simplest possible case: a single memristor. The corresponding configuration space is one-dimensional and the segregation is easy to visualize regardless of which initial condition is used. This might be much harder to do for higher dimensional spaces.[3]

We sampled the initial state space by using $S = 6$, 11, and 21 values uniformly separated in the one-dimensional interval $[x_{\min}, x_{\max}]$. The boundaries were always included. The analysis of Figs. 28.2, 28.3 and 28.4 indicates that the segregation of states does indeed happen. The signals depicted were spawned automatically by the algorithm. In all figures memristance values are driven into different regions of the configuration space by the task and the background signals, regardless of the initial condition. Every initial memristance value progressively "locks on" to a signal. In addition to varying the number S of possible resistive values at $t = 0$, we also gradually increased the number of features n_{slices} from 5, to 10, and finally 15. The respective number of integration steps n_{steps} was 20, 40, and 60. In all graphs the turning power was set to TP $= 10$. The signals are classified nearly perfectly: $\delta = 0$ over the set $f = \{1, 0, 0\}$.

In Fig. 28.2 the signal searching algorithm allows for only relatively simple pulses since $n_{\text{slices}} = 5$. Indeed, visual inspection of the signals in panels b, d, and f confirms

---

[2]For instance, a typical use case of the code would be: the user could select $\tau = 7\,$s digital pulse of $V_{\text{values}} = 7$, $n_{\text{slices}} = n_{\text{steps}} = 10$, and TP $= 20$ (e.g. in volts), specify $f = \{1, 0, 0\}$, and instruct the code to run for 10 min, and see what signals were best recognized by her chosen system over time. The code will return three signals. The first one can be recognized well against the remaining two.

[3]Already for a two dimensional configuration space this might be problematic if the initial conditions are sampled uniformly in the whole region.

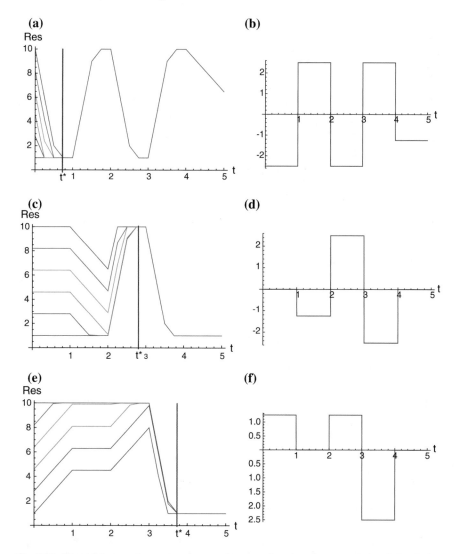

Fig. 28.2 The single memristor network: memristance values (panels **a, c,** and **e**) and signals (u_α, panels **b, d,** and **f**) selected by the algorithm. The *first row* (the **a–b** panel pair) and the *bottom two rows* (panel pairs **c–d** and **e–f**) correspond to the signal being recognized and the background signals respectively. The number of allowed features in signal sampling algorithm is $n_{\text{slices}} = 5$, and the number of starting points is $S = 6$. The detailed list of parameters used in the random search is given in the appendix

this. The signal being recognized (panel b) steers the network's states (panel a) towards a larger value of approximately $x \approx 7$, while other background signals steer the initial memristance values towards opposite values of $x \approx 1$ (panels c and e).

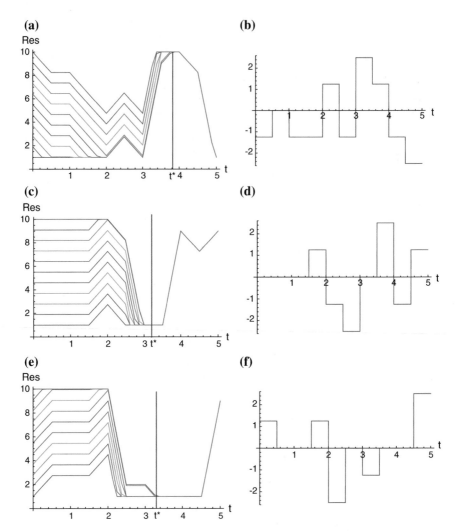

Fig. 28.3 The single memristor network: memristance values (panels **a**, **c**, and **e**) and signals (panels **b**, **d**, and **f**) selected by the algorithm. Same as in Fig. 28.2 with a difference that the number of allowed features in signal sampling algorithm is higher ($n_{\text{slices}} = 10$). The detailed list of parameters used in the random search is given in the appendix

Since $S = 6$, we observe 6 different colored lines in the figures each representing a different history that begins from one of the possible initial starting points.

Once the signals have been identified, it is possible, but not always easy, to understand why these particular signals have been selected: The alternating spikes of b cause the network to "lock on" them: all lines in panel a merge together quite quickly. Conversely, the strong concluding negative pulses of d and f push the network to its final resting states of $x \approx 1$. To quantify the time evolution property of a signal, we

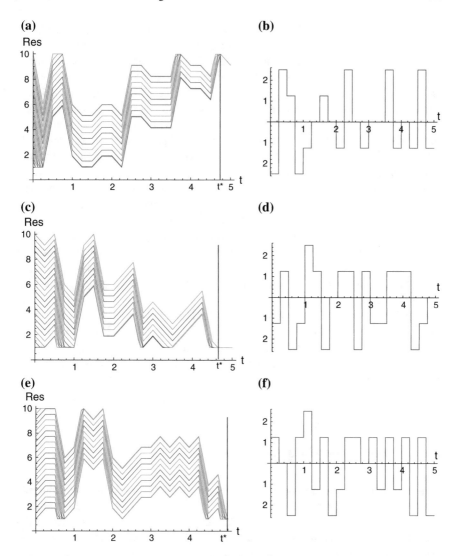

Fig. 28.4 The one memristor network: The signals being classified (panels **b**, **d**, and **f**) and the configuration space (panels **a**, **c**, and **e**). Same as in Figs. 28.2 and 28.3 but with yet higher number of features $n_{\text{slices}} = 15$ and more possible starting states $S = 21$, explaining the density of lines

introduce t^* as the time after which the network "locks on" completely: e.g. $t^* = 0.8$ (panel a), 2.9 (c), and 3.8 (e). In the examples all initial memristance values are merging, since the signals have relatively large turning power of TP $= 10$.

In the intermediate case of more complex and feature-rich input signals ($S = 11$ and $n_{\text{slices}} = 10$, as shown in Fig. 28.3), all key features of Fig. 28.2 are clearly visible again. There are neatly separated concluding memristive values between (a) at $x \approx 1$ and (c, e) at $x \approx 10$. However, the lock-on time is larger, in particular for the recognized function ($t^\star = 3.9$). As in Fig. 28.2 it is always the largest positive (as in d and f) or negative (b) concluding signal features which seem to be definitive.

Figure 28.4 depicts the results of the analysis of the most complex signal search. There are 21 starting conditions (histories) as well as 15 different regions in the signal functions. Although there is still a clear separation between the ending values of (a) at $x \approx 9$ and (c, e) at $x \approx 1$, the convergence is now happening quite late, at almost the last moment; $t^\star = 4.8$, 4.7, and 4.9 for signals (b, d, f) respectively. At this number of features, it becomes difficult to determine why the first signal pulse is recognized so well against the second two.

The analysis of Figures 28.2, 28.3 and 28.4 provides hints with respect to the main question posed in this work. The signals that can be efficiently recognized by a single memristor have to be strong enough to "lock" the value of the memristance onto the input signal before the signal ends (to ensure that $t^\star < \tau$). The initial values of the memristance have to be "forgotten" in order to be driven by a signal progressively. This progressive convergence is especially visible in Fig. 28.4. Two classes of signals are distinguished, the ones that are predominantly positive, or negative, towards the end of the signal duration.

Finally, we have already seen that strong signal segregation is possible with even the simplest possible memristive system, which is encouraging. This is evidence to support the claim that memristors function well, *naturally*, as a binary classifier, and are a candidate building block for hardware instantiations of reservoir computing principles. In the following sections the tolerance to increasing initial conditions will be further explored for multi-memristor networks.

28.5.2 Multi-memristor Networks

We programmed in 17 different pre-defined network topologies for in depth analysis; their exact configurations are shown in Figs. 28.5 and 28.6. Although these topologies are pre-set, the user is allowed to define parameters for the operation of the individual memristor model devices. Once defined, this "model" is then propagated along the whole network. Previous work by Burger and Teuscher has evaluated the effect of device variation in crossbar (i.e. mesh) memristor networks similar to ours, and concluded that a certain degree of variation actually improves computational performance [12]. At the moment such variation is not considered.

28.5.2.1 Small Networks (Fig. 28.5)

We explored the set of smaller networks, defined as consisting of 4 nodes and less. Using an arbitrary indexing system, the network indices (NIs) are as follows: NI = 1–2 point to the one-memristor systems, NI = 3–6 are variants of two memristors in series, NI = 7–10 are variants of two memristors in parallel, NI = 11 is a case of

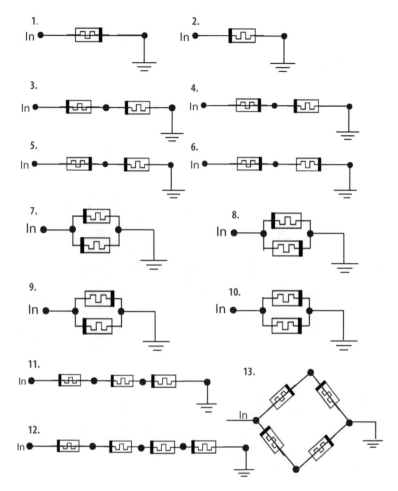

Fig. 28.5 Illustration of the smaller memristor reservoirs used in our numerical studies. An arbitrary indexing system was used (in the range 1–13), as explained in the text. In general, networks with larger numbers of memristors appear later in the list

three memristors in series and similarly NI = 12 with four in series, and finally NI = 13 corresponds to a network with four memristors in parallel (equivalent to the '1-Xbar' or diamond motif depicted in Fig. 28.6).

Figure 28.7 shows the recognition error as a function of the network index NI. At each data point the signal searching algorithm was executed with typically hundreds of iterations. Further, for each network several analyses were performed with varying S. The coloured lines connect data points belonging to different networks with a given number of initial conditions used. The figure suggests that, in general, the one-memristor and two memristor-in-parallel systems outperform the other systems dramatically: the effective TP is larger (smaller) for memristors coupled in parallel

(a)　　　　*'Diamond'/ [1-Xbar]: 4 memristors*　　　　　**(b)**　　　　*2-Xbar: 12 memristors*

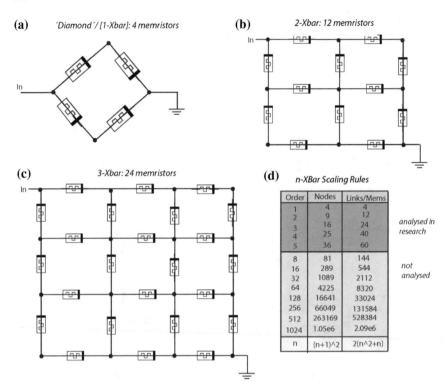

(c)　　　　*3-Xbar: 24 memristors*　　　　　**(d)**　　　*n-XBar Scaling Rules*

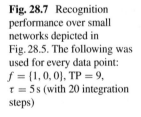

Order	Nodes	Links/Mems	
1	4	4	
2	9	12	
3	16	24	*analysed in research*
4	25	40	
5	36	60	
8	81	144	
16	289	544	*not analysed*
32	1089	2112	
64	4225	8320	
128	16641	33024	
256	66049	131584	
512	263169	528384	
1024	1.05e6	2.09e6	
n	$(n+1)^2$	$2(n^2+n)$	

Fig. 28.6 Some of larger memristor networks used in numerical studies are depicted in panels **a**, **b** and **c**. These illustrate the key principles of how the networks are constructed. Nodes to be connected are placed on a regular two-dimensional grid (the n-Xbar architecture). Only nearest neighbours are connected. Panel **d** depicts the how the number of nodes and links scales with an Xbar index n. Due to the rapidly increasing number of components only a subset of these networks has been analysed

Fig. 28.7 Recognition performance over small networks depicted in Fig. 28.5. The following was used for every data point: $f = \{1, 0, 0\}$, TP = 9, $\tau = 5$ s (with 20 integration steps)

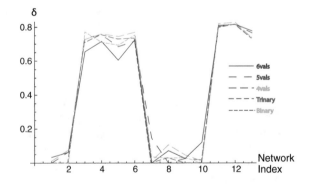

(series). Furthermore, increasing the number of initial conditions is not deleterious, as curves with different colours roughly follow each other. Due to the combinatorial increase in the number of initial conditions possible for each network, this space could not be sampled uniformly. Each initial condition has been chosen at random.

28.5.2.2 Larger Networks (Fig. 28.6)

Some of the larger networks were studied in the final part of our work. The first three are visible in panes a, b, and c of Fig. 28.6. The diamond (a), alternately '1-XBar', is by far the most important because it is ubiquitous: the feed-forward diamond shape is a frequently occurring design pattern not only in biological neural networks, but electronic circuit architectures [13]. In general, n-degree cross-bar networks are often realized experimentally.

28.5.2.3 Larger Networks: Legendre Patterns

As a calibration test for our larger networks we investigated a given pre-defined set of patterns in order to investigate some specific aspects of the classification. In particular, in reservoir computing performance evaluation, the ability of a network to recognize linear and strongly non-linear inputs alike is an important criterion; there is often considered to be a direct trade-off between non-linear mapping capability and raw memory capacity of the system [14]. The Legendre Polynomial set provides a perfect litmus test, since it combines both relatively simple functions at the lower orders with rather complex heavily-nonlinear ones at the higher orders. The first ten functions were chosen (Fig. 28.8).

We tested this training set on all the larger memristor model reservoirs from 1-Xbar listed in Fig. 28.6d (4 memristor links), up through 5-Xbar (60 memristor

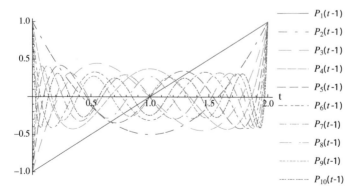

Fig. 28.8 The first ten Legendre Polynomials normalized over the time interval $0 \leq t \leq \tau = 2$ (in the unit of seconds)

Fig. 28.9 Representation of reservoir recognition performance with gradually increasing initial conditions. Panel **a** the case of the diamond network; **b** the 3-XBar; **c** the 5-Xbar. The *colors* and line styles match with Fig. 28.8. Five pools of initial conditions have been used, with increasing number of initial conditions S, as indicated in the legends. Each set is indexed by the ICS variable that marks the abscissa of the graphs

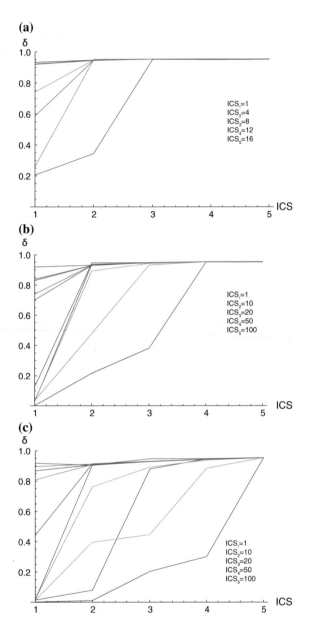

links). Every network is run as a reservoir being optimized to select every one of the Legendre polynomials: the task vectors f were chosen as $\{1, 0, 0, 0, 0, 0, 0, 0, 0, 0\}$, $\{0, 1, 0, 0, 0, 0, 0, 0, 0, 0\}$, and so on.

Results of the numerical analysis are shown in Fig. 28.9 for the diamond network (panel a), the 3-Xbar (b), and 5-Xbar (c). Each individual curve describes the recognition error of a *particular recognition task*. (The colors and line styles match the ones used in Fig. 28.8 to indicated which Legendre polynomial is being referred to.) The horizontal axis refers to various sets of initial conditions. These are indexed using a progressive indexing system, where the sets are sorted according to the number of initial conditions used. The set index *ICS* ranges from 1 (relatively small subset of initial conditions) to 5 (relatively high subset). Precisely, these indices represent $S = 1, 10, 20, 50, 100$ initial conditions for 2-Xbar up to 5-Xbar. Since the diamond network (1-Xbar) only has a total set of 16 such conditions, the indices in the diamond case represent $S = 1, 4, 8, 12, 16$.

The recognition error (δ) that is computed per initial condition, saturates towards an upper limit (less than one). This average recognition error does not scale up with the number of initial conditions. The error per initial condition is bound. It seems, however, that the initial condition-bound deterioration takes longer on the larger network- but *only for some tasks* (e.g. the solid line at the bottom of panel C, which corresponds to the first Legendre Polynomial). In fact, this first polynomial is recognized well relative to the other tasks in all three test networks.

28.5.2.4 Large Networks: Varying Turning Power

We investigated how a given network's recognition error varies over a spectrum of turning power values, as shown in Figs. 28.10 and 28.11. These graphs provide an *overall* snapshot of recognition performance for a given network (at a given functional target f). In each case (model network), we consider the same random subset of initial conditions to ensure that we can draw analogous conclusions about overall recognition performance.

For the diamond network we ran a series of super-simulations across the turning power spectrum, and with a range of increasing simulation times (number of iterations) to allow for more signal search at each slice (Fig. 28.10). In general, the turning power of the sampled signals quickly allows the network to begin to generate functions that are sufficiently strong to achieve a lock onto the signal. Roughly, around $TP = 20$ the network reaches a reduced-error regime with $\delta = 0.1 - 0.2$. Yet, there is lot of noise noise (variation), with many peaks of poor performers, and valleys of excellent performers that approach perfect recognition. As one might expect, the curves smooth out and become less noisy as the algorithm has more time to search the signal space properly. The black graph in the figure, with the largest time of 8 min, is a near average of all the others.

We also ran turning power gradient simulations for several other large networks from 2-XBar to 5-Xbar. The simulation results are displayed in Fig. 28.11. Again, the recognition error decreases with increasing turning power, as is expected. However,

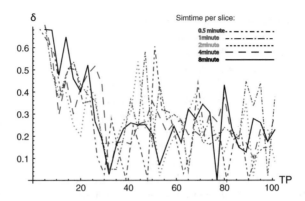

Fig. 28.10 The dependence of the recognition error on the turning power TP. Each line represents a different super-simulation of the diamond network. For every value of TP the pattern sampling algorithm has been run, with gradually increasing simulation time (and equivalently the number of iterations during the random search procedure). To make for an accurate benchmarking of δ, in all cases S = 8

Fig. 28.11 Same as in Fig. 28.10 but for different networks. Each line represents a separate memristor reservoir exposed to the iterative signal-searching algorithm under a gradually easing turning power constraint. In all cases S = 50

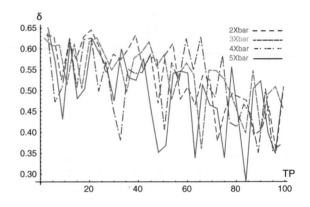

it does not reduce as observed in the nearly perfect recognition ($\delta = 0$) achieved with some of the smaller network model system cases and the diamond network. There seems to be a "floor" to recognition potential: $\delta > \delta_{min}$ with $\delta = 0.375$ for 2-Xbar, $\delta_{min} = 0.36$ (3-Xbar), 0.35 (4-Xbar), and 0.28 (5-Xbar). Most likely these values exist since only finite parts of the vast functional space of the input signals can be sampled.

28.6   Conclusion

We have suggested a formal approach to address an inverse pattern recognition problem: given a defined memristor network, which patterns can it recognize best? This is a special version of the more general implementation and the natural computability problems. As such, it should be easier to address.

A rigourous algorithmic approach has been formulated to solve the inverse pattern recognition problem. The problem has been solved in a rigorous mathematical way since the solution is presented as a minimization problem on the space of input patterns with appropriate constraints. The algorithm is generic and easily implementable in any computer language.

The algorithm works by sampling the infinite space of input signals and identifying the signals that can be classified well. The quality of the recognition is expressed as a recognition error δ where $\delta = 0$ describes a perfect recognition. Several key challenges have been addressed in the process. We believe that these are rather generic for similar types of problems. The greatest challenges are to (i) meaningfully sample the infinite configuration space of input signals, (ii) compare different input signals, and (iii) analyse the signals that the algorithm produces (signal semantics). To address these issues the turning power (TP) concept has been suggested, and formalized as a functional on the space of input signals: any input signal can be assigned a turning power value which can be used to rigorously assign a semantics for the signal.

The examples of single memristor response illustrates the remarkably complex concept of configuration space separation. By increasing the number of features n_{slices} progressively, more complicated functions (signals) are introduced into the total pool of base functions. For more complex signals the fading memory of the network may be overwhelmed if the network is too simple. Due to the bipolar properties of the signals at the end of their duration, classification was possible.

The analysis of the recognition of individual Legendre polynomials by larger networks provided two key results. As expected, recognition deteriorates as the total number of initial conditions increases. However, the recognition error saturates towards an upper limit (less than one). This implies that these networks, and likely memristor crossbars in general, have the potential to perform real-time computational tasks.

The analysis of how the recognition error varies with the turning power for larger networks revealed that larger crossbar networks with more nodes tend to perform better recognition. In particular, the (largest) 5-Xbar network seems to out-perform the others for some turning power values.

The results are encouraging, and our work can be improved in several ways. Larger sets of base signals should be explored to obtain better statistics to understand classes of task and background signals (rather than individual signals), though this is rather hard to do due to the standard computer hardware limitations. The sampling algorithm is rather elementary, genetic algorithms could be utilized. One future challenge is to gain enough understanding to generalize and expand the turning power concept. In particular, assigning some semantics to a spawned set of signals seems to be rather challenging. In this context, the turning power concept is a first attempt to formalize how a signal should be analyzed. In cases of off-line computing, it would be interesting to understand if and to what extent the choice of an initial condition can be used as a training parameter, and how the initial conditions can be selected by applying a particular set of input patterns. For on-line computation, which is very likely the most challenging, understanding of how to sample the initial conditions in the most efficient way is desired. These and other generic questions are left for future analysis.

Acknowledgments This work has been supported by the European Commission under contract FP7-FET-318597 SYMONE, and by Chalmers University of Technology.

Appendix

Parameters Used in Simulations

For the sake of completeness and reproducibility, the full set of parameters used in the simulations is given here. In all network examples each memristor was defined by the same set of parameters: $a = 1$ ohm volt^{-1} sec^{-1}, $\beta = 10$ ohm volt^{-1} sec^{-1}, $x_{min} = 1$, $x_{max} = 10$. The memristor threshold voltage V_{th}, and the so called training voltage V_{tr} are both in volts. The specific choice of parameters that are unique for each figure are listed below. The quantities specify time duration are always given in seconds unless specified otherwise. The turning power is always given in volts. The number of initial conditions used can be inferred from $S = n_{bits} + 1$:

- Figure 28.2: $\tau = 5$, $n_{bits} = 5$, $n_{slices} = 5$, $n_{steps} = 20$, $V_{values} = 5$, $V_{th} = 1$, $V_{tr} = 2.5$, $f = \{1, 0, 0\}$, TP $= 10$, simulation time (simtime) $= 1$ min. Iterations completed, 480, the best recognition error $\delta = \{0., 1.38 * 10^{-16}, 1.38 * 10^{-16}\}$.
- Figure 28.3: $\tau = 5$, $n_{bits} = 10$, $n_{slices} = 10$, $n_{steps} = 40$, $V_{values} = 5$, $V_{th} = 1$, $V_{tr} = 2.5$, $f = \{1, 0, 0\}$, TP $= 10$, simtime $= 1$ min. Iterations completed: 137, $\delta = \{4.3 * 10^{-16}, 2.2 * 10^{-16}, 2.2 * 10^{-16}\}$.
- Figure 28.4: $\tau = 5$, $n_{bits} = 20$, $n_{slices} = 20$, $n_{steps} = 80$, $V_{values} = 5$, $V_{th} = 1$, $V_{tr} = 2.5$, $f = \{1, 0, 0\}$, TP $= 10$, simtime $= 5$ min, $\delta = \{0., 2.64 * 10^{-16}, 2.64 * 10^{-16}\}$.
- Figure 28.6: $\tau = 5$, $n_{slices} = n_{steps} = 40$, $V_{values} = 5$, $V_{th} = 1$, $V_{tr} = 2.5$, $f = \{1, 0, 0\}$, TP $= 9$, simtime $= 15$ min, $n_{bits} = \{1, 2, 3, 4, 5\}$, progressively. Iterations per hardware vary from 1000 in one memristor, low nbits, to around 50 in diamond, high n_{bits}.
- Figure 28.9: $\tau = 2$, $n_{bits} = 1$, $n_{steps} = 20$. Please recall, this input is set and not searched so no further parameters required by pattern recognition algorithm; $V_{th} = 1$, $V_{tr} = 2.5$.
- Figure 28.10: $\tau = 5$, $n_{bits} = 1$, $f = \{1, 0\}$, $n_{slices} = 5$, $n_{steps} = 15$, $V_{values} = 5$, $V_{th} = 2$, $V_{tr} = 4$. Simtime varies 0.5, 1, 2, 4, 8 and for each iterations per slice are approximatively 20, 50, 80, 200, 500. Initial conditions for each set at $S = 8$.
- Figure 28.11: $\tau = 5$, $n_{bits} = 1$, $n_{slices} = n_{steps} = 10$, $f = \{1, 0\}$, $V_{th} = 2$, $V_{tr} = 4$, $V_{values} = 5$, simtime $= 2$ min. Initial conditions for each set at $S = 50$. Iterations vary from approx 150 per slice (2-XBar), to approx 20 per slice (5-Xbar).

References

1. Sienko, T., Adamatzky, A., Rambidi, N.G., Conrad, M.: Molecular Computing. MIT Press, Cambridge (2005)
2. Chalmers, D.J.: Does a rock implement every finite-state automaton? Synthese **108**, 309–333 (1996)
3. Zoran, K.: A perspective on Putnam's realizability theorem in the context of unconventional computation. Int J Unconv Comput **11**, 83–102 (2015)
4. Maass, W., Natschläger, T., Markram, H.: Real-time computing without stable states: a new framework for neural computation based on perturbations. Neural Comput **14**(11), 2531–2560 (2002)
5. Jaeger, H., Haas, H.: Harnessing nonlinearity: predicting chaotic systems and saving energy in wireless communication. Science **304**(5667), 78–80 (2004)
6. Kulkarni, M.S., Teuscher, C.: Memristor-based reservoir computing. In: Proceedings of the 2012 IEEE/ACM International Symposium on Nanoscale Architectures (NANOARCH), pp. 226–232 (2012)
7. Goudarzi, A., Stefanovic, D.: Towards a calculus of echo state networks. Procedia Comput. Sci. **41**, 176–181 (2014)
8. Forcada, M.L., Carrasco, R.C.: Learning the initial-state of a 2nd-order recurrent neural-network during regular-language inference. Neural Comput. **7**, 923–930 (1995)
9. Pershin, Y.V., Di Ventra, M.: Experimental demonstration of associative memory with memristive neural networks. Neural Netw. **23**(7), 881–886 (2010)
10. Pershin, Y.V., Di Ventra, M.: Solving mazes with memristors: a massively parallel approach. Phys. Rev. E **84**, 046703 (2011)
11. Zoran, K., Göran, W.: A generic simulator for large networks of memristive elements. Nanotechnology **24**, 384007 (2013)
12. Burger, J., Teuscher, C.: Variation-tolerant computing with memristive reservoirs. In: Proceedings of the 2013 IEEE/ACM International Symposium on Nanoscale Architectures (NANOARCH), pp. 1–6. IEEE (2013)
13. Milo, R., Shen-Orr, S., Itzkovitz, S., Kashtan, N., Chklovskii, D., Alon, U.: Network motifs: simple building blocks of complex networks. Science **298**(5594), 824–827 (2002)
14. Butcher, J.B., Verstraeten, D., Schrauwen, B., Day, C.R., Haycock, P.W.: Reservoir computing and extreme learning machines for non-linear time-series data analysis. Neural Netw. **38**, 76–89 (2013)

Chapter 29
Self-Awareness in Digital Systems: Augmenting Self-Modification with Introspection to Create Adaptive, Responsive Circuitry

Nicholas J. Macias and Lisa J.K. Durbeck

Abstract The question of augmenting self-modification with introspection to create flexible, responsive digital circuitry is discussed. A specific self-configurable architecture—the Cell Matrix—is introduced, and features that support introspection and self-modification are described. Specific circuits and mechanisms that utilize these features are discussed, and sample applications that make use of these capabilities are presented. Conclusions are presented, along with comments about future work.

29.1 Introduction

"Your visions will become clear only when you can look into your own heart. Who looks outside, dreams; who looks inside, awakes." [6]

The relevance of Dr. Jung's quote to the human experience seems clear: one must be self-aware to achieve their full potential to "awake" rather than merely to dream. But what might this mean in different contexts: say, in the context of human-made systems?

Self-awareness has been defined by Stephen Franzoi as "...a psychological state in which one takes oneself as an object of attention" [5]. Christopher Jamison further differentiates this from simple introspection, stating "Introspection is only looking at me, whereas self-awareness involves considering how I interact with the world around me" [7]. Under these definitions, it may be interesting to ask if human-made systems can be, in some sense, self-aware.

N.J. Macias (✉)
Clark College, Vancouver, WA, USA
e-mail: nmacias@clark.edu

L.J.K. Durbeck
Cell Matrix Corporation, Newport, VA, USA
e-mail: lisa@cellmatrix.com

© Springer International Publishing Switzerland 2017
A. Adamatzky (ed.), *Advances in Unconventional Computing*,
Emergence, Complexity and Computation 22,
DOI 10.1007/978-3-319-33924-5_29

Looking at digital electronics, one may ask if such circuitry has the necessary mechanisms for any form of self-awareness. In analyzing this question, it may be useful to classify digital circuits as falling into five different classes.

The first class (Class I systems) contains pure digital logic: collections of logic gates, wired together to perform certain operations. In this class, digital circuits can perform algorithmic computations: simple state machines exhibit this behavior.

Class II digital circuitry is circuitry whose structure can be changed. For example, fuse-programmed PLAs have a given initial configuration, but their behavior can be modified (once) by a programming operation that changes the connectivity of the internal elements [1].

The third class extends the notion of configurable hardware to *re-configurable* hardware: devices whose wiring can be modified multiple times. A field-programmable gate array (FPGA) exemplifies this class of hardware systems [8]. In this class, circuitry can be tailored to a particular situation. For example, in a software-defined radio (SDR) system, the hardware which encodes/decodes an audio signal according to one standard can be re-wired to change that coding as the device moves to a different geographic region (where a different coding standard is used) [3]. While Class III circuitry is in some ways more powerful than circuitry from Classes I and II, it requires an additional component, typically a CPU/memory system (e.g. a PC) to generate configuration information for modifying the device. This works well in many situations, but the PC which is generating the configuration strings is fundamentally separate from the hardware that's being modified (Fig. 29.1). In this sense, the PC is not "taking itself as an object of attention," nor is the FPGA. Rather, the PC is considering the FPGA. This misses Franzoi's criteria for self-awareness.

The fourth class of digital circuitry adds the ability of circuitry to *self-modify*. An example of this would be an FPGA containing a multitude of programmable planes, each configured for a different situation, and a small supervisor which chooses which plane to use based on sensory input. Continuing the example above, circuits for different types of cellphone codings could be stored in the device, and a monitor could analyze incoming signals, determining the coding, and selecting the proper

Fig. 29.1 FPGA being configured by an external device. Here the source and object of the configuration are fundamentally different from each other

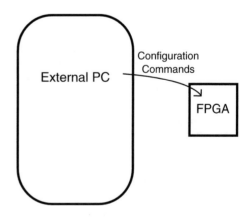

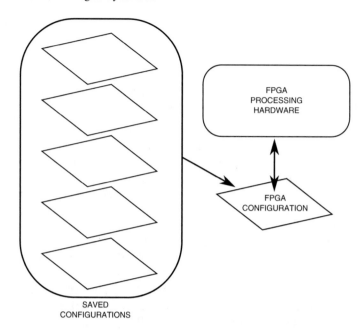

Fig. 29.2 A multi-plane FPGA. Several configurations can be developed and saved in a configuration memory, and then swapped into the actual configuration memory being used by the FPGA

hardware configuration plane (Fig. 29.2). In Class IV, the device is (in some sense) considering itself as an object of attention, since it effectively chooses how to change its digital circuitry based on the results of other pieces of digital circuitry inside itself. This situation is thus arguably one step closer to self-awareness. There is still a key element missing though: the ability to introspect, in the sense of analyzing oneself and making decisions that haven't been pre-wired. Choosing one of several pre-configured circuits based on which one of several possible results comes from an analysis algorithm doesn't "feel" the same as performing a general analysis and synthesizing new circuitry on-the-fly.

With Class V digital circuitry, introspection becomes a central element in the behavior of digital circuits. Circuitry processes information about its own configuration in the same way that it processes other data, and the results of that processing can be used to change or create new digital circuits as easily as changing an output from one value to another. Circuitry in this class is thus able to consider itself as an object of attention, including analyzing how its own circuitry is interacting with other connected circuits, and making decisions about changing its own configuration in response.

This paper discusses Class V circuitry, and describes specific examples of how this combination of introspection and self-modification fosters the implementation of unique and powerful circuits.

29.2 Background

The two key characteristics of Class V reconfigurable devices are:

1. the ability of circuits to *analyze* other circuits configured within the device; and
2. the ability of circuits to *modify* other circuits configured within the device.

These two features mean that the system can analyze and modify itself. The system used for this research is the Cell Matrix [14]. The Cell Matrix was developed in 1986, in part as a way to explore these two aspects of digital circuit design. One of the original goals was to build-in self-modification as an intrinsic piece of the architecture. By also allowing circuits to analyze other circuits, it was hoped that systems could be designed that would, for example, read from a library of sub-circuits and synthesize circuits elsewhere in the matrix by placing and connecting those sub-circuits. A second requirement for the system was that it should be extremely fine-grained: composed of a collection of simple *cells* that, while limited in their individual behavior, could be assembled into arbitrarily-complex circuits. In support of this idea, the third requirement of the system was a high degree of scalability: in particular, connecting two matrices together on their edges should result in a larger matrix, without requiring any changes to either of the pre-existing matrices.

Figure 29.3 shows a simplified view of a single cell within a Cell Matrix. This cell is a 4-sided device, with each side having a single input and a single output. The device is purely combinatorial: the values of the four outputs are determined completely by the contents of a 64-bit *truth table* memory, as shown in Table 29.1.

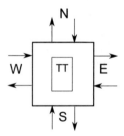

Fig. 29.3 Simplified view of a Cell Matrix cell. The four outputs are determined by the four inputs, which specify a single row in the 16×4 truth table ("TT")

Table 29.1 Truth table for a simplified 4-sided cell

N_{in}	S_{in}	W_{in}	E_{in}	N_{out}	S_{out}	W_{out}	E_{out}
0	0	0	0	D_3	D_2	D_1	D_0
0	0	0	1	D_7	D_6	D_5	D_4
0	0	1	0	D_{11}	D_{10}	D_9	D_8
0	0	1	1	D_{15}	D_{14}	D_{13}	D_{12}
...	...						
1	1	1	1	D_{63}	D_{62}	D_{61}	D_{60}

Each of the four outputs are completely determined by the combination of the four inputs

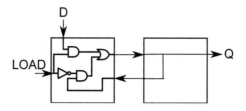

Fig. 29.4 D flip-flop implemented with two cells. The cell on the *left* echoes its eastern input (*LOAD* = 0) or copies its western input (*D*) to its eastern output (*LOAD* = 1). The eastern cell (on the *right*) re-circulates its western input to its western output, while also copying that input to its eastern output (*Q*)

By loading the appropriate pattern of 1s and 0s into the truth table memory, a cell can be configured to perform any four combinatorial functions of four inputs (i.e., wires, 1-bit adder, 2-1 selector, basic logic gates, etc.). The combination of 64 bits $(D_0, D_1, D_2, \ldots, D_{63})$ thus completely defines the behavior of a cell. Collections of cells are used to implement more complex function, including those with a sense of state. For example, Fig. 29.4 shows a D flip-flop implemented with two cells. When *LOAD* = 1 the D input is fed into the cell on the right, whereas when *LOAD* = 0 the output from the cell on the right is recirculated back by the cell on the left.

The cell shown in Fig. 29.3 does not include a mechanism for loading bits into the cell's truth table. The addition of a second set of lines—the C (or "control") lines—adds this capability. Figure 29.5 shows a complete cell with these extra lines included. Each side now has two inputs (C and D) and two outputs (C and D). The truth table doubles in size, as shown in Table 29.2.

Now, the combination of 128 bits $(D_0, D_1, D_2, \ldots, D_{127})$ completely specifies the behavior of the cell. While the cell has 8 outputs, those output values are still

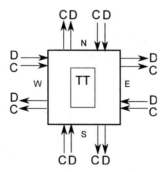

Fig. 29.5 A more-complete view of a Cell Matrix cell, showing both C and D inputs and outputs. The D inputs are used to select a row from the cell's 128-bit (16×8) truth table, in order to generate its 8 outputs. The C inputs are used to place the cell in configuration mode ("C mode"), wherein its truth table can be read or written by a neighbor

Table 29.2 Truth table for a complete 4-sided cell

DN_{in}	DS_{in}	DW_{in}	DE_{in}	CN_{out}	CS_{out}	CW_{out}	CE_{out}	DN_{out}	DS_{out}	DW_{out}	DE_{out}
0	0	0	0	D_7	D_6	D_5	D_4	D_3	D_2	D_1	D_0
0	0	0	1	D_{15}	D_{14}	D_{13}	D_{12}	D_{11}	D_{10}	D_9	D_8
0	0	1	0	D_{23}	D_{22}	D_{21}	D_{20}	D_{19}	D_{18}	D_{17}	D_{16}
0	0	1	1	D_{31}	D_{30}	D_{29}	D_{28}	D_{27}	D_{26}	D_{25}	D_{24}
...	...										
1	1	1	1	D_{127}	D_{126}	D_{125}	D_{124}	D_{123}	D_{122}	D_{121}	D_{120}

Each of the eight outputs are completely determined by the combination of the four inputs

selected based on only 4 inputs (the D inputs); the C inputs have a special function, and are not involved in the selection of a truth table row.

Instead of being inputs into the truth table evaluation, the C inputs are used to control the writing and reading of a cell's truth table, as follows:

- If all of a cell's C inputs are 0 then the cell is said to be in D (or "data") mode. In this mode, the cell's outputs are derived from the truth table, using a row determined by the cell's 4 D inputs.
- If any of a cell's C inputs are 1 then the cell is said to be in C (or "control") mode. In this mode, the cell's outputs on each side depend on whether or not that side's C input is 1 or 0:

  - if the C input is 0 on a given side (called an "inactive side"), then that side's D output is 0, and its D input is ignored;
  - if the C input is 1 on a given side (called an "active side"), then that side's D output is determined by a bit of the cell's current truth table. Additionally, that side's D input is OR'd with the D inputs on all other active sides, and the resulting bit will replace the truth table bit that's being presented on the D output(s).

In the case of two adjacent cells ("SOURCE" and "TARGET," as shown in Fig. 29.6), SOURCE can interrogate TARGET's truth table as follows:

1. SOURCE asserts its own C output to its east, which asserts TARGET's C input from its west;

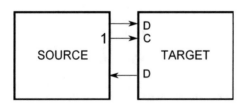

Fig. 29.6 An example of cell analysis. The cell on the *left* ("SOURCE") is examining the truth table of the cell on the *right* ("TARGET") by controlling TARGET's C input and monitoring TARGET's D output

2. TARGET will present a bit of its truth table to its D output on its west; and
3. SOURCE reads its D input from the east, which contains that truth table bit.

Cells map inputs to outputs asynchronously: that is, as soon as a cell's inputs change, its outputs will change according to the cell's truth table, without regard to any sort of synchronizing clock. In contrast, there is a system-wide clock used for C mode operations. Specifically:

- when a cell enters C mode, an internal (per cell) bit counter is reset to 0;
- the truth table bit indexed by this counter is presented to the appropriate D output(s);
- on the rising edge of the system clock, the appropriate D inputs are OR'd, and the resulting bit is saved internally;
- on the falling edge of the system clock, the saved bit it loaded into the truth table at the position indexed by the bit counter; the bit counter is incremented; and the new bit now indexed by the bit counter is sent to the appropriate D output(s).

This process continues as long as the cell remains in C mode.

In this way, the SOURCE cell in Fig. 29.6 can completely read the TARGET cell's truth table by simply asserting its own C output, and repeatedly reading its D input each time the system clock ticks low. Assuming it outputs 0's on its D output, the TARGET cell's truth table will be filled with all 0's after 128 clock ticks. Figure 29.7 shows a slightly different configuration by which the SOURCE cell can non-destructively read the TARGET cell's truth table. Figure 29.8 shows a further modification, in which a third cell ("DEST") is also placed in C mode, and configured to be a clone of the SOURCE cell.

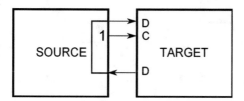

Fig. 29.7 Non-destructive read of a cell. The SOURCE cell re-circulates the truth table as it's read from the TARGET cell, so that TARGET's truth table is unchanged by the reading

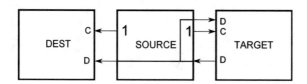

Fig. 29.8 Non-destructive cell replication. SOURCE reads TARGET's truth table, and simultaneously refreshes TARGET's truth table while also configuring DEST's truth table to be a copy of TARGET's. This effectively makes DEST a clone of TARGET

To create a Cell Matrix, cells are tiled in a 2-dimensional array, with each cell connected to four adjacent cells (one on each side), as shown in Fig. 29.9. This allows each cell to be potentially interrogated and/or modified by four neighbors; and also allows each cell to potentially interrogate and/or modify those same neighbors. It's important to note that *this is the full extent of the direct control any cell has over other cells within the matrix*. Of course, to be useful, cells must be able to interact with more than just a few immediately-adjacent cells. This takes place via intermediary cells. For example, one cell X may communicate directly with a neighboring cell Y, and may configure Y to allow X to control one of Y's neighbors Z (Fig. 29.10). This type of multi-cell control is typical of circuits built on the Cell Matrix. Several specific examples are presented in the next section.

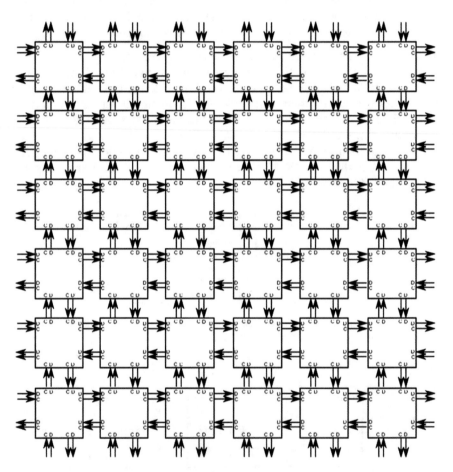

Fig. 29.9 A two-dimensional 6 × 6 Cell Matrix. Each cell communicates only with its immediate neighbors. Edge cells are accessible from outside the matrix

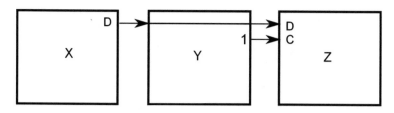

Fig. 29.10 Non-adjacent cell configuration. Cell X is using its neighbor (Y) to configure non-adjacent cell Z

The limitation of nearest-neighbor communication is a key benefit to the architecture, since it allows the system to scale nicely. Connecting two matrices along their edges results in a larger matrix. This has important implications for manufacturing large Cell Matrices, especially with techniques that employ self-assembly [15, 16].

While a 2D matrix is more-easily explained and analyzed than a 3D one, the architecture itself is agnostic to dimensionality and interconnection topology. One can also implement, for example, a 2D matrix with 3-sided cells, or a 3D matrix with 4-sided cells. The most common 3D implementation is using 6-sided cells (cubes) [11], though research is ongoing using 8- and 14-sided cells [18]. Higher dimensionalities are also possible, though scalability of the matrix is generally impaired at higher dimensions, as the matrix generally can't be extended without modifying the interconnection network in already-assembled sub-matrices.

29.3 Approach

In this section, several examples are presented of Cell Matrix circuitry that can be used for introspection and modification of digital circuits.

29.3.1 Multi-channel Wires

The first example of a modification circuit is a *multi-channel wire* [10, 14]. This is a circuit that can be used to establish and exercise control over a remote area of cells. Figure 29.11 shows a wire with three channels:

- The Program Channel ("PC"), which is used to control a target cell ("*")'s D input;
- the Control Channel ("CC"), which is used to control a target cell's C input; and
- the Break Channel ("BRK"), which is used to carry additional C or D information.

Each channel consists of a number of single cells. Figure 29.12 shows the details of the cells comprising the PC and CC channels. As can be seen, sending a 1 or 0 down PC causes the same value to be sent to the target cell's D input. The PC is

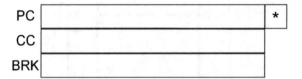

Fig. 29.11 A three-channel wire, used for controlled access to a remote cell ("*"). Each channel is a 1 × n collection of n single cells. The PC is used to control *'s D input, and CC is used to control *'s C input. BRK is an auxiliary line, generally used to break or retract the wire

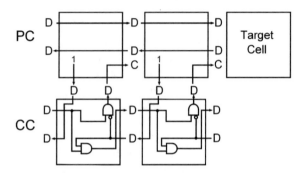

Fig. 29.12 Details of the cells comprising the PC and CC. A wire of length 2 is shown. The PC always transmits data to the D input of the Target Cell. The CC transmits a signal from cell to cell; at the end of the channel, the CC also transmits a 1 to the PC head cell's southern D input, which the PC head cell transmits to the target cell's C input

usually bi-directional, so that the target cell's D output is delivered through the PC to its own western output.

Similarly, sending a 1 or 0 down the CC drives the target cell's C input. The operation of this channel is more complicated than that of the PC. The CC employs a feedback signal to determine where the head of the wire is. Each PC/CC pair works together to route a signal to the previous pair's CC cell. Any CC cell receiving this signal simply routes its western D input to its eastern D output; but in the absence of this signal, the CC cell routes its western D input to its *northern* D output, which the PC then uses to drive the target cell's western C input. The final effect of this mechanism is that in a line of PC/CC cells (running west-to-east), the easternmost pair acts differently from the other cells, working together to control the target cell's mode. It's assumed here that the cell to the south of the target cell is empty (or at least is sending a 0 to its western D output). If that is not the case, then this cell must be *pre-cleared* before extending the wire.

By building a chain of PC and CC cells, placed side-by-side as in Fig. 29.12 (which shows a wire of length 2), one set of cells can thus control a non-adjacent cell's C and D inputs and read its D output. This allows circuity in one region of the matrix to interrogate and modify remote cells, provided a multi-channel wire exists between the source and the target cell. This leaves the question: how does one construct a multi-channel wire?

(a)   **(b)**   **(c)**

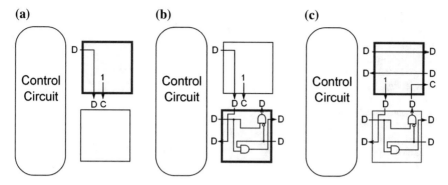

Fig. 29.13 Construction of an initial PC/CC pair. In each sub-figure, the cell being configured is indicated in bold. First **a** the control circuit configures the northern cell so that it can be used in **b** to configure the southern cell as a CC cell. In **c** the northern cell is reconfigured, this time as a PC cell. After these steps, a wire of length 1 has been built

The answer is that **a multi-channel wire can be used to construct a multi-channel wire**. By beginning with access to a single pair of cells, one can build an initial PC/CC pair in three steps, as shown in Fig. 29.13:

1. in (a), the top cell is configured so as to allow configuration of the lower cell;
2. in (b), the lower cell is configured as a CC; and
3. in (c), the top cell is configured as a PC.

Note that only access to the northernmost cell's C and D inputs is required. Since that cell is used to configure the cell to its south, no direct access is required to that southern cell.

This is called a "wire-building sequence," or simply a "sequence." A sequence is generally any collection of configuration operations, typically where certain cells are alternately configured and then used to configure other cells. Later, sequences of sequences ("super-sequences") will be used for more complex configuration operations, such as configuring 2D or 3D regions of cells.

Given an initial wire of length 1, the above steps can be repeated, since the control circuit *now has access to the non-adjacent target cell* (Fig. 29.14). Once access is available to that new target cell, it can of course be used to configure yet another PC/CC pair, and so on. In this way, wires of arbitrary length can be built, thus allowing access to remote regions of the matrix. Note that the key to performing this wire extension is the feedback path in the CC cell. Following the third step in the above sequence, the PC sends a 1 into the CC, which then feeds it back to the previous stage's CC, which thus ceases acting as the final cell ("head cell") in the wire, and simply routes its western D input to its eastern D output. This is the basic mechanism by which wire extension is possible.

Figure 29.15 shows a circuit for turning a wire. In this case, a wire that was extending from west to east is now turned to the south. This 2 × 2 circuit works effectively the same as the simple PC/CC pair, but the feedback path is modified

Fig. 29.14 Once a wire of length 1 has been constructed, the Control Circuit can use it to configure a non-adjacent Target Cell. By repeating the steps shown in Fig. 29.13, the Control Circuit can extend the wire to a length of 2. This extension can be repeated as many times as needed to make wires of arbitrary length

Fig. 29.15 Circuit for turning a 2-channel wire. The PC and CC control the Target Cell. The wire can now continue to extend to the south

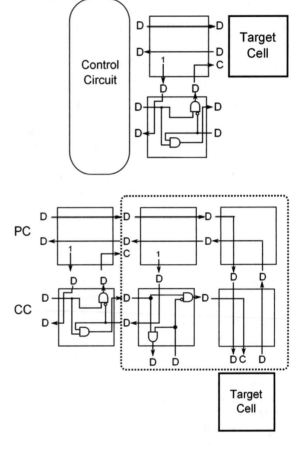

slightly, so as to allow the feedback 1 signal to be presented following the final configuration step. Constructing a corner is a 12-step process: details can be found in [10, 14].

The use of a BRK line is sometimes useful for interrupting a wire, as shown in Fig. 29.16. In this example, a wire of length 4 has been constructed, granting access to the target cell marked "*". However, two cells of the BRK line have also been constructed adjacent to the 2-channel wire. The leftmost cell simply routes D from west to east; but the second cell of the BRK line routes its western D input to its northern C output. Thus, if a 1 is sent down the BRK line, it will clear the second cell of the control channel. The effect of this is to interrupt the feedback from the second PC cell, thus causing the first CC cell to become the new head CC cell (which in turn causes the first PC cell to become a head cell as well). This effectively causes the cell marked "+" to become the new target cell of the wire. This mechanism can thus be used to restore control to a prior location after a wire has been extended. This may be used, for example, in a bootstrap super-sequence [11].

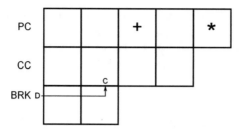

Fig. 29.16 Example of a break line. The PC/CC pair initially give access to the target cell ("*"). By asserting the BRK line, the second CC cell can be cleared, changing the location of the target cell to the cell marked "+"

Fig. 29.17 A breakable wire. The BRK channel is dynamic: whatever its length, the rightmost cell will route its western D input to its northern C output, thus clearing the CC cell above it. This changes the wire's target cell to be the cell directly above the end of the BRK line. The target cell can then be configured to clear the last cell on the BRK line, thus retracting it

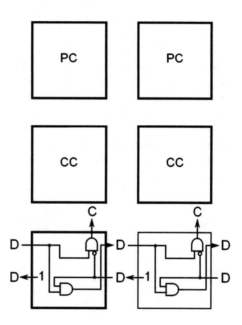

In some cases, rather than simply breaking a wire, it may be desirable to back a wire up one step. Figure 29.17 shows a *retractable wire*: a circuit to allow a wire to be retracted, one cell at a time. This is useful for synthesizing a circuit in reverse: extending a wire, configuring the target cell, then backing up one step and configuring the new target left in the wire's wake, and so on.

The key to a retractable wire is the dynamic BRK line. Instead of a line of cells that simply route data from one side to the other, each cell of this BRK line actively determines whether it is the last cell in the channel: if not, it simply routes data from west to east; but if it sits at the end of the wire—i.e., if there is no BRK cell to its right—then it routes data from the west into the C output to the north.

The wire is extended in the usual way: the target cell (in front of the easternmost PC cell) is configured to build a new BRK cell, then a new CC cell, and finally a new PC cell. This allows the wire to be extended as far as desired. To back the wire up, the BRK signal is activated to clear the head CC cell, which moves the target cell back one spot. The new target is then configured to clear the current head BRK cell. This allows the target to be configured, but leaves the BRK line ready to clear the previous stage of the wire, thus allowing further retraction of the wire.

29.3.2 Multi-headed Wires

The above circuits are used to configure a single target cell at a time. However, modified versions allow parallel configuration of multiple target cells, as shown in Fig. 29.18. In this figure the PC and CC inputs drive the D and C inputs, respectively, on a series of target cells (marked "*"). Such a wire (with multiple heads) is called a "Medusa Wire," and allows multiple targets to be configured simultaneously. Of course, since there is only a single PC/CC pair delivering information to each target cell, all target cells will be configured identically. However, *identical circuitry built near each target cell can be used to effect non-identical synthesis* by using techniques such as local indexing (discussed at the end of this section).

A Medusa wire is not easily extended in the same manner as a single-headed multi-channel wire. However, a multi-channel wire *below* the Medusa wire can be used to synthesize the Medusa. And once built, the Medusa wire allows configuration of a line of identical circuits. Since the Medusa wire itself is a line of identical circuits, it follows that *the Medusa wire can be used to build another Medusa wire* (directly below the original). Similarly, that Medusa can be used to build a 3rd copy, and so on (as shown in Fig. 29.19). A column of additional control circuitry is also built

Fig. 29.18 A multi-headed Medusa wire. The PC and CC control access to the D and C inputs of multiple target cells (each labeled "*")

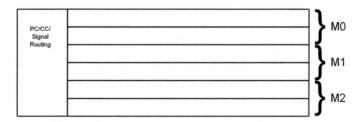

Fig. 29.19 Medusa wires built from Medusa wires. In this example, $M0$ was built first (sequentially), and then used to build $M1$ *in parallel*. $M1$ was then used to build $M2$ in parallel. This process can be repeated for more-efficient configuration of a region of cells

along the left edge of the assembly, to route the PC/CC signals to all Medusa wires. While the first wire—assuming it contains n heads—takes $O(n)$ steps to build, the second wire takes $O(1)$, as does the third, fourth, and so on. This allows n^2 heads to be configured in $O(n)$ steps.

By themselves, these heads have limited purpose if they're only used to build more Medusa wires. In practice, this 2D plane of Medusa heads might be used to configure a 2D plane of target cells above the Medusa plane (up the Z axis in a 3D Cell Matrix, for example). This utilizes a modified Medusa head circuit, which can be switched between multiple operating modes.

Alternatively, a scheme such as shown in Fig. 29.20 can be used. In (a), a single-headed multi-channel wire has been built, and as it was extended, it was used to synthesize the pieces of a Medusa wire ("MED0"). MED0 is a multi-headed wire that configures regions to its north or south, depending on the value of a direction signal. In (b), MED0 is used to synthesize a 1D collection of 2D target circuits to its

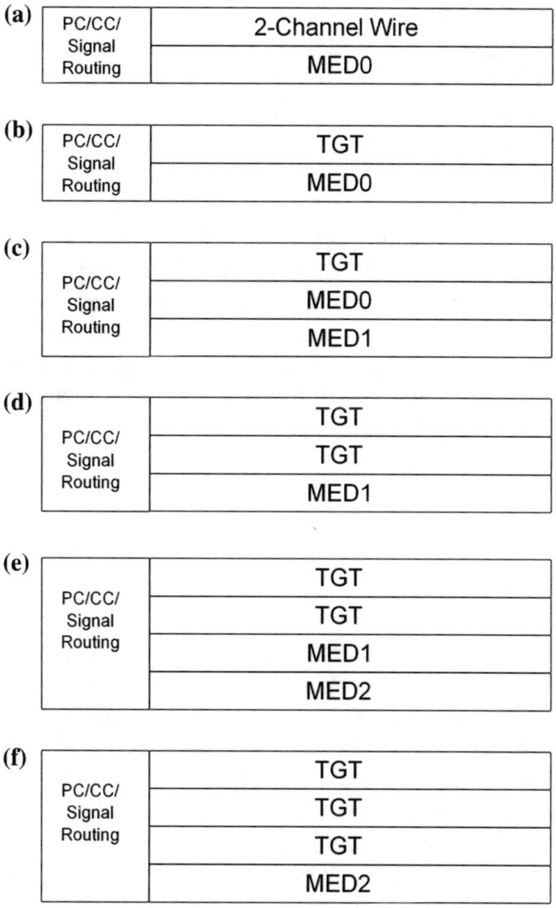

Fig. 29.20 Parallel build of target circuits and new Medusa wires using Medusa wires: **a** a standard two-channel wire has been used to build Medusa wire MED0; **b** MED0 configures the region to the north into a row of desired target circuits; **c** MED0 configures the region to the south into a new Medusa wire MED1; **d** MED1 configured a second target region; **e** a third Medusa (MED2) is configured; and **f** MED2 configures a third target region

north (for example, pieces of a finite-element analysis system [19]). In (c) MED0 is directed south to configure a new Medusa wire ("MED1") to its south. In (d), MED1 configures regions to its north, creating another row of target circuits. In (e), MED1 configures a new wire "MED2" to its south, and in (f) MED2 configures a third row of target circuits to its north. This process repeats, configuring (in parallel) new rows of target cells, leading again to an efficient initialization of n^2 regions of target cells in $O(n)$ time steps. For simplicity, the target regions are shown as being thin, but in practice taller regions can be configured by placing the Medusa wires further to the south.

29.3.3 Analysis of Cells: Intrinsic Operations

The above techniques and circuits are useful for synthesis or modification of circuits within the Cell Matrix. Another useful task is the *analysis* of circuits or states within the Cell Matrix. The simplest example of this is responding to outputs from neighboring cells. Since each cell's outputs are directly connected to the inputs of any neighboring cells, each cell automatically responds to its neighbors outputs according to its own truth table configuration.

A more interesting example is a cell that reads another cell's truth table, by asserting the target cell's C input and reading that cell's D output. This is also a basic cell-level operation within the matrix. Once read, a cell's truth table can be stored (perhaps in another cell's truth table memory, or in a larger structure composed of many cells acting in concert as, say, a shift register). Truth tables that have been read can be compared to templates (useful for reverse-engineering dynamically-built circuits); analyzed for patterns (to examine the results of, say, random generation of configuration strings); rotated (to build a new circuit that faces a different direction); combined with other truth tables (to breed two configuration strings in a genetic algorithm [17]); and so on.

29.3.4 Circuits and Techniques for Testing Single Cells

Moving away from intrinsic cell-level operations, one can create test circuits and test patterns. Figure 29.21 shows a simple arrangement whereby one cell—called the source ("SRC")—may test a target cell ("*"). A basic test is to load the truth table for the equation *DataWestOut* ← *WestIn* (abbreviated $DW = W$) into the target cell, and then examine the target's D output as the SRC sends 0's and 1's into the target. The expected behavior is that the target echoes back whatever value the SRC sends. This can detect errors where the target's DW_{out} is stuck-at-1 or stuck-at-0 (or where there's a problem configuring the target at all).

Next, the target cell be configured with the equation $DW = !W$ (e.g. an inverter), and the above test repeated. This test can detect shorts between the target's DW_{in} and

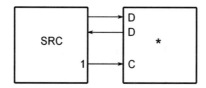

Fig. 29.21 Simple cell-testing arrangement. "SRC" is able to load test patterns into "*," send values to *'s D input, and monitor its response on *'s D output

DW_{out}. Additional tests can be used to check other aspects of the target's D mode behavior (though each of these tests is also to some degree checking the target's configurability).

C mode can be more-thoroughly tested by loading different bit patterns into the target's truth table memory, and immediately reading back those patterns (without letting the target leave C mode). If the test pattern reads correctly, then the SRC knows that the target's memory can successfully store the given test pattern (and can also successfully enter C mode, index bits in the truth table, and so on). By using different patterns (all 0's; all 1's; alternating 1's and 0's; alternating groups of 1's and 0's, arranged so that each physical row of the truth table's memory stores complementary values; and so on), different possible failure modes can be tested.

As an example, consider Fig. 29.22 which shows a potential layout for the 128 bits of a truth table inside a cell. Here, the memory is physically arranged as a 16 × 8 array of single-bit memories. For this example, assume there is a defect involving the 6th cell on the 3rd and 4th rows, whereby their outputs are shorted together. This defect will cause those cells to report the same value as each other. If we attempt to load the memory with the pattern

11111111100000000011111**1**1100000**0**0001111111100000000011111111100000000

the boldfaced bits will be read-back as the same value as each other, which will indicate to the testing circuit that there is a problem in the memory storage subsystem.

29.3.5   Testing Regions of Cells

After exercising a cell's capabilities *sufficiently-enough* to determine that the cell is usable, that cell can then be used to test additional cells. This is a process similar to bootstrapping a 2D region of cells: first one cell is tested; then a second cell; then these cells are configured as a wire, and used to test a 3rd and 4th cell, which are used to extend the wire, and so on. As more cells are tested (and found to be good), more control can be exercised over the region being tested [4].

In any defect-testing system, there is always a concern that the test circuitry itself may be defective. In the case of an introspective, self-modifying system such as the Cell Matrix, these concerns are largely obviated, since the test circuitry is not a

Fig. 29.22 Example of
memory defects inside a
16 × 8 truth table. Here,
there is a short between the
3rd and 4th rows, in the 6th
cell on each row. This can't
be detected by loading all 0's
or all 1's, but would show up
when loading a repeating
pattern of 8 0's followed by
8 1's

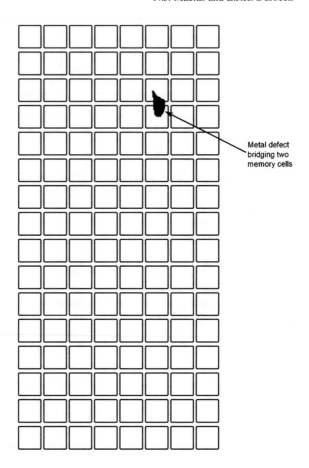

Metal defect
bridging two
memory cells

pre-existing, hardwired system, but rather is constructed, cell by cell, *as each cell is
subjected to fault testing and found to be working correctly*. Thus, if a cell is found
to be defective, it is simply not used in constructing the test circuit (of course, there
are extreme edge cases, such as if every cell along the outer perimeter of the matrix
is defective, in which case there is no way to reach any internal cells). If a wire is
being built, and encounters a defective cell, the wire can be backed-up two steps, and
a corner built to move down past the defective cell, and so on.

Testing can also be done from different sides, as shown in Fig. 29.23. Since wires
can be bent and turned, a cell being tested can be approached from different sides,
to allow testing of the I/O lines on all sides of a cell.

Note that it is generally not feasible to test a system for *every* possible failure
mode. For example, there are 2^{128} possible ways to arrange a 4-sided cell's memory,
and numerous possible present states inside a cell's circuitry. While these are too
numerous for exhaustive testing, it may be possible to test many of the states in
which a cell is expected to actually operate. For example, due to the layout of cells

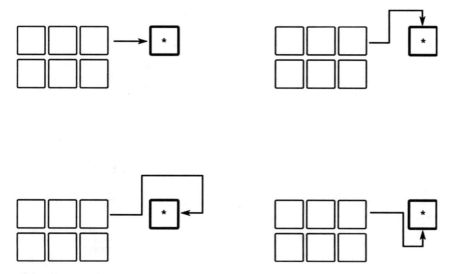

Fig. 29.23 Interrogating a cell from different sides. By making a series of turns, a multi-channel wire can access a target cell ("*") from different sides. This provides a greater testing capability, since I/O lines on each side can be examined

within a circuit, it may not be possible to interrogate the cell from (say) its eastern edge. In that case however, it's likely that the circuit that's being built may not actually use the cell's eastern edge, in which case an I/O error associated with that side may not be relevant to the expected operation of the circuit being constructed.

29.3.6 Isolating Defects

The above techniques are primarily concerned with **fault detection**: identifying when one or more cells are defective and may potentially impair the behavior of a circuit built with those cells. An additional consideration arises given the possibility that a defective region of cells could inadvertently affect nearby cells in an undesirable way. Since cells can configure other cells, and cause those cells to configure still other cells, it's theoretically possible (though unlikely under normal circumstances) that a defective cell could alter the configuration of nearby (or even remote) non-defective cells. To prevent such a situation, one can construct circuits (*from non-defective cells*) that isolate defect-free regions of cells from regions containing defects. One such circuit is called a "Guard Cell." A guard cell employs a second cell, as shown in Fig. 29.24. The middle cell is the guard cell, but it is controlled by the cell on its left, which is simply placing the guard cell into C mode. Under the Cell Matrix architecture, when a cell is in C mode, it is unable to assert any of its C outputs. Since the guard cell is built from a non-defective cell, the guard cell is guaranteed to

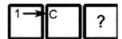

Fig. 29.24 Example of a *guard cell*. The two cells on the *left* work together to prevent the *rightmost* cell ("?") from being able to affect the cells on its *left*. Since the *middle* cell is always in C mode, it can never assert its own C outputs, and thus can't be used by cell "?" to configure other cells

Fig. 29.25 A line of guard cells acts collectively as a *guard wall*, which protects one region of the matrix (to the *left* of the wall shown here) from the behavior of another region (immediately to the *right* of the wall shown here)

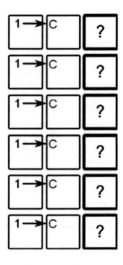

be outputting $C = 0$ on all sides. Thus it is impossible for the defective cell (marked "?") to use the guard cell to configure other cells.

A line of guard cells constitutes a "Guard Wall" (Fig. 29.25); and a closed perimeter of guard cells (Fig. 29.26) creates a "Guard Ring" that isolates a region of cells from the rest of the matrix.

29.3.7 Auto-Location Within the Matrix

Another use of introspection is in determining ones location within a collection of identical cells. Since parallel-configuration techniques such as Medusa wires lead to the creation of a number of identical sub-circuits, these circuits don't intrinsically have any identifying information that can be used to self-identify as being distinct from other circuits. Such self-identification is an important first step in subsequent differentiation of the matrix into heterogeneous components.

Figure 29.27 shows a line of identical circuits, each containing an increment unit. Each circuit receives—either in parallel or as a serial stream—an integer from the circuit on the left; adds one to it; and sends the incremented value to the circuit on its right. Assuming the leftmost circuit is receiving all 0's from its left (either because it is adjacent to a region of unconfigured cells, or because it is at the edge of the

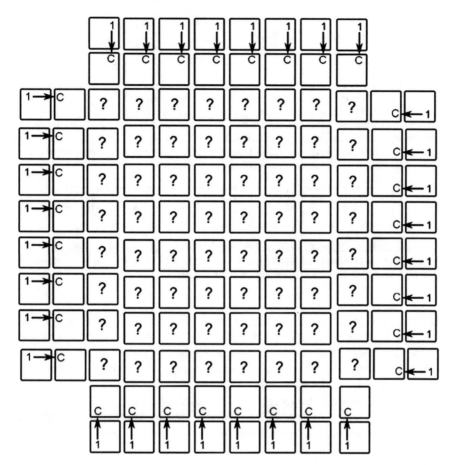

Fig. 29.26 A closed ring of guard cells creates a *guard ring*, which isolates the behavior of cells (labeled "?" here) inside the ring from the rest of the matrix

matrix), that circuit's X value will be 0, while the circuit to its right will see $X = 1$; the next circuit to the right will see $X = 2$; and so on. This generated value of X effectively tells each circuit what column it sits in. Similar circuitry can be used in the Y direction to allow cells to determine which row they occupy in a 2D collection of identical circuits. Given such positional information, circuits can then differentiate into pieces of a larger target circuit according to the shared circuit map. The circuits can also dynamically determine an optimal configuration for a target circuit based on available resources [13].

The above scheme can be simplified to let cells determine whether or not they are adjacent to an edge of the matrix. Circuits can read from (as in Fig. 29.28 for example) the north, and output a signal (1) to the south. If placed top-to-bottom, each circuit will read a 1 from the north, except the circuit placed at the very top of the matrix. Repeating this in different orientations, circuits can in this way identify

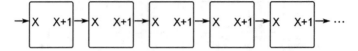

Fig. 29.27 A line of identical circuits that can locate themselves within the matrix. Each circuit (which is a collection of cells operating as a single unit) receives an integer from the *left*, increments it, and outputs the incremented value to the *right*. The leftmost circuit will be the only one receiving a 0 (assuming it's either at the edge of the matrix, or is adjacent to a region of cells that is not participating in this behavior). Every circuit will receive a unique value of X from its neighbor (provided enough bits are used in transmitting the integer). This allows circuits to identify themselves with a unique index relative to all other circuits in the collection

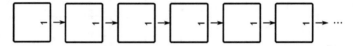

Fig. 29.28 Detecting the northern edge of the matrix. Even though each cell is configured identically, only one cell (the northernmost one) will receive a 0 on its northern input. This allows cells to identify when they are located at the top edge of the matrix

the boundaries of the matrix. A different scheme is to simply build a multi-channel wire, constructing a simple feedback test circuit in front of the wire as it extends (Fig. 29.29). While this arrangement is normally used for fault detection, it will also indicate where the end of the matrix is, since attempts to configure a feedback cell ("F") will fail beyond the edge of the matrix. By repeating this analysis on multiple rows, one may determine the location (and shape) of the matrix's edge. Note that this edge may be the physical edge of the matrix, but if a region of cells are defective, this will define the *effective* edge of the matrix.

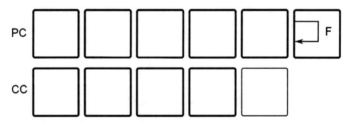

Fig. 29.29 Another example of edge detection in the matrix. A two-channel wire is grown to the right. At each step, a feedback cell ("F") is configured; a test ping is sent; and an echo is detected. The echo detection will fail when the wire has reached the rightmost edge of the matrix (since there is no cell available for building the feedback cell). This allows a circuit to determine the horizontal width of the matrix. A similar technique can be used to determine the matrix's height

29.4 Applications

The above sections have covered some basic underlying mechanisms that support self-analysis and self-modification. This section presents examples of digital circuits that utilize introspection to drive self-modification. These specific applications themselves are not necessarily "killer apps" for this technology. The goal is to present a set of examples illustrating ways in which this technology can be used to achieve unique, potentially interesting behaviors.

Four examples are presented:

1. An overflow-proof counter. This illustrates the use of self-modification to handle specific, expected conditions.
2. An autonomous circuit-scrubbing system. This system employs extensive self-modification, but also utilizes self-analysis of circuit configurations, to develop a system that can **repair** soft errors.
3. A system for parallel synthesis of circuits, including detection and avoidance of defective regions. This illustrates the use of introspection and self-modification to **work around** hard errors.
4. A system for detection of mis-oriented cells in a Cell Matrix: a system that **adapts-to and works-with hard errors**.

29.4.1 An Overflow-Proof Counter

Figure 29.30 shows a simple ripple-carry counter based on negative edge-triggered toggle flip flops. The counter is driven by the clock input on the right, and as each stage flips from 1 to 0, it toggles the next stage to its left. The configuration shown is a 3-bit counter, which counts from 000 to 111 before overflowing and returning to 000.

Figure 29.31 shows the same circuit extended by adding a fourth flip flop to the left. Adding this additional stage changes the circuit to a 4-bit counter. Such an extension requires no additional changes to the pre-existing circuit; the change is made only on the leftmost edge, by feeding the current counter's MSB into the new stage's clock input.

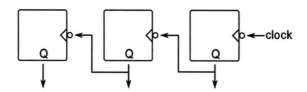

Fig. 29.30 Basic 3-bit ripple-carry counter. Each T flip flop toggles when the prior bit changes from 1 to 0. After 8 ticks, this counter overflows from 111 to 000

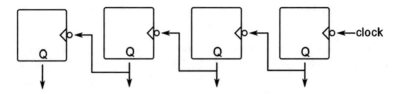

Fig. 29.31 By simply adding a 4th flip flop to the *left* of the MSB, the counter is now extended to a 4-bit counter, and can count to 1111 before overflowing to 0000

When implemented on a Cell Matrix, a toggle flip flop requires only three cells, using the configuration show in Fig. 29.32. The middle cell includes a pair of 1-bit multiplexers which select either their 0 or 1 input (based on the value of the S input).

The top and bottom cells provide a feedback path, while the middle cell receives the incoming clock and alternates the bit that's circulated among the cells. The bottom cell delivers the flip flop's Q output, which is also sent from the middle cell's left edge. This allows a new stage to be added by simply placing it adjacent to the prior rightmost stage (assuming those cells aren't already being used for some other purpose).

Fig. 29.32 3-cell implementation of a T flip flop. The clock is applied to the middle cell's eastern D input. The flip flop's output (Q) is sent to the bottom cell's southern D output. It's also copied to the middle cell's western D output, where it can be applied to the clock of another T flip flop positioned directly to its left. The "S" blocks in the middle cell are one-bit multiplexers

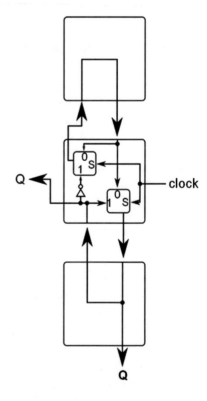

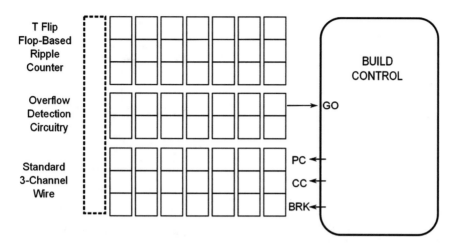

Fig. 29.33 An expanding counter. In addition to an *n*-stage ripple counter (comprised of *n* T flip flops), there is a set of circuitry located below each flip flop. This circuitry determines if the two MSBs of the count are 11, and reports that fact to the main control circuit (via Build Control's "GO" input); it also routes build commands from Build Control to the edge of the flip flop collection, allowing Build Control to construct a new flip flop (in the *dashed* region). This allows the counter to grow in size dynamically, as necessary to prevent overflow

Figure 29.33 shows a toggle flip flop-based counter with additional hardware to allow it to automatically build more stages as the count increases. Below each flip flop is an *overflow-detection* circuit with three functions:

1. it contains an edge detection circuit to determine if it is located below either of the two leftmost flip flops;
2. it contains simple logic to determine—if it is located below either of the two leftmost flip flops—whether each of those flip flops is currently outputting 1; and
3. it contains a piece of a two-channel wire which can be used to configure cells to the left of the leftmost flip flop.

Figure 29.34 shows the details of this overflow-detection circuit.

The logic in step 2 above is fed through the collection of circuits to the state machine located to the right of the flip flop assembly. When the *GO* input is triggered (raised from 0 to 1), the state machine begins generating supersequences. Those are sent down the two-channel wire (function 3 described above), and cause the synthesis of a new flip flop and overflow-detection circuit to the left of the current assembly. Once the flip flop has been built, the wire itself is reconfigured to extend it another stage and re-build a new corner at the leftmost edge, thus preparing for the next future build operation.

Since the build is triggered when the leftmost two bits are both 1, it happens when the counter has reached approximately 75 % of its maximum value. As long as the build can be completed in less time than it takes the counter to count the remaining 25 % of its maximum value, the build will finish before the counter overflows, i.e., a

Fig. 29.34 Overflow-
detection circuitry. The
lower cell determines if it is
at the leftmost-edge of the
collection, and outputs a 1 to
the north if so. The upper
cell uses this 1 signal, along
with the counter bit from
above and from the right, to
determine if an overflow is
going to occur soon. The
resulting "OFLOW" signal is
routed from west to east
through these same cells (via
the OR gate). This upper cell
also routes the current stage's
data to the western cell

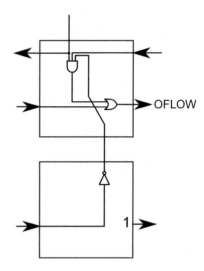

new flip flop will be present before it needs to operate in order to maintain the correct count. This circuit thus implements an *overflow-proof* counter.

Of course there are some details to be noted here. The first is that this is a ripple counter, and may not be suitable for counting to extremely large values. The design, however, can be easily modified to create a synchronous output. Likewise, preset and clear inputs can be incorporated, as well as other embellishments; the basic design of the expanding element remains essentially the same.

This scheme assumes there are available cells on the left of the counter. One may ask, if there are free cells there, why not just pre-configure them to be counters, rather than waiting until they are needed? The answer is that one may not know ahead of time how large a counter is needed. Until those extra cells on the left are used for the counter, they are available for use by other circuits. This raises another question: what happens if the circuit tries to expand, but some other circuit is using the needed cells? The answer is that something must be relocated: either the circuit currently using the needed cells, or the counter itself. This may seem a daunting managerial task, but it is in many ways no different from how a typical memory management system works in a modern operating system. This raises the possibility of implementing a *hardware management system*, perhaps incorporating constructs similar to virtual memory, paging, swapping, and so on.

This technique thus makes good use of self-modification to respond to detected conditions in the environment (in this case, the approach of an overflow condition). In this case, "introspection" is limited to analysis of the system's state. The next example introduces a circuit that uses self-analysis of its own memory configuration.

29.4.2 Triple Redundancy with Autonomous Defect Scrubbing[1]

A simple defect detection and correction technique is to use *triple redundancy* [9], where three independent copies of a digital circuit operate simultaneously. Assuming defects are relatively rare, i.e., if there is a high likelihood that only one circuit will fail at a time, one can compare the states of the three copies, and in the case of a discrepancy, the majority state is taken to be correct. While relatively simple to implement, a triple-redundant system is only useful for detecting a single defect. Once one of the three copies is corrupt, the system now has only two pristine copies, which does not allow determination of a majority state if one of those remaining circuits becomes corrupt.

Some errors are transient—for example, a momentary voltage spike in the output of a logic gate—and will disappear shortly after they appear. Their *effect*, however, may be longer-lasting, particularly if they occur in part of a memory circuit. If they occur in the configuration memory of a reconfigurable device, their effect may become permanent (until the system is reconfigured). While reconfigurable systems are potentially well-suited to using triple redundancy (since, by employing a sufficiently-large device, there may be enough space to make three copies of the target circuit, and to incorporate the necessary mechanisms for voting), they suffer from this particular vulnerability in the event of a *configuration memory upset* [2].

On a substrate that allows analysis of configured circuitry, one can implement a circuit that does more than simply compare three copies of a circuit. One can actually compare three copies *of the circuit's configuration memory*, and, upon determining that there is a bit error somewhere, *correct that error* to restore the circuit to pristine condition.

Figure 29.35 shows a system that uses these concepts to mitigate memory upset-related faults. $C1$ $C2$ and $C3$ are three copies of the target circuit. Each block labeled "Wire Head" contains the circuitry for a small multi-channel wire, which can be used to build longer multi-channel wires. The control circuit ("CTRL") directs these wire-building operations, extending these wires across the top of each circuit ($C1$–$C3$). As the wires extend, the configuration memory of the cells along the top of each circuit is read and conveyed to the control circuit. The control circuit performs a bit-by-bit comparison of the three configuration strings, checking to see whether all three strings agree. In the case of a discrepancy, the majority bit value is taken to be correct. As the wires extend across the circuits, the control circuit stores these (corrected) configuration strings in the block labeled "Storage," until the wires reach the rightmost edge of the circuits. At this point, the storage circuit contains a (presumably) pristine copy of the configuration memory for the top row of the circuits. Next the wires are de-constructed (with the end of the wire moving from the

[1]This work was supported by DOE/LANL under subcontract 90843-001-04 with the Regents of the University of California.

Fig. 29.35 Overview of triple-redundancy system. The CTRL circuit guides the construction of three wires which scan three copies (C1–C3) of the target circuit. By comparing the truth tables in each copy's cells, temporary upsets in the truth table memories can be detected and corrected by the CTRL circuit

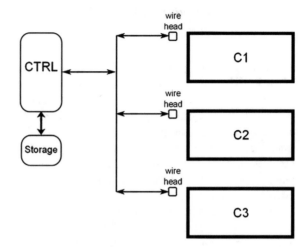

rightmost edge back to the initial position at the Wire Head), and the stored configuration information is used to re-build perfect circuitry in its wake.

Following the above operations, the controller builds wire from each wire head to the south, turns counter-clockwise 90 degrees, and begins to move across the next rows of circuits $C1–C3$, again reading, voting, saving, and finally re-building their configuration memories. This process is repeated, until the entire circuits have been "scrubbed" using this technique. In the end, any upsets in the configuration memory of a circuit should have been corrected.

Figures 29.36a–d illustrate the first row of this build process. The process illustrated in these figures is repeated in all three circuits $C1–C3$ (from Fig. 29.35). In Fig. 29.36a, the circuit is intact, and the three cells marked "*" are about to be examined via a set of three three-channel wires. In Fig. 29.36b three cells in the upper-left of the circuit have been read, and are now a part of the extended wire. In Fig. 29.36c, the wire has been extended most of the way across the top of the circuit. Figure 29.36d shows the state of the circuit after the entirety of the top three rows have been read. Those rows are now completely overwritten with the wire itself, but the configuration of the original cells are stored in the repair circuit's storage elements ("Storage" in Fig. 29.35). Moreover, if there were errors in any of the configuration memories that were read, they should have been corrected before those configurations were stored.

In Fig. 29.37a, the wire has been retracted one step to the left, and the configurations of the original cells have been repaired in the wire's wake. In (b), more of the top rows have been repaired, as the wire continues to retract. Finally, in (c), the entire top three rows have been restored to their intended configuration. Moreover, since this operation occurs in parallel in each of the three circuits $C1$ $C2$ and $C3$, all three copies are repaired simultaneously.

In Fig. 29.38, the wire head has moved down three cells, and is ready to repeat the above steps on the next three rows of the circuit. This repeats until the entire 2D circuit has been repaired.

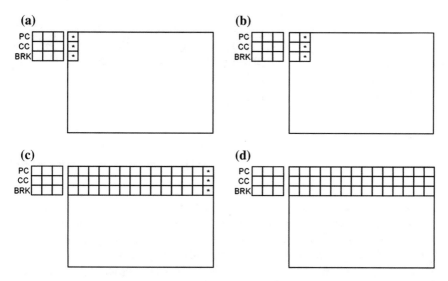

Fig. 29.36 Initial steps of cell repair. In figure **a** the cells marked "*" are about to be read. In **b** the original cells have been replaced with the wire, which is now ready to read the new cells marked "*". If **c** most of the top three rows have been read and replaced with wire. In **d** the entire top three rows have been read and replaced with wire. The overwritten cells' definitions are stored in the block marked "Storage" in Fig. 29.35

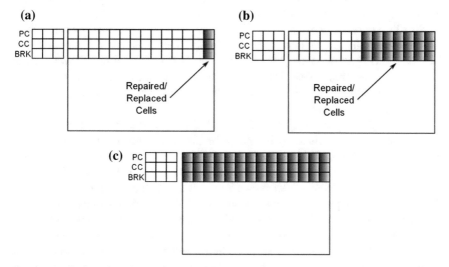

Fig. 29.37 Retraction of the wire. In figure **a** the *shaded* cells have been replaced with the original (if undamaged) or repaired versions of the original cells. In **b** the wire has been retracted further, with more of the original circuit replaced/repaired. In **c** the entire top 3 rows have been repaired/replaced

Fig. 29.38 After the top row has been replaced/repaired, the initial wire moves south 3 spots, and the above process repeats. This continues until the entire 2D region has been replaced/repaired

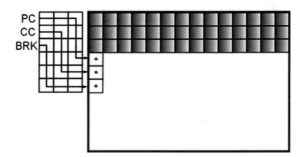

The above process works fine, provided that there is not an upset *in the same bit position* of the configuration memory for two or more cells *in the same position of the circuit.* For a 4-sided cell, each cell's configuration memory has 128 bits. Figure 29.39 shows a condition under which this scrubbing process will *not* work. In the indicated cell in circuit $C1$, one particular bit has an upset. If the same bit in the same cell in, say, circuit $C2$ also has an upset, then the majority vote on that bit will fail. Provided that condition does not occur, the scrubbing process will be able to restore the correct configuration.

In any system where circuitry is used to test or repair other circuitry, one must deal with the possibility that the test circuit itself may incur a defect. This present strategy, while not immune from this consideration, presents a relatively small opportunity for such critical defects: the wire heads, control and storage circuits must be defect-free. Of these, only the storage circuits have a size that increases with target circuit size: however, this is an order $O(\sqrt{n})$ situation: for an $n \times n$ target circuit, we need to protect a storage area whose size is $O(n)$ (since we are only processing one row of the target circuits at a time). For an extension of this technique to 3D circuits, the critical region's size is $O(\sqrt[3]{n})$.

One remaining issue is that while the scrubbing operation is underway, all 3 circuits are unusable. This may be acceptable, if the circuits are used infrequently, and the scrubbing is used to keep them clean in-between runs. An alternative approach is to keep 6 copies of the circuit, in two sets (A and B) of 3, and to alternate between a live circuit and a circuit-under-repair. A and B are each a complete fault-tolerant triple-redundant implementation of the target circuit, complete with its own copy of this scrubbing circuitry. While copy A is being used as a live circuit (with one of its three circuits pre-chosen to be the actual live circuit), copy B is being scrubbed. Once copy B is clean, it can become the live circuit, and copy A can be scrubbed. This alternation continues as long as the circuits are being used, and scrubbing occurs whether or not there are any errors present.

This technique thus combines self-analysis and self-modification to provide a reliable method of detecting and correcting defects in the configuration memories of a target circuit's cells. By continually detecting and repairing errors, the configuration memories are maintained in pristine condition; and by keeping them in this condition, they continue to be usable for detecting and repairing errors.

Fig. 29.39 A condition from which the system cannot recover. If a defect occurs in exactly the same bit of the same cell in two of the three copies of the circuit, the scrubbing mechanism will incorrectly modify the third copy to also be defective

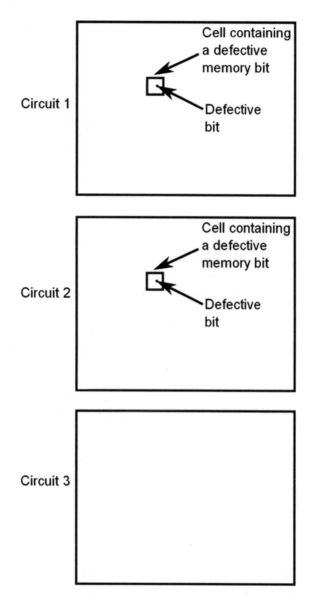

29.4.3 Fault-Tolerant Detection and Isolation of Defects[2]

The Cell Matrix does not include any pre-existing circuitry for configuring a region of cells with a given set of configurations. While this may seem like a disadvantage of the architecture, its advantage over a hardwired configuration mechanism is that it

---

[2]This work was supported by NASA under contract NAS2-01049.

Fig. 29.40 Configuration of
a single cell with a
multi-channel wire

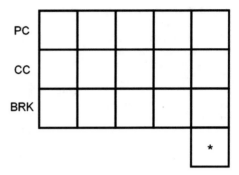

offers a great deal of flexibility in how the system is configured. This section explores
two particular ways in which this flexibility can be used:

1. custom bootstrap circuits can be built for performing massively-parallel configurations; and
2. regions can be tested for defects before being configured.

Recall that a cell's configuration can be written by any adjacent cell (if that
adjacent cell asserts one of the target cell's C inputs). Figure 29.40 shows a simple
configuration example. "*" indicates the target cell which is being configured. This
circuit employs a multi-channel wire: PC is used to send data to the target cell's
D input; CC is used to assert the target cell's C input; and BRK is used to convey
a secondary data signal. Circuitry located near the beginning of the wire (the end
labeled "PC" "CC" and "BRK") is responsible for generating bitstreams into the PC
and CC inputs so-as to configure the target cell appropriately.

Using this setup, the target cell can be configured to configure other cells in the
vicinity, which allows the wire to be extended, bent, retracted, broken, and generally
manipulated so as to allow the configuration of a **region** of cells near the initial target.
Let us call the time required to configure a particular region of cells τ.

Figure 29.41 shows a similar setup, except that the channel now has two heads,
and thus configures two target cells in parallel. In the setup shown, since the heads
are situated 5 cells apart from each other, it's assumed a region no wider than 5 cells
is being configured below each head. Note that aside from the extra propagation
delay (arising from the longer path to the second target), **configuring two regions
like this takes the same time as configuring one region**.

Figure 29.42 shows a further-modified setup (for simplicity, the multi-channel
wire is now represented with a single line). In this case, there are multiple target
cells (each at the end of an arrowhead). If there are n such heads, then n regions will
be configured in parallel. While the configuration of these targets is thus relatively
efficient ($O(1)$ for configuring n regions), one must consider the cost of building the
wire itself, which is $O(n)$. Thus, to configure n regions is still an $O(n)$ process.

However, once an entire row of targets can be configured in parallel, it's possible
to configure n regions in time τ. Those configurations can be used to create a new

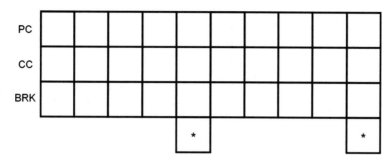

Fig. 29.41 A two-headed wire. The two target cells labeled "*" can each be configured at the same time

Fig. 29.42 A multi-headed wire. Each target cell "*" is configured at the same time

multi-headed wire below the initial wire, which can then be used to configure another n regions in time τ. Thus, in time τ we can configure one region; after $n\tau$ we can configure n regions; after $(n + 1)\tau$ we can configure $2n$ regions, and so on. After $(2n - 1)\tau$ we will have configured n^2 regions (Fig. 29.43).

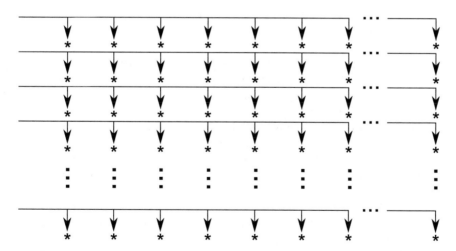

Fig. 29.43 Once an *initial row* has been configured (sequentially), it can be used to configure a *second row in parallel*. That row can configure a *3rd row*, and so on, with each row being configured in a fixed amount of time (independent of the length of the row). The entire region can thus be configured in $O(n^2)$ time

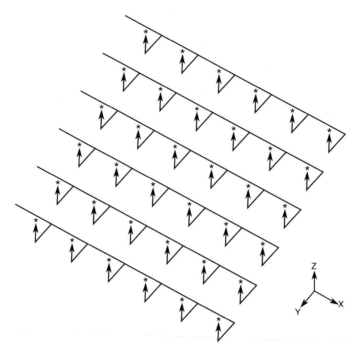

Fig. 29.44 Three-dimensional Medusa circuit. After configuring a plane of identical circuits, those circuits can configure a second plane in a fixed amount of time (independent of the dimensions of that plane). The entire 3D region can thus be configured in $O(n^3)$ time

This process can be further extended to support parallel 3D bootstrapping. If each configured region is built to allow configuration of cells in the Z axis, the circuit now allows configuration of n^2 regions in one additional timestep τ (Fig. 29.44). Target cells are again denoted with "*".

After another timestep τ, this newly-configured collection of n^2 circuits can be replicated again in the Z axis, creating another n^2 regions. Continuing for another $O(n)$ steps, and entire 3D collection of n^3 regions can be configured in $(3n - 2)\tau$ timesteps. Configuring n^3 regions thus takes a total time of $O(n)$ steps. Table 29.3 summarizes these results.

This is one example of a "*Medusa System.*" In general, a Medusa system is one that employs multiple heads to configure multiple regions in parallel, particularly with the goal of accelerating the configuration as it runs. The above example is a first introduction to a Medusa system, but a modified algorithm is actually more useful. By utilizing a different (but still accelerating) configuration order, we can incorporate fault detection and isolation into a parallel bootstrap system. Figure 29.45 shows the first 5 steps of this modified Medusa scheme. In this (and the next two figures), each square represents a collection of cells corresponding to a multi-cell region to be configured. At the end of the entire build process, all regions will be configured identically.

Table 29.3 Number of regions configured versus total configuration time

Number of Configured Regions	Total Time
1	τ
2	2τ
3	3τ
…	…
n	$n\tau$
$2n$	$(n+1)\tau$
$3n$	$(n+2)\tau$
$4n$	$(n+3)\tau$
…	…
n^2	$(2n-1)\tau$
$2n^2$	$(2n)\tau$
$3n^2$	$(2n+1)\tau$
…	…
n^3	$(3n-2)\tau$

The initial configurations proceed with linear time complexity, whereas the middle set of configurations accelerates to $O(\sqrt{n})$. The final set of configurations is $O(\sqrt[3]{n})$, which is also the time complexity for the entire process

The build begins with the configuration of a single region. The next 4 steps proceed as follows:

1. the first region configures a second region to its east;
2. those two regions configure two regions to their south (resulting in 4 configured regions);
3. the two westernmost regions configure two regions to their west (resulting in 6 configured regions); and
4. the three northernmost regions configure three regions to their north (resulting in 9 configured regions).

Continuing this way, after each pair of successive builds, the number of configured regions grows from k^2 to $(k+1)^2$. This is thus also an $O(\sqrt{n})$ process (and can again be extended into 3D for an $O(\sqrt[3]{n})$ process). Whereas the prior Medusa system grows first in one dimension, and then in the second, this system grows symmetrically from the middle, extending outward evenly in all directions. This is essential for incorporating fault handling into the system.

To detect faults, test patterns are applied in the region(s) which are being configured. Figure 29.46 shows a sequence of steps for parallel, fault-tolerant configuration of a 3×3 collection of multi-cellular regions. Steps which don't result in any new configurations (i.e., testing regions to the west from a westernmost cell) are not shown. In Fig. 29.46, the regions at the end of the arrowheads are being tested. Assuming the tested regions are found to be operating correctly, they are configured

Fig. 29.45 First 5 steps of
an increasingly-parallel build
system. In each step, the
arrow(s) show where the next
build will occur. Note that all
steps require the same
amount of time to complete

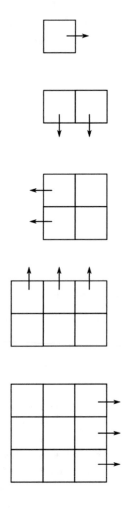

as new additions to the Medusa system, and are subsequently used for further parallel test/configuration operations.

Figure 29.47 shows how this process runs in the presence of a fault. Here, a 3 × 3 layout of multi-cell regions is shown. The shaded regions have already been tested and configured; arrows indicate which adjacent regions are being tested (and, if found to be defect-free, configured). The region marked with a "*" contains a defect in at least one of its cells. Dark lines surrounding that defective region indicate *guard walls*, which are circuits within the adjacent defect-free regions that act to isolate the defective region from non-defective regions.

The following description concentrates on the detection and isolation of defective regions. To make use of this process for the construction of useful target circuitry, the target circuit is either built from elements contained within the testing circuitry

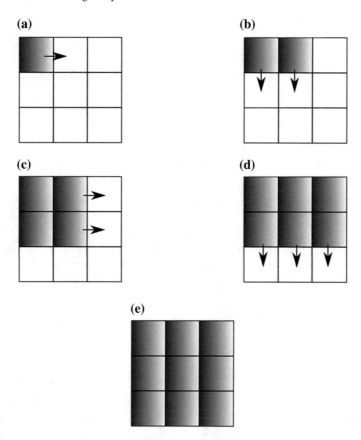

Fig. 29.46 Sequence of steps for parallel fault detection. Each square is a multi-cellular region. *Shaded* areas represent already-tested-and-configured regions. *Arrows* point to regions being tested and configured. In this example, there are no defects present in the region

itself, or a post-testing synthesis step reconfigures the collection of regions, using the test results to avoid defective cells.

The process begins with one initial region (in the upper-left corner) being configured. That region then tests the region of cells to its east ((a) in Fig. 29.47). Upon finding that region defect-free, it is configured as another piece of the collective test-circuit. Those two regions then test—in parallel—the regions to their south as shown in (b). The region on the left passes the tests, but the region on the right contains a defect, and fails one or more of the tests. A guard wall is activated in response to this, as illustrated by the heavy line to the north of the defective region in (c). The next tests—to the west and north—are not shown, as there are no cells to test in those directions.

In (c), non-defective regions now test to the east, resulting in the discovery of one good region (in the upper-right of the 3 × 3 area), as well as re-discovery of the defective region in the middle, causing a guard wall to the west of that region

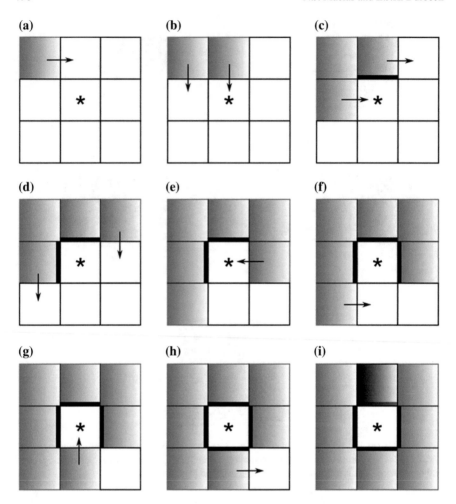

Fig. 29.47 Sequence of steps for parallel fault detection and isolation. Each square is a multi-cellular region. *Shaded* regions are already tested and configured. *Arrows* indicate which regions are being tested and configured. The region marked "*" contains a defect, which the system attempts to detect and isolate. *Bold* lines indicate guard walls, which work to isolate the defective region. Steps **a–i** show how the algorithm proceeds

to be activated (the heavy line in (d)). In (d), testing occurs to the south, resulting in configuration of two additional defect-free regions (seen in (e)). Note that the middle region of the upper row will attempt to re-test the defective region, but the guard-wall blocks all signals from entering or leaving that region, and thus those tests fail quickly.

In step (e), testing to the west *does* take place, with the only effect being the activation of a guard wall to the east of the defective region. The results of this are seen in (f), where the defective region is now surrounded on three sides by guard walls. (f) also shows testing to the east, resulting in discovery of a 7th defect-free region, which in (g) will test to the north, and thus activate the final guard-wall around the defective region as shown in (h). Finally, one more test to the east leads to the final configuration shown in (i). As can be seen, the 8 defect-free regions have been identified and used in the establishment of a guard-wall surrounding the defective region in the middle.

As described above, this process as shown is only testing the 3×3 collection of regions, but in practice this is a preamble to subsequent synthesis of a parallel target circuit using these pre-tested regions. Further steps could then be used if differentiation of the regions is desired, as described in [12].

29.4.4 Detection of and Adaption to Cell-Level Orientation in a Disoriented Matrix[3]

One area of research for the physical assembly of a Cell Matrix is to employ *self-assembly* of individual cells into a larger 2D or 3D structure [15]. This raises a potential issue of *cellular disorientation*:

- each cell (in the 2D/4-sided case) has a notion of north, south, east and west;
- circuits are made from collections of cells communicating with each other;
- this means each cell needs to know how it is oriented relative to other cells.

For example, Fig. 29.48 shows a set of 4 cells being used to make a short wire. Each cell passes input from its west to its east, thereby making a 4-cell wire. In this case, each cell is oriented normally. In Fig. 29.49, the four cells are oriented randomly. While each has been configured to act as a wire from west to east, the cells are unable to act together to pass information from one cell to another. Given this orientation, three of the cells would need to be programmed differently, as shown in Fig. 29.50. The necessary equations are shown below each cell.

As described earlier, determining a cell's orientation is relatively easy. First, program the cell with a feedback circuit that copies west to west; then send a 1 out your own eastern output, and listen for an echo. If you receive it, then the target cell is oriented normally; if not, configure it to copy north to north, and try again. Knowing the configuration that results in an echo tells you the cell's orientation. Provided a cell's orientation can be determined, it is then easy to modify the cell's configuration string to restore the intended behavior. Thus, via introspection and appropriate modification of configuration instructions, a disoriented cell can be *effectively* re-oriented, so that it can be used as if it were oriented normally.

Of course, this only works for a target cell *directly adjacent* to a correctly-functioning set of cells. However (as usual), if an adjacent target cell can be effectively re-oriented, then it can be used to re-orient non-adjacent cells, by building and using

[3]This work was supported by the Cross-Disciplinary Semiconductor Research (CSR) Program award G15173 from the Semiconductor Research Corporation (SRC).

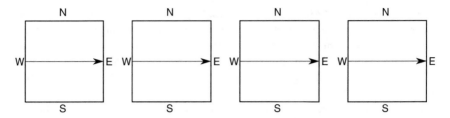

Fig. 29.48 Four cells, each configured to pass a single bit of data from west to east

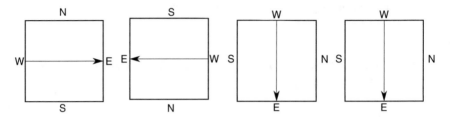

Fig. 29.49 Four cells, each configured to pass a single bit of data from west to east. In this case, the cells are disoriented, so the structure is not able to pass data from west to east

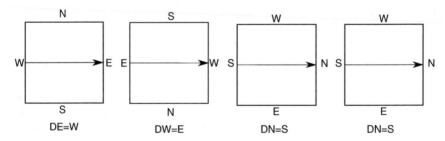

Fig. 29.50 By taking into account the disorientation of the cells, they can be configured differently to restore the data-passing capability of this circuit. Each cell's configuration is listed below the cell

multi-channel wires. The controlling circuit needs to be more complex, to allow for generation of test sequences and rotation of desired truth tables to match true orientation. Aside from that change, the build process for configuring a region of cells is essentially unchanged.

The situation is different though for parallel configuration. Consider the situation in Fig. 29.51. Here, a Medusa wire is setup to configure n cells (T_0 through T_{n-1}) in parallel. But unless all of these cells have the same orientation, there is no single configuration string that can be sent through the wire to configure all cells with the same effective configuration. For example, if T_0 is rotated 90 degrees clockwise and T_1 is rotated 90 degrees counter-clockwise, then the configuration strings sent into their respective D inputs must be adjusted accordingly. Similarly, to determine the orientation of a cell, test patterns must be fed to the cell and its behavior studied.

Fig. 29.51 A Medusa wire being used to configure *n* cells in parallel. To configure all target cells with the same *behavior*, different configurations need to be loaded into each cell based on its orientation

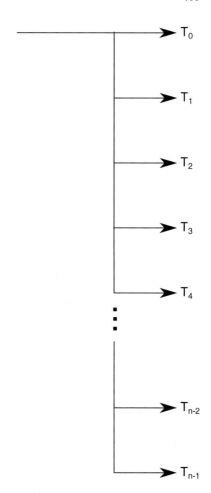

Since different cells may have different responses, each must be studied individually; yet to maintain parallelism, a set of cells should be tested in parallel. How can these seemingly conflicting requirements be reconciled?

The answer, again, is that the mechanisms of introspection and modification are purely local to each cell, meaning that the circuits to determine and respond to each target cell's orientation can be built near each target cell $T_0, T_1, \ldots, T_{n-1}$. This allows a universal set of instructions to be sent through the Medusa wire to each head, but also allows each head to act differently based on its local observations.

To accomplish this, each head is augmented with additional control circuitry as shown in Fig. 29.52. Three main channels are transmitted:

1. the PC, which delivers D inputs for configuring the target cell, as well as data to be used in testing a cell's orientation. The PC also returns data output from the target cell;

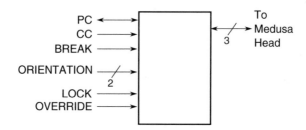

Fig. 29.52 Additional circuitry configured at each head of the Medusa wire. PC, CC and BREAK are the usual channels of the 3-channel wire. ORIENTATION carries two bits that code the current orientation being tested; LOCK is used to record the current value on the ORIENTATION lines; and OVERRIDE causes the CC to be asserted regardless of the ORIENTATION lines' values

2. the CC, which delivers the C input to control the target cell's mode (C or D); and
3. the BREAK line, which is an auxiliary data line that can be used to interrupt a wire and return it to a previous/shorter state.

In addition to these channels, there are three extra sets of lines sent to the control circuit:

4. an ORIENTATION wire, which carries two bits of information representing the current orientation being tested or configured;
5. a LOCK signal used to indicate that a test pattern is being delivered; and
6. an OVERRIDE signal, used to set the target's C input regardless of the state of the ORIENTATION lines.

Figure 29.53 shows how these lines work together to solve the simultaneous reorientation problem for multiple heads. Asserting the OVERRIDE line causes the CC line to be passed to the target cell's C input, thus allowing the target to be configured regardless of its orientation. This is used to load a feedback test pattern (e.g. $DW = W$) into the target cell. Next, the ORIENTATION lines are set to 00 (indicating a rotation of 0 degrees from "normal" orientation), and a 1 is then sent through the PC channel. While maintaining those signals, the LOCK line is momentarily pulsed high. Assuming the target cell is oriented normally, the two D flip flops will load the current orientation (00); otherwise, the flip flops' values will remain unchanged.

By cycling through all four possible test patterns ($DW = W$, $DN = N$, $DE = E$ and $DS = S$), the flip flops will be loaded with the orientation that resulted in feedback detection, and thus indicate the orientation of the target cell. This completes the introspection phase of the system. Next, this information is used to configure the target cell with the desired configuration. This is accomplished by asserting the CC line and sending the desired configuration four times: once for each possible orientation. As each configuration is sent, the ORIENTATION lines are set to indicate the orientation corresponding to the current configuration being sent. In the unique case where the ORIENTATION lines match the saved orientation, the flip flops' MATCH output will be asserted, which will allow the CC signal to drive the target cell's C input.

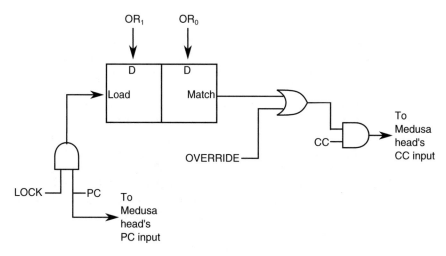

Fig. 29.53 Circuitry for simultaneous determination and correction of head cell orientation. LOCK is used to load the 2-bit orientation latch with the current orientation (OR_1 and OR_0) provided PC is receiving a 1 (echo) from the target cell. If the orientation matches the latched value, the MATCH output is asserted, which causes CC to be routed to the Medusa head. The OVERRIDE signal also causes CC to be routed to the Medusa head

Since the control circuitry in Fig. 29.53 is local to each head of the Medusa wire, the saved orientation information is correct for each target cell, and after sending the four possible configuration strings, each target cell will have been configured as desired. While this now takes $4\times$ longer than configuring a normally-oriented matrix, the process is still $O(\sqrt{n})$. The $4\times$ factor can be eliminated (in exchange for a $4\times$ increase in space utilization) by making the Medusa Wire's PC channel four times as wide, so that the configuration bitstream for all four possible orientations are sent in parallel. The control circuit can be modified to act as a 4-1 selector to route one of those four PC sub-channels into the target cell's D input.

Thus, by combining analysis and self-modification with local hardware to preserve and utilize local information, a high degree of parallelism can be maintained despite the need to perform different operations at each Medusa head.

29.5 Conclusions and Future Work

The ideas of introspection and self-modification have been discussed, with respect to a particular self-configurable architecture called the Cell Matrix. Native mechanisms for support of introspection and self-modification have been described, and these have been used to develop circuits and methodologies for various low-level behaviors, including wire building, parallel configuration, defect detection and isolation, orientation-determination, and self-location within the matrix. These have been used as building blocks to create more-complex systems that utilize introspection and self-modification in a variety of ways. The simplest example demonstrates circuit synthesis in response to system state; whereas the most complex uses parallel differentiation in response to local introspection to tailor circuit synthesis as a way to work with disoriented cells.

The techniques and examples presented herein are starting points for more-sophisticated systems. It is hoped that this work will be extended in multiple directions, including work in higher-dimensional spaces; extension to other self-modifying, introspective architectures; and creation of larger-scale end-products that exhibit fault detection/avoidance/adaption, parallel synthesis and operation, and self-analysis. Eventually, these may not be ends unto themselves, but will simply be pieces of a larger picture: a picture of artificial systems that utilize self-awareness to afford resiliency, adaption and growth as part of their normal operation.

References

1. Birkner, J., Chua, H.T., Chan, A.K.L., Chan, A.: Programmable logic array with added array of gates and added output routing flexibility. US Patent 4,758,746, 19 July 1988
2. Ceschia, M., Violante, M., Sonza Reorda, M., Paccagnella, A., Bernardi, P., Rebaudengo, M., Bortolato, D., Bellato, M., Zambolin, P., Candelori, A.: Identification and classification of single-event upsets in the configuration memory of sram-based fpgas. IEEE Trans. Nucl. Sci. 50(6), 2088–2094 (2003)
3. Dubowski, S.: Software defined radio aims to bury gsm-cdma hatchet. Netw. World Can. 12(17) (2002)
4. Durbeck, L.J., Macias, N.J.: Defect-tolerant, fine-grained parallel testing of a cell matrix. In: ITCom 2002: The Convergence of Information Technologies and Communications, pp. 71–85. International Society for Optics and Photonics (2002)
5. Franzoi, S.: Social psychology (1996)
6. http://quotes.lifehack.org/quote/carl-jung/your-vision-will-become-clear-only-when/. Accessed 30 June 2015
7. Jamison, C.: Finding Happiness: Monastic Steps for a Fulfilling Life. Liturgical Press (2009)
8. Kuon, I., Tessier, R., Rose, J.: Fpga architecture: survey and challenges. Found. Trends Electron. Des. Autom. 2(2), 135–253 (2008)
9. Lyons, R.E., Vanderkulk, W.: The use of triple-modular redundancy to improve computer reliability. IBM J. Res. Dev. 6(2), 200–209 (1962)
10. Macias, N.J.: Circuits and sequences for enabling remote access to and control of non-adjacent cells in a locally self-reconfigurable processing system composed of self-dual processing cells. US Patent 6,297,667, 2 Oct 2001
11. Macias, N.J.: Self-Modifying Circuitry for Efficient, Defect-Tolerant Handling of Trillion-element Reconfigurable Devices. Ph.D. thesis, Virginia Polytechnic Institute and State University (2011)
12. Macias, N.J., Athanas, P.M.: Application of self-configurability for autonomous, highly-localized self-regulation. In: Second NASA/ESA Conference on Adaptive Hardware and Systems, 2007. AHS 2007, pp. 397–404. IEEE (2007)
13. Macias, N.J., Durbeck, L.J.K.: Self-assembling circuits with autonomous fault handling. In: Proceedings. NASA/DoD Conference on Evolvable Hardware, 2002, pp. 46–55. IEEE (2002)
14. Macias, N.J., Durbeck, L.J.K.: Self-organizing computing systems: songline processors. In : Advances in applied self-organizing systems, pp. 211–262. Springer (2013)

15. Macias, N.J., Pandey, S., Deswandikar, A., Kothapalli, C.K., Yoon, C.K., Gracias, D.H., Teuscher, C.: A cellular architecture for self-assembled 3d computational devices. In: 2013 IEEE/ACM International Symposium on Nanoscale Architectures (NANOARCH), pp. 116–121. IEEE (2013)
16. Patwardhan, J.P., Dwyer, C., Lebeck, A.R., Sorin, D.J.: Circuit and system architecture for dna-guided self-assembly of nanoelectronics. In: Foundations of Nanoscience: Self-Assembled Architectures and Devices, pp. 344–358 (2004)
17. Sekanina, L., Dvok, V.: A totally distributed genetic algorithm: From a cellular system to the mesh of processors. In: Proceedings of 15th European Simulation Multiconference 2001, pp. 539–543 (2001)
18. Udall, J., Teuscher, C., Macias, N.: Truncated Octohedron Circuits. private communication (2015)
19. Zienkiewicz, O.C., Leroy Taylor, R.: The Finite Element Method, vol. 3. McGraw-Hill, London (1977)

Chapter 30
Looking for Computers in the Biological Cell. After Twenty Years

Gheorghe Păun

Abstract This is a personal, in a great extent autobiographical, view on natural computing, especially about DNA and membrane computing, having as a background the author work in these research areas in the last (more than) two decades. The discussion ranges from precise (though informal) computer science and mathematical issues to very general issues, related, e.g., to the history of natural computing, tendencies, questions (deemed to remain questions, debatable) of a, say, philosophical flavor.

30.1 Preliminary Cautious and Explanations

The previous title needs some explanations which I would like to bring from the very beginning.

On the one hand, it promises too much, at least with respect to my scientific preoccupations in the last two decades and with respect to the discussion which follows. It is true that there are attempts to use the cell as it is (bacteria, for instance) or parts of it (especially DNA molecules) to compute, but a research direction which looks more realistic, at least for a while, and which has interested me, is to look in the cell for *ideas* useful to computer science, for *computability models* which, passing from biological structures and processes to mathematical models, of a computational type, can not only ensure a better use of the existing computers, the electronic ones, but they can also return to the starting point, as tools for biological investigations.

This is the English version of the Reception Speech I have delivered on October 24, 2014, at the Romanian Academy, Bucharest, and printed by the Publishing House of the Romanian Academy in December 2014. Some ideas and some paragraphs of the text have appeared, in a preliminary version, in the paper Gh. Păun "From cells to (silicon) computers, and back", published in the volume *New Computational Paradigms. Changing Conceptions of what is Computable* (B.S. Cooper, B. Lowe, A. Sorbi, eds.), Springer, New York, 2008, 343–371.

G. Păun (✉)
Institute of Mathematics of the Romanian Academy, 1-764, 014700 Bucharest, Romania
e-mail: gpaun@us.es

© Springer International Publishing Switzerland 2017
A. Adamatzky (ed.), *Advances in Unconventional Computing*,
Emergence, Complexity and Computation 22,
DOI 10.1007/978-3-319-33924-5_30

Looking to the cell through the mathematician-computer scientist glasses, this is the short description of the present approach, and in this area it is placed the personal research experience which the present text is based on.

On the other hand, the title announces already the autobiographical intention. Because a Reception Speech is a synthesis moment, if not also a career summarizing moment, it cannot be less autobiographical than it is, one uses to say, any novel or poetry volume. And, let us not forget, the life in the *purity and signs world* (a syntagma of Dan Barbilian-Ion Barbu, a Romanian mathematician and poet) of mathematics assumes/imposes a great degree of loneliness, as acad. Solomon Marcus reminded us in his Reception Speech (2008), [51], while the loneliness (it is supposed to) make(s) us wiser, but it also moves us farther from the "world-as-it-is", so that at some stage you no longer know how much from a mathematician belongs to the "world" and how much belongs to mathematics. That is why we can consider that a mathematician is autobiographical both in his/her theorems and in the proofs of his/her theorems, as well as in the models (s)he proposes.

Looking back in time, I find that I am now at the end of two periods of two decades each, the second one completely devoted to "searching computers in the cell", while the first period was almost systematically devoted to preparing the tools needed/useful to this search. The present text describes mainly the latter of these two periods.

30.2 Another Possible Title

For a while, I had in mind also another title, much more general, namely, *From bioinformatics to infobiology*. It was at the same time a proposal and a forecast, and the pages which follow try to bring consistency to this forecast. Actually, the idea does not belong to me, in several places there were discussions about a new age of biology—the same was predicted also for physics—based on using the informational-computational paradigms, if not also based on further chapters of mathematics, not developed yet. The idea is not to apply computer science, be it theoretical or practical, to biology, but to pass to a higher level, to a systematic approach to biological phenomena in terms of computability, with the key role of information being understood. Attempts which illustrate this possibility, also advocating for its necessity, can be found in many places, going back in time to Erwin Schrödinger and John von Neumann. In a recent book, [47], Vincenzo Manca also pleads for "a new biology", which he calls *infobiotics*, starting from the observation that *the life is too important to be investigated only by biologists*. I would reformulate in more general terms: *the life is too important and too complex to be investigated only by the traditional biology*—with the important emphasis that exactly the biologists are called to not only benefit, but also to provide consistency to infobiology. Together with the computer scientists and, more plausibly and more efficient, borrowing from the computer scientists ideas, models, techniques, making them their own ideas, models and techniques and developing them. There is here also a plead for multi-trans-interdisciplinarity (starting with the higher education), but also a warning: this is not only possible, but, it seems, this is also at the right time, on the verge to become urgent.

30.3   The Framework

Having in mind the title before and looking for an "official" enveloping area, the first syntagma which appears is *natural computing*—with the mentioning, however, that it covers a very large variety of research areas, including the bioinformatics and also moving towards infobiology. For an authoritative description, let us consider the four volumes handbook [65]. From the beginning of the Preface, we learn that *Natural Computing is the field of research that investigates human-designed computing inspired by nature as well as computing taking place in nature, that is, it investigates models and computational techniques inspired by nature, and also it investigates, in terms of information processing, phenomena taking place in nature.* The generality is obvious, adding to the desire to identify in nature (important: not only in biology) ideas useful to computer science, a position which, as I have already said, although it is not completely new, if it is systematically applied, it can lead to a new paradigm in biological research and in other frameworks too: the informational approach, hence surpassing the traditional approach, the chemical–physical one.

The idea was formulated also in other contexts: the computational point of view (to the information processing one adds the essential aspect of computability) can also lead to a new physics—among others, this is the forecast of Jozef Gruska, an active promoter of quantum computing. On the same idea is grounded also the collective volume [79]. Many chapters have exciting-enthusiastic titles: *Life as Evolving Software, The Computable Universe Hypothesis, The Universe as Quantum Computer*, etc. There also is a chapter-long Preface, by sir Roger Penrose, not always fully agreeing with the hypotheses from the book.

Actually, also the *Handbook of Natural Computing* mentioned before includes the quantum computing among the covered domains. Here is its contents (the main sections, without specifying the chapters): *Cellular Automata, Neural Computation, Evolutionary Computation, Molecular Computation, Quantum Computation, Broader Perspective—Nature-Inspired Algorithms, Broader Perspective—Alternative Models of Computation.* There is some degree of "annexationism" here (for instance, cellular automata are not too much related to the biological cells), but let us mention that the section devoted to the molecular computation covers DNA computing, membrane computing, and gene assembly in ciliates, the former two areas being exactly what we are interested in here.

30.4   The Popularity of a Domain

Even remaining only at the editorial level and at the level of conferences (without considering also the research projects, hence the financial support), one can say that there is a real fashion of natural computing—more general, of unconventional computing, more restricted, of bioinformatics.

Here are only a few illustrations. Springer-Verlag has a separate series of books dedicated to natural computing monographs, named exactly in this way, it also has a journal, *Natural Computing*. A typical journal is *International Journal of Unconventional Computing*, published by OCP Science, Philadelphia, USA. There is an international conference, *Unconventional Computing*, which became, in the last year, slightly pleonastic, *International Conference on Unconventional Computation and Natural Computation*. *BIC-TA*, that is, *Bio-Inspired Computing—Theory and Applications*, is another conference of a real success, at least in what concerns the number of participants, a meeting whose format I has established, together with colleagues from Spain and China, in 2005, and which is organized since then each year, in China or in the neighboring countries—this can explain the massive participation, as the Chinese researchers are very active in this area.

We have reached the closest upper envelope of the area discussed here: the computability inspired from biology. It is important to note that the term "bioinformatics" (bio-computer science) has a double meaning, with, one can say, a geographical determination. In the "pragmatic West", it mainly covers the computer science applications to biology (in the "standard" scenario, one goes from problems towards tools, without too much theory). In Europe, both directions of influence are taken into consideration, from biology towards computer science and conversely. Although it is just natural that both these two research directions should be developed together, in collaboration, the reality is not always so. In search of solutions for current questions, some of them really urgent, for instance from the biomedical area, mathematics and computer science often provide tools prepared and developed in other areas. The typical example is that of differential equations, with a glorious history in physics, astronomy, mechanics, meteorology, and which are "borrowed" to biology, not always checking their adequacy. I will return to this issue, of a great importance for promoting new tools for biology.

"The European strategy", of constructing a mathematical theory which looks for applications after it is developed, has its appeal and advantages—but also its traps. Being an European, being a mathematician, I have been especially attracted by this strategy, but, in time, I became more and more interested by "reality", by applications.

30.5   What Means to Compute?

Let us come back to the title, with the fundamental question concerning the definition of the notions of computation and computer. This is a question of the same type as "what is mathematics?", with many different answers, none of them complete, none of them fully agreed. If information processing is a computation, then we can see computations everywhere. With a very important detail, hidden in the previous formulation: *we can see*. We, the human beings. Otherwise stated, an observer, which interprets a process as being a computation. I do not want to push the discussion as far as asking questions of the form "does a tree which falls in the water of a lake, in the middle of an uninhabited forest, produce any noise, taking into account that there is nobody there to hear it?"—I mention the fact that this question was the topic of a paper accepted some years ago by a conference on unconventional computing, that is why I recall it—and, on the other hand, I also do not want to involve God in this issue, the omnipresent, omniscient, omnipotent God, considered as an universal observer (at least, not for observing computations, maybe only for noticing noises in desert forests…).

A somewhat exaggerated but rather suggestive example is that of a drop of liquid which falls freely in the air. During its falling down, the drop instantaneously "solves" on its surface, by the form it takes, complex differential equations. Is this a computation? I would not go so far. Similarly with what happens continuously in the cells of a leaf or of the human body, at the biochemical or even at the informational level.

The idea of a computation as a process considered so by an observer is not at all new. One of the conclusions of the John Searle book [68] is exactly this—a computation is not an intrinsic property of a process, but it is *observer–relative*.

A very suggestive formulation of the role of the observer in considering a process as being a computation belongs to Tommaso Toffoli, and it appears in a paper with a statement-title, [72]: "We've just seen that it is not useful to call *computation* just any nontrivial yet somewhat disciplined coupling between state variables. We also want this coupling to have been *intentionally* set up for the purpose of predicting or manipulating—in other words, for *knowing* or *doing* something. This is what shall distinguish bona-fide computation from other intriguing function–composition phenomena such as weather patterns or stock–exchange cycles. But now we have new questions, namely, 'Set up by whom or what?', 'What is it good for?', and 'How do we recognize intention?'

Far from me to want to sneak animistic, spiritualistic, or even simply anthropic considerations into the makeup of computation! *The concept of computation must emerge as a natural, well-characterized, objective construct, recognizable by and useful to humans, Martians and robots alike*" (my emphasis, Gh.P.).

Toffoli's questions should be remembered and discussed, but they move us far from our subject. Let us return to John Searle, namely, to a more technical reading of the idea of implying an observer in the definition of a computation. This was the approach of Matteo Cavaliere and Peter Leupold, both of them my students in the PhD school in Tarragona, Spain, the former one being my first PhD student there. They have published a series of papers with this subject, I cite here only a recent one, by Peter Leupold, [44]. Actually, in the Cavaliere–Leupold approach there appear two observers, one of them—we can call it observer of the first order— following a simple process and "translating" the steps of the process in an external language, and the second observer, closer to the Searle–Toffoli observer, interpreting as a computation the results of the activity of the first observer. Cavaliere and Leupold consider a series of process–observer (of the first order) pairs which, separately, have a reduced (computing) power, but which, together, lead to the computing power of Turing machines from the point of view of the external observer.

30.6   The Turing Machine

Let us start also from another direction, from the meaning given by mathematics to the notion of computation. Already from the thirties of the previous century we have a definition of what is computable, the answer Alan Turing gave to the question "what is mechanically computable?", formulated by David Hilbert at the beginning of the twentieth century. "Mechanically", i.e., "algorithmically" in our today reformulation. There were many proposed answers (I recall only the recursive functions and the lambda-calculus), given by great names of mathematics–computer science (I recall here only Alonzo Church, Stephen Kleene, Emil Post), but the solution given by Turing, what we call now *Turing machine*, has been accepted as the most convincing one (a fact certified even by the highly exigent Gödel). This is now in computer science the standard model of an algorithm (I have not said *definition*, because we have only an intuitive understanding of the idea of an *algorithm*, but we can say that in this way we have a definition of what is *computable*).

Without entering into details, I mention only that Hilbert's problem was more general. It started from the algorithmic resolution of diophantine equations, those with integer coefficients (the tenth problem in Hilbert's 1900 list as published in 1902), but in its later (in 1928) formulation Hilbert was saying that "the *Entscheidungsproblem* [the decision problem in the first order logic] would be solved if we would have a procedure which, for any logical expression we would decide through a finite number of operations whether it is satisfiable... *Entscheidungsproblem* should be considered the main problem of the mathematical logic". At this general level, Gödel theorems answer negatively Hilbert's program. Negative answers gave also Church and Turing, while Hilbert tenth problem was solved—also negatively—in 1970, by Yurii Matijasevich (after many efforts of several mathematicians: Julia Robinson, Hilary Putnam, Martin Davis). Turing not only gives a negative answer, moreover, he not only defines "the frontiers of computability", but he also produces an example of a problem placed behind these frontiers, a problem which is not algorithmically solvable, *the halting problem* (there is no algorithm, hence a Turing machine, which, taking as input an arbitrary Turing machine, can tell us, in a finite number of steps, whether the given input machine halts or not when starting from an arbitrarily given initial data). To the halting problems reduce, directly or indirectly, most if not all undecidability results obtained after that.

The Turing machine is so important for computer science, including the natural/unconventional computability, that it is worth discussing it a little bit more.

30.7 Some More Technical Details

It is interesting to note that when he defined his "machine", Turing explicitly started—he states this at the beginning of the paper—from the attempt to abstract the way a human being computes, reducing to the minimum the resources used and the operations made. In this way, in the end one obtains a "computer" which consists of a potentially infinite *tape*, bounded to the left, divided in *cells* where one can write *symbols* from a given finite *alphabet*; these symbols can be read and rewritten by a *read-write head*, which can "see" only one cell, can read the symbol written there, can change it, then it can move to the neighboring left or right cell or it can stay in the same place; the activity of the read-write head is controlled by the finitely many *states* of a *memory*. Thus, we get *instructions* of the form $s_1 a \rightarrow s_2 bD$ with the following meaning: in state s_1, with the head reading symbol a, the machine passes to state s_2, modifies a to b (in particular, a and b can be identical), and moves the read-write head as indicated by D. One starts with the tape empty, with the machine in a special initial state s_0; one writes the initial data on the tape (for instance, two numbers which have to be multiplied), one places the head on the first cell of the tape (the leftmost one), and one follows the instructions of the (e.g., multiplication) "program" until one reaches a *final state* and the machine *halts*, no further instruction can be applied. The contents of the tape at that moment is the result of the computation.

Extremely reductionistic, but this is the most general model of an algorithmic computation—because no previous definition of what is computable is known, this assertion is only a hypothesis, called the *Turing–Church thesis*. However, what made Turing machine so attractive were not only the *simplicity* of its definition and its *power* (it was proved that the Turing machine can simulate any other computing model), but also its *robustness* (the computing power is not changed if we add further ingredients to the architecture or to the functioning, such as further tapes, if we infinitely prolong the tape also to the left, if we consider non-deterministic computations, etc.), and, mainly, the existence of *universal Turing machines*: there exists a fixed Turing machine TMU which can simulate any particular Turing machine TU, in the following sense. If a code of the machine TM (let us denote it by $code(TM)$) is placed on the tape of TMU together with an input x of TU, then TMU will provide the same result as that provided by TU when starting from input x. A little bit more formally (but still omitting some details—e.g., codifications), we can write $TMU(code(TM), x) = TM(x)$. And Turing proved that there are universal Turing machines, [74].

This is the "birth certificate" of the today computers, consequently called of Turing–von Neumann type (in forties, when he has participate in the designing of the first programmable electronic computers, von Neumann was influenced by Turing ideas).

A couple of things deserve to be mentioned: the code of machine TM is the program to be executed/simulated on TMU, starting from the data x; the instructions of TMU form the "operating system" of our "computer"; the data and the programs are written in the same place, on the tape of the universal Turing machines (in the "computer memory")—from here it follows the possibility to process programs in the same way as we process data, hence the vulnerability of programs to computer viruses.

Several details are important from the point of view of natural computing. The work of the Turing machine is sequential, in each time unit one performs only one instruction. In many places in nature, if not in most of them, in particular, in biology, the processes develop in parallel, which is a very appealing feature for computer science, but these processes are not necessarily synchronized, which, in turn, raises difficulties for computer science.

There also are further differences between Turing machines, the "biological computers", and the electronic computers, but we will discuss these differences later.

For the time being we keep in mind that in what follows *to compute* has the meaning suggested by Turing machines: there are an input and an output, between them there is an algorithm which bridges inputs and outputs, and the result of a computation is obtained in the moment when the machine halts. Very restricted, but precise. With such a framework at hand, we can look around for computations, moreover, we can investigate them in a well developed context, the computability theory—actually, a set of several theories, such as automata theory, formal language (grammar) theory, complexity theory and others.

30.8   Computer Science and Mathematics

This is maybe the place to remind a debate which motivated many discussions and points of view, often biased, concerning the relation between computer science and mathematics. Discussions of this kind have appeared in the higher education (in the sixties-seventies of the last century, at the time of sputniks and hydroelectrical plants, we had many faculties of "mathematics–mechanics", now mechanics was replaced by computer science), the issue is often debated in mass media. Actually, the context is larger, sometimes it is put in question the relation of mathematics with other sciences, with school education, with the society. There are persons who are proud of the fact that they "were not good in math". It was even expressed the opinion that mathematics is a luxury, a "national fetish" (this expression has recently appeared in the title of a Romanian newspaper article), in short, that one makes too much fuss of mathematics and one teaches too much mathematics. This opinion is getting more and more popular, supported also by the ubiquitous penetration of computers ("we no longer need to know the multiplication table, the computer knows it for us").

Of course, there is a problem with the mathematical education. *What, how much*, and, mainly, *how*?—and there also are further questions; we can find them, often also together with solutions, in the papers dedicated in the last years by professor Solomon Marcus to education. The problem cannot be solved from bottom up, the mathematicians involved in research and in higher education should consider it—this is, for instance, the opinion of Juraj Hromkovic, from ETH Zürich, [38], based on the practical activity in this respect carried out in the institute where he works (among others, this activity was materialized in mathematical school books of a new type). In general, the mathematicians should enter public debates and plead for their discipline, mainly they are guilty if the domain loses its popularity. It is true that for a mathematician mathematics is a great game, which, like any game, has an intrinsic rewarding, in the very development of the game, therefore it is natural that the interest for "popularization" is low among mathematicians, but the persons who are proud of their mathematical infirmity, be it real or only claimed, are always much more visible, more vocal, and the danger which comes from this is obvious.

Having in mind only the relation between mathematics and computer science, let us mention that the theoretical computer science, placed at the intersection of the two domains, is often considered by computer scientists as a part of mathematics, and by mathematicians as a part of computer science. Sometimes, theoretical computer science has problems even inside computer science—as it happens also with other theoretical branches of science with a strong practical dimension. Of course, all these are false problems by themselves, but they can have unpleasant practical consequences.

Being of the same opinion, I cite here an authoritative voice, that of Edsger W. Dijkstra, one of the classics of computer science, in fact, of the practical computer science: it is sufficient to remind that during sixties he has worked for implementing the Algol language in the Amsterdam Mathematical Center, and, furthermore, he was the promoter of structured programming, well-known among the software practitioners. (Maybe it is good to add here that the first four years after graduation I have intensively written computer programs, in Cobol and Fortran, realizing even the programs for computing the salaries of the workers in a large Bucharest factory—I remember, therefore, what practical computer science means...)

"The end of computer science?", asks Dijkstra, ironically-rhetorically, already in the title of his note [21], which starts with the following phrase: "In academia, in industry, and in the commercial world, there is a widespread belief that computing science as such has been all but completed and that, consequently, computing has matured from a theoretical topic for the scientists to a practical issue for the engineers, the managers, and the entrepreneurs." Then, it adds: "This widespread belief, however, is only correct if we identify the goals of computer science with what has been accomplished and forget those goals that we failed to reach, even if they are too important to be ignored."

Much more explicit is Dijkstra in the speech he delivered in May 2000 at a symposium (*In Pursuit of Simplicity*) organized at the Austin-Texas University, on the occasion of his retirement. The title of the speech is also relevant, [22]. I recall a couple of aphoristic phrases: "What is theoretically beautiful tends to be eminently useful." "In the design of sophisticated digital systems, elegance is not a dispensable luxury but a matter of life and death, being a major factor that decides between success and failure." "These days there is so much obsession with application that, if the University is not careful, external forces, which do make the distinction [between theory and practice], will drive the wedge between *theory* and *practice* and may try to banish the *theorists* to a ghetto of separate departments and separate buildings. A simple extrapolation will tell us that in due time the isolated practitioners will have little to apply; this is well-known, but has never prevented the financial mind from killing the goose that lays the golden eggs. The worst thing with institutes explicitly devoted to applied science is that they tend to become institutes of second-rate theory."

The plead to place us under the spell of Leibniz is obvious, because, Leibniz said it, "the symbols direct the reason", and, after having a language where "all reason truths will be reduced to a kind of calculus", "the errors will only be computation errors". (Leibniz program, continued and formulated in more precise terms by David Hilbert, cannot be realized, on the one hand, mathematics is too exact-rigorous while the reality is too complex and nuanced to can transform everything in formal computations, on the other hand, Gödel theorems proved that even the Hilbert program is not realizable.)

Of course, the mathematics–computer science relationship is much more complex, but we cannot explore it further here. I close the discussion returning to the starting point: the today computers, programmable, of Turing–von Neumann type, are born from the Turing universality theorem from 1936. It is interesting to note (and comfortable for Dijkstra position) that, by means of a vote through Internet, in 2013, looking for the most important scientific and technological British discovery, the first place was won, surprisingly for our pragmatic times, by the Turing machine and Turing universality theorem, which were placed ahead of the steam engine, the telephone, the cement, the carbon fiber and other similarly important things.

30.9 Does Nature Compute?

Having in mind the computability in the sense of Turing, the previous question becomes more restrictive, but the discussion above provides us the borderlines in between which we have to look for the answer: yes, nature computes at least at the level of... humans, and yes, nature computes whenever there is a process which can be interpreted as a computation by a suitable observer. Opinions which are placed closer to the former or the latter of these limits can be easily found, I cite here only one from the very permissive extreme, even passing over the borderline, because the observer is not mentioned anymore.

At the beginning of Chap. 2 ("Molecular Computation") of the collective volume [29] M. Gross says: "Life is computation. Every single living cell reads information from a memory, rewrites it, receives data input (information about the state of its environment), processes the data and acts according to the results of all this computation. Globally, the zillions of cells populating the biosphere certainly perform more computation steps per unit of time than all man made computers put together."

In what follows, I adopt a more conservative and, at the same time, more productive position: bearing in mind the mathematical definition of computability, more precisely, the Turing approach, let us look around, especially in biology, in search of ideas, data structures, operations with them, ways to control the operations, "computer" architectures, which can suggest (1) new computability models, (2) ways to

better use the existing computers, (3) possibilities of improving the existing computers at the hardware level, maybe even (4) new types of computers, based on biological
materials. It should be noticed the increased ambition from a point to the next one.
It is worth remembering that DNA computing started from the very beginning from
the attempt to compute in a test tube, thus directly addressing the fourth goal in the
list above.

We mainly had here in mind the goals of computer science, but the first objective
also covers the second direction of research mentioned in the preface of the *Handbook
of Natural Computing*, the investigation of processes taking place in nature in terms
of computability, and this research direction should be explicitly and separately
emphasized, especially for pointing to a "side effect" of this approach, namely the
return to biology, delivering models useful to the biologist.

At this moment, DNA computing was not too much useful to practical computer
science, it was useful to biology and much useful to nano-technology, suggesting new
research questions. Membrane computing has significant applications to computer
science and biology, with higher promises in the latter area, including biomedicine
and ecology among the application directions.

A detail: "the goal of computer science" also covers the theoretical interest, which
is not supposed to necessarily lead to applications, in the restricted meaning of the
term. Let us think, for instance, to ciliates. In the division process, when passing
from the micronuclear genes to the macronuclear genes, these unicellular beings
complete complex operations of list processing, and they are doing this since millions
of years, much before the computer scientists gave name and investigated these data
structures. Of course, the ciliates are not thinking to computations when doing this,
but we, the humans, can build beautiful theories starting from their activity, including
computability models, sometimes equivalent in power with Turing machines. Details
and references can be found in the monograph [23].

30.10 An Eternal Dilemma

The previous discussion inevitably pushes us towards the long debate concerning
the relation between invention and discovery. The bibliography is huge, I cite here
only the book [49]. How much is invention and how much is discovery in computer science—with particularization to natural computing? I do not try to provide
an answer, there are as many answers as many view points, personal experiences,
philosophical positions. The models we work with are of a mathematical nature, the
Platonic point of view ensures us that everything is discovery, because mathematics
itself is a revealed reality. Yes, but it is already agreed that notions, concepts, theories,
and models are inventions, the theorems are discovered, the proofs are invented. We
can continue the alternating sequence by adding that the applications are discovered.
Therefore, the models are considered inventions.

However, I would like to introduce a nuance. The models are based on structures
which already exists, but they have not received yet a name. Moreover, differently

from a wall which can be discovered both by an archaeologist who knows what he is searching for, but also by a fruit trees farmer who digs the soil with other goals than finding the basement of an old church, a computability model can be "seen" in a cell only by a computer scientist who has already in mind computability models. For instance, the processes called by biologists symport and antiport exist, they function since long ages in their ingenious ways, but they *compute* only for a mathematician who is looking for a computing model based on passing "objects" from a cell compartment to another one. "Computing by communication"—I have searched for a while something like that, having the intuition that it exists, and I had the solution when a biologist (Ioan Ardelean) told me about symport and antiport operations. This was a model mostly discovered than invented. Actually, a discovery which was not done by bringing to light the discovered object, but by means of superposing the intuition of a model over a piece of reality. The imagined model, similar to other existing computability models, was actualized during the dialogue between reality and the formal framework. I can say that this is at the same time invention and discovery.

30.11 Another Endless Discussion

I am not continuing with other similarly delicate questions, always of interest in spite of any given answers. (For instance, providing us the opportunity to ask how much *art* and how much *science* is in computer science, Donald Knuth entitled an impressing editorial project, planned to have a dozen of volumes, *The Art of Programming*.) However, I touch here another very sensitive topic, with which I was confronted sometimes in the form of the newspaper question (but not completely nonsensical): "During your research in the cell area, have you ever met God?" Of course, the expected answer is something different from "yes" or "no", and similarly obvious is that, if we take the question seriously, we will get lost on the slippery sands of personal options, beliefs, metaphors.

If God is the order, the organization, the good and the beautiful, Spinoza's God, visible in the harmony of the Universe laws, as Einstein would say, then yes, I meet Him continuously, both in cells and outside them. Furthermore: in the title of the book [45] Mario Livio asks *Is God a Mathematician?* I answer in the style of Plato: no, God is not a mathematician, He is mathematics itself (the "grammar of the world")—hence, again, I meet Him every moment.

If, however, God is what the Book proposes to me, then I go in line with Galileo Galilei, who, in a letter sent to don Benedetto Castelli, on December 21, 1613, said (I recall it following [14]): "God has written two books, the Bible and the Book of Nature. The Bible is written in the language of men. The Book of Nature is written in the language of mathematics. That is why the language of the Bible is not suitable for speaking about nature. The two books must be studied independently from each other." And Galileo added: the Book of Nature teaches us "how the Sky/Heaven goes", while the Bible teaches us "how to go to the Sky/Heaven".

After centuries of separation—mainly dogmatical, from both sides—, alternating with attempts, most of them pathetical, of reconciliation of science with religion, the words of Galileo can look too simple or opportunistic, but they cut in an efficient way a continuously regenerated Gordian knot. Let me mention also a more sophisticated, but somewhat symmetrical position, of Francis S. Collins, [13], not only contemporary with us, but also connected to the topic of these pages, as he was the director of National Human Genome Research Institute, one of the leaders of the famous Human Genome Project. The syntagma "language of God" was used also by Bill Clinton, in 2001, when he has announced the completion of "the most important, most wondrous map ever produced by humankind", the map of the around three billions of "letters" of the "book of life". Even if the title seems to suggest this, Collins is neither a creationist, nor an adept of the intelligent creation, but he is an "evolutionary deist" and the conclusion of his book is that "the God of Bible is also the God of the genome" (p. 222 in the Romanian version), while "science can be a form of religiosity" (p. 240). This is a very comfortable positioning, but, in what follows, I remain near Galileo.

30.12 The Limits of Today Computers

The fashion of natural computing and especially of the computing inspired from biology does not have only the internal motivation, of the numerous research directions explored in the last decades and proved to be theoretically interesting and at least promising if not directly useful in practice, but it has also an external motivation, related to the limits of the current computers, some of them rather visible. Indeed, the computers are the twentieth century invention with the widest impact, with implications in all components of our life, from communication to the functioning of the financial system, from the health system to the army, from the numerous gadgets around us to Internet. In spite of all these—actually, just because of that—the computers which we have now have limits which we reach often (with the mentioning that also here, like in most things, there is something bad in the good and something good in the bad: powerful computers can be used both in positive ways, but also for bad goals, such as breaking security systems and cryptographic protocols on which, for instance, the protected communication is based.) Let us however think positively and note that there are many tasks which the today computers cannot carry out, but which we would like to have performed.

The processors become continuously faster and more compact, the memory storage larger and larger. Sure, but how much this tendency will last? It was much invoked the so-called *Moore law*, stated in 1965 by Gordon A. Moore, co-founder of Intel Corporation, with respect to the number of transistors which can be placed on an integrated circuit, extended then to the cost of information unit stored, formulated sometimes even in the form "in each year, the computers become two times smaller, two times faster, and two times more powerful". Exponential in all the three directions, thus tending fast towards the quantum limit in the dimension of processors.

Even at the more technical level, confirmed for a couple of decades, of doubling the capacity of processors, the law—actually, only an observation, followed by a forecast—has been adjusted several times, with the doubling/halving moved first at one year and a half, then at two years, then at three years. Still, it is not too bad, but one cannot continue too much even at this pace.

In fact, the real problem is a different one. Progresses are made continuously at the technological level, but the current computers have intrinsic limits, which cannot be overcome only by means of technological advances. The computer recognizes fingerprints, but not human faces, it plays chess at the level of the world champion, but (on the standard board, not on reduced boards) it plays GO only at the level of a beginner, it proves propositional calculus theorems, but cannot go over this level (and definitely cannot distinguish trivial and non-interesting theorems from theorems which deserve to be collected). All these and many more, mainly because these computers are… of Turing–von Neumann type. That is, sequential. Uniprocessor. (It also has other weaknesses, less restrictive in the current applications—for instance, it is a considerable energy consumer.) It computes whatever can be computed, but this is true in principle, at the *competence* level. There is here also a historical aspect. In the beginning, we were interested in what it can be computed, in the frontiers of computability, of algorithmic decidability. All these are important mathematical questions, but in applications it is of a direct relevance the *performance*, the resources needed for a given computation, what we can compute now and here, in specified conditions. How much electricity consumes a computer and how much space it needs are no longer questions of current interest, as they were in sixties (and still are in special frameworks, such as in cosmos and robotics), but the time we have to wait before receiving the answer to a given problem or the result of a computation is a crucial aspect in any application. And, I already mentioned it, in this respect not the technological promises are crucial, but the mathematical limits, the borderline between feasible and non-feasible.

30.13   A Great Challenge: The Exponential Complexity

A powerful theory was developed dealing with this subject, the computational complexity theory. Since the very beginning, it has defined as tractable the problems which can be solved in a polynomial time with respect to the size of the problem. (An example: consider a graph—a map with localities and roads among them—with n nodes. Which is the time necessary for an algorithm to tell us whether or not the graph contains a Hamiltonian path, i.e., a path which visits all nodes, passing only once through each of them? If this time is bounded by a polynomial in n, then we say that the algorithm is of a polynomial complexity.) The problems of an exponential complexity, those which need a time of the type 2^n, 3^n, etc. for an input of size n were considered intractable. The former class was denoted by **P**, the latter one with **NP**, with the abbreviations coming from "polynomial" and "non-deterministically polynomial", respectively: a problem belongs to **NP** if we can decide in a polynomial time whether a proposed solution for it is indeed a solution or not (otherwise stated, we "guess" a solution, then we check whether it is correct; more technically, the solution is found by a non-deterministic Turing machine, one which has several possible transitions at a computation step and we rely on the fact that it always chooses the right continuation, without exhaustively checking all possibilities). For precise details the reader can consult [55].

Let us recall that in the class **NP** there is a subclass, of "the most difficult problems in **NP**", the **NP**-*complete* problems: a problem is of this type if any other problem in **NP** can be reduced to it in a polynomial time. Consequently, if an **NP**-complete problem could be solved in a polynomial time, then *all* problems in **NP** could be solved in a polynomial time. The problems used in cryptography are in most cases **NP**-complete.

A beautiful theory, which, however, in its basic version has three weaknesses: (1) it cannot tell us yet whether or not $\mathbf{P} = \mathbf{NP}$, whether or not polynomial solutions can be found also for the problems which are now supposed to be of an exponential complexity, (2) the theory does not take into account such "details" as the coefficients and the degree of the polynomials and which, at the practical level, can have a crucial influence on the computation time, and (3) the theory takes into consideration the extreme cases, it is of the *worst case* type, it counts the steps of computations which solve the most difficult instances of a problem, while the reality is placed in most cases in the middle, near the "average". Here is an example with a practical relevance: the linear programming problem is in **P**, because the ellipsoid algorithm of Leonid Khachiyan (1979) solves the problem in a polynomial time, but this algorithm is so complex that practically, in most cases, it is less efficient than the old simplex algorithm, proposed during the Second World War, considered one of the most important ten algorithms ever imagined, but which is, theoretically, of an exponential complexity.

For these reasons, the complexity theory was refined and diversified (average complexity, approximate algorithms—these algorithms have a direct connection with natural computing), while the definition of tractability was carefully redefined.

Anyway, the general feeling was transformed in a slogan: *the Turing–von Neumann computers cannot solve in a reasonable time problems of an exponential complexity.*

The interest for the $\mathbf{P} = \mathbf{NP}$ problem is enormous. On the one hand, most of the (no-trivial) practical problems are in the class **NP** and are not known to be in **P**, hence they are (considered) intractable, cannot be efficiently solved, on the other hand, most of the cryptographic systems in use are based on problems of an exponential complexity, hence solving them in a polynomial time would lead to breaking these systems. The problem whether **P** is or not equal to **NP** was already formulated in 1971 (by Stephen Cook), and in the year 2000 it was included by Clay Mathematical Institute, Cambridge, Massachusetts, in the list of the seven "millennium problems", with a prize of one million dollars for a solution.

While the importance of this problem for the theoretical computer science cannot be overestimated, it is not clear which would be the practical consequences of a solution, whichever this will be. There were many discussions on this topic—see, for instance, [15]. If a proof of the strict inclusion of **P** in **NP** will be obtained, as most computer scientists (but not all of them!) believe, then almost nothing will be changed at the level of the practical computer science. If the equality will be proved in a non-constructive manner, or the proof will be a constructive one, but in a non-feasible manner (polynomial solutions to problems in **NP** will be found, but with polynomials of very high degrees or with very large coefficients), then the practical consequences will not be significant (but a race will start for ad-hoc solutions, having the time estimated by polynomials with reasonable degrees and coefficients). If, however, a "cheap" passage from **NP** to **P** would be found, then the consequences for the practical computer science will be spectacular—in the good sense, excepting the cryptography, where the consequences will be dramatic.

At the level of the software there is one further problem, which I recall here in the formulation of Edsger W. Dijkstra (from [21]): "Most of our systems are much more complicated than can be considered healthy, and are too messy and chaotic to be used in comfort and confidence. The average customer of the computing industry has been served so poorly that he expects his system to crash all the time." The lack of robustness of the complex software systems is today a concern of the same interest as it was in the year 2000.

In order to illustrate the fact that not by means of technological progresses one can face the exponential complexity, let us examine a simple case: let us consider a problem of exponential complexity of the range of 2^n, for instance, a graph problem, which can be solved on a usual computer, say, for graphs with 500 nodes, in approximately one quarter of hour; let us suppose that the technology provides us a computer which is 1000 times faster than the ones we have, which is a totally non-trivial advance, not very frequently met. Using the new computer, we will solve the same problems as before in about one second (around 15 min means approximately 1000 s), but if we try to address the same problem for graphs with more than 500 nodes, the progresses are negligible: with the new computer we will solve in a quarter of hour only problems for graphs with at most 510 nodes. The simple reason for that is the fact that 2^{10} is already bigger than 1000. If the problem were of complexity 3^n, then we will stop around 506 nodes...

30.14 Promises of Natural Computing

In order to cope with the exponential complexity, but also for other reasons which I will mention later, computer science has imagined several research directions, most of them also related to the natural computing, even to bio-inspired computing: (1) looking for massively parallel computers, (2) looking for non-deterministic computers/computations, (3) looking for approximate/probabilistic solutions to computationally hard problems.

All these three research directions were explored already in the framework of the "standard" computer science, both at the theoretical level and at the technology level, the electronic one. Multiprocessor computers are available since several years—but without reaching the massive parallelism which is supposed to solve complex problems. If a large number of processors are put together, there appear other problems, some of them technological (e.g., high temperature dissipation), others, maybe more important, theoretical, concerning the synchronization of the processors. A distinct research area deals with the synchronization complexity—see, for instance, [37]. One of the conclusions of this theory says that, for a large number of processors, the synchronization cost (measured by the number of bits necessary to this aim) becomes larger than the cost of the computation itself, which suggests to get rid of synchronization, but then other problems appear, as we are not accustomed to use asynchronous computers.

Even less used we are to construct and utilize "non-deterministic computers". In exchange, the last of the three ideas mentioned above is rather attractive, and in this respect of a great help is the "brute force" of existing computers. The approach is useful especially in addressing complex optimization problems: exploring randomly the candidate solutions space, for a large enough time, with a sufficiently high probability we will reach optimal or nearly optimal solutions. Approximate solutions, possibly found with a known probability of being optimal.

Here it enters the stage, with great promises, the natural computing. From now on I will only refer to the one having a biological inspiration.

In a cell, a huge number of "chemical objects" (ions, simple molecules, macro-molecules, DNA and RNA molecules, proteins) evolve together, in an aqueous solutions, at a high degree of parallelism, and, at the same time, of non-determinism, in a robust manner, controlled in an intricate way, successfully facing the influences coming from the environment, and getting in time very attractive characteristics, such as adaptation, learning, self-healing, reproduction. Many other details are of interest, such as the reversibility of certain processes or the energy efficiency, with the number of operations per Joule much bigger than in the case of the electronic processing of information (erasing consumes energy, that is why the reversible computers are of interest; see, e.g., [5, 43]).

It seems, therefore, that during millions of years of evolution nature has polished many processes (and material supports for them) which wait to be identified and understood by the computer scientists, in order to learn new computability methods and paradigms, maybe for constructing computers of a new kind. And, the computer scientists have started to work since a long time…

Here are a few steps on this road, very shortly: *Genetic algorithms*, as a way to organize the search through the space of candidate solutions, imitating the Darwinian evolution, in order to solve optimization problems. Generalization to *evolutionary computing* and *evolutionary programming*. *Neural networks*, trying to imitate the functioning of the human brain, also used for finding approximate solutions, especially for pattern recognition problems. A little bit later, *DNA computing*, which has proposed a new hardware, massively parallel, based on using the DNA molecules as a support for computations. Even younger, *membrane computing*, taking as the starting point the biological cell itself and cell populations.

In turn, the evolutionary computing, in general, the area of approximative algorithms inspired from biology, is spectacularly ramified, in the most diverse (in certain cases, also picturesque) directions: *immune computing, ant colony algorithm, bee colony algorithm, swarm computing, water flowing computing, cultural algorithm, cuckoo algorithm, strawberry algorithm*—and it is highly probable that in the meantime further algorithms have been proposed...

It is important to note that all the above mentioned branches of natural computing, with the exception of DNA computing, are meant to be implemented on the usual computer, in the aim of having a better use of it; one proposes new types of software/algorithms, not to change the computers architecture or new types of hardware.

30.15   Everything Goes Back to Turing

In a certain sense and to a certain extent, the whole history of theoretical computer science is related to biology, it has searched and has found inspiration in biology. I have already mentioned that, in 1935–1936, when he has defined the machine which bears now his name, Turing tried to imitate the way the humans are computing.

After one decade, McCulloch, Pitts, Kleene have founded the theory of finite automata starting from the modeling of neurons and of neural networks. Later, the same starting point led to what is called today neural computing.

It is interesting to note that the beginnings of this research area can be identified in unpublished papers of the same Allan Turing. We have here an interesting case which can illustrate the influence of psychology and sociology on the development of science, telling about uninspired group leaders and about researchers interested more in their research than in the publication of the obtained results. Specifically, in 1948, Turing has written a short paper, called "Intelligent machinery", which has remained unpublished until 1968, because his boss from the London National Physical Laboratory, ironically, named sir Charles Darwin, the grandson of the famous biologist with the same name, has written on the corner of the first page of the paper "schoolboy essay", thus preventing the publication.

"In reality, this farsighted paper was a manifesto of the field of artificial intelligence. In the work (...) the British mathematician not only set out the fundamentals of connectionism but also brilliantly introduced many of the concepts that were later to become central to AI, in some cases after reinvention by others."—I have cited from [18]. Among others, Turing paper introduces two types of "neural networks", with the neurons randomly connected. This was proposed as a first step towards an intelligent machine, one of the key features of these networks being that of learning, of getting trained for solving problems. This is neural computing *avant la lettre*, with the main ideas rediscovered later, without referring to Turing. Details about Turing "unorganized machines" can also be found in [69, 70]. Furthermore, at the address http://www.AlanTuring.net one can find details about Turing unpublished manuscripts and about the recent efforts to reintroduce them in circulation.

The same Turing, in the same year 1948, has proposed the "genetic or evolutionary search", the first ideas of the evolutionary computing developed later, a domain which contains now several powerful branches, (re)launched during the years: evolutionary programming (L.J. Fogel, A.J. Owens, M.J. Walsh), genetic algorithms (J.H. Holland), evolutionary strategies (I. Rechenberg, H.P. Schwefel), all three initiated in the sixties, genetic programming (J.R. Koza, the years 1990). The first experiment of computer "optimization through evolution and recombination" was carried out in 1962, by Bremermann. Details can be found in [24].

It would not be completely surprisingly if among Turing manuscripts we would discover also ideas related to DNA computing—let us remember that Turing died in June 1954, and the paper where J.D. Watson and F.H.C. Crick described the double helix structure of the DNA molecule was published one year before, [75].

It is worth mentioning that two other concepts with a high career in computer science come from Turing, thus supporting the assertion that "everything starts with Turing". First, Turing himself raised the question whether or not one can compute… more than the Turing machine, imagining Turing machines with oracles, which is a much investigated topic in the current computer science. Then, Turing can be considered not only a founder of artificial intelligence, but also a forerunner of what is called now *artificial life*: in the last years of his life, Turing was interested in morphogenesis, in modeling the evolution from the genes of a fertilized egg to the structure of the resulting animal.

30.16 An Encouraging Example: The Genetic Algorithms

Before passing to the DNA and membrane computing, topics which I will describe in more details, let us spend some time discussing a branch of natural computing inspired from biology which is, at the first sight, surprisingly efficient. This is the *genetic algorithms* area, used for solving complex optimization problems for which there do not exist deterministic optimal algorithms or these algorithms are not efficient. The implicit slogan can look confusing: *if you do not know where to go, then go randomly*—with the mentioning that the "randomness" here is directed, the "random walk" is done "like in nature, in species evolution".

Everything is a metaphorical imitation of some elements from the Darwinian evolution. Let us assume we have a two variables function (we can suggestively represent it as a ground surface, with valleys and hills) for which we need to find the maximum (one of them, if there are several). If we cannot analytically address the problem, then we can choose to walk randomly through the definition domain, looking for the greatest value of the function. To this aim, we represent the domain points as "chromosomes", binary strings of a constant length, we choose (randomly or through other methods) a given number of starting points, and we compute the function value for all of them. Then we pass to "evolution": we take two by two the "chromosomes" and we recombine them (by crossover), that is, we cut them at a specified position and then we recombine the fragments, the prefix of one "chromosome" with the suffix of the other one and conversely. In this way, we obtain two new "chromosomes", describing two new individuals of the next "generation". We repeat this procedure for a specified number of times, we select the best solution obtained so far, and we stop.

Nothing guarantees that in this way we reach the solution of the problem, that, for instance, we do not get stuck in a local maximum, without being able to escape, but, and this is the (pleasant) surprise, in a large number of practical applications, this strategy works. Sure, there are a lot of variations of the previous scenario, it is even said that the monographs in this area are a sort of "cooking books", collections of recipes, lists of ingredients and suggestions of improvements of the algorithms: besides recombination, similar to the biological evolution, one also uses the local mutation operation, the passing from a generation to another one can be done in many ways, the "chromosomes" population can be distributed, we can evolve it locally, communicating in a way or another among regions, there are several halting criteria, and so on and so forth.

We have here at work the brute force of the computers and the evolutionary metaphor—with results, I repeat and stress it, unexpectedly good: non-intuitive solutions, rapid initial convergence, in many cases succeeding to avoid local maxima. The only "explanation" for these good results is the "bio-mystical" one: genetic algorithms are so good because they involve ingredients which nature has polished for many millions of years in the species evolution.

All these induce, at the same speculative level, a rather optimistic conclusion: if the genetic algorithms are so useful, in spite of the lack of any mathematical argument for their usefulness, let us try to imitate biology also in other aspects, with a great probability that, if we are similarly inspired to extract the right ideas, to obtain other fruitful suggestions for improving the use of the existing computers and, maybe, for imagining computers of other kinds, more efficient.

However, this optimism should be cooled down by the observation that a famous result in the area of evolutionary computing, in the area of approximate optimization algorithms in general, is the so-called *no free lunch theorem*, [77], which, informally, says that any two methods of approximate optimization are equally good, in average, over all optimization problems. "Equally good" can be also read "equally bad", for each method there are problems for which the method does not provide satisfactory solutions.

30.17 A Coincidence

Before passing to DNA computing, an autobiographical intermezzo. In April 1994 I was in Graz, Austria, attending a conference, and there I got a copy of the paper by Tom Head, from State University of New York at Binghamton, USA, soon after that a friend and collaborator, [33]. It was a revelation. I was then after twenty years of formal language theory research and I immediately felt that it is open there a large area of application of what I have done before. It is true, I should had a similar revelation earlier, namely when I have read professor Solomon Marcus paper [48], but probably it was too early, at that time I had not passed yet through the twenty years of preparation for natural computing which will be shortly described in a forthcoming section. In his paper, Tom Head introduces a formal operation with strings which formalizes the operation of recombination of DNA molecules. He calls it *splicing*, and I will call it in the same way, thus distinguishing it from the recombination operation from the genetic algorithms. The two operations are related, but they are not identical. Still being in Graz, I have imagined a sort of grammar based on the splicing operation, in fact, a variant simpler than that of Tom Head and closer to the string operations in language theory. The paper emerged in this way [57] has consecrated the splicing version I have proposed. After a few weeks, I was in Leiden, The Netherlands, where I have written a paper [61] together with Grzegorz Rozenberg and Arto Salomaa, the latter one from Turku, Finland, the place where I have spent after that several years, initially devoted to DNA computing and then to membrane computing. As usually well inspired, G. Rozenberg gave to our paper the title "Computing by splicing". Because, starting then, we have named H systems the computing devices based on splicing, thus reminding the name of the one who has introduced (invented or discovered?...) the respective operation, we have sent the paper, in manuscript, to Tom Head. He has immediately replied, by phone, asking us rather excited: have you known that right now it was carried out a successful experiment of computing with DNA?! No, we did not know—this was only a coincidence, which I place in the category of significant coincidences.

30.18   The First Computation in a Test Tube

Tom Head was talking about Leonard Adleman experiment, reported in the autumn of 1994, [1]. Speculations about the possibility of using DNA molecules for computing were made already in seventies of the last century (Ch. Bennett, M. Conrad, even R. Feynman, with his much invoked phrase "there is plenty of space at the bottom", referring to physics but also extended to biology). Adleman has confirmed these expectations, solving in a laboratory the problem whether a Hamiltonian path exists or not in a given graph (I have mentioned it in a previous section). The problem is known to be **NP**-complete, hence among the most difficult intractable problems, of an exponential complexity (we assume that **P** is not equal to **NP**), but Adleman solved it in a number of steps which is linear with respect to the size of the graph. It is true, these steps are biochemical operations, performed by making use of a massive parallelism, even of non-determinism, all these made possible by the characteristics of the DNA molecules and the related biochemistry.

In short, millions of copies of one stranded sequences of nucleotides, codifying the nodes and the edges of the graph, were placed in an aqueous solution. Then, by decreasing the temperature of the solution, these sequences annealed, forming double stranded molecules, corresponding to the paths in the graph. Because there were used sufficiently many copies of the initial sequences, with a high probability we obtain in this way *all* paths in the graph. From them, the paths were selected which pass through all nodes, and this was done by usual laboratory procedures: gel electrophoresis for separating the molecules according to their length, then selection through denaturation and amplification by PCR of the paths passing through all nodes (hence Hamiltonian).

This procedure assumes a number of biochemical operations which is linear with respect to the number of the nodes in the graph. The problem is **NP**-complete, hence this is an extraordinary achievement—and the consequences were accordingly sound. Already in the next year, 1995, it was organized in Princeton a meeting with the title "DNA Computing", which became an international conference which is still continuing. However...

30.19   Pro and Against Arguments

Adleman experiment was a historical achievement, the proof that *it is possible*. However, the experiment has considered a graph with only 7 nodes, for which the problem can be solved by a simple visual inspection. In comparison, at the beginning of the nineties, the computers were already able to solve the Hamiltonian path problem for graphs with several hundred nodes, sufficient for current practical applications (in the meantime, the progresses continued).

Moreover, the solution was obtained by means of a space–time trade-off, the number of molecules used was exponential with respect to the number of nodes. Juris Hartmanis, an authoritative name in computer science, after expressing his enthusiasm (Hartmanis compares computer science with physics, [31]: while the latter progresses by means of crucial experiments, the former progresses by means of proofs that something can be done, by *demos*; Adleman has produced such a *demo*!), has computed, [32], the quantity of ADN which is necessary in order to apply Adleman's procedure for a graph with 200 nodes and he has found that the weight of the ADN would be greater than the weight of the Earth...

From a practical point of view, DNA computing is, to a certain extent, in the same point even now. Numerous experiments, but all of them always dealing with "toy problems", a lot of theory, a lot of lab experience gained in dealing with DNA molecules, with results of interest for the general lab technology (just one example: an improved version of PCR, the Polymerase Chain Reaction, called XPCR, was proposed, [26]), but the domain has moved towards nano-technology, no computability practical applications were reported (unless if, and this is plausible, there were applications in cryptography which are still classified).

However, the list of possible advantages of using DNA molecules for computing is large: a very good efficiency as a data support, with one bit at the level of a nucleotide; energy efficiency; parallel and non-deterministic behavior, two dreams of computer science (with the mentioning that the non-determinism also brings problems, for instance, providing false solutions); a very developed laboratory technology; robustness, predictability, reversibility of certain processes.

30.20   The Marvelous Double Helix

The DNA molecule has surprising properties at the informational and computational level. Let us remind that, formulated in "syntactic" terms, we have two strings of letters A, C, G, T, the four nucleotides, placed face to face, in Watson–Crick complementary pairs, always A being paired with T and C with G. The two strings are oriented, in opposite directions with respect to each other; the biochemists indicate the directionality by marking one end of a string with $3'$ and the other end with $5'$. There already appear here a surprise, first pointed out in [66]: the structure of the DNA molecule "hides", in a codified manner, the computing power of Turing machines! The formulation above is not precise, it however corresponds to the following observation. Already in 1980, it was proved, [25], that *any language whose strings can be recognized by a Turing machine can be written as the image of a specified fixed language, let us denote it with $TS(0, 1)$, by means of a sequential transducer.*

The previous language is the so-called "twin-shuffle" over 0, 1 (hence the used notation). Shuffle is the operation of mixing the letters of the two words, without changing their ordering (exactly as in the case of shuffling two decks of playing cards of different back colors). Here we shuffle the letters of two "twin" words, one string of symbols 0, 1 and the second string identical with this one, but changing the "color" of each symbol (for instance, we can add an upper bar or a prime to each symbol in order to get the twin string). In turn, the sequential transducers are the simplest transducers, with a finite memory and with a head which scans the string from left to right. Let us note that we work with four symbols, let us say 0, 1 and their pairs $0'$, $1'$. Exactly the number of the nucleotides, four. Let us also note that $TS(0, 1)$ is a fixed language. Given an arbitrary language, if it is recognized by a Turing machine, then it can be obtained from this unique language $TS(0, 1)$, only the transducer depends on the language.

The nice and significant surprise is that the language $TS(0, 1)$ can be obtained by means of "reading" the DNA molecules, in the following way: let us walk along the two Watson–Crick complementary sequences, from the left to the right, advancing randomly along the two strands, and associating with the four nucleotides A, C, G, T symbols 0, 1 according to the following rule: $A = 0, G = 1, T = 0', C = 1'$. Collecting all these strings over 0, 1, $0'$, $1'$, for all readings of all DNA molecules, we get a set which is exactly $TS(0, 1)$!

Consequently, any language which can be defined by a Turing machine can be obtained by translating these readings of the DNA molecules by means of the simplest transducer, the sequential one, with a finite memory. The transducer depends on the language, it "extract" from $TS(0, 1)$ the result of the computations of a Turing machine. The power is there, what we have to do is only to make it visible. (In a certain sense, we have again the coupling of a simple process, the "reading" of the DNA molecule, and an observer of the first order, a simple one, the sequential transducer, like in the papers of M. Cavaliere and P. Leupold mentioned before, with the result reaching the highest level of computability, the power of Turing machines.)

Two questions arise in this framework. For instance, we mentioned the different orientation of the two strands of the DNA molecule, but in the previous reading we pass along the two strands in the same direction, from the left to the right. There is no problem, the reading of the double stranded DNA molecules can proceed in opposite directions and the result is the same. Second: nature is redundant, are all the four nucleotides (the four symbols $0, 1, 0', 1'$) necessary in order to cover, in the sense discussed above, the power of Turing machines? No, three symbols are sufficient— but not two! Proofs for all these results can be found in the monograph [62].

Speaking about computations and redundancy, let us remember that a large part of the DNA molecule is "residual", it does not codify genes and we do not exactly know what it is used for. We can then speculate: if in the cell, at the genetic level, one performs computations (the viruses are strings of nucleotides, hence their identification is a parsing operation, hence a computation), and these computations are supposed to be complex, why not?, even of the level of Turing machines, then we need a "workspace", a "tape" which in the end remains empty in most of its length, with the result placed in a finite part of it (at the beginning in the case of the Turing machine tape). Can then the DNA without an apparent usefulness be the workspace for complex computations, which we cannot yet understand?

30.21 Computing by Splicing

In his experiment, Adleman has not used the splicing operation, but the biochemical ingredients specific to the splicing have been used in many other cases: restriction enzymes, which cut the DNA molecules in well specified contexts, ligases which glue back the nucleotides thus repairing the strands, recombination on the basis of the "sticky ends" of the molecules with the strands of different lengths, hence with nucleotides which do not have their Watson–Crick pairs.

I do not recall biochemical details or mathematical details concerning the splicing operation. In short, two molecules (represented as simple strings, because the nucleotides of a strand are precisely identified by their complementary nucleotides placed on the other strand) are cut in two parts each, in the middle of a context specified by a pair of substrings, and the fragments obtained are recombined crosswise, thus obtaining two new strings. Starting from an initial set of strings and applying this operations repeatedly (with respect to a given finite set of contexts, hence of *splicing rules*), we obtain a computing device, a language generator, similar to a grammar. We obtain an *H system*. A large part of the monograph *DNA Computing. New Computing Paradigms* cited before is dedicated to the study of these systems: variants, extensions, generative power, properties.

Always when a new computing model is introduced, the first question to clarify concerns its power, in comparison with the automata theory and language theory classifications—the Turing machine and its restrictions, the Chomsky grammars, the Lindenmayer systems. Let us only note that the two "poles" of computability are the power of Turing machines, through the Turing–Church thesis the maximal level of algorithmic computability, and the power of the finite automata, the minimal level. In terms of grammars and languages, the maximal class is that of unrestricted Chomsky grammars and of recursively enumerable languages, while the minimal one corresponds to regular grammars and languages.

The H systems with a finite number of starting strings and a finite number of splicing rules generate only regular languages. This is not sufficient as computing power, moreover, a "computer" of this level cannot have (convenient) universality properties, hence it cannot be programmable.

Interesting and attractive enough is the fact that, adding a minimal control on the splicing operation, with many controls of this kind suggested by the area of regulated rewriting or coming from biology (example: associate a promoter, a symbol, with each splicing rule and the rule is applied only to strings which contain that symbol; a variant—the symbol does not appear, it acts as an inhibitor), then we obtain H systems which are equivalent with the Turing machine. The proof is constructive, therefore we "import" in this way from the Turing machines the existence of the universal machine, which means that we get an universal H system, a programmable one.

Unfortunately, so far, no such universal "computer" based on splicing was realized in a laboratory. The passage from the natural case, with an uncontrolled splicing operation (thus with the power under the power of the finite automaton), to the controlled case was not yet done in a laboratory and it is not clear whether it can be realized in the near future. The construction of the universal computer based on splicing has to still wait...

30.22 An Important Detail: The Autonomous Functioning

Let us not forget that a universal, programmable computer should work autonomously, that is, after starting a program, the computer continues without any external control. This is completely different from the usual DNA computing experiments, where the human operator (or a robotic operator) controls the whole process. For instance, in the case of the 1994 experiment, Adleman was, in fact, the "computer", he has only used the DNA molecules as a support for the computation, while the computation complexity counted the lab steps performed by the biochemist, not the internal steps, the DNA operations, performed in parallel.

There are, however, promising progresses towards the implementation of autonomous computations, the key-word, very much promoted in the last years, being *self-assembly*. Remarkable achievements in this direction has obtained Erik Winfree and his group from Caltech, Pasadena, USA, and his approach is worth mentioning also because it starts (pleasantly enough for the discussion concerning the usefulness of mathematics for computer science) from an old chapter of theoretical computer science, the domino calculus of Wang Hao, developed in the beginning of sixties of the last century. In short, square dominos, with the edges colored (marked), can be used for computing (by placing the dominos adjacently, in such a way that the neighboring dominos have the contact edges of the same color), thus simulating the work of a Turing machine. We obtain once again a computing model which is universal.

Erik Winfree has constructed "dominos" from DNA molecules, with the edges marked with suitable sequences of nucleotides, he has left them free in a solution, such that the dominos glued together according to the Watson–Crick affinity of the nucleotides "coloring" the edges. The approach worked well, the experiments were successful—but everything has remained once again, in Hartmanis terms, at the level of a *demo*. It is important to underline that this time it was not addressed a given problem, as in Adleman case and as in most of the experiments reported in the DNA computing literature, but it was implemented in a laboratory a Turing machine, hence an universal computing model—that is why this *demo* is perhaps farther reaching than that of Adleman (however, Adleman was the first one…).

There also are other attempts to obtain autonomous "computers" in a laboratory. I mention here only the simulation of a finite automaton, an achievement of a team from Weizmann-Rehovot and Tehnion-Haifa, Israel, [3], with the mentioning that one deals with a finite automaton with only two states. Again, only a *demo*…

30.23   What Means to Compute *in a Natural Way*?

The *DNA Computing* monograph has also chapters dedicated to other ways of computing, inspired from the DNA biochemistry, for instance, by insertion and deletion of substrings (in given contexts), by means of a "domino game" with DNA molecules which are coupled on the basis of the Watson–Crick complementarity, a model different from the Wang Hao one.

Splicing, insertion–deletion, prolongation of strings. In membrane computing we use the multiset processing. The evolution itself is mainly based on recombination/splicing, the local mutations appear only accidentally. In contrast, the existing computers and almost all theoretical models of computing use the string rewriting operation. One works locally, on strings of arbitrary length. This observation is valid for automata, grammars, Post systems, Markov algorithms. All these operations, both the rewriting and the "natural" ones (the splicing only with an additional control), so different among them, lead to computability models of the same power, that of the Turing machine.

The question is obvious: what means to compute *in a natural way*? With many continuations: Why computer science has not considered (with rare exceptions) also other operations different from rewriting? Can we devise (electronic) computers based on "natural operations" (for instance, using the splicing or other forms of recombination)? When Hilbert has formulated the question "what is mechanically computable?", he probably had in mind formal logical systems, where the substitution is a central inference rule, and Turing has proposed an answer in the same language. Were we influenced in this way to think in the same terms when we have designed the first computers? I have never heard that the engineers have said that we cannot imagine, maybe also construct, computers based on different operations.

It remains the question whether or not such new types of computers would be better than the existing computers or not. Theoretically, they will have the same power, hence the differences should be looked for on different coordinates: computational efficiency, easiness of use, learning possibilities and so on.

I said above that the H systems are either of the power of finite automata or equivalent with Turing machines. Similar situations are met in the membrane computing. Can we then say that the classes of automata and grammars which lie in between finite automata and Turing machines—and there are many such classes investigated in the theoretical computer science—are not "natural"? In some sense, this is the case. For instance, the context-free languages have a definition which has a mathematical-linguistic motivation, while the context-sensitive languages have a definition with a motivation coming from the complexity theory (it refers to the space needed for generating or recognizing the strings of a language).

30.24 Let Us Pass to the Cell!

In spite of the theoretical achievements, of numerous successful experiments (however, dealing with problems of small dimensions) and of the continuous progresses in what concerns the lab techniques, the DNA computing has not confirmed the enthusiasm of the twenty years ago, after the announcement of the Adleman experiment—if not having, as I have suggested before, application in cryptography which will be declassified only after several decades. There are elements which can support this assumption. For instance, during the first DNA computing conference, Princeton, 1995, a communication was presented, [7], which described a possibility to break Data Encryption Standard, DES, the system used by the American administration, using DNA, in four months. Next year, the subject was discussed by a team containing also Adleman, and the proposed DNA experiment was supposed to can break DES in five days, provided that the lab operations would be done by robots. A further paper of this kind was presented in 1997, the year when DES was broken also with electronic computers and then abandoned.

Anyway, at some years after Adleman experiment it was clear that one cannot go essentially further, it was necessary to have one more innovative idea, one more "breakthrough" in order to make an essential step towards applications (towards a "killer-app", as the Americans use to say), and one of the "explanations" of this situation was the fact that DNA molecules behave better in vivo (more predictable, more robustly) than in vitro. The suggestions is just natural: let us go to the cell!

30.25 The Fascinating Cell

The cell is really fascinating for a mathematician-computer scientist. I am sure that this is true also for biologists. The smallest entity which is unanimously considered *alive*. The topic is not trivial: at the middle of years 1980, at the Santa Fe Institute for complexity studies a new research vista was initiated, under the name of *artificial life*, as an extension of artificial intelligence, aiming to investigate the life per se, to simulate it on non-biological supports, on computer and in mathematical terms. The starting point was, of course, the attempt to have a definition for what we intuitively call *life*, but the progresses have not gone too far: all definitions either left out something alive, or they ensured that, for instance, the computer viruses are alive (they have "metabolism", self-reproduction etc.). Let us also remember that already Erwin Schrödinger has a book whose title asks *What is Life?*, [67].

The cell passes this test. It is an extraordinarily small "factory", with a complex, intricate and efficient internal structure, where an enormous number of agents interact, from ions to large macromolecules like that of DNA, and where informational processes are carried out at each place and in each moment. Some cells live alone (I am not saying "isolated"), as unicellular organisms, other cells form tissues, organs, organisms.

It is a topic of interest the one concerning the role of the cells in making possible the life itself. I am only citing the reference book [2] the paper [34], also important for what follows because it proposes the slogan "life means surfaces inside surfaces", referring to the membranes which define the inner structure of the cells, and I end with a paragraph from [40]: "The secret of life, the wellspring of reproduction, is not to be found in the beauty of Watson–Crick pairing, but in the achievements of collective catalytic closure."

I am adding also a suggestive equation-slogan, which acad. Solomon Marcus has launched during one of the first Workshops on Membrane Computing, the one in Curtea de Argeş, 2002, [50]:

$$Life = DNA\ software + membrane\ hardware.$$

30.26 The Membrane. From Biology to Computability

We have thus arrived to a fundamental ingredient—the membrane. One can speak very much about it, and the biologists and the experts in bio-semiotics have done it. The cell itself exists because it is separated from the neighboring environment by a membrane. Not only metaphorically, any entity exists because it is delimited by a "membrane", actual or virtual, from the world around.

The (eukaryotic) cell also has a number of membranes inside: the one which encloses the nucleus, the complicated Golgi apparatus, vesicles, mitochondria. From a computational point of view, the main role of these membranes is to define "protected reactors", compartments where a specific biochemistry takes place. There also are other features-functions of the biological membranes which are important for membrane computing: in membranes are placed protein channels which allow the selective communication among compartments; on membranes are bound enzymes which control many of the biochemical processes which take place around them; the membranes are useful also for creating reaction spaces small enough so that the molecules swimming in solution can get in contact so that they can react. It is said that when a compartment is too large for the local biochemistry to be efficient, nature creates new membranes, in order to obtain small enough "reactors" (so that, by Brownian movement, the molecules collide sufficiently frequent and react) and for creating new "reaction surfaces".

I stress the fact that I look here to the cell, its structure, and processes inside it through the glasses of the mathematician-computer scientist, ignoring many biochemical details (for instance, the structure itself of the membranes) and interpreting the selected ingredients according to the goal of this approach: to define a computing model.

Let us give some details, starting with the essential role of membranes in communication. If, in the biological cell or in the model we are going to define, the compartments delimited by membranes evolve separately, then we will not have one "reactor", but a number of neighboring "reactors", evolving independently. However, the membranes ensure the integration. The polarized molecules or those of great dimensions cannot pass through the (phospholipid, with a polarized "head" and two hydrophobic "legs") molecules of the membranes, but they can pass across a membrane through the protein channels embedded in it. This passage is selective and sometimes it is done against the gradient, from a smaller concentration to a higher concentration. A very interesting case is that of the simultaneous passage through a protein channel of two or more molecules: the respective molecules cannot pass separately, but they can do it together, either in the same direction (*symport*) or in opposite directions, one molecule entering the respective compartment and the other one going out, simultaneously (*antiport*). An important chapter of membrane computing is based on these operations and the interest comes from the particularity of this process: there is no rewriting, but only object transport across the borders defined by the membranes, there is no erasing, but only communication. *Computing by communicating* (objects). We can formulate also in this context the question *what means to compute in a natural way?*

We can read in many places about the informational processes taking place in a cell, in most cases with the involvement of membranes, too.

"Many proteins in living cells appear to have as their primary function the transfer and processing of information, rather than the chemical transformation of metabolic intermediates or the building of cellular structures. Such proteins are functionally linked through allosteric or other mechanisms into biochemical 'circuits' that perform a variety of simple computational tasks including amplification, integration, and information storage."

This is the abstract of the paper [8]. In their turn, in [30], one interprets the cytoskeleton as an automaton, while in [46] one constructs a whole theory starting from the informational aspects of the cell life. About the bio-semiotics of the cell has elaborated in many places Jesper Hoffmeyer, already mentioned at the previous pages. I am citing here only his paper [35].

30.27 A Terminology-History Parenthesis

Before passing to a quick description of membrane computing, let me point out a few preliminary things.

First, about the name of the domain. I have called it *membrane computing*, starting from the role of the membrane in the life of the cell, in its architecture and functioning, but the choice was not the best one. "Cellular computing" was probably the most "marketable" choice, but I have discarded it as being too comprehensive.

Then, the name of the models: in the first papers, I used "membrane systems", but soon those who started to investigate these models have called them "P systems", continuing the line of other computing devices having a name (H systems are the closest ones). In the beginning this induced to me some public discomfort, for instance, during conferences, but the letter P soon became autonomous, completely neutral to me.

The domain has grown very rapidly and it is still active after more than sixteen years since its initiation. I have sometimes asked myself which were the explanations, and what I can do for enhancing the growth. Several aspects concurred to the interest for the membrane computing: the favorable context (the natural computing "fashion" mentioned in the beginning); the right moment, on the one hand, with respect to the DNA computing (which, in some sense, is covered and generalized by membrane computing), on the other hand, with respect to the theoretical computer science in general and the formal language theory in particular.

There are several things to be mentioned here. After four decades since the introduction of Chomsky grammars, the formal language theory became "classical" enough and got somewhat retired from the front research (almost completely in USA), even if there still exist specialized conferences (for instance, about finite automata and their applications) or more general conferences (DLT—Developments in Language Theory). Membrane computing appeared as a continuation and an extension of formal language theory: the main investigation objects are no longer the strings of symbols and the languages, but (I anticipate) the multisets of symbols and the sets of multisets. Strings, without taking into account the ordering of the symbols, more technically speaking, strings "seen" through the Parikh application, the one which tells us the number of occurrences of each symbol in a given string. The consequence was that a large number of researchers in formal language theory became interested in the new research area.

A comprehensive information about the membrane computing area can be found at the domain website from the address http://ppage.psystems.eu, hosted in Vienna (it is the successor of a page which has functioned for many years in Milan, Italy, at the address http://psystems.disco.unimib.it).

Of course, it has counted very much the "sociology" of the domain. A *community* was soon created, and this is very important, not only in science, but in culture in general. They have contributed to that the seniors mentioned above, the yearly conferences (started in 2000, with the first three editions organized in Curtea de Argeș, Romania, where the meeting returned for the tenth edition and where I intend to also organize the twentieth edition) as well as a series of meetings which I would like to specially emphasize, one of a unusual format, which I have organized for the first time in Tarragona, Spain, in 2003. After that, it took place every year in Seville, also in Spain. Because it had to have a name, I called it "Brainstorming Week on Membrane Computing". One week when researchers interested in membrane computing work together, far from the current preoccupations, teaching or bureaucratic tasks. A very fruitful idea was to collect in advance open problems and research topics and to circulate them among the participants before the meeting in Seville, then addressed, in collaboration, during the Brainstorming. Very useful meetings—in the website of membrane computing one can find the yearly volumes, with the papers written or only started during the Brainstorming.

Very useful was, of course, the Internet. The first paper, [58], has waited more than one year before it was published, but, because I was in Turku, Finland, in the autumn of 1998, I made the paper available on Internet, in the form of an internal report of TUCS, Turku Center for Computer Science (*Report No.* 208, 1998, www. tucs.fi). Until 2000, when the journal paper has appeared, there were written some dozens of papers, making possible the organization of the first meeting dedicated to this topic, in Curtea de Argeș.

30.28   A Quick View on Membrane Computing

Let us not forget: we want to start from the cell and to construct a computing model. The result (the one proposed in the fall of 1988) is something of the following form. We look to the cell and we abstract it until we only see the *structure* of the hierarchically arranged *membranes*, defining *compartments* where *multisets of objects* are placed (I am using a generic term, abstract, free of any biochemical interpretation); these objects evolve according to given *reactions*. A multiset is a set with multiplicities associated with its elements, hence it can be described by a string; for instance, *aabcab* describes the multiset which contains three copies of *a*, two of *b*, and one of *c*. All permutations of the string *aabcab* describe the same multiset. The reactions, in their turn, are described by multiset "rewriting" rules, of the form $u \rightarrow v$, where u and v are strings which identify multisets. Initially (in the beginning of a computation), in the compartments of our system we have given multisets of objects. The evolution rules start to be applied, like biochemical reactions, in parallel, simultaneously, making evolve all objects which can evolve—and thus the multisets change. Using a rule $u \rightarrow v$ as above means to "consume" the objects indicated by u and to introduce the objects indicated by v. We have to notice that the objects and the rules are localized, placed in compartments, the rules in a given compartment are applied only to objects from that compartment. Certain objects can also pass through membranes. We proceed by applying rules until (like in the case of a Turing machine) we get stuck, no rule can be applied, and then the computation halts. The result of a halting computation is "read", for instance, in the form of the number of objects placed in a compartment specified in advance.

Processing of multisets (of symbols), in parallel, in the compartments defined by a hierarchical structure of membranes—this is the short description of a "P system". A distributed grammar, working with multisets of symbols—this is the direct connection with the formal language theory.

The working site starting here looks endless.

First, one can introduce a large number of variations of P systems, with a mathematical, computer science, biological motivation, or motivated by applications.

From the point of view of mathematics, the models should be minimalistic, they have to contain the smallest number of ingredients. For computer science, a computing model is good to be as powerful as possible, in the best case universal, equivalent with the Turing machine, and as efficient as possible, in the best case able to solve **NP**-complete problems in polynomial time.

Biology and applications provide a long list of alternatives, starting with the way of arranging the membranes (hierarchical, as in a cell, or placed in the nodes of an arbitrary graph, as in tissues and other populations of cells), the types of objects (symbols as before, strings or even more complex data structures, such as graphs or bidimensional arrays), the form of the evolution rules (also dependent on the type of objects), the strategies of applying them, the way of defining the result of a computation.

I have mentioned before the multiset rewriting rules. They can be *arbitrary, non-cooperative* (with the left hand multiset consisting of a single object, which corresponds to context-free rules in Chomsky grammars), or, an intermediate case, *catalytic* (of the form $ca \rightarrow cv$, where c is a catalyst, an object which assists object a in its transformation to the multiset v). Then, we have the *symport* and *antiport* rules, which move objects from a compartment into another one (example: the antiport rule $(u, out; v, in)$, associated with a membrane, moves the objects indicated by u from this membrane to the surrounding compartment and the objects indicated by v in the opposite direction). Very important are the rules which *divide* membranes, because they increase, even exponentially, the number of membranes in the system. Many other types of rules were investigated (for instance, with a control on their application—with promoters, inhibitors, etc.), but I do not mention them here, the presentation would become too technical for the intentions of this text.

If the objects in the compartments of a system are strings, then they evolve by means of operations specific to strings: rewriting, insertion and deletion, or, in order to make the model more uniform from a biological point of view, by the splicing operation from the DNA computing.

An interesting situation is that when we work with symbol objects, hence with numbers, but the result of a computation is "read" outside the system, in the form of the string of the objects which are expelled from the system. It is worth noticing the qualitative difference between the internal data structure, the multiset, and the external one, the string, which carries out positional information.

In turn, the applications need a completely different strategy of constructing the models—far from minimalistic, but adequate to the modeled piece of reality; this time not the computing power is of interest, but the evolution in time of the system. I will come back to applications.

Over this small jungle of models one superposes the investigation program of the classic computer science: computing power, normal forms, descriptional complexity, computational complexity, simulation programs, etc., etc.

30.29   Classes of Results (and Problems)

Of course, I will not recall precise theorems, but I will only mention the two main classes of results in membrane computing and their general form.

Computational completeness/universality: most of the classes of P systems considered so far are equivalent with Turing machines, they are computationally complete. Because the proofs are constructive, in this way one also brings to membrane computing the universality property in the sense of Turing (that is why we speak about computational completeness and universality as they would be synonymous). In most cases, this result is obtained for systems of a reduced, particular form, with a small number of membranes. For instance, cell-like P systems with only two membranes, using catalytic rules (hence not of the general form) can compute whatever the Turing machines can compute.

An important detail: *two* catalysts are sufficient. It is an open problem whether the P systems with only one catalyst are universal. The conjecture is that the answer is negative, but the proof still fails to appear. This is one of the most interesting types of open problems in membrane computing (many of them still open): identifying the precise borderline between universality and non-universality.

Efficiency: the classes of P systems which can grow (exponentially) the number of membranes can solve **NP**-complete problems in a polynomial time. The idea is to generate, in a polynomial time, an exponential working space and then to use it, in parallel, for examining the possible solutions to a problem. Membrane division helps, similarly the membrane creation, similarly other operations. Like in the case of the Adleman experiment, we have again a space–time trade-off, but in our case the space is not provided in advance, but it is created during the computation, through "mitosis" or by means of other "realistic" biological operations.

There are also in this area open problems concerning the borderline between efficiency and non-efficiency, but more difficult to be stated in plain words.

Interesting is a somewhat unexpected fact. Using rules of the form $a \rightarrow aa$, applied in parallel, we can produce an exponential number of copies of a in a linear number of steps. (In n steps, we get 2^n copies of a.) However, such an exponential working space is not of any help in solving high complexity problems in a feasible time—this is what the so-called *Milan theorem*, from Claudio Zandron PhD thesis, says. If these objects are localized, placed in an exponentially large number of membranes, then the situation is different. Otherwise stated, not only the size of the working space matters, but also its structure, the possibility to apply different rules in different compartments. This is a subtle aspect, which I do not know whether it has been met also in other frameworks.

For details, the reader is refereed to the monograph [59] and, especially, to the handbook [63].

30.30   Significations for Computer Science and for Biology

A computing model which has the same power as the Turing machine is a good thing, such a computer is universal not only in the intuitive sense, but it is also programmable. Moreover we have here a distributed, parallel computer, with a great degree of non-determinism, controlled in various biologically inspired ways.

Let us, however, observe the similarities and the differences between a usual computer program, a set of instructions of a Turing machine, and a set of evolution rules of a P system. In the programming languages, the programs consist of precisely ordered instructions, perhaps labeled and addressed by means of these labels. In the case of the Turing machine, the sequence of instructions to be applied is determined by the states of the machine and by the contents of the tape. In the cell case, the reactions are potential, their set is completely unstructured, and their application depends on the available molecules. The evolution rules are just waiting for the data to which they can be applied, there is a competition between rules with respect to the objects to process.

The differences are visible and they suggest once again the question *what means to compute in a natural way?*, adding now the question whether we can work with programs in the form of completely unstructured sets of instructions.

On the other hand, in the first moment, it is expected that the biologist reaction to results of the type of the equivalence with the Turing machine is indifference, a raising of the shoulders. Another domain, another language, another book… But: if the cell is so powerful from a computational point of view, then, according to an old result, the Rice theorem ("all nontrivial problems—having both instances with a positive answer and instances with a negative answer—about a computing model equivalent with Turing machines are algorithmically undecidable"), no nontrivial question about the cell can be solved in an algorithmic way, by means of a program. The biologists formulate every day such questions: How a cell, a cell population, an organ or an organism evolves in time? Is there a substance which gets accumulated over a given threshold, in a given compartment? What happens if we add a multiset of molecules (a medicine), does the state of an organ improves (from specified points of view)?—and so on. If a model of the cell would be decidable, then we could find the answer to such questions by (algorithmically) examining the model, at a given initial state. But, because this is not possible (cannot be done in principle, not only we cannot do it now, here), what remains to do are the laboratory experiment (expensive and time consuming), the computer experiment (cheap, fast, but with the relevance depending on the quality of the model), and, theoretically, the non-algorithmic, ad-hoc, approach.

The previous paragraphs can be seen also as a plead for biology to learn new languages, in particular, the language of theoretical computer science, thus having the possibility of raising problems and of finding solutions which cannot appear, cannot be even formulated in the previous language. This would be an essential step towards infobiology.

30.31 Three Novel Computer Science Problems

In the continuation of the discussion about the significance for computer science, let us point out a remarkable fact: natural computing in general and membrane computing in particular raise theoretical questions which were not considered in the framework of the classical computer science. Here are three questions of this kind, all three pertaining to complexity theory.

Like in the case of Adleman, most experiments of DNA computing started from an instance of a problem and constructed a "computer" associated with that instance. The standard complexity theory does not allow such an approach, it asks for *uniform* solutions, for programs/algorithms which start from the problem (and its size) and solve all instances of the problem. The idea is that during the programming stage one can already work on solving the problem, so that one can then pretend that the solution was found faster than it was the case in reality. That is why, also for the uniform solutions one limits the time allowed for programming, for constructing the algorithm. Let us then place a bound also on the programming time in the case when we start from an instance, so that we cannot cheat here either. The relationship between uniform solutions and semi-uniform (with a limited time for programming) solutions is not clarified yet, in spite of its importance for the natural computing. In membrane computing there were reported a series of related results—see, for instance, recent papers by Damien Woods (Caltech, USA), Niall Murphy (Microsoft Research, Cambridge, UK), Mario J. Pérez-Jiménez (Seville University, Spain).

Second: in DNA computing and in many models in membrane computing, at least part of the steps of a computation are of a non-deterministic type, but in the end the experiment/computation provides a unique result. The idea is to organize the computation in such a way that it is *confluent*, with two variants: either the system evolves non-deterministically for a while, then it "converges" to a unique configuration and then it continues in a deterministic way, or the system "converges logically", it gives the same result irrespective of how it evolves. Again, the complexity theory lacks a study of these situations, of the cases intermediate between determinism and non-determinism.

Finally, the biology provides situations where extended resources wait for external challenges which activate a suitable portion of the resource. The examples of the brain and of the liver, from which we use at any given time only part of the huge number of available cells, are the most known. We can then imagine "computers"—for instance, neural networks—with an arbitrarily large number of cells/neurons, but containing only a limited quantity of information (not to hide there the solution of a problem); after introducing a problem in the system, one activates the necessary number of cells/neurons for solving it. There is no theory dealing with this strategy (of using *pre-computed resources*). How the pre-computed working space should look in order to contain only "a limited quantity of information", how this information can be defined and measured, when a system with pre-computed resources is acceptable/honest, it cannot hide the solution of a problem in its structure?

Natural computing not only motivates the improvement of old results in computer science, but it also makes necessary new developments, which were not imagined before.

30.32 About the Tools Used in Membrane Computing

In order to stress once again the relationships between various branches of theoretical computer science which, at the first sight, look far from each other, and the fact that membrane computing, the natural computing in general, use many old techniques and results, let me remind some details from my personal experience.

In the first universality proof for P systems I have used the result of Yuri Matijasevich mentioned also before, of characterizing the sets of numbers computed by Turing machines as solutions of diophantine equations. I have, however, soon realized that a simpler proof can be obtained starting from the characterization of the same sets of numbers with the help of the matrix grammars. The initial paper was published in this form. In this context it appears the necessity of improving some old results in this area. After a while, also the matrix grammars were replaced, the proofs are now based mainly on register machines, investigated already in the sixties.

A technique even older was useful in the first universality proof for H systems, namely the way of functioning of Post systems, which were introduced at the beginning of the years 1940. Adapted to the splicing operation, this has led to a technique called *rotate-and-simulate*, which has become almost standard for H systems and their variants.

In the first years of my research activity, I was much interested in matrix grammars and I have concluded this research with a monograph (published in Romanian, in 1981), extended after a while to a book, [19], in collaboration with Jürgen Dassow, from Magdeburg, Germany, dedicated to all restrictions in the derivation of context-free grammars. The same happened with other domains which were useful in the membrane computing; the Marcus contextual grammars and the grammar systems are the most important of them.

In mathematics and computer science it is not possible to say in advance whether and when a subject or a result will be useful...

30.33 Spiking Neural P (SNP) Systems

A class of P systems inspired from the brain structure and functioning deserves to be separately discussed. It was introduced later than other models, [39], but it seems that it will get earlier hardware implementations useful to computer science (details about this possibility can be found in [78]).

In a few words, such a system consists of "neurons" linked through "synapses" along which circulate electrical impulses, produced in the neurons by means of specific rules. Like in the case of the real neurons (see, for instance, [52]), the communication among neurons is done by means of identical electrical impulses, *spikes*, for which the frequency is relevant, codifying information. Otherwise stated, important is the distance in time between spikes. In each moment, the axons are a sort of "bar codes", sequences of 0 and 1 which move from a neuron to another one. Obviously, the model ignores many neuro-biological details, but even at this reductionistic level we can formulate a series of questions concerning the relevance for computer science. In a certain sense, the SNP systems use the time as a support of information. The distance between two events, two spikes here, codifies a number. Can we construct a computer with such a "memory"? I mention the question only as a speculation—provocative at the theoretical level.

A result which deserves to be recalled refers to the search of SNP systems which are universal in the Turing sense, that is, they can be programmed in such a way to simulate any other SNP system. From the equivalence with the Turing machine, it follows immediately that such a system exists. The problem of interest concerns the number of neurons of an "universal brain" of this kind, able to simulate any computation in any particular system. This number is not at all too large. In [56] one uses 50–80 neurons, depending on the type of rules for producing spikes, but these numbers were subsequently decreased. In newspaper terms, we can say that "there are computationally universal brains consisting of only a few tens of neurons". From here we can either infer that a computing model of the form of SNP systems is very powerful, actually, that the neurons of these systems are *too powerful*, or that the Turing computability level is not very high—or both these conclusions. Of course, the human brain does not function as a Turing machine—but the computational paradigm was useful, to a certain extent, in modeling the brain functioning.

30.34 About Implementations

The DNA computing started by the definition of the splicing operation, in 1987, but about the possibility of using DNA molecules for computing there were discussions already one decade before. However, the domain became popular after Adleman experiment in 1994. An example was thus created, so that the question whether or not there are implementations of P systems is both natural and frequent. It is understood that one speaks about implementations on a biological substrate. The answer is negative. There were some attempts, but no successful experiment was reported.

An experiment of this kind was designed in the group of professor Ehud Keinan (with well known research both in chemistry and biology) from the Technion Institute in Haifa, Israel, where I have spent one week in 2006, exactly with this purpose. Two main related problems were identified from the beginning: finding a P system plausible to be implemented in a laboratory and, of course, finding the biochemical techniques necessary. We did not intend to solve an **NP**-complete problem, we have not found a reasonable one, but we have looked for a system whose behavior was illustrative for membrane computing (compartments, multisets, parallel processing), and we have chosen a system generating numbers in the Fibonacci sequence. The lab implementation seemed to be only a time issue—as well a question of money, for buying the laboratory equipments and the...DNA molecules. The plan was to simulate the membranes by means of the micro-chambers of a reconfigurable lab installation, with the objects being DNA molecules.

The first experiments did not succeed, then the...sociology of science struck again: the two PhD lady students who were in charge with this experiment moved to USA. In the meantime, an USA patent has appeared, on the name of Ehud Keinan, for implementing a P system, but using another technique, based on three non-miscible liquids placed in a common space. As far as I know, it is about a "theoretical implementation", no successful experiment was reported.

The question which naturally arises is whether or not such an experiment would bring something useful from the point of view of applications. Recalling a saying of Benjamin Franklin, "it is impossible to say what will become a newborn baby", but, having in mind the case of DNA computing, it is highly possible that this will only be a *demo*, at the level of simple calculations.

Completely different is the situation of implementations on an electronic hardware. There are several promising implementations on a parallel hardware (on NVIDIA graphic cards, in Seville, Spain), on a hardware especially designed for membrane computing (Madrid–Spain and Adelaide–Australia), on networks of computers, even on web. All these succeed in a great extent to capture the essential characteristics of P systems, the parallelism. Having in mind the parallelism, I do not call implementations, but *simulations* the cases when one uses standard sequential computers.

On the other hand, both the simulation programs and, still more, the implementations are useful in applications.

30.35 Applications

Membrane computing confirms an observation already made in several situations: when a mathematical theory, starting from a piece of reality, is sufficiently developed at the abstract, theoretical level, there are high chances to find applications not only in the domain which has inspired it, but in other areas too, some of them far away, at the first sight, from the reality from where the theory emerged (but having a common deep structure). It is, very convincingly, the present case.

It was just natural to return to the cell. Biology needs tools and models, the cell is not easy to model. It was stated that, after completing the human genome reading, the main challenge for the bioinformatics is the modeling of the cell, [73]. I have already mentioned that many of the models currently used in biology are based on differential equations. In many cases they are adequate, in many cases not. Differential equations belong to the mathematics of the continuum, they are appropriate to very large populations of molecules, uniformly stirred. However, in a cell, many molecules can be found in small numbers, therefore the approximation of the finite through the infinite, as necessary for applying differential equations, can lead to wrong results. This makes necessary the discrete models, in particular, the P systems, which also have other characteristics which are attractive for the biologist: they come from biology, hence they are easily understandable, which is an aspect which should not be underestimated; furthermore, P systems are algorithmic models, directly programmable in order to simulate them on the computer; can be easily extended, are scalable, adding new components, of any type, does not change the simulation program; their behavior is emergent, cannot be predicted by just looking to the components.

There are many applications of membrane computing in biology and biomedicine. From the individual cell, the applications passed to populations of cells (e.g., of bacteria) and then to... ecosystems. I cite here only [11]—the last two authors are biologists, experts in the ecology of the bearded vulture and animal protection from Lleida, Spain. Of course, the ecosystem is a metaphoric cell, while the "molecules" are the vultures, goats, wolves, hunters, all these in discrete quantities, small known numbers, with no possibility to be modeled with the instruments of the continuous mathematics. Other ecosystems which were investigated concern Panda bears in China and the zebra mussel from the water basins of the Spanish hydroelectrical plants.

So far, plausible applications. Not so expected are the applications in computer graphics (but in this respect we have a previous example, that of Lindenmayer systems), cryptography (in the organization of the attack against certain cryptographic systems), approximate optimization (distributed evolutionary computing, with the distribution organized like in a cell; the number of papers in this area is very large, the topic being popular in China, and the results are surprisingly and pleasantly good—with the mentioning that the famous no free lunch theorem should cool down also here the enthusiasm), economic modeling (a metaphorical extension similar to that to ecosystems), robot control.

These two last areas of applications are part of a potentially larger one, based on the use of the so-called *numerical P systems*, where, in a cell-like framework there evolve numerical variables, not molecules; the evolution is done by means of certain *programs*, consisting of a *production function* and a *repartition protocol*. The inspiration comes from economics, [60]. The systems of this kind compute functions of several variables, in a parallel way, and this computation is rather efficient, that is why it is expected that this somewhat exotic class of P systems will find further applications.

Details about applications can be found in the webpage of membrane computing, in the mentioned *Handbook*, as well as in the collective volumes [12, 27].

30.36   Doubts, Difficulties, Failures

During ceremonies like the today one [delivering a Reception Speech in the Academy] or with the occasion of periodical reports, it is not usual, even not appropriate, to also speak about difficult moments, even if this would be instructive for the reader and useful for the domain.

On the other hand, the hesitations and the doubts are continuously a component of the researcher life. For instance, I can compile a long list of moments where my expectations were of a certain type and the results were different.

This happened starting with the mathematical results. For instance, in the beginning I did not believe that the catalytic P systems are universal, furthermore, that they are universal even in the case of using only two catalysts. Similarly, for a while I have expected to find a class of systems for which the number of membranes induces an infinite hierarchy (of the classes of sets of computed numbers). In exchange, almost always the universality is obtained with only one or two membranes. One membrane means no structure of the system, a trivial architecture. Of course, we can see here the positive fact: the (catalytic) processing of multisets is powerful enough in order to simulate a Turing machine.

Because I have in mind the case of the DNA computing, I do not count as a failure the fact that there are no biological implementations of P systems (although such an event would have a great publicity impact), but I still wait for an implementation on a parallel or a dedicated hardware having a "commercial" value. Such an implementation is necessary and, I believe, it is also possible. For instance, some years ago, a team of biologists and computer scientists from Nothingham, Sheffield (UK), and Seville have tried to simulate on a computer the communication among bacteria, modeling the so-called *quorum sensing*. The simulation programs were able to deal with hundreds of bacteria, the biologists wanted to pass to populations of thousands of bacteria. My expectation is that the implementations, for instance, on NVIDIA cards, will reach soon this level of magnitude requested by the biologists.

Concerning the applications in general, although they were not of interest at the beginning of membrane computing, at some moment it was clear that the domain cannot pass over a certain level of development and notoriety without "real" applications. For a while, there were applications, but of the *postdiction*, not of the *prediction* type. The frequent scenario is the following: we take a biological phenomenon, discussed in a paper or in a book, we formalize it as a P system, we write a simulation program (or we take one available—at this moment we also have a specialized programming language, *P-lingua*, realized in the Seville University), we perform experiments with data from the paper or the book, and if the results are similar to those obtained in a laboratory or through other methods, we are happy. Postdiction, nothing new for the biologists, we only get more confidence in the new model and we can tune it with real data. In order to pass over this stage it is necessary to have a biologist in the team, who should come with a research question, with hypotheses which need to be checked. In turn, the computer scientist should come with sufficiently versatile models and with sufficiently efficient programs, in order to cope with the complexity of biological processes. After sixteen years, the bibliography of membrane computing applications (among them, comfortably many being of the prediction type) is rather large—see the references from the previous section (we add the recent paper [6])—although still we need biologists who have to come towards the computer scientists, maybe to learn membrane computing or, at least, to learn to use the instruments which the computer scientists have already realized (and tested).

I said before that I was continuously interested in forming a community—initially, this was an intuitive desire, later it became conscious, as this was a way to stabilize the domain against the dynamics of the groups. This looks as an external aspect, but we do not have to ignore the influence of the psycho-sociology on science, especially in the case of young branches. A group which is broken can mean a group less (it depends where its members land, whether they continue or not the research activity) or the apparition of several new groups, in new places. I have been the witness of both these two types of consequences. Fortunately, at the present time, the membrane computing community has dimensions which provide it with a comfortable inertia—which, however, does not mean that membrane computing will not get dissolved into infobiology, it already works for that…

30.37 At the Frontier of Science-Fiction

The main promise of natural computing is a better use of the existing computers, pushing forward the frontier of feasibility, by providing solutions, perhaps approximate, to problems which cannot be solved by means of traditional techniques. The DNA computing came with a more ambitious goal, that of providing a new type of hardware, of "biological chips", "wet processors", efficient not only in computational terms, but also in what concerns the energy consumption, or making plausible very attractive features, self-healing, adaptation, learning. Biology can suggest also new computer architectures or ideas for implementing other dreams of computer science, such as the parallel computation, the unsynchronized one, the control of distributed processes, the reversible computation and so on.

All these are somewhat standard expectations, but there also are some ideas which point out to the science of tomorrow, if not directly to science-fiction.

One of these directions is that which aims to *hypercomputability*, to "compute the uncomputable", to pass beyond the "Turing barrier". The domain is well developed, there are over one dozen basic ideas which lead to computability models stronger that the Turing machine—while physics does not forbid any one of these ideas, moreover, it even suggests ideas which look genuinely SF, like, e.g., the use of an internal time of the model which contains cycles or of a bidimensional time. It is true, Martin Davis [20] considers all of them tricks by which the computing power is introduced in the model from the very beginning, in disguise, and then one proves that the model passes beyond the Turing machine (for example, one considers real numbers, which can codify, in their infinite sequence of decimals, all possible computations), but there also are some ideas which look more realistic that others.

One of them is that of *acceleration*, already discussed several decades ago, not only in computer science: R. Blake (1926), H. Weyl (1927), B. Russell (1936), have imagined processes which need one time unit (measured by an external clock) for performing the first step, half of a time unit for the second step (the process "learns"), and so on, at every step, half of the time needed by the previous step. In this way, in two time units (I insist: external, measured by the observer) one performs infinitely many (internal) steps. Such an accelerated Turing machine can solve the halting problem, hence it is more powerful than a usual Turing machine.

Let us now remember the observation that nature creates new membranes in order to get small reactors, where the reactions are enhanced, because of the higher possibilities of molecules to collide. Consequently, *smaller is faster*. The biochemistry in an inner membrane is faster than in the surrounding membrane. Let us push the speculation to the end and assume that the "life" in a membrane is twice faster than in the membrane containing it. Exactly the acceleration we have mentioned above. One can prove [10] that, exactly as in the case of the accelerated Turing machine, an accelerated P system (able to repeatedly create inner membranes) can decide the halting problem.

Hypercomputability might seem to be only a mathematical exercise, but it is estimated that passing beyond the Turing barrier could have more important consequences than finding a proof, even an efficient one, of the $\mathbf{P} = \mathbf{NP}$ equality; see, for instance, [17].

Let us get closer to the laboratory. I have mentioned the lab implementation of a finite automaton with an autonomous functioning. A finite automaton can parse strings. The genes are strings, the viruses are strings (of nucleotides). A hope of medicine is to cure illnesses by editing genes, to eliminate viruses by identifying them and then cutting them in pieces. A more efficient idea than to introduce medicines in out body is to construct a "machinery" which can recognize and edit the necessary sequences of nucleotides, genes or viruses. To this aim, we need a carrying vector, to bring the gene editor in the right place. The identification of that place can be done by an automaton, possibly a finite one, while the vector can be a sort of nano-carrier which can be also built from DNA molecules. In short, un nano-robot, suitably multiplied, which can move from a cell to another one, curing what it is necessary to be cured. A pre-project of such a nano-robot was presented in [4]. In a great extent, it was the same team which has implemented the autonomous finite automaton mentioned before.

There still are many things to be done, the possibility to have our body continuously scanned by a gene repairing robot is not at all close to us. (Such a robot can also have malevolent tasks, it can be a weapon—one can open here a discussion about the ethics of research, but there are sufficiently many debates of this type, even in bio-computer science. Also Francis S. Collins speaks about bioethics in *The Language of God*, the book mentioned several pages before.) On the other hand, there are numerous nano-constructions made of DNA, "motors", "robots", etc. The nano-technology based on DNA biochemistry is spectacularly developed. I cite, as a reference, the paper [64].

It is worth mentioning here also an observation made by Jana Horáková and Jozef Kelemen in [36], with respect to the evolution of computers, somewhat in parallel with the evolution of the idea of a robot: from organic to electromagnetic, then to electronic, and in the end tending to return to organic.

Further speculations? Without any limits, starting from facts with a solid scientific background. In the extreme edge, one can mention Frank Tipler, with his controversial eternal life, in informational terms, which is nothing else than artificial life at the scale of the whole universe [71]. In any case, we have to be conscious that all these are plans for tomorrow formulated today in the yesterday language, to cite a saying of Antoine de Saint-Exupéry. The progresses in bioengineering can bring surprises which we cannot imagine in this moment.

30.38 Do We Dream Too Much?

Let us come down on the Earth, to the reality, to the natural computing as we have it now and how it is plausible to have it in the near future, adopting a lucid position, even a skeptical one, opposed to the enthusiasm from the previous section and to the enthusiasm of many authors. (I am not referring here to newspaper authors, which too often use big words when talking about bioinformatics.)

In order to promote an young scientific branch, the enthusiasm is useful and understandable—but natural computing is no longer an young research area. Let us oppose here to the previous optimism a more realistic position, starting from the differences, many and significant, between computer science and biology, from the difficulties to implement bio-ideas in computer science and computations in cells: the goal of life is life, not the computations, we, the computer scientists, see everywhere computations and try to use them for us; in a certain sense, life has unbounded time and resources, it affords to make experiments and to discard the results of unsuccessful attempts—all these are difficult to extend to computers, even if they are based on biomolecules. Similarly, life has a great degree of redundancy and non-determinism. Then, the biological processes have a high degree of complexity, moreover, they seem to mainly use the mathematics of approximations, probabilities, fuzzy sets, all of which are difficult to be captured in a computing model, not to speak about the difficulty to implement them.

Still more important: we perhaps dream too much even from the theoretical point of view. First, the space–time trade-off does not redefine the complexity classes, at most it can enlarge the feasibility space (see again Hartmanis remarks about Adleman experiment).

Then, there is a theorem of Michael Conrad [16] which says that three desired characteristics of a computer, *programmability* (universality), *efficiency*, and *evolvability* (the capacity to adapt and learn), are contradictory, there is no computer which can have all these three features at the same time. We can interpret this result as a general *no free lunch* theorem for the natural computing.

A similar theorem of limitation of "what can be done in principle" belongs to Robin Gandy, a student and collaborator of Turing, which offers general mathematical arguments to Martin Davis: the hypercomputability is a difficult thing to reach (see, for instance, [28]). Gandy wanted to free the Turing–Church thesis of any anthropic meaning (in Turing formulation, the thesis says that "everything which can be computed by a human being can be computed by a Turing machine"). To this aim, he has defined a general notion of a "computing machine", described by four properties formulated mathematically and which any "computer", an actual or a theoretical one, should possess. Then, Gandy proved that any machine having these properties can be simulated by a Turing machine.

Passing from theoretical computer science to applications, let me notice that there are visible limitations also in this respect. I am even convinced that, if one will make lists with the properties the models and the simulations we would like to have (adequacy, relevance, accuracy, efficiency, understandability, programmability, scalability and so on), then impossibility theorems similar to Arrow, Conrad, Gödel theorems will be proved concerning the modeling and the simulation of the cell—the very task which M. Tomita formulated.

30.39   Everything Is New and Old All Are…

(The title of this section reproduces a verse from a poem by Mihai Eminescu, the national poet of Romania.)

In spite of what was said above, there is a more and more visible interest in the modeling of the cell. Actually, a dedicated research direction was proposed, the *systems biology*, with several programmatic papers, published in high visibility journals, such as *Science* and *Nature*. The main promotor was H. Kitano [41, 42], which has in mind a general model of the cell, meant to be simulated on a computer and then used, in relation also with other computer science and biological instruments, in such a way "to transform biology and medicine in a precise engineering". The goal is important and probably feasible in a medium-long term, but the insistence with which one speaks about "systems biology" as about a novel idea made Olaf Wolkenhauer to ask already in the title of his paper [76] whether this is not only "the reincarnation of systems theory applied in biology". The paper recalls the efforts in this respect made in the years 1960, with the disappointments appeared at that time, due, among others, to the limits of the computers (but also to the limits of biology: let us remember that the Singer-Nicolson model of the membrane as a "fluid mosaic" dates only from 1972). But, besides the computing power, it is possible that something else was missing, which is perhaps missing even today, both in computer science and in biology. The last paragraph from Olaf Wolkenhauer paper invokes the name of Mihailo Mesarovic, a classic of systems theory, which, in 1968, said: "in spite of the considerable interest and efforts, the application of systems theory in biology has not quite lived up the expectations... One of the main reasons for the existing lag is that systems theory has not been directly concerned with some of the problems of vital importance in biology". His advice for biologists, continues Olaf Wolkenhauer, is that such a progress can only be obtained by means of a stronger direct interaction with the systems theory researchers. "The real advance in the applications of systems theory to biology will come about only when the biologists start *asking questions* which are based on the system-theoretic concepts rather than using these concepts to represent in still another way the phenomena which are already explained in terms of biophysical or biochemical principles... then we will not have *the application of engineering principles to biological problems*, but rather a field of *systems biology* with its own identity and in its own right [54]."

Mesarovic words can be taken as a motto of infobiology in favor of which the whole present text pleads.

The transformation of biology and medicine in "a precise engineering" can be also related with the current difficulties to understand what is life, materialized, among others, in the current limits of the artificial intelligence and artificial life. One says, for instance, that up to now the computers are good in IA, the intelligence amplification, but not equally good in AI, artificial intelligence. Still less progresses were made in what concerns the artificial life. In terms of Rodney Brooks, [9], this suggests that "we might be missing something fundamental and currently unimagined in our models of biology". Computers are good in crunching numbers, but "not good at modeling living systems, at small or large scale". The intuition is that life is more than biophysics and biochemistry, but what else it is can be something which we cannot imagine today, "some aspects of living systems which are invisible to us right now". "It is not completely impossible that we might discover some new properties of biomolecules or some new ingredient". An example of such a "new stuff", R. Brooks says, can be the quantum effects from the microtubules of the neural cells, which, according to Penrose, "might be the locus of consciousness at the level of the individual cell" (citation from R. Brooks).

A similar opinion was expressed by another great name of the artificial intelligence, John McCarthy, [53]: "Human-level intelligence is a difficult scientific problem and probably needs some new ideas. These are more likely to be invented by a person of genius than as part of a Government or industry project".

Anyway, the progresses related to the collaboration between computer science and biology should not be underestimated. If we do it, then we take a risk which has struck big names of science and cultures. I close with a funny example of this kind, some statements (dated around 1830) of the French philosopher Auguste Comte: "Every attempt to employ mathematical methods in the study of biological questions must be considered profoundly irrational and contrary to the spirit of biology. If mathematical analysis should ever hold a prominent place in biology—an aberration which is happily almost impossible—it would occasion a rapid and widespread degeneration of that science."

Thanks to God, the philosopher was wrong—but we needed about two hundred years to see that...

30.40   (Provisory) Last Words

I hope that this quick description was convincing in showing that the way from biology to computer science and back to biology is intellectually fascinating and useful to both sciences.

A few things should be remembered: (i) in all its history, computer science tried to learn from biology, (ii) and this effort brought important benefits to computer science and equally to biology; (iii) the progresses in this area should not be underestimated, (iv) but, in general, it is plausible that we expect too much (and too fast) from the computer science-biology symbiosis, (v) because we ignore the essential differences between the two universes, the inherent limits of computability and the fact that biology is not a mathematically formalized science, (vi) with the mentioning that possibly one needs a new mathematics in order to model and simulate life and intelligence; finally, (vii) let me anticipate a new age of biology, beyond the today bioinformatics and the today natural computing, and let me also propose a name for it, *infobiology*.

Should we wait two further decades in order to see it taking shape?

From an intellectual point of view, during the forty years which I have told about here I have lived around academician Solomon Marcus, a "big tree" which invalidates the phrase ("In the shadow of big trees not even the grace is growing.") by which Constantin Brancusi motivated his decision to refuse to work under the guidance of Rodin: professor Solomon Marcus never puts shadow on his numerous students and collaborators, but on the contrary. I repeat, in order to stress it: on the contrary. I witness this and I dedicate to him this discourse, thanking him once again.

Acknowledgments Thanks are due to an anonymous referee, who has carefully read the text and also pointed to us paper [6].

References

1. Adleman, L.M.: Molecular computation of solutions to combinatorial problems. Science **226**, 1021–1024 (1994)
2. Alberts, B., Johnson, A., Lewis, J., Raff, M., Roberts, K., Walter, P.: Molecular Biology of the Cell, 4th edn. Garland Science, New York (2002)
3. Benenson, Y., Paz-Elizur, T., Adar, R., Keinan, E., Livneh, Z., Shapiro, E.: Programmable and autonomous computing machine made of biomolecules. Nature **414**, 430–434 (2001)
4. Benenson, Y., Shapiro, E., Gill, B., Ben-Dor, U., Adar, R.: Molecular computer. A 'smart drug' in a test tube. In: Pre-proceedings of DNA Computing Conference, p. 49. Milano, invited talk, extended abstract (2004)
5. Bennett, C.H.: Logical reversibility of computation. IBM J. Res. Dev. **17**, 525–532 (1973)

6. Bollig-Fischer, A., Marchetti, L., Mitrea, C., Wu, J., Krüger, A., Manca, V., Draghici, S.: Modeling time-dependent transcription effects of HER2 oncogene and discovery of a role for E2F2 in breast cancer cell-matrix adhesion. Bioinformatics **30**, 3036–3043 (2014)

7. Boneh, D., Dunworth, C., Lipton, R.: Breaking DES using a molecular computer. In: Proceedings of DNA Based Computers. DIMACS Workshop, pp. 37–66, Princeton (1995)

8. Bray, D.: Protein molecules as computational elements in living cells. Nature **376**, 307–312 (1995)

9. Brooks, R.: The relationship between matter and life. Nature **409**, 409–411 (2001)

10. Calude, C., Păun, Gh: Bio-steps beyond Turing. BioSystems **77**, 175–194 (2004)

11. Cardona, M., Colomer, M.A., Pérez-Jiménez, M.J., Sanuy, D., Margalida, A.: Modeling ecosystems using P systems: The bearded vulture, a case study. Proceedings of WMC, Edinburgh, UK. LNCS, vol. 5391, pp. 137–156. Springer, Berlin (2008)

12. Ciobanu, G., Păun, Gh, Pérez-Jiménez, M.J. (eds.): Applications of Membrane Computing. Springer, Berlin (2006)

13. Collins, F.S.: The Language of God. A Scientist Presents Evidence for Belief. Simon & Schuster Inc., New York (2006)

14. Constantinescu, E.: God Does not Play Dice. MajestiPress Publishing House, Arad (2008). (in Romanian)

15. Cook, S.: The importance of the P versus NP question. J. ACM **50**, 27–29 (2003)

16. Conrad, M.: The price of programmability. In: Herken, R. (ed.) The Universal Turing Machine: A Half-Century Survey, pp. 285–307. Kammerer and Unverzagt, Hamburg (1988)

17. Copeland, B.J.: Hypercomputation. Mind. Mach. **12**, 461–502 (2002)

18. Copeland, B.J., Proudfoot, D.: Alan Turing's forgotten ideas in computer science. Sci. Am. **280**, 77–81 (1999)

19. Dassow, J., Păun, Gh: Regulated Rewriting in Formal Language Theory. Springer, Berlin (1989)

20. Davis, M.: The myth of hypercomputation. In: Teuscher, C. (ed.) Alan Turing: The Life and Legacy of a Great Thinker, pp. 195–212. Springer, Berlin (2004)

21. Dijkstra, E.W.: The end of computer science? Commun. ACM **44**, 92 (2001)

22. Dijkstra, E.W.: Under the spell of Leibniz's dream. Inf. Process. Lett. **77**, 53–61 (2011)

23. Ehrenfeucht, A., Harju, T., Petre, I., Prescott, D.M., Rozenberg, G.: Computation in Living Cells. Gene Assembly in Ciliates. Springer, Berlin (2004)

24. Eiben, A.E., Smith, J.E.: Introduction to Evolutionary Computing. Springer, Berlin (2003)

25. Engelfriet, J., Rozenberg, G.: Fixed point languages, equality languages, and representations of recursively enumerable languages. J. ACM **27**, 499–518 (1980)

26. Franco, G., Giagulli, C., Laudana, C., Manca, V.: DNA extraction by XPCR. In: Proceedings of DNA Computing Conference, Milano. LNCS, vol. 3384, pp. 104–112. Springer, Berlin (2004)

27. Frisco, P., Gheorghe, M., Pérez-Jiménez, M.J. (eds.): Applications of Membrane Computing in Systems and Synthetic Biology. Springer, Berlin (2014)

28. Gandy, R.: Church's thesis and principles for mechanisms. In: Barwise, J., et al. (eds.) The Kleene Symposium, pp. 123–148. North-Holland, Amsterdam (1980)

29. Gramss, T., Bornholdt, S., Gross, M., Mitchel, M., Pellizzari, Th (eds.): Non-Standard Computation. Wiley-VCH, Weinheim (1998)

30. Hameroff, S.R., Dayhoff, J.D., Lahoz-Beltra, R., Samsonovich, A.V., Rasmussen, S.: Models for molecular computation: conformational automata in the cytoskeleton. IEEE Comput. **25**, 30–39 (1992)

31. Hartmanis, J.: About the nature of computer science. Bull. EATCS **53**, 170–190 (1994)

32. Hartmanis, J.: On the weight of computation. Bull. EATCS **55**, 136–138 (1995)

33. Head, T.: Formal language theory and DNA: an analysis of the generative capacity of specific recombinant behaviors. Bull. Math. Biol. **49**, 737–759 (1987)

34. Hoffmeyer, J.: Surfaces inside surfaces. On the origin of agency and life. Cybern. Hum. Knowing **5**, 33–42 (1998)

35. Hoffmeyer, J.: Semiosis and living membranes. Seminário Avançado de Comunicaçao e Semiótica. Biosemiótica e Semiótica Cognitiva, Sao Paolo, Brasil (1998)

36. Horáková, J., Kelemen, J.: Capek, Turing, von Neumann, and the 20th century evolution of the concept of machine. In: Proceedings of the International Conference in Memoriam John von Neumann, pp. 121–135. Budapest Polytechnic (2003)
37. Hromkovic, J.: Communication Complexity and Parallel Computing. Springer, Berlin (1997)
38. Hromkovic, J.: Why is mathematics useful and how it should be taught? Curtea de la Argeş (2014). www.curteadelaarges.ro (in Romanian)
39. Ionescu, M., Păun, Gh, Yokomori, T.: Spiking neural P systems. Fundamenta Informaticae **71**, 279–308 (2006)
40. Kauffman, S.: At Home in the Universe. Oxford Univ, Press (1995)
41. Kitano, H.: Systems biology: a brief overview. Science **295**, 1662–1664 (2002)
42. Kitano, H.: Computational systems biology. Nature **420**, 206–210 (2002)
43. Landauer, R.: Irreversibility and heat generation in the computing process. IBM J. Res. Dev. **5**, 183–191 (1961)
44. Leupold, P.: Is computation observer-relative? In: Sixth Workshop on Non-Classical Models of Automata and Applications, Kassel, Germany (2014)
45. Livio, M.: Is God a Mathematician?. Simon & Schuster Inc., New York (2009)
46. Loewenstein, W.R.: The Touchstone of Life. Molecular Information, Cell Communication, and the Foundations of Life. Oxford University Press, Oxford (1999)
47. Manca, V.: Infobiotics. Information in Biotic Systems. Springer, Berlin (2013)
48. Marcus, S.: Linguistic structures and generative devices in molecular genetics. Cahiers de Linguistique Thèorique et Appliquée **11**, 77–104 (1974)
49. Marcus, S.: Invention or Discovery. Cartea Românească Publishing House, Bucharest (1989). (in Romanian)
50. Marcus, S.: Bridging P systems and genomics: a preliminary approach. In: Proceedings of WMC-CdeA. LNCS, vol. 2597, pp. 371–376. Springer, Berlin (2002)
51. Marcus, S.: The Loneliness of a Mathematician. The Publishing House of the Romanian Academy, Bucharest (2008). (in Romanian)
52. Maass, W.: Networks of spiking neurons: the third generation of neural network models. Neural Netw. **10**, 1659–1671 (1997)
53. McCarthy, J.: Problems and projection in CS for the next 49 years. J. ACM **50**, 73–79 (2003)
54. Mesarovic, M.D.: System theory and biology – view of a theoretician. In: Mesarovic, M.D. (ed.) System Theory and Biology, pp. 59–87. Springer, New York (1968)
55. Papadimitriou, C.H.: Computational Complexity. Addison-Wesley, Reading (1994)
56. Păun, A., Păun, Gh: Small universal spiking neural P systems. BioSystems **90**, 48–60 (2007)
57. Păun, Gh: On the splicing operation. Discrete Appl. Math. **70**, 57–79 (1996)
58. Păun, Gh: Computing with membranes. J. Comput. Syst. Sci. **61**, 108–143 (2000)
59. Păun, Gh: Membrane Computing. An Introduction. Springer, Berlin (2002)
60. Păun, Gh, Păun, R.: Membrane computing and economics: numerical P systems. Fundamenta Informaticae **73**, 213–227 (2006)
61. Păun, Gh, Rozenberg, G., Salomaa, A.: Computing by splicing. Theor. Computer Sci. **168**(2), 321–336 (1996)
62. Păun, Gh, Rozenberg, G., Salomaa, A.: DNA Computing. New Computing Paradigms. Springer, Berlin (1998)
63. Păun, Gh, Rozenberg, G., Salomaa, A. (eds.): The Oxford Handbook of Membrane Computing. Oxford University Press, Oxford (2010)
64. Reif, J.H., LaBean, T.H., Sahu, S., Yan, H., Yin, P.: Design, simulation, and experimental demonstration of self-assembled DNA nanostructures and motors. In: Proceedings of the Workshop on Unconventional Programming Paradigms, UPP04, Le Mont Saint-Michel (2004)
65. Rozenberg, G., Bäck, T., Kok, J.N. (eds.): Handbook of Natural Computing, vol. 4. Springer, Berlin (2012)
66. Rozenberg, G., Salomaa, A.: Watson-Crick complementarity, universal computations, and genetic engineering. Technical Report 96-28, Leiden University, The Netherlands (1996)
67. Schrödinger, E.: What is Life? & Mind and Matter. Cambridge Univ. Press, Cambridge (1967)
68. Searle, J.: The Rediscovery of the Mind. MIT Press, Cambridge (1992)

69. Teuscher, C. (ed.): Alan Turing. Life and Legacy of a Great Thinker. Springer, Berlin (2003)
70. Teuscher, C., Sánchez, E.: A revival of Turing's forgotten connectionist ideas: exploring unorganized machines. In: French, R.M., Sougne, J.J. (eds.) Proceedings of Connectionist Models of Learning, Development and Evolution Conference, Liége, Belgium, 2000, pp. 153–162. Springer, Berlin (2001)
71. Tipler, F.: The Physics of Immortality. Doubleday, New York (1994)
72. Toffoli, T.: Nothing makes sense in computing except in the light of evolution. J. Unconv. Comput. **1**, 3–29 (2005)
73. Tomita, M.: Whole-cell simulation: a grand challenge of the 21st century. Trends Biotechnol. **19**, 205–210 (2001)
74. Turing, A.M.: On computable numbers, with an application to the *Entscheidungsproblem*. In: Proceedings of the London Mathematical Society, Series 2, vol. 42, pp. 230–265 (1936) (with an erratum in vol. 43, pp. 544–546 (1936))
75. Watson, J.D., Crick, F.H.C.: A structure for deoxyribose nucleic acid. Nature **171**, 737–738 (1953)
76. Wolkenhauer, O.: Systems biology: the reincarnation of systems theory applied in biology? Brief. Bioinf. **2**(3), 258–270 (2001)
77. Wolpert, D.H., Macready, W.G.: No free lunch theorems for optimization. IEEE Trans. Evol. Comput. **1**(1), 67–82 (1997)
78. Xu, Z., Cavaliere, M., An, P., Vrudhula, S.: The stochastic loss of spikes in spiking neural P systems: Design and implementation of reliable arithmetic circuits. Fundamenta Informaticae **134**(1–2), 183–200 (2014)
79. Zenil, H. (ed.): A Computable Universe. Understanding and Exploring Nature as Computation. World Scientific, Singapore (2013)

Chapter 31
Unconventional Computing: A Brief Subjective History

Cristian S. Calude

Abstract In this chapter we present a few stages of the evolution of the emerging area of unconventional computing from a personal perspective.

31.1   Introduction

In 1994 John Casti[1] and I started talking about the eventual decay of Moore's law and the advance of new models of computation, which we called *unconventional*.[2]

The famous open problem NP versus P informally asks whether every problem whose solution can be quickly verified by a Turing machine can also be quickly solved by a Turing machine.[3] By mid 1990s there was a wide spread belief that the problem will be solved in the negative before the end of the century. This motivated the imperious need of "fast" ways to solve NP problems (quickly checkable) not in P (quickly solvable), a computational challenge unlikely, if not impossible, to succeed using Turing computability.

The third reason was the Turing barrier derived from the Church-Turing Thesis: all computations are extensionally equivalent to Turing machines. Is it possible to design new models of computation capable of transgressing Turing's barrier?

The need to have a conference on that subject appeared obvious. The first conference of the new series called *Unconventional Models of Computation* was held

---

[1] At that time at the Santa Fe Institute.

[2] The earliest written reference to the term which I have is from an email sent by Seth Lloyd to John Casti Sat on 27 Jul 1996 17:12:41 in which Seth, answering an email from John, lists some researchers in "unconventional and non-Turing models of computation".

[3] The problem was formulated in an equivalent form by Kurt Gödel in a letter to John von Neumann in 1956. Its current mathematical formulation was given by Steven Cook in 1971 using the classes NP and P defined via polynomial time computability, a debatable model of "feasible computation", see also [18].

C.S. Calude (✉)
Department of Computer Science, University of Auckland, Auckland, New Zealand
e-mail: cristian@cs.auckland.ac.nz

© Springer International Publishing Switzerland 2017   855
A. Adamatzky (ed.), *Advances in Unconventional Computing*,
Emergence, Complexity and Computation 22,
DOI 10.1007/978-3-319-33924-5_31

in 1998; 13 further conferences followed. Journals in the area of unconventional computing started to appear in 2002.

31.2 Moore's Law and Turing's Barrier

Moore's law is a famous empirical trend stating that the number of transistors in a dense integrated circuit doubles approximately every two years.[4] It was proposed in 1965 by Gordon Moore, co-founder of Intel, in [22]; hence in 2015 we mark its 50th anniversary.

In the IT world the law informally translates into the prediction that the silicon chips that power servers, PCs, phones, and wifi gadgets can run faster and consume less power roughly doubling every two years or so. The law is so accurate that it is used in the semiconductor industry to guide long-term planning and to set targets for research and development.

Various analyses concluded that the law will run out of steam, i.e. the improvements of conventional ways of manufacturing microprocessors, graphics chips and other silicon components will hit a wall: drastically new ideas will be required. Predictions are notoriously difficult and this case is no exception. Indeed, 2000 was a failed estimation; other dates, 2013 and 2015, suffered the same fate. At the 2015 IEEE international Solid-State Circuits Conference (http://isscc.org)[5] organised in San Francisco, USA on February 22–26, Intel engineers discussed the challenges of moving from current 14 nm chips to the 10 nm manufacturing node and even smaller. Moore's law is not (yet) dead!

The Church-Turing Thesis[6] states that a function on the natural numbers is computable in an informal sense (i.e. computable by a human being using a pencil-and-paper method, ignoring any resource limitations) if and only if it is computable by a Turing machine. Identifying an informal concept with a mathematical one makes the "thesis" mathematically unprovable; however, in principle, it can be disprovable. There are many studies on the Church-Turing Thesis, see for example [19].

The major challenge posed by the Church-Turing Thesis is *whether there are any possibilities of going beyond Turing's barrier, i.e. whether one can develop tools to compute (in some meaningful way) uncomputable functions.* This problem is both theoretical as well as practical. There is an active community devoted to this problem—called hypercomputation—see [26]. Davis [16, 17] argues strongly against hypercomputation on the ground that it is physically unfeasible. Curiously enough, there exists a practical model of hypercomputation (see [2]) which has a (yet) unknown computational capability (in particular, it is not know whether it can solve the Halting Problem).

---

[4]The period is often quoted as 18 months.

[5]One of the top academic conferences on chip design.

[6]Named after American mathematician Alonzo Church and his British Ph.D. student Alan Turing.

31.3 Unconventional Computation or Unconventional Computing?

The adjective unconventional means "not conforming to accepted rules or standards". The name *unconventional computation* was chosen for its "neutrality": it included both major new computing trends at the time when it was coined, in 1994—biological/molecular/DNA and quantum physics models—and gives none of them any preference. The year 1994 was an exceptional time for both molecular and quantum computation: Adleman's molecular computation paper appeared in *Science* [6] and Shor's quantum algorithm for factoring quickly large integers was published in the same year [23]. They both generated an immense interest across disciplines.

The adjective unconventional has also shortcomings. For example, it is time-dependent: what is unconventional today might become conventional tomorrow. Also, various researchers may have different opinions regarding whether a model of computation is unconventional or not. For example, a model using a finite automaton seems hardly unconventional now as it was in 1994: however, if the use of the automaton is essentially different from the standard ones, then one could argue that the model of computation is unconventional.

The initial title of the book [8] was *The Human Face of Computation*, a title coined by Solomon Marcus, one of its contributors. Then, in a private exchange of emails, Joseph Sifakis, another contributor, argued that "computing" is a better term than "computation". An interesting discussion, involving also Solomon Marcus, followed.

Computation refers to any type of calculation or use of computing technology in information processing [24], while computing denotes any goal-oriented activity requiring, benefiting from, or creating algorithmic processes, e.g. through computers [25]. Computing connotes the use (or study) of computers. Computation connotes calculation (not necessarily by computer and not necessarily via mathematical operations, though this is technically what a computer does). Computation seems to be more theoretical; computing tends to be more general.[7] Also, as Marcus noted, "Computing has only three syllables while computation has four; musically, the former is better than the latter". So the title of the book has been changed!

The initial title of the conference *Unconventional Models of Computation* has been changed in 2005 to *Unconventional Computation*[8] and again in 2012 to the current *Unconventional Computation and Natural Computation*.[9] Here the term "computation" has (till now?) survived.

---

[7]See more at [14].

[8]To emphasise both theoretical and practical studies.

[9]UCNC page: https://www.cs.auckland.ac.nz/research/groups/CDMTCS/conferences/uc/uc.html.

31.4   Unconventional Models of Computation 1998 (UMC'98)

The first conference in the series was organised in Auckland New Zealand on 6–9 January 1998 by the Centre for Discrete Mathematics and Theoretical Computer Science (https://www.cs.auckland.ac.nz/research/groups/CDMTCS/) in Auckland and the Santa Fe Institute (http://www.santafe.edu).

   The website of the conference appears in Fig. 31.1. Its proceedings [10] were published by Springer and distributed to participants during the conference, see Fig. 31.2. Pictures from the conference can be found at UMC98 pictures (https://www.cs. auckland.ac.nz/research/groups/CDMTCS/conferences/umc98/umcphotos.html).

          UMC-98

First International Conference on
UNCONVENTIONAL MODELS OF COMPUTATION

5-9 January 1998
Auckland, New Zealand

The First International Conference on Unconventional Models of Computation was organised by the Centre for Discrete Mathematics and Theoretical Computer Science, NZ, and the Santa Fe Institute, USA. The proceedings volume has been published in the DMTCS Series of Springer-Verlag, Singapore. Click here for some photos.

This web page contains a number of pointers to information that may be of interest in the context of this conference.

- Call for Participation (old)
- Programme
- Electronic Conference Registration (old)
- Registration of Interest (old)
- Accommodation Information
- Homepage of the Centre
- Information about Auckland
- Handy Restaurant Guide
- Handy Cafe Guide
- Handy Pub and Wine Bar Guide
- Links to UMC-Related Web Pages
- NZ Herald presentation
- InfoTech Weekly presentation: Scientists Ponder Future of Computers
- K. Svozil unconventional review (to appear in the EATCS Bull. 64(1998).)
- K. Gh. Paun review (to appear in the EATCS Bull. 64(1998).)

Fig. 31.1 UMC'98 website (https://www.cs.auckland.ac.nz/research/groups/CDMTCS/conferences/umc98/)

The conference attracted a lot of attention. Various reports were published in prestigious international publications like *Nature* [9], *Complexity* [11] and the *Bulletin of the European Association for Theoretical Computer Science* (64 (1998), presentations by G. Păun (https://www.cs.auckland.ac.nz/research/groups/CDMTCS/conferences/umc98/paun.pdf) and K. Svozil (https://www.cs.auckland.ac.nz/research/groups/CDMTCS/conferences/umc98/karl.pdf)) as well as in local media, New Zealand Herald (https://www.cs.auckland.ac.nz/research/groups/CDMTCS/docs/nzherald.pdf) and InfoTech Weekly. Bob Doran's Computer History Time Line (https://www.cs.auckland.ac.nz/historydisplays/TimeLine/TimeLineMain.php) which includes computing history displays in the Computer Science Department at the University of Auckland also marks the UMC'98 event (https://www.cs.auckland.ac.nz/historydisplays/TimeLine/TimeLine4.5/4.5.13-CDMTCS).

Fig. 31.2 UMC'98 Proceedings

31.4.1 Scope

By mid 1990s both areas of biological and quantum computing had their own specialised conferences. The scope of UMC'98 was *to bring together* researchers from as many as possible areas of unconventional computing and encourage/stimulate interaction. The organisers paid special attention to "merge" in the programme papers from different areas, to avoid the creation of clusters of specialised mini-conferences and to stimulate interdisciplinarity.

31.4.2 Preparation

The conference took almost four years to prepare. As mentioned above, discussions had started in 1994. The title included the word "model" because of the theoretical nature of the domain at that stage: unconventional computation was on paper only!

In 1998 the area of unconventional computation included biological/molecular/ DNA and quantum models of computation, neural networks, cellular automata, reversible computation, genetic algorithms, hyper-Turing machines or any other model of computation going beyond Turing's barrier and the meaning and relevance of Church-Turing Thesis for the physics of computation.

31.4.3 Invited Speakers

The list of invited speakers consisted of a rather "unconventional" mixture of[10]

- well-known eminent researchers, H. Jeff Kimble, John. H. Reif and Arto Salomaa,
- and young rising stars, Martyn Amos, Artur Ekert, Seth Lloyd, and Christopher Moore,

with slightly more from the latter category (all well-known researchers today).

31.4.4 Opposing "Philosophies"

In the final session we discussed the goals of biological and quantum computing. In both areas "computers" were understood "by default" to be universal, i.e. equivalent in power to a universal Turing machine.

Regarding the future, two interesting, divergent "philosophies" for quantum computing emerged:

- for the Europeans the main interest was in using the new framework of quantum computing to prove new impossibility results in quantum physics, in particular the impossibility of building a "real" quantum computer, while

---

[10]In alphabetical order.

- the Americans didn't care much whether quantum computers can be built or not, they were determined to build and commercialise them.

Today the idea of universality has been by and large abandoned. Various proposals to solve "quickly" NP-hard problems have also been discarded because of practical unfeasibility.

The European preference can be illustrated with the work on locating "value indefinite observables" [1, 3] which led to a new form of the Kochen-Specker theorem [20] and a better understanding of the unpredictability of quantum mechanics [4, 5].

In the area of quantum computing a new emerging trend called *quassical computing*[11] combines classical and quantum computing by taking advantage of their complementary capabilities.[12] The example of the adiabatic quantum machines D-Wave produced by the Canadian company D-Wave Systems confirms the American prediction. D-Wave One (2011) operates on a 128-qubit chipset while D-Wave Two (2013) works with 512 qubits [15].

31.5 Unconventional Computation and Natural Computation 2015 (UCNC'2015)

After Brussels, Belgium (December 2000), Kobe, Japan (October 2002), Sevilla, Spain (October 2005), York, UK (September 2006), Kingston, Canada (August 2007), Vienna, Austria (August 2008), Ponta Delgada, Portugal (September 2009), Tokyo, Japan (June 2010), Turku, Finland (June 2011), Orléans, France (September 2012) Milano, Italy (June 2013) and London, Ontario, Canada (July 2014), the 14th Unconventional Computation and Natural Computation (http://ucnc15.wordpress. fos.auckland.ac.nz) conference returns to Auckland, New Zealand and was held from 31 August to 4 September 2015.

The UCNC conference series is overseen by a Steering Committee which includes Thomas Back (Leiden University, The Netherlands), Cristian S. Calude (University of Auckland, New Zealand), as founding chair, Lov K. Grover (Bell Labs, Murray Hill, New Jersey, USA), Nataša Jonoska (University of South Florida, USA), as co-chair, Jarkko Kari (University of Turku, Finland), as co-chair, Lila Kari (University of Western Ontario, Canada), Seth Llloyd (Massachusetts Institute of Technology, USA), Giancarlo Mauri (University of Milano-Bicocca, Italy), Gheorghe Păun (Institute of Mathematics of the Romanian Academy, Romania), Grzegorz Rozenberg (Leiden University, The Netherlands), as emeritus chair, Arto Salomma (University of Turku, Finland), Tommaso Toffoli (Boston University, USA), Carme Torras

---

[11]A term coined by N. Allen [7].

[12]An early example of quassicality appears in [21] where a method to enhance Grover's quantum search by incorporating it into classical procedural algorithms is presented. See also [12].

(Institute of Robotics and Industrial Informatics, Barcelona, Spain), and Jan Van Leeuwen (Utrecht University, The Netherlands).

The list of main areas of interest of the conference has expanded considerably: molecular (DNA) computing, quantum computing, optical computing, chaos computing, physarum computing, computation in hyperbolic spaces, collision-based computing; cellular automata, neural computation, evolutionary computation, swarm intelligence, nature-inspired algorithms, artificial immune systems, artificial life, membrane computing, amorphous computing; computational systems biology, genetic networks, protein-protein networks, transport networks, synthetic biology, cellular (in vivo) computing. computations beyond the Turing model and philosophical aspects of computing.

31.6   Unconventional Computing Journals

As in any new area of computer science, conferences dominate in the beginning, then, after maturing, new international refereed specialised journals are founded. For unconventional computing this process started in 2002:

- Springer journal Natural Computing (http://www.springer.com/computer/theoreti cal+computer+science/journal/11047) founded in 2002 by G. Rozenberg.
- Elsevier journal Theoretical Computer Science C—Theory of Natural Computing (http://www.journals.elsevier.com/theoretical-computer-science) founded in 2004 by G. Rozenberg.
- Old City Publishing journal International Journal of Unconventional Computing (http://www.oldcitypublishing.com/journals/ijuc-home/) founded in 2005 by A. Adamatzky.
- IGI Global publisher journal International Journal of Nanotechnology and Molecular Computation (http://www.igi-global.com/journal/international-journal-nanote chnology-molecular-computation/1117) founded in 2009 by B. MacLennan.

31.7   Unconventional Computing: An Area in Full Expansion

An interesting article titled "Unconventional Computing" [13] commenting on recent achievements in the field was written by V. Cerf, the Chief Internet Evangelist at Google. Among them is the IBM TrueNorth (http://www.research.ibm.com/articles/brain-chip.shtml), the one million neuron brain-inspired processor. The chip consumes just 70 mW and is capable of 46 billion synaptic operations per second, per watt, "literally a synaptic supercomputer in your palm".

The advance in neural chips, Watson-like systems and quantum computers require strategies, solutions and programming styles very different from those used in conventional computing (see, for example [12]). Academia[13] is not anymore the sole place for such studies: commercial companies, from those with a well-established research track-record like IBM and Lockheed Martin, to new ones, like D-Wave Systems or TDK-Headway Technologies, and to IT giants like Google and Microsoft, have ambitious programmes.

International projects like Truce (www.truce-project.eu), Training and Research in Unconventional Computing, a Coordination Action, supported by the Future and Emerging Technologies (FET) programme within the ICT theme of the Seventh Framework Programme for Research of the European Commission, play an important role.

Acknowledgments I am grateful to Andy Adamatzky, Elena Calude, John Casti, Michael Dinneen, Bob Doran, Grzegorz Rozenberg, Karl Svozil and Garry Tee for conversations on the topic of the paper and personal recollections.

This work was supported in part by Marie Curie FP7-PEOPLE-2010-IRSES Grant RANPHYS.

References

1. Abbott, A.A., Calude, C.S., Conder, J., Svozil, K.: Strong Kochen-Specker theorem and incomputability of quantum randomness. Phys. Rev. A **86**, 062109 (2012)
2. Abbott, A.A., Calude, C.S., Svozil, K.: A quantum random oracle. In: Barry Cooper, S., van Leeuwen, J. (eds.) Alan Turing: His Work and Impact, pp. 206–209. Elsevier Science, Amsterdam (2013)
3. Abbott, A.A., Calude, C.S., Svozil, K.: Value-indefinite observables are almost everywhere. Phys. Rev. A **89**, 032109 (2013)
4. Abbott, A.A., Calude, C.S., Svozil, K.: A variant of the Kochen-Specker theorem localising value indefiniteness. Report CDMTCS-478, Centre for Discrete Mathematics and Theoretical Computer Science, University of Auckland, Auckland, New Zealand (2015)
5. Abbott, A.A., Calude, C.S., Svozil, K.: A non-probabilistic model of relativised predictability in physics. Information **6**, 773–789 (2015)
6. Adleman, L.M.: Molecular computation of solutions to combinatorial problems. Science **226**, 121–1024 (1994)
7. Allen, N.: Personal communication to C.S. Calude, 7 Nov 2014
8. Calude, C.S. (ed.): The Human Face of Computing. Imperial College Press, London (2015)
9. Calude, C.S., Casti, J.L.: Parallel thinking. Nature, pp. 549–551, 9 April 1998
10. Calude, C.S., Casti, J., Dinneen, M.J. (eds.): Unconventional Models of Computation. Springer, Singapore (1998)
11. Calude, C.S., Casti, J.L., Gibbons, P.B., Lipponen, M.: Unconventional models of computation: a conventional report. Complexity **3**(4), 8–11 (1998)
12. Calude, C.S., Calude, E., Dinneen, M.J.: Adiabatic quantum computing challenges. ACM SIGACT News **46**(1), 40–61 (2015)
13. Cerf, V.G.: Unconventional computing. Commun. ACM **57**(10), 7–7 (2014)
14. "Computing" vs. "computation" (2015). http://english.stackexchange.com/questions/79262/computing-vs-computation. Accessed 3 Mar 2015
15. D-Wave. D-Wave overview: A brief introduction to D-Wave and quantum computing (2013). http://www.dwavesys.com/sites/default/files/D-Wave-brochure-102013F-CA.pdf

---

[13]From specialised centres, like International Center of Unconventional Computing (http://uncomp.uwe.ac.uk/people.html) in Bristol, UK, to groups scattered all around the globe.

16. Davis, M.: The myth of hypercomputation. In: Teuscher, C. (ed.) Alan Turing: Life and Legacy of a Great Thinker, pp. 195–212. Springer, Berlin (2004)
17. Davis, M.: Why there is no such discipline as hypercomputation. Appl. Math. Comput. **178**, 4–7 (2006)
18. Davis, M.: Interview with Martin Davis. Not. Am. Math. Soc. **55**, 560–571 (2008)
19. Jack Copeland, B.: The Church-Turing thesis. In: Zalta, E.N. (ed.) The Stanford Encyclopedia of Philosophy (2008). Fall 2008 edition
20. Kochen, S.B., Specker, E.: The problem of hidden variables in quantum mechanics. J. Math. Mech. **17**, 59–87 (1990). Reprinted in E. Brikhäuser Verlag, Basel, Specker. Selecta (1967)
21. Lanzagorta, M., Uhlmann, J.K.: Hybrid quantum-classical computing with applications to computer graphics. In: ACM SIGGRAPH 2005 Courses, SIGGRAPH '05. ACM, New York, NY, USA (2005)
22. Moore, G.E.: Cramming more components onto integrated circuits. Electronics pp. 114–117 (1965), 19 April 1965. Reprinted in: Proceedings of the IEEE, vol. 86, no. 1, pp. 82–85 (1998)
23. Shor, P.W.: Algorithms for quantum computation: discrete logarithms and factoring. In: Proceedings of the 35th Annual Symposium of on Foundations of Computer Science, Santa Fe, NM, Nov. 20–22, 1994. IEEE Computer Society Press (1994). arXiv:quant-ph/9508027
24. Wikipedia. Computation (2015). http://en.wikipedia.org/wiki/Computation. Accessed 27 Feb 2015
25. Wikipedia. Computation (2015). http://en.wikipedia.org/wiki/Computing. Accessed 27 Feb 2015
26. Wikipedia. Hypercomputation (2015). http://en.wikipedia.org/wiki/Hypercomputation. Accessed 27 Feb 2015

Index

© Springer International Publishing Switzerland 2017
A. Adamatzky (ed.), *Advances in Unconventional Computing*,
Emergence, Complexity and Computation 22,
DOI 10.1007/978-3-319-33924-5